Electroweak Effects at High Energies

ETTORE MAJORANA
INTERNATIONAL SCIENCE SERIES
Series Editor:
Antonino Zichichi
European Physical Society
Geneva, Switzerland

(PHYSICAL SCIENCES)

Recent volumes in the series:

Volume 15 **UNIFICATION OF THE FUNDAMENTAL PARTICLE INTERACTIONS II**
Edited by John Ellis and Sergio Ferrara

Volume 16 **THE SEARCH FOR CHARM, BEAUTY, AND TRUTH AT HIGH ENERGIES**
Edited by G. Bellini and S. C. C. Ting

Volume 17 **PHYSICS AT LEAR WITH LOW-ENERGY COOLED ANTIPROTONS**
Edited by Ugo Gastaldi and Robert Klapisch

Volume 18 **FREE ELECTRON LASERS**
Edited by S. Martellucci and Arthur N. Chester

Volume 19 **PROBLEMS AND METHODS FOR LITHOSPHERIC EXPLORATION**
Edited by Roberto Cassinis

Volume 20 **FLAVOR MIXING IN WEAK INTERACTIONS**
Edited by Ling-Lie Chau

Volume 21 **ELECTROWEAK EFFECTS AT HIGH ENERGIES**
Edited by Harvey B. Newman

Electroweak Effects at High Energies

Edited by

Harvey B. Newman

California Institute of Technology
Pasadena, California

Plenum Press • New York and London

Library of Congress Cataloging in Publication Data

Europhysics Study Conference on Electroweak Effects at High Energies (1st: 1983: Erice, Sicily)
Electroweak effects at high energies.

(Ettore Majorana international science series: Physical sciences; v. 21)
"Proceedings of the First Europhysics Study Conference on Electroweak Effects at High Energies, held Feb. 1–12, 1983, at the Ettore Majorana Center for Scientific Culture, Erice, Sicily, Italy"—T.p. verso.
Includes bibliographical references and index.
1. Electroweak interactions—Congresses. 2. Particles (Nuclear physics)—Congresses. 3. Lepton-nucleon scattering—Congresses. 4. Electrons—Scattering—Congresses. 5. Collisions (Nuclear physics)—Congresses. I. Newman, Harvey B. II. Series.
QC794.8.E44E97 1983 539.7′54 84-26444
ISBN 0-306-41904-1

Proceedings of the First Europhysics Study Conference on Electroweak Effects at High Energies, held February 1–12, 1983, at the Ettore Majorana Center for Scientific Culture, Erice, Sicily, Italy

A Division of Plenum Publishing Corporation
233 Spring Street, New York, N.Y. 10013

Printed in the United States of America

In memory of

J.J. Sakurai,

whose keen vision in the face of orthodoxy helped us maintain the separation between pure speculation, and firmly established experimental fact.

—H. Newman

PREFACE

The first Europhysics Study Conference on Electroweak Effects at High Energies was held at the "Ettore Majorana" Centre for Scientific Culture in Erice, Sicily from February 1 - 12, 1983. The conference was attended by 61 physicists from 11 countries.

The conference was sponsored by the European Physical Society, the Italian Ministry of Public Education, the Italian Ministry of Technological Research, the Sicilian Regional Government and the California Institute of Technology.

CONFERENCE FORMAT

The Study Conference followed a new intensive format in which the state of our knowledge of the electroweak interaction, and the relation of the electroweak sector to Grand Unified and Superunified Theories was reviewed in some depth. During the two week conference, 54 experimental and theoretical talks were presented, and four evening discussion sessions were held.

The Erice surroundings, the wide-ranging conference program, and the fact that nearly all of the participants were directly involved in recent major experimental or theoretical developments, led to animated and very friendly discussions. Participants had the rare opportunity to view most of the major trends in high energy physics in a short interval of time, and to discuss and contemplate the trends in the uniquely peaceful yet stimulating atmosphere which is an Erice tradition.

The conference program placed special emphasis on the most

recent experimental results and prospects for future experiments bearing on:

(1) the structure of the weak neutral and charged currents,
(2) the nature and mass spectrum of the gauge bosons, and
(3) the possible compositeness of quarks and leptons.

Parts of the conference also related electroweak phenomena to schemes of dynamical symmetry breaking, grand unification and supersymmetric models which include the gravitational interaction. Recent experimental developments in the fields of proton decay, neutrino and neutron-antineutron oscillations and magnetic monopoles were presented, and their theoretical implications were discussed.

THE J. J. SAKURAI MEMORIAL FELLOWSHIP AND LECTURE

The conference was touched with sadness by the untimely passing of Prof. J. J. Sakurai, who was scheduled to give a key lecture summarizing the state of our knowledge in the electroweak sector. His wit and keen insight were missed by all at Erice.

In honor and memory of his many fundamental contributions, which extended from his early work on vector Yang - Mills theories of hadronic physics to his later work on electroweak unification and composite models, the conference organizers decided to establish the J. J. Sakurai Memorial Fellowship. The fellowship was ultimately awarded to Prof. Sakurai's last student, Pisin Chen, who reported on their joint work (continued with F. M. Renard) on the final day of the meeting. Prof. Roberto Peccei delivered an excellent Memorial Lecture on the Structure of the Fundamental Interactions during the same session, in which he put the essence of J. J. Sakurai's contributions in perspective.

CONFERENCE HIGHLIGHTS

The discovery of the W boson by UA1 and UA2 at the CERN $P\bar{P}$ collider (reported by G. Salvini and G. Goggi) made the conference one of the most exciting. It was clear to us at the conference that the discovery of the Z°, or its failure to appear at the prescribed mass, was imminent. (Its appearance, in agreement with the standard electroweak theory of Glashow, Weinberg and Salam, was first reported by the UA1 collaboration in May, 1983). This was the dawning of the era of direct experimental tests of electroweak unification.

The conference was also marked by a number of other events of great significance which were first presented at, or just prior to the conference. These included:

(1) The lower bound from the IMB experiment on the proton lifetime of 6.5 x 10^{31} years. The Kolar Gold Field and NUSEX Candidates for proton decay were also presented, corresponding to lifetimes of 1 - 2.5 x 10^{31} years (as summarized by M. Rollier).

(2) The emergence of clear narrow two and possibly three jet structures in high E_T $P\bar{P}$ events (presented by G. Goggi for UA2).

(3) The discovery of the B mesons at CESR by the CLEO collaboration (presented by D. Kreinick). Data from CUSB and CLEO on the semileptonic momentum spectrum and in B decays showed that the branching ratio for b quarks $(b \to u)/(b \to c)$ is less than 5%.

(4) The measurement of the s-dependence of electroweak interference in $e^+e^- \to \mu^+\mu^-$, in agreement with the standard electroweak theory of Glashow, Weinberg and Salam. The results were summarized by A. Boehm, on the basis of contributions from eight PETRA and PEP experiments. Asymmetry results for $e^+e^- \to \tau^+\tau^-$ are also in agreement with the standard model.

(5) The absence of further "monopole events" seen in Blas Cabrera's improved superconducting loop detector. This made the hypothesis that the "St. Valentine's Day Event" (February 14, 1982) was caused by the passage of a magnetic monopole look increasingly doubtful. The new limit on the monopole flux of 7 x 10^{-11} cm^{-2} sec^{-1} sr^{-1} obtained after 125 days of running with the new detector is still far above the Parker bound of 10^{-15}, which is based on the fact that a larger flux of monopoles would erase the galactic magnetic field.

Monopole bounds, and the status of present and future experiments designed to reach these astrophysical limits, were thoroughly reviewed (with style) by D. Groom and R. Peccei.

The impact of the recent CESR data, and data on dimuon production by neutrinos, was discussed in reviews of the charge current sector (the weak mixing angles and matrix elements) by G. Altarelli, L. L. Chau and K. Kleinknecht.

Results on $\sin^2\theta_W$ from νN scattering by the CCFRR, CDHS, and CHARM collaborations were presented in talks by Borgia, Geweniger and Rapidis, with the CCFRR result (0.24) showing some indication of being higher than the CHARM result (0.21). Given the small statistical and systematic errors in the data (±0.013 and $\lesssim$±0.01

typically), radiative corrections (typically -0.01) and QCD corrections arising from Q^2 evolution of the proton structure (±0.01) now play a significant role.

Prior to the UA1 and UA2 results on the W and Z masses (the latter after the conference), the most precise result on $\sin^2\theta_W$ in purely leptonic interactions was obtained from the CHARM neutrino electron scattering data. This field was reviewed by K. Winter. The question of residual uncertainty in the background from coherent pizero production was discussed by Winter, J. Morfin and Y. Suzuki. Suzuki also presented preliminary data from his experiment at BNL which uses an almost totally sensitive detector, and which gives a large signal-to-background ratio in νe scattering for the first time.

In looking forward to the next generation of experiments at e^+e^- and ep colliders, we included a special "study session" on beam polarization. The basic physics concepts and techniques were very ably presented by B. Montague for e^+e^- storage rings. J. Buon concentrated on schemes for attaining longitudinal polarization (at LEP for example), and D. Barber presented the state of the art of polarization measurements as practiced at DESY. C. Prescott explored the precise measurements of the weak neutral couplings of leptons and quarks, with emphasis on the SLC, which could be made possible by the use of longitudinal beam polarization. K. Mess discussed the analogous physics possibilities at ep colliders such as HERA, where both the charged and neutral current couplings may be measured. Both Prescott and Mess showed that longitudinally polarized beams could allow experiments to detect a new heavy gauge boson up to the 0.5 TeV range.

The profound importance of "radiative corrections" emerged in a variety of sessions throughout the conference. It was therefore appropriate to explore the status of the subject in lectures by the world's experts: Berends, Gaemers, and Gastmans. Key points in Berends' and Gaemers' lectures were the effects higher order corrections have on the relationship between gauge boson masses and charge asymmetries, and the basic neutral current coupling constants. Gastmans presented the major advances of the "Calkul" collaboration in analytic calculations of higher order graphs in QED and QCD, which may play a pivotal role in the precise comparison of experiment and theory in the coming years.

On the eve of (apparent) vindication of the Standard Model, the conference was a time of intense speculation on whether the known electroweak phenomena (the W mass, the consistency with $\sin^2\theta_W$ values in lower energy νN experiments, the asymmetries in e^+e^- and μN experiments) could be described equally well by alternative models. The idea of Sakurai and Bjorken that global

$SU(2)_L$ symmetry and γ-Z mixing is a viable alternative to the standard scheme was fully developed by Sakurai and Chen. The possibility that the gauge bosons are bound states of more fundamental constituents was explored by Fritzsch and Schildknecht. Further motivation for composite models in explaining the heirarchy of lepton and quark masses, and the mass generation mechanism was given by Fritzsch.

Proponents of composite models held forth the exciting prospect that the "standard" W and Z are no more than the lowest bound states in a spectrum, and that a continuum might even open up at accelerator energies starting in the $\sqrt{s}$ = 1-10 TeV energy range.

Theoretical speculation was also fueled by the non-observation of proton decay or magnetic monopoles (even though the Parker bound was still to be breached by experiment). The IMB result clearly put the minimal SU(5) GUT in difficulty, and GUT's based on larger gauge groups, and more complicated spontaneous symmetry-breaking schemes, were presented with renewed vigor by Marshak and Shafi.

In the context of non-minimal GUT's, Roos applied a systematic analysis of experimental data in the charged and neutral current sectors to set limits on the couplings and mixing angles of hypothetical "Mirror Leptons" (V + A) with ordinary leptons and quarks.

Felix Boehm presented a review of experiments on neutrino oscillations, indicative of mass differences between neutrino flavors, and neutron-antineutron oscillations which would indicate baryon number non-conservation ($\Delta B = 2$). There is no conclusive evidence for either phenomenon. Recent reactor-based searches at Gosgen set new restrictive upper limits in the Δm^2 vs $\sin^2 2\theta$ plane for neutrino oscillations. The current limit on the $\Delta B = 2$ lifetime of the neutron from IMB may be extended to the 10^{32} year level by proposed experiments at Grenoble, LAMPF and ORNL. These and other experimental tests of GUT's were also discussed by Al Mann.

The remaining mysteries of nature, left unexplained either in the context of the standard electroweak theory, or the minimal SU(5) GUT, were reviewed from the theoretical standpoint, and approaches were proposed in the talks by Peccei, Shafi, and Nanopoulos. Starting from SO(10) as the Grand Unified symmetry group, Shafi gave possible solutions the strong CP problem, an explanation for the dark matter in the universe (axions), the mechanism for large scale density fluctuations in the universe, and the lack of an observed monopole flux. Recent theoretical work by Nanopoulos and collaborators, based on local N=1 Supersymmetric GUT's, shows possible solutions to the problems of flavor changing neutral current suppression, the nonobservation of proton decay, the gauge hierarchy and fermion mass generation problems, and an alternative solution to the strong CP problem.

Murray Gell-Mann, in his unique style, gave a sweeping review of the main approaches to connecting supergravity to observation. He explored various attempts to account for all the known particles in multiplets of supergravity, and in extended supergravity theories. He raised the possiblity that the supersymmetries actually apply only to the sub-constituents of the known particles, called "haplons", and presented the consequences in terms of the charge, color and family quantum numbers of the haplons. He also reviewed recent work on theories containing extra dimensions, which provide the mechanism for the symmetry breaking, and hence the "supergap" in mass between each particle and its supersymmetric partner. The extra dimensions may be made to "roll up into a little ball" with a scale inaccessible to present day experiments, but which may have observable consequences on the nature of space-time as probed by very high energy experiments.

In closing the conference, we returned to the key question of possible future scenarios which may face the high energy experimentalist. Whether we find ourselves at the edge of the "Great Desert", amidst a spectrum of W's and Z's, in a new continuum where quark and lepton subconstituents are produced, or in the SUSY world of gauge boson, quark and lepton partners -- there is much to be learned from the immediate past. The new principles of detector design introduced by UA2 and especially UA1 will serve experimenters well if refined and extended for the next collider generation. The need for superb lepton and jet identification and high resolution energy measurement, and the crucial role of (nearly) hermetic calorimetry, were made clear in the early days of the W's discovery. In the future, the particular signatures of E_T single (or multi-) leptons, and associated jets and/or missing E_T, may provide a high degree of sensitivity in detecting new physics phenomena. In following these guidelines in future detector design, experimenters will be well-placed to explore any one of the imagined scenarios, discussed during the conference, which are energetically accessible to accelerators.

. . . Until nature proves once again that her subtleties transcend our imagination.

H. Newman
California Institute of Technology

May, 1984

ACKNOWLEDGEMENTS

The success of the conference depended on extraordinary efforts in planning and organizing the meeting, and preparing the proceedings, on the part of many people.

First and foremost thanks go to Prof. Antonino Zichichi for establishing the Erice "Ettore Majorana" Centre as a scientific and cultural resource for all nations. I wish to thank Prof. Zichichi personally for the opportunity to chair this meeting.

Prof. Zichichi's staff at Erice carried forward the meeting organization with great professional skill. Particular thanks go to Dr. A. Gabriele and Miss N. Zaini. The entire Erice staff could not have been more accommodating in arranging both the conference sessions, and the social events, in the warmest atmosphere of friendship.

The intensive conference schedule made the task of organizing and exchanging written information very difficult. Dr. Manfred Steuer, as General Scientific Secretary for the conference, did an outstanding job and exhibited infinite patience. Dr. Ling-Lie Chau invested enormous effort in helping me review the sessions, for which I am most grateful. I would also like to thank Drs. G. Bella, R. Brown, P. Chen, P. Grosse-Wiesmann, H. Maxeiner, J. Morfin, P. Rapidis, J. Salicio, and Y. Suzuki for their work as Scientific Secretaries.

Mrs. JoAnn Anderson and Mrs. Margaret Brennan were instrumental in completing the final stages of the manuscript of the proceedings for submission to the publisher.

Special thanks go to Mrs. June Bressler, who invested uncounted hours in preparing the conference in all details, from poster to sessions to proceedings. Her work at Caltech over a two year period, and her personal attention to each of the participants' needs before and during the conference, were indispensible.

Finally, to my wife Lynda, go the warmest thanks of many participants as well as myself, for her help in arranging the social events and her unique role in bringing all the conference participants together in the spirit of lasting friendship.

H. Newman

CONTENTS

I. EXPERIMENTS AT THE CERN $\bar{P}P$ COLLIDER

II. THE WEAK NEUTRAL CURRENT: FIXED TARGET EXPERIMENTS

A. LEPTON NUCLEON SCATTERING

B. NEUTRINO ELECTRON SCATTERING

C. PLENARY SESSION ON NEUTRINO ELECTRON SCATTERING

III. ELECTRON POSITRON INTERACTIONS

A. EXPERIMENTS AT PETRA AND PEP

B. PLENARY SESSION

IV. THE WEAK CHARGED CURRENT

V. POLARIZED BEAMS

VI. ELECTRON PROTON COLLISIONS

VII. STANDARD ELECTROWEAK THEORY AND PHENOMENOLOGY: A VIEW TOWARD HIGHER ENERGIES

VIII. EXTENSIONS AND GENERALIZATIONS OF THE STANDARD ELECTROWEAK PICTURE

IX. COMPOSITE MODELS

X. SPECIAL TOPICS IN GAUGE THEORIES

XI. MONOPOLES

XII. PROTON DECAY

XIII. SELECTED TOPICS: ON THE PATH TO TESTS OF GRAND UNIFIED THEORIES

XIV. SUPERSYMMETRY AND SUPERGRAVITY: INTERMEDIATE MASS SCALES

XV. SPECIAL SESSIONS

HADRON JETS AND LARGE TRANSVERSE MOMENTUM ELECTRONS AT THE SPS $\bar{p}p$ COLLIDER

Presented by G. Goggi[5]
The UA2 Collaboration

M. Banner[6], R. Battiston[7,9], Ph. Bloch[6], F. Bonaudi[2], K. Borer[1], M. Borghini[2], J-C. Chollet[4], A.G. Clark[2], C. Conta[5], P, Darriulat[2], L. Di Lella[2], J. Dines-Hansen[3], P-A. Dorsaz[2], L. Fayard[4], M. Fraternali[5], D. Froidevaux[2], J-M. Gaillard[4], O. Gildemeister[2], V.G. Goggi[5], H. Grote[2], B. Hahn[1], H. Hänni[1], J.R. Hansen[3], T. Himel[2], V. Hungerbühler[2], P. Jenni[2], O. Kofoed-Hansen[3], M. Livan[5], S. Loucatos[6], B. Madsen[3], P. Mani[1], B. Mansoulié[6], G.C. Mantovani[7], L. Mapelli[2], B. Merkel[4], M. Mermikides[2], R. Mollerud[3], B. Nilsson[3], C. Onions[2], G. Parrour[2,4], F. Pastore[2,5], H. Plothow-Besch[2,4], J-P. Repellin[4], A. Rothenberg[2], A. Roussarie[6], G. Sauvage[4], J. Schacher[1], J-L. Siegrist[2], H.M. Steiner[8,2], G. Stimpfl[2], F. Stocker[1], J. Teiger[6], V. Vercesi[5], A. Weidberg[2], H. Zaccone[6], and W. Zeller[1]

ABSTRACT

A preliminary analysis of the UA2 data collected at the SPS $\bar{p}p$ collider during the 1982 run (20 nb^{-1}) is presented, with particular emphasis on large transverse momentum hadron jets and on electrons having the configuration expected from the decay of electroweak bosons. The data provide very strong evidence of two-jet dominance in events having a large transverse energy released in the central region. Four isolated electrons condidates having a transverse momentum in excess of 20 GeV/c are observed, with no significant transverse momentum balance within the UA2 acceptance. These events are consistent with the hypothesis $W \to e\nu$.

[1]Laboratorium für Hochenergiephysik, Universität Bern, Sidlerstrasse 5, Bern, Switzerland.
[2]CERN, 1211 Geneva 23, Switzerland.
[3]Niels Bohr Institute, Blegdamsvej 17, Copenhagen, Denmark.
[4]Laboratoire de L'Accélérateur Linéaire, Université de Paris-Sud, Orsay, France.
[5]Dipartimento di Fisica Nucleare e Teorica, Università di Pavia and INFN, Sezione di Pavia, Via Bassi 6, Italy.
[6]Centre d'Etudes nucléaires de Saclay.
[7]Gruppo INFN del Dipartimento di Fisica dell'Università di Perugia, Italy.
[8]On leave from Lawrence Berkeley Laboratory, USA.
[9]Also at Scuola Normale Superiore, Pisa, Italy.

1 - INTRODUCTION

The major objective of the UA2 experiment at the CERN $\bar{p}p$ Collider is the search for the electroweak intermediate vector bosons Z^o and W through their electronic decay modes [1]

$$\bar{p}p \rightarrow W^{\pm} + X \qquad W^{\pm} \rightarrow e^{\pm}\nu \qquad (1)$$

$$\bar{p}p \rightarrow Z^{o} + X \qquad Z^{o} \rightarrow e^{+}e^{-} \qquad (2)$$

For this reason the UA2 apparatus is designed to detect and identify electrons over its whole acceptance. Despite their small branching fractions the electronic decay modes

$$B\ (W \rightarrow \ell\nu) \qquad \simeq 8\%\ \text{per lepton type} \qquad (3)$$

$$B\ (Z^{o} \rightarrow \ell^{+}\ell^{-}) \qquad \simeq 3\%\ \text{per lepton type} \qquad (4)$$

provide the strongest signature against background contamination and can be detected with superior resolution by means of compact calorimeters. In addition, hadron detection is also implemented over the whole acceptance of the apparatus, for the study of hard scattering processes of hadron constituents yielding large transverse-momentum hadronic jets.

Both kinds of processes become accessible at the center of mass energy, $\sqrt{s}$ = 540 GeV, provided by the SPS $\bar{p}p$ Collider[2]. Theoretical expectations[1] for this energy region, which can be uniquely explored by this facility, predict in fact cross sections of a few nb for the production of electroweak bosons, and an increase of about four orders of magnitude with respect to ISR energies for the yield of large-p_T jets.

The results presented in this report correspond to a selected set of data collected during the runs at the end of 1982.

The data were recorded using triggers sensitive to events with high transverse energy deposited in the UA2 calorimeters.

In Section 2 we first give a description of the experimental apparatus and its performance. Preliminary results on the production of large-p_T jets are given in Sect. 3. The evidence for the production of large transverse momentum electrons, consistent with the hypothesis $W \to e\nu$, is discussed in Sec. 4. The search for electron pairs originating from reaction(2) in then briefly discussed in Sect. 5.

2 - DETECTOR AND DATA TAKING

2.1 - Detector

The UA2 detector[3-5] is essentially a highly segmented calorimeter having cylindrical symmetry around the beam axis and tower structure pointing to the interaction region. In addition, the forward-backward regions are equipped with toroidal magnetic spectrometers.

A plan view of the detector is shown in Fig. 1. The detector acceptance extends from $\theta = 20^{o}$ to $\theta = 160^{o}$ to the beams, covering about 3.5 rapidity units in the central region. This pseudorapidity range is optimal for the detection of both large transverse momentum binary parton collisions and of the decay products of electroweak bosons.

The cylindrical chambers of the vertex detector cover 2π radians in ϕ and from 20° to 160° in θ, with an outer radius of about 35 cm. The chamber assembly consists of four multiwire proportional chambers (MWPC) with analog read-out on helicoidal cathode strips, two JADE-type drift chambers with six staggered sense wires per drift cell and read out including multihit and charge division electronics, and 24 scintillation counters arranged as a cylindrical hodoscope. A fifth MWPC is preceded by 1.5 radiation lengths of tungsten and is used for shower localization in front of the electromagnetic central calorimeter.

The central calorimeter, covering ± 1 unit in rapidity around 0, is segmented into 240 cells, each covering 15° in ϕ and 10° in θ, and is built in a tower structure pointing to the centre of the interaction region. The cells are segmented longitudinally into a 17 radiation length thick electromagnetic compartment (lead-scintillator)

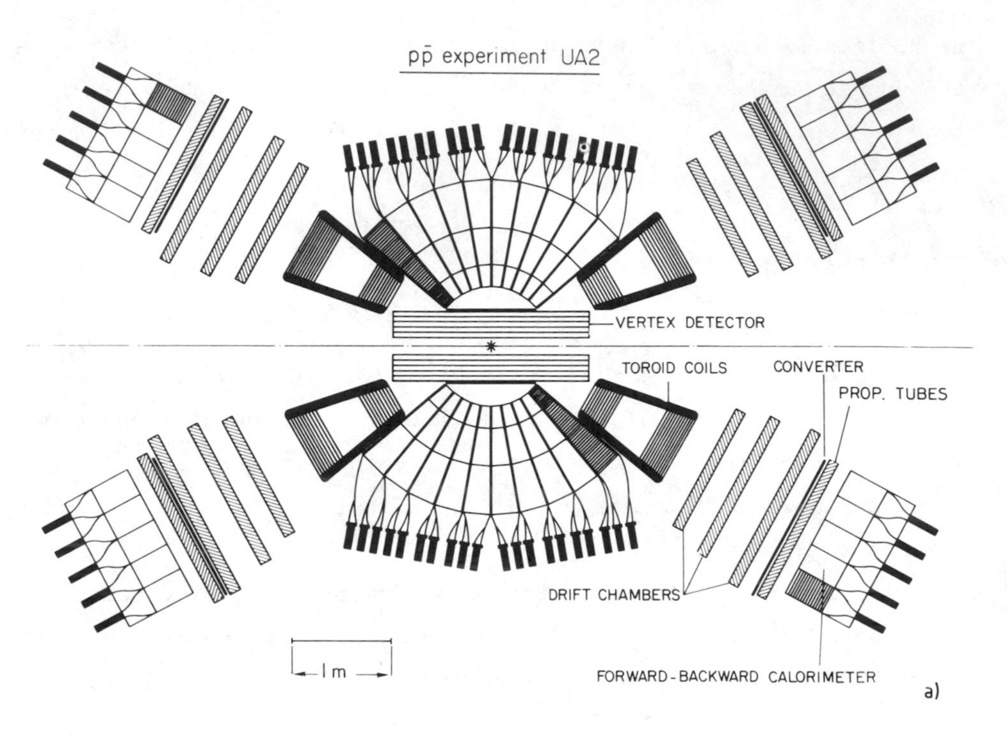

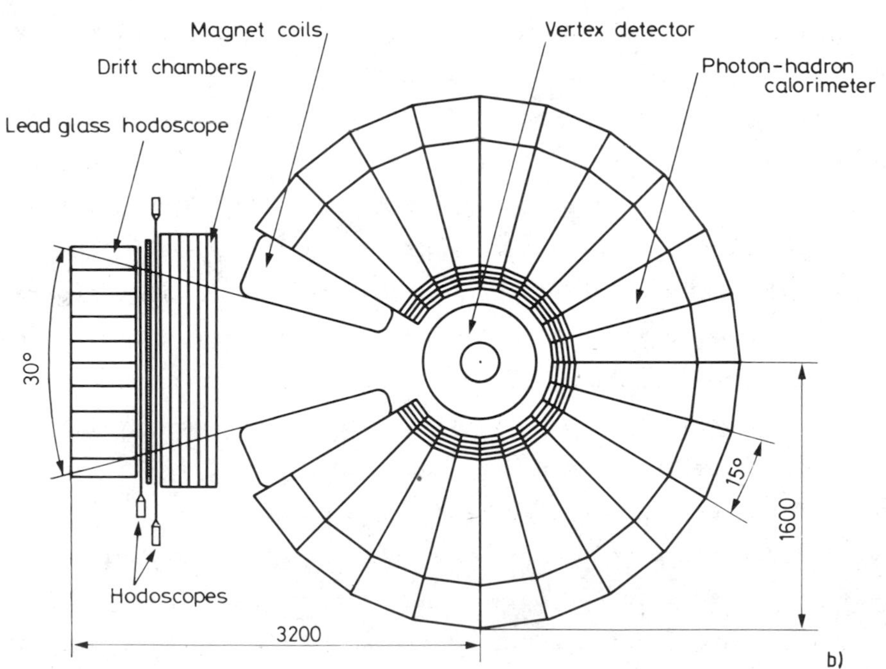

Figure 1. Schematic detector assembly :
a) longitudinal cut along the beam,
b) transverse cut normal to the beam.

followed by two hadronic compartments (iron-scintillator) of two absorption lengths each, giving a total of 4.5 absorption lengths (Fig. 2). The light from each compartment is collected by two BBQ-doped light guide plates on opposite sides of the cell.

All calorimeters, including the forward-backward modules, have been calibrated in a 10 GeV/c beam from the CERN PS using incident electrons and muons. The calibration has since been tracked with a Xe light flasher system. In addition, the response of the electromagnetic compartments is checked regularly by accurately positioning a Co^{60} source in front of each cell and measuring the direct current from each photomultiplier. The systematic uncertainty in the energy calibration for the data discussed here is less than ±2% for the electromagnetic calorimeter and less and ±3% for the hadronic one.

The response of the calorimeter to electrons, single hadrons and multi-hadrons (produced in a target located in front of the calorimeter) has been measured at the CERN PS and SPS machines using beams from 1 to 70 GeV. In particular, the longitudinal and transverse shower development and the effect of particles impinging near the cell boundaries have been studied in detail.

The energy resolution for electrons is measured to be $\sigma_E/E = 0.14/\sqrt{E}$ (E in GeV). In the case of hadrons, σ_E/E varies from 32% at 1 GeV to 11% at 70 GeV (approximately like $E^{-\frac{1}{4}}$). The resolution for multi-hadron systems of more than 20 GeV is similar to that of single hadrons.

The polar angular regions of 20° to 37.5° and 142.5° to 160° are each equipped with toroidal magnets. The coils define 12 forward and 12 backward azimuthal sectors covering 80% of 2π radians in ϕ. The average field integral along the particle trajectories is 0.38 Tm. Each spectrometer sector (Fig. 3) is instrumented with 9 drift chamber planes. Together with the vertex detector these chambers allow a measurement of the electron charge from reaction (2) up to 60 GeV/c transverse momentum.

The drift chambers are followed by a 1.4 radiation length thick iron-lead converter and four layers of multitube proportional chambers (MTPC) which define two coordinate axes at an angle of 77°

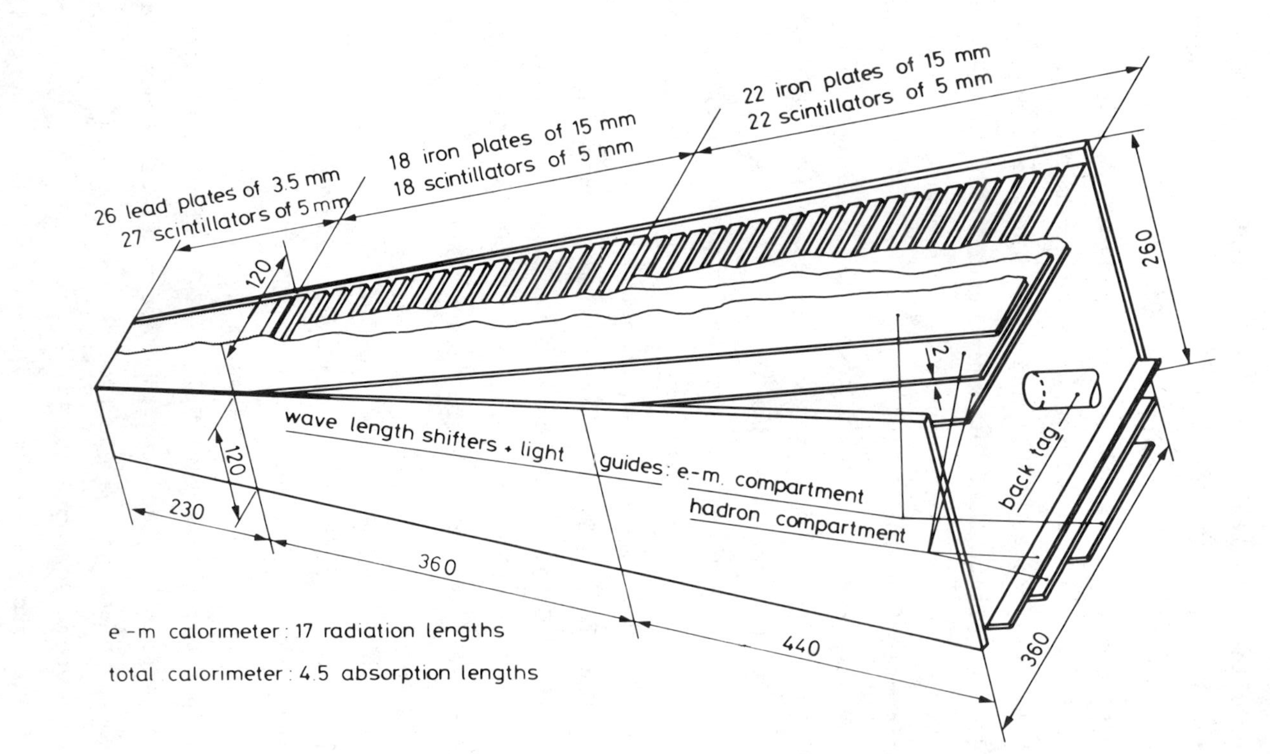

Figure 2. Exploded view of a central calorimeter cell.

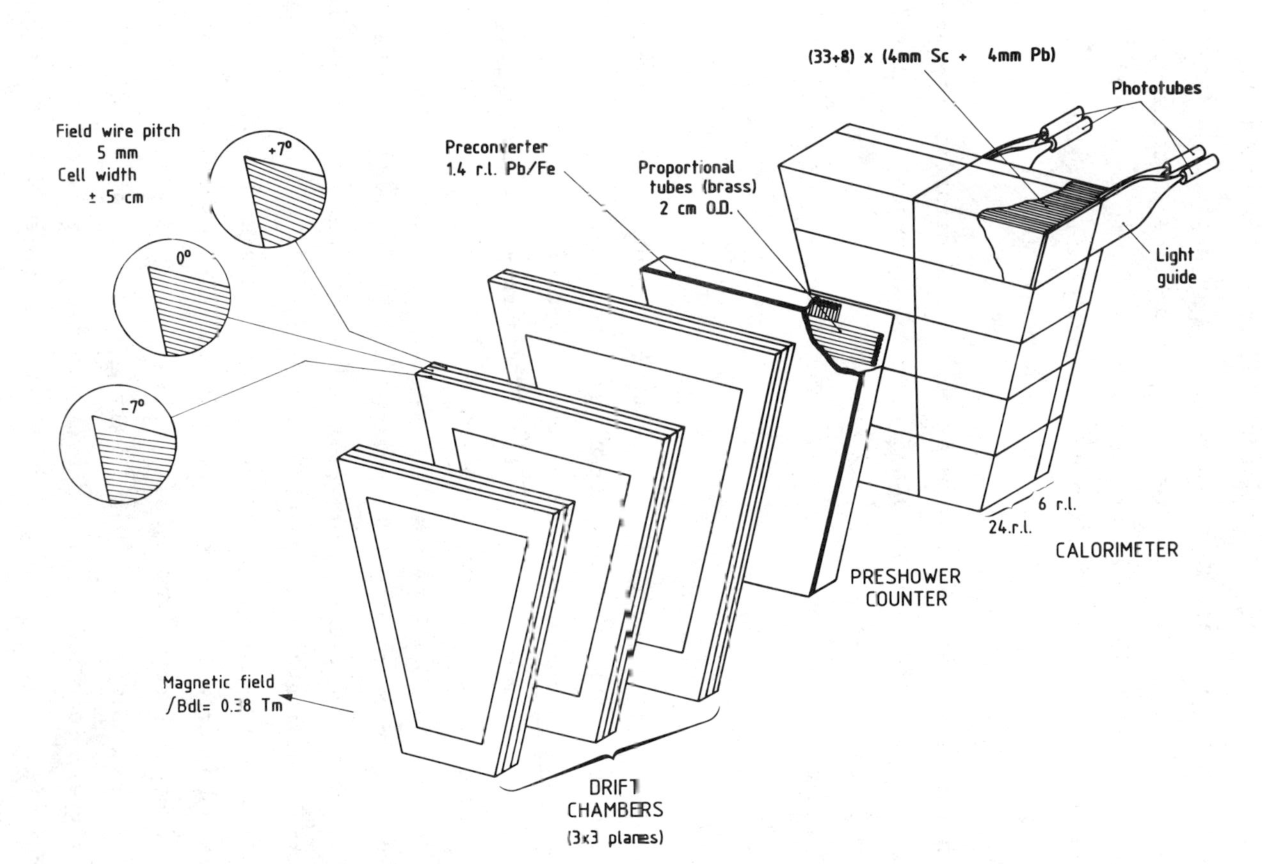

Figure 3. Schematic view of one of the 24 sectors of the forward detectors.

between them. Resolutions of 5 mm (rms) have been obtained in each projection. Electromagnetic calorimeter sectors finally complete the forward-backward detectors. Each calorimeter sector is divided into ten cells covering 15° in ϕ and 3.5^{o} in θ. The cells are lead-scintillator sandwiches. They are segmented in depth into two compartments (24 and 6 radiation lengths), to achieve good electron-hadron separation. The light of each compartment is collected by two photomultipliers via BBQ doped wavelength-shifting light guides. The energy resolution for electrons measured in test beams is $\sigma_E/E = 0.15/\sqrt{E}$ (E in GeV).

In this stage of the experiment no central calorimenter units were mounted in a 60° azimuthal wedge, allowing space for a large angle magnetic spectrometer ($\Delta\phi \times \Delta\theta \simeq 30^{o} \times 70^{o}$) instrumented with drift chambers, lead-glass cells and scintillator hodoscopes(4) (Fig. 1b).

2.2 - Data Taking

The data discussed in this report were recorded using triggers sensitive to events with large transverse energy in the central and forward calorimeters. They were of three types :

- The ΣE_T triggers required a total transverse energy (ΣE_T) measured in the central calorimeter (electron and hadron cells linearly added) in excess of $\sim$ 35 GeV.
- the W trigger required the presence of at least one quartet (2×2) of electron calorimeter cells (central or forward) in which the measured transverse energy exceeds $\sim$ 8 GeV.
- the Z^{o} trigger required the presence of two such quartets, each having a transverse energy in excess of $\sim$ 3.5 GeV, and azimuthally separated by $\Delta\phi \geqslant 60^{o}$.

In addition the ΣE_T and W triggers (but not the Z^{o} trigger) required a coincidence with two signals obtained from scintillator arrays covering scattering angles $0.47 < \theta < 2.84^{o}$ on both sides of the collision region. This additional condition is satisfied by nearly all non-diffractive collisions(6).

Early signals measured in these scintillator arrays were used to tag background events induced by beam halo particles.

3 - LARGE TRANSVERSE ENERGY HADRONS

3.1 - Data Reduction

In the present Section the analysis is restricted to events satisfying the total transverse energy trigger (ΣE_T). The energy in each cell is the sum of the energies of the three compartments, whenever at least one of them exceeds a minimal threshold of 150 MeV. Adjacent cells containing at least 400 MeV are joined into clusters, which are then split if they contain local maxima separated by "valleys" more than 5 GeV deep. The results obtained are largely insensitive to the exact values of these parameters.

The errors resulting from the fact that the dependence of the calorimeter response upon impact and energy are not taken into account are of the order of $\sim$ 10%. They can be reduced to a negligible level after proper correction but only uncorrected data are presented in this Section.

A small background contamination of beam halo particles interacting in the detector survives the time of flight selection in the small angle scintillator arrays. It is easily recognised from an abnormally large energy fraction measured in the hadronic compartments and is rejected from the event sample.

3.2 - Two-Jet Events

Fig. 4 shows the distribution of the observed events as a function of their total transverse energy ΣE_T, measured in the central calorimeter. A departure from an exponential behaviour is clearly seen when ΣE_T exceeds $\sim$ 60 GeV.

Fig. 5 illustrates that this departure corresponds to the emergence of two-jet configurations at large values of ΣE_T. It shows the dependence upon ΣE_T of h_1 and h_2, the mean values of the fractions $E_T^1/\Sigma E_T$ and $(E_T^1 + E_T^2)/\Sigma E_T$, where E_T^1 and E_T^2 are the largest

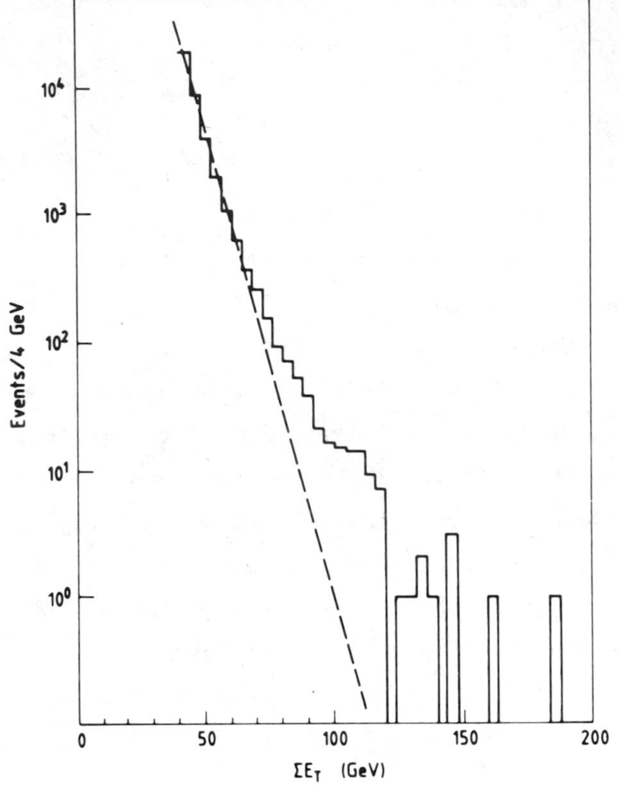

Figure 4.
Observed (uncorrected) ΣE_T distribution in the central calorimeter. The line is an exponential eye-fit to the data at low ΣE_T.

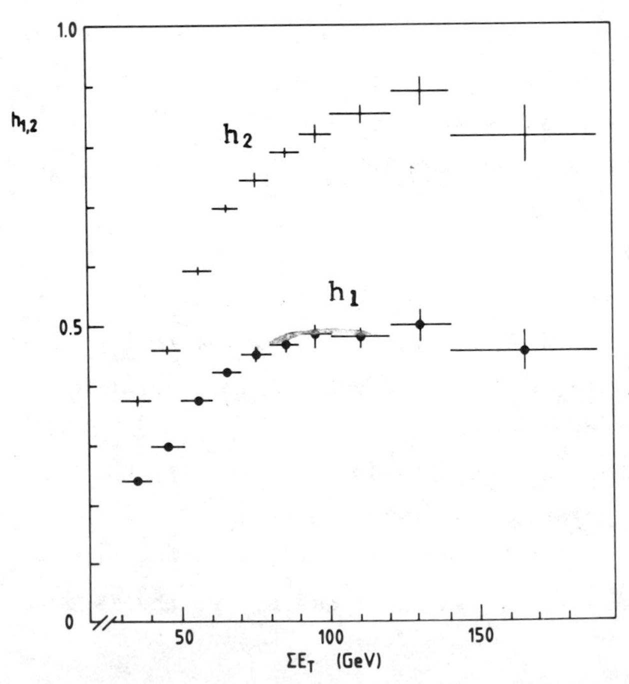

Figure 5.
Two-jet dominance : the dependence of h_1 and h_2 upon ΣE_T (see text).

and second largest transverse energies of the clusters in an event, respectively. An event containing only two jets of equal transverse energies would have $h_1 = 0.5$ and $h_2 = 1$. Immediate evidence for two-jet dominance is also obtained from a simple inspection of the energy distribution in the θ-ϕ plane : examples are shown in Figures 6a to d.

The azimuthal separation between the two clusters having the largest transverse energies ($E_T^{12} > 15$ GeV) in events having $\Sigma E_T >$ 80 GeV is observed to peak near 180° (Fig. 7).

In the present data (Fig. 8) 400 events (compared to 3 in the 1981 data(3)) contain a large E_T cluster (jet) having a transverse energy in excess of 40 GeV.

Finally we show in Fig. 9 the uncorrected invariant mass (M_{j-j}) distribution of the two-jet systems and their transverse momentum distribution for $M_{j-j} > 80$ GeV/c^2. While the invariant mass distribution shows no significant structure, it is not inconsistent with the presence of $\sim$ 20 events in the 80 ± 10 GeV/c^2 region from hadronic decays of the electroweak bosons. Such decays are expected to proceed mainly via $q\bar{q}$ pairs.

4 - SEARCH FOR $W \to e\nu$ DECAYS

Large transverse momentum electrons are expected to originate from the decay $W^{\pm} \to e^{\pm}\nu$, whith a p_T spectrum having a Jacobian peak at $p_T^e \simeq M_W/2$. This structure is smeared by the W transverse momentum, which may be similar to that of a two-jet system of a same mass (see Fig. 9), but remains clearly detectable. This is because the smearing of the Jacobian peak does not induce significant correlations between the transverse momentum of the decay electron and that of the W (or of its associated recoil particles). We shall therefore search for large transverse momentum electrons which are not accompanied by other particles at small angle to the electron momentum. This simplified approach will strongly reduce a possible two-jet background and should not affect seriously the $W \to e\nu$ signal.

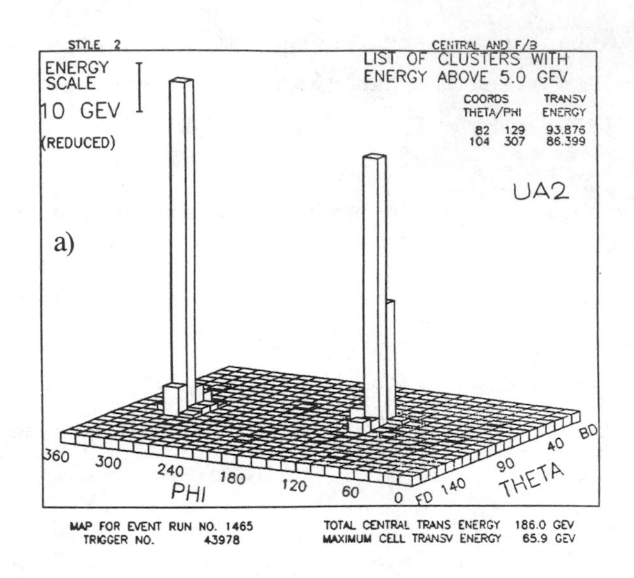

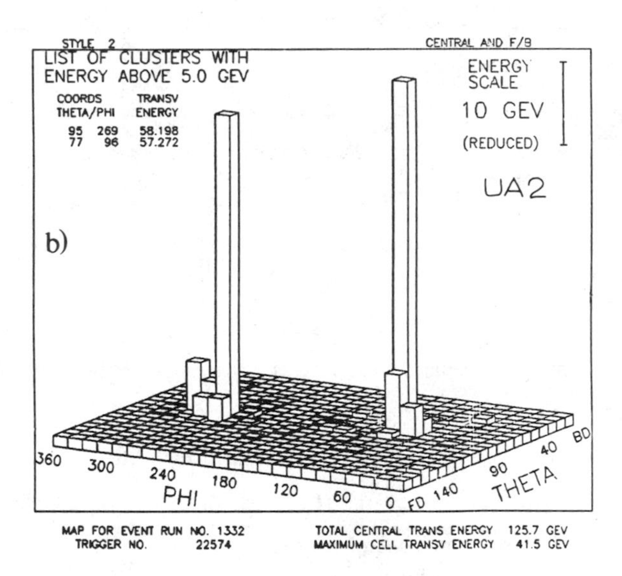

Figure 6 a to d.
Typical θ-ϕ distributions of the transverse energy in large ΣE_T events.

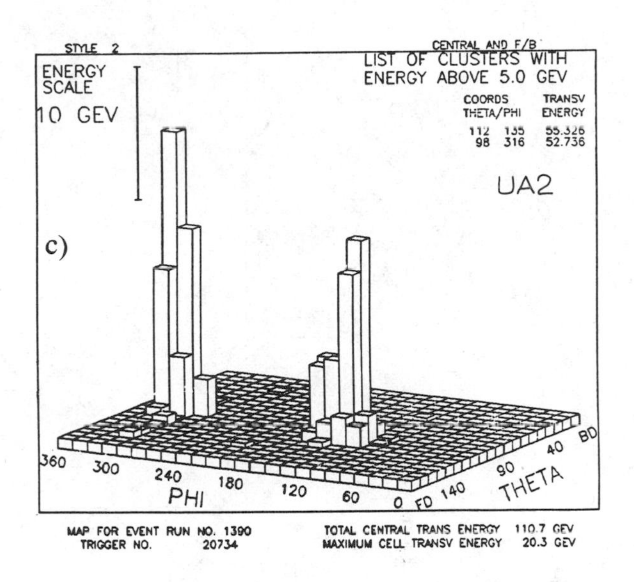

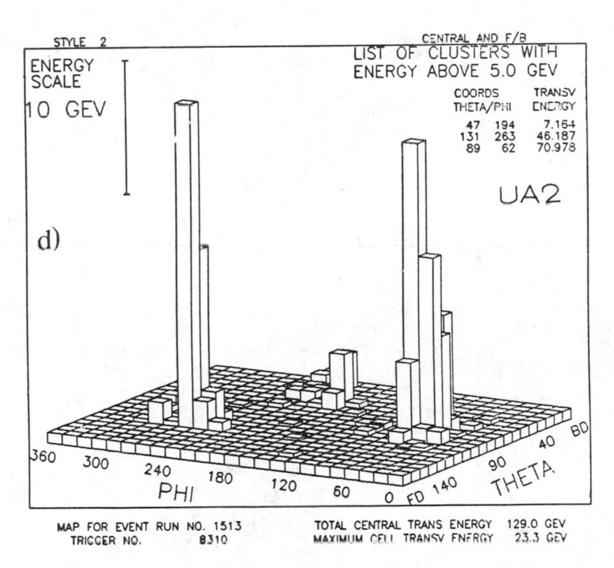

Figure 6 a to d.
Typical θ-ϕ distributions of the transverse energy in large ΣE_T events.

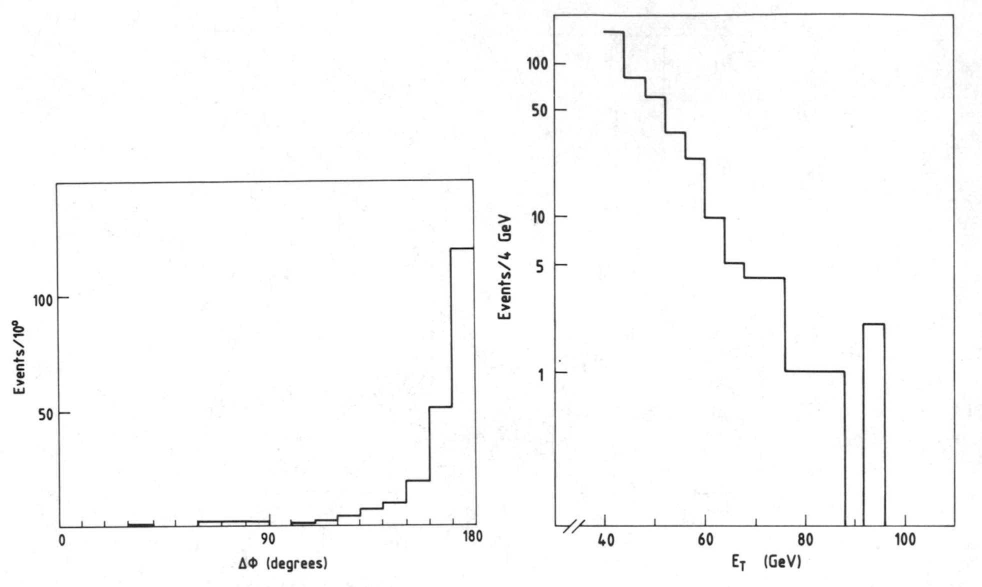

Figure 7. Azimuthal separation $\Delta\phi$ between two jets, each having $E_T > 15$ GeV, in events with $\Sigma E_T > 30$ GeV.

Figure 8. Observed (uncorrected) transverse energy distribution of jets having $E_T > 40$ GeV.

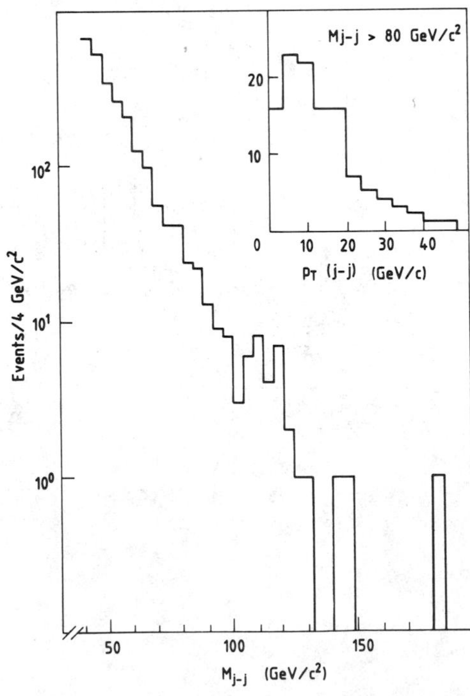

Figure 9. Two-jet invariant mass (M_{j-j}) distribution (uncorrected). The observed (uncorrected) transverse momentum distribution of two-jet systems having $M_{j-j} > 80$ GeV/c^2 is shown in the insert.

However it precludes any search for large transverse momentum electrons among jet fragments.

We shall deal separately with the central and forward regions, in which different experimental methods are used. In both cases the initial event sample is that of W-triggers, which correspond to a total integrated luminosity of 19 nb^{-1} and 16 nb^{-1} respectively.

4.1 - Search for $W \to e\nu$ in the central region.

We first search for energy clusters in the central calorimeter which have a configuration consistent with that expected from an isolated electron : the cluster must be contained in a 2 × 2 cell quartet of the electromagnetic calorimeter, the leakages in the associated hadronic cells (first compartment) and in the adjacent electromagnetic cells must not exceed 10% each. In addition we exclude clusters having their centroid in a cell located at a boundary of the calorimeter acceptance. These straightforward requirements leave only 363 events in which such a cluster has a transverse energy E_T in excess of 15 GeV (including transverse and longitudinal leakages), with only seven events above E_T = 30 GeV. The E_T distribution is shown in Fig. 10a.

This sample is further reduced by simply requiring that only one track measured in the vertex detector points to the energy cluster. If σ_θ and σ_ϕ are the angles between the track and the line joining the event vertex to the cluster centroid, we require that $\sigma^2 = |(\sigma_\theta/10^o)^2 + (\sigma_\phi/15^o)^2| < 1$, a condition always satisfied by single electrons ($< \sigma^2 > \simeq 0.13$, with 95% having $\sigma^2 < 0.4$).

With the further condition that no other charged particle track be present in a cone of 10° half aperture around the candidate track, 96 events survive, with the E_T distribution shown in Fig. 10b.

We then require that the track develops an early shower in the tungsten converter, with an associated charge cluster in the outermost vertex chamber (C5) as expected in the case of electrons. Fig. 11a shows a clear peak in the distribution of d^2, the square of the distance between the track and the closest charge cluster centroid in C5. The charge of the cluster (for $d^2 < 500$ mm^2) is shown in Fig. 11b. The condition $Q_5 < 4$ mip, satisfied by more than 90% of electrons in the energy range from 20 to 60 GeV, reduces the event sample from 96 to 35 (Fig. 10c). Although electrons always satisfy the condition $d^2 < 50$ mm , no cut is applied in order to evaluate the background contamination. The additional requirement that no conversion cluster be present in C5 in the same 10° half-aperture around the track selects isolated electron candidates. The E_T and d^2 distributions of the remaining 10 events are shown in Fig. 10d and 11c, respectively.

As a final selection criterion we use the high segmentation of the central calorimeter to check if the shape of the energy cluster is consistent with that expected from an electron impinging along the direction of the observed track. To this purpose we define a quality fractor f as follows

$$f = \frac{1}{E} \sqrt{\Sigma (E_i - \tilde{E}_i)^2} \qquad (5)$$

where E is the measured cluster energy, E_i is the energy measured in cell i, $\tilde{E}_i$ is the energy predicted from test beam data for cell i under the assumption that the observed charged particle track is an electron of energy E. Fig. 11d shows the f distribution for the sample of 10 events.

Three events appear to satisfy the condition $f < 0.05$. Such condition is satisfied by more than 95% of single isolated electrons, as verified with electron beams from the CERN SPS (see also Fig. 11d).

The d^2 distribution for the three surviving events is shown in Fig. 11e. These events all have $d^2 < 50$ mm^2, as expected for charge clusters in C5 which are associated with a track. The E_T distribution for these events, which represent our final sample of electron candidates, is shown in Fig. 10e.

Backgrounds from standard sources contribute a negligible contamination to this event sample :

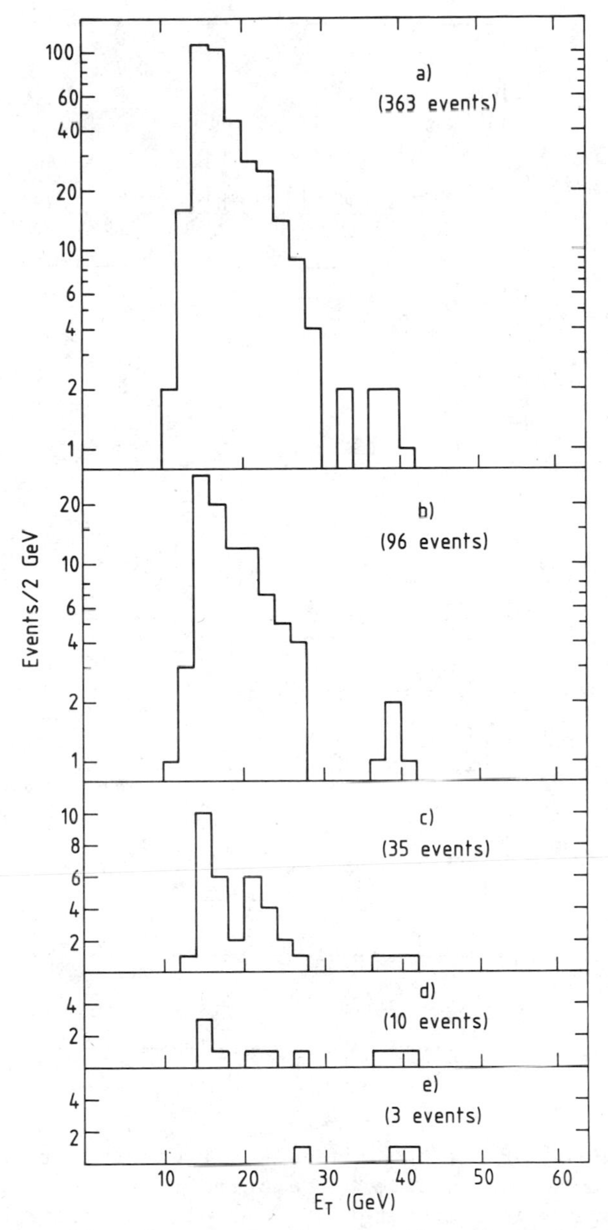

Figure 10. Transverse energy distribution of the event samples in the central calorimeter :

a) After requirements on energy cluster ;
b) After association with a track ;
c) After association of the track with a shower in the tungsten converter ;
d) After requiring only one such shower ;
e) After further cuts on the quality of the track-energy cluster matching. This is the final electron sample.

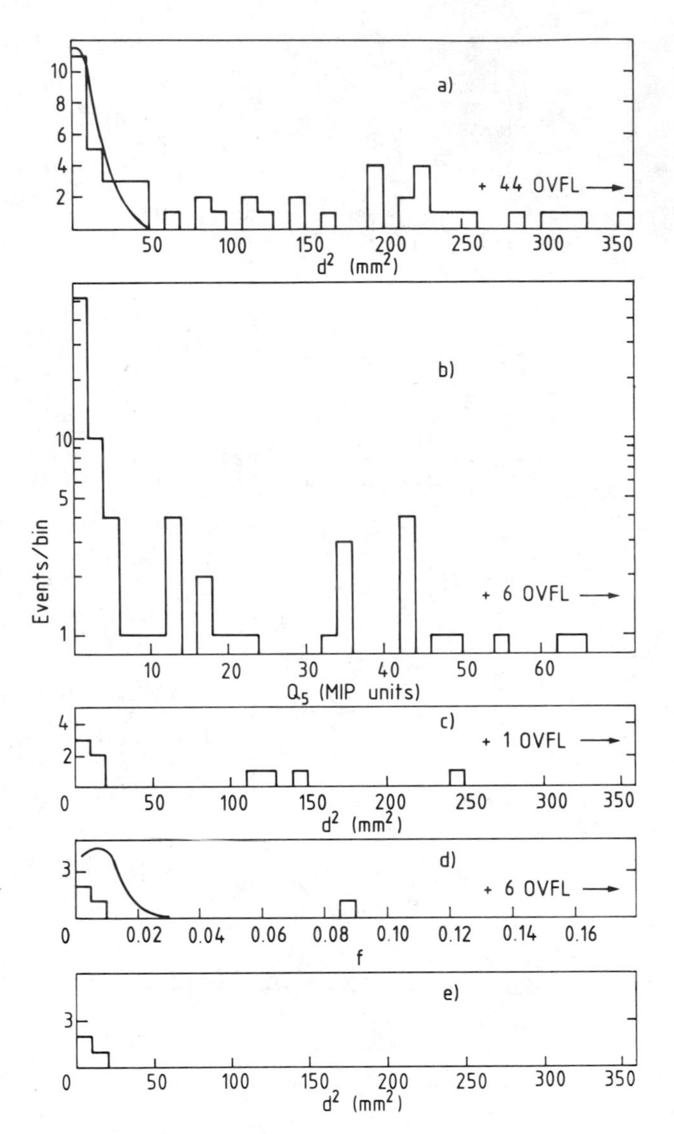

Figure 11.

a) Distribution of d^2, the distance between the track and the shower in the tungsten converter. The curve represents test beam data.

b) Distribution of the charge Q_5 (MIP units), observed in chamber C_5 after the tungsten converter.

c) Distribution of d^2 for the sample of 10 events satisfying the isolation criteria.

d) Distribution of f, the quality parameter for the energy cluster shape, as defined in Eq. (5). The curve represents test beam data.

e) Distribution of d^2 for the three electron candidates.

a) single, isolated high-p_T π^0 mesons undergoing Dalitz decay or high-p_T photons converting in the vacuum pipe. Out of the 363 events of the original sample, only one event above E_T = 25 GeV satisfies the requirements of having no charged track and one C5 cluster with $f < 0.05$. Taking into account the conversion probabilities and the Dalitz decay branching ratio, the background from this source amounts to less than 0.04 events for E_T > 25 GeV.

b) single high-p_T charged hadrons interacting in the converter and depositing a large fraction of the energy in the electromagnetic calorimeter. The measured rejection factor is 1/400, so that 3/400 $\simeq$ 0.01 events are expected for E_T > 25 GeV.

c) overlap between a charged particle and one or more high-p_T photons. In these events the charge cluster in C5 is not correlated to the track, giving rise to a flat background in both d^2 and f variables (see Figs. 11c,d). An extrapolation in the (d^2,f) plane yields a background of approximately 0.1 events for E_T > 25 GeV.

In conclusion the total background contribution to the three electron candidates amounts to less than 0.2 events.

4.2 - Search for $W \rightarrow e\nu$ in the forward regions

The search for $W \rightarrow e\nu$ decays in the forward regions follows the same general guidelines as in the central region. The full data sample recorded corresponds in this case to a total integrated luminosity of 16 nb^{-1}.

The following selection criteria are applied to the sample of W triggers to select possible electron candidates :

- a transverse energy deposition exceeding 15 GeV must be measured in one of the forward sectors. Depending upon the position of the track impact, it must be contained in one or at most two cells. A total of 761 events are selected from the full data sample.

- there must be a track in the spectrometer corresponding to the position of the energy cluster and matching a track in the vertex detector.
- the proportional tubes associated with the selected track must give signals. The quality of the matching between the track, the preshower counter cluster and the energy deposition in the calorimeter must fulfill selection criteria illustrated in Figs. 12a to c.
- the momentum p measured in the magnetic spectrometer must be consistent with the energy E measured in the calorimeter. Specifically we require $\frac{1}{p}$ and $\frac{1}{E}$ to be equal within three standard deviations (Fig. 12d).
- in addition to the above selection criteria we require that the electron candidate is isolated : the energy measured in adjacent calorimeter cells (including the contribution of track momenta) must not exceed 3 GeV, no other track must point to the same cell as the electron candidate and no other signal facing this cell must be measured in the preshower counter.

Four events survive all these conditions. Fig. 13a shows the E_T spectrum of selected clusters ; Fig. 13b shows the distribution of R, the ratio of the energy deposited in the second section of the calorimeter to that deposited in the first section. For comparison the distribution of the fractional leakage in the back calorimeter compartment is shown in Fig. 14.

Out of the four events, three satisfy the electron criterion, $R < 0.02$, whereas one has $R > 0.1$, as expected for hadrons. This is further confirmed by Fig. 13c, which shows the charge distributions of the clusters measured in the MTPCs for the two classes of events separately. The large charges associated with the events having $R < 0.02$ are consistent with an electron behaviour. The transverse energy distribution of the 3 electron candidates is shown in Fig. 13d.

Backgrounds to this sample are expected to originate from the same sources as for the central calorimeter, but for photon

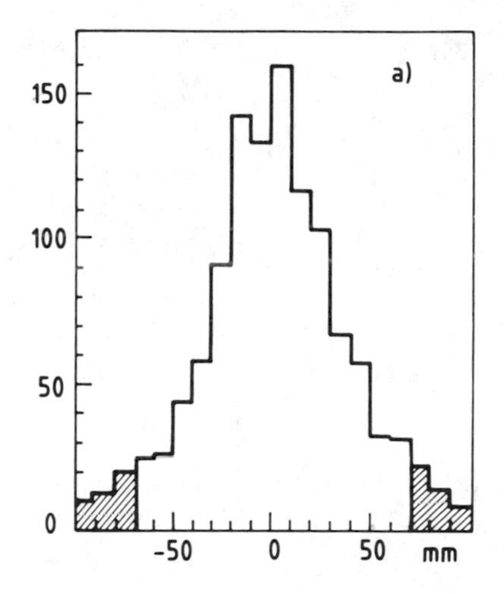

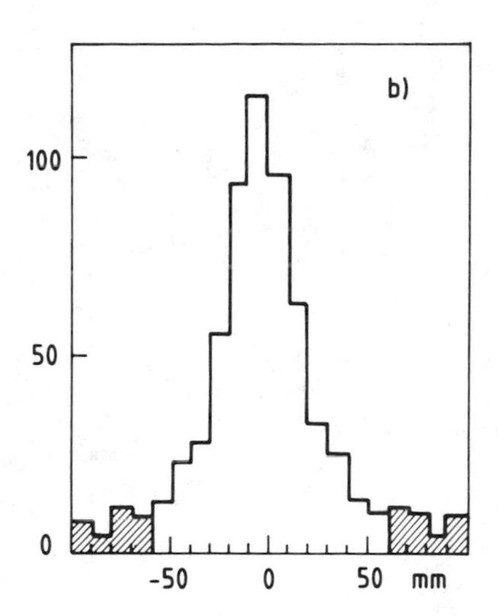

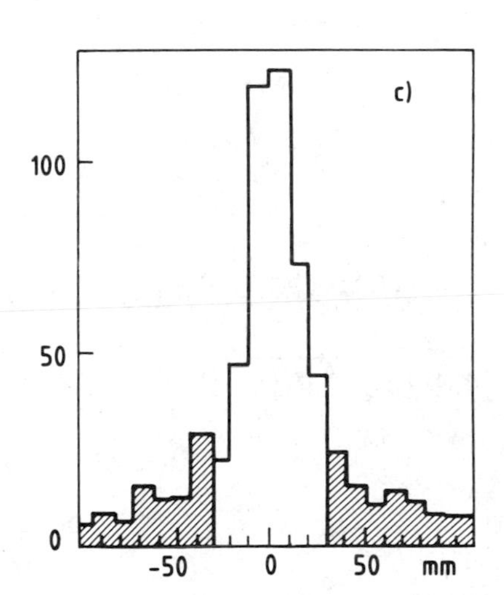

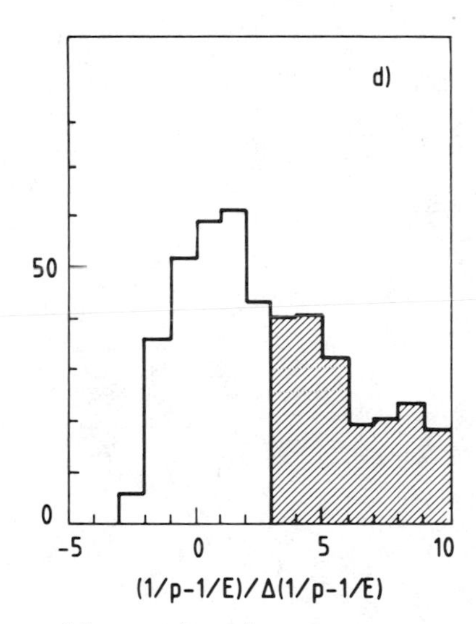

Figure 12. Matching quality criteria.

a) Between the position in the calorimeter (measured from phototube ratio) and the preshower cluster (measured along the magnetic field) ;

b,c) Between the track measured in the drift chambers and the preshower cluster (measured along the magnetic field and normal to it respectively) ;

d) Between the momentum measurement in the drift chambers and the energy measured in the calorimeter for events having $E_T > 10$ GeV.

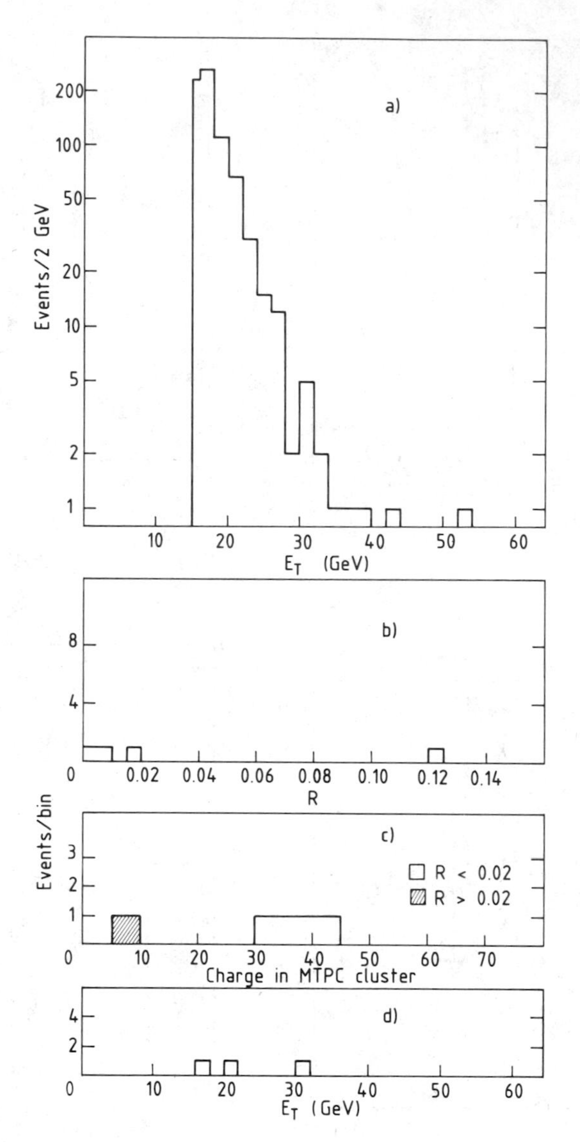

Figure 13.

a) E_T-distribution of the sample of 761 events selected from energy requirements in the forward calorimeters ;
b) Distribution of R, the ratio between the energy deposited in the hadron veto section of the calorimeter and the cluster energy ;
c) Distribution of the charge of the track-associated cluster measured in the MTPCs after the lead-iron converter (MIP units) ;
d) Transverse energy distribution for the final sample of electron candidates.

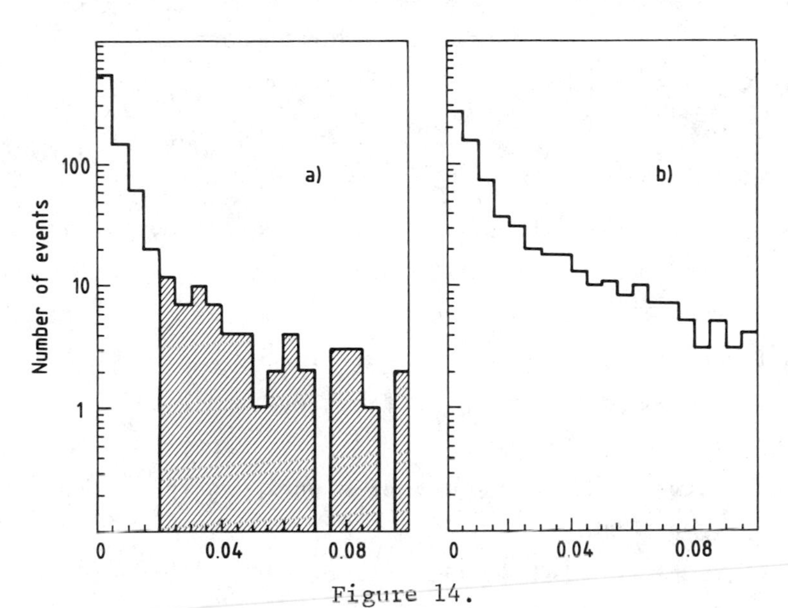

Figure 14.
Leakage into the hadron veto
compartment of the forward calorimeters.

a) For events associated with a signal in the preshower counter ;
b) For events not associated with a signal in the preshower counter.
The cut used in the analysis rejects events in the cross-hatched region ($\sim$ 2%).

conversions which are resolved by the magnetic field. The overall background contamination, estimated from an extrapolation in Fig. 13b, amounts to 0.2 events for $E_T > 15$ GeV.

4.3 - Missing transverse momentum

A distinctive feature of the decay $W \to e\nu$ is given by the presence of an undetected neutrino having transverse momentum roughly balancing that of the electron. In order to detect a possible transverse momentum unbalance we evaluate the quantity

$$p_T^{miss} = \frac{\vec{p}_{tot} \cdot \vec{p}_T^{\,el}}{E_T^{el}} \qquad (6)$$

where $\vec{p}_{tot}$ is given by the vector sum over all momenta (or directed energy vectors) measured by the whole detector.

The ratio p_T^{miss}/E_T is shown in Fig. 15a for the electron candidates found both in the central calorimeter and in the forward detectors. The events with $p_T^{miss}/E_T \simeq 1$ are consistent with reaction (1).

The E_T distribution of the four events with $p_T^{miss}/E_T > 0.8$ is given in Fig. 15b.

Finally, Fig. 15c shows the cell energy distribution in θ and ϕ for the highest E_T event. The only significant energy flow within the detector acceptance is carried by the electron candidate. The other three events have the same spectacular configuration.

5 - SEARCH FOR $Z^o \to e^+e^-$ DECAYS

A search for electron pairs having an invariant mass in excess of 40 GeV/c^2 has been conducted by selecting events of the Z-trigger sample in which two quartets of electromagnetic cells (central and forward regions were considered together) were found to have transverse energies E_T^1 and E_T^2 such that $E_T^1 + E_T^2 > 30$ GeV.

Although the Z-trigger did not require a coincidence with the small angle scintillator arrays, all events of the selected sample

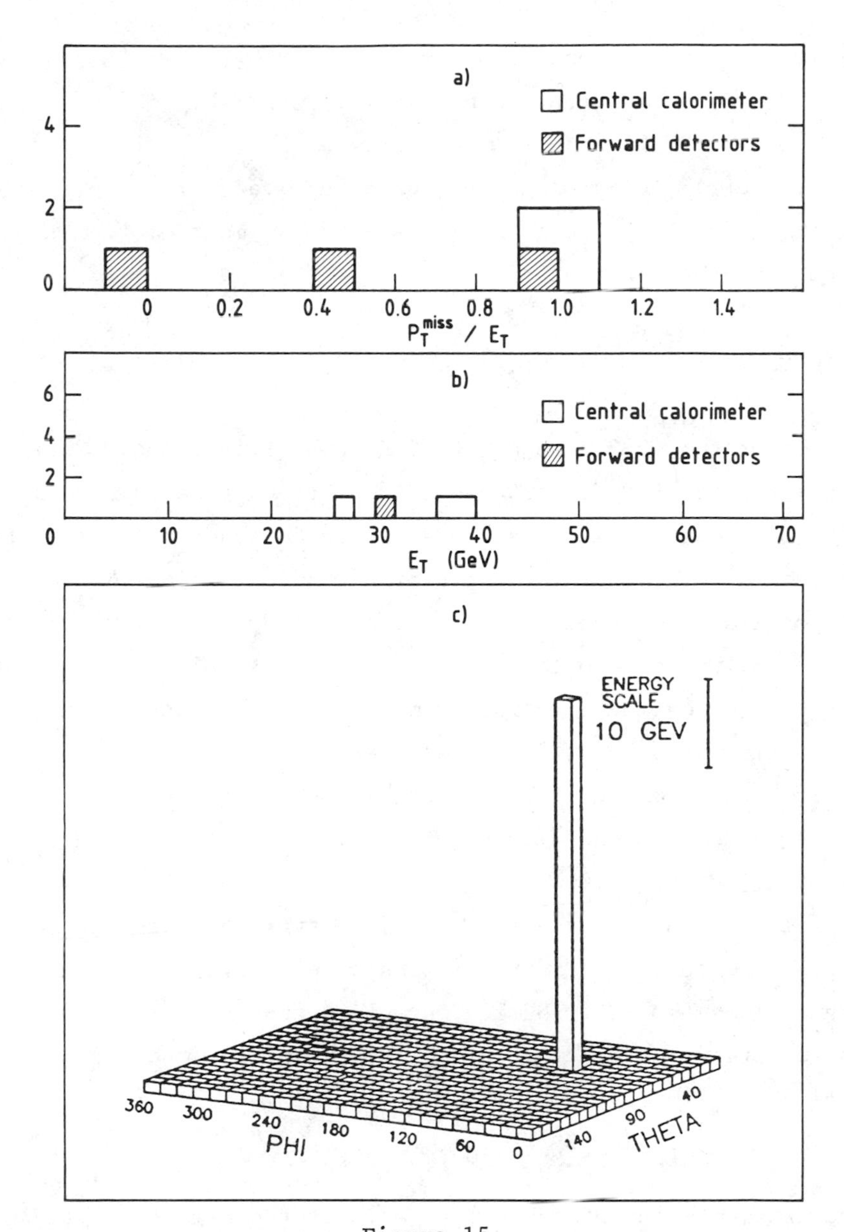

Figure 15.

a) Distribution of p_T^{miss}/E_T, the ratio between the total missing transverse momentum and the transverse energy of the electron candidate ;

b) Transverse energy distribution of the electron candidates observed in events with large missing transverse momentum ($p_T^{miss}/E_T > 0.8$) ;

c) The cell energy distribution as a function of polar angle θ and azimuth ϕ for the electron candidate with the highest E_T value.

in which this coincidence was not satisfied have been observed to result from sources other than $\bar{p}p$ collisions (cosmic rays, beam halo, etc...).

Of the 257 events selected, none survived very simple selection criteria. All were scanned visually and most of them show typical two-jet configurations.

6 - CONCLUSIONS

During the 1982 runs at the CERN $\bar{p}p$ Collider the UA2 detector collected about 400 events having total transverse energy in excess of 40 GeV. These events display a clear dominance of two-jet configuration, with a smooth mass distribution up to about $M_{jj} \simeq 180$ GeV. This confirms the spectacular collimated hadronic structures first observed by the UA2 collaboration[3].

The application of straightforward selection criteria to identify electrons isolates four events with $E_T > 15$ GeV.

Both the event configuration, single electrons with strong transverse momentum unbalance, and their rate of occurrence are consistent with reaction (1).

A fit to the W mass, assuming W decay kinematics and V-A coupling, turns out to be rather insensitive to the particular choice of the W momentum distribution, provided that $p_T(W) \ll M_W c$. Using Gaussian distributions for both $p_L(W)$ and $p_T(W)$, the result of the fit gives

$$M_W = 80^{+10}_{-6} \quad \mathrm{GeV}/c^2$$

where the quoted error takes into account the effect of varying the widths of the Gaussians within the limits of reasonable production models.

ACKNOWLEDGEMENTS

We gratefully acknowledge the remarkable performance of the Staff of the $\bar{p}p$ project and of the operation crews of the relevant CERN accelerators. We thank the UA4 Collaboration who let us use their small angle scintillator hodoscopes.

We are deeply indebted to the technical staffs of the Institutes Collaborating in UA2. Financial support from the Danish Natural Science Research Council to the Niels Bohr Institute group and from the Schweizerischer National fonds zur Forderung der wissenschaftlichen Forschung to the Bern group are acknowledged.

REFERENCES

[1] L.B. Okun' and M.B. Voloshin, Nucl. Phys. B120 (1977) 459.
C. Quigg, Rev. Mod. Phys. 94 (1977) 297.
J. Kogut and J. Shigemitsu, Nucl. Phys. B129 (1977) 461.
R.F. Peierls, T. Trueman and L.L. Wang, Phys. Rev. D16 (1977) 1397.
F.E. Paige, BNL - 27066 (1979).
F. Rapuano, Lett. Nuovo Cim. 26 (1979) 219.
R. Horgan and M. Jacob, CERN 81-04, p. 65 (1981).

[2] The Staff of the CERN proton-antiproton project, Phys. Lett. 107B (1981) 306.

[3] M. Banner et al., Phys. Lett. 118B (1982) 203.

[4] M. Banner et al., Phys. Lett. 115B (1982) 59.
M. Banner et al., Phys. Lett. 121B (1983) 187.
M. Banner et al., Inclusive charged particle production at the CERN $\bar{p}p$ Collider, to be published in Phys. Lett.

[5] A.G. Clark, Proceedings of the Int. Conf. on Inst. for Colliding Beam Phys., SLAC, 1982.
The UA2 Collaboration, First Results from the UA2 experiment, presented at the 2nd Int. Conf. on Physics in Collisions, Stockholm, Sweden, 2-4 June 1982.

[6] R. Battiston et al., Phys. Lett. 117B (1982) 126.

RECENT RESULTS FROM UA1: FUTURE PERSPECTIVES

G.Salvini
Dipartimento di Fisica, Università di Roma, Roma, Italy
INFN - Sezione di Roma, Italy

This talk is mainly devoted to the recent results of the UA1 collaboration, with one experiment which is based on a 4Π solid angle detector placed in the CERN Superprotonsynchrotron used as a Proton-Antiproton Collider.. All the measurements I shall report have been performed at a centre of mass energy of 540 GeV.

I shall proceed in the following order:

1 - A short historical introduction

2 - The UA1 experiment. Experimental disposition and a summary of previous results

3 - The observation of the $W^{\pm}$ boson with the UA1 experiment. Some open experimental and theoretical problems

4 - Future perspectives: objectives and machines.

In Table 1 are reported the authors of the UA1 collaboration who took part in the observation of the $W^{\pm}$ bosons.

Table 1

EXPERIMENTAL OBSERVATION OF ISOLATED LARGE TRANSVERSE ENERGY ELECTRONS WITH ASSOCIATED MISSING ENERGY AT $\sqrt{s}$ = 540 GeV.

UA1 Collaboration, CERN, Geneva, Switzerland.

G.Arnison, A.Astbury, B.Aubert, C.Bacci, G.Bauer, A.Bezaguet, R.Böck, T.J.V.Bowcock, M.Calvetti, T.Carroll, P.Catz, P.Cennini, S.Centro, F.Ceradini, S.Cittolin, D.Cline, C.Cochet, J.Colas, M.Corden, D.Dallman, M.DeBeer, M.Della Negra, M.Demoulin, D.Denegri, A.Di Ciaccio, D.Di Bitonto, L.Dobrzynski, J.D.Dowell, M.Edwards, K.Eggert, E.Eisenhandler, N.Ellis, P.Erhard, H.Faissner, G.Fontaine, R.Frey, R.Frühwirth, J.Garvey, S.Geer, G.Ghesquière, P.Ghez, K.L.Giboni, W.R.Gibson, Y.Giraud-Héraud, A.Givernaud, A.Gonidec, G.Grayer, P.Gutierrez, T.Hansl-Kozanecka, W.J.Haynes, L.O.Hertzberger, C.Hodges, D.Hoffmann, H.Hoffmann, D.J.Holthuizen, R.J.Homer, A.Honna, W.Jank, G.Jorat, P.I.P.Kalmus, V.Karimäki, R.Keeler, I.Kenyon, A.Kernan, R.Kinnunen, H.Kowalski, W.Kozanecki, D.Kryn, F.Lacava, J.-P.Laugier, J.-P.Lees, H.Lehmann, K.Leuchs, A.Lévêque, D.Linglin, E.Locci, M.Loret, J.-J.Malosse, T.Markiewicz, G.Maurin, T.McMahon, J.-P.Mendiburu, M.-N.Minard, M.Moricca, H.Muirhead, F.Muller, A.K.Nandi, L.Naumann, A.Norton, A.Orkin-Lecourtois, L.Paoluzi, G.Petrucci, G.Piano Mortari, M.Pimiä, A.Placci, E.Radermacher, J.Ransdell, H.Reithler, J.-P.Revol, J.Rich, M.Russenbeek, C.Roberts, J.Rohlf, P.Rossi, C.Rubbia, B.Sadoulet, G.Sajot, G.Salvi, G.Salvini, J.Sass, J.Saudraix, A.Savoy-Navarro, D.Schinzel, W.Scott, T.P.Shah, M.Spiro, J.Strauss, K.Sumorok, F.Szoncso, D.Smith, C.Tao, G.Thompson, J.Timmer, E.Tscheslog, J.Tuominiemi, S.Van der Meer, J.-P.Vialle, J.Vrana, V.Vuillemin, H.D.Wahl, P.Watkins, J.Wilson, Y.G.Xie, M.Yvert, and E.Zurfluh.

Aachen - Annecy(LAPP) - Birmingham - CERN - Helsinki - Queen Mary College, London - Paris(College de France) - Riverside - Rome University and INFN - Rutherford Appleton Lab. - Saclay(CEN) - Vienna Collaboration.

1 - A SHORT HISTORICAL INTRODUCTION -

1.1 - Some important steps forward -

The experiments at CERN with the SPS used as a proton antiproton collider at 540 GeV (experiment UA1, UA2) have their main justification in the search for the famous intermediate vector boson (IVB) of the so-called Standard Theory. They are in a sense the second important step in the physics of matter colliding with antimatter, which always had Europe among the protagonists.

The first act was the physics of e^+e^- colliders. Let me go through some main historical steps.

- It is in the Institute of Rome, that Enrico Fermi in 1933 (half a century ago) wrote an historical work of a few pages: The first theory of weak interactions.[1]

- In March 1960 Bruno Touschek, professor in the University of Rome, proposed at Frascati[2] the first e^+e^- ring ADA (Anello di Accumulazione). (I was then director of Frascati, and I am proud for having favoured and encouraged as much as possible the starting of this development). Physicists and technicians of the Laboratories of Frascati started immediately for a realization which is one of the fastest in the history of machines: ADA was ready in February 1961.

- 1961-1963 are the beautiful years with ADA in Orsay and the intense Italo-French collaboration[3].

- Preparation of ACO starts in 1964. In 1966 the physics of ACO and the companion Russian VEP begin.[3] The ρ , ω, ϕ Vector Mesons, on the theoretical line of the still partially true Vector Dominance, were thoroughly studied (it is the occasion here to remember our unforgettable inspirer J. Sakurai)[(4)].

- Construction of ADONE, the e^+e^- 3 GeV colliding ring, started in 1963. ADONE was ready in 1969.

- 1970: the physics of ADONE begins. The expected electromagnetic desert[4] is filled with hadronic trees. It is the multihadronic production, in part expected after the results from SLAC[5]. Unfortunately nature was not generous to us: our maximum highest programmed energy was 3 GeV c.m., but J/ψ has a mass of 3070 $\pm$ .1 MeV/c^2 and the τ lepton has a mass of 1784 $\pm$ 3 MeV/c^2.

- 1973 and after: it is the physics of CEA, SPEAR, DORIS, PETRA, PEP, i.e. the beginning of the revolution of our times, and I have not to recall it here. Let me only recall that December 1974 issue of Phys. Rev. Letters, announcing J/ψ from more than two sources, with the Editorial hailing the newcomer[6].

But the second act of matter-antimatter (from e^+e^- to $p\bar{p}$ colliders) had already started during the sixties.

- In 1966 G.I.Budker[7] proposed electron cooling, a method to collimate antiprotons in order to realize proton-antiproton collisions. I remember the interest for his proposal (I first heard it at Saclay, on September 1966), in the eyes of the ADA builders and in me. In the same period Budker exposed his very advanced ideas during a visit in Frascati.

- 1966-68: the stochastic cooling is proposed by S.Van der Meer[8]. Both methods for cooling antiprotons became successful in these last ten years, as we know.

- 1976: We arrive at the recent times of proton-antiproton with the proposal of C.Rubbia, D.Cline, P.McIntyre[9] to use big SPS machines as colliders.

- In 1977 a complete project started at CERN to use CERN SPS as a collider. This has been an operation of many engineers, physicists and technicians, with C.Rubbia and S.Van der Meer among the protagonists.

- Preparation of four experiments, UA1, UA2, UA4, UA5, started in 1978, and their basic structure and role shall be recalled at this meeting, together with the recent results.

- In the middle of 1981 we had the first indication that proton antiproton was not a dream: the p-$\bar{p}$ interactions were clearly coming and well separated from the beam gas interactions.

1.2 - Beginning of the experiments -

In the last part of 1981 we had the first proton-antiproton events, with a still low luminosity ($\sim 10^{26}$ cm^{-2} s^{-1}); but this was already enough to get some hundred thousands events, and to publish altogether more than 10 original scientific papers, between UA1, UA2, UA4, UA5.[10] It is amusing to recall that the huge experiments we are going to describe pretended, at the beginning, to be rather small. The first sketched UA1 project for the study of $p\bar{p}$ interactions and for the production of W, Z was considering the possibility of keeping the experiment within the maximum dimensions of the already existing tunnel of the SPS, by using a dense high field superconducting magnet, and a condensed Tungsten or Uranium calorimeter.

The next was an enlargement of the tunnel up to a width of 8 meters, but it was clearly too small. Then Giorgio Brianti offered us 12 meters, and we found it slightly insufficient. Then he arrived at 20 m, not one cm more! O.K., Giorgio! We shall wallow in it! Go and see, now! We are using all this space, and not many centimeters are left.

But let's go now to the final UA1 experiment, reasons, experimental dispositions, previous and recent results.

2 - THE UA1 EXPERIMENT. DISPOSITION, SUMMARY OF PREVIOUS RESULTS

The ambition of UA1 is to succeed to study the interactions

(1) $p + \bar{p} \rightarrow$ anything at 540 GeV c.m.

all over the 4Π solid angle, with the maximum of information on the energy and momentum of the emitted particles, both charged (by wire chambers and calorimeters when possible) and neutral (by calorimeters when possible).

Our experimental disposition is rather obviously the same for any possible interactions (1), including IVB research. It is given in fig.1, and we shortly recall it in the following.[11]

The basic structure is a magnetic dipole whose free space is 3x3x6 cubic meters with an horizontal field B of .7 tesla. The free space is filled with the central detector, i.e. an ensemble of drift chambers whose electronics can reconstruct the charged tracks of the events, the vertex and the impact point on the surrounding calorimeters, yielding a bubble-chamber quality picture.

Momentum precision for high-momentum particles is dominated by a localization error inherent to the system ($\leqslant$ 100 μm) and the diffusion of electrons drifting in the gas. This results in a typical relative accuracy of ±20% for 1 m long track at p = 40 GeV/c, and in the plane normal to the magnetic field. The precision, of course, improves considerably for longer tracks.

The central section of electromagnetic and hadronic calorimetry (see fig.1) has been used in the present investigation to identify electrons over a pseudorapidity interval $|\eta| < 3$ with

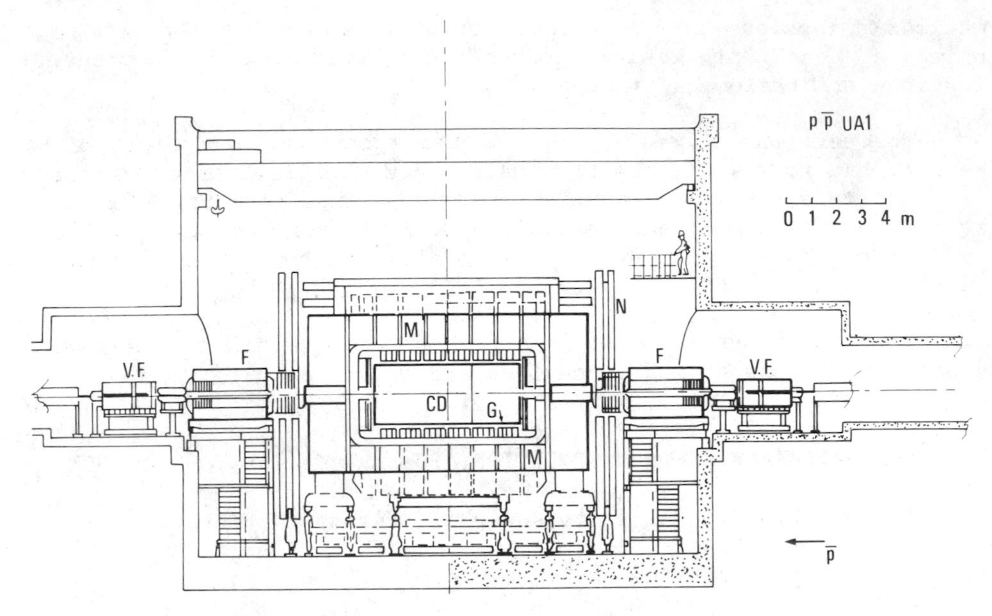

Fig.1 The UA1 detector: schematic cross section in the vertical plane containing the beam. M, Magnet. CD, Central Detector. G., Gondolas. F, Compensating Magnets and Forward Calorimeters. VF, Very Forward Calorimeters; N, Muon Detectors.

full azimuthal coverage. Additional calorimetry, both electromagnetic and hadronic, extends to the forward regions of the experiment, down to 0.2°.

The central electromagnetic calorimeters consist of two different parts:

i) 48 semicylindrical modules of alternate layers of scintillator and lead (gondolas), arranged in two cylindrical half-shells, one on either side of the beam axis with an inner radius of 1.36 m. Each module extends over approximately 180° in azimuth and measures 22.5 cm in the beam direction.

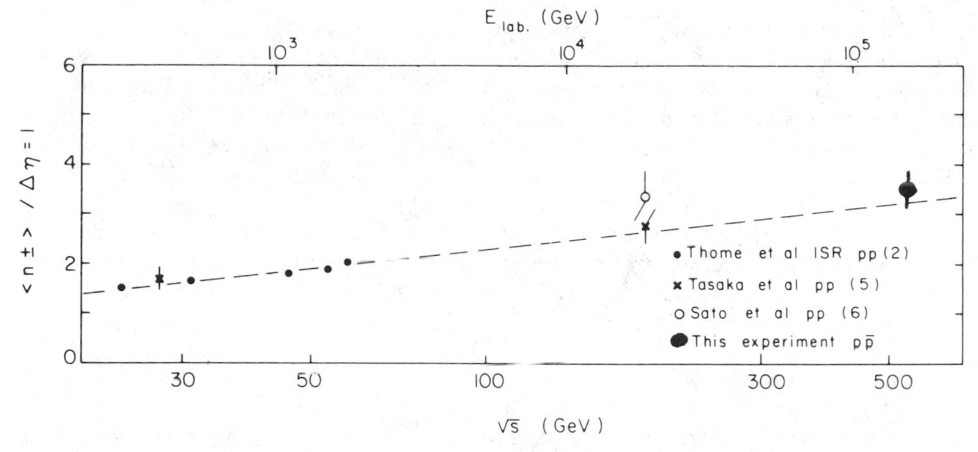

Fig.2 The mean charged particle multiplicity per unit of η for $|\eta| < 1.5$ as a function of center-of-mass energy for events with at least one charged particle in the fiducial region. The line is the linear fit of Thomé et al. to their data (Nucl.Phys. B129:365 (1977).

ii) 64 petals of end-cap electromagnetic shower counters (bouchons), segmented four times in depth, on both sides of the central detector at 3 m distance from the beam crossing point. The position of each shower is measured with a position detector located inside the calorimeter at a depth of 11 radiation lenghts, i.e. after the first two segments. It consists of two planes of orthogonal proportional tubes of 2x2 cm^2 cross-section and it locates the centre of gravity of energetic electromagnetic showers to ± 2 mm in space. The attenuation length of the scintillator has been chosen to match the variation of sin θ over the radius of the calorimeters so as to directly measure in first approximation $E_T = E \sin \theta$ rather than the true energy deposition E, which can be determined later, using the position detector.

Electromagnetic showers are identified by their characteristic transition curve, and in particular by the lack of penetration in the hadron calorimeter behind them.

The emission of one (or more) neutrinos can be signalled only by an apparent visible energy imbalance of the event (missing energy).

We shall report later (§ 3) the results on the $W^{\pm}$ Intermediate Vector Boson, which only could be achieved with an integral luminosity L larger than 10^{34} cm^{-2}. But a number of important results had been obtained already since 1981, with a total luminosity L dt $\sim 10^{32}$ cm^{-2}. We briefly recall them in the following:

- Angular distribution of produced charged particles.[12] One convenient quantity to express this distribution is to give the number m of particle per unit of pseudorapidity η , being $\eta = -\ln \tan \theta/2$. In fact one should foresee m~const for $-2 < \eta < 2$ ($\sim 15 < \theta < 165°$). UA1 has found $m = 3.3 \pm 0.2$. (see fig. 2).

- Transverse momentum spectra for charged particles.[13] The results (see fig.3) have been compared with data at ISR energies and with the predictions of a QCD model. The charged particle spectrum shows a clear dependence on charged track multiplicity.

- Comparison with cosmic ray results. A search for events having the characteristics of cosmic ray Centauros[14] has been made, using information on charged particle multiplicities and transverse momenta from our central detector image chamber, together with energy deposition in our calorimeters. No such events have been found in 48.000 low bias events.

- Production of jets.[15] UA1 collaboration has measured the total transverse energy spectrum up to E_T = 130 GeV in the pseudorapidity range $|\eta| > 1.5$. Using two different algorithms, we have looked for localized depositions of transverse energy (jets). For $\Sigma E_T > 40$ GeV, the fraction of events with two jets increases with ΣE_T; this event structure is dominant for $\Sigma E_T > 100$ GeV. We have measured the inclusive jet cross section up to E_T (jet) = 60 GeV, and the two-jet mass distribution up to 120 GeV/c^2. The measured cross sections are compatible with the predictions of hard scattering models based on QCD. (See fig. 4).

It is clear that the jet structure (which was observed first by UA2)[16] may be the fact which promises the proton antiproton collider a brilliant future among the machines of the future: this shall be discussed again in § 4.

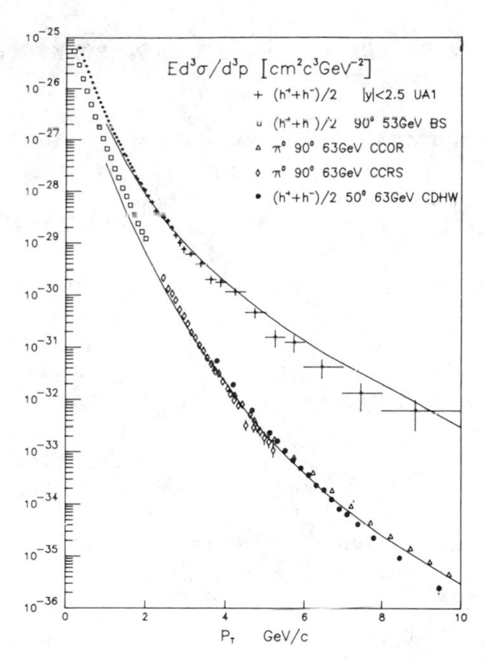

Fig.3 The invariant cross-sections as a function of transverse momentum for: UA1 data at $\sqrt{s}$ = 540 GeV for charged hadrons in the rapidity interval $|\eta|$ <2.5. ISR data at $\sqrt{s}$ = 63 GeV for π^0 at 90° and charged hadrons at 90° and 50°. QCD predictions at $\sqrt{s}$ = 540 GeV and $\sqrt{s}$ = 63 GeV for π^0 at 90°.

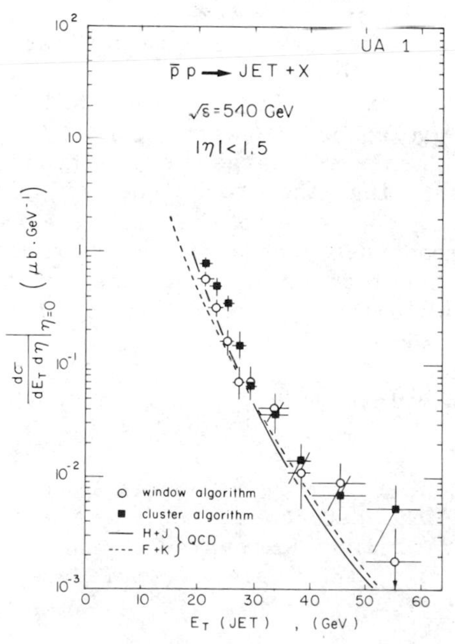

Fig.4 Differential jet cross section $(d\sigma / dE_T d\eta)\ |\eta=o$ for jets found by the window algorithm (open circles) and the cluster algorithm (black squares). See ref. 15.

3 - THE OBSERVATION OF THE $W^{\pm}$ IVB AT UA1. SOME OPEN EXPERIMENTAL AND THEORETICAL PROBLEMS -

The results reported here have been published by the authors reported in Table I.

It is generally postulated that the beta decay, namely (quark) $\to$ (quark) + $e^{\pm}$ + ν is mediated by one of two charged Intermediate Vector Bosons (IVBs), W^{+} and W^{-} of very large masses.

Properties of IVBs become better specified within the theoretical frame of the unified weak and electromagnetic theory and of the Weinberg-Salam model.[17] The mass of the IVB is precisely predicted[18]:

$$M_{W^{\pm}} = 82 \pm 2.4 \text{ GeV}/c^2$$

for the presently preferred[19] experimental value of the Weinberg angle $\sin^2 \theta_W = 0.23 \pm 0.01$. The cross-section for production is also reasonably well anticipated (see fig. 5).

$$\sigma\,(p\bar{p} \to W^{\pm} \to e^{\pm} + \nu) \cong 0.4 \times 10^{-33} k \text{ cm}^2,$$

where k is an enhancement factor of ~1.5, which can be related to a similar well-known effect in the Drell-Yan production of lepton pairs. It arises from additional QCD diagrams in the production reaction with emission of gluons. In our search we have reduced the value of k by accepting only those events which show no evidence for associated jet structure in the detector.

We report now the results of two searches made by UA1 collaboration on data recorded at the CERN SPS proton-antiproton collider: one for isolated large-E_T electrons, the other for large-E_T neutrinos using the technique of missing transverse energy. Both searches converge to the same events, which have the signature of a two-body decay of a particle of mass ~80 GeV/c^2. As we shall see, the topology as well as the number of events fits well the hypothesis that they are produced by the process $\bar{p} + p \to W^{\pm} + X$, with $W^{\pm} \to e^{\pm} + \nu$.

3.1 - Electron identification -

The performance of the detectors with respect to hadrons and electrons has been studied extensively in a test beam as a function of the energy, the angle of incidence, and the location of impact. The fraction

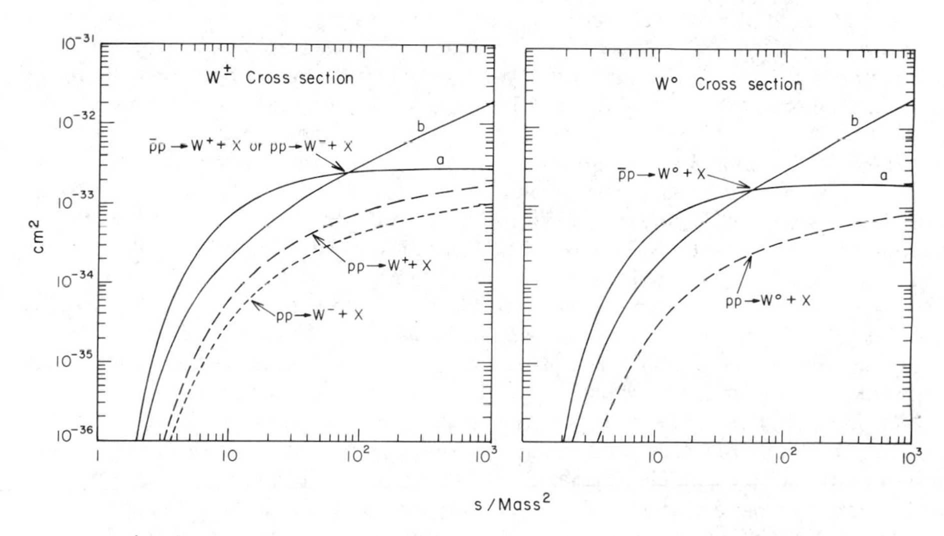

Fig.5 $W^{\pm}$ and Z° production cross section in $p\bar{p}$ collisions from quark parton model as a function of the mass M. $\sqrt{s}$ is the center of mass energy. Curve a) assumes validity of scaling[11]; curve b) is non scaling, i.e. keeping into account the Q^2 dependence of the structure functions as in J.F.Owens and E.Reya, Phys.Rev. D17:3003 (1978).

of hadrons (pions) delivering an energy deposition E_c below a given threshold in the hadron calorimeter is a rapidly falling function of energy, amounting to about 0.3% for $p \simeq 40$ GeV/c and $E_c < 200$ MeV. Under these conditions, 98% of electrons are detected.

3.2 - Neutrino identification -

The emission of one (or more) neutrinos can be signalled only by an apparent visible energy imbalance.

In order to permit such a measurement, calorimeters have been made completely hermetic down to angles of 0.2° with respect to the direction of the beams. (In practice, 97% of the mass of the magnet is calorimetrized). It is possible to define an energy flow vector $\overrightarrow{\Delta E}$, adding vectorially the observed energy depositions over the whole solid angle. Neglecting particle masses and with an ideal calorimeter response and solid-angle coverage, momentum conservation requires $\overrightarrow{\Delta E} = 0$. We have tested this technique on minimum bias and jet-enriched events for which neutrino emission ordinarily does not occur. The transverse components ΔE_y and ΔE_z exhibit small residuals centred on zero with an r.m.s. deviation well described by the law $\Delta E_{y,z} = 0.4 \ (\Sigma_i |E_T^i|)^{1/2}$, where all units are in GeV and the quantity under the square root is the scalar sum of all transverse energy contributions recorded in the event. The distributions have Gaussian shape and no prominent tails. The longitudinal component of energy ΔE_x is affected by the energy flow escaping through the 0° singularity of the collider's beam pipe and it cannot be of much practical use. We remark that, like neutrinos, high-energy muons easily penetrate the calorimeter and leak out substantial amounts of energy. A muon detector, consisting of stacks of eight planes of drift chambers, surrounds the whole apparatus and has been used to identify such processes, which are occurring at the level of 1 event per nanobarn for $\Delta E_{y,z} \geqslant 10$ GeV.

3.3 - Data-taking and initial event selections -

The present work is based on data recorded in a 30-day period during November and December 1982. The integrated luminosity after subtraction of dead-time and other instrumental inefficiencies was 18 nb^{-1}, corresponding to about 10^9 collisions between protons and antiprotons at $\sqrt{s} = 540$ GeV.

For each beam-beam collision detected by scintillator hodoscopes, the energy depositions in all calorimeter cells after fast digitization were processed, in the time prior to the occurrence of the next beam-beam crossing, by a fast arithmetic processor in order to recognize the presence of a localized electromagnetic energy deposition, namely of at least 10 GeV of transverse energy either in two gondola elements or in two bouchon petals. In addition, we have simultaneously operated three other trigger conditions: i) a jet trigger, with $\gtrsim$ 15 GeV of transverse energy in a localized cluster of electromagnetic and hadron calorimeters; ii) a global E_T trigger, with $>$ 40 GeV of total transverse energy from all calorimeters with $|\eta| < 1.4$; and iii) a muon trigger, namely at least one penetrating track with $|\eta| < 1.3$ pointing to the diamond.

In total, 9.75×10^5 triggers were collected, of which 1.4×10^5 were characterized by an electron trigger flag.

Event filtering by calorimetric information was further perfected by off-line selection of 28,000 events with $E_T > 15$ GeV in two gondolas, or $E_T > 15$ GeV in two bouchon petals with valid position-detector information. These events were finally processed with the central detector reconstruction. Of these events there are 2125 with a good quality, vertex associated charged track of $p_T > 7$ GeV/c.

3.4 - Search for electron candidates -

We now require three conditions in succession in order to ensure that the track is isolated, namely to reject the debris of jets:

i) The fast track ($p_T > 7$ GeV/c) as recorded by the central detector must hit a pair of adjacent gondolas with transverse energy $E_T > 15$ GeV (1106 events).

ii) Other charged tracks, entering the same pair of gondolas, must not add up to more than 2 GeV/c of transverse momenta (276 events).

iii) The ϕ information from pulse divison from gondola phototubes must agree within 3 σ with the impact of the track (167 events).

Next we introduce two simple conditions to enhance its electromagnetic nature:

iv) The energy deposition E_c in the hadronic calorimeters aimed at by the track must not exceed 600 MeV (72 events).

v) The energy deposited in the gondolas E_{gon} must match the measurement of the momentum of the track p_{CD}, namely $|1/p_{CD} - 1/E_{gon}| < 3\ \sigma$.

At this point only 39 events are left, which were individually examined by physicists on the visual scanning and interactive facility Megatek. The surviving events break up cleanly into three classes, namely 5 events with no jet activity, 11 with a jet opposite to the track within a 30° angle in ϕ , and 23 with two jets (one of which contains the electron candidate) or clear e^+e^- conversion pair. A similar analysis performed on the bouchon has led to another event with no jets. The classes of events have striking differences. We find that whilst events with jet activity have essentially no missing energy (fig. 6), the ones with no jets show evidence of a missing transverse energy of the same magnitude as the transverse electron energy (fig. 7a), with the vector momenta almost exactly balanced back-to-back (fig. 8). In order to assess how significant the effect is, we proceed to an alternative analysis based exclusively on the presence of missing transverse energy.

3.5 –Search for events with energetic neutrinos –

We start again with the initial sample of 2125 events with a charged track of $p_T > 7$ GeV/c. We now move to pick up validated events with a high missing transverse energy and with the candidate track not part of a jet:

i) The track must point to a pair of gondolas with deposition in excess of $E_T > 15$ GeV and no other track with $p_T > 2$ GeV/c in a 20° cone (911 events).

ii) Missing transverse energy imbalance in excess of 15 GeV.

Only 70 events survive these simple cuts, as shown in fig. 9. The previously found 5 jetless events of the gondolas are clearly visible. At this point, as for the electron analysis, we process the events at the interactive facility Megatek:

iii) The missing transverse energy is validated, removing those events in which jets are pointing to where the detector response is limited, i.e. corners, light-pipe ducts going up and down. Some very evident, big secondary interactions in the beam pipe are also removed. We are left with 31 events, of which 21 have $E_c > 0.01\ E_{gon}$ and 10 events in which $E_c < 0.01\ E_{gon}$.

iv) We require that the candidate track be well isolated, and there is no track with $p_T > 1.5$ GeV in a cone of 30°, and that $E_T < 4$ GeV for neutrals in neighbouring gondolas at similar ϕ angle. Eighteen events survive: ten with $E_c \neq 0$ and eight with $E_c = 0$.

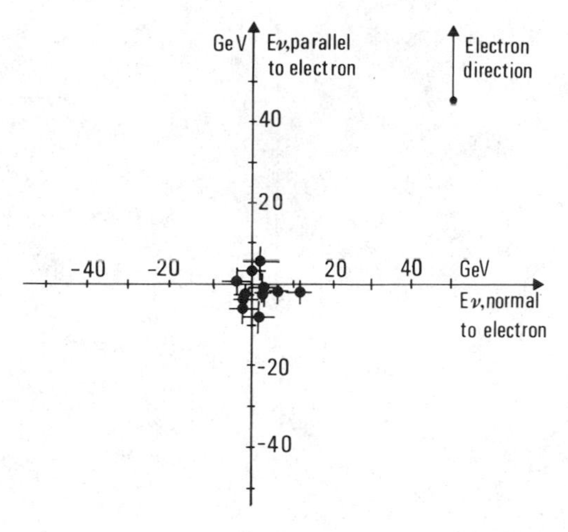

Fig.6 The missing transverse energy (E_{ν}) is plotted vectorially against the electron direction for the events yelded by the electron search. Those reported here are the events with jets.

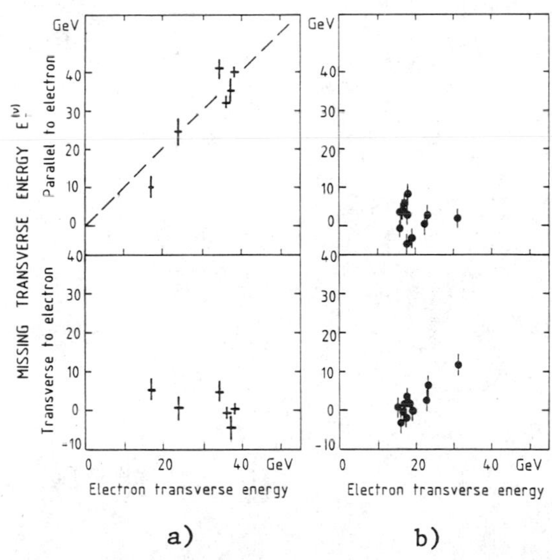

Fig.7a The missing transverse energy is of the same magnitude than the electron transverse energy in the events without jets. This is not the case for events with jets (Fig.7b).

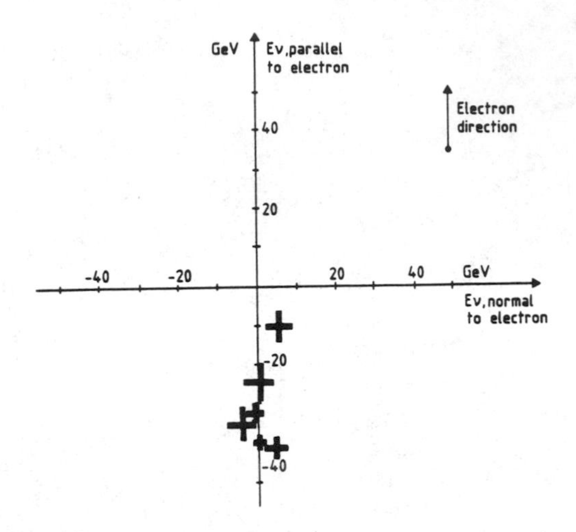

Fig.8 As in Fig.6, but those reported here are the events without jets.

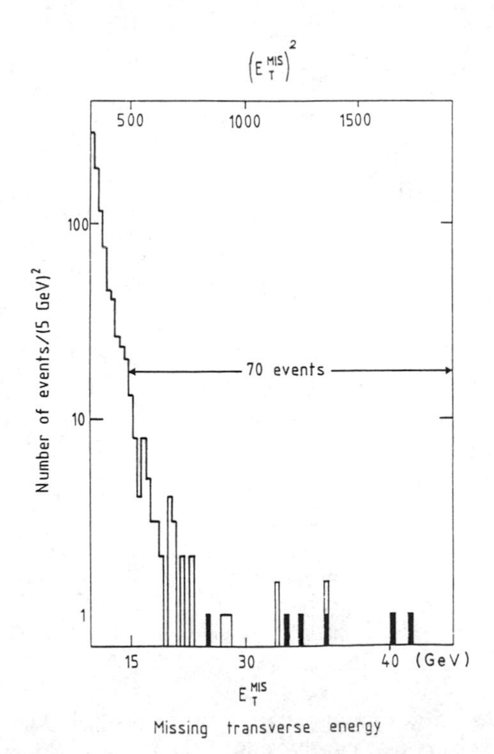

Fig.9 Distribution of the square of the missing transverse energy for events surviving the cuts requiring association of the central detector isolated track and a struck gondola in the missing energy search.

The events once again divide naturally into the two classes: 11 events with jet activity in the azimuth opposite to the track, and 7 events without detectable jet structure. If we now examine E_c, we see that these two classes are strikingly different, with large E_c for the events with jets (fig. 10) and negligible E_c for the jetless ones (fig 11). We conclude that whilst the first ones are most likely to be hadrons, the latter constitute an electron sample.

We now compare the present result with the candidates of the previous analysis based on electron signature. We remark that five out of the seven events constitute the previous sample (fig. 11). Two new events have been added, eliminated previously by the test on energy matching between the central detector and the gondolas. Clearly the same physical process that provided us with the large-p_T electron delivers also high-energy neutrinos. The selectivity of our apparatus is sufficient to isolate such a process from either its electron or its neutrino features individually. If (ν_e, e) pairs and (ν_τ, τ) pairs are both produced at comparable rates, the two additional new events can readily be explained since missing energy rise equally well from ν_e and ν_τ. Indeed, closer inspection of these events shows them to be compatible with the hypothesis, for instance, $\tau^- \to \pi^- \pi^0 \nu_\tau$ with leading π^0. However, our isolation requirements on the charged track strongly biases against most of the τ decay modes.

3.6 - Detailed description of the electron-neutrino events -

The main properties of the final sample of six events (five gondolas, one bouchon) are given in table 2 and marked A through F. Another event, G, is probably a τ candidate.[2,3] One can remark that both charges of the electrons are represented. The successive energy depositions in the gondola samples are consistent with test beam findings. All but event D have no energy deposition in the hadron calorimeter; event D has a 400 MeV visible, 1% energy leakage beyond 25 radiation lengths. Test beam measurements show that this is a possible fluctuation. Multiplicity of the events is widely different: event F (fig. 12) has a small charged multiplicity (14), whilst event A (fig. 13) is very rich in particles. Event B is the bouchon event, and it has a number of features which must be mentioned. A 100 MeV/c track emerges from the vacuum chamber near the exit point of the electron track, which might form a part of an asymmetric electron pair with the candidate. The initial angle between the two tracks would then be 11°, not incompatible with this hypothesis once Coulomb scattering and measurement errors of the two tracks are taken into account. There is also some activity in the muon detector opposite to the electron candidate; the muon track is unmeasurable in the central

Table 2

MAIN PARAMETERS OF ELECTRON EVENTS WITH A LARGE MISSING TRANSVERSE ENERGY

Run, event	Properties of the electron track										Calorimeter information					General event topology				
											Electromagnetic energy deposition									
	E_T (GeV)	E (GeV)	p (GeV/c)	Δp a)	Q	dE/dx I/I_0	y b)	Track No.	Length (m)	Sagitta (mm)	Sample 1 (GeV)	Sample 2 (GeV)	Sample 3 (GeV)	Sample 4 (GeV)	E_{had} (GeV)	E_{tot} (GeV)	Missing E_T (GeV)	Δφ c) (deg.)	Charged tracks	$\sum\lvert E_T\rvert$ (GeV)
A 2958 1279	24	39	33.8	+6.3 -4.6	-	1.22 ±0.2	+1.1	36	1.36	1.7	3	34	2	0.2	0	278	24.4 ± 4.6	179	65	81
B 3522 214	17	46	47.5	+8.2 -6.1	-	1.37 ±0.16	+1.7	18	1.64	1.5	2	32	10	0.5	0	296	11.6 ± 4.0	219	49	60
C 3524 197	34	45	21.6	+21.8 -7.2	-	1.37 ±0.3	-0.8	26	1.25	2.11	1	30	14	0.2	0	367	41.3 ± 3.6	187	21	68
D 3610 760	38	40	33.4	+33.0 -11.1	-	1.64 ±0.34	+0.3	9	0.98	0.75	3	9	26	2.2	0.4	111	40.0 ± 2.0	181	10	47
E 3701 305	37	37	56.2	+121.3 -22.8	+	1.54 ±0.28	-0.1	12	0.95	0.4	1	18	17	0.9	0	363	35.5 ± 4.3	173	39	87
F 4017 838	36	70	53.1	+6.6 -5.3	-	1.30 ±0.26	+1.4	3	2.01	2.0	19	48	3	0.3	0	177	32.3 ± 2.4	179	14	49
G 3262 1108	40	40	6.7	+1.9 -1.2	-	1.23 ±0.28	0.0	21	0.85	3.0	2	22	15	0.9	0	218	33.4 ±2.9	172	21	63

a) Including 200 μm systematic error.

b) y is defined as positive in the direction of outgoing $\bar{p}$.

c) Angle between electron and missing energy (neutrino).

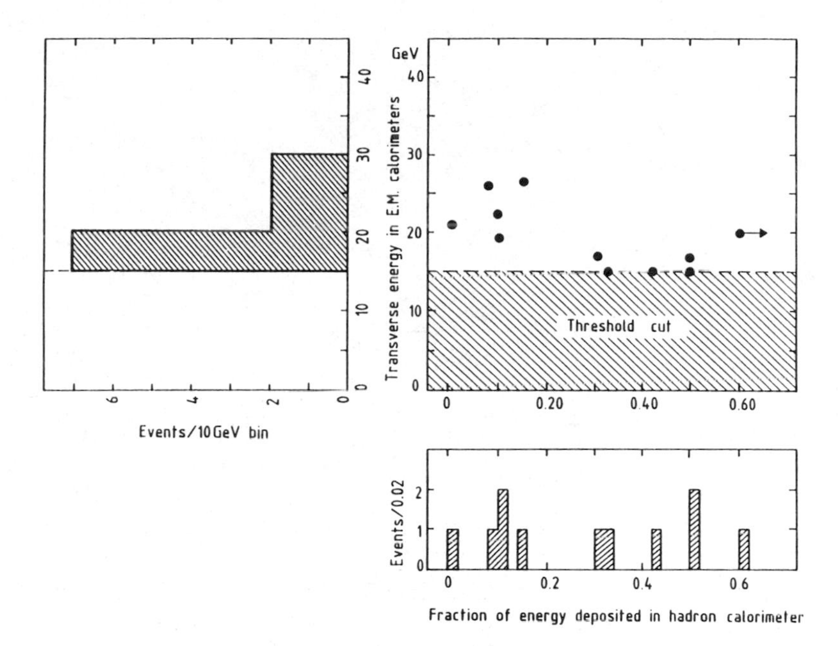

Fig.10 A plot of the transverse energy in the e.m. calorimeters versus the fraction of energy deposited in the hadron calorimeters for events which survive the missing energy search. With jets.

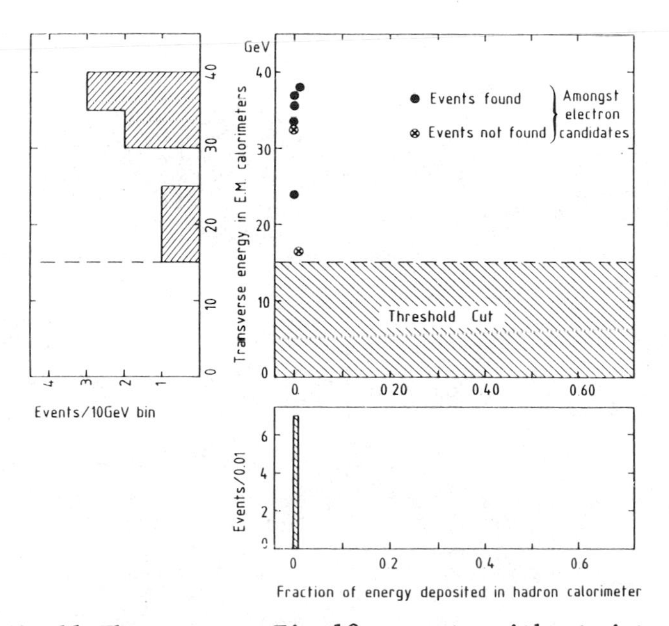

Fig.11 The same as Fig.10, events without jets;

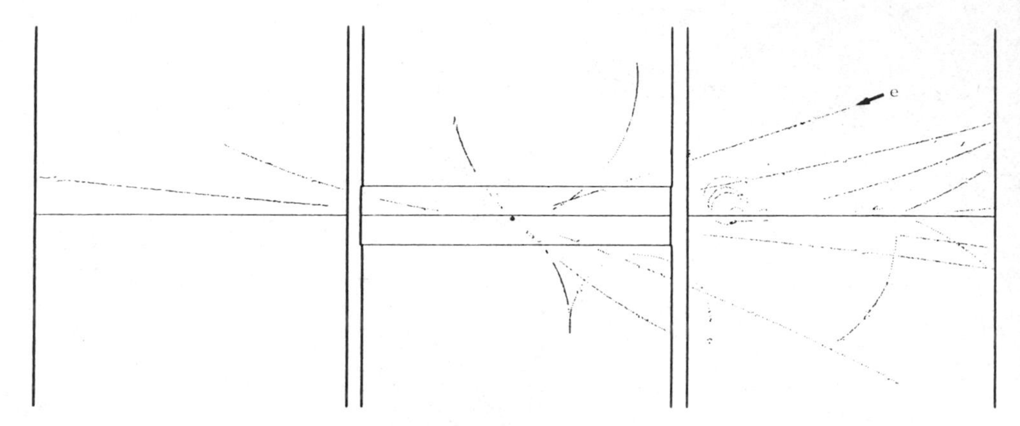

Fig.12 Digitization from the Central Detector for the tracks in two of the events which have a well measured high p_T electron. In this image, low multiplicity.

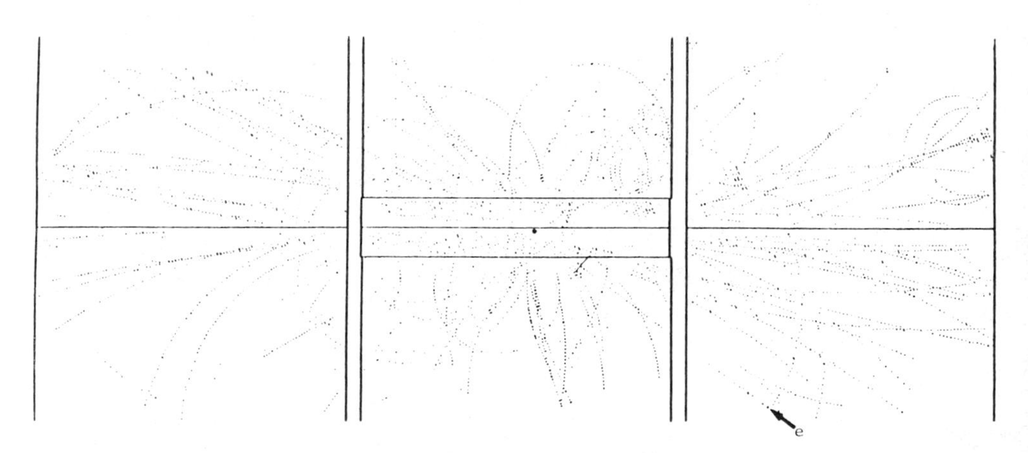

Fig.13 As in Fig.12, a case of high multiplicity (65 associated tracks).

detector. For these reasons we prefer to limit our final analysis to the events in the gondolas, although we believe that everything is still consistent with event B being a good event.

3.7 - Background evaluations -

I cannot report the results of the accurate analysis we did on this point.

I only say here that we have been unable to find a background process capable of simulating the observed high-energy electrons. Thus we are led to the conclusion that they are electrons. Likewise we have searched for backgrounds capable of stimulating large-E_T neutrino events. Again, none of the processes considered appear to be even near to becoming competitive.

3.8 - Comparison between events and expectations from W decays -

The simultaneous presence of an electron and (one) neutrino of approximately equal and opposite momenta in the transverse direction (fig. 14) suggests the presence of a two-body decay, $W \to e + \nu_e$. The main kinematical quantities of the events are given in Table 2. A lower model-independent bound to the W mass m_W can be obtained from the transverse mass, $m_T^2 = 2p_T^{(e)} p_T^{(\nu)} (1 - \cos \phi_{\nu e})$ remarking that $m_W \geqslant m_T$ (fig. 15). We conclude that:

$$m_W > 73 \text{ GeV}/c^2 \text{ (90\% confidence level).}$$

A better accuracy can be obtained from the data if one assumes W decay kinematics and standard V-A couplings. The transverse momentum distribution of the W at production also plays a role. We can either extract it from the events or use theoretical predictions.[20]

As one can see from fig. 16, there is good agreement between two extreme assumptions of a theoretical model and our observations. By requiring no associated jet, we may have actually biased our sample towards the narrower first-order curve. Fitting of the inclusive electron spectrum and using full QCD smearing gives $m_W = 74 {}^{+4}_{-4}$ GeV/c^2. The method finally used is the one of correcting, on an event-to-event basis, for the transverse W motion from the $(E_\nu - E_e)$ imbalance, and using the Drell-Yan predictions with no smearing. The result of a fit on electron angle and energy and neutrino transverse energy with allowance for systematic errors, is

$$m_W = 81^{+5}_{-5} \text{ GeV}/c^2,$$

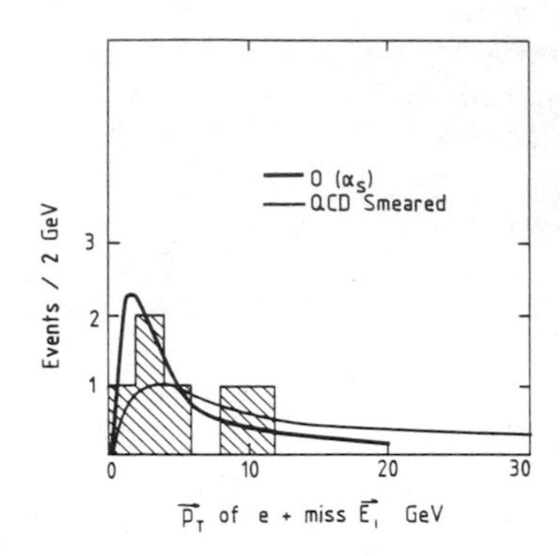

Fig.14 The missing transverse energy component parallel to the electron, plotted versus the transverse electron energy for the final six electron events without jets.

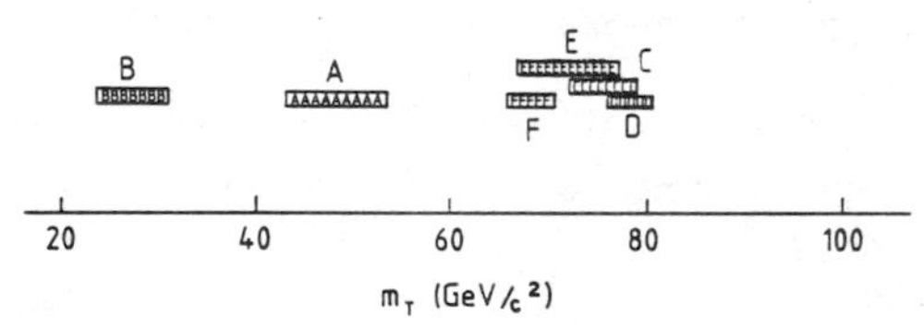

Fig.15 Distribution of the transverse mass derived from the measured electron and neutrino vectors of the six electron events.

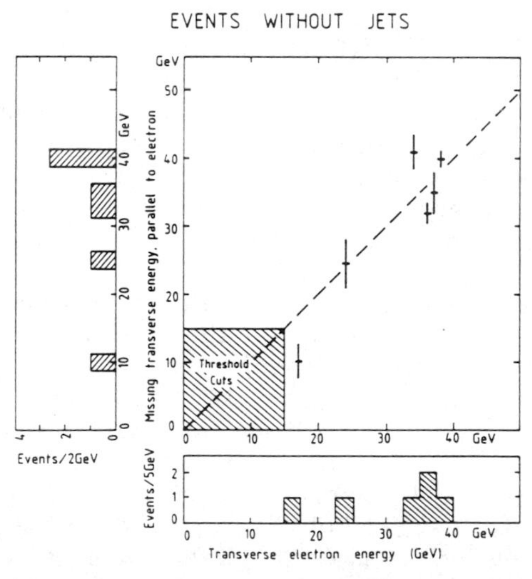

Fig.16 The transverse momentum distribution of the W derived from our events, using the electron and missing-energy vectors. This is compared with the theoretical predictions of Ref.20 for W production without [O(α_s)] and with QCD smearing.

in agreement with the expectation of the Weinberg-Salam model.

We find that the number of observed events, once detection efficiencies are taken into account, is in agreement with the cross-section estimates based on structure functions, scaling violations, and the Weinberg-Salam parameters for the W particle.

3.9 - One open problem -

As we already said at the Rome meeting, there is one point which is not clear yet, and in case gives an evidence which is not in favour of the interpretation of our events as being W^+ or W^-. This is the asymmetry between the lepton emitted to the left and to the right of the interaction point. Let's try to be more specific.[21]

If we call θ the angle in the W rest frame between the positon (electron) and the antiproton (proton), the theoretical expectation for the e^+ (e^-) produced in $W \rightarrow e +$ must be a typical $(1 + \cos\theta)^2$ distribution as long as we assume V-A coupling. This means that $e^+(e^-)$ should be preferentially emitted (see Table 2) in the direction of positive (negative) rapidity,where rapidity is defined as positive in the direction of the outgoing antiproton.

This is not the case for our electrons; they move rather in the opposite direction. The wrong asymmetry is not seriously pronounced for events C, D, E, F, G. It is rather strong for event A, which results to be rather improbable (~1:10) having negative charge.

This is a point which should be explored soon,and it is clear that it is very important to look to this asymmetry in the future 20-30 W events. The conservation of events with clearly disfavored asymmetry should compel us to unexpected conclusions on the validity of the so called standard theory. This comment opens our considerations on the immediate and remote future of high energy with colliders.

4 - FUTURE PERSPECTIVES -

I present here some personal opinions which also arise from a number of discussions we already had at CERN, and particularly inside the UA1 community. I shall proceed with the following order:

4.1 The obvious near future;
4.2 A touch of philosophy;

4.3 The increasing importance of the proton-antiproton way;
4.4 One important point: the resolving power of the detectors;

4.1 - The obvious near future -

A number of important objectives shall be in the hands of UA1 and UA2 - and only in their hands - during 83 and 84. One is to bring to a higher statistics (40-100?) the W cases, and to verify mass and decay asymmetry of the W. This is a must. The second point is to get some Z° events. We should have more or less 0.5 Z° events in our present 20 nb^{-1}. With the expected luminosity in April-June 83, we could have at the end, hopefully, three to five Z° in our hands. It is clear that the mass M_{Z° from the leptonic decay $Z^\circ \rightarrow e^+ + e^-$ and Z° $\mu^+ + \mu^-$ is direct and straightforward. Apart from some unexpected subtleties that mother nature could present to us, we expect that the mass of the Z° can be estimated with an error lower than 5%. If they exist, of course.*

This collection of events through the leptonic decays of the IVB's will also be of great help when we shall try to identify the W's and Z°'s decaying in two hadronic jets among the many jet pairs of high mass produced in $p\bar{p}$ through ordinary QCD processes. This identification is certainly not easy, but it is clear that, if successful, it will put at our disposal hundred of events with IVB's at the end of '84.

Another not remote goal is the search for new heavy flavours, among which and first comes the Top quark. Search for heavy flavours was of course one of our goals since the beginning[11]. All our present apparatus UA1 will be heavily engaged, when looking for the Top: the forward part, to analyse that region which more or less properly we call the diffractive region; the central part of UA1; the apparatus for the identification of the muons, for we can expect an high number of muons from the semileptonic decay of the heaviest flavours.

The search for the Higgs is obviously very important, and probably very hard[11]. The search for the Higgs was clearly already stated in our first proposal of 1977. The fact is that the present Weinberg-Salam theory cannot live its normal or renormalized life without these scalars. But it will be very difficult to verify the existence of the Higgs or to deny it, considering that their mass can vary from 20 to 200 GeV[22], and their decay modes are not well fixed yet.

Someone said that the Higgs are imposed by coherence with the local theories of our time, in the same way that Ether was imposed at the end of last century, to maintain the Galileian coherence of

* (Note added in proof -- they do! H.B.N.).

space and time. Should our evidence of W, Z etc be confirmed, and the existence of the Higgs excluded (something, I repeat, which will in case be established only within several years from now), this will produce some basic changes in our present understanding.

4.2 - A touch of philosophy -.

Let me dedicate to this point not more than a few seconds. To what do we pretend, beyond the great success of the standard theory? It has been a fantastic success, which took 16 years, from the theoretical proposal[17] (1967) to the experimental verification.[17] Why is not this enough? We are also having here in Erice a continuous moaning of the theoreticians on the fact that we are still far from a real unification, that too many parameters (someone said more than twenty) are still independent or introduced ad hoc in the standard theory; and there is already some demand for new accelerating machines of high energy to look for bigger brothers of the existing particles. One could ask whether this request is an obvious consequence of our society (machine builders insist for new machines, theoreticians for new theories, etc) or there is a real necessity for the development of new ideas and frontiers. Personally, I think that we are only at the beginning. But the braking or the encouragement shall also descend for the attitude of other scientists, outside the high energy field. Everyone of them is now probably willing to acknowledge our splendid results; but we must convince them of our future aims and the necessity to go ahead and beyond. In other words we must accept the toil of explaining and defending our work in front of our society, accepting the fact that they at the end are our judges. In particular, we must know what we really want in high energy physics, from now to the end of this century. How many of us are ready to express clearly our desires and programs?

4.3 - Importance of the proton antiproton collider -

A few important points descend from our experimentation (UA1 and UA2):

- The proton antiproton collider is a clean machine,[24] the best one could imagine or desire: when it gets its optimal conditions, the ratio (real events)/(background, beam-gas collisions etc) is very high, even with a minimum bias trigger.

- The close collisions have a jet structure which becomes more and more evident when the transverse energy increases[16]. This is very impressive and significant. The definition of a jet and its structure are not completely understood yet, but one can perhaps say that it will be possible in future and especially at higher energies to intepret the $p\bar{p}$ ring as a quark-antiquark collider.

- The analysis of the $p\bar{p}$ events can be made in reasonable times (within weeks), when the detectors are well chosen and prepared: this has been demonstrated by the results presented at Rome in January, one month or so after the data taking.

- The luminosity is increasing, still with a collider which was originally only a protosynchrotron. The limit of 6×10^{28} cm^{-2} s^{-1} has been already reached, and more than 10^{29} was reached during 1983.

All this means that the physics at an energy larger than one TeV in the center of mass will become possible in future, and therefore the $p\bar{p}$ collider is going to be a unique instrument for future research at the highest energies. In particular, this new way will overtake the energy limits of e^+e^- colliders. In conclusion, it seems that the physics of $p\bar{p}$ collider is not a temporary episode, but it is a new way which came to stay, and perhaps to become dominant in future years.

But there is at present an obvious limit to the physics with $p\bar{p}$ colliders, at least in comparison with the e^+e^- colliders: this is the quality of detectors, a point which we must not underestimate.

4.4 – The problem of the detectors –

Let's take the inclusive detection of the Z°. Contrary to the exclusive reaction $e^+e^- \to Z°$, reaction

$$p + \bar{p} \to Z^\circ + X$$

allows a measurement of the mass of Z° with an error which depends on the detectors of the particles. Just to give numbers, one can expect that the measurement of M_{Z° may be obtained by e^+e^- colliders (LEP) with a precision of $\pm$ .3 GeV, i.e. 3 per thousand of the mass of the Z°. In fact the precision of M_{Z° is closely related to the precision in energy of the e^+e^- colliding beams. But in the case of $p\bar{p}$ colliders, the measurement of M_{Z° depends on the detectors measuring the momentum of the decay particles e^+e^- or μ^+ μ^-.

The best we can hope to get today, with the present detector of UA1, by combined use of the calorimeters and the central

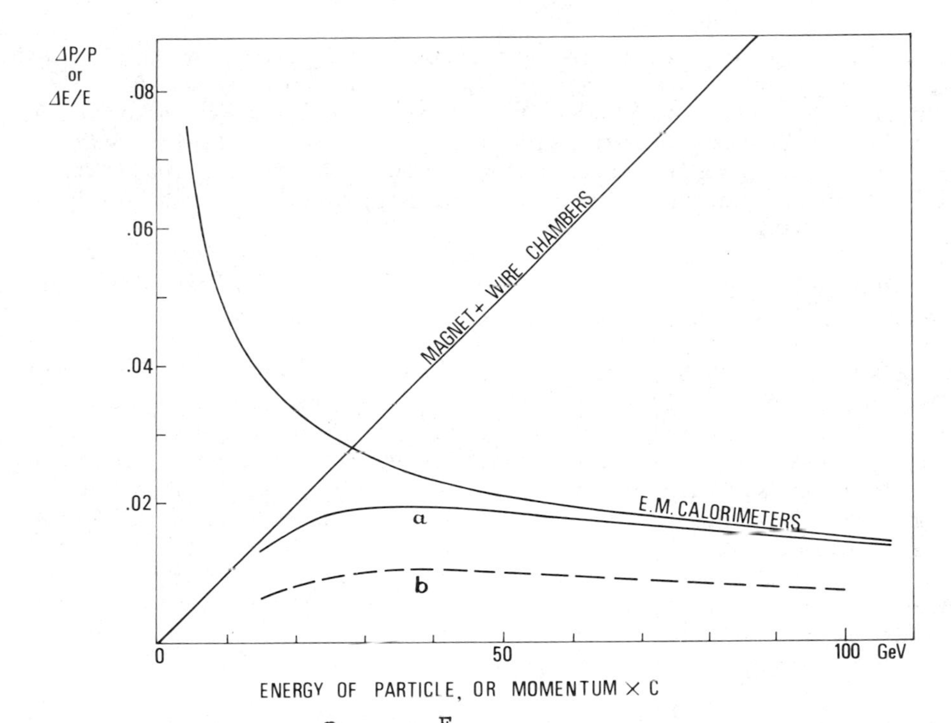

Fig.17 Resolution $\frac{\Delta p}{p}$ or $\frac{\Delta E}{E}$ for a Central Detector with magnetic field, and for a calorimeter measuring electrons, respectively. Obviously the percentage resolutions improve (i.e. decrease) in opposite directions. Line a) is an added contribution of the two resolutions. Line b) is what we should have to compete with Lep I and Lep II.

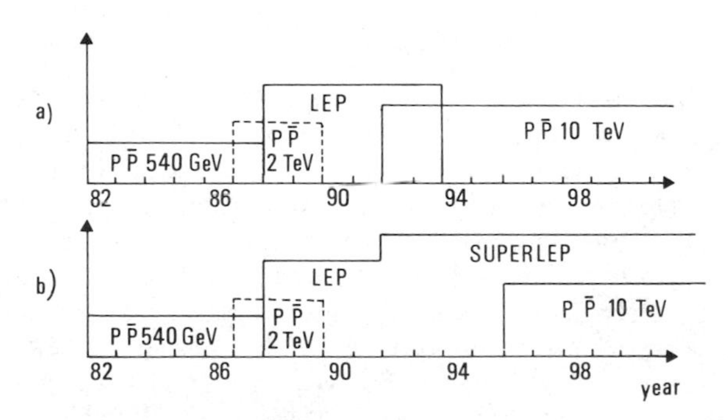

Fig.18 a), b) Two possible scenarios of future research. The two scenarios shall depend strongly on the future resolution of the particle detectors: development of fig. 18a) will be possible only with detectors arriving to a precision in energy and momentum better than 1%.

detector, is a precision case by case of the order of ± 2 GeV in M_{Z°. In fig. 17 we have reported the resolutions $\frac{\Delta p}{p}$, $\frac{\Delta E}{E}$ for a Central Detector with magnetic field, and for a calorimeter measuring electrons. Obviously the resolutions improve (i.e. decrease) in opposite directions. Line a) is a qualitative combination of the two resolutions. Line b) is what we should have to compete with LEP I and II.

This makes clear that LEP may have many advantages in the detailed study of the Z°, respect to the $p\bar{p}$ collider.

There are already definite plans in the future use of $p\bar{p}$: the 2 TeV Fermi Lab collider, and the possible use of the tunnel for LEP at CERN as a multiTeV $p\bar{p}$ collider.[24] It is clear anyway that a relevant point shall be the condition of the detectors used in these colliders. This progress seems to me rather important or necessary. What has been obtained until now descends from the strict collaboration of the users with the physicists of solid state and with the physicists and engineers working to the computers. The activity in the European laboratories to invent new detectors and to improve and simplify the software connected to them is probably an important part of the future of our research.

Just as a naive guess, I have presented in fig. 18a and in fig. 18b two possible scenarios of future research. Suppose that in the abscissa we put some undefined quantity, like the number of experimental physicists employed in the researches. As we see the two scenarios shall depend strongly, in my opinion, on the future resolution of the particle detectors.

In the optimistic case (fig. 18a) of detectors arriving to a precision better that 1%, I incline to believe that electron machines shall not go beyond the 120 GeV center of mass level, and superLEP of 200 GeV or so could never be built. The tunnel of LEP will rather be occupied in the 90's from the $p\bar{p}$ multiTeV $p\bar{p}$ collider.

Let me close with a comment strictly connected with the previous points. As we also saw in this meeting, the theoretical proposals for new particle and states is continuous and high. This is not strange, it is an obvious consequence of the fact that we are working for the "greatest unification or synthesis", an objective which never looked so close, since the end of last century. But I think that we should make a request to our theoretical friends: each proposal they do of new states and particles should be accompanied by the analysis of how one could observe it: this is something like giving a "realistic operative definition" of the quantity which is suggested or proposed to observe.

The limits we have today in energy resolution and in absolute energy are an important part, not a secondary part of the landscape of knowledge we wish to explore. It is interesting to observe that this contact with reality has been always the quality of the greatest theoreticians we had in our times. They were conscious of the danger that our explorations may trespass on dreamland.

References

1. E. Fermi, "Tentativo di una teoria dell'emissione dei raggi 'Beta'", Ricerca Scientifica, IV, Vol.2: 491-495 (1933).

2. G. Bernadini, G.F. Corazza, G. Ghigo, B. Touschek, Nuovo Cimento, 18:1293 (1960).

3. J. Perez-y-Jorba, Proceedings of the 4th International Symposium on Electron and Photon Interactions at High Energies, Daresbury Nuclear Physics Laboratory: 213 (1969).

4. J. Sakuray, Proceedings quoted in 3.: 91.

5. R.E. Taylor, Proceedings quoted in 3.: 251.

6. J.J. Aubert et al., Phys.Rev.Lett. 33:1404 (1974).
J.E. Augustin et al., Phys.Rev.Lett. 33:1406 (1974).
C. Bacci et al., Phys.Rev.Lett. 33:1408 (1974).

7. G.I. Budker, Proceedings of the International Symposium on Electron Positron Storage Rings, Saclay 26-30 September (1966).

8. Staff of the CERN $p\bar{p}$ project, Phys.Lett. 107B:306 (1981).

9. C. Rubbia, P. Mc Intyre and D. Cline, Proceedings International Neutrino Conference, Aachen (1976) p.683 (Vieweg, Braunshweig, 1977).

10. We do not report here the list of published papers of UA1 experiments. They are regularly reported mostly in Phys. Lett., to start from 1981.

11. A. Astbury et al., "A 4π solid angle detector for the SPS used as a proton antiproton collider", CERN/SPSC/78-06 30/Jan 1979. M. Calvetti et al., Nucl. Instr. and Methods, 176:255 (1980).

12. UA1 Collaboration, G. Arnison et al., Phys. Lett. 107B:320 (1981), paper presented to the XXI International Conference on High Energy Physics.

13. UA1 Collaboration, G. Arnison et al., Phys.Lett. 118B:167 (1982).

14. UA1 Collaboration, G. Arnison et al., Phys.Lett. 122B:189 (1983).

15. UA1 Collaboration, G. Arnison et al., submitted for publication in Phys.Lett.

16. UA2 Collaboration, M. Banner et al., Phys.Lett. 118B: 203 (1982).

17. S.Weinberg, Phys.Rev.Lett. 19:1264 (1967).
A. Salam, Proc. 8th Nobel Symposium, Aspenäsgarden, p. 367 1968, Almqvist and Wiksell, Stockholm, 1968.

18. A. Sirlin, Phys.Rev. D22:971 (1980).
W.J.Marciano and A.Sirlin, Phys.Rev. D22:2695 (1980).
F. Antonelli, M. Consoli, G. Corbò, Phys.Lett. 91B:90 (1980).
F. Antonelli, M. Consoli, G. Corbò, O. Pellegrino, Nucl. Phys. B183:195 (1981).
F.Antonelli, L.Maiani, Nucl. Phys. B186:269 (1981).

19. For a review, see M.Davier, Proc. 21sr Int. Conf. on High Energy Physics, Paris, p. C3-471, 1982; J.Phys. (France), No. 12, t. 43, 1982.

20. F. Halzen and D.M. Scott, Phys. Lett. 78B:318 (1978).
P. Aurenche and F. Lindfors, Nucl.Phys. B185:301 (1981).
F. Halzen, A.D. Martin, D.M. Scott and M. Dechantsreiter, Phys.Lett. 106B:147 (1981).
M. Chaichian, M. Hayashi and K.Yamagishi, Phys.Rev. D25:130 (1982).
A. Martin, Proc. Conf. on Antiproton-Proton Collider Physics, Madison, 1981, AIP Proceedings No. 85:216 (1982).
F. Halzen, A.D. Martin and D.M. Scott, Phys.Rev. D25:754 (1982).
V.Barger and R.J.N. Phillips, University of Wisconsin preprint MAD/PH/78 (1982).

21. F. Halzen, A.D. Martin, D.M. Scott, quoted in ref. 20.

22. N. Cabibbo, L. Maiani, G.Parisi, R.Petronzio, Nucl. Phys. B158:295 (1979).

23. As reported in Table I, and submitted to Phys.Lett.

24. Proceedings of the 3rd Topical Workshop on Proton Antiproton Collider Physics, Rome, CERN Report 83-04 (1983).

RESULTS ON WEAK CURRENT FROM CHARM COLLABORATION[1]

Bruno Borgia

Istituto Nazionale di Fisica Nucleare
Roma, Italy

ABSTRACT

The fine grained calorimeter of the CHARM collaboration constitutes a well suited instrument to measure fundamental quantities of neutral current interactions as the coupling constants to quarks and leptons. Results derived from neutrino interactions on nuclei and on electrons are presented.

PARAMETERS OF CHARM DETECTOR

The full description of the detector can be found in the published article[2]. I will recall some of the most relevant features. The apparatus consists of two parts:

a) the target-calorimeter made of 78 identical units
b) the muon spectrometer composed of toroidal magnet and drift chambers.

Each of the 78 calorimeter units is composed of a marble absorber ($3x3m^2x8cm$), a plane of 128 proportional drift tubes ($3x3x400\ cm^3$ each) and a plane of scintillator hodoscopes (20 counters, $15x3x300\ cm^3$ each).

This structure has the following physical properties:

a) the target is isoscalar
b) the hadronic and electromagnetic showers have approximately the same length

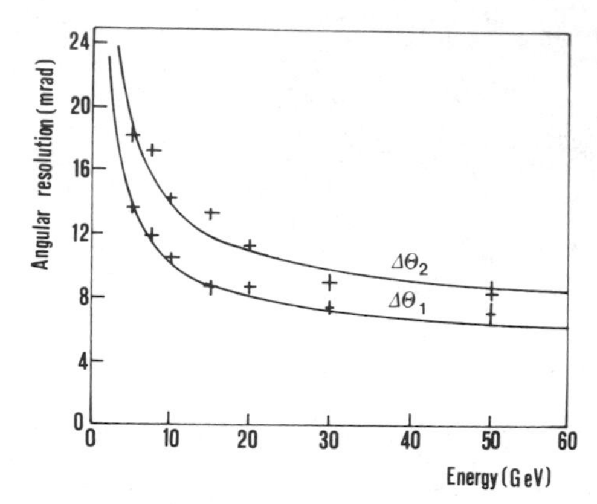

Fig. 1. Angular resolution for electromagnetic showers in the close ($\Delta\theta_1$), and far ($\Delta\theta_2$) projections.

c) the energy resolution follows the empirical law

$$\frac{\sigma(E)}{E} \simeq \frac{K}{(E/M_p)^{1/2}}$$

with K = 0.53 for hadronic showers
= 0.20 for e.m. showers

d) the shower direction can be measured accurately. For e.m. showers the angular accuracy is described in fig.1. $\Delta\theta_1$ is the angular resolution for the projected view where the proportional tubes give the coordinate closer to the event vertex.

e) the transverse profile of the shower can be measured by scintillators and by proportional tubes. The two width estimators are used to discriminate between hadronic and e.m. showers. Fig. 2 shows some results from test measurements.

SEMILEPTONIC INTERACTIONS

The cross section of neutrino scattering on an isoscalar target can be expressed in terms of chiral couplings of quarks and in terms of quark content of the nucleon. The inclusive total cross section provides the most direct information of the quantities $u_L^2+d_L^2$ and $u_R^2+d_R^2$ or on ρ, the relative overall strength of NC interactions with respect to the CC interactions, and $\sin^2\theta_W$. The differential cross section $d\sigma/dy$ (y is the inelasticity) provides additional information on the quark content of the nucleon and under a more restrictive hypothesis, also the first determination of the coupling constant of the strange quark.

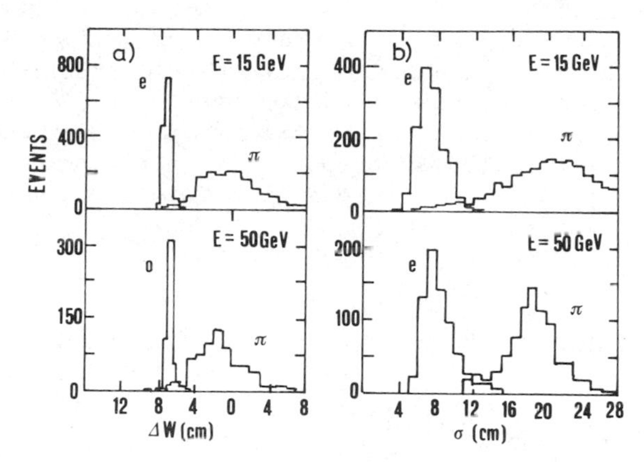

Fig. 2. Distributions of the width of electron and pion shower, a) as measured by scintillator, b) as measured by the proportional tubes.

Total cross section

The most precise determination of $\sin^2\theta_W$ is obtained by the total cross section measurement either using the explicit expressions in terms of quark parton distributions or using the Paschos-Wolfenstein equation derived by isospin invariance.

Data[3] were collected in the neutrino narrow band beam (NBB) of Cern produced by the decay of a pion and kaon beam of known momentum (200±9 GeV/c). A total of 9200 neutrino and 3800 antineutrino induced events were selected among all interactions with the following main criteria: a) interaction vertex has to lie inside the fiducial volume (65 tons in the neutrino beam, 37 tons in the antineutrino beam); b) the shower energy has to be greater than 2 GeV.

Classification in to CC or NC interactions is based on the identification of a track associated with the interaction vertex, without kink or interaction and having a range more than 1.0 GeV/c. The two classes of events were corrected for various effects mixing NC and CC events and for background:

i) Events induced by neutrinos produced by hadrons before the momentum slits were evaluated using special runs with the momentum slits closed.

ii) Cosmic ray events were evaluated in runs with the SPS off.

iii) Events lost or misclassified by the automatic track recognition program were determined by eye scan of a sample of rejected events and using a semi-Monte Carlo method.

iv) The decay in flight of pions and kaons can produce a muon track satisfying the CC criteria. This correction was determined from the analysis of dimuon events.

v) The NBB contains electron-neutrinos. Events induced by this background were calculated using the kaon branching ratios.

Applying the above corrections we obtain the events listed in table 1.

Table 1: Events for shower energy > 2 GeV

	Antineutrino		Neutrino	
	NC	CC	NC	CC
Raw events	1126±34	2751±53	2361±49	6503±81
Correction (%)	15.8±2.4	8.6±1.4	12.8±1.0	1.1±0.3
Corrected events	948±43	2514±64	2059±54	6433±84

On the derivation of the experimental values

$$R = \frac{\sigma\,(\nu_\mu N \to \nu_\mu X)}{\sigma\,(\nu_\mu N \to \mu^- X)} \quad ; \quad \bar{R} = \frac{\sigma\,(\bar{\nu}_\mu N \to \bar{\nu}_\mu X)}{\sigma\,(\bar{\nu}_\mu N \to \mu^+ X)}$$

doesn't enter any absolute or relative beam flux measurement,

while in the ratio $r = \frac{\sigma\,(\bar{\nu}_\mu N \to \mu^+ X)}{\sigma\,(\nu_\mu N \to \mu^- X)}$

it is necessary the knowledge of the relative antineutrino/neutrino fluxes. We used for this purpose the measurements of the muon fluxes in the iron shield following the decay tunnel.

We then obtain the following results:

$R = 0.320 \pm 0.010 \qquad \bar{R} = 0.377 \pm 0.020$

and $r = 0.470 \pm 0.017$
for shower energies above 2 GeV

Following the method of reference[4], the weak mixing parameter can be derived from R, by QCD model calculation:

$$\sin^2\theta_W = 0.220 \pm 0.013\ \text{(stat)} \pm 0.004\ \text{(syst)}$$

The above result assumes $\rho = 1$.

Using both R and $\bar{R}$, the parameters $\sin^2\theta_W$ and ρ can be simultaneously determined, see fig. 3, giving

$$\rho = 1.027 \pm 0.023; \qquad \sin^2\theta_W = 0.247 \pm 0.038$$

The alternative of using the Paschos-Wolfenstein relation[5] must again assume $\varrho = 1$:

$$\frac{R - r\bar{R}}{1 - r} = \varrho^2 \left(\frac{1}{2} - \sin^2\theta_W\right)$$

Then $\quad \sin^2\theta_W = 0.230 \pm 0.023$

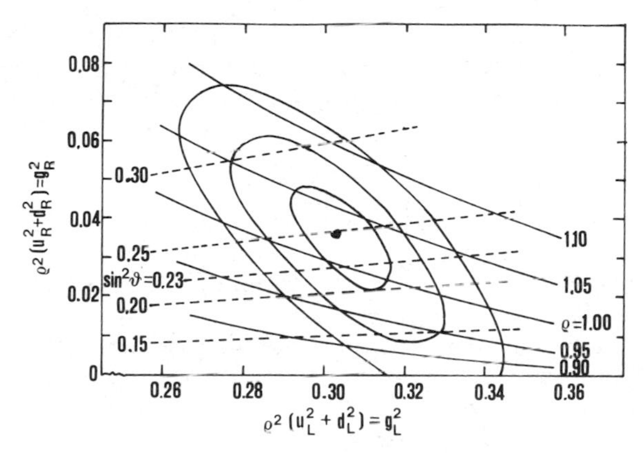

Fig. 3. Best fit and confidence limits (39%, 87%, 99%) on the chiral coupling constants with lines of constant ϱ and $\sin^2\theta_W$.

We estimate a systematic error of ± 0.008 due the uncertainties in scaling violations in neutrino and antineutrino interactions and in the knowledge of the neutrino and antineutrino energy spectra.

Above, in all derivations radiative corrections have been neglected. This subject will be discussed below.

Theoretical corrections and uncertainties

The error reached by the CHARM experiment is small enough to consider radiative corrections. Several authors[6] have contributed to this field computing also numerically the correction for several experiments.

From the beginning, one should note that the Fermi constant

G_F has a logarithmic correction resulting into a negative renormalization of the parameter . Therefore the assumption $\rho = 1$ in the derivation of $\sin^2\theta_w$ should be substituted with the corrected value.

Sirlin, Marciano[7], and Wheater, Llewellyn-Smith[8] reach very similar values in the actual computation of the corrected $\sin^2\theta_w$:

$$\sin^2\theta_w(m_w) = 0.210\pm0.004 \text{ (theor.uncert.)}$$
$$(m_t = 18 \text{ GeV}, m_H = m_z) \quad \text{reference 7}$$

$$= 0.208$$
$$(m_t = 20 \text{ GeV}, m_H = 200 \text{ GeV}) \quad \text{reference 8}$$

where $\sin^2\theta_w$ is defined in the $\overline{MS}$ scheme at the mass scale m_w.

From Paschos, Wirbel[9] we quote the radiative corrected value of the mixing angle derived by the Paschos, Wolfenstein equation:

$$\sin^2\theta_w = 0.225$$

Here $\sin^2\theta_w$ is defined in terms of the renormalized coupling constants, independent of q^2. In the same article uncertainties due to the hadronic model are discussed, leading to an estimate of $\sin^2\theta_w = +0.005$ due to the charm development.

Uncertainties in $\sin^2\theta_w$, due to the QPM and QCD model calculations used to express the cross sections ratio R, have been recently analyzed by Llewellyn-Smith[10] with the following conclusions: (a) higher twist corrections in isoscalar targets are smaller than 0.005, (b) uncertainties from the present knowledge of KM matrix elements and strange quark distribution in the nucleon give an error of ±0.01. However this uncertainty can be reduced by further experimental information.

Differential cross sections

As mentioned above, the differential cross section for neutral current interaction can provide additional information on the coupling constants and on the quark content of the nucleon. The results of the CHARM experiment[11], based on the same data used for the total cross section analysis, for the first time yield the differential cross section fully unfolded.

While in CC events the incoming neutrino energy appears in the final state as $E_\nu = E_\mu + E_h$, in NC events the muon energy is replaced by the outgoing neutrino energy. Therefore to compute $y = E_h/E_\nu$ one must relay on the knowledge of the beam energy.

The neutrino spectrum in the narrow band beam can be computed starting from the parameters of the parent beam and from the two body kinematics of the parent particle, pion or kaon, decay. As a result, at a given radius the neutrino energy will have a twofold ambiguity, spread out by the beam resolution. The unfolding procedure adopted in our analysis is briefly described below.

Assuming a total cross section linear on E_ν, we parametrize the differential cross section with a linear combination of bell-shaped functions (B-splines) $b_i(y)$ and coefficients a_i.
The measured event distribution as function of E_h and radius r can be expressed as

$$\frac{d^2N}{dE_h dr} = \sum_i a_i \int b_i(E_h/E_\nu) F(E_\nu, r) dE$$

where $F(E_\nu, r)$ is the computed neutrino spectrum. A maximum likelihood fit to the experimental distribution determines the coefficients a_i. The differential cross section is then obtained in terms of a_i and $b_i(y)$. The resulting unfolded distributions are shown in fig. 4. The CC distributions, with radiative corrections included, are obtained exactly by the same method as the NC distributions.

The value of the result can be appreciated by expressing the differential cross sections without neglecting the strange quark content of the nucleon as[12]:

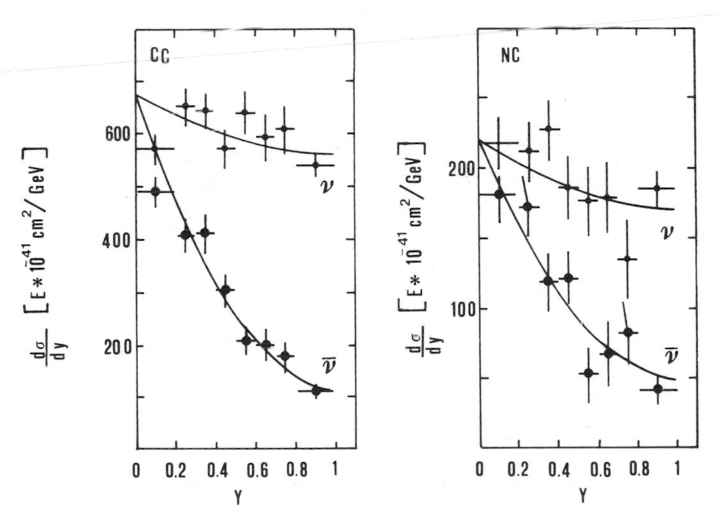

Fig. 4. The differential cross sections after unfolding. The curves correspond to a two parameter fit (see reference 11).

$$\left(\frac{d\sigma}{dy}\right)^{CC}_{\nu} = B\left[(1-\alpha_\nu) + \alpha_\nu(1-y)^2\right]$$
$$\left(\frac{d\sigma}{dy}\right)^{CC}_{\bar\nu} = B\left[\alpha_{\bar\nu} + (1-\alpha_{\bar\nu})(1-y)^2\right]$$

$$\left(\frac{d\sigma}{dy}\right)^{NC}_{\nu,\bar\nu} = B\left\{ g^2_{L,R}\left[(1-\alpha_{\bar\nu}) + \alpha_\nu(1-y)^2\right] + \right.$$

$$g^2_{R,L}\left[\alpha_\nu + (1-\alpha_{\bar\nu})(1-y)^2\right] +$$

$$\left. g^2_s(\alpha_{\bar\nu} - \alpha_\nu)\left[1 + (1-y)^2\right]\right\}$$

A simultaneous fit of all five unknowns $\alpha_\nu, \alpha_{\bar\nu}$, g^2_L, g^2_R, g^2_s is precluded by the form of the NC distributions where in the last term appears the product $g^2_s(\alpha_{\bar\nu} - \alpha_\nu)$. Therefore in order to obtain the value of the chiral coupling g^2_s to the strange quark, we constrain in the NC cross section the quantity $(\alpha_{\bar\nu} - \alpha_\nu) = 0.086 \pm 0.050$ as obtained from the CC data alone. We also express g^2_L and g^2_R in terms of $\sin^2\theta_W$ accordingly to the Glashow, Salam, Weinberg model. The fitted values of the free parameters are given in table 2.

The coupling constant g^2_s can be compared to the analogous coupling constant g^2_d as derived from the fitted electroweak mixing angle giving $g^2_s/g^2_d = 1.39 \pm 0.43$.

The error includes the systematic uncertainties. Fig. 5 shows the domain of g^2_s and $\sin^2\theta_W$ allowed by this fit.

The equality of the coupling constants of the strange and down quark is assumed in the GIM mechanism, and our result is consistent with this hypothesis. The "universality" of coupling constants of the neutral weak current among the quark families is beginning to be tested.

Further information derived from the differential cross sections can be found in reference[11].

Table 2: Result of the fit with $(\alpha_{\bar\nu} - \alpha_\nu) = 0.086 \pm 0.050$

$\alpha_{\bar\nu}$	$\sin^2\theta_W$	g^2_s
0.18±0.02	0.23±0.02	0.26±0.06

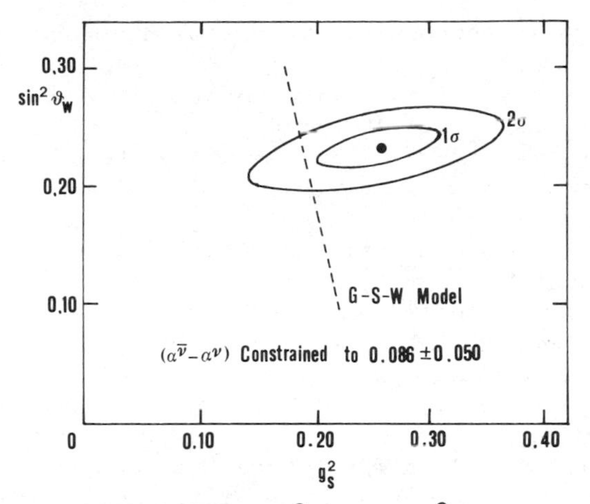

Fig. 5. The allowed domain of g_s^2 and $\sin^2\theta$. The 1σ and 2σ contours are statistical only.

LEPTONIC INTERACTION

The process $\nu_\mu + e^- \rightarrow \nu_\mu + e^-$ or the counter part initiated by antineutrino has the distinctive advantage over the corresponding processes on nuclei or nucleons to be free from details of the quark parton model, as the electron is a free "point like" particle. On the other end, since its cross section is small, being $\sigma(\nu_\mu e)/\sigma(\nu_\mu N) \sim m_e/M_p$, a high statistics experiment has never been performed until now.

Defining $y = E_e/E_\nu$, the differential cross section can be written as

$$\left(\frac{d\sigma}{dy}\right)_{\nu,\bar{\nu}} = \frac{G_F^2 m_e E}{2}\left[(g_V \pm g_A)^2 + (g_V \mp g_A)^2(1-y)^2 - \frac{m_e}{E_\nu}\,y(g_V^2 - g_A^2)\right]$$

where in the standard model $g_A = -1/2$ and $g_V = 2\sin^2\theta_W - 1/2$.

The CHARM detector was exposed to the wide band neutrino beam (WBB) of the Cern 400 GeV SPS. We have collected 1.3 million neutrino and 1.4 million antineutrino interactions[13].

Events were selected with the following criteria, in order to isolate e.m. showers at small angle:

a) Events with single track with length $l > 8\ x_o$ are rejected

b) θ_{SH} must be less than 100 mrad

c) Estimators of shower width must be less than 1.6 cm for scintillator based parameter and less than 9 cm for a proportional tubes parameter

d) In the first proportional tube plane following the interactin vertex, only 1 hit must be present. This requirement rejects interactions on nuclei with backscattering.

The efficiency to detect e.m. showers is $(85\pm5)\%$ and $(71\pm10)\%$ respectively for selection criteria c) and d). The combined efficiency is $(61\pm9)\%$ while the efficiency for hadronic showers is less than $2 \cdot 10^{-3}$.

Only events with a shower energy E between 7.5 GeV and 30 GeV have been retained, to reduce background due to electron-neutrino interactions.

We then obtain 267 neutrino induced events and 665 antineutrino events.

Fig. 6a and 6b show the events plotted versus $E^2\theta^2$. This variable was choosen in order to have a convenient representation of electron-neutrino background. Two sources of background are assumed:

(a) Elastic and quasi-elastic charged current events induced by the ν_e and $\bar{\nu}_e$ components of the beam. (CC $\overset{(-)}{\nu}_e$ in fig.6)

(b) Neutral current events with γ or π^o in the final state produced by coherent scattering of ν_μ and $\bar{\nu}_\mu$ on nuclei. (NC $\overset{(-)}{\nu}_\mu$ in fig.6).

Distribution of background (a) was determined experimentally from our data on elastic and quasi elastic interactions induced by muon neutrinos and replacing the muon energy and angular resolutions with the electron ones. The cross section of this process above $E_\nu = 7.5$ GeV is function of only $P_T^2 \simeq E^2\theta^2$ and not of the neutrino energy.

Distribution of background (b) was calculated using the predicted differential cross section[14] for coherent π^o production by neutrinos on nuclei.

The decomposition and normalization of the two backgrounds was performed by the analysis of the energy deposited in the first scintillator plane following the vertex (E_F). In fact while a large fraction of showers initiated by electrons after one-half radiation length will appear still as a single particle, almost all showers due to photons will emerge with two or more minimum ionizing particles. Fig. 7 shows the result of the test with electrons of 15 GeV, dotted line, compared to the E_F distribution of events, continuous line, with $E^2\theta^2 > 0.54\ GeV^2$ where the π^o coherent production by neutrinos dominates.

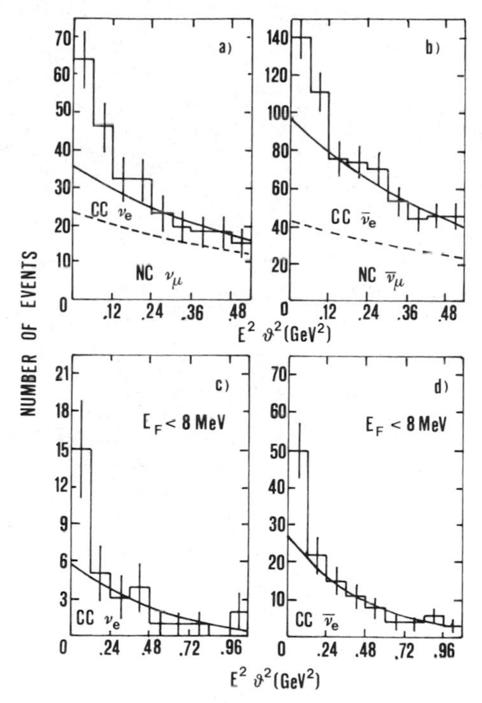

Fig. 6. Distributions of the candidate events as a function of $E^2\theta^2$. (a) for neutrinos, (b) for antineutrinos. (c) and (d) show events with the additional cut E_F<8 MeV.

To take advantage of this feature, the events were subdivided into two regions:

i) $E^2\theta^2 < 0.12$ GeV^2 where 90% of the signal is contained (forward region)

ii) $0.12 < E^2\theta^2 < 0.54$ GeV^2 where almost only the background is present (reference region).

The respective ratios forward/reference of backgrounds (a) and (b) are computed according to the prescriptions discussed

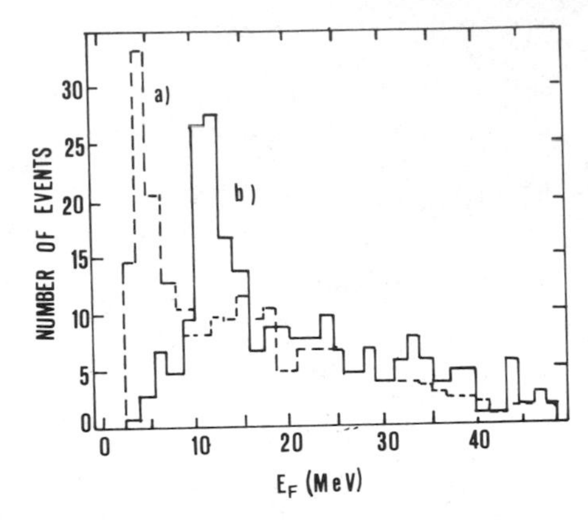

Fig. 7. Energy deposition in the first scintillator plane following the shower vertex: a) showers induced by 15 GeV electrons traversing an average 1/2 marble slab; b) photon induced showers produced by neutrinos in energy-angle range where π^0 coherent processes dominate.

above. The background normalization is obtained by applying a cut $E_F = 8$ MeV as the fraction of events from background (b) with $E_F <$ 8 MeV will be negligible.

Three equations can be written, relating the events in forward region, events in reference region with $E_F < 8$ MeV and events in reference region with $E_F > 8$ MeV to the three unknowns: signal, background (a), background (b), all other quantities being computed or measured. The result of this analysis is summarized in table 3.

Table 3: "Signal" and "Background" in forward region

	Signal	CC '$\bar{\nu}$'$_e$	NC '$\bar{\nu}$'$_\mu$
Neutrino	46 ± 12	20 ± 7	44 ± 7
Antineutrino	77 ± 19	95 ± 14	79 ± 14

An alternative method of analyzing the data can however be performed. Applying the cut $E_F < 8$ MeV to all candidates, the only background left is of the type (a). This background is then normalized in the reference region and extrapolated in the forward region (see fig. 6c and 6d).

We then obtain 10 ± 4 events to attribute to the signal in the neutrino and 25 ± 8 events in the antineutrino beam. These numbers are in good agreement with the signal derived by the first method, taking into acount the efficiency for detecting electrons with $E_F < 8$ MeV, i.e. 0.32 ± 0.05.

Before computing the total cross sections and their ratio, the full composition of the WBB is taken into account to assign the proper number of events to each neutrino type present in the beam.

Finally the absolute cross sections are obtained by normalizing the event numbers to the muon-neutrino and antineutrino interactions, the cross section of which is known. The normalization can also be performed with respect to the quasi elastic muon neutrino events. Different corrections apply to the two methods, thus providing an estimate of systematic errors.

Because the events have the kinematical cut $7.5 \lesssim E < 30$ GeV, the visible cross section σ_{vis} has to be corrected in order to obtain the total cross section σ_{tot}. In the standard model the inelasticity y depends on $\sin^2\theta_W$, therefore the relationship between σ_{vis} and σ_{tot} is a function of $\sin^2\theta_W$. We then obtain the following values:

$$\sigma(\nu_\mu e)/E_\nu = 2.1 \pm 0.55(\text{stat}) \pm 0.49(\text{syst}) \times 10^{-42} \text{cm}^2/\text{GeV}$$

$$\sigma(\bar{\nu}_\mu e)/E_\nu = 1.6 \pm 0.35(\text{stat}) + 0.36(\text{syst}) \times 10^{-42} \text{cm}^2/\text{GeV}$$

$$R_{exp} = \frac{N(\nu_\mu e)}{N(\bar{\nu}_\mu e)} \cdot F = 1.37^{+0.65}_{-0.44}$$

where F = 2.09±0.15 is the energy-weighted flux ratio of $\bar{\nu}_\mu$ and ν_μ.

In the cross section ratio, the systematic errors due to uncertainties in the background, efficiencies and normalization procedure partly cancel out. The main sources are due to background subtraction (±10%) and to the normalization (±7%). The final result from R on the weak mixing angle is then

$$\sin^2\theta_W = 0.215 \pm 0.040(\text{stat}) \pm 0.015(\text{syst})$$

Fig. 8 shows the relationship between R and $\sin^2\theta_W$.

CONCLUSION

A tremendous effort is under way since the pioneering work of Reines in 1956 to complete the experimental knowledge of the neutrino interactions. Only recently experiments are approaching the accuracy needed to confirm or contradict the theoretical framework given by the "standard model".

Along this way, CHARM provides still one of the best result on the weak mixing angle.

In the semileptonic sector, the weak mixing angle is

$$\sin^2\theta_W = 0.220 \pm 0.015$$

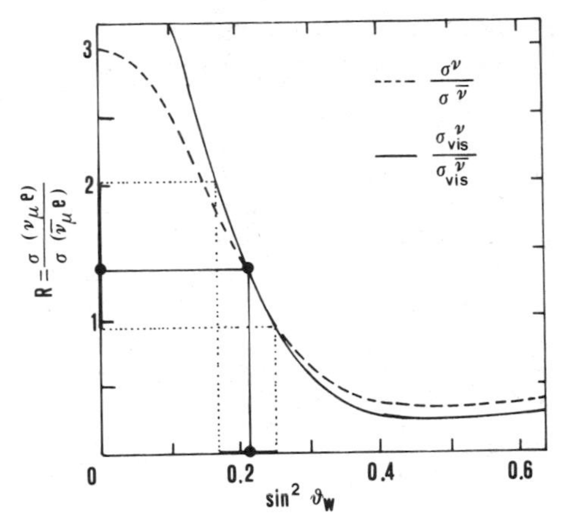

Fig. 8. Ratio of muon neutrino-electron to antineutrino-electron cross sections as a function of $\sin^2\theta_W$. The full curve represents the expected ratio in the energy range 7.5-30 GeV. The dashed curve is the expected R in case of full acceptance.

with ρ fixed to 1, while in the two parameter fit to the data we obtain the remarkable result on

$$\rho = 1.027 \pm 0.023 \quad \text{and} \quad \sin^2\theta_W = 0.247 \pm 0.038$$

A stringent test[10] of the theory will require a factor 3 on the $\sin^2\theta_W$ error and at least a factor 2 on the error of ρ. This achievement not only seems feasible[15], at least for $\sin^2\theta_W$, but also theoretically meaningful, as the QPM and QCD calculations seems to be under control[10].

While the $\sin^2\theta_w$ derived from semileptonic interactions has an error comparable to the theoretical corrections due to second order terms, the value of $\sin^2\theta_w$ obtained in purely leptonic interaction has a greater experimental error by a large factor.

Still the CHARM result is by far the best value today:

$$\sin^2\theta_w = 0.215\pm0.040\pm0.015$$

where the larger error is of statistical nature. A new experiment has been proposed[16] by the CHARM collaboration to reduce the error down to ±0.005.

The simplest Glashow, Weinberg, Salam theory with one Z°, one Higgs doublet fits our data and all present experimental results on weak interaction physics, but still plenty of room remains for surprises.

REFERENCES

1. The CHARM collaboration: F.Bergsma, J.Dorenbosch, M.Jonker, C.Nieuwenhuis and F.Udo (Amsterdam); J.V.Allaby, U.Amaldi, G.Barbiellini, L.Barone, A.Capone, G.Cocconi, W.Flegel, W.Kozanecki, K.H.Mess, M.Metcalf, J.Meyer, R.S.Orr, J.Panman, A.M.Wetherell and K.Winter (CERN); J.Aspiazu, V.Blobel, F.W.Buesser, P.D.Gall, H.Grote, B.Kroeger, E.Metz, F.Niebergall, K.H.Ranitzsch and P.Staehelin (Hamburg); P.Gorbunov, E.Grigoriev, V.Kaftanov, V.Khovansky and A.Rosanov (Moscow), A.Baroncelli, R.Biancastelli, B.Borgia, C.Bosio, F.Ferroni, E.Longo, P.Monacelli, F. de Notaristefani, P.Pistilli, C.Santoni, L.Tortora and V.Valente (Rome).
2. A. N.Diddens et al., Nucl. Inst. Meth. 178:27 (1980).
3. M. Jonker et al., Phys. Lett. 99B:265 (1981).
4. J. Kim et al. Rev. Mod. Phys. 53:211 (1981).
5. E.A. Paschos, L. Wolfenstein, Phys. Rev. D7:91 (1973).
6. In addition to references 7,8 and 9, the calculations of A.De Rujula, R.Petronzio, A.Savoy-Navarro, Nucl.Phys. B154:394 (1979) were used to correct our differential cross section measurements. A general presentation of the subject is contained in F.Antonelli and L.Maiani, Nucl.Phys. B186:269 (1981). A complete list of higher order calculation can be found in the review by M.A.B.Bég, A.Sirlin, Phys.Rep. 88:1 (1982).
7. A. Sirlin, W.J. Marciano, Nucl. Phys. B189:442 (1981).
8. J.F.Wheater, C.H. Llewellyn-Smith, Nucl. Phys. B208:27 (1982).
9. E.A. Paschos, M.Wirbel, Nucl. Phys. B194:189 (1982).
10. C.H. Llewellyn Smith, in Proc. of Workshop on SPS Fixed Target Physics, CERN 83-02: 180 vol. II
11. M.Jonker et al., Phys. Lett. 102B:67 (1981)
12. L.M.Sehgal, Proc. Intern. Neutrino Conf. 1978, Purdue Univ. pag. 263.

13. M. Jonker et al., Phys. Lett. 105B:242 (1981).
    M. Jonker et al., Phys. Lett. 117B:272 (1982).
14. K. S. Lackner, Nucl. Phys. B153:526 (1979).
    S. S. Gershtein et al. Preprint IHEP 81-9 (1981).
15. J. Panman, in Proc. of Workshop on SPS Fixed Target Physics, CERN 83-02:146 vol. II.
16. K. Winter, in Proc. of Workshop on SPS Fixed Target Physics, CERN 83-02:170 vol. II.

RESULTS ON WEAK NEUTRAL CURRENTS FROM THE CDHS COLLABORATION

Christoph Geweniger

Institut für Hochenergiephysik

Universität Heidelberg, F.R.G.

INTRODUCTION

The results of an experiment to measure the neutral (NC) to charged current (CC) cross section ratios for inclusive neutrino nucleon scattering

$$R_\nu = \frac{\sigma(\nu+Fe \rightarrow \nu+X)}{\sigma(\nu+Fe \rightarrow \mu^-+X)} \quad \text{and} \quad R_{\bar\nu} = \frac{\sigma(\bar\nu+Fe \rightarrow \bar\nu+X)}{\sigma(\bar\nu+Fe \rightarrow \mu^+X)}$$

are presented. The experiment has been carried out by the CERN-Dortmund-Heidelberg-Saclay collaboration[1]. Preliminary results from the part of the data have been reported previously[2]. One of the aims of this experiment is to obtain a precise determination of the electroweak mixing parameter $\sin^2\theta_W$. Model calculations to derive $\sin^2\theta_W$ from R_ν will be discussed. Such calculations also relate R_ν to $R_{\bar\nu}$, so that a measurement of the two quantities can be used to test the model.

In the original version of this talk a limit on right handed weak currents, obtained by the CDHS collaboration, was discussed. Since the presented material is already published[3], it will be omitted here.

EXPERIMENTAL DETAILS

The experiment is very similar to the first neutral current experiment of this collaboration[4]. The data were taken in the years 1978/79 with the CDHS detector[5] exposed to the CERN 200 GeV narrow band neutrino beam. The effective neutrino and antineutrino spectra relevant to the conditions of this experiment are shown in Fig. 1. The detector is a magnetized iron cylinder, 3.85 m in diameter, with the axis parallel to the beam axis. Longitudinally it is subdivided into 19 modules consisting of iron slabs with a total thickness of 0.75 m. The slabs are interleaved with planes of scintillators for hadron calorimetry, and drift chambers are inserted between the modules for muon tracking. The scintillator structure allows one to measure the position of the shower.

The main idea of the experiment is to measure only the hadronic showers produced in neutrino interactions, in order to treat NC and CC interactions alike, as much as possible. In particular, no pattern recognition or track reconstruction for muons of CC events is performed. In this way, when forming the ratio of NC events to CC events, many sources of systematical errors are eliminated.

The detector records neutrino-iron interactions producing hadronic showers of energies above the trigger threshold of about 3 GeV. For each event the hadronic energy E_H, the interaction vertex, and the event length L are measured. L is defined as the thick-

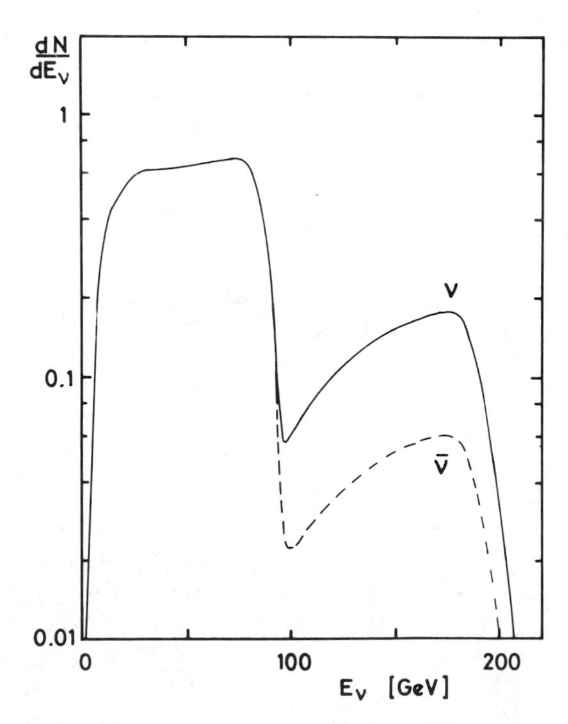

Fig. 1. Neutrino and antineutrino spectra of the CERN 200 GeV narrow band beam valid for the cuts of the neutral current experiment. The abscissa is given in arbitrary units.

ness of iron between the shower vertex and the last downstream signal of the event seen in the detector. It is measured by counting successive scintillator planes and drift chambers with hits, starting from the shower vertex; no position information perpendicular to the detector axis is required.

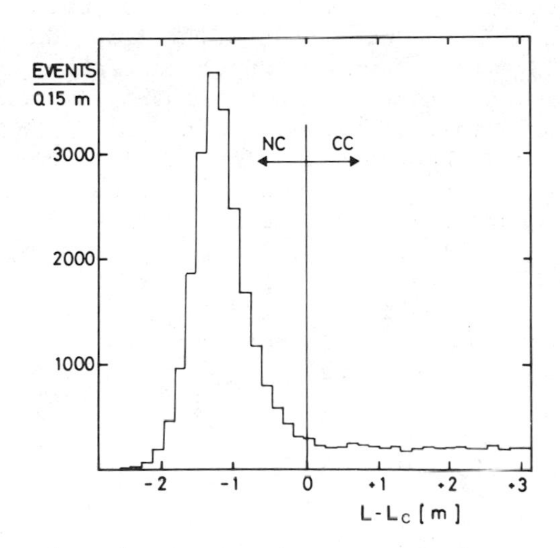

Fig. 2. Event length of neutrino events given in meters of penetrated iron relative to L_C. Only the short events are shown.

A cut on the event length L is used to separate NC and CC events. The length of a NC event is given by the length of the hadronic shower, the length of a CC event in general by the range of the final state muon, which for most CC events is more penetrating than the accompanying shower. The cut L_C is experimentally determined by the requirement, that 99% of all showers have $L < L_C$. L_C depends logarithmically on E_H and is typically at 2 meters of iron. In Fig. 2 neutrino events are plotted versus $\Delta L = L-L_C$. A large peak of NC events and the beginning of a long tail of CC events is seen.

Further cuts are applied to the hadron energy, with $E_H > 10$ GeV, and to the fiducial volume. The fiducial volume is restricted in order to reject possible background and to have well contained events. The lateral distance of a shower vertex from the edge of the detector is larger than 45 cm, and there are at least 6 m of iron after the vertex for a good measurement of L. The effective target mass is about 260 tons.

CORRECTIONS

The two event classes of NC and CC events as defined by the cut L_C have to be corrected for misidentified events. 1% of the NC events have $L > L_C$ by definition of L_C (long NC events). A few percent of the CC events have a muon whose track, projected onto the detector axis, is shorter than L_C (short CC events). The short CC events consist mostly of high-y events ($y \sim E_H/E_\nu$) and a small fraction of events with a muon leaving the detector at the side. The correction is determined by means of Monte Carlo generated CC events which are normalized to measured events with $(L_C + 0.75\text{ m}) < L < 5.1$ m. This corresponds to an extrapolation in y from $\langle y \rangle \sim 0.85$ to $\langle y \rangle \sim 0.95$. In this region of y the differential cross sections $d\sigma/dy$ for neutrinos and antineutrinos are almost flat, and the correction is not very sensitive to uncertainties in the structure functions.

Further corrections are due to background events coming from three sources:

1. Cosmic rays. A small correction is caused by cosmic ray events, which is reliably determined from events taken at times without beam spill.

2. ν_e contamination of the beam from K_{e3} decay. CC events from ν_e interactions in the detector are misidentified as NC events, since the final state electron produces an electromagnetic shower with $L < L_C$. These events have to be removed from the NC sample, where they are entered with their total energy, and to be added to the CC sample corrected for the cut in E_H. The correction is calculated using the known beam properties and K decay characteristics, and assuming e-μ universality for the cross sections.

3. Wide-band neutrino background. This background originates from the decay of particles not passing the momentum selection of the narrow-band beam line. Part of this background is due to low energy particles which decay directly after the production target, before they are swept by the first magnet. It has been measured by closing the momentum defining collinator in the beam line. A raw ratio NC/CC being four times bigger than the ordinary events has

been found. After the collimator the beam passes through small amounts of matter and enters into the decay tunnel. Relatively few secondaries are produced which, however, have a very long decay path. This background has only been calculated, but it accounts for about half of the total background in the neutrino setting and for about 20% in the antineutrino setting, with a rather large systematical uncertainty.

Table 1. Summary of analysis for R_ν and $R_{\bar{\nu}}$

	Neutrino	Antineutrino
Total events	62 000	21 000
NC/CC (raw)	0.407	0.470
Correction factors:		
1) Cosmics	0.990	0.976
2) WBB	0.982	0.876
3) Short CC	0.815	0.917
4) Long NC	1.013	1.014
5) ν_e (K_{e3})	0.918	0.956
Overall corr. (1-5)	0.737	0.760
R_ν, $R_{\bar{\nu}}$	0.300	0.357

RESULTS

The results of the analysis are summarized in Table 1. Starting from a raw ratio NC/CC as obtained by the cuts on E_H, fiducial volume, and L, correction factors are applied to this ratio in the same order as listed in Table 1. The final results for $E_H > 10$ GeV and averaged over the energy spectra shown in Fig. 1 are:

$$R_\nu = 0.300 \pm 0.005 \pm 0.005 \quad (\pm 0.007)$$

$$R_{\bar{\nu}} = 0.357 \pm 0.0095 \pm 0.0115 \quad (\pm 0.015)$$

The errors are given in the sequence: statistical, systematical, and combined error, respectively. In Table 2 measurements of R_ν on isoscalar targets are compared.

Table 2. R_ν results for isoscalar targets.

Experiment	Target	E_H cut [GeV]	R_ν	Ref.	(Year)
CDHS	Fe	12	0.293± 0.010	4	(77)
HPWF	CH	4	0.30 ± 0.04	6	(78)
CITF	Fe	12	0.28 ± 0.03	7	(78)
ABCLOS (BEBC)	Ne	15	0.32 ± 0.03	8	(78)
CHARM	marble	2	0.320± 0.010	9	(81)
IMSTT (15')	D_2	5	0.30 ± 0.03	10	(82)
ABBPPST (BEBC)	D_2	5	0.34 ± 0.03	11	(82)
CDHS (this expt.)	Fe	10	0.300± 0.007	-	(83)

THEORETICAL INTERPRETATION

In the framework of the standard model of electroweak interactions the mixing parameter $\sin^2\theta_W$ can be determined from R_ν. Using a quark parton model for the nucleon and the predicted NC couplings of the quarks, the neutrino cross section ratio for NC and CC interactions can be calculated as a function of $\sin^2\theta_W$. Such a model calculation for the conditions of this experiment has been performed by Kim et al.[12] For $R_\nu = 0.300 \pm 0.007$, and including radiative corrections[13,14], the model yields

$$\sin^2\theta_W = 0.228 \pm 0.012.$$

The error reflects the experimental error of R_ν. The model uncertainty is not explicitly given in Ref. 12 for the determination of $\sin^2\theta_W$ from R_ν alone. One of the authors, however, estimates it to about ± 0.003[15]. Some improvement of the model of Ref. 12 is possible. In order to check the above result, we have performed a new model calculation. The main differences are discussed in the following.

1. More precise nucleon structure function data have become available since the publication of Ref. 12. New measurements of the structure functions of valence quarks[16], antiquarks[16] and strange quarks[17] have been used as input to the quark parton model.

2. Structure functions depending on the hadronic energy E_H only have been directly determined from the data, rather than inte-

grating parametrizations of structure functions over x and Q^2. This method avoids, in particular, the extrapolation to $Q^2 = 0$, which is usually not well defined in parametrized fits to structure function data at large Q^2.

3. Kinematical effects due to the mass of the charm quark have been taken into account. They were ignored in Ref. 12. For instance, the strange sea component of the nucleon is kinematically suppressed in CC interactions due to the production of (heavy) charm quarks, whereas there is no suppression in NC interactions leading to (light) strange quarks in the final state. We have used the slow rescaling model of Barnett[18] to include the mass effects in the caluclation, assuming an effective charm mass of 1.5 GeV.

4. Radiative corrections have been incorporated directly into the model, since detailed formulae for neutrino and antineutrino interactions with quarks and antiquarks have recently been published[14]. This allows for a correct treatment of the experimental cuts and easy variations of the model. Previous numerical results [13,14] on radiative corrections were obtained under some simplifying assumptions.

The new model calculation yields a preliminary result from this experiment:

$$\sin^2\theta_W = 0.232 \pm 0.012.$$

This is very close to the result quoted above. The model uncertainty has also been investigated. It is completely dominated by uncertainties related to the charm mass effect. For a given flavour content of the nucleon the relative amount of charm production in NC interactions is fixed, for CC interactions, however, it is controlled by the Kobayashi-Maskawa (KM) matrix. Therefore, the fraction of quarks, which may be suppressed in CC interactions, is uncertain to the extent, that the KM matrix elements describing transitions to charm are known. Llewellyn Smith[19] estimates an error of $\Delta\sin^2\theta_W = \pm\ 0.008$ coming from this source. In addition, the suppression itself depends on the effective charm mass m_c. Using $m_c = (1.5 \pm 0.3)$ GeV or a running mass[20] with $m_c^c(m_c) = 1.27$ GeV[21] yield an uncertainty of about $\Delta\sin^2\theta_W = \pm\ 0.004$.

Other effects are less severe than the charm mass effect (see also Ref. 19). The total uncertainty is of the order of ±0.01. However, this is expected to be reduced in the future, in particular when the KM matrix elements will be known better. Then a more model independent determination of $\sin^2\theta_W$ proposed in Ref. 19 will further reduce the theoretical uncertainties.

The results of this experiment on the electroweak mixing parameter are compared in Table 3 to other precision experiments. The agreement between the different determinations of $\sin^2\theta_W$ is rather good even though the actual values range from 0.208 to 0.244. There is also good agreement with the SU(5) prediction $\sin^2\theta_W = 0.215 \pm 0.006$[14].

Table 3. Determinations of $\sin^2\theta_W$ from precision experiments. Radiative corrections are included, errors are experimental only.

Experiment		Ref./Year	$\sin^2\theta_W$	Model
CDHS	R_ν=0.293±0.010	4/77	0.244±0.017	This calc. (prelim.)
SLAC	eD asymmetry	22/79	0.215±0.015	Ref. 12, 13
Charm	R_ν=0.320±0.010	9/81	0.208±0.014	Ref. 9, 14
CDHS	R_ν=0.300±0.007	-/83	0.232±0.012	This calc. (prelim.)

$R_{\bar\nu}$ is rather insensitive to $\sin^2\theta_W$, given the actual value of this parameter. However, the above model calculations can be used to relate $R_{\bar\nu}$ to R_ν. This relation is tested in Fig. 3. Data from the CHARM and CDHS experiments are shown. The results of the earlier experiments[4,9] have been adjusted to the conditions of this experiment, in order to allow for a direct comparison. The agreement among the experiments and between the data and the prediction is satisfactory.

CONCLUSIONS

A new precision measurement of the neutral to charged current cross section ratios R_ν and $R_{\bar\nu}$ for inclusive neutrino and antineutrino scattering off an isoscalar target has been reported. From the experimental result two values of the electroweak mixing parameter $\sin^2\theta_W$ have been determined: $\sin^2\theta_W = 0.228 \pm 0.012$ using existing model calculations[12,13,14] and $\sin^2\theta_W = 0.232 \pm 0.012$ using an improved calculation presented above. The latter calculation is still preliminary. Both results include radiative corrections and are in very good agreement with each other. The earlier model calculation ignored the kinematical effects related to charm production. These effects lead to a rather large uncertainty in the determination of $\sin^2\theta_W$ from R_ν, which in total is of the order of ±0.01. This uncertainty is expected to be reduced in the future[19].

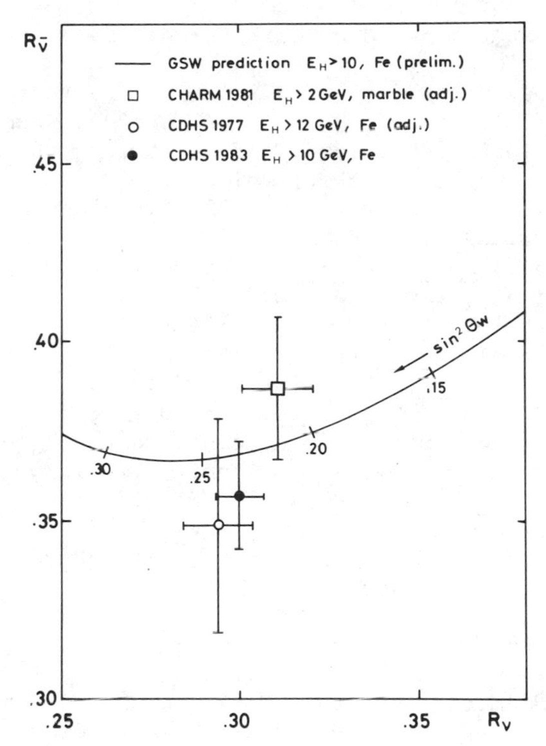

Fig. 3. Comparison of R_ν and $R_{\bar{\nu}}$ with the prediction of the standard model. The curve is a preliminary calculation for the conditions of this experiment including radiative corrections. Other data are adjusted to these conditions.

ACKNOWLEDGEMENTS

I wish to thank Harvey Newman for organizing this stimulating and enjoyable conference. The warm hospitality experienced at the Centre "Ettore Majorana" is gratefully appreciated.

REFERENCES

1. H. Abramowicz, J.G.H. de Groot, J.T. He, J. Knobloch, J. May, P. Palazzi, A. Para, F. Ranjard, T.Z. Ruan, A. Savoy-Navarro, D. Schlatter, J. Steinberger, H. Taureg, W. von Rüden, H. Wahl, J. Wotschack and W.M. Wu. (CERN, Geneva, Switzerland)
   P. Buchholz, F. Eisele, H.P. Klasen, K. Kleinknecht, H. Lierl, D. Pollmann, B. Pszola, B. Renk and H.J. Willutzki. (Institut für Physik der Universität Dortmund, F.R.G.)
   F. Dydak, T. Flottmann, C. Geweniger, J. Królikowski and

K. Tittel. (Institut für Hochenergiephysik der Universität Heidelberg, F.R.G.)
P. Bloch, B. Devaux, C. Guyot, J.P. Merlo, B. Peyaud, J. Rander, J. Rothberg and R. Turlay. (DPhPE, CEN-Saclay, France).

2. C. Geweniger, in: Proc. Intern.Conf. Neutrino '79, Bergen 1979, ed. A. Haatuft and C. Jarlskog, Vol. 2 p. 392.
3. H. Abramowicz et al., Z. Phys. C12:225 (1982).
4. M. Holder et al., Phys. Lett. 71B:222 (1977).
5. M. Holder et al., Nucl. Instrum. Methods 180:429 (1981).
6. P. Wanderer et al., Phys. Rev. D17:1679 (1978).
7. F.S. Merritt et al., Phys. Rev. D17:2199 (1978).
8. P.C. Bosetti et al., Phys. Lett. 76B:505 (1978).
9. M. Jonker et al., Phys. Lett. 99B:265 (1981).
10. T. Kafka et al., Phys. Rev. Lett. 48:910 (1982).
11. P.H.A. van Dam, in: Proc. Intern. Conf. Neutrino '82, Balatonfüred 1982, ed. A. Frenkel and L. Jenik, Vol.II p. 51.
12. J.E. Kim et al., Rev. Mod. Phys. 53:211 (1981).
13. A. Sirlin and W.J. Marciano, Nucl. Phys. B189:442 (1981).
14. J.F. Wheater and C.H. Llewellyn Smith, Nucl. Phys. B208:27 (1982)
15. P. Langacker, cited in Ref. 13.
16. H. Abramowicz et al., Z. Phys. C17:283 (1983).
17. H. Abramowicz et al., Z. Phys. C15:19 (1982).
18. R.M. Barnett, Phys. Rev. Lett. 36:1163 (1976).
19. C.H. Llewellyn Smith, in: Proc. Workshop on SPS Fixed-Target Physics in the Years 1984-1989, CERN Yellow Report 83-02, Vol. II p. 180.
20. O. Nachtmann and W. Wetzel, Nucl. Phys. B187:333 (1981).
21. J. Gasser and H. Leutwyler, Physics Reports 87:76 (1982).
22. C.Y. Prescott et al., Phys. Lett. 84B:524 (1979).

MEASUREMENT OF $\sin^2\theta_W$ IN SEMILEPTONIC ν Fe AND $\bar{\nu}$ Fe INTERACTIONS

Presented by P.A. Rapidis

R.E. Blair[1], A. Bodek[5], R.N. Coleman[5], D. Edwards[4], H.T. Edwards[4], O.D. Fackler[6], H.E. Fisk[4], Y. Fukushima[4], K.A. Jenkins[6], B.N. Jin[4], Q.A. Kerns[4], T. Kondo[4], D.B. McFarlane[1], W.L. Marsh[5], F.S. Merritt[2], R.L. Messner[1], D.B. Novikoff[1], M.V. Purohit[1], P.A. Rapidis[4], P.G. Reutens[2], F.J. Sciulli[3], S.L. Segler[4], M.H. Shaevitz[3], R.J. Stefanski[4], D.E. Theriot[4], and D.D. Yovanovitch[4]

We report results from neutral current scattering of muon neutrinos and antineutrinos on iron nuclei by the CCFRR experiment at Fermilab. Our objective is the measurement of $\sin^2\theta_W$ from inclusive semileptonic neutrino-nucleon interactions. Such a measurement, even though it is subject to more corrections than the analogous measurement in neutrino-electron scattering (e.g. due to the presence of a strange quark sea in the nucleon), has a high degree of statistical accuracy and may afford the most precise test of the standard model of electro-weak interactions. Furthermore, the recent observations[1] of the Z^0 and $W^\pm$ in proton-antiproton collisions give such a measurement added topical interest. The comparison of $\sin^2\theta_W$ determined from the masses of the Z^0 and $W^\pm$ and from neutrino interactions, i.e. from different processes and energy regions, may hold some surprises. Finally, the grand-unified theories make definite predictions[2] about the value of $\sin^2\theta_W$ and the lifetime of the proton; $\sin^2\theta_W$ measurements test the validity of these theories.

---

[1]California Institute of Technology, Pasadena, CA 91125
[2]University of Chicago, Chicago, IL 60637
[3]Columbia University, New York, NY 10027
[4]Fermi National Accelerator Laboratory, Batavia, IL 60510
[5]University of Rochester, Rochester, NY 14627
[6]Rockefeller University, New York, NY 10021

The experiment was performed in Fermilab's dichromatic neutrino beam and utilized a large detector consisting of an iron-scintillator target calorimeter followed by a steel toroidal muon spectrometer. Since the detector and beamline have been described elsewhere[3] we will only give a short description.

Briefly, 400 GeV protons impinge on a BeO target and produce secondary particles which are momentum and sign selected by a point-to-parallel beam channel. The secondary particles include e, π, K, and p's with an rms momentum spread of 10% and an rms angular

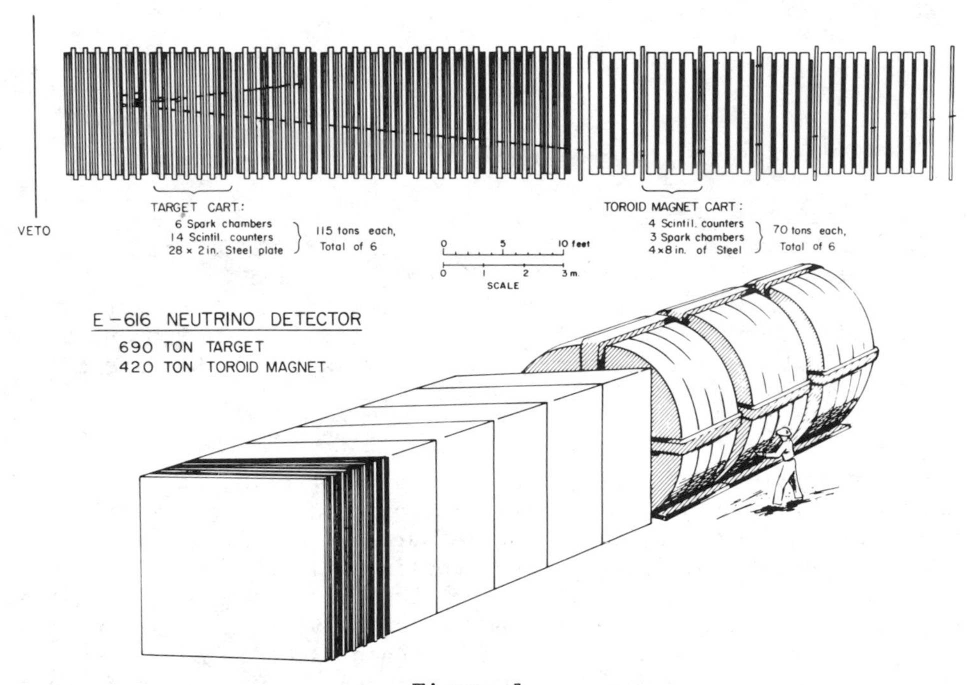

Figure 1

width of approximately 0.15 mrad. These secondaries travel through a 340m long evacuated decay space where some of the π's and K's decay and produce neutrinos and muons. A beam dump at the end of the decay region and a 900m long steel and earth shield absorb the surviving secondaries and the decay muons. Instrumentation in the beam line measures the intensity and particle composition of the secondary beam and allows us to calculate the neutrino flux at the end of the shield where our detector is located. Data were taken at five settings of the secondary beam momentum with both polarities.

The detector (Fig. 1) is a hadron calorimeter made of 690 tons of steel plates interspersed with spark chambers (every 20cm of steel) and liquid scintillation counters (every 10cm of steel).

It is followed by a steel toroidal spectrometer instrumented with spark chambers (every 80cm of steel) and scintillation counters (every 20cm of steel). Hadron energies and muon emission angles are measured in the calorimeter, and muon momenta are reconstructed in the spectrometer. The hadronic shower energy is measured with an rms resolution of

$$\Delta E_H = 0.89 \sqrt{E_H}(\text{GeV}) \quad .$$

The sparks in a hadronic shower can be used to determine the transverse position of the event vertex with an energy independent rms resolution of 1.7 inches. Even though the vertex for charged current events with a final state muon can be determined much more precisely, the charged current and neutral current events were treated in an identical fashion and only the vertex determined by the hadronic shower information was used. Calibration data taken with pions of known energy show that the maximum length for hadronic showers in our calorimeter for energies up to 200 GeV is 210 cm of steel.

Three types of event triggers were installed :

i. The muon trigger requiring a particle to penetrate through part of the calorimeter and the upstream part of the spectrometer; no hadronic energy requirement was imposed.

ii. The penetration trigger requiring the deposition of at least 4 GeV of energy in the calorimeter and the presence of a penetrating particle (muon) travelling through 160 cm of steel or more.

iii. The hadronic energy trigger requiring only an energy deposition of 12 GeV in the calorimeter.

Events of the first two triggers were used for the study of charged current interactions[4] and of trigger and reconstruction efficiencies for events that also satisfied the hadronic energy trigger. From such studies we have determined that for hadronic energies greater than 18 GeV the hadronic trigger is 100% efficient. The events used in this analysis were required to satisfy the hadronic energy trigger, have a minimum hadronic energy of 20 GeV, and have a vertex within a 40 inch radius from the center of the detector.

The traditional technique[5] of determining $\sin^2\theta_W$ is based on the study of the cross-section ratios

$$R_\nu = \frac{\sigma_\nu^{NC}}{\sigma_\nu^{CC}} \quad , \text{ and } \quad R_{\bar{\nu}} = \frac{\sigma_{\bar{\nu}}^{NC}}{\sigma_{\bar{\nu}}^{CC}} \quad .$$

The precision of the determined $\sin^2\theta_W$ depends on the precision with which one can measure R_ν and $R_{\bar{\nu}}$. In determining these ratios one has to estimate the number of neutral current events in the presence of backgrounds due to misidentified charged current events.

Neutral current events are identified through the study of the length distribution of events where the length of the event refers to the distance over which energy is deposited in the target calorimeter. Charged current interactions, besides producing a relatively short hadronic shower at the event vertex, are accompanied by a final state muon that travels many meters in our steel target, and therefore, have a long event length. Neutral current interactions give rise to events with short lengths since the final state contains only hadrons. The major problem in determining R_ν and $R_{\bar{\nu}}$ arises from charged current events where the muon is produced at a large angle and leaves the calorimeter from the side after a short distance. These events can be easily misidentified as neutral current events. Such backgrounds are approximately 20% of the neutral current interactions (Fig. 2a).

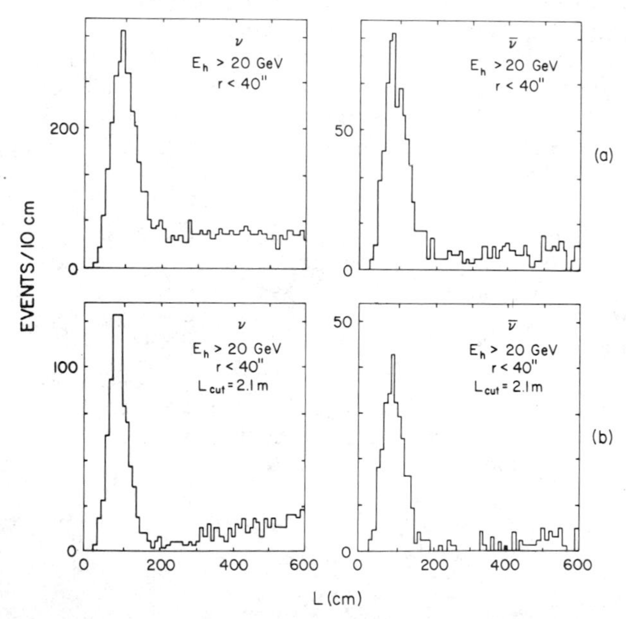

Figure 2. Length distribution for (a) events satisfying the trigger , hadronic energy, and radius requirements, (b) events that in addition satisfy the requirement $y_{NC} < y_{cut}$. The data shown here and in Figures 4 and 5 were obtained for a secondary beam momentum of 200 GeV/c. They correspond to approximately 25% of the full data sample.

A different method has been proposed by Paschos and Wolfenstein[6], who suggest using the ratios :

$$R_- = \frac{\dfrac{d^2\sigma_\nu^{NC}}{dxdy} - \dfrac{d^2\sigma_{\bar{\nu}}^{NC}}{dxdy}}{\dfrac{d^2\sigma_\nu^{CC}}{dxdy} - \dfrac{d^2\sigma_{\bar{\nu}}^{CC}}{dxdy}} = \rho^2 \left(\frac{1}{2} - \sin^2\theta_W \right) \quad \text{and}$$

$$R_+ = \frac{\dfrac{d^2\sigma_\nu^{NC}}{dxdy} + \dfrac{d^2\sigma_{\bar{\nu}}^{NC}}{dxdy}}{\dfrac{d^2\sigma_\nu^{CC}}{dxdy} + \dfrac{d^2\sigma_{\bar{\nu}}^{CC}}{dxdy}} = \rho^2 \left(\frac{1}{2} - \sin^2\theta_W + \frac{10}{9} \sin^4\theta_W \right) ,$$

These ratios are valid over any region of x and y; using them allows us to restrict our attention to particular regions in y where the separation between neutral current interactions and charged current interactions is clear. Furthermore, R_- being in essence the ratio of two non-singlet structure functions, is largely independent of the quark sea component of the target nucleons.

The following procedure was used to limit the y region in our analysis. A scan of the data was used to determine the minimum event length in the target, L_{cut}, necessary for the muon to be recognized efficiently. L_{cut} was determined to be 210 cm, a range sufficient to contain all neutral current events, as verified by our calibration data.

Figure 3a is a schematic representation of a neutrino event in the target portion of our detector. A muon may not have sufficient range for identification in the target if its angle relative to the neutrino direction is greater than θ_{max}, which depends on the vertex position of the interaction. Charged current events of this type can be removed by a cut on $y_{NC}=E_{vis}/E(r)$ by employing the kinematical relations between the neutrino energy, the lepton scattering angle and y. E_{vis} is the observed hadron energy of the interaction, while E(r) is the neutrino energy predicted[3] from the dichromatic beam at a given radius. The cut on the event sample requires that

$$y_{NC} < y_{cut} \equiv \frac{E(r)\theta^2_{max}}{2M_p + E(r)\theta^2_{max}} ,$$

where $\theta_{max} = \arctan \{ (R-r)/L_{cut} \}$, with R = 60 inches.

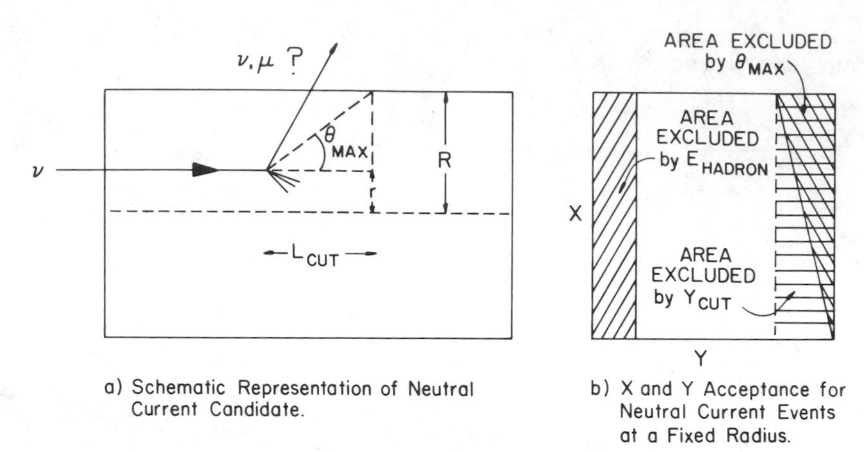

a) Schematic Representation of Neutral Current Candidate.

b) X and Y Acceptance for Neutral Current Events at a Fixed Radius.

Figure 3

Figure 3b shows schematically, for a fixed radius, the regions in x and y excluded by the y_{cut} and the minimum energy requirement. Figure 4 is a scatter plot of events in the y_{NC}-radius plane.

Figures 2 and 5 show the event length distribution with and without the y_{cut}. After the cut on y_{NC} is applied to all events, the length cut at 210 cm gives a much better separation of neutral current and charged current events. Since our event sample is large, this loss of events is not a very severe penalty.

Events with $y_{NC} < y_{cut}$ and $L < L_{cut}$ were assumed to include all neutral current events. The remaining charged current contamination in this region is estimated using a Monte Carlo calculation normalized to the number of events with length between 210 and 310 cm. The shape of this calculation agrees well with the data. The contamination due to charged currents is reduced to 5.5% for neutrinos and 2.3% for antineutrinos. To assure that the charged and neutral current events came from the same x-y regions, the charged current events were required to pass the identical criteria as the neutral current events (after correcting for the muon ionization contribution to the hadronic energy) with the exception that the length requirement was $L > L_{cut}$.

In applying the y_{NC} cut to the data shown in Figs. 2 and 4, we used the expected neutrino energy for neutrinos originating from pion decay. The use of this pion neutrino energy to calculate y_{NC} removes the majority of events due to neutrinos originating from kaon decays since pion neutrinos have roughly half the energy of kaon neutrinos. The remaining background from kaon neutrino neutral current events with small hadronic energy was subtracted using Monte Carlo calculations. The number of such kaon neutrino events surviving the y_{cut} is rather small. Table I shows the number of events at various stages of the analysis.

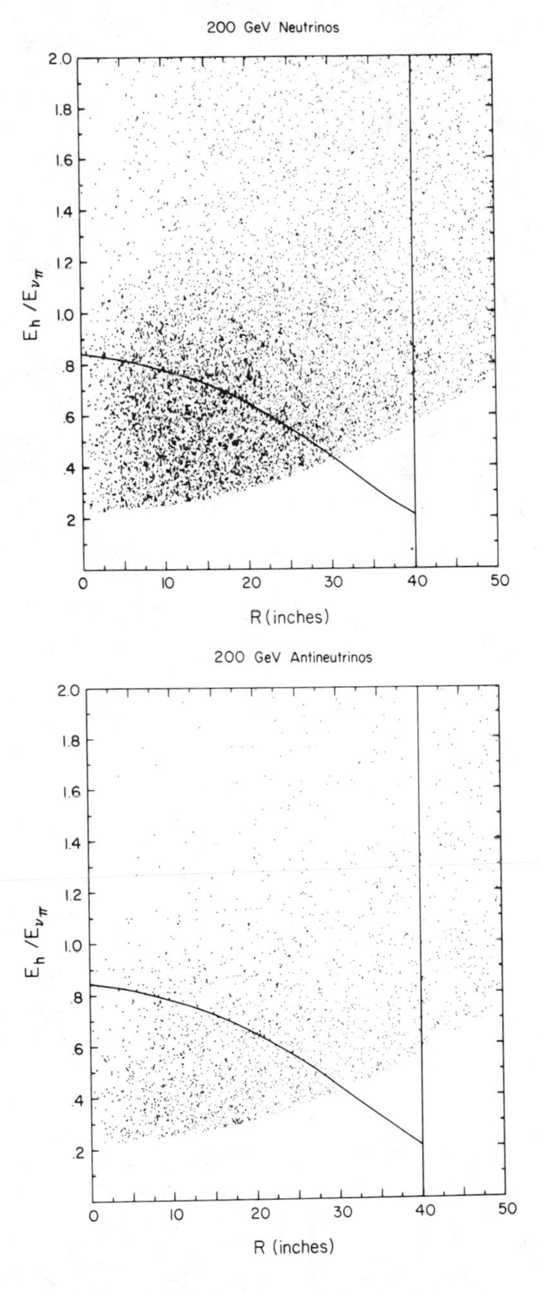

Figure 4. $y=E_H/E_\nu$ versus radius distribution for the data shown in Figure 2(a). The low y behaviour of the data reflects the 20 GeV energy requirement. The superposed lines indicate the $y_{NC} < y_{cut}$ and the $r < 40$ inches cuts.

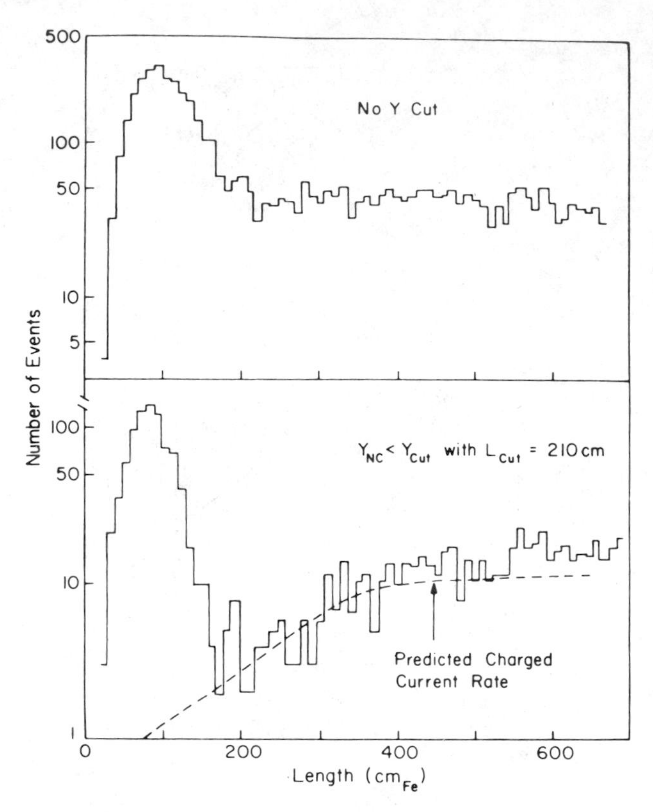

Figure 5. Length distribution for neutrino events (see Fig. 2). The charged current Monte Carlo prediction is also indicated.

TABLE I

	Number of Events	
	Neutrinos	Antineutrinos
Hadronic Energy Trigger, $r<40$ inches, $E_H > 20$ GeV	32316	7455
$y < y_{cut}$	13950	3760
Pion Neutrino Events	10340	3381
$L < L_{cut}$	2514	961
$L > L_{cut}$	7826	2420
Kaon Neutrino Events	3610	379
$L < L_{cut}$	1029	131
$L > L_{cut}$	2581	248

Table II shows the remaining corrections to the neutral current event sample for pion neutrino events. Due to the limited hadron energy of the pion neutrino events the contamination due to electron neutrino charged current interactions originating from neutrinos produced in K_{e3} decays, which our apparatus and analysis techniques misidentify as neutral current events, has been substantially reduced by the use of the y_{cut}. The final event population was determined by applying this correction as well as corrections due to wide band neutrinos, cosmic ray contamination, neutral current events where a π or a K in the hadronic shower decays and gives a muon causing them to be misidentified as charged current events, and other minor corrections.

Rather than calculating cross sections, in order to determine the ratios R_+ and R_-, we calculated quantities proportional to the cross sections, i.e. the number of events divided by the integrated flux of incident neutrinos or antineutrinos.

TABLE II

Corrections to Neutral Current Events		
	Neutrinos	Antineutrinos
Events	10340	3381
E_H Cut	20 GeV	20 GeV
WBB and Cosmic Rays	-1.3%	-5.0%
CC Background	-5.5%	-2.3%
K_{e3} Background	-1.1%	-0.5%
π/K Decay in Shower	+0.1%	+0.1%
Other Corrections	+0.2%	+0.4%
(ν_k Allowance)	(-9.4%)	(-5.2%)
Total	-7.6%(-17.3%)	-7.3%(-12.5%)

Events with hadronic energies larger than the maximum possible energy that a pion neutrino can have can be uniquely identified as kaon neutrino events. We analyzed this kaon neutrino event sample in a completely analogous manner.

Combining both the pion and kaon neutrino results we find that

$$R_+ = 0.246 \pm 0.016 \qquad \text{and} \qquad R_- = 0.308 \pm 0.007 \quad ,$$

Figure 6 shows these ratios, for the pion neutrino sample, as functions of the neutrino energy. The ratios show no significant dependence on the neutrino energy or on y (Ref. 7).

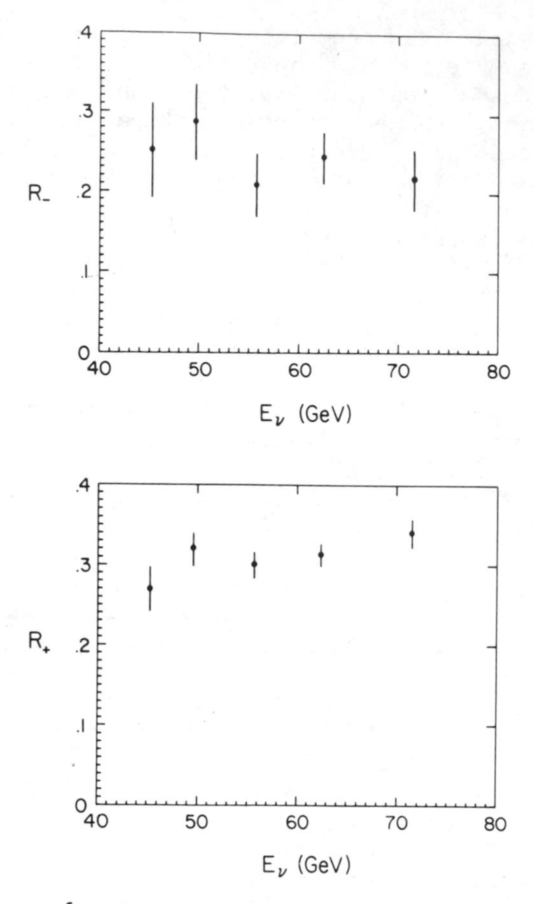

<u>Figure 6.</u> R_- and R_+ versus mean neutrino energy for the pion data sample after corrections.

Before obtaining $\sin^2\theta_W$ it is necessary to include additional corrections for the contribution of the strange and charmed sea quarks (to R_+), the small difference between the neutrino and antineutrino energy spectra, and the small excess neutron content of our iron target. The sea correction, for the results reported here, has been evaluated in a Monte-Carlo calculation utilizing Buras-Gaemers[8] structure functions and does not include the effects of charm excitation (slow rescaling). These corrections amount to +2.8% for R_- and +4.1% for R_+.

Using these corrected ratios we obtain the following values for ρ and $\sin^2\theta_W$:

$$\rho = 1.0054 \pm 0.024 \text{ and}$$

$$\sin^2\theta_W = 0.253 \pm 0.026 \quad .$$

Figure 7 shows the confidence regions in the ρ $\sin^2\theta_W$ plane. If we

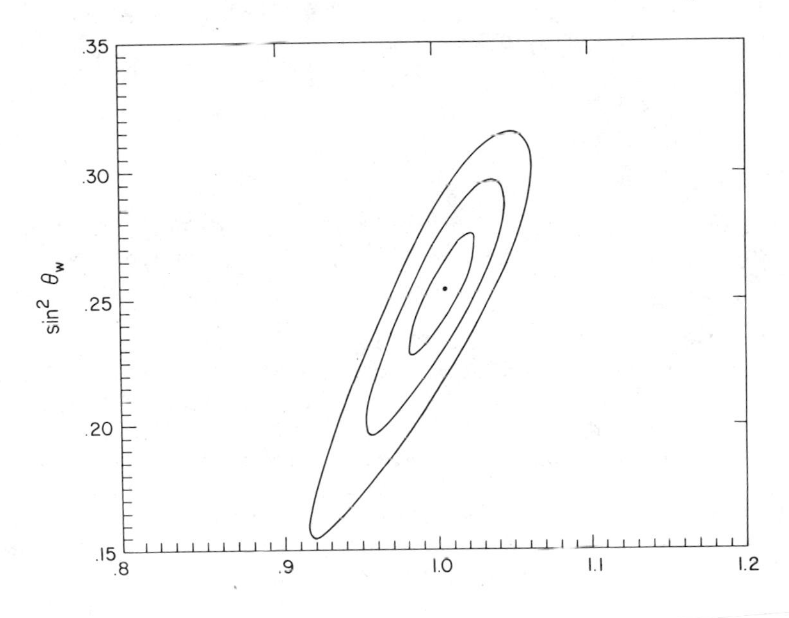

Figure 7. 1σ contours in the $\sin^2\theta_W$ - ρ plane. The innermost contour corresponds to the 39.4% confidence level.

fix the parameter ρ to be exactly equal to 1, we then obtain for

$$\sin^2\theta_W = 0.248 \pm 0.012 \quad \text{(statistical error)} .$$

These results are preliminary for the following two reasons : i. The corrections due to the strange/charmed sea were evaluated using Buras-Gaemers structure functions and do not include slow rescaling. We are currently evaluating these corrections using structure functions extracted[4] from our experiment. ii. Radiative corrections[9] have not been included. We are also evaluating radiative corrections at this time. Preliminary results from this program indicate that $\sin^2\theta_W$ decreases by approximately 0.01. We have only quoted statistical errors; a careful evaluation of systematic errors is also under way. Our preliminary estimate of our systematic uncertainties is 0.015 in the value of $\sin^2\theta_W$.

Our value for $\sin^2\theta_W$, while consistent with other reported[10] values from neutrino-nucleon scattering, tends to be higher than the world average. The same is also true if we compare our measurement to the values of $\sin^2\theta_W$ obtained from the measurements of the masses of the $W^{\pm}$ and Z^0 in $\bar{p}p$ collisions. Our measured value of $\sin^2\theta_W$ is higher than 0.212, the value predicted by SU(5) GUTS (assuming that $\Lambda_{\overline{MS}}$= 0.24 GeV).

REFERENCES

1. G. Arnison et al., Phys. Lett. 122B, 103 (1983); G. Arnison et al., Phys. Lett. 126B, 398 (1983); M. Banner et al., Phys. Lett. 122B, 476 (1983); P. Bagnaia et al., Phys. Lett. 129B, 130 (1983).
2. W.J. Marciano and A. Sirlin, Phys. Rev. Lett. 46, 163 (1980).
3. B.C. Barish et al., Proceedings of the 9th SLAC Summer Institute of Particle Physics, Stanford, 1981, A. Mosher ed., p. 641 (1982); R.E. Blair, Ph. D. thesis (1982), California Institute of Technology (unpublished); R.E. Blair et al., Fermilab Report Pub-83/26-Exp,1983 (to be submitted to Nucl. Instr. and Methods).
4. D.B. McFarlane et al., contributed paper to International EPS Conference, Brighton, England, July 1983 ; R.E. Blair et al., Phys. Rev. Lett. 51, 343 (1983).
5. For a review see:J.E. Kim et al., Rev. Mod. Phys. 53, 211 (1981).
6. E.A. Paschos and L. Wolfenstein, Phys. Rev. D7, 91 (1973). P.Q. Hung and J.J. Sakurai, Phys. Lett. 63B, 295 (1976).
7. M.H. Shaevitz et al., Proceedings of Neutrino-81, International Conference on Neutrino Physics and Astrophysics, Hawaii, 1981, R.J. Cence ed., Vol. I,p. 311; R. Messner, Proceedings of 16th Rencontre de Moriond, 1981, J. Tran Thanh Van ed., Vol.I,p. 341.
8. A.J. Buras and K.J.F. Gaemers, Nucl. Phys. B132, 249 (1978).
9. A. Sirlin and W.J. Marciano, Nucl. Phys. B189, 442 (1981); J.F. Wheater and C.H. Llewellyn Smith, Nucl. Phys. B208, 27 (1982); E.A. Paschos and M. Wirbel,Nucl. Phys. B194, 189(1982).
10. See the contributions by C. Geweniger and B. Borgia in these proceedings; also Ref. 5.

ELECTROWEAK EFFECTS IN DEEP INELASTIC MUON SCATTERING

G. Smadja

DPhPE, CEN-Saclay
91191 Gif-sur-Yvette Cedex, France

I - CHARGE ASYMMETRY IN $\mu+$ AND $\mu-$ SCATTERING

The Electroweak Currents

The interference between weak and electromagnetic amplitudes generates charge and polarization asymmetries which can be used to measure neutral current parameters.

The weak neutral current amplitude is written as :

$$\frac{G_F}{\sqrt{2}} \; \bar{\mu} \, \gamma_\mu \, (v_\mu - a_\mu \gamma_5) \mu \; \bar{q} \, (v_q - a_q \gamma_5) q$$

The interference between photon and Z° exchange at first order will contribute to the cross section an amount given by A. Love et al. and other authors[1] :

$$(1) \qquad \frac{d^2\sigma^{\pm}_{\gamma z}}{dQ^2 d\nu} = \frac{G_F}{\sqrt{2}} \frac{\alpha}{Q^2 \nu} \left\{ \begin{array}{l} (1 + (1-y)^2) \; G_2(x) \; (-v_\mu \pm \lambda a_\mu) \\ \quad + (1 - (1-y)^2) \; xG_3(x) \; (\pm a_\mu - b v_\mu) \end{array} \right\}$$

where λ is the beam polarization, x and y the usual deep inelastic variables.

a_μ and v_μ are here functions of the weak isospin assignments of muons :

$$(2)\quad \begin{cases} v_\mu = I_3^L + I_3^R - 2\,Q\,\sin^2\theta_w = -\frac{1}{2} + 2\,\sin^2\theta_w \\ a_\mu = I_3^L - I_3^R = -\frac{1}{2} \end{cases}$$

The functions G_2 and G_3, can be expressed within the parton model in terms of the vector and axial vector couplings of quarks

$$(3)\quad \begin{cases} G_2(x) = 2x \sum_i v_i Q_i (q(x) + \bar{q}(x)) \\ xG_3(x) = -2x \sum_i a_i Q_i \ (q(x) - \bar{q}(x)) \end{cases}$$

The charge asymmetry which is measured is $B = \dfrac{\sigma_+(\lambda) - \sigma_-(-\lambda)}{\sigma_+(\lambda) + \sigma_-(-\lambda)}$

where the simultaneous flip of charge and polarization should be noted.

From (1) and (2) one derives

$$(4)\qquad B = -\frac{1}{\sqrt{2}}\,\frac{G}{2\pi\alpha}\,Q^2\,g(y)\,[a_\mu A - |\lambda|\,v_\mu\,A]$$

with
$$g(y) = \frac{1 - (1-y)^2}{1 + (1-y)^2}$$

$$a_\mu = -\,1/2$$

$$v_\mu = -\,1/2 + 2\,\sin^2\theta_w \qquad (\sin^2\theta_w = 0.23)$$

$$A^{n+p} = \frac{6}{5}\,(a_d - 2\,a_o) = -\,1.8$$

so that the expected value is

$$(5)\qquad \boxed{B = -\,1.51\cdot 10^{-4}\ g(y)\ Q^2\ \ \mathrm{GeV}^2}$$

The result has been derived while accepting the whole of the standard model :

- quark description of deep inelastic interactions

- lepton and hadron couplings are computed with $I_3^R=0$.

In muon scattering, the dominant contribution to B is the axial-axial coupling $a_\mu A$ in formula (3), but some $\lambda|v_\mu|A$ contribution is present.

This should be contrasted with other asymmetry measurements :

- at SLAC, the polarization asymmetry in e D_2 scattering measured by Prescott et al.[2]

$$A(\lambda) = -\frac{G}{\sqrt{2}}\frac{Q^2}{2\pi\alpha}\lambda(-a_e V + V_e A)$$ contains only "parity violating"

couplings aV, vA.

- on the other hand, the forward backward asymmetry in $(e^+e^- \to \ell^+ \ell^-)$ scattering :

$$A^{FB} = \frac{3}{2} a_e a_\mu \frac{G_F}{8\pi\alpha\sqrt{2}} S$$

is nearly pure axial-axial contribution and does not give a determination of $\sin^2\theta_w$ away from the Z^o mass.

Higher order corrections

Radiative corrections to the first order γZ^o interference have been computed in the one loop approximation by Bardin et al.[3]. These large corrections, shown on Fig. 1 as a function of $g(y)Q^2$, result in a substraction of 30 to 50 % to the first order asymmetry.

The magnitude of the radiative corrections is a weakness of the interpretation of charge asymmetry measurements. The comparison with the observed asymmetry may however also be considered as an indirect check of higher orders of the electroweak theory : it turns out that weak electromagnetic terms almost cancel out at the 1 loop level, and second order QED dominates[4].

Charge Asymmetry Measurement by the BCDMS Collaboration

The data. The asymmetry measurement of Argento et al.[5] has recently been published. The BCDMS spectrometer, shown on Fig. 2 has a good

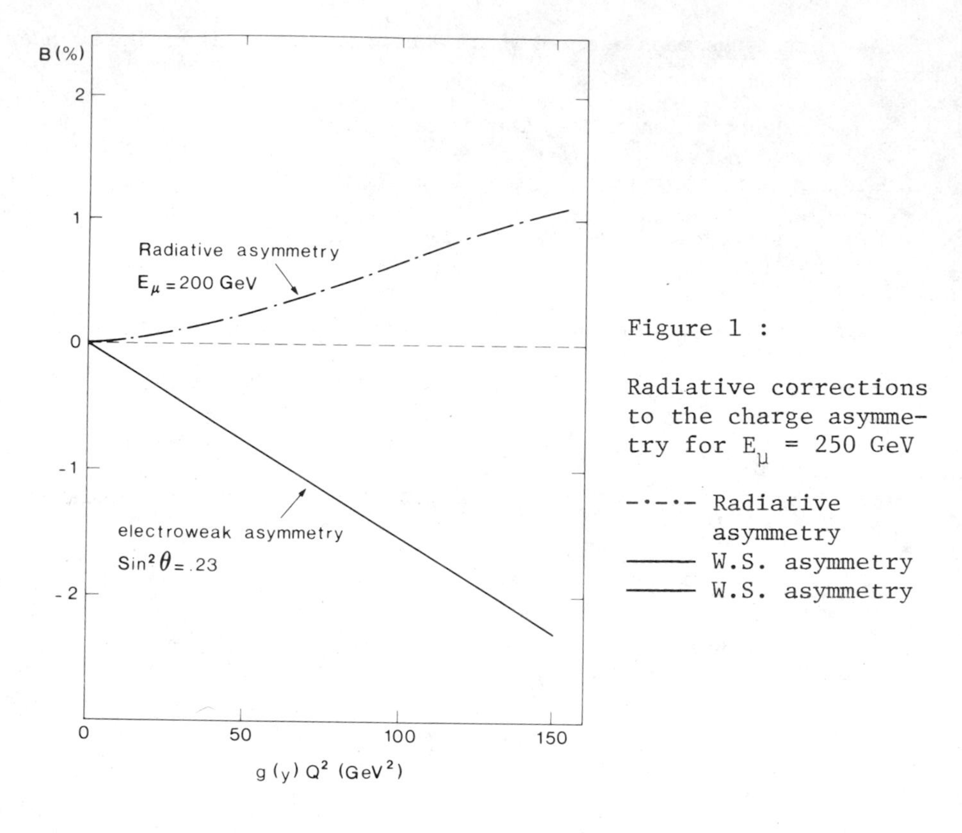

Figure 1 :

Radiative corrections to the charge asymmetry for E_μ = 250 GeV

-·-·- Radiative asymmetry
——— W.S. asymmetry
——— W.S. asymmetry

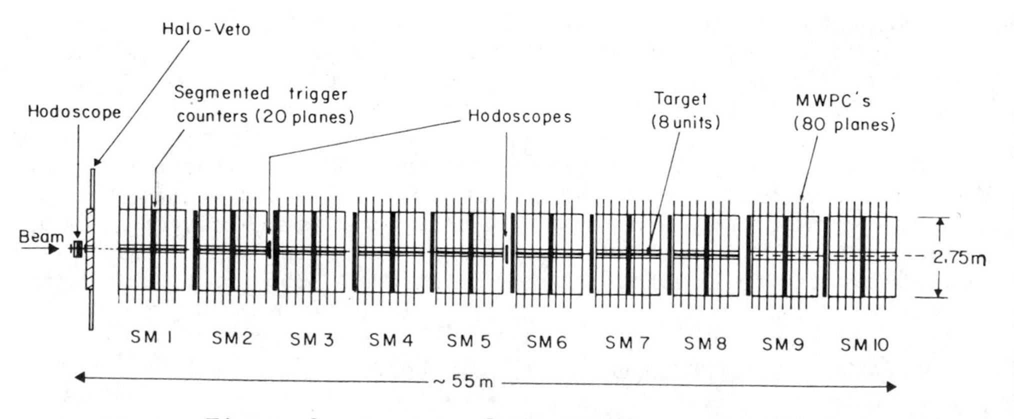

Figure 2 : Layout of the BCDMS spectrometer

luminosity, a flat acceptance at large Q^2, and an azimuthal symmetry. The data have been taken at two beam energies, 120 and 200 GeV, with $\mu+$ polarizations of -0.66 and -0.81 respectively.

After a selection of good runs on the basis of stability criteria, $3 \cdot 10^6$ events are obtained within the acceptance cuts, in a Q^2 range which extends from 40 to 180 GeV^2 at 200 GeV, and from 15 to 70 GeV^2 at 120 GeV.

Corrections. The differential cross sections for $\mu+$ and $\mu-$ data are corrected for the efficiencies in all the elements of the detector : counter efficiencies, chamber efficiencies, triggering efficiency. The size of the correction to B as a function of the variable $g = g(y)\ Q^2$ is shown on fig. 3 for a typical data taking period.

Asymmetry measurement. The observed asymmetry corrected for radiative effects

$$B(g) = \frac{\sigma_+(g) - \sigma_-(g)}{\sigma_+(g) + \sigma_-(g)}$$

is shown on Fig. 4 as a function of $g(y)\ Q^2$, and compared to the straight line fits expected from formula (4).

The systematic uncertainty on the normalization asymmetry is $4 \cdot 10^{-3}$, and it has not been used in the determination of $\sin^2\theta$. The slope measurements at 120 and 200 GeV from straight line fit of the form $B = a + bg(y)Q^2$ are summarised in Table I. An improved measurement of the normalization asymmetry, say with an error of $2 \cdot 10^{-3}$, is attempted and may reduce by 30 % the statistical error on b given in Table I.

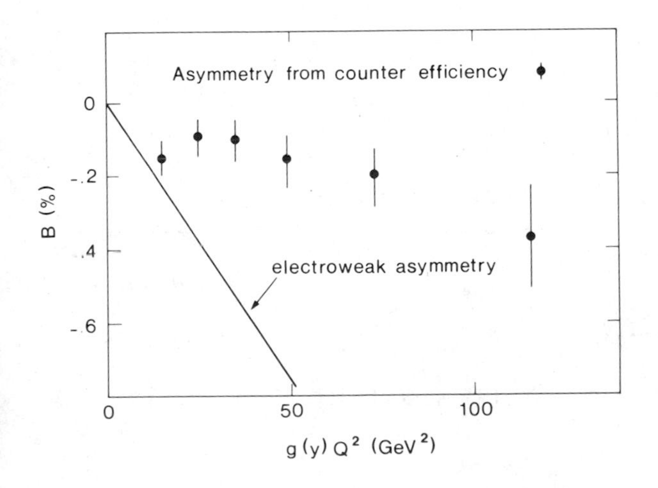

Figure 3 :

Correction to the asymmetry measurement from counter efficiency

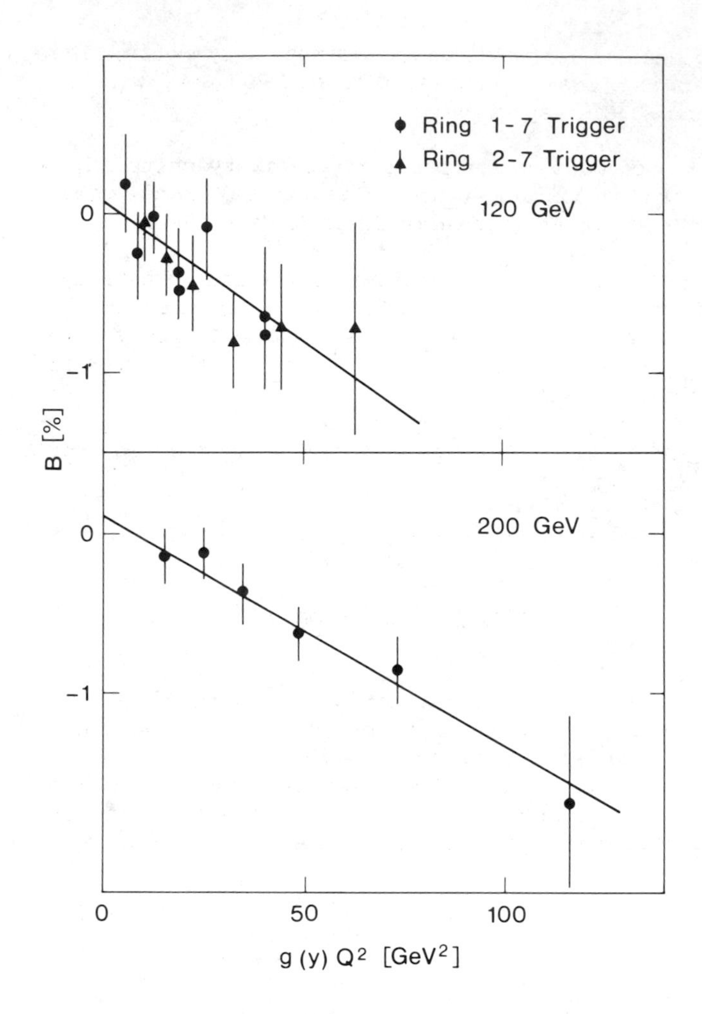

Figure 4 :

Charge asymmetry as a function of $g(y)Q^2$ at 120 and 200 GeV.

Table I - Slope measurement

E (GeV)	a	b (GeV^{-2})	W-S $\sin^2\theta = 0.23$
120	$(0.06\pm0.17)10^{-2}$	$(-1.74\pm0.75)10^{-4}$	$-1.53\ 10^{-4}$
200	$(0.15\pm0.17)10^{-2}$	$(-1.47\pm0.37)10^{-4}$	$-1.51\ 10^{-4}$

Systematics. The point of this reminder of old data is to emphasize the control of the apparatus needed for such measurements. The main sources of systematics are given in Table II. In particular :

- the relative beam calibration between $\mu+$ and $\mu-$ was known to $\Delta E/E = 6 \cdot 10^{-4}$ thanks to a set of Hall probes.

- the spectrometer field reversal was checked to be good to $\Delta B/B = 2 \cdot 10^{-4}$.

The resulting effect of these uncertainties on the asymmetry is shown on Fig. 5.

Table II - Main systematics Δb (10^{-6} GeV^2)

E	120 GeV	200 GeV
Beam momentum	18	8
BCDMS spectrometer	3	3
Beam phase space	8	8
Energy loss	5	5
δ rays	1	1
π+K+ charm decays	3	5
Halo contamination	20	20
$\sqrt{\Sigma(\Delta b_i)^2}$	29	24

It can be concluded from Table II that a new generation of asymmetry experiments (at FNAL ?) should make a special effort to reach a better control of the relative beam calibration, and of its phase space. Background events should also be kept to less than 10^{-4}.

Checks on Systematics and Final Results. The estimates given in Table II can be checked directly by a comparison of different partitions of the data :

i) in time :the asymmetry B can be measured in each of 8 data taking period.

ii) in space : different subdivisions of the detector along the target give the same asymmetry.

iii) in other kinematical variables : (Q^2, x, y, θ) no inconsistency is found in any case, which leads to the final result:

120 GeV	$b = (-1.74 \pm 0.75 \pm 0.29)10^{-4}$ GeV^2
200 GeV	$b = (-1.47 \pm 0.37 \pm 0.24)10^{-4}$ GeV^2

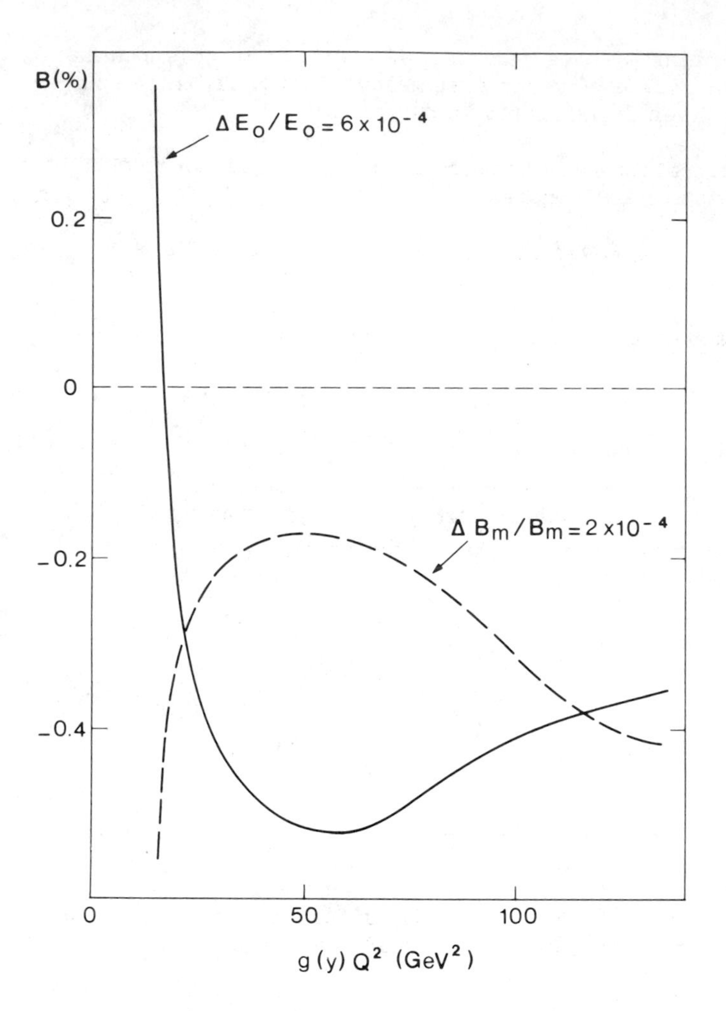

Figure 5 :

Effect of a calibration error on the beam momentum or on the spectrometer field.

or in terms of $\sin^2\theta_W = (-b/5.1 - 0.062)$ (at 200 GeV)

$$\sin^2\theta = 0.23 \pm \underset{stat}{0.07} \pm \underset{syst}{0.04}$$

If we assume $\sin^2\theta = 0.23$, equations (2) and (4) can be used to compute $I_3(\mu)$, as suggested by M. Klein et al.[6] :

$$I_3^R(\mu) = 0 \pm 0.06 \pm 0.04$$

which rules out 1/2 : the muon does not belong to a weak isospin doublet with a heavy partner, whatever its mass, as pointed out by Sakurai[7].

Conclusions. Such measurements can be repeated at higher energies at FNAL, in a larger Q^2 range, but this determination of $\sin^2\theta$ will not easily compete with ν measurements. The result $I_3^R = 0$ is an interesting way to present the asymmetry measurement : if one wants to restore the Left-Right Symmetry, one is forced to introduce a family of heavy right handed leptons as emphasized by J. Maalampi et al.[8] within mirror lepton models.

II - OTHER ELECTROWEAK EFFECTS

Polarization Asymmetry.

The polarization asymmetry A is defined as

$$A^{\pm} = \frac{\sigma^{\pm}(\lambda_1) - \sigma^{\pm}(\lambda_2)}{\sigma^{\pm}(\lambda_1) + \sigma^{\pm}(\lambda_2)}$$

$$A^{\pm} = - K \frac{\lambda_1-\lambda_2}{2} (v_\mu A_o\ g(x) \pm a_\mu V)\ Q^2$$

$K = (G/\sqrt{2})\ (1/2\pi\alpha) = 1.79\ 10^{-4}\ GeV^2$ has already appeared in the expression of the charge asymmetry B.

$$A_o = \frac{6}{5} (a_d - 2\ a_v) = - 1.8$$

$$V = \frac{6}{5} (2\ V_v - V_d) = \frac{6}{5} (\frac{3}{2} - \frac{10}{3} \sin^2\theta)$$

so that

$$(6) \quad \boxed{A^{\pm} = 1.79 \frac{\lambda_2-\lambda_1}{2} \left(\pm \frac{3}{4} - \frac{3}{4} g(y) + \sin^2\theta\ (\mp \frac{5}{3} + 3\ g(y)\right)}$$

good sensitivity to $\sin^2\theta$ is only obtained with $\mu-$. Assuming $\sin^2\theta = 0.23$

$$A\pm = 1.79\ (\Delta\lambda/2)\ [\pm 0.44 - 0.07\ g(y)]\ Q^2$$

A is about 3 times smaller than B as computed from equation (5).

Four data taking conditions are necessary to obtain the polarization asymmetry, with $\mu+$ and $\mu-$ beams. Given the practical range of λ : $|\lambda_1| = 0.9$, $|\lambda_2| = 0.1$, 8 to $10 \cdot 10^6$ events are needed in total to reach an adequate accuracy corresponding to $3 \cdot 10^{18}$ protons on target at the CERN SPS operated at 450 GeV.

For each polarity, one obtains a statistical error $\Delta_1 \sin^2\theta = 0.016$. Assuming a factor of 3 improvement on the present systematical uncertainties of BCDMS[5], this would add a contribution $\Delta_2 \sin^2\theta = 0.016$. Finally, an uncertainty of 0.04 on the polarization change $\Delta\lambda$ (twice better than the present estimate of BCDMS) would contribute an extra $\Delta_3 \sin^2\theta = 0.01$.

The A± measurements would give both a_μ and v_μ in an independent fashion, but it is not the safest method to measure $\sin^2\theta$ accurately. The effect would be twice larger at the FNAL Tevatron, easing the needed control over systematics.

Charged Right handed Currents

General features. There are theoretical arguments which motivate the search for right handed currents, as described by J.C. Pati and A. Salam[8].

The $\mu+$ is left handed in forward π decays, and therefore well adapted to the search for right handed currents. This was pointed out earlier by K. Winter[9]. As the helicity is conserved by V, A couplings, two cases should be distinguished :

- If there is no ν_R, $\bar{\nu}_L$, or if they are of such mass as to lie above the kinematically accessible range, the right handed currents can only couple to the quark vertex. Muon experiments cannot compete with the CDHS upper limit for quark couplings :

$$\bar{\sigma}_R / \sigma_L < 0.009 \qquad (90\ \%\ \text{C.L.})$$

by H. Abramowicz et al.[10].

- ν_R, $\bar{\nu}_L$ exist but are not produced in π/K decays (for instance because $m_{\nu_R} > m_\pi$). As neutrino beams single out left handed ν states in the final state, no wrong helicity particles can be created by V, A currents. On the other hand, $\bar{\nu}_L$ could be produced in deep inelastic muon scattering if its mass were in the few GeV range.

A specific model. If we assume a specific left right symmetric model $SU(2)_L \times SU(2)_R \times U(1)$ of J.C. Pati and A. Salam[8] and discussed by other authors[11] where the mass eigenstates W_1, W_2 of mass M_1, M_2 are linear combinations of the chiral bosons W_L, W_R

$$W_1 = W_L \cos\xi - W_R \sin\xi$$
$$W_2 = W_R \sin\xi + W_L \cos\xi$$

The contribution of W_L and W_R to the amplitude T can be separated : $T = T_L + T_R$

$$T_L \;\alpha\; \left\{ (V-A)_\ell (V-A)_n \left[\frac{\cos^2\xi}{M_1^2} + \frac{\sin^2\xi}{M_2^2} \right] \right.$$

$$\left. - \; (V-A)_\ell \; (V+A)_n \; \sin\xi \; \cos\xi \left[\frac{1}{M_1^2} - \frac{1}{M_2^2} \right] \right\}$$

$$T_R \;\alpha\; \left\{ (V+A)_\ell \; (V+A)_n \left[\frac{\cos^2\xi}{M_2^2} + \frac{\sin^2\xi}{M_1^2} \right] \right.$$

$$\left. + \; (V+A)_\ell \; (V-A)_n \; \sin\xi \; \cos\xi \left[\frac{1}{M_1^2} - \frac{1}{M_2^2} \right] \right\}$$

$$\sigma_{\mu^+}(\lambda) = \frac{1+\lambda}{2} \; \sigma^R_{\mu^+} + \frac{1-\lambda}{2} \; \sigma^R_{\mu^-}$$

The second contribution is non zero only above eventual thresholds, and is not present in deep inelastic neutrino interactions. In terms of quark distributions

$$\sigma^R_{\mu^+} = \frac{G^2ME}{\pi} \left[q(1-y)^2 + \bar{q} \right] \left[\cos^2\xi + \sin^2\xi \left(\frac{M_1}{M_2} \right)^2 \right]^2$$

$$+ \frac{G^2ME}{\pi} \left[q + \bar{q} \; (1-y)^2 \right] \sin^2\xi \; \cos^2\xi \left(1 - \frac{M_1^2}{M_2^2} \right)^2$$

$$\sigma^L_{\mu^+} = \frac{G^2ME}{\pi} \left[q(1-y)^2 + \bar{q} \right] \left(\cos^2\xi \left(\frac{M_1}{M_2} \right)^2 + \sin^2\xi \right)^2$$

$$+ \frac{G^2ME}{\pi} \left[q + \bar{q} \; (1-y)^2 \right] \sin^2\xi \; \cos^2\xi \left(1 - \frac{M_1^2}{M_2^2} \right)^2$$

In particular, the y distributions of left and right handed cross sections are the same, so that the only handle to separate them is the absolute normalization as a function of the polarization λ.

Rates and errors. We consider a run of 100 days ($5\cdot10^5$ good spills) with $2\cdot10^7$ μ/spill on a 10 m lead target.

$$\sigma_o = \sigma_R^{\bar{\nu}} = 0.3 \text{ E } 10^{-38} \text{ cm}^2$$

is the normal background of right handed antileptons, corresponding to 4 10^4 events. After applying kinematical cuts, one would be left with about 2 10^4 events.

For a $\mu+$ beam of polarization $\lambda = -0.8$: $\sigma(\lambda) = [0.1 + 0.9(M_1/M_2)^4]\sigma_o$.

If the normalization for the standard contribution were known to better than 2 % then $M_1/M_2 < 1/4$ (90 % C.L.). Else, one needs an addditional measurement with $\lambda = -0.1$.

$$\sigma(\lambda=-0.1) = [0.45 + 0.55 \, (M_1/M_2)^4] \, \sigma_o \quad .$$

The normalization cancels out in the ratio of both measurements, but one loses accuracy : $M_1/M_2 < 0.35$ (90 % C.L.). The uncertainty on normalization lowers the mass limit on W_R from 320 to 230 GeV. The range covered by a charge current measurement at the CERN SPS is therefore somewhat marginal. It could complement other attempts reviewed by Maalampi et al.[11] which are insensitive to massive ν_R . The mass limit is directly obtained from event rates, without additional model dependent assumptions. Let us point out that limits quoted by Hera study groups[12] are barely higher.

III - LEPTON NUMBER CONSERVATION

Neutrino Oscillations

The quark mixing matrix of Kobayashi Maskawa does not imply a priori any new interaction ; it only expresses the difference between flavour and weak interaction eigenstates. S.E. Eliezer et al.[13] have pointed out that for massive neutrinos, lepton number violating interactions could mix neutrinos and be observed as neutrino oscillations.

Borrowing the notations of P.H. Frampton and P. Vogel[14], we assume that there exists a Majorana mass matrix mixing chiral neutrinos ν_L , ν_R :

$$\mathcal{L}_M = \nu_L^T M_L C \nu_L + \nu_R^T M^R C \nu_R + \text{h.c.}$$

The neutrino mass eigenstates φ_A of mass M_A are self conjuguate Majorana particles :

$$\varphi_A = \varphi_A^L + \varphi_A^R$$

$$\varphi_A^L = U_A \tilde{\nu}_L \qquad \tilde{\nu}_L = \begin{pmatrix} \nu_L \\ \\ \nu_L^c \end{pmatrix} \qquad \tilde{\nu}_R = \begin{pmatrix} \nu_R^c \\ \\ \nu_R \end{pmatrix}$$

$$\varphi_A^R = U_A^* \tilde{\nu}_R$$

ν_L interacts in the standard weak Lagrangian

$$\mathcal{L}_W = g_L \bar{\nu}_L \gamma_\mu \ell_L^- W_L + \text{h.c.} \ldots$$

The probability to produce a lepton l_j from an initial ν_i is

$$|A_{ij}|^2 = | < \nu_j | \nu_i > |^2 = | \sum_A \exp [it \, (p_A^2 + m_A^2)^{1/2}] \, U_{Aj}^* \, U_{Ai} |^2$$

Charged Leptons

Neutrinos mixing generates lepton number violating transitions between charged leptons.

S.M. Bilenki et al.[15] compute

$$R_\mu = \frac{\mu \to e\gamma}{\mu \to e\nu\nu} = \frac{3}{32} \frac{\alpha}{\pi} \left(\frac{\Delta M^2}{M_W^2} \right) \sin^2\theta \cos^2\theta$$

where ΔM^2 is the difference of the squared masses of the neutrino mass eigenstates. Let us recall that according to J.D. Browman et al.[16] $R_{exp} < 2 \; 10^{-10}$. Within another model involving charged heavy leptons W.J. Marciano and Sanda[17] obtain a similar result, the mass difference between charged heavy leptons replacing ΔM^2.

In pion decay, one considers the channels $\pi^0 \to \mu + e-$, $\pi+ \to \mu + \nu_e$, $\pi+ \to e + \nu_\mu$. A specific model for π decay is needed to relate $R_\pi = (\pi \to \mu\nu_e)/(\pi \to \mu\nu_\mu)$ to R_μ, along the arguments discussed by D.A. Bryman et al.[18]. Experimentally, the neutrino and muon beam lines are well adapted to an investigation of $\pi \to \nu_\mu e$, on which no good limit exists at present. The ratio of fluxes $\pi \to \nu_\mu \; / \; \pi \to \nu_e$ cannot give an accuracy of 10^{-3} on R_π. On the contrary, a neutrino detector can tag the ν_μ in correlation with charged particles at the end of the decay tunnel, and a limit of $R_\pi < 10^{-3}$ is easily achievable.

If the neutrino mass difference is smaller than 0.1 eV, this correlation method would be quite competitive with other experiments on lepton number conservation.

ACKNOWLEDGEMENTS

In preparing these notes, I have largely benefitted from previous manuscripts of A. Benvenuti, R. Voss and G. Vesztergomby. Numerous conversations with all members of the BCDMS Collaboration have also been quite helpful.

REFERENCES

[1] A. Love, G.G. Ross, D.V. Nanopoulos
Nucl. Phys. B49 (1972) 513

E. Derman
Phys. Rev. D7 (1973) 2755

S.M. Berman and J.R. Primack
Phys. Rev. D9 (1974) 2171

R. Cahn and F.J. Gilman
Phys. Rev. D17 (1978) 1313

[2] C.Y. Prescott et al.
Phys. Letters 77B (1978) 347, PL 84B 524 (1979) 524

[3] D.Y. Bardin, P.Ch. Christova and D.M. Fedorenko
Nucl. Phys. B175 (1980) 435

[4] D.Y. Bardin
Private communication

[5] A. Argento et al.
Phys. Letters 120B (1983) 245

[6] M. Klein, T. Riemann and J. Savin
Phys. Letters B85 (1979) 385

[7] J.J. Sakurai
XVII Rencontre de Moriond, les Arcs 1982,
Vol. I, p. 561

[8] J.C. Pati and A. Salam
Phys. Rev. D10 (1974) 275

[9] K. Winter
Phys. Letters 67B (1977) 236

[10] H. Abramowicz et al. (CDHS Coll.)
Z Phys. C 12 (1982) 225

[11] R. W. Mohapatra and Pati
Phys. Rev. D11 (1975) 566, PR 11 (1975) 2558

D. Wyler and L. Wolfenstein
preprint CERN TH 3435 (1982)

J. Maalampi, K. Mursula and M. Roos
Nucl. Phys. B207 (1982) 233

J. Maalampi and K. Mursula
preprint CERN TH 3476 (1982)

[12] Study on the proton electron storage ring HERA
reports ECFA 80/42 and DESY HERA 80/01

[13] S.E. Eliezer and D. Ross
Phys. Rev. D10 (1975) 3088

S.M. Bilenki and B. Ponte Corvo
Phys. Letters 61B (1976) 248

[14] PH. Frampton and P. Vogel
Phys. Rep. 02 (1982) 339

[15] S.M. Bilenki, S.T. Petrov and B. Ponte Corvo
Phys. Letters 67B (1977) 309

[16] J.D. Browman et al.
Phys. Letters 42 (1979) 556

[17] W.J. Marciano and A.I. Sanda
Phys. Letters 67B (1977) 303

[18] D.A. Bryman, P. Depommier and C. Leroy
Phys. Rep. 88 (1982) 151

A SURVEY OF $\nu_\mu e$ PHYSICS WITH EMPHASIS ON RECENT FERMILAB RESULTS

Jorge G. Morfín

Fermilab
Batavia, IL 60505
U.S.A.

From both an experimental and an historical point of view it is particularly appropriate to summarize the development of $\nu_\mu e^-$ physics at this time. Historically, it was ten years ago last week that the announcement (see Figure 1) of the first $\nu_\mu e^-$ event was sent from Aachen to the other members of the Gargamelle Collaboration. The event, shown in Figure 2, is of a single electron identified via its characteristic bremsstrahlung and curvature. The significance of this event far exceeds its visual impact. With a background[1] of less than .03 events, it became the first solid indication for the existence of the weak neutral current. On the experimental front, the investigation of the $\nu_\mu e$ interaction is about to enter a new phase, having graduated from experiments yielding 2-3 events to those which will be analyzing hundreds of events. With these high statistics experiments it should be possible to study the differential as well as the total crosssections of $\nu_\mu e$ and $\bar{\nu}_\mu e$ scattering. Before reviewing the increasingly sophisticated methods with which the experimentalists have studied $\nu_\mu e$ scattering, let's briefly recall how the theoretical interpretation has evolved.

PHENOMENOLOGY OF $\nu_\mu e$ SCATTERING

The theory of $\nu_\mu e$ scattering has been covered[2] many times. It is a purely leptonic neutral current interaction not complicated by a (relatively) poorly known hadronic component. Let me here directly introduce the vector and axial vector coupling constants g_V and g_A in the effective Lagrangian.

$$L_{eff} = \frac{G}{\sqrt{2}} (\bar{\nu}_\mu \gamma_\alpha (1+\gamma_5) \nu_\mu)(\bar{e} \gamma_\alpha (g_V + g_A \gamma_5) e) \qquad 1)$$

It is the accurate determination of these two quantities- g_V and g_A - which has been the goal of the last decade of experiments. These constants appear as measurable quantities in the cross sections

$$\frac{d\sigma}{dy} = \frac{G^2 m_e}{2\pi} E_\nu \left[(g_V + g_A)^2 + (g_V - g_A)^2 (1-y)^2 \right] \qquad 2)$$

with $y = E_e/E_\nu$. The $\bar{\nu}$ cross section is obtained by changing the sign of g_A $(+ \to -)$ in the above formula.

III. PHYSIKALISCHES INSTITUT
der Rhein.-Westf. Technischen Hochschule Aachen
PROF. DR. H. FAISSNER

51 AACHEN, den 19.1.73

Institutsgebäude: Jägerstraße
Telefon: 4222464-65/34552
Fernschreiber 08 32704

Dear Colleagues,

Enclosed you will find photographs of a single electron event found here at Aachen. The main vertex is at x = -163.7, y = 6.9, z = 13.9. The measured momentum (by curvature) is P = ·359 ± ·035 and the angle θ = ·025 radians. This event thus qualifies as a leptonic neutral current candidate according to our previously defined constraints.

We have found the use of photographs to be extremely helpful in the analysis of the leptonic neutral current question. We would propose then that photographs of all events having a single electron from the main vertex be collected at the next collaboration meeting. These would include:

1) Any other leptonic neutral current candidates
2) All ν_e and $\bar{\nu}_e$ events
3) Any event classified as having a $\mu^\pm$ candidate and an $e^\pm$ candidate.

Sincerely,

(For the Aachen Group)

Jorge G. Morfín

Figure 1. Letter sent to the Gargamelle Collaboration announcing the discovery of the first $\bar{\nu}_\mu e$ event.

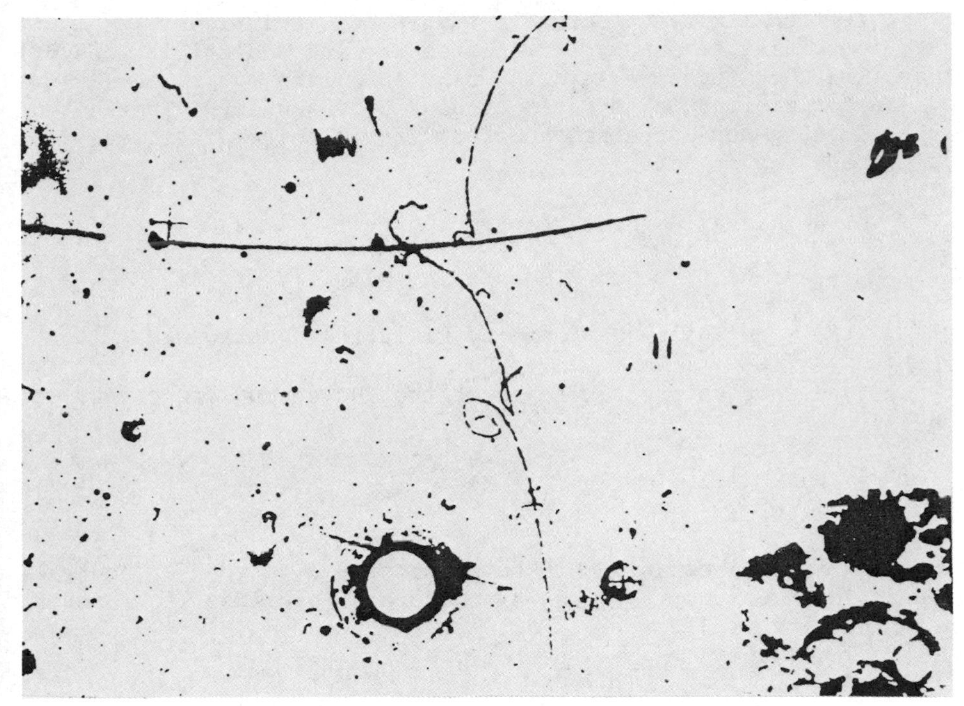

Figure 2. One view of the first $\nu_\mu e$ event found by the Gargamelle Collaboration.

Most experimental results of $\nu_\mu e$ scattering are presented as contours in the g_V - g_A plane. Note that if equation 2 is integrated over y to give the total cross sections σ and $\bar{\sigma}$, they each describe an ellipse in the g_V - g_A plane. Because of this quadratic dependence of $\sigma(\bar{\sigma})$ on the coupling constants, there is a four-fold ambiguity in the values of g_V and g_A no matter how accurately the total cross sections are measured. Even with measurements of $d\sigma(\bar{\sigma})/dy$, which the new high statistics experiments can (in principle) provide, the ambiguity still remains twofold.

We need not wait for these measurements of the y distribution to reduce the number of possible solutions. The interaction of ν_e and $\bar{\nu}_e$ with electrons is also described with g_V and g_A. However, since $\bar{\nu}_e e$ scattering also has a charged current contribution, the differential cross section becomes

$$\frac{d\bar{\sigma}(\bar{\nu}_e e)}{dy} = \frac{G^2 m_e}{2\pi} E_\nu \left[(g_V - g_A)^2 + (g_V + g_A + 1)^2 (1-y)^2\right] \qquad 3)$$

This introduces an ellipse in the g_V - g_A plane oriented like the $\bar{\nu}_\mu e$ ellipse, but with an offset toward the $-g_V$, $-g_A$ quadrant.

To further limit the possible number of solutions we must leave the realm of purely leptonic scattering and utilize one of the many elegent legacies left to us by the late J.J. Sakurai, the "factorization" hypothesis of Hung and Sakurai.[3] They noted that the most general neutral current formulation involves ten coupling constants:

g_V and g_A: ν-e scattering

α, β, γ and δ: ν-quark scattering (I=0,1;V and A)

$\tilde{\alpha}$, $\tilde{\beta}$, $\tilde{\gamma}$ and $\tilde{\delta}$: e-quark parity violating scattering

They derived an equality, in terms of the above coupling constants to be;

$$C_\nu^2 = 2g_A \frac{\alpha}{\tilde{\alpha}} = 2g_A \frac{\gamma}{\tilde{\gamma}} = 2g_V \frac{\beta}{\tilde{\beta}} = 2g_V \frac{\delta}{\tilde{\delta}} \qquad 4)$$

This established a relationship between the three types of neutral current interactions νe, νq, and eq, and provides a further constraint on g_V and g_A

$$\frac{g_V}{g_A} = \frac{(\alpha + \gamma/3) \quad (\tilde{\beta} + \tilde{\delta}/3)}{(\tilde{\alpha} + \tilde{\gamma}/3) \quad (\beta + \delta/3)} \qquad 5)$$

With the introduction of factorization, the allowed values of g_V and g_A become limited as in Figure 3.

This then is the most general model-independent way of fixing values of the initial coupling constants g_V and g_A. By introducing factorization we increased the number of coupling constants involved in the interpretation to ten which we quickly reduced-by expression 5) (and the assumption that $C_\nu^2 = 1$) -to seven. Further reductions in the number of constants is possible but only at the cost of model independence. Assuming general SU(2) ⊗ U(1) introduces two further constraints[4] which reduces the overall number of constants to five: ρ, the ratio of charged current(CC) to neutral current(NC) coupling "strengths"; T^u_{3R}, T^d_{3R}, T^e_{3R}, the right handed isospin assignments of the u-quark, d-quark and electron; and Θ, the electro-weak mixing angle. In terms of this new set of coupling constants our initial pair of constants can be expressed as

$$g_V = \frac{1}{2}\,(-1 + 2T^e_{3R} + 4\sin^2\Theta) \qquad 6)$$
$$g_A = \frac{1}{2}\,(-1-2\,T^e_{3R})$$

A further and final reduction in the number of involved constants brings us to the minimal SU(2) X U(1) model otherwise known as the "standard" or Weinberg-Salam model [5]. In this model, the strengths of NC and CC interactions are equal (ρ=1) and all right handed components are iso-singlets ($T^{u,d,e,}_{3R} = 0$). Thus we have just one remaining constant, the electro-weak mixing angle commonly referred to as the Weinberg angle, Θ_W. Expression 6) reduces to the "standard" representation

$$g_V = -\frac{1}{2} + 2 \sin^2\Theta_W$$
$$g_A = -\frac{1}{2} \qquad 7)$$

DEVELOPMENT OF EXPERIMENTAL TECHNIQUES

Turning now to the experimental study of ν_μe scattering, we will see that the sophistication of the experimentalists' interpretation of their results followed, naturally, the increasing statistical power of the experiments.

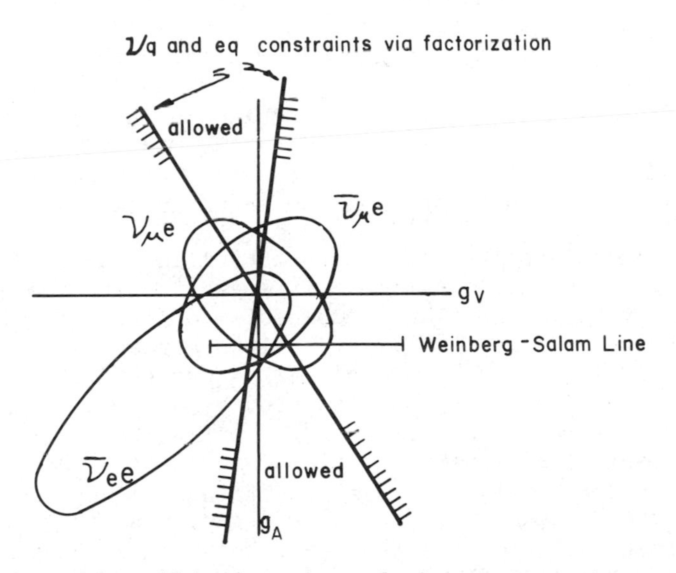

Figure 3. How the allowed region of the $g_V - g_A$ plane is reduced by the various experimental inputs.

Referring back to expression 2) it quickly becomes apparent why $\nu_\mu e$ experiments are so difficult. The quantity $G^2 m_e/2\pi$ corresponds to a cross section of $\sim 4.3 \times 10^{-42}$ cm^2 so that the rate of these purely leptonic events is down by three orders of magnitude compared to the semileptonic NC and CC interactions. This implies that not only is it difficult to acquire substantial statistics but that backgrounds, coming from NC or CC interactions, have much higher cross sections than the signal interaction. Fortunately, kinematics proves to be an indispensable aid in reducing the background to managable size. In particular an electron resulting from $\nu_\mu e$ scattering will subtend a very small angle with respect to the ν direction. All experiments have made use of this fact in various forms.

At the presentation of the first results[6] and interpretation, based on $\sim$ 1 (after subtraction) $\bar{\nu}_\mu e$ event and zero $\nu_\mu e$ events found by February 1973, direct comparison was made with the minimal SU(2) X U(1) model of Weinberg and Salam. The resulting limit of $\sin^2\Theta_W < 0.9$ was not a particularly bold statement - it was, however, a beginning! At the completion of the Gargamelle PS Freon experiment, with 2.6 $\bar{\nu}_\mu e$ and $\sim$ 0.7 $\nu_\mu e$ events, the limits[7] on the mixing angle were $0.1 < \sin^2\Theta_W < 0.4$.

The next experiment to study $\nu_\mu e$ scattering was the Aachen-Padova spark chamber. In general, electronic detectors will yield higher statistics but have a more difficult time separating signal from background whereas bubble chambers have limited statistics but good signal/background separation. The results of $\bar{\nu}_e e$ scattering were combined with the Aachen-Padova results (based on 9.6 $\bar{\nu}_\mu e$ events and 11.5 $\nu_\mu e$ events) by Sehgal[8] reducing the allowed regions to two areas referred to as the g_V dominant and g_A dominant solutions as shown in Figure 4.

It was at this point in time that Fermilab experiments began contributing to the world sample of $\nu_\mu e$ events with the 15' Bubble Chamber results of a Brookhaven-Colombia collaboration. In an exposure using a heavy (64%) Ne/H mixture they determined their angular resolution to be $\sim$ 4mr and $\Delta E/E$ to vary from 10% at 2 GeV to $\sim$ 15% at 20 GeV. Results presented by N. Baker[9] gave limits of $\sin^2\Theta_W = 0.20^{+0.16}_{-0.08}$. Factorization is now used to combine: 1) the SLAC e-D results; 2) ν semileptonic NC results: 3) Gargamelle and Aachen-Padova $\bar{\nu}_\mu e$ results; 4) $\bar{\nu}_e e$ results and 5) the Colombia-BNL results to yield Figure 5. The allowed region has been reduced to the g_A dominant solution and is completely consistent with the minimal Weinberg-Salam model.

With this Fermilab 15' experiment, the era of significant contributions from Bubble Chambers to the study of $\nu_\mu e$ physics came to an end, and electronic detectors, with improved

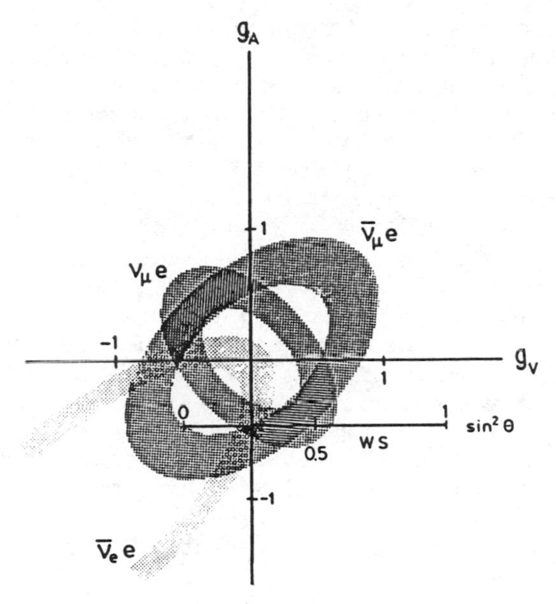

Figure 4. Allowed region of $g_V - g_A$ using the Gargamelle, Aachen Padova $\bar{\nu}_\mu e$ and the Reines $\bar{\nu}_e e$ results.

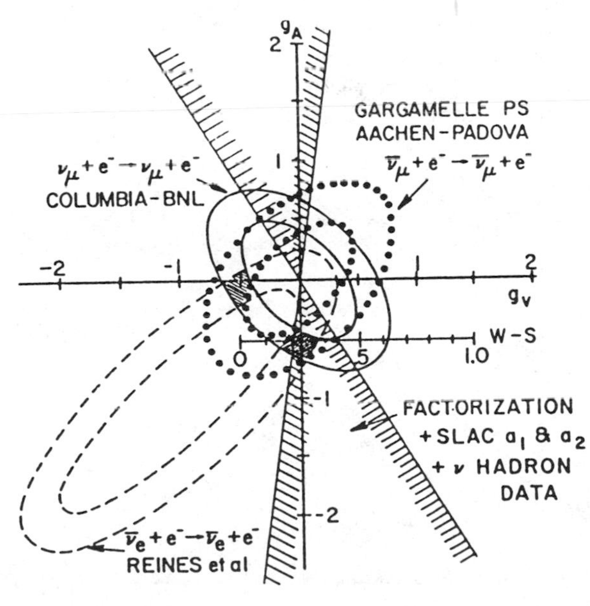

Figure 5. Allowed region of $g_V - g_A$ using input of Figure 4($\bar{\nu}$) and the Brookhaven - Colombia result and introducing the constraints from factorization.

resolutions and higher statistics, took over the lead. The results from the CHARM collaboration's experiment at CERN and the new dedicated experiment at Brookhaven will be described by others[10,11] at this conference. I will concentrate on the two Fermilab experiments beginning with the high resolution detector of the VPI-Maryland-NSF-Oxford-Peking collaboration. The apparatus consisted of 49 modules each consisting of $\sim$ 1 radiation thick Al plate, 1 MWPC and 1 layer of plastic scintillation counter. The resolution in energy was determined to be < 8%. The angular resolution is illustrated in Figure 6, taken from Reference 12., which can be assumed to be a distribution of $\Delta\Theta = \Theta$ (measured) - Θ(real). The authors quote the angular resolution as $\pm$ 5 mr (FWHM), which is an overestimate, and independent of energy. In fact the distribution of Figure 6 demonstrates $\langle\Delta\Theta\rangle$ = 0.36 mr with a σ of 2.65 mr. This is far better angular resolution than any other detector except, perhaps, the new Brookhaven detector and allows this collaboration to make excellent use of kinematics to separate signal from background. The final sample of 40 $\nu_\mu e^-$ events, when interpreted in the minimal model, corresponded to $\sin^2\Theta_W = 0.25^{+0.05}_{-0.03}$. Incorporating these results in the full $g_V - g_A$ plane analysis yields Figure 7.

The second major Fermilab electronic detector experiment initially dedicated to the study of $\nu_\mu e$ scattering is experiment E-594 a Fermilab, MIT, Michigan State and Northern Illinois collaboration with participants as shown in Figure 8a. The detector, shown in Figure 8b, is a fine-grain calorimeter consisting of 608 flash tube planes (4×10^5 cells) interspersed amongst planes of iron shot and sand to give interaction mass. After every 16 flash tube planes there is a proportional tube plane used

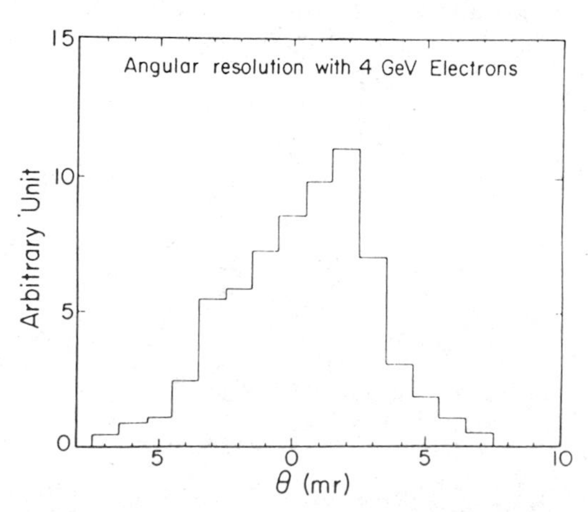

Figure 6. Measured angular resolution of UPI-Maryland - NSF-Oxford Peking detector at Fermilab.

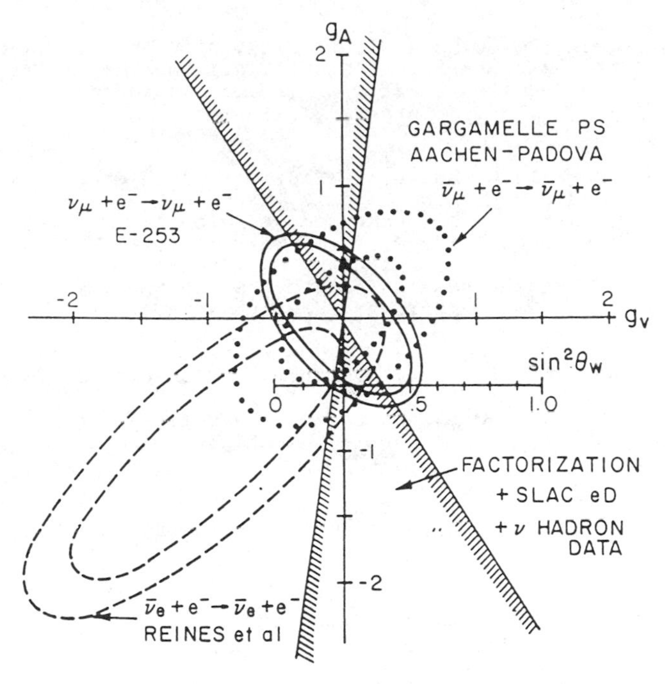

Figure 7. As in Figure 5 with the VPI -- Collaboration results repalcing the BNL - Colombia results.

in triggering and energy determination of very high energy (> 75 GeV) electrons. The calorimeter was followed by a muon spectrometer consisting of three 24' and two 12' magnetized toroids with four double planes of proportional tubes to trace the muon trajectory through the toroids. In a calibration run employing electrons between 5 and 75 GeV and hadrons with 10 to 125 GeV energy, the resolution of the detector was determined to be as shown in Figure 9.

In a 1981 engineering run,[F1)] dedicated to bringing the detector up and developing triggers, $\sim 3 \times 10^{18}$ protons were directed to the Fermilab wideband neutrino beam. The resultant neutrino flux yielded 58.3K electron triggers defined as a shower with length less than 21 radiation lengths, no muon and a minimum deposited energy of $\sim$5 GeV. It was determined that this trigger was 90% efficient in detecting electrons. There were also triggers to record conventional neutral and charged current neutrino events for normalization purposes. Direct use of the fine grained nature of the calorimeter was made by examining the density of showers. Defining the density (ρ) as the number of hit

(a)

D. Bogert, R. Burnstein*, R. Fisk, S. Fuess, J. Morfín,
T. Ohska, M. Peters†, L. Stutte, J.K. Walker, H. Weerts
Fermi National Accelerator Laboratory
Batavia, Illinois

J. Bofill, W. Busza, T. Eldridge, J.I. Friedman,
M.C. Goodman, H.W. Kendall, T. Lyons, R. Magahiz,
T. Mattison, A. Mukherjee, L. Osborne, R. Pitt,
L. Rosenson, A. Sandacz, M. Tartaglia, R. Verdier,
S. Whitaker, G.P. Yeh
Massachusetts Institute of Technology
Cambridge, Massachusetts

M. Abolins, R.Brock, A. Cohen, J. Ernwein*, D. Owen,
J. Slate
Michigan State University
E. Lansing, Michigan

F.E. Taylor
Northern Illinois University
DeKalb, Illinois

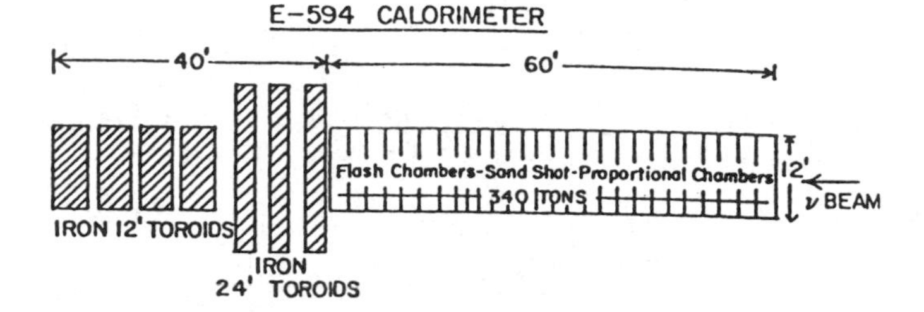

(b)

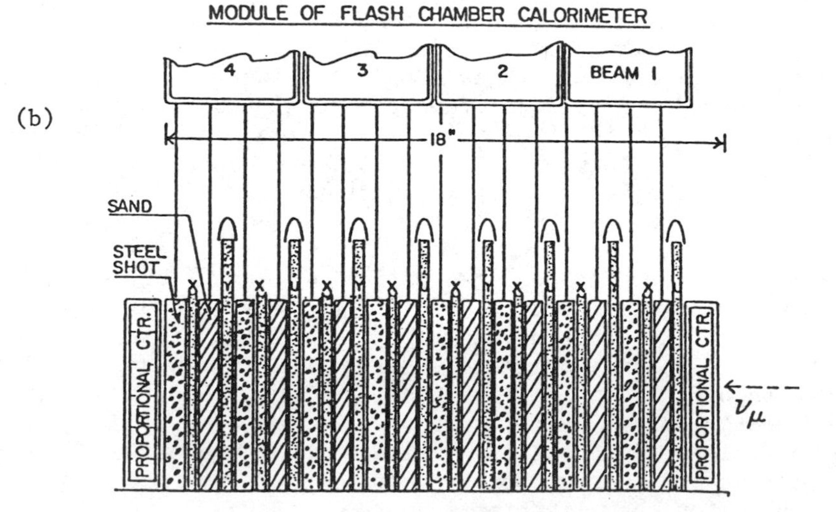

Figure 8. a) Participants of Experiment 594

b) The E-594 fine-grained Flash Tube Detector

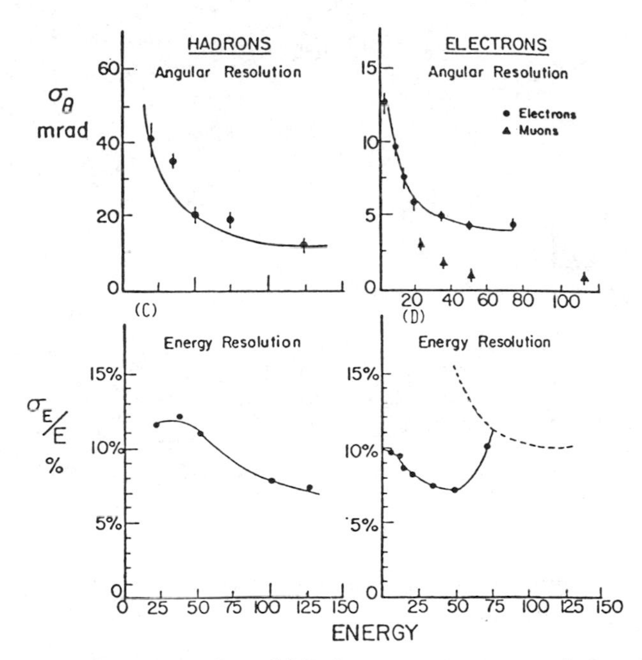

Figure 9. Resolutions of E-594 detector measured in the Fermilab Calibration beam.

cells in a cone of given volume and normalizing ρ to be ~ 1 for electrons (< 1 for hadrons) a cut at $\rho = 0.78$ was found to be about 80% efficient in rejecting hadrons and reduced the number of candidates to 13K. These candidates were each visually scanned by professional Fermilab scanners who, with 95% scan efficiency, reduced the number to $\sim$ 1000 candidates. Physicists then examined each candidate and further reduced the number to 300 events. To improve the signal to background ratio, a fiducial volume cut ($r <$ 110cm) and energy cut (5 GeV $< E_e <$ 30 GeV) were introduced. For further analysis, the kinematic variable $E_e\theta_e^2$ was chosen since this variable is limited to $<2m_e$ for true $\nu_\mu e$ scatters. The finite energy and angular resolutions of our detector indicated that we would contain 95% of the $\nu_\mu e$ events within $0 < E_e\theta_e^2 < 6$ MeV so that Figure 10 presents the remaining candidates in 6 MeV bins.

There is an obvious peak in the first bin and we need now separate the signal from the background which has survived all previous cuts. The backgrounds which, in principle, will contribute are:

1. $\nu_e + N \rightarrow e + X$; where E_X is small

2. $\nu_\mu + N \rightarrow \nu_\mu + X$; where the X state is dense

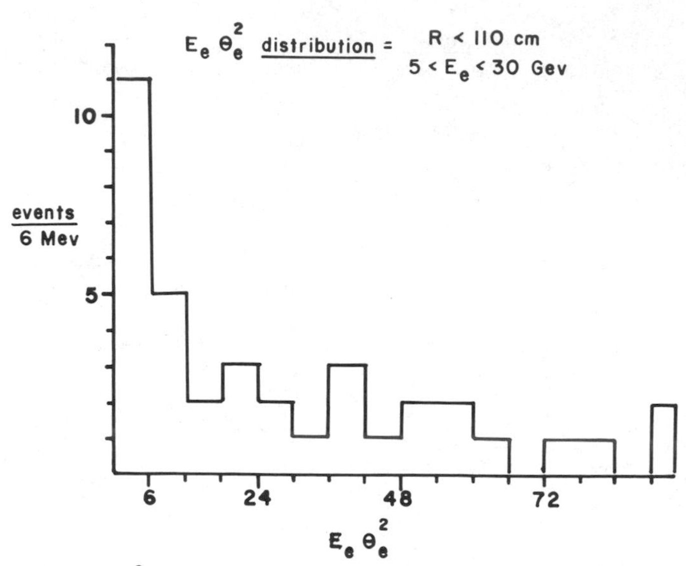

Figure 10. The $E\Theta^2$ distribution of electron candidates surviving the various cuts.

3. $\nu_\mu + N \rightarrow \nu_\mu + N + \pi^0$; Coherent π^0 production

Recent theoretical[13)] and experimental [14)] results indicate that the third background - ν induced coherent π^0 production - is much larger and more forward peaked than previously assumed. This would imply that all experiments which are unable to consistently distinguish e^+e^- pairs from single electrons should carefully (re)evaluate this background for the given experimental conditions. Within the kinematical cuts of this experiment the expected <u>relative shapes</u> of the three types of backgrounds are shown in Figure 11. The absolute contribution of each source was determined using the relative absolute cross sections of coherent π^0 production[13)] and standard CC + NC cross sections, the relative ν_e and ν_μ energy spectra, and shape of the observed $E\Theta^2$ distribution at large $E\Theta^2$ where source 2. is dominant. We find that of the 11 events in the first 6 MeV $E\Theta^2$ bin, 2.4 events are due to source 2; 1.0 event is due to source 1 and 0.9 event is due to neutrino induced coherent π^0 production. This leaves a signal of 6.7 $\pm$ 3.6 events which we attribute to $\nu_\mu e$ scattering.

Thus, although the 1981 engineering run did not yield sufficient statistics to add, significantly, to the world sample

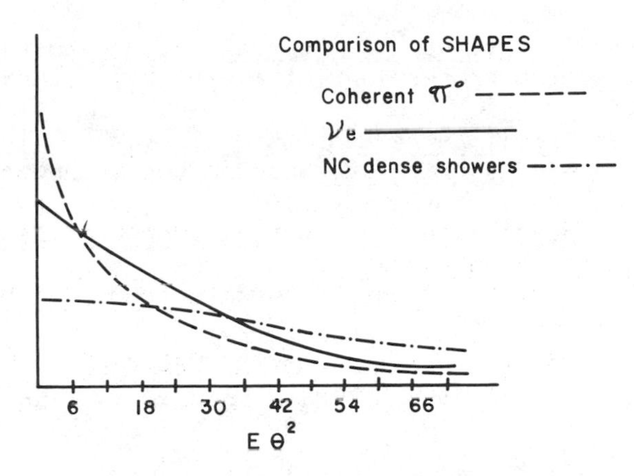

Figure 11. The relative shapes of the three background sources. The curves are all normalized to the same number of events.

of $\nu_\mu e$ events, it did enable us to demonstrate the capability of this fine-grained calorimeter. It furthermore allowed us to emphasize the importance of correctly accounting for the background coming from neutrino induced coherent π^0 production. Had we neglected to account for the very forward peaked distribution of the coherently produced π^0's, we would have underestimated the background by ∿20%.

REFERENCES

1. F.J. Hasert et.al, Phys. Lett. 46B, 121(1973)
2. L.W. Mo, Review on Purely Leptonic Interactions of Weak Neutral Currents, Proceedings of the Neutrino 1980 Conference, Erice, 1980 and references therein.
3. P.Q. Hung and J.J. Sakurai, UCLA Report UCLA 79/TEP/9; Proc. of Neutrino 79, Vol. 1, 267 (1979).
4. J. Bernabeu and C. Jarlskog, Phys. Lett. 69B, 71 (1977)
5. S. Weinberg, Phys, Rev. Lett, 1264(1967); A. Salam in Elementary Particle Theory, ed. by N.J. Svartholm, 367(1968)
6. J.G. Morfín, Purely Leptonic Neutral Currents, submitted to Deutsche Physikalichses Gesellschaft, 21 February 1983, available as PITHA-69(1973), III Phys. Inst. RWTH Aachen, Germany.
7. J. Blietschau et al., Nucl. Phys. B114, 189 (1976) J. Blietschau et al. Phys. Lett. 73B, 232(1978)

8. L.M. Sehgal, Status of Neutral Currents in Neutrino Interactions, in Proceedings of Neutrino 78, Purdue University 253 (1978).
9. N.J. Baker, Preliminary Results On $\nu_\mu + e^- \rightarrow \nu_\mu + e^-$, in Proceeding of Neutrino 82, Balotonfüred, Hungary, Vol. 2, 1 (1982)
10. B. Borgia, Results on Weak Neutral Currents from the CHARM Collaboration, Proceedings of this conference.
11. Y. Suzuki, Neutrino Electron Results from BNL, Proceedings of this Conference.
12. T.A. Numamaker et al., An Electromagnetic Shower Detector using Proprotional Wire Chambers with Cathode Plane Delay-Line Readout; Virginia Poly. Inst. Preprint VPI-HEP-80/3, February 1980.
13. D. Rein and L. Sehgal, to be published in Nuclear Physics available as III Phys. Inst, RWTH Awachen preprint.
14. E. Isiksal D. Rein and J.G. Morfin, T.H. Aachen preprint PITHA 82/83 submitted to Phys. Rev. Lett.; H. Faissner et al., to be published in Phys. Lett.

FOOTNOTE

F1. At the time of the 1981 engineering run approximately 2/3 of the calorimeter was instrumented and the two 24' proportional planes were not yet in operation.

PROGRESS IN AN EXPERIMENT TO MEASURE ELASTIC $\nu_\mu e \to \nu_\mu e$ SCATTERING

Presented by Y. Suzuki

Physics Department, Osaka University
Toyonaka, Osaka 560, Japan

K. Abe, L.A. Ahrens, K. Amako, S.H. Aronson, E. Beier, J. Callas, P. Clarke, P.L. Connolly, D. Cutts, D.C. Doughty, R.S. Dulude, L.S. Durkin, T.E. Erickson, R.S. Galik, B.G. Gibbard, S.M. Heagy, D. Hedin, B. Hughlock, M. Hurley, S. Kabe, R.E. Lanou, Y. Maeda, A. Mann, M.D. Marx, J.T. Massimo, T. Miyachi, M.J. Murtagh, S.J. Murtagh, Y. Nagashima, F.M. Newcomer, T. Shinkawa, E. Stern, Y. Suzuki, S. Tatsumi, S. Terada, A. Thorndike, R. Van Berg, P. Wanderer, D.H. White, H.H. Williams, and T. York.

Brookhaven National Laboratory, Brown University, KEK (Japan), Osaka University, University of Pennsylvania, SUNY/Stony Brook, Tokyo (INS)

The formalism and present status of $\nu_\mu e \to \nu_\mu e$ pure leptonic neutral current interaction has been thoroughly discussed by Dr. Morfin[1] and by Dr. Winter[2] at this conference. We report here on entirely new results from an experiment at Brookhaven National Laboratory.

The experiment is designed primarily to measure elastic neutral current interactions:

$$\overset{(-)}{\nu}_\mu + e \to \overset{(-)}{\nu}_\mu + e \qquad (1)$$

$$\overset{(-)}{\nu}_\mu + p \to \overset{(-)}{\nu}_\mu + p \qquad (2)$$

in the low energy region.

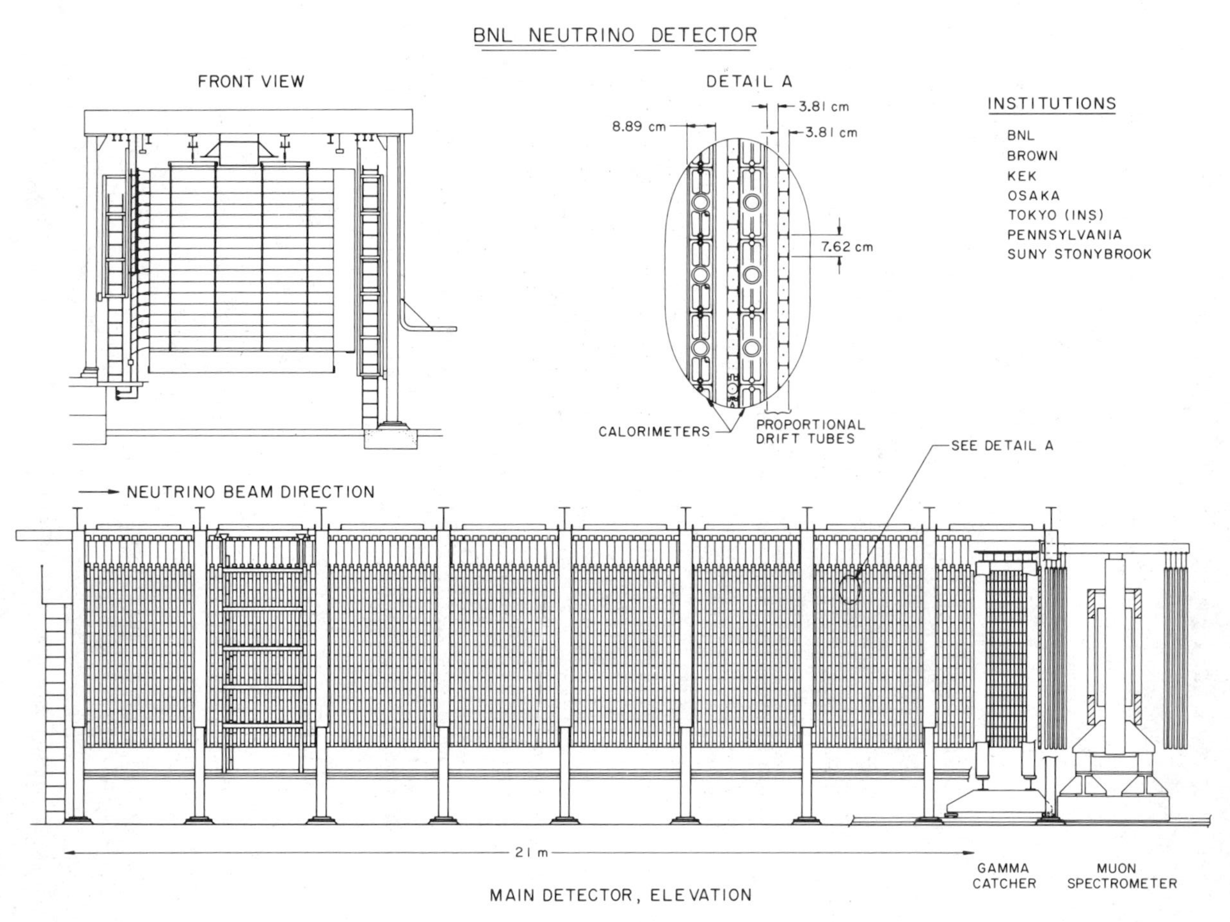

Fig. 1 Schematic view of the experimental apparatus.

Table 1. Module Properties

CALORIMETER (LIQUID SCINTILLATOR)
Active Area 4.22 x 4.90 m^2 Thickness 7.9 cm
Weight (Liquid & Acrylic) 1.35 metric tons
16 cells/module 2 Amperex 2212A phototubes/cell
1 Pulse Height Measurement/2 Time Measurements Per Tube Readout

PROPORTIONAL DRIFT TUBES (PDT)
Active Area 4.2 x 4.2 m^2
Thickness (x and y) 7.6 cm 54 x wires 54 y wires
1 Pulse Height Measurement/2 Time Measurements Per Wire Readout

The apparatus, shown in Fig. 1, consists of 112 basic modules followed by a gamma catcher and a spectrometer. Each module, 4m x 4m in area, consists of one plane of a liquid scintillator calorimeter slab segemented into 16 cells in vertical direction and two planes (X and Y) of proportional drift tubes, 54 tubes for each coordinate. The details of one module are given in Table 1. The main detector is followed by a gamma-catcher, which consists of ten modules of liquid scintillator slabs with one radiation length of lead between each slab. This allows good total energy measurements of electron and photon events occuring in the downstream portion of the detector. Following the gamma catcher is a magnetic spectrometer which aids in measuring $\nu_\mu(\bar{\nu}_\mu)$ contamination by $\bar{\nu}_\mu(\nu_\mu)$. Total weight of the detector is about 200 metric tons and it is almost (85%) fully active. The fine segmentation of the detector and the moderately longer radiation length enable us to make multiple measurement of dE/dX by both calorimeters and PDTs. So we can make good separation between electrons and photons as well as pion and proton identification.

The detector electronics is operated in a triggerless mode. With an approximately 1.4 second repetition rate, the primary beam from the Alternating Gradient Synchrotron of the Brookhaven National Laboratory is extracted in twelve radio-frequency buckets. Each bucket separated by 220 nanoseconds with an intensity

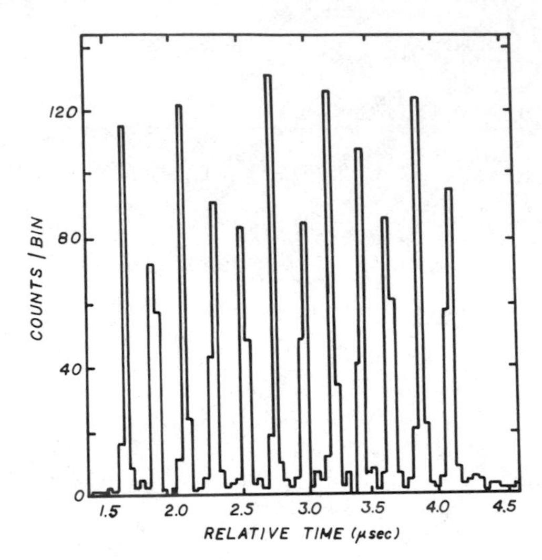

Fig. 2. Time distribution of the events. This shows clear twelve R.F. bucket structure of the neutrino beam.

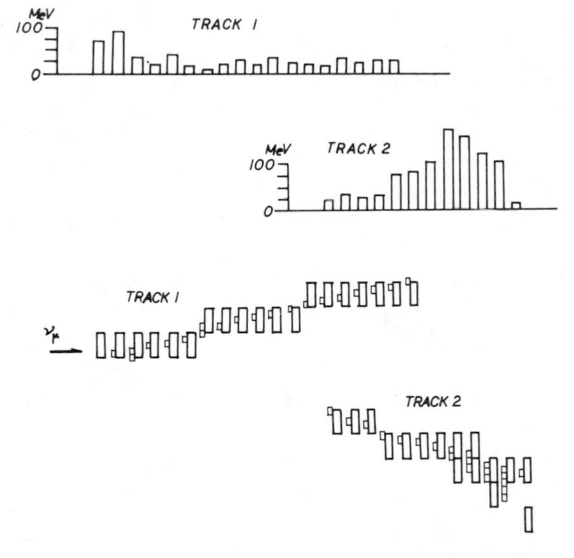

Fig. 3. A neutrino interaction in the detector. Large boxes are the calorimeter cells and small boxes are PDT cells (y-coordinate). Two tracks are in a same R.F. bucket. The plots show the energy deposits in the calorimeter along the tracks.

approaching 1 x 10^{13} protons on target (POT) per pulse into a horn-focussed wide band beam. A 10 μsec electronics gate covers the ν-beam, PDT drift time and μ-decay. Through the bucket structure of the ν-beam and the timing measurement of the calorimeter, time clustering can separate events happening in different buckets, which is collected throughout the gate. Fig. 2 shows the 12 R.F. bucket structure of the ν-beam seen by events. This confirms that the out-of-time background is negligible.

An example of the dE/dX patterns for muons, hadrons and photons/electrons in the detector is shown in Fig. 3. The two tracks are in the same time cluster. The dE/dX patterns of the first track reveals that there is some hadronic energy deposit around the vertex and a minimum ionizing particle runs through 18 modules of the calorimeter. The second track behaves very differently showing the characteristics of the electromagnetic showers.

Data runs completed so far are listed in Table 2. Data taken from December 1981 through February 1982 having 8.8 x 10^{18} protons on target with full detector were used for the analysis presented here. The data contains 1.25 M beam bursts. One neutrino interaction per burst was expected.

The major background to the $\nu_\mu e \to \nu_\mu e$ signal comes from $\nu_\mu n \to \nu_\mu n \pi^\circ$ where the neutron and one of the photons from the π° were missed and another photon was detected in the very forward direction and from $\nu_e n \to e^- p$ where the proton missed detection. Other reactions involving charged currents and/or charged hadrons are easily suppressed via energy deposition in both the calorimeters and PDTs.

We first applied simple software filters to remove events which contained muons and/or hadrons and large angle events. The filter reduced the amount of the data to about 30K events. According to a Monte Carlo calculation based on the program EGS, 94% of the $\nu_\mu e \to \nu_\mu e$ signals were passed. The first 3/4 of the

Table 2. Summary of the Completed Runs

Run	ν-beam (POT)	$\bar{\nu}$-beam (POT)	Fiducial Mass
June-July 1981	6.3 x 10^{18}	5.0 x 10^{17}	HALF
Dec-Feb. 1982	8.8 x 10^{18}	9.1 x 10^{18}	FULL

events has been scanned by physicists. The scanning eliminated events which had extra tracks at the vertices, large angle events ($\geq 15°$), side entering events, front entering events and ones which had additional gammas or neutrons in the same time cluster. The number of events which resulted from the scanning was 1114. We then applied cuts; they were fiducial cuts of about 60% of the total volume and the total visible energy requirement greater than 150 MeV. The total visible energy is about 70% of the true energy. In the remainder of this text, the visible energy will be used as energies for electrons and photons. Events which satisfied those requirements were 744. The events were sorted into two categories, single forward e/γ showers with separated upstream energy deposits and those without separated upstream energy deposits. The events with separated upstream energy deposits are mostly gammas which come from $\nu_\mu p \rightarrow \nu_\mu p \pi^\circ$ and $\nu_\mu n \rightarrow \nu_\mu n \pi^\circ$, in either case the protons or neutrons are seen upstream of the vertex where one of the gammas from the pi-zero is converted. These events were used as controlled samples including gamma rays from which we see the dE/dX distribution of the first few modules of calorimeters and PDTs, the angular and $E\theta^2$ distributions.

Five hundred fifty two (552) events were categorized as single forward e/γ showers without separated upstream energy deposits. These consisted mostly of the $\nu_\mu e$ candidates, gamma background originating either from neutral current resonance pi-zero production or from coherent pi-zero production and electron background from $\nu_e n \rightarrow e^- p$ interaction in cases where protons were not visible in the detector.

In Fig. 4a we show the dE/dX distributions for muon tracks for a single calorimeter cell and a single PDT cell. The mean energy deposited by minimum ionizing particles is 13 MeV for a calorimeter cell and 8 keV for a PDT cell. The full width at half maximum of the energy deposition is 60% for both calorimeters and PDTs. In Fig. 4b is the dE/dX distributions of a first calorimeter cell and a first X and Y PDT cell for the controlled gamma samples after the vertex. There is a clear difference of the peaks between minimum ionizing particlesand the controlled gamma samples. In Fig. 4c is the dE/dX distributions of those for the events without separated upstream energy deposits.

We have used a simple algorithm to reduce gamma background by using the dE/dX in the calorimeter and the PDTs. We define a function of the dE/dX of the first two X and Y PDT planes $\langle \Delta EPDT \rangle$, as follows:

$$\langle \Delta EPDT \rangle = \frac{\min(XPDT1, YPDT1) + \min(XPDT2, YPDT2)}{2}$$

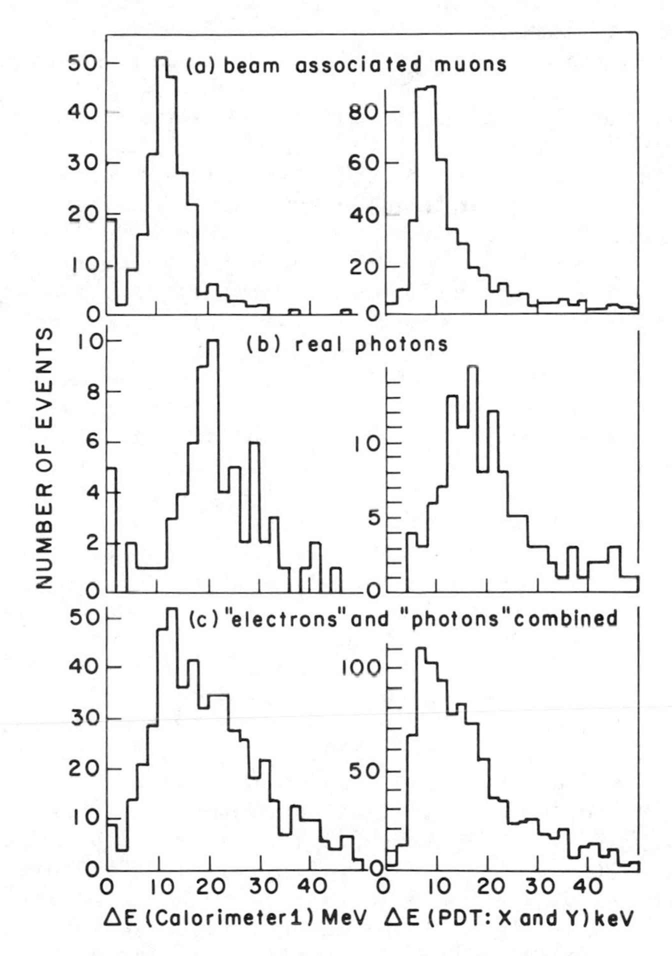

Fig. 4. The dE/dX distributions for both calorimeters and PDTs immediately following the vertices; for a) beam associated muons, b) controlled gamma samples and, c) "electrons" and "photons" combined.

where XPDTi represents dE/dX of the ith XPDT plane. The events with energy deposit larger than 20 MeV in the first calorimeter cell after the vertex and the <ΔEPDT> greater than 12 keV were defined as "gammas". Using the EGS Monte Carlo calculation we estimate that 95% of the $\nu_\mu e$ signals pass this separation. About 50% of the gamma events are estimated to pass the separation by the Monte Carlo calculation and the controlled gamma samples.

We measured the angles for the events without separated upstream energy deposits and retained the events having angles less than 180 mrad. After this procedure we are left with 292 events. Applying the e/γ separation, 186 events passed as "electrons" and 106 events were failed and assigned as "gammas".

Most of the background comes from the $\nu_\mu N \to \nu_\mu N\pi^\circ$ process, either by resonant or coherent production, where N represents a nucleon or a nucleus. The energy of the gammas from the background process is expected to be less than 300 MeV. The gamma background is insignificant above 300 MeV. The background from the process $\nu_e n \to e^- p$, which is caused by ν_e contamination in the ν_μ beam at the level of 0.5%, contributes mostly in the higher energy region due to the high energy tail of the ν_e spectrum. In the region where the energy is less than 1500 MeV and the angle less than 180 mrad, the $e^- p$ background is estimated to be less than 10% of the $\nu_\mu e$ signals; furthermore, the $\nu_\mu e$ signals and the ep background do not overlap in this kinematic region. Taking into account these facts, we evaluate the background from ν_e to be negligibly small in the $\nu_\mu e$ signal region. Seventy five percent of the $\nu_\mu e$ signals are in the energy region between 150 MeV and 1500 MeV by the Monte Carlo calculation using the Standard Model[3] with $\sin^2\theta_W = 0.23$. The kinematical limit of the angle changes rapidly up to $\approx$ 80 mrad below 300 MeV. We have divided the data into the three energy regions: 150 MeV $<$ E $<$ 300 MeV, (Region 1); 300 MeV $<$ E $<$ 1500 MeV, (Region 2); and 1500 MeV $<$ E, (Region 3). Region 1 and 2 consist mainly of the $\nu_\mu e$ signals and the gamma background. In the region 1 the signal to background ratio is worse than that in the region 2, and the angular distribution of the signal is broader. Region 3 is mainly $\nu_e n \to e^- p$ background with very little signal.

Fig. 5 shows the θ^2 distributions in the region 2 of the "electrons", the "gammas" and the controlled gamma samples. The distributions of the "gammas" and the controlled gamma samples are flat up to 0.025 $(\text{rad})^2$ of θ^2 which indicates the gamma background is spatially uniform enabling us to estimate the $\nu_\mu e$ signals

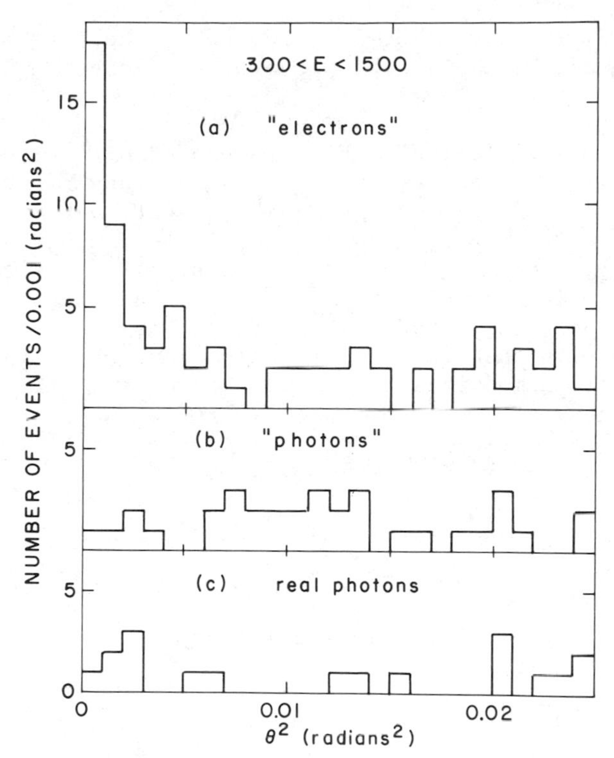

Fig. 5. The θ^2 distributions in Region 2 (300 MeV < E < 1500 MeV); for a) "electrons", b) "photons", and c) controlled gamma samples.

simply by extrapolating the flat background in the large θ^2 region into the small θ^2 region. The absence of a large forward peak in the "gamma" samples suggest that contribution from coherent π° production is not a difficulty in this experiment.

In the low energy region, shown in Fig. 6, the θ^2 distribution has a broad bump instead of a sharp peak in the forward direction. This is due to the fact that the angle of the signal can go up to ≈100 mrad in this energy region. Note however the θ^2 distribution of the "gammas" is still flat. Fig. 7 shows the θ^2

distribution in the region 3, E > 1500 MeV: there is no forward peak (the signal, were it to be present, should be at $\theta^2 < 0.002$ $(rad)^2$).

The kinematical variable, $E\theta^2$, is expected to have a much more limited range of values for $\nu_\mu e$ than for any of the background reactions. In our energy range and with our present resolution, the $\nu_\mu e$ signal should be entirely contained in $E\theta^2 \lesssim 3$. Conversely, the backgrounds are more smoothly distributed out to very large values of $E\theta^2$. The $E\theta^2$ distributions are given in Fig. 8. The number of events estimated in the signal region is consistent with that which is extracted by using the θ^2 distribution. Table 3 summarizes preliminary results on the $\nu_\mu e \rightarrow \nu_\mu e$ interactions.

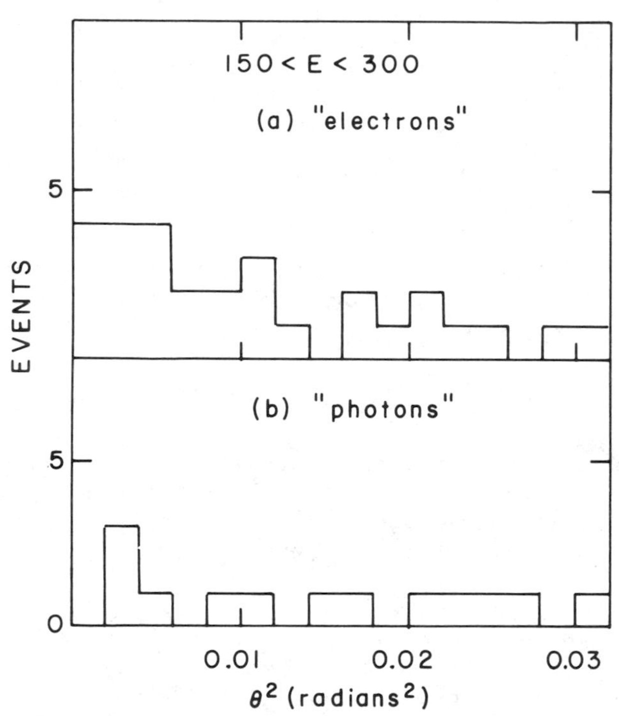

Fig. 6. The θ^2 distributions in Region 1 (150 MeV < E < 300 MeV); for a) "electrons", and b) "photons".

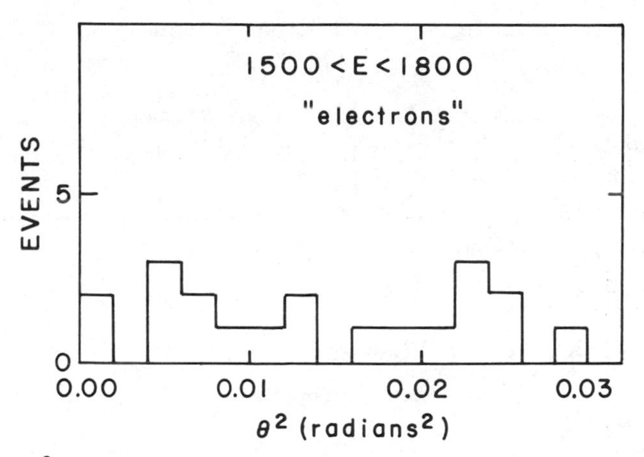

Fig. 7. The θ^2 distribution in the energy region between 1500 MeV and 1800 MeV for "electrons".

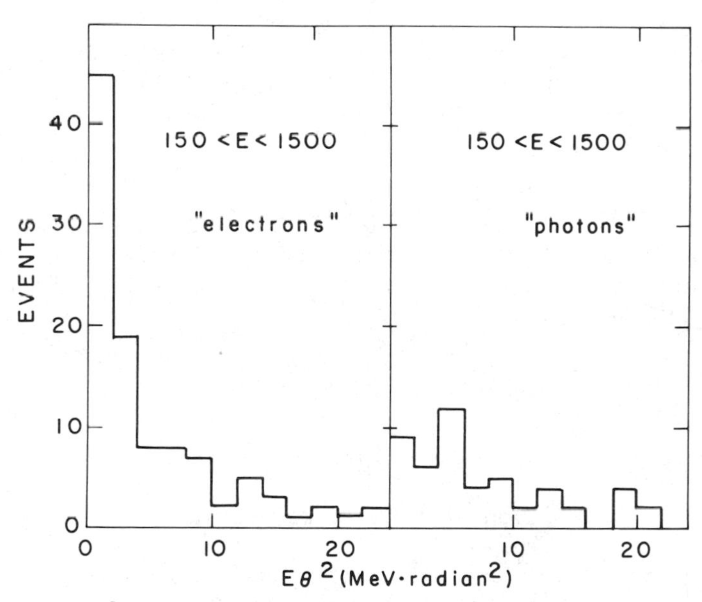

Fig. 8. The $E\theta^2$ distributions in the energy region between 150 MeV and 1500 MeV for "electrons" and "photons".

Table 3. Preliminary Results

Processed	Signal & Background	Estimated Background	Signal
June-July 1981[4]	23	≈10	13
Dec-Feb 1982 (≈3/4 are analyzed)			
150 < E < 300 MeV	19	≈6	13
300 < E < 1500 MeV	39	≈10	29
	58	≈16	42
TOTAL			55

We are making measurements with a charged particle beam to simulate the behavior of particles in the apparatus. When this is complete we expect to have a better understanding of the energy and angular resolution and the efficiency of the filter. Efforts to obtain the cross section normalization are under way. Another 2000 hours of running was recently approved in order to achieve more than 120 events for each sample of $\bar{\nu}_e$ and ν_e interactions. The running is scheduled to start at the end of August of this year.

This research is supported in part through the Japan-USA Cooperative Research Project on High Energy Physics under the Japan Ministry of Education, Science and Culture and the U.S. Department of Energy.

REFERENCES

1. J.G. Morfin, Talk given at this Conference.

2. K. Winter, Talk given at this Conference.

3. S. Weinberg, Phys. Rev. Lett. 19, 1264 (1967); A. Salam, in Elementary Particle Theory, edited by N. Svartholm (Stockholm 1968), p. 367.

4. R.E. Lanou, Proc. of International Conf. on Neutrino Physics & Astrophysics, Balatonfured, Hungary, June 24-19, Vol. II. pp. 9-19, A. Frenkel, L. Jenik, Editors, Budapest, 1982.; K. Abe et al., presented at the Eighth International Workshop on Weak Interactions and Neutrinos, Sept. 5-11, 1982, Javea, Spain.

NEUTRINO-ELECTRON SCATTERING

Klaus Winter

CERN, Geneva, Switzerland

1. INTRODUCTION

The first experimental observation of a weak neutral-current phenomenon was on the scattering of $\bar{\nu}_\mu$ on electrons ten years ago by the Gargamelle Collaboration at CERN[1]. Further observations, in particular on semileptonic neutral-current neutrino interactions[2] and on the scattering of polarized electrons on deuterium[3], have by now lent strong support to a unified gauge theory of the weak and the electromagnetic interaction as proposed by Weinberg and Salam. The main goal of pursuing the study of neutrino scattering on electrons is to determine the coupling constants of the leptonic weak neutral current, thus avoiding the ambiguities inherent in the use of hadronic targets. Until now, however, the extremely low cross-section has drastically limited the number of observed events, as illustrated in Tables 1 and 2. Taking all experiments together, about 100 events have now been observed in each channel.

At this conference we have heard reports on recent results on the ratio of $\sigma(\nu_\mu e)/\sigma(\bar{\nu}_\mu e)$, obtained by the CHARM collaboration at CERN[4], on progress from a beautiful new experiment[5] at the AGS in Brookhaven and on preliminary results from the flash-tube detector at Fermilab.

The axial-vector coupling constant g_A^e and the vector coupling constant g_V^e can be determined by a phenomenological, model-independent analysis. Assuming that leptons are point-like and that the outgoing neutrino is identical to the incoming neutrino, we obtain for the differential cross-section,

Table 1. Summary of past activities on $\sigma(\nu_\mu e)$

Experiment		$\nu_\mu e$	Background	σ 10^{-42} cm^2/GeV
1973 GGM PS		1	0.3 ± 0.1	< 3(90% c.l.)
Aachen/Pad		11	3	1.1 ± 0.6
GGM SPS	64K	9	0.5 ± 0.2	$2.4 \pm^{1.2}_{0.9}$
Col-BNL 15' BC	84K	8	0.5 ± 0.5	1.8 ± 0.8
LM O		40	12	1.4 ± 0.3
CHARM	10^6	46 ± 12	64 ± 10	2.1 ± 0.55(stat) ± 0.49
World average	$\sin^2\theta = 0.24 \pm^{0.06}_{0.04}$			1.55 ± 0.26

Table 2. Summary of past activities on $\sigma(\bar{\nu}_\mu e)$

GGM PS		3	0.4 ± 0.1	$1.0 \pm^{2.1}_{0.9}$
A-P		8	1.7	2.2 ± 1.0
CHARM SPS	10^6	77	146	1.6 ± 0.35(stat) ± 0.36
World average	$\sin^2\theta = 0.37 \pm 0.05$			1.58 ± 0.30

$$\frac{d\sigma}{dy}(\nu_\mu e) = \frac{\sigma_0}{2} E_\nu \left[\left(g_V^e+g_A^e\right)^2 + \left(g_V^e-g_A^e\right)^2(1-y)^2 + \frac{m_e}{E_\nu}\left(g_V^{e^2}-g_A^{e^2}\right)\right]\rho^2 \quad (1a)$$

and

$$\frac{d\sigma}{dy}(\bar{\nu}_\mu e) = \frac{\sigma_0}{2} E_\nu \left[\left(g_V^e-g_A^e\right)^2 + \left(g_V^e+g_A^e\right)^2(1-y)^2 - \frac{m_e}{E_\nu}\left(g_V^{e^2}-g_A^{e^2}\right)\right]\rho^2 \quad (1b)$$

where $\sigma_0 = G^2 m_e/\pi = 8.6 \times 10^{-42}$ cm^2. ρ is an overall factor related to the coupling at the $\nu_\mu Z^0 \nu_\mu$ vertex. If $\rho = 1$, $M_W = M_Z \cos\theta$ and the coupling of the neutral and charged current is of universal strength, $G_F^{NC} = G_F^{CC}$.

In the standard model $\rho = 1$ if one doublet of Higgs bosons is sufficient to describe the spontaneous symmetry breaking. Recently, C. Jarlskog and F. Yndurain[7] have estimated deviations of ρ from one due to the possible existence of lepton doublets with large mass differences between the neutral (neutrino) and charged component. They find

$$\varepsilon = |\rho^2 - 1| \geq \frac{G_F}{8\sqrt{2}\pi^2} \sum_{\ell=1}^{N_e} m_\ell^{\,2}$$

for N_ℓ new families beyond the (τ, ν_τ). Inserting the present experimental limit of

$$\varepsilon < 3\%,$$

derived from a recent analysis of semileptonic neutral current data[8], and assuming $m_\ell \gtrsim 20$ GeV, as suggested by searches for new heavy leptons at PETRA[9], a limit on the number of new lepton generations can be derived, giving

$$N_\nu < 737.$$

Combining this analysis with the present limit on modifications to the photon propagator in the reaction $e^+e^- \to \mu^+\mu^-$ due to heavy leptons

$$\sigma = \sigma^{(0)} (1+\delta) \; .$$

With

$$\delta = \frac{2\alpha s}{15\pi} \sum \frac{1}{m_\ell^{\,2}} < 3\% \text{ at } s = 1600 \text{ GeV}^2$$

the authors conclude that $N_\nu < 134$. At the large e^+e^- colliders a measurement of the cross-section of the reaction

$$e^+e^- \to \gamma + \text{nothing}$$

10 GeV above the Z^0 pole can give a sensitivity of $\Delta N_\nu \sim 0.2$.

2. Electron-Neutrino-Electron Scattering

The cross-section of the scattering of antielectron-neutrinos on electrons has been measured by Reines et al.[10]. It gives important constraints on the neutral current coupling of the electron, as well as a sensitive test of their generation universality. Both neutral and charged currents contribute to the reaction cross-section. Averaging over the unobserved electron helicities λ and λ'

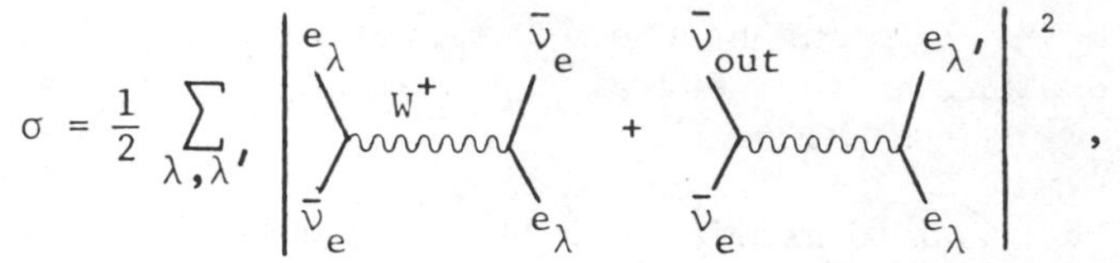

$$\sigma = \frac{1}{2} \sum_{\lambda,\lambda'} \left| \text{[diagram: } e_\lambda, \bar{\nu}_e \xrightarrow{W^+} \bar{\nu}_e, e_\lambda \text{]} + \text{[diagram: } \bar{\nu}_e, e_\lambda \to \bar{\nu}_{out}, e_{\lambda'} \text{]} \right|^2 ,$$

where the interference term is given, in the context of the Weinberg-Salam model, by

$$\left[T^3_L(e) + \sin^2 \theta \right] \cong -\frac{1}{2} + 0.23 ,$$

and should therefore be negative. Table 3 shows the predicted values in terms of σ^{CC}_{V-A} for two energy regions and for three assumed cases, destructive, constructive and no interference, the latter corresponding to a situation in which $\nu^{in} \neq \nu^{out}$. The experimental results of Reines et al.[10] favour the destructive interference case but cannot reject, in a compelling way, the other two cases.

It seems, therefore, that a new experiment, using modern technology would be required.

Observation of this interference term would prove that the outgoing and incoming neutrinos are identical; that the neutral current couples to left-handed electrons, solving the ambiguity under exchange of g^e_V and g^e_A in Eq. (1); and that the neutral current is helicity-conserving in the same way as the charged current[11], and hence that its space-time structure is given by V and A contributions only.

Table 3. Cross-section for $\bar{\nu}_e e \to \bar{\nu}_e e$ for three cases

Case	$1.5 \leq T_e \leq 3$ MeV	$3.0 \leq T_e \leq 4.5$ MeV
Destructive	$0.83\ \sigma^{CC}_{V-A}$	$1.2\ \sigma^{CC}_{V-A}$
Constructive	2.2	2.8
No interference ($\nu^{in} \neq \nu^{out}$)	1.5	2.0
Exp (Reines)	0.87 ± 0.25	1.7 ± 0.44

3. Coupling of the weak neutral electron current

The effective Lagrangian of neutrino-electron interactions and of electron-hadron interactions involves the weak neutral electron current

$$J_\mu^e = \bar{e}\gamma_\mu(g_V^e+g_A^e\gamma_5)e \ . \qquad (2)$$

It is determined by two coupling constants, g_A^e and g_V^e . The study of neutrino-electron scattering at high energy leaves a four-fold ambiguity, one sign ambiguity and an ambiguity under the exchange of g_V and g_A. Combining these results with those from $\bar{\nu}_e e \to \bar{\nu}_e e$ scattering[10] and from $e^+e^- \to \mu^+\mu^-$ at PETRA[12] a unique solution emerges, the one predicted by the standard model of electroweak interactions, with

$$g_A^e = -0.523 \pm 0.035$$

and[13]

$$\sin^2\theta = 0.215 \pm 0.043.$$

Figure 1 summarizes the present situation. Using the value of g_A^e determined from neutrino-electron scattering one can deduce, from measurements of the angular asymmetry in $e^+e^- \to f^+f^-$, which is proportional to the product $g_A^e \cdot g_A^f$, the axial vector coupling of the other leptonic neutral currents. Table 4 summarizes these results.

Table 4. Neutral current coupling of the leptonic currents

Current	Process	g_A^f	I_3^R
ee	νe	-0.523 ± 0.035	0.02 ± 0.03
$\mu\mu$	$e\bar{e} \to \mu\bar{\mu}$	-0.50 ± 0.08	0.01 ± 0.08
$\tau\tau$	$e\bar{e} \to \tau\bar{\tau}$	-0.39 ± 0.12	-0.11 ± 0.12

The SU(2) structure is clearly confirmed by these measurements, the left-handed leptons transform as doublets and the right-handed leptons as singlets. An independent way of resolving the ambiguities has been achieved before[14], using results of an experiment performed at SLAC on the helicity asymmetry of electron-deuteron scattering and of experiments on neutrino-hadron scattering, under the assumption of factorization.

The overall coefficient ρ^2, multiplying the theoretical expressions of the cross sections of the neutrino-electron scattering, determines the relative strength of the neutral and of the charged current coupling. In the standard model $\rho = 1$, provided the Higgs fields form an isospin doublet. This simplest version of the theory is already confirmed by the present measurement[15] giving

$$\rho = 1.12 \pm 0.16 \ .$$

At a higher level of accuracy small deviations of ρ from one are expected as discussed in the introduction. It would therefore be important to determine $|\rho^2-1|$ to $\sim \pm 1\%$.

The most precise determination of the value of $\sin^2\theta$ in the leptonic sector has been obtained by the CHARM Collaboration[15], making use of the direct relation between the ratio of $\sigma(\nu_\mu e)$ and $\sigma(\bar{\nu}_\mu e)$ and $\sin^2\theta$.[16]

$$R = \frac{\sigma(\nu_\mu e)}{\sigma(\bar{\nu}_\mu e)} = 3 \, \frac{1-4\sin^2\theta + (16/3)\sin^4\theta}{1-4\sin^2\theta + 16\sin^4\theta} \ .$$

In the vicinity of $\sin^2\theta = \frac{1}{4}$ this relation gives $\Delta\sin^2\theta \sim \frac{1}{8}\,\Delta R$ and, hence, a very precise determination of the mixing angle. The detection efficiency cancels in the ratio and many systematic uncertainties are reduced. Using their upgraded detector, incorporating now measurements of the shower barycentre in two orthogonal projections behind each target plate, the CHARM Collaboration will further improve the accuracy of this result. The plans for further measurements at CERN, FNAL and BNL are summarized in Table 5.

Are further, improved measurements of neutrino-electron scattering required? The electroweak theory predicts the Born term ($Q^2 \sim 0$) of the mass of the Z^0 as a function of $\sin^2\theta$

$$M_{Z^0} = \frac{37.4 \text{ GeV}}{\sin\theta\cos\theta}$$

Second order corrections shift M_{Z^0} by 5 GeV[17], e.g. for $\sin^2\theta = 0.220 \pm 0.015$, as determined by the CHARM Collaboration from semi-leptonic neutral current interactions[18]. Wheater and Llewellyn-Smith[17] predict

$$M_{Z^0} \text{ (Born)} = 90 \pm 2.3 \text{ GeV}$$

$$M_{Z^0} \text{ (physical)} = 94.6 \pm 2.3 \text{ GeV}$$

and hence a shift of $\sim$ 5 GeV. A precise experimental determination of this shift would constitute a decisive test of the underlying electroweak gauge theory. A significant test could be claimed if the present error on M would be reduced to 0.7 GeV corresponding to an error on $\sin^2\theta$ of $\pm$ 0.005[19].

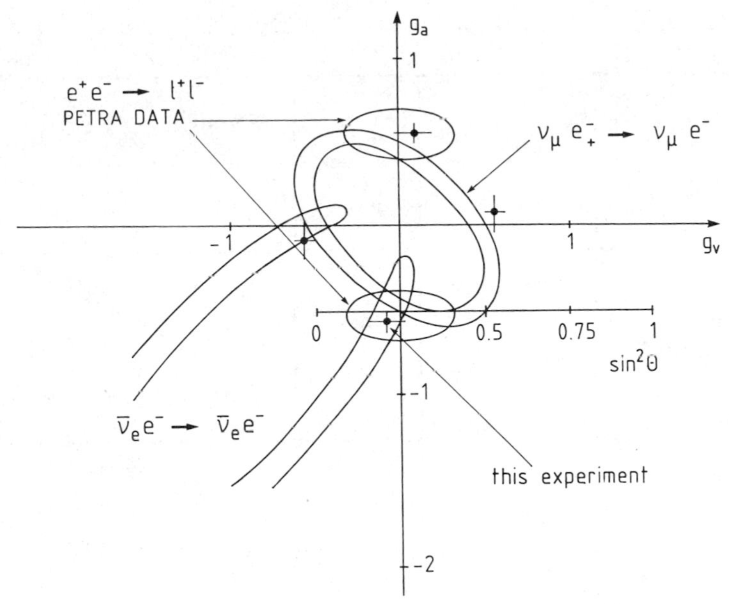

Fig. 1 Results of various leptonic reactions determining g_A^e and g_V^e

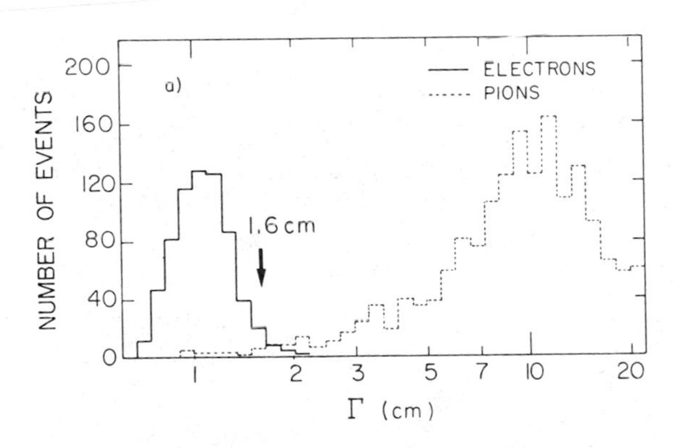

Fig. 2 Distribution of the width Γ of electron and pion induced showers as measured by the lateral energy deposition profile in the scintillator planes of the CHARM I detector

Table 5. Summary of present and future activities on $\nu_\mu e$ scattering

Group	$\sigma(\theta)$ E = 15 GeV	S/B	Events	$\Delta\sin^2\theta$ (stat)
Flashtube det. FNAL 200 tons	$\sim$ 8 mrad	1/1	$\sim$ 10 $\nu_\mu e$	
CHARM 1982 70 tons	$\sim$ 12 mrad	1/2	77 $\bar{\nu}_\mu e$ 46 $\nu_\mu e$	$\pm$ 0.04
CHARM 1983	$\sim$ 8 mrad	1/1	$\sim$ 60 $\nu_\mu e$ $\sim$ 60 $\bar{\nu}_\mu e$	Total $\sim$ $\pm$ 0.02
BNL 1982			80 $\nu_\mu e$ 25 $\bar{\nu}_\mu e$	
BNL 1982+1983 80 tons	$\frac{12\text{ mrad}}{\sqrt{E}}$	1/0.25	Total 140 $\nu_\mu e$ 90 $\bar{\nu}_\mu e$	Total $\pm$ 0.017

Can this accuracy be achieved in experiments? Llewellyn-Smith[20] has discussed the theoretical uncertainties of the quark model picture of the nucleon and its effects on attempts to measure $\sin^2\theta$ in semi-leptonic neutral current interactions. These uncertainties do not exist in the leptonic reactions of neutrino-electron scattering. Here only the experimental problems have to be solved, namely:

1) event rate requiring a large fiducial tonnage and high selection efficiency over a wide window of electron energies;

2) background, dominantly due to quasi-elastic electron-neutrino scattering and to coherent π^0 production[15] has to be reduced by efficient e/π discrimination and precise measurements of the shower direction;

3) monitoring of the relative flux of the different beam components ν_μ, $\bar{\nu}_\mu$, ν_e and $\bar{\nu}_e$ is required to determine the ratio of $\sigma(\nu_\mu e)/\sigma(\bar{\nu}_\mu e)$.

4. Study of a new detector for neutrino-electron scattering

A tentative design of a new dedicated ($\nu_\mu e$) detector has been discussed by the CHARM Collaboration. It is based on the principle of a fine-grain target-calorimeter and on the accumulated experience of the CHARM Collaboration with instrumentation of this type. The main concern of this study was the question: can the present technique used for $\nu_\mu e$ studies be sufficiently improved to match the aim of $\Delta \sin^2\theta = 0.005$ or is a new technique required?

The accuracy of shower direction measurements depends mainly on three contributions

$$\sigma^2(\theta) \sim \underbrace{(\Delta\theta_{\sqrt{N}})^2}_{\text{statistics}} + \underbrace{(\Delta\theta_V)^2}_{\text{vertex}} + \underbrace{(\Delta\theta_{SMEAR})^2}_{\text{smearing}}$$

- on the sampling frequency and on the method used to count the number of shower particles N;
- on the plate thickness and grain size of the calorimeter near the vertex;
- on the lateral shower sampling used to determine the barycentre of the shower.

An optimization is required if limited space is available.

The limiting accuracy is given by the Z number of the target material,

$$\sigma(\theta) \sim \frac{Z}{\sqrt{E}} \text{ const.}$$

The mean Z of the present marble target is 13, with $MgCO_3$ (dolomite) a value of ~ 8 could be achieved. In these light targets with a nuclear absorption length of $\Lambda_{abs} \sim 4$ radiation lengths, electromagnetic and hadronic showers have very similar longitudinal profiles[21]. The CHARM Collaboration has developped a new method to discriminate between electromagnetic and hadronic showers in low Z materials based on the characteristic difference of their lateral profiles (see figure 2). A discrimination by a factor of ~ 100 has been achieved[21]. Figure 3 shows the resolution in shower direction measurements achieved with the present CHARM detector and with the upgraded version.

The structure of the new, dedicated detector is sketched in fig.4. It consists of 330 modules of 3.5×3.5 m^2 surface area, each composed of a 4 cm thick target plate (marble) and of a plane of streamer tubes with 1 cm wire spacing, read out by crossed cathode strips of 2 cm spacing in two orthogonal projections. Using analog electronics the centroid position of a track or a shower can be reconstructed with $\pm$ 2 mm accuracy. Simulating this structure using Monte Carlo methods (EGS) we find an angular resolution of $\sigma(\theta) \sim 16$ mrad$/\sqrt{E/\text{GeV}}$ corresponding to ~ 4 mrad at 15 GeV and an energy resolution of $\sigma(E)/_E \sim 20\%/\sqrt{E/\text{GeV}}$. Hence, we expect a reduction of background proportional to $\sigma^2(\theta)$ by a factor of ~ 9.

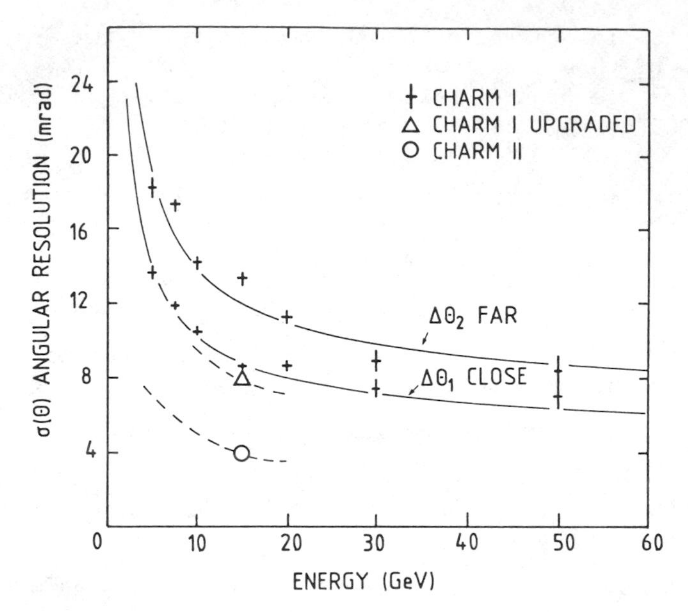

Fig. 3 Angular resolution of electron shower direction as a function of electron energy. In the CHARM I detector close and far refer to the projection following the target plate in which the event started (close) and the next plate (far). In the upgraded CHARM I detector additional planes of streamer tubes have been installed and the resolution is now the same in both projections

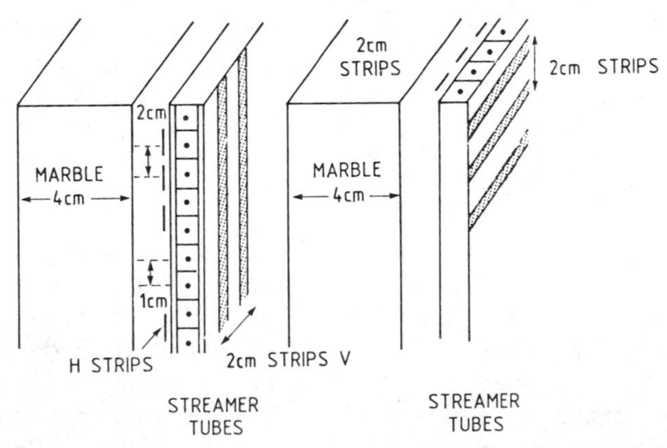

Fig. 4 Schematic sketch of a new dedicated CHARM II detector

Figure 3 shows this estimate of $\sigma(\theta)$ compared to the present CHARM I detector. The choices made of analog versus digital readout of strips and of the strip width are illustrated in figures 5a and 5b. This configuration has a fiducial target weight of $\sim$ 320 tons, whereas the CHARM I detector has 70 tons.

Monitoring of the beam composition has been studied in detail for the CHARM I experiments using quasielastic $\nu_\mu(\bar{\nu}_\mu)$ events. This method can be extended to lower neutrino energies by restricting the value of Q^2 to less than 0.01 GeV^2 to equalize the cross sections for the neutrino and the antineutrino induced reactions. Figure 6 shows the measurements for quasielastic reactions[22] at low Q^2. At $E_\nu > 10.8$ GeV a peak appears at small Q^2 because of the inverse μ decay reaction, $\nu_\mu e \to \mu^- \nu_e$. The CHARM II detector will have good angular resolution for measurement of the muon direction and, combined with a muon spectrometer, will therefore directly monitor the flux ratio $\phi(\nu_\mu)/\phi(\bar{\nu}_\mu)$ to $\pm 2\%$, required to evaluate the ratio R and the neutrino spectra.

Electron neutrino-electron scattering has a ten time higher cross section than $\nu_\mu e$ scattering. The beam contains approximately 2% electron neutrinos and antineutrinos. Aiming at a 5% measurement of R requires a knowledge of the $(\nu_e + \bar{\nu}_e)$ flux to better than $\pm$ 10%. I think that this can be achieved by making dE/dx measurements in the plane following the vertex, at angles outside the forward peak. The CHARM Collaboration has also developed a method to recognize inclusive $\nu_e N \to eX$ events with $y < 0.6$[21].

I summarize this study in Table 6 which gives the selection criteria, the angular resolution, the rate and background for $\bar{\nu}_\mu e$ scattering which would be obtained for 10^{19} protons on target.

Table 6. Rates and background of $\bar{\nu}_\mu e$ scattering in CHARM I and in CHARM II for 10^{19} protons

Selection	CHARM I	CHARM II
Rate for $7.5 < E_e < 30$ GeV	154 ± 27	909 ± 33
ε_e (1 hit)	61%	90%
$\sigma(\theta)$ at $\bar{E}_e = 15$ GeV	12 mrad	4 mrad
Signal/background	1/2	1/0.22
Rate for $2.5 < E_e < 30$ GeV		1364 ± 40

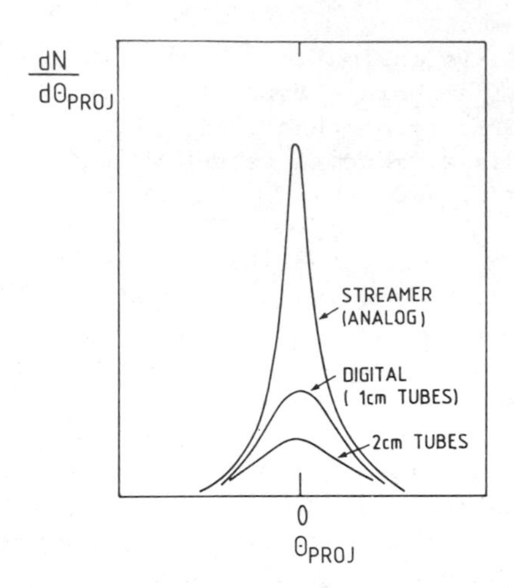

Fig. 5a Shower profile sampling by detectors with analog and with digital measurements

The same statistics for $\nu_\mu e$ would be obtained with $5 \cdot 10^{18}$ protons. Hence, we expect to measure

$$R = \frac{N(\nu_\mu e)}{N(\bar{\nu}_\mu e)} xF = \frac{1364 \pm 40}{1364 \pm 40} (F \pm 2\%) = \overline{R} \pm 5\%,$$

corresponding to $\Delta\sin^2\theta = \pm 0.005$.

An additional physics result may be obtained on the value of ρ, derived from a measurement of the ratio of cross sections for $\nu_\mu e \rightarrow \nu_\mu e$ and for $\nu_\mu e \rightarrow \mu^- \nu_e$. The flux monitoring is now internal and, apart from systematic uncertainties due to background subtraction, electron detection efficiency and spectrum determination, the statistical error of ρ would be $\pm$ 2%.

The main conclusion that we draw from this study is that the experiment seems to be feasible using the presently developed fine-grain calorimeter technqiue. Of course, many details have to be investigated further and it is not the aim of this study to be at the level of a proposal.

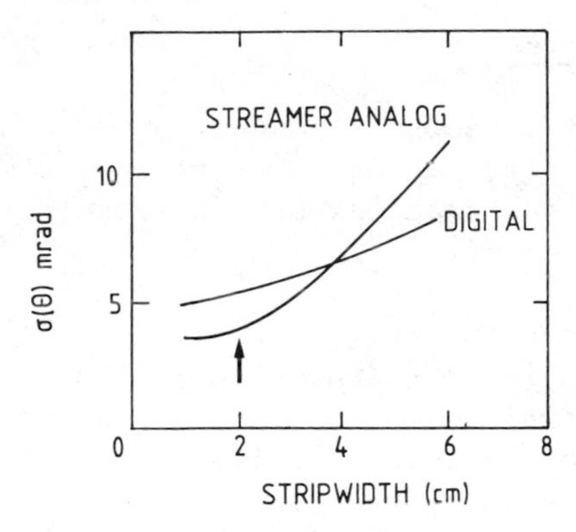

Fig. 5b Resolution of the shower direction as a function of the strip width

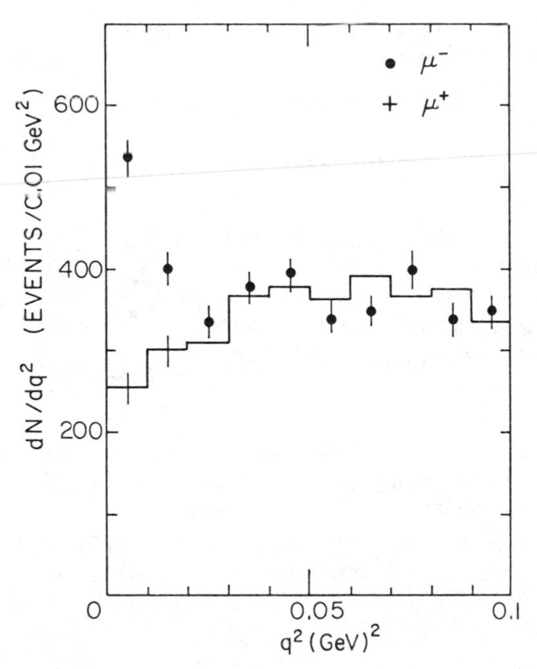

Fig. 6 Observed Q^2 dependence of quasielastic $\nu_\mu(\mu^-)$ and $\bar{\nu}_\mu(\mu^+)$ events

5. Concluding remarks

With the recent advances in experiments showing helicity-asymmetry in electron-deuteron scattering and forward-backward angular asymmetry in $e^+e^- \rightarrow \mu^+\mu^-$ our understanding of the weak neutral current has progressed.

Neutrino-electron scattering is a field that is expected to progress rapidly, following the demonstration that experiments at high energy, using fine-grain electronic calorimeters, can detect these rare events and separate them reliably from the dominant semileptonic reactions.

Future results at the Z^0 pole and on neutrino-electron scattering together will provide an important test of the electroweak standard model.

Acknowledgements

It is a great pleasure to thank the organizers of the Study Conference on Electroweak Interactions for their kind and warming hospitality and for having created a stimulating atmosphere of discussions.

In preparing these notes I have enjoyed many discussions with my colleagues of the CHARM Collaboration whom I wish to thank sincerely.

References

1) F.J. Hasert et al., Phys. Letters 46B (1973) 121.
J. Blietschau et al., Nucl. Phys. B114 (1976) 189.
2) For a recent review see, for example, F. Niebergall in Proceedings of the International Conference Neutrino '82, Balatonfüred, Hungary 1982 (Budapest, 1982) Volume II, p.62.
3) C.Y. Prescott et al., Phys. Letters 77B (1978) 347 and 84B (1979) 524.
4) Report by B. Borgia (CHARM Collaboration), these proceedings.
5) Report by R.E. Lanou, these proceedings.
6) Report by J. Morfin, these proceedings.
7) C. Jarlskog and F.J. Yndurain, Phys. Letters 120B (1983) 361.
8) M. Jonker et al., (CHARM Collaboration), Phys. Letters 99B (1981) 265.

9) PLUTO Collaboration, Ch. Berger et al., Phys. Letters 99B (1981) 489
MARK-J Collaboration, D.P. Barber et al., Phys. Rev. Letters 45 (1980) 1904
TASSO Collaboration, R. Brandelik et al., Phys. Letters 99B (1981) 163
JADE Collaboration, W. Bartel et al., Phys. Letters 123B (1983) 353.

10) F. Reines et al., Phys. Rev. Letters 37 (1976) 315.

11) M. Jonker et al., (CHARM Collaboration), CERN-EP/82-207 and Z. f. Physik C. (in print).

12) See e.g. A. Böhm, DESY 82-084, December 1982, and M. Davier, LAL 82/93, Oct. 1982.

13) F. Bergsma et al., (CHARM Collaboration), Phys. Letters 117B (1982) 272.

14) See e.g. review of K. Winter, in Proc. of the 1979 Int. Symposium on Lepton and Photon Interactions at High Energies, (FNAL, Batavia, 1979) p. 258.

15) See reference 13).

16) A precision measurement of $\sin^2\theta$ by the ratio of $\sigma(\nu_\mu e)/\sigma(\bar{\nu}_\mu e)$ has been discussed by K. Winter in New Aspects of Subnuclear Physics (Plenum Press, New York and London 1980) p.218.

17) J. Wheater, C.H. Llewellyn-Smith, Phys. Letters 105B (1981) 486.

18) See reference 8).

19) K. Winter, in Workshop on Weak Interactions, Javea (Spain) September 1982.

20) Ch. Llewellyn-Smith, contribution to the SPS fixed target Physics Workshop at CERN, Dec. 1982 and Nucl. Phys. (to be published)
J. Panman, Contribution to the SPS Fixed Target Physics Workshop at CERN, Dec. 1982 (to be published).

21) A.N. Diddens et al. (CHARM Collaboration) Nucl. Instr. Methods 178 (1980) 27
M. Jonker et al. (CHARM Collaboration) Nucl. Instr. Methods 200 (1982) 183.

22) F. Bergsma et al., (CHARM Collaboration), Phys. Letters 122B (1983) 465.

RESULTS FROM PLUTO

Gideon Bella

Department of Physics and Astronomy
Tel Aviv University
Tel Aviv, ISRAEL *

ABSTRACT

Results on the reaction $e^+e^- \to \mu^+\mu^-$ are presented. The data were taken with the PLUTO detector at the c.m. energy of 34.7 GeV. In the integrated luminosity of 44000 nb^{-1} taken, a forward backward asymmetry of $-(12.0\pm3.2)\%$ is observed for $|\cos\theta|<1$ which is in agreement with the standard model.

1. INTRODUCTION

Electron-positron collisions are a very convenient tool to investigate electroweak as well as strong interactions. The main process studied is the one photon annihilation where the electron and positron produce a virtual time-like photon which convert into a pair of fermions (see Fig. 1a). A similar diagram where the photon is replaced by the neutral intermediate vector boson Z (see Fig. 1b) is also present. The interference between these two diagrams allows to test the standard model of electroweak interactions.

The PLUTO detector which is operated by a collaboration of several institutions (1) is described in the next section. The high luminosity obtained from PETRA at a high c.m. energy, thanks to the implementation of mini β quadrupoles, enabled us together with the other PETRA detectors to obtain a significant electroweak effect, which will be described in sections 3-7.

* Partially supported by the Israeli Academy of Sciences and Humanities - Basic Research Foundation.

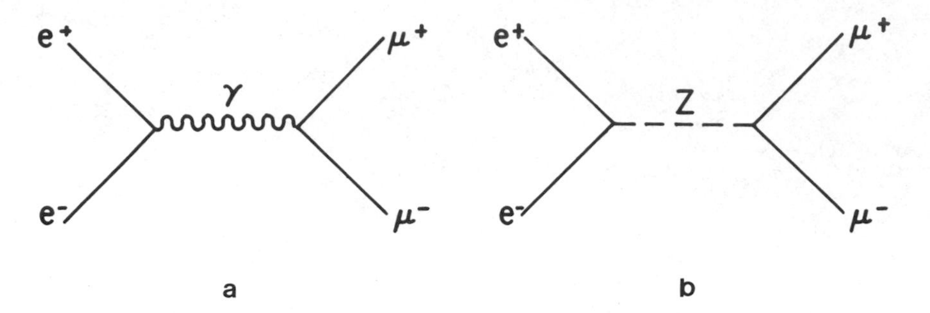

Figure 1: Annihilation diagrams mediated by a) photon (pure QED)
b) Z (weak neutral curren

2. The PLUTO Detector

The PLUTO detector (Fig. 2) has been described elsewhere (2). It consists of a central detector and two forward spectrometers which were added in 1981 in order to improve its capability to detect two photon reactions (3). For the sake of completeness, let us mention here the main components of the central detector which was used for the analysis described in this report.

a) The magnetic field of 1.65 Tesla is provided by a superconducting solenoid.
b) The inner detector consists of 13 layers of proportional chambers.
c) Inside the magnetic field volume there are barrel and endcap lead scintillator shower counters which give also timing information.

3. Electroweak Effects in e^+e^-

The differential cross section which is derived from the diagrams of Fig. 1 can be written in the following form:

$$\frac{d\sigma}{d\Omega} = \frac{\alpha^2}{4s} \; R \; (1 + \cos^2\theta \; + \frac{8}{3} \; A \; \cos\theta \;) \tag{1}$$

where

$$R = Q_f^2 \; - 2Q_f v_e v_f \chi + (v_e{}^2 + a_e{}^2)(v_f{}^2 + a_f{}^2)\chi^2 \tag{2}$$

is the ratio between the total cross section of fermion pair $f\bar{f}$ production and the total pure QED cross section to produce point-like leptons with unit charge. A is given by

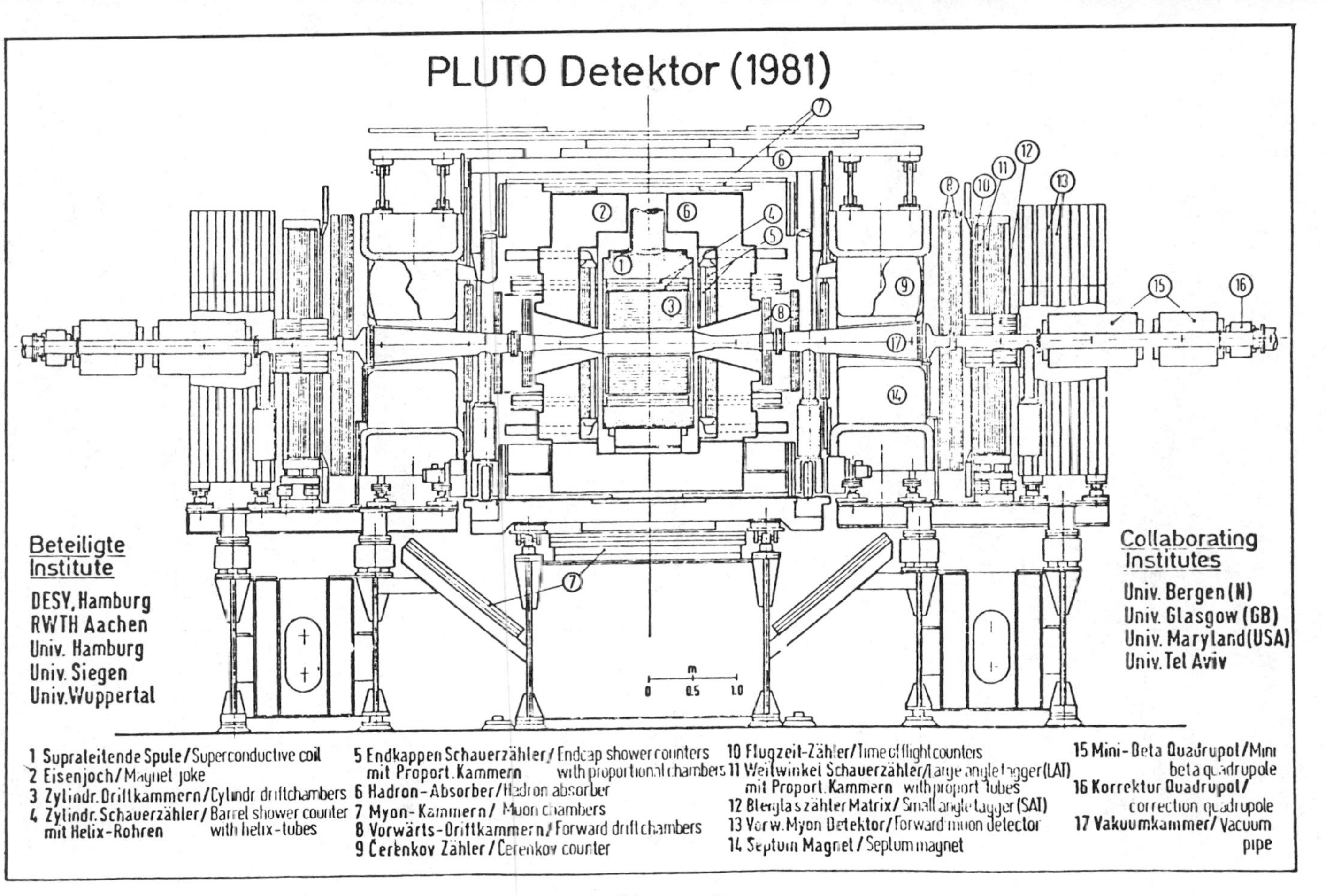
PLUTO Detektor (1981)
Beteiligte Institute
DESY, Hamburg
RWTH Aachen
Univ. Hamburg
Univ. Siegen
Univ. Wuppertal
Collaborating Institutes
Univ. Bergen (N)
Univ. Glasgow (GB)
Univ. Maryland (USA)
Univ. Tel Aviv
m
0 0.5 1.0
1 Supraleitende Spule / Superconductive coil
2 Eisenjoch / Magnet joke
3 Zylindr. Driftkammern / Cylindr. driftchambers
4 Zylindr. Schauerzähler / Barrel shower counter mit Helix-Rohren with helix-tubes
5 Endkappen Schauerzähler / Endcap shower counters mit Proport. Kammern with proportional chambers
6 Hadron-Absorber / Hadron absorber
7 Myon-Kammern / Muon chambers
8 Vorwärts-Driftkammern / Forward driftchambers
9 Čerenkov Zähler / Cerenkov counter
10 Flugzeit-Zähler / Time of flight counters
11 Weitwinkel Schauerzähler / Large angle tagger (LAT) mit Proport. Kammern with proport tubes
12 Bleiglaszähler Matrix / Small angle tagger (SAT)
13 Vorw. Myon Detektor / Forward muon detector
14 Septum Magnet / Septum magnet
15 Mini-Beta Quadrupol / Mini beta quadrupole
16 Korrektur Quadrupol / correction quadrupole
17 Vakuumkammer / Vacuum pipe

Figure 2

$$A = \frac{3}{2} a_e a_f \chi (- Q_f + 2 v_e v_f \chi)/R \tag{3}$$

and corresponds to the forward backward asymmetry, namely

$$A = \frac{\sigma(\cos\theta>0) - \sigma(\cos\theta<0)}{\sigma(\cos\theta>0) + \sigma(\cos\theta<0)} \tag{4}$$

Q_f is the charge of the final fermion in units of positron charge. v_e and v_f are the weak vector charges of the electron and the final fermion respectively. a_e and a_f are the corresponding weak axial vector charges. Finally χ is given by,

$$\chi = g \frac{m_z^2 s}{s - m_z^2} \tag{5}$$

where $g = \frac{G_F}{8\sqrt{2}\pi\alpha} = 4.49 . 10^{-5}\ \mathrm{GeV}^{-2}$ with G_F being the Fermi coupling constant which is known from weak interactions at low energies and m_z is the mass of the Z boson.

So far our formulae were model independent. In order to compare them with experimental data we will use the standard model to fix the weak charges a and v and m_z. In Table 1 we listed the standard model values for a and v for the four different types of fermions. In addition, the Z mass is then given by

$$m_z = \sqrt{\frac{\pi\alpha}{\sqrt{2} G_F}} \cdot \frac{1}{\sin\theta_w \cos\theta_w} = \frac{37.3\ \mathrm{GeV}}{\sin\theta_w \cos\theta_w} = 89\ \mathrm{GeV} (\sin^2\theta_w = 0.228)$$

From m_z we can calculate through Eq. 5 a value $\chi = - 6.38\%$ for the c.m. energy of $\sqrt{s} = 34.7$ GeV. The expected values for A and R shown

Table 1

fermion	a	v	v for $\sin^2\theta_w = 0.228$
$\nu_e \nu_\mu \nu_\tau$	1	1	1
e, μ, τ	-1	$-1 + 4\sin^2\theta_w$	- 0.088
u,c,t	1	$1 - \frac{8}{3}\sin^2\theta_w$	0.392
d,s,b	-1	$-1 + \frac{4}{3}\sin^2\theta_w$	- 0.696

Table 2

Fermion	A	R
μ,τ	- 9.5%	1.003
u,c	-14.2%	0.446
d,s,b	-27.2%	0.115

in Table 2 were calculated using Eqs. 2 and 3. So far, no clear evidence has been reported on electroweak effects in $e^+e^-\rightarrow$hadrons,(4). The main reason for this situation is the difficulty in identifying flavors and charges of quarks because they fragment into hadrons. One can however try to tag heavy flavors using inclusive muon production or identify heavy flavored mesons, like D^*, through their decay. Even with these methods the results have not been so far very significant in spite of the relatively large values expected for the asymmetry.

The total hadronic cross section, which does not require flavor identification, can be calculated by adding the contributions of the different quarks and multiplying them by the color factor 3. The expected electroweak result of R_h = 3.709 at $\sqrt{s}$ = 34.7 GeV is higher than the pure QED expectation by only 1.1% (mainly due to the fact that unfortunately v_e <<1). This must be compared with the first order QCD correction of ≈ 5% and with the present systematical errors on R_h which are larger or equal to 3%. Hence,it is presently impossible to establish the electroweak effect from R_h and it is only possible to set limits on the value of $v_e v_f$ and thus, on $\sin^2\theta_w$. The same holds for $\mu^+\mu^-$ and $\tau^+\tau^-$ total cross sections. For electrons, there is an additional contribution to the cross section coming from the exchange of photon in the t-channel (5). This complicates the equations and reduces the electroweak effect considerably.

In conclusion we see that at present, the only practical way to establish the existence of electroweak effect is the study of μ pair or τ pair asymmetry in the differential cross section. From these two channels the μ pair is simpler because muons are easier to identify.

4. Systematic Effects in the Determination of Muon Pair Asymmetry

In determining the μ pair asymmetry there are mainly three types of systematic effects which must be considered:
a) radiative corrections;
b) background contributions;
c) detector effects.

The diagrams contributing to radiative effects at the order of α^3 are plotted in Fig. 3. The vertex correction (Fig. 3a) and the vacuum polarization (Fig. 3c) have no effect on the angular distribution of the outgoing muons. On the other hand, the interference between the two photon exchange diagrams (Fig. 3b) and the lowest order diagram (Fig. 1a) is totally antisymmetric and gives a positive

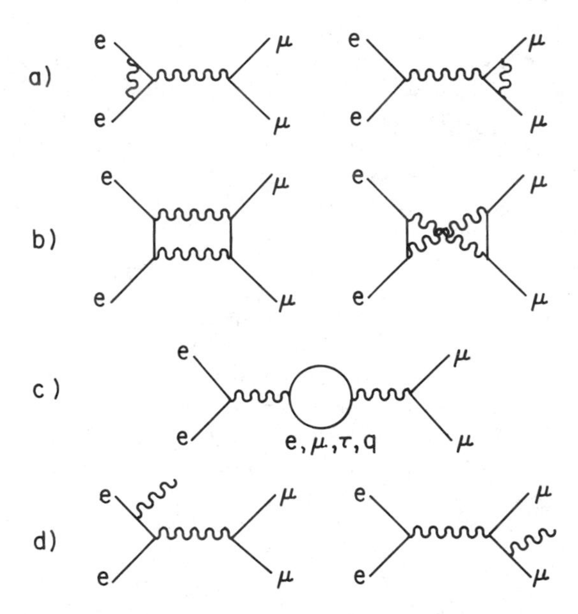

Figure 3: Diagrams contributing to radiative corrections of the order α^3
a) vertex corrections, b) two photon exchange, c) vacuum polarization, d) soft and hard photon radiation.

contribution to the forward backward asymmetry. Part of it is cancelled out by the negative contribution coming from the interference between the initial state and the final state radiation (Fig. 3d). This depends also on the experimental cuts which determine the fraction of hard photon events introduced in the final sample.

In addition to the pure QED diagrams there are also weak radiative contributions, where any one of the virtual photons can be replaced by Z, and also other diagrams which include virtual $W^{\pm}$ bosons. In particular one should mention initial state radiation with Z exchange. The invariant mass of the Z in this case is smaller than $\sqrt{s}$ and hence leads to a smaller asymmetry.

In order to take into account the effects of radiative corrections one can use the Monte Carlo generator introduced by Berends, Kleiss and Jadach (6) which includes all the pure QED and most of the weak radiative corrections to the power α^3. The effect of the diagrams not taken into account in Ref.6 is expected to be negligible at PETRA energies (4).

The second type of systematics which must be carefully considered in the analysis of μ pair asymmetry is the background contributions. The most dangerous is a possible contamination of Bhabha events in the μ pair sample. For $|\cos\theta|<0.75$ the Bhabha cross section is larger than the μ pair cross section by a factor of more than 20 and at the same time its asymmetry is, $A_{ee} = +\ 77.6\%$ due to the t-channel photon exchange. Much less important is the contamination of cosmic ray muons which do not have any asymmetry.

The third type of systematics is related to detector effects. The outgoing muons have high momenta and their tracks are almost straight lines. Consequently, forward and backward scattered events look almost the same in the detector and have also the same acceptance. On the other hand, the probability to assign to a given muon the wrong charge may not be negligible. In so far as the charge misassignment is completely random, it would tend to reduce the asymmetry. Nevertheless, this effect would be negligible as far as the charge misassignment probability is less than 10% per track. Here one should note that the random assumption may well be wrong and in that case its effect on the asymmetry determination can be significant. For instance, if the whole inner detector is twisted by a small amount around the beam pipe it will tend e.g. to increase the momenta of both muons in the case of forward scattering and to decrease it in the opposite case. Consequently, the probability of charge misassignment would be different for both cases and a large effect on the asymmetry result will follow.

5. Muon Pair Selection Criteria and Background Considerations

In order to obtain the μ pair sample collected with the PLUTO detector, the following criteria were required:

a) The event must have exactly two tracks which have a successful fit to the vertex in the $r\phi$ co-ordinates and momenta larger than 5 GeV. This cut suppresses cosmic muons and two photon events of the type $e^+e^- \to e^+e^-e^+e^-$, $e^+e^-\ \mu^+\mu^-$.

b) Both tracks must have a common vertex along the beam line (z coordinate) which lies within 40 mm from the expected z vertex. This cut was introduced to remove cosmic muons.
c) The angle θ between the track and the beam line (averaged over both tracks) must satisfy $|\cos\theta|<0.752$. This cut ensures more than 8 hits per track in the detector and therefore reduces the probability for charge misassignment.
d) The acollinearity angle between both tracks must be smaller than 10°. This cut suppresses τ lepton and two photon events.
e) In order to reject Bhabha events, the total energy measured in the barrel and endcap shower counters must be smaller than 2 GeV.
f) For the final reject of the left over cosmic ray events, timing cuts were also applied.
The timing information is obtained from the barrel and endcap shower counters which provide also an overall trigger signal. After requiring that this signal is within a certain time interval, the difference in time between the lower and upper barrel hits of the remaining events is plotted (Fig. 4).

The distribution obtained shows that cosmic ray muons can be separated out very clearly. For the endcaps, the timing resolution was not that good and some cosmic background was left in the sample. A residue of the cosmic background was left in the sample where only one track gave timing information due to inefficiency of the counters.

After all the cuts, a total number of 1550 events was left for final physics analysis which corresponds to an integrated luminosity of 44000 nb^{-1}. In this sample the background sources were estimated to be :

a) Cosmic muons (4.2%)
b) $ee \to \tau\tau$ (3.0%)
c) $ee \to ee\ \mu\mu$ (0.5%)
d) $ee \to ee$ (<0.2%)

The left over cosmic background had to be statistically subtracted. The τ pair background is expected to give the same asymmetry as the μ pairs and therefore, no correction for it was necessary. The corrections due to the third and fourth background sources are negligible.

6. The Effects of Radiative Corrections and Charge Misassignment

In order to minimize systematic effects due to charge misassignment, both tracks of each event were fitted to one long track. This fit improves the momentum resolution significantly and the resulting 1/P distribution is plotted in Fig. 5. The number of events with small 1/P values (i.e. those events which might have a wrong charge assignment) is very small. A scatter plot of 1/P vs. $\cos\theta$ (Fig. 6) looks flat and rules out any possibility of a large twist deformation in our detector. Small twist deformation which

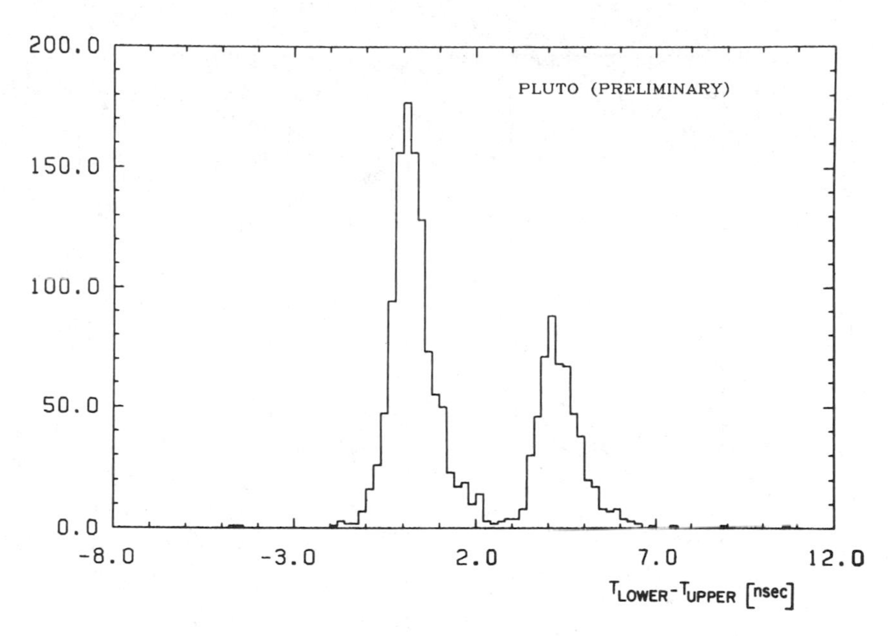

Figure 4: The difference in time between the lower and upper barrel hits of events after a preliminary cosmic rejection (see text). The genuine μ pairs are peaked around 0 while the cosmic muons are peaked around 4.2 nanoseconds. The cut was made at 2.8 nanoseconds.

would leave that distribution more or less flat, would change the asymmetry only by 0.2% as confirmed by Monte Carlo events and by software deformations of the real events. The only disadvantage of this common long track fit is for events where hard photons are radiated. Since these events obviously fit poorly a single track, they will more often have a wrong charge assignment and at the same time these events have a preference to go backwards (see also section 4). This would reduce the observed asymmetry. The presence of these events is observed in the longs tails of the 1/P distribution (Fig.5) which are seen more clearly in a logarithmic scale (Fig.7).

The corresponding Monte Carlo distribution (solid line in Fig.7) does follow the data very closely. For the Monte Carlo simulation, the event generator based on Ref. 6 was used followed by a complete detector simulation and the complete μ pair analysis including the common track fit procedure. The observed asymmetry for these Monte Carlo events was found to be $-(2.9 \pm 0.4)\%$, while the input asymmetry to the event generator program was taken to be -7.9% according to the

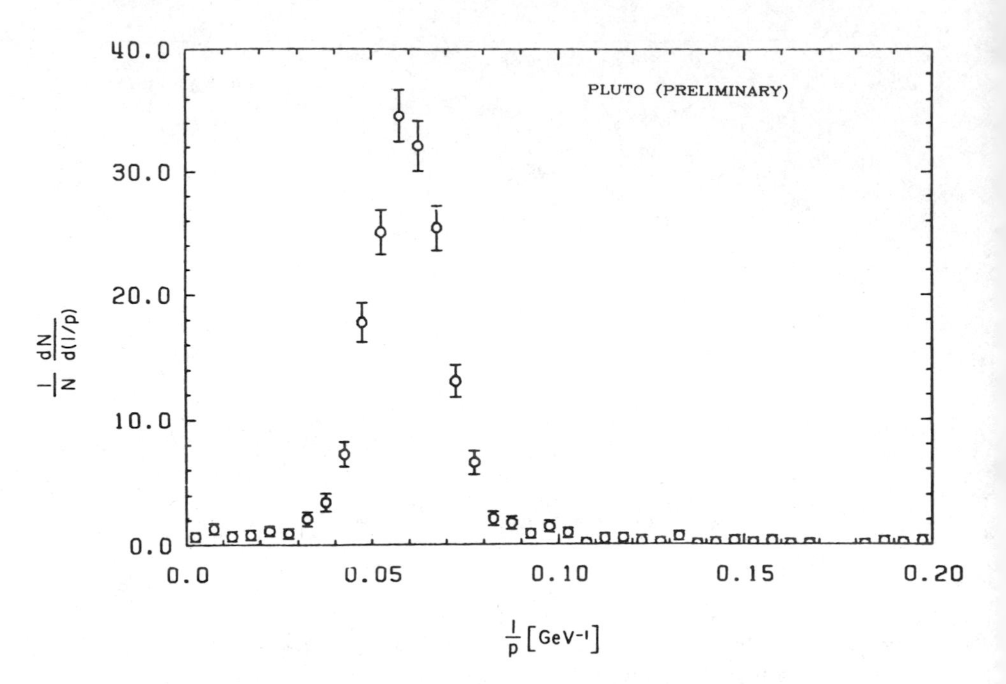

Figure 5: 1/P distribution for μ pair sample using a common long track fit. The resolution obtained is :$\sigma(P)/P = 1.3\% \cdot P$.

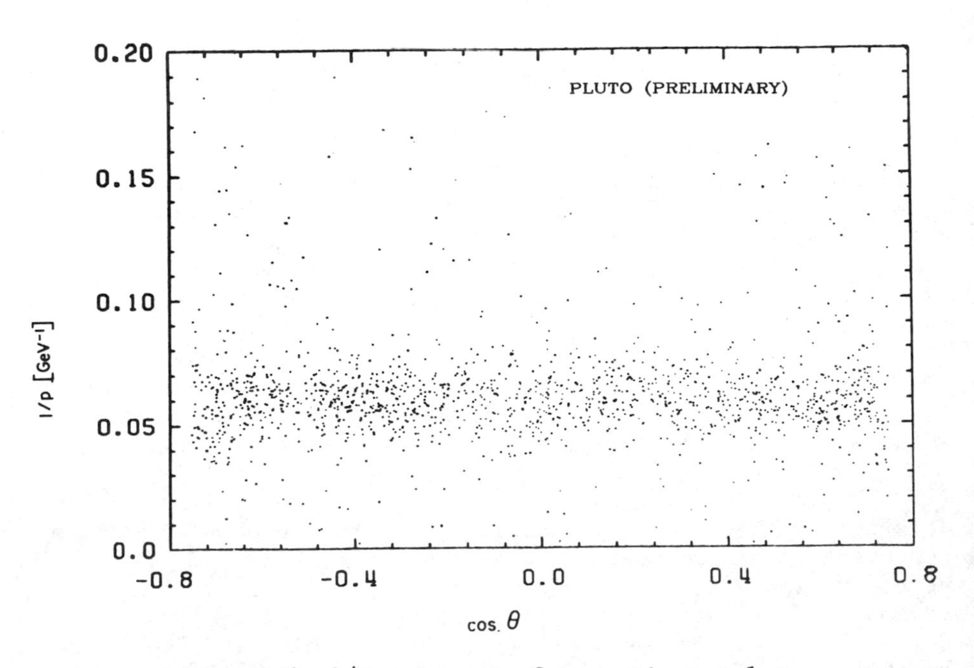

Figure 6: 1/P vs. $\cos\theta$ for μ pair sample.

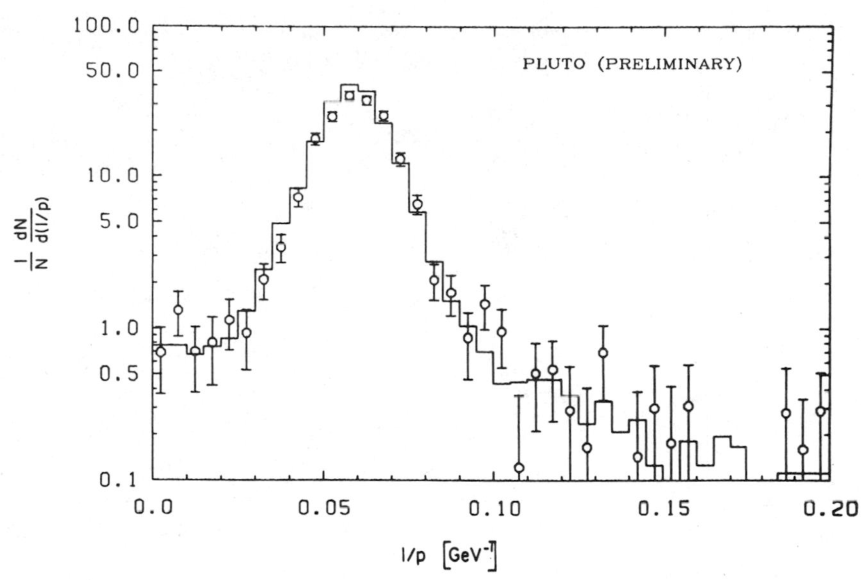

Figure 7: 1/P distribution for μ pair sample (as in Fig.5 but with logarithmic scale). The solid line represents Monte Carlo events.

standard model for lowest order QED + electroweak effect. This difference of -5% is partially due to purely radiative effects (-2.6%) and the rest (-2.4%) comes from the systematic effect due to the common long track fit.

7. Results on Muon Pair Asymmetry

The $\cos\theta$ distribution of the raw data, after subtracting the cosmic ray background bin by bin, was corrected for radiative effects and charge misassignment as described in the last section. The corrected distribution is shown in Fig. 8.

A fit to a pure QED curve of the form $a(1+\cos^2\theta)$ (see dotted line) gives a poor χ^2 (24.3 for 7 degrees of freedom) and does not describe the data. On the other hand, a fit to a curve with an additional $\cos\theta$ term (Eq.1) results in a good χ^2 of 5.2 for 6 degrees of freedom. The asymmetry for the full polar angle range was calculated to be,

$$A(|\cos\theta|<1) = -(12.0 \pm 3.2)\%.$$

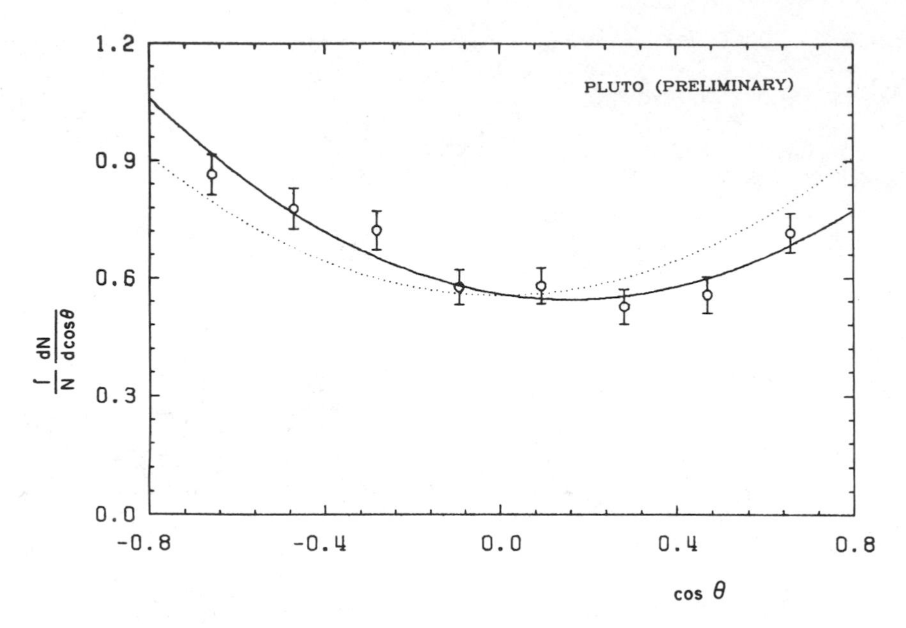

Figure 8: cosθ distribution for μ pairs. The dotted line represents the result of a fit to a pure QED prediction while the solid line includes also the electroweak term.

The statistical error also includes the uncertainty due to the cosmic background subtraction. The systematical error is negligible compared with the statistical one. It consists of ~0.2% due to uncertainty of the detector deformation and ~0.4% due to the finite Monte Carlo sample statistics. The value obtained for A is within errors in good agreement with the standard model prediction of -9.5% and with the latest experimental results from the other PETRA experiments (7).

For the experimental limits in the polar angle of $|\cos\theta|<0.752$ we evaluated the asymmetry parameter directly from the number of events in the forward and backward hemispheres (Eq. 4) with a result that $A=-(10.5 \pm 2.8)\%$, to be compared with the standard model prediction of -7.9%.

Acknowledgement

I am indebted to all members of the PLUTO collaboration for thei help in preparing this talk. I would like to thank G. Alexander for a careful reading of the manuscript. During most of my stay at DESY I was supported by the Minerva Gesellschaft für Forschung mbH.

REFERENCES

1. The present members of the PLUTO collaboration are:
Ch.Berger, H.Genzel, W.Lackas, J.Pielorz, F.Raupach, W.Wagner
I. Phys. Institut der RWTH Aachen, Germany.
L.H.Flølo, A.Klovning, E.Lillenstöl, J.M.Olsen
University of Bergen, Norway.
J.Burger, Ch.Dehne, A.Deuter, A.Eskreys, G.Franke, Ch.Gerke,
U.Jacobs, G.Knies, B.Lewendel, U.Maurus, J.Meyer, U.Michelsen,
K.H.Pape, U.Timm, G.G.Winter, S.T.Xue, M.Zachara, P.Waloschek,
W.Zimmermann
Deutsches Elektronen Synchrotron DESY, Hamburg, Germany.
P.J.Bussey, S.O.Cartwright, J.B.Dainton, B.T.King, C.Raine,
J.M.Scarr, I.O.Skillicorn, K.M.Smith, J.C.Thomson
Department of Natural Philosophy, University of Glasgow, UK.
O.Achterberg, L.Boesten, D.Burkart, K.Diehlmann, V.Hepp, H.Kapitza,
B.Koppitz, M.Kruger, W.Lührsen, M.Poppe, H.Spitzer, R. van Staa,
II Institut für Experimentalphysik, Universität Hamburg, Germany.
C.Y.Chang, R.G.Glasser, R.G.Kellogg, S.J.Maxfield, R.O.Polvado,
B.Sechi-Zorn, J.A.Skard, A.Skuja, A.J.Tylka, G.E.Welch, G.T.Zorn,
University of Maryland, USA.
M.Gaspero, B.Stella
Rome University, Italy.
F.Almeida, A.Bäcker, F.Barreiro, S.Brandt, K.Derikum, C.Grupen,
H.J.Meyer, H.Müller, B.Neumann, M.Rost, K.Stupperich, G.Zech
Universität Siegen, Germany.
G.Alexander, G.Bella, Y.Gnat, J.Grunhaus
Tel Aviv University, Israel.
H.J.Daum, H.Junge, K.Kraski, C.Maxeiner, H.Maxeiner, H.Meyer,
D.Schmidt , Universität Wuppertal, Germany.
2. L.Criegee and G.Knies, Phys. Rev. 83 (1982) 153.
3. PLUTO Collaboration,
A Proposal to Study $\gamma\gamma$ Interactions with the Detector PLUTO at PETRA, Internal Report PLUTO - 79/01 (PRC 79/06) June 1979 (+ Appendix, February 1980).
4. A.Bohm, Weak Neutral Currents in e^+e^- Experiments, talk given at this Conference.
5. R.Budny, Phys. Lett. 55B (1975) 227.
6. F.A.Berends, R.Kleiss and S.Jadach, Radiative Corrections to Muon Pair and Quark Pair Production in Electron-Positron Collisions in the Z_o Region, Preprint Instituut Lorentz,Leiden.
7. P.Grosse-Wiesmann:Results from CELLO
J.Salicio: Results from Mark J
B.Naroska: Results from JADE
G.Mikenberg: Results from TASSO
Talks given at this Conference.

RESULTS FROM MAC*

George B. Chadwick

for the MAC Collaboration - Colorado, Frascati, Houston, Northeastern, SLAC/Stanford, Utah, Wisconsin Universities

Stanford University, Stanford, California 94305

ABSTRACT

The MAC detector has been exposed at PEP to 40 pb^{-1} luminosity of e^+e^- collisions. The detector is described and recent results of a continuing analysis of hadronic cross section, lepton pair charge asymmetry, Bhabha process, two photon final state and radiative μ pairs are given. New results on "flavor tagging" of hadronic events with an inclusive μ, and some searches for new particles are presented.

INTRODUCTION

The MAC detector (MAgnetic Calorimeter) endeavors to capture all possible observable energies from e^+e^- collision events at PEP, the SLAC storage ring. The large mass needed for this makes the detector an excellent muon identifier, and so the iron of the calorimeter is magnetized to provide a muon momentum measurement. To maximize capture at reasonable cost requires the charged particle tracking chamber to be small. Loss of resolution here is partially compensated for by decreased particle decay probability. The detector thus specializes in the following areas: total hadronic cross section, lepton pair properties, "flavor tagging" with muons, and new particle searches.

The detector has collected an integrated luminosity of ~40 pb^{-1}, most of which data is used in the analyses presented.

* Work supported by the Department of Energy, contract number DE-AC03-76SF00515.

THE DETECTOR

Figure 1 shows end and side views of the detector. Going out radially from the beam pipe there is first a central drift (CD) chamber of 10 layers in a ~1 m diameter solenoid of 5.7 kG field. Next are six sextants of shower chambers (SC), constructed of lead plates and proportional chamber layers wired into 192 azimuthal cells with layers grouped into 3 sections. An axial coordinate is obtained by current division. Scintillator trigger counters (TC) follow, then comes the central hadron calorimeter (HC) of 2.5 cm steel plates for a total thickness ~1 m, interleaved with proportional chambers wired like the SC, including current division. As noted above, the iron is excited toroidially to ~17 kG by the coils shown. Finally, four layers of drift chamber tubes (MO) are mounted on the iron, which can locate muon tracks exiting the iron and provide a momentum determination.

The endcaps (EC) are made of iron plates, also magnetized, sandwiched with proportional chambers. The segmentation is coarser here than for the central part of the detector, except for a small angle region used for shower detection. The endcaps are followed by six layers of drift tubes (MO) for muon detection: each layer is set at a 60° rotation to the preceding one to allow spatial reconstruction.

The trigger counters (TC) of scintillator, already mentioned, are positioned after the SC, providing a hexagonal "barrel" enclosure, and two end walls of scintillator are set within the EC

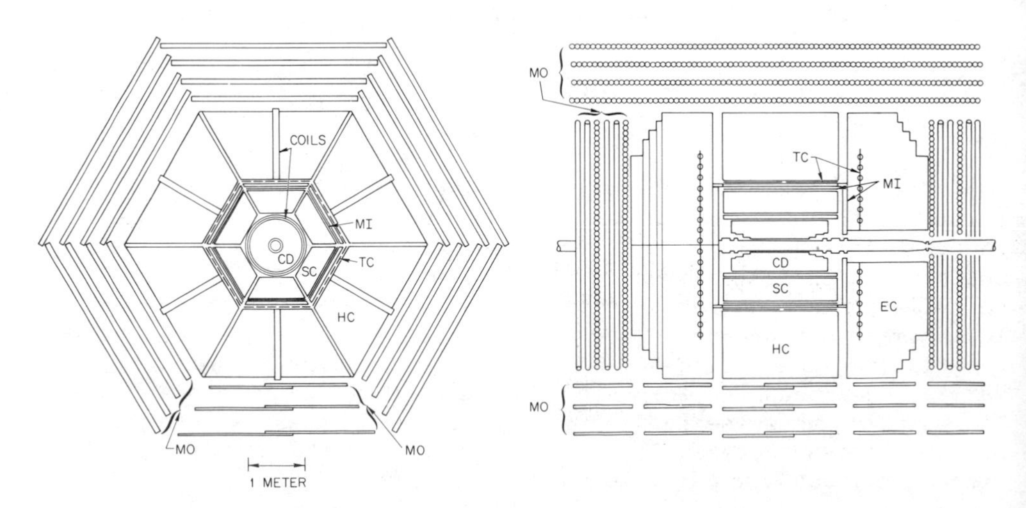

Fig. 1. MAC detector end and side views. Components are: central drift chamber (CD), shower chamber (SC), trigger/timing scintillation counters (TC), central and endcap hadron calorimeters (HC,EC), and inner and outer muon drift chambers (MI,MO).

steel, as shown in Fig. 1. These provide time of flight measurement as well as forming a part of the hardware trigger. The resolution properties of the detector are described in Table I. More details can be found in Ref. 1.

Table I. Resolution of the various sections of the MAC detector.

Section	Resolution	Comments
Central Detector	$\Delta p/p = 0.06 p_{\perp}$	200μ in space
Shower Chamber	$\Delta E/E = 0.2/\sqrt{E}$	Gaussian part
	$\sigma_{\phi} = 0.8^{o}$	segmentation
	$\sigma_{\theta} = 1.3^{o}$	current division
Endcap SC	$\Delta E/E = 0.45/\sqrt{E}$	
	$\sigma_{\phi} = 2^{o}$	cathode strips
	$\sigma_{\theta} = 1.5^{o}$	anode wire groups
Central and Endcap Hadron Calorimeter	$\Delta E/E = 0.87/\sqrt{E}$	
Outer Muon Drift Chambers	$\Delta p/p = 0.3$	multiple scattering limited

HADRONIC TOTAL CROSS SECTION

The hadronic total annihilation cross section is expressed as a ratio R between hadron production and the purely QED production of muon pairs. The latter has a cross section

$$\sigma_0 = \sigma(e^+e^- \to \gamma \to \mu^+\mu^-) = 4\pi\alpha^2/3s \quad .$$

At $\sqrt{s}$ = 29 GeV, our running energy, σ_0 = 0.103 nb.

Experimentally

$$R = \frac{1}{\sigma_0} \cdot \frac{N_{h,corr}}{A} \cdot \frac{1}{\int L\,dt}$$

where $N_{h,corr}$ is the number of purely hadronic events corrected for detector efficiency, background subtraction, and radiative processes. A is the apperture and $\int L\,dt$ is the integrated luminosity.

For hadron production purely by known quark pairs, with no final state interactions, the theoretical expectation for the basic process is

$$R_0 = 3 \sum_f \left(\frac{q_f}{q_e}\right)^2 = \frac{11}{3}$$

where q_e is the electron charge and q_f is the quark charge for flavors $f = u,d,s,c,b$. A correction for gluon radiation is also postulated,[2] giving

$$R = R_0(1 + \alpha_s/\pi + \ldots) = 1.06\ R_0$$

for $\alpha_s = 0.17$, the most commonly quoted value.[3,4] This implies $R = 3.90$.

Experimentally, the ratio $N_{h,corr}/A$ is determined as a unit by a Monte Carlo procedure in which the computer code of Berends and Kleiss[5] is used to generate quark pairs. This procedure takes into account all radiative processes to order α^3. The quarks are then subjected to hadronic fragmentation using the Lund Monte Carlo procedure.[6] Finally, each event is subjected to a Monte Carlo simulation of the detector response, and the cuts applied to real data are used on the simulation.

The term $\int L\,dt$ is determined in three independent ways: a) by the corrected number of central section Bhabha events (polar angles 55°–125°); b) by endcap Bhabhas (15°–25°); and c) by a separate luminosity monitor at ~30 mr.

The first data cuts used were based on the energy flow, a vector for each calorimeter cell of the detector, $(E_i, \theta_i\ \phi_i)$. These cuts were:

1. $E_{vis} = \sum |E_i| > 12$ GeV
2. $E_{vis,\perp} = \sum |E_i| \sin\theta_i > 7.5$ GeV
3. Imbalance $I = |\sum \vec{E}_i| / E_{vis} < 0.65$.

In the second set of cuts, at least 3 good CD tracks were required ($\sum |p_i| > 2$ GeV/c, fit to a common vertex). The vertex position $|Z_{vertex}|$ had to be <5 cm, and the total of all tracks >4. With these cuts defining $N_{h,corr}$, the corresponding A is found to be 1.08.

The major corrections applied, and the estimated systematic errors were as follows:

1) virtual $\gamma\gamma \rightarrow$ hadrons, 1% ± 0.5%
2) $\tau \rightarrow 6$ prongs, 0.5% ± 0.3%
3) other (pileup etc.), 0.5% ± 0.5% .

The final numbers were $N_{h,corr} = 10{,}870$ events, $\int L\,dt = 24.6\ \mathrm{pb}^{-1}$, giving

$$R = 3.97 \pm 0.04 \pm 0.12 \pm 0.12 \quad .$$

Here the uncertainties are respectively statistical, experimental systematic, and theoretical uncertainty in higher order (α^4) corrections. The overall uncertainty is 4.2%. This result compares favorably with previously reported values.[7]

The agreement with simple QCD is striking; the 6% α_s correction seems beautifully confirmed. It is clear that the production of a charge 2/3 top quark pair is ruled out, and the presence of a charge 1/3 quark is very unlikely. At this energy there is no detectable effect on R from the weak interactions.

MUON PAIR PRODUCTION CHARGE ASYMMETRY

In the standard model, muon pairs are expected to show a detectable asymmetry in the angular distribution of μ^+ with respect to the initial e^+ beam direction, owing to interference between annihilation through γ and through Z^o intermediate states:

$$\frac{d\sigma}{d\cos\theta} = \frac{\pi\alpha^2}{2s}\left[R_{\mu\mu}(1+\cos^2\theta) + B\cos\theta\right]$$

$$R_{\mu\mu} = 1 + 2g_V^2\chi \approx 1 \qquad \text{(to order } \alpha G\text{)}$$

$$B = 4g_A^2\,\chi$$

$$\chi = \frac{G_F}{2\sqrt{2}\,\pi\alpha}\;\frac{s}{s/M_Z^2 - 1}$$

where the symbols have their well known meanings. The charge asymmetry

$$A_{\mu\mu} = \frac{N_F - N_B}{N_F + N_B} = \frac{3B}{8R_{\mu\mu}}$$

is predicted this way to be -6.3% at $\sqrt{s}$ = 29 GeV. Using a recent modification of the standard Monte Carlo procedure which includes radiation on the Z^o diagrams,[8] the expected asymmetry becomes -6.0%.

Experimentally, detector fields were reversed periodically. The data was selected as follows:

1. Two CD track found, vertex constrained, and associated with minimum ionizing tracks in shower and hadron calorimeters and/or muon track in the outer drift system.
2. CD momenta satisfy $|p_1| + |p_2| > 8$ GeV/c.
3. $|z_{vertex}| < 5$ cm, $|x_v| < 0.4$ cm, $|y_v| < 0.2$ cm.
4. Scintillator time difference $-10 < \Delta t < 4$ nsec.
5. Acollinearity angle $<10^o$.
6. Opposite signs of momenta (CD or OD).

We have only recently incorporated the OD muon detectors into the analysis. This has helped to reduce charge sign ambiguities and identify some feed-in of forward Bhabha events, which are highly asymmetric. Background estimates are:

1. Cosmic rays, $< 0.5\%$
2. $ee\mu\mu$, $\sim 1.8\%$
3. Bhabhas , $< 0.3\%$
4. $\tau\tau$ events , 2.5%

The final sample had 3067 events, with 130 attributed to background. Integrated luminosity was 39.9 pb^{-1}. For this data the observed asymmetry was corrected for an asymmetry from QED processes[8] of +2.8%, to give

$$A_{\mu\mu} = - 0.076 \pm 0.018 \pm 0.003 \quad .$$

For $M_Z = 90$ GeV, this implies via the standard model

$$g_A^e \, g_A^\mu = 0.31 \pm 0.08 \quad .$$

If the expected value of 0.25 is used as input, the asymmetry implies

$$M_Z > 40 \text{ GeV}$$

at 90% confidence level.

When all detector efficiencies are included, the normalization of the reaction gives

$$R_{\mu\mu} = 0.99 \pm 0.02 \pm 0.05 \quad .$$

This result implies

$$g_V^e \, g_V^\mu = 0.03 \pm 0.16 \quad .$$

Both results are in excellent agreement with measurements from other detectors (see paper of A. Böhm, this conference).

The analysis of τ pair processes for charge asymmetry remains as reported at the Paris Conference.[4]

BHABHA SCATTERING

The elastic scattering reaction has its charge asymmetry overwhelmed by the intensity of the t-channel process. However the angular distribution is modified by electroweak effects, even if $\sin^2\theta_W = 0.25$. This process is so prolific that very small errors are possible, and an understanding of the data provides an excellent assurance that one understands the detector response.

Events were selected with two vertex constrained CD tracks, collinear to 10°, and having a total energy in the shower counters of over 0.5 $E_{c.m.}$. Corrections were made for tracking efficiencies, energy cut efficiencies, and radiative corrections with the standard Monte Carlo techniques. In these events no external charge identification is available, and so the small CD double sign reversal ambiguity puts enough forward high cross section events into the backward hemisphere to spoil the beauty of the distribution. Fortunately, little actual sensitivity is lost in the fit to a folded distribution.

In Fig. 2, the angular distribution is shown as a ratio to the pure QED prediction. The solid line shows the best fit for a variation of $\sin^2\theta_W$. The broken line shows the charge for a 95% confidence level variation, in either direction. The result is

$$\sin^2\theta_W = 0.24 \pm 0.08 \quad .$$

The statistical level of data from this reaction has reached a point at which a significant test of "preon" models of composite leptons can be made. Details may be found elsewhere,[9] but briefly, composite structure would imply constituent exchange, which would induce an effective contact interaction of the form

$$\frac{g^2}{2\Lambda^2}\eta_{ij}\,\bar{\psi}_i\,\gamma_\mu\,\psi_i\,\bar{\psi}_j\,\gamma_\mu\,\psi_j$$

where i,j are over left and right handed spinors. $g^2/4\pi$ is taken equal to unity (making Λ a "mass scale" rather than an actual mass) and η's are zero or ± 1. Interference with the normal exchange process would produce a deviation from QED of a quite marked character, for masses in the usual range of testing.

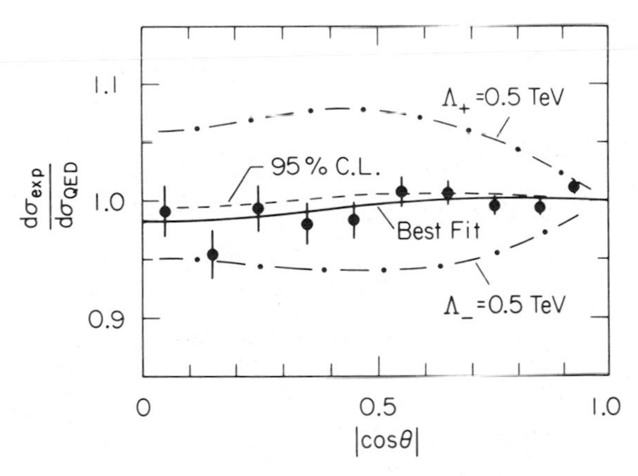

Fig. 2. Angular distribution of the reaction $e^+e^- \rightarrow e^+e^-$ compared with QED. BEST FIT is from varying $\sin^2\theta_W$ and 95% C.L. is from such a variation. The dot-dashed lines are meant to show the sensitivity of the data to preon mass.

The dot-dash lines of Fig. 2 show a prediction for $\Lambda_{RR} = 0.5$ TeV, which would produce a wildly unacceptable fit. At the 95% confidence level, actual fits produce the limits:

$$\eta_{LL} = 1, \quad \eta_{RR} = \eta_{RL} = 0 : \Lambda_{LL} > 1.2 \text{ TeV}$$
$$\eta_{RR} = \eta_{LL} = \eta_{RL} = 1 \quad : \Lambda_{VV} > 2.5 \text{ TeV}$$
$$\eta_{RR} = \eta_{LL} = 1, \quad \eta_{RL} = -1 : \Lambda_{AA} > 1.3 \text{ TeV}$$

where V and A stand for vector and axial vector coupling combinations.

TWO PHOTON FINAL STATE

The process

$$e^+e^- \rightarrow \gamma\gamma$$

shows no effects from weak interactions, and hence provides the cleanest test of QED theory. When the experimental angular distribution is compared with QED, deviation from agreement are parameterized as the mass of an excited or "ortho" electron which can be exchanged. The cross section becomes

$$d\sigma/d\Omega = (d\sigma/d\Omega)_{QED} \; (1 + s^2 \sin^2\theta/2\Lambda^4)$$

where Λ is the e^* mass.

We have found our data to be consistent with QED. At the 95% confidence level

$$A = M_{e*} > 55 \text{ GeV} \quad .$$

RADIATIVE MUON PAIRS

The interest in the reactions

$$e^+e^- \rightarrow \mu\mu^* \rightarrow \mu\mu\gamma \tag{1}$$

is twofold. First, it provides a search for an "orthomuon," i.e. an excited state. Secondly, it provides an excellent test of our understanding of radiative corrections. Thus we win either way.

Process (1) is described by[10]

$$\frac{d\sigma}{d\Omega} = \lambda^2 \alpha^2 \; \frac{(s - M)^2}{s^3} \left[(s + M^2) - (s - M^2) \; \cos^2\theta\right]$$

which is quite distinguishable from the usual Bremsstrahlung. Here λ is a factor modifying the usual coupling of pairs.

If such a state as μ^* exists, we would also expect to see the process

$$e^+e^- \to \mu^*\mu^* \to \mu\mu\gamma\gamma \qquad (2)$$

with an equally distinctive distribution

$$\frac{d\sigma}{d\Omega} = \frac{\alpha^2}{4s} |F(s)|^2 \beta\left[1 + \cos^2\theta + (1 - \beta^2) \sin^2\theta\right] \quad .$$

Therefore we study both final states.

The $\mu\mu\gamma$ event sample was selected by the following criteria:

1. Two tracks, satisfying muon criteria, with collinearity $> 10^o$.
2. $E_\gamma > 1$ GeV.
3. Acceptable 4-constraint kinematic fit.

The $\mu\mu\gamma\gamma$ selection used the above, plus an extra requirement of $>10^o$ between any tracks or showers.

The invariant $\mu\gamma$ mass for reaction (1) is shown in Fig. 3(a). Superimposed is the QED prediction (Berends-Kleiss[5]), in excellent agreement. In Fig. 3(b) the angular distribution is shown, with the QED prediction. A strong charge asymmetry, well explained, is evident, of around -20%. Its source is interference between initial and final state real radiation. At large angles and energies the two sources become comparable, but have different intermediate state C parities.

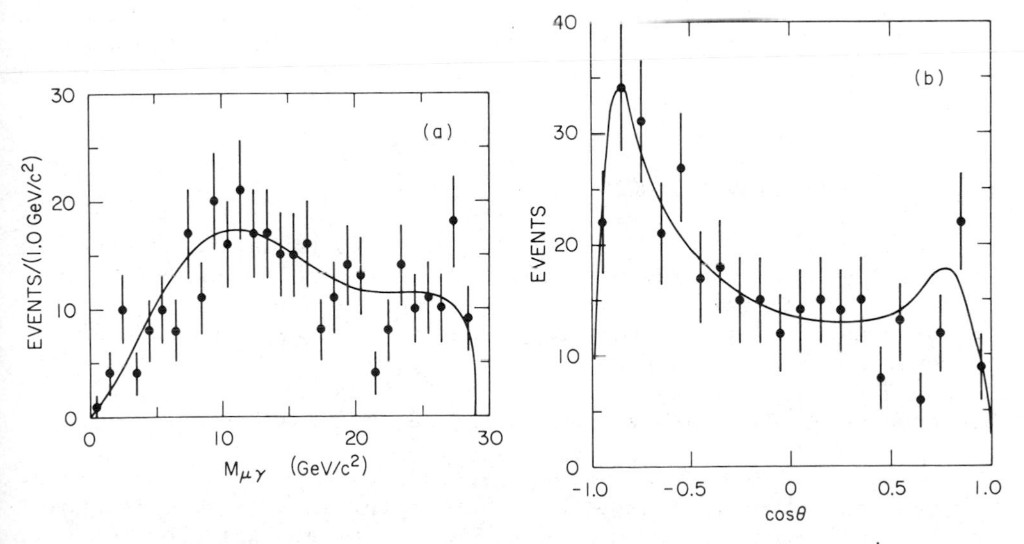

Fig. 3. (a) Invariant mass of $\mu\gamma$ system in the reaction $e^+e^- \to \mu^+\mu^-\gamma$. The solid curve is the absolute QED prediction; (b) angular distribution of the same reaction, along with the QED prediction, both showing a strong charge asymmetry.

In Fig. 4(a) we show a scatter plot for reaction (2) $\mu\gamma$ masses. If a $\mu^*\mu^*$ pair was being made, the area enclosed by broken lines would have an enhanced population, which is not found. Figure 4(b) shows the 95% confidence level mass limits assigned as a function of $M_{\mu\gamma}$ and effective couplings assumed.

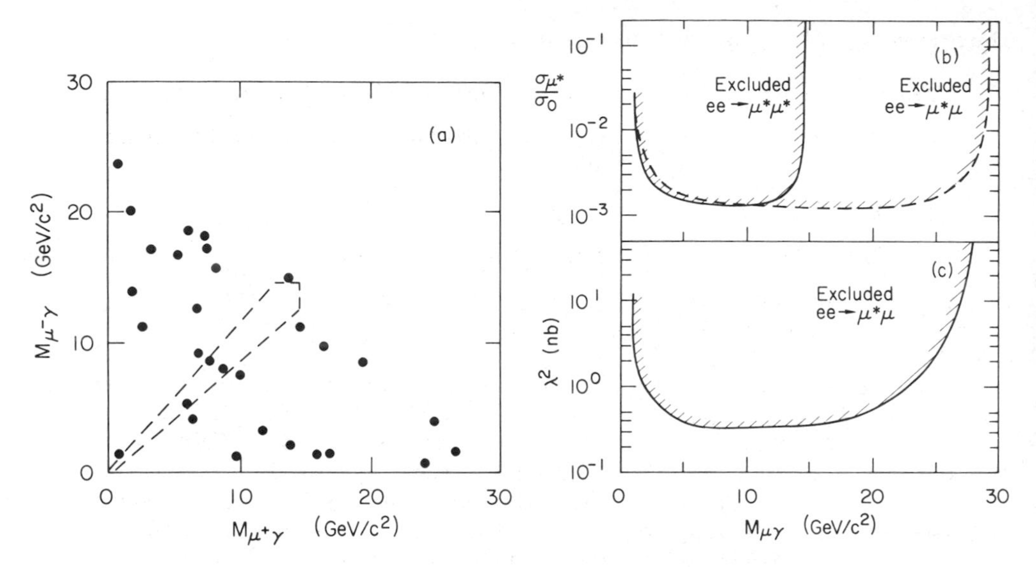

Fig. 4. (a) Scatter plot of invariant $\mu\gamma$ masses in the reaction $e^+e^- \to \mu^+\mu^-\gamma\gamma$. Each event appears twice. The broken line indicates the search area for an excess of events due to excited muon production; (b) mass limits (95% confidence level) for the excited muon state corresponding to assumed parameters of the production process given in the text.

INCLUSIVE MUON FLAVOR TAGGING

With the MAC detector surrounded by drift tubes, it is possible to identify muons in hadronic events, measure their momentum, and deduce their initial angles. A large momentum transverse to the jet axis carried by the muon is indicative of a massive quark semi-leptonic decay. The main background is from muonic decay of π or K mesons and "punch-through" of hadrons. Our estimates of these backgrounds in the selected sample are 20% and 10% respectively.

A total of 240 events were selected from a 27 pb^{-1} data sample. The muon momentum transverse to the event thrust axis, $p_\perp$, is shown in Fig. 5(a). Superimposed upon the histogram are the expected distributions from b and c quark decays and from background. This decomposition was arrived at by first deducing a quark fragmentation spectrum,[11] then fitting the data assuming a semi-muonic decay fraction. The Ali[12] Monte Carlo procedure was used.

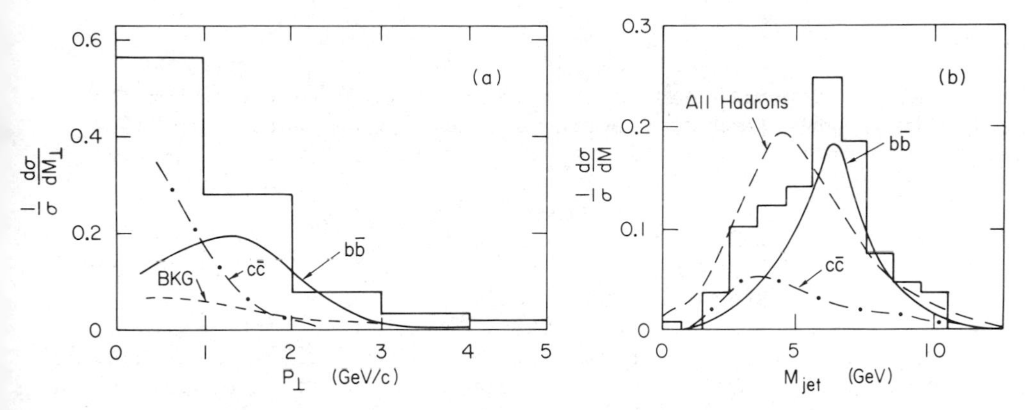

Fig. 5. (a) Muon momenta perpendicular to the thrust axis in hadronic annihilation events with an identified "inclusive muon." Curves indicate the contributions from $b\bar{b}$ (solid), $c\bar{c}$ (broken) and background (dot-dash) production processes as obtained in an overall fit; (b) jet mass, defined in the text, for above events with $p_\perp >$ 1 GeV/c. Fit determined contributions are again indicated.

For the jet opposite the one with the muon, a "jet mass" is calculated:

$$M_{jet} = E_{beam}\sqrt{1 - T_{1/2}^2}$$

where $T_{1/2}$ is the thrust of particles in the jet hemisphere. Figure 5(b) shows the normalized distribution for $p_\perp > 1$ GeV/c, the b quark selection cut. Superimposed is the jet mass found for all hadronic events (broken line) which peaks noticeably lower. We note parenthetically that for $p_\perp < 1$ GeV/c, the jet mass agrees with the "all hadron" curve. The solid lines show the deduced $b\bar{b}$ and $c\bar{c}$ contributions. A cut for $M_{jet} > 5$ GeV urther enhances the b signal.

The charge asymmetry for quark pairs is expected to be larger than for μ pairs, i.e.

$$A_{ff} = \frac{3\chi}{2Q_f}\, g_A^e\, g_A^f$$

because for $b\bar{b}$, $Q_f = 1/3$. We naively expect $A_{bb} \approx -19\%$. Unfortunately, with the above cuts, only 64 events remain, and we find

$$A_{bb} = 0.06 \pm 0.13$$

which is not a significant test. Further data and deeper analysis will be needed to use this promising technique.

SUMMARY

MAC has now produced results which in each test confirm the validity of the Standard Electroweak Model. Its future now lies in the search for phenomena outside that model.

REFERENCES

1. W. T. Ford, SLAC-PUB-2894 (1982).
2. M. Dine and J. Sapirstein, Phys. Rev. Lett. 43:668 (1979).
3. D. P. Barber et al., Phys. Lett. 89B:139 (1979); R. Brandelik et al., Phys. Lett. 94B:459 (1980); D. Schlatter et al., Phys. Rev. Lett. 49:521 (1982).
4. D. M. Ritson, Contribution to XXI International Conference on High Energy Physics, Paris, 1982 (SLAC-PUB-2986, Oct. 1982).
5. F. A. Berends and R. Kleiss, Nucl. Phys. B177:237 (1981); Nucl. Phys. B178:141 (1981).
6. T. Sjostrand, University of Lund Report LUND LU TP 82-3, 82-7 (1982, unpublished).
7. See e.g. K. H. Mess and B. H. Wiik, DESY 82-011 (1982).
8. F. A. Berends, R. Kleiss and S. Jadach, Nucl. Phys. B202:63 (1982).
9. E. J. Eichten, K. D. Lane and M. E. Peskin, FERMILAB-PUB-83/15-THY (1982).
10. F. E. Low, Phys. Rev. Lett. 14:238 (1965).
11. W. T. Ford et al., to be published.
12. A. Ali et al., Phys. Lett. 93B:155 (1980).

RESULTS FROM CELLO ON ELECTROWEAK INTERACTIONS

Presented by Paul Grosse Wiesmann+)
CELLO Collaboration

Max-Planck-Institut für Physik und Astrophysik
München, Western Germany

The CELLO collaboration working at the e^+e^- storage ring PETRA has performed several experiments on electroweak interactions. Results are presented on

- lepton pair production and
- search for new particles.

CELLO[1] is a large solid angle detector with a solenoidal spectrometer, electromagnetic calorimetry and muon identification. Fig. 1 shows the detector and Table 1 summarizes its major properties.

The data sample comprises 11 pb^{-1} at $\sqrt{s}$ = 34 GeV collected between Spring 1980 and Summer 1981.

I. Leptonic Reaction

Due to its high energy, PETRA offers the possibility to study the interference between electromagnetic and weak neutral currents[2].

+) Now at University Karlsruhe, Western Germany

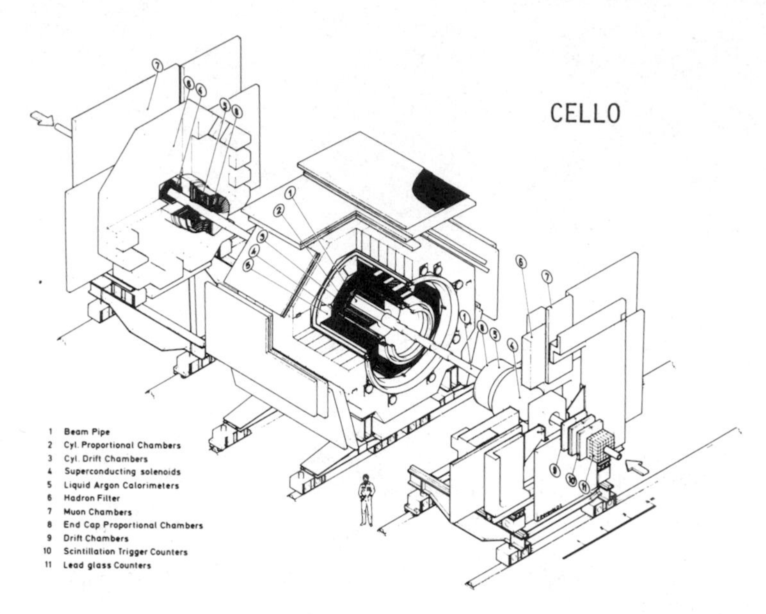

Fig. 1: The CELLO-Detector at PETRA

Table 1: CELLO-Detector Properties

COLLABORATION	DESY, Hamburg KfK und Universität, Karlsruhe MPI, München LAL, Université de Paris XI, Orsay Université de Paris VI CEN, Saclay
MAGNET	1.3 Tesla superconducting thin (0.49X_0) solenoid 0.75 m inner bore radius 3.8 m length
TRACKING	5 cylindrical proportional chambers with cathode strips 7 cylindrical drift chambers 8 planar end-cap chambers 2 cylindrical layers of drift tubes (vertex detector installed end 1982) Tracking down to θ = 150 mrad $\sigma_p/p = 2\%\ p \sin\theta$ ($p > 2$ GeV, $\theta > 30°$) σ_z = 0.4 mm centroid measurement on cathode strips
EM CALORIMETRY	20 lead-liquid argon modules (20X_0) down to θ = 130 mrad 5 layers in depth for shower sampling $\sigma_E/E = 13\%/\sqrt{E}$ σ_θ = 4 mrad
MUON DETECTION	32 planar proportional chambers (covering 92% of 4π) $\sigma_{position}$ = 6 mm $p_{cut\text{-}off}$ = 1.4 GeV
FORWARD DETECTORS	Pb-glass blocks + scintillators 50 θ < 100 mrad
TRIGGERS	Charged-track trigger in $r\phi$ (p_T > 200 MeV) and rz projections Calorimetric triggers

The experiments mainly concentrate on purely leptonic reactions, since they have clean experimental signatures and the theoretical predictions are unambiguous.

Taking into account the annihilation through γ and Z^0, the differential cross section for lepton pair production is

$$\frac{d\sigma}{d\Omega} = \frac{\alpha^2}{4s} \left| R\cdot(1 + \cos^2\theta) + A \cos\theta \right| \tag{1}$$

$$R = 1 + 2 v_e v_1 \chi + (v_e{}^2 + a_e{}^2)(v_1{}^2 + a_1{}^2)\chi^2$$

$$A = 4 a_e a_1 \chi + 8 v_e v_1 a_e a_1 \chi^2$$

$$\chi = \frac{G_F}{8\pi\alpha\sqrt{2}} \frac{s}{s/M_{z^0}{}^2-1} ;$$

G_F: Fermi Constant; M_{z^0}: Z^0 mass

θ: angle to the beam axis

where v_e and a_e (v_1, a_1) are the vector and axial vector coupling constants of the incoming electron (produced lepton). In the standard model $v = (4 \sin^2\theta_w - 1)$ and $a = -1$. Universality requires $v_e = v_1$ and $a_e = a_1$. In the PETRA energy range, the pure weak term is small compared to the interference term. Therefore, in our measurements, the total cross section is sensitive to the vector couplings while the forward-backward charge asymmetry is sensitive to the axial vector couplings. Extrapolating v, a and $\sin^2\theta_w$ from the low q^2 data, one expects, in lepton pair production, a measurable charge asymmetry of $\sim -10\%$ and practically no effect in the total cross section. In Bhabha scattering[3], the situation is complicated by additional t-channel diagrams. On the other hand, only the electron couplings v_e and a_e are involved.

I.1 Bhabha Scattering

Bhabha scattering has the clean signature of collonear, high momentum charged particle tracts pointing to electromagnetic showers in the calorimeter and no hit in the muon chambers. In the CELLO detector, the process is measured from $\Theta = 30°$ to $150°$ with the magnetic spectrometer and the central calorimeter[4]. The basic cuts

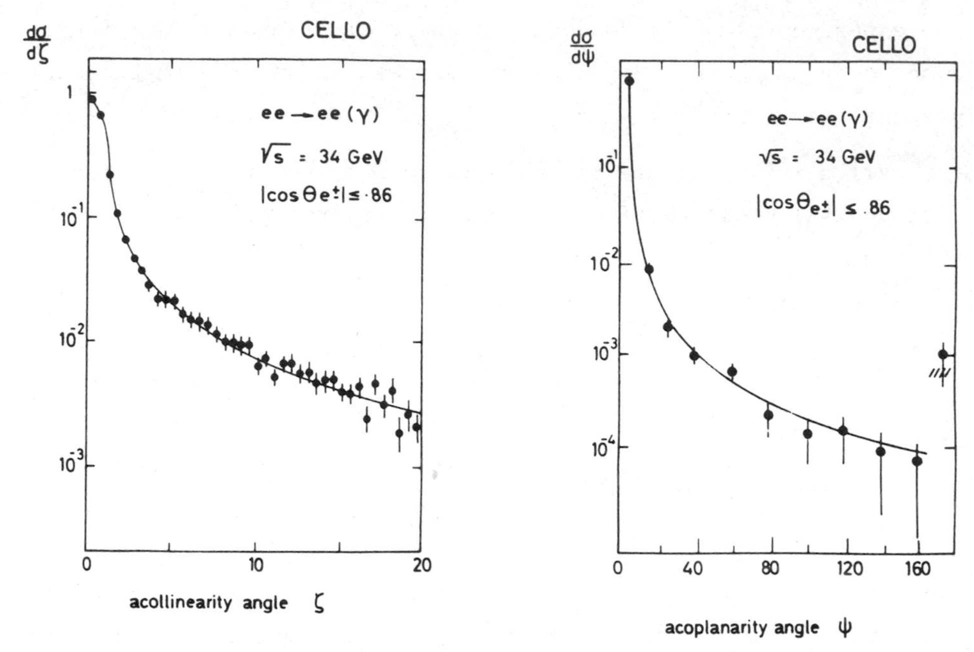

Fig. 2: The acollinearity and acoplanarity distribution for Bhabha scattering. The solid line is the QED prediction.

are requirements for vertex constrained charged particles, shower energy greater than $1/3\ \sqrt{s}$ and acollinearity less than 250 mrad. A unique charge determination is possible in more than 99% of the events. Inefficiencies are determined by relaxing the charged particle or the shower criteria; the experimentally determined losses are well below 1%. Monte Carlo techniques are used only for radiative corrections. Since background and losses are negligible, a systematic point to point error of 1% is achieved. The radiative corrections[5] are checked by the acollinearity and acoplanarity angle distributions. Fig. 2 shows the agreement of our measurement with the prediction of the order α^3 QED Monte Carlo calculation.

In Fig. 3, the measured differential cross section is compared with the QED expectation. Within the mainly statistical errors there is no deviation. The modification caused by an electroweak interference term is indicated. The data mainly constrain the vector coupling $v^2 = v_e \cdot v_e$ and are, therefore, sensitive to $\sin^2\theta_W$.

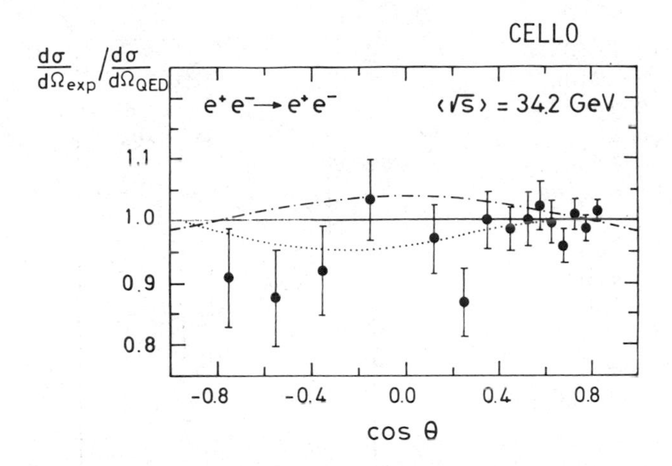

Fig. 3: The differential cross section for Bhabha scattering normalized to the QED prediction. The dotted line represents the best fit; the dotted-dashed line corresponds to the vector-dominated solution in neutrino electron scattering.

I.2 Muon- and Tau-Pair Production

Muon-pair production[6] also gives collinear high momentum particles in the magnetic spectrometer. It is distinguished from Bhabha scattering by requiring the energy deposition of minimum ionizing particles in the calorimeter and at least one track with an associated hit in the muon chambers. Due to the insufficient timing information, a background of cosmic-ray particles remains in the final data sample. For studying the total cross section and angular asymmetry in μ-pair production, the cosmic background is subtracted statistically using the information of the vertex distribution along the beam axis. After applying all corrections, including radiative effects, we determine an asymmetry, extrapolated over the full azimuth angle, of -6.4% ± 6.4%. The expectation of the standard model is -9.2%.

τ-pairs at PETRA energies[7] have the distinctive signature of almost collinear, low mass, low multiplicity jets with missing energy and a small acoplanarity angle due to the unobserved neutrinos. Our selection is sensitive to all τ-decay topologies. With the fine grained liquid argon lead calorimeter, the dominant two charged prong events are clearly distinguished from Bhabha scattering

Fig. 4: The differential cross section for τ-pair production at $\sqrt{s}$ = 34.2 GeV. The solid line shows the best fit with an asymmetry of -10.3%. The dashed line is the QED expectation.

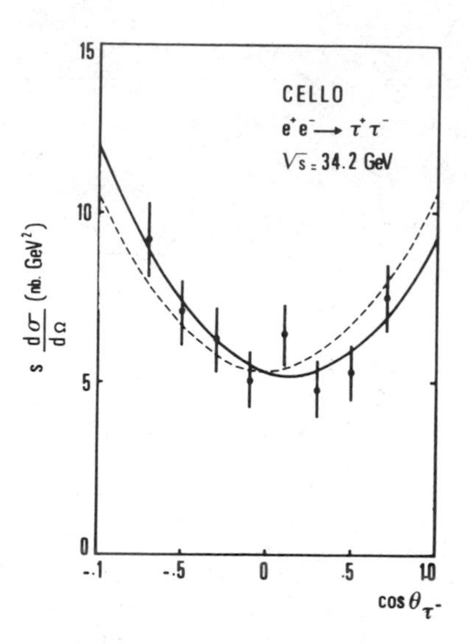

events. To eliminate confusion with two photon processes, we require a minimum momentum for the charged particles and exclude (e, e) and (μ, μ) final states. Altogether, 94% of the τ-decay combinations are used, so that a significant statistical accuracy is reached.

Fig. 4 shows the differential cross section obtained with 434 events. The angular distribution shows a clear $1 + \cos^2\theta$ dependence (seen for the first time, when published!) with a forward-backward charge asymmetry of $-10.3\% \pm 5.2\%$. The total cross section is measured to be $1.03 \pm .05 \pm .07$ in units of the pointlike cross section.

A maximum contamination of 8 multihadron annihilation events and 9 two photon scattering events is estimated. Event candidates are finally accepted only after a visual scan.

Fig. 5 compares the acollinearity and acoplanarity distributions with the Monte Carlo expectations based on V-A charged weak current decay and radiative corrections.

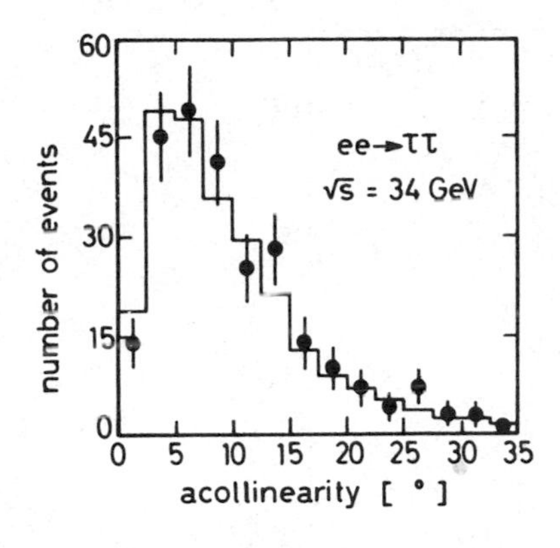

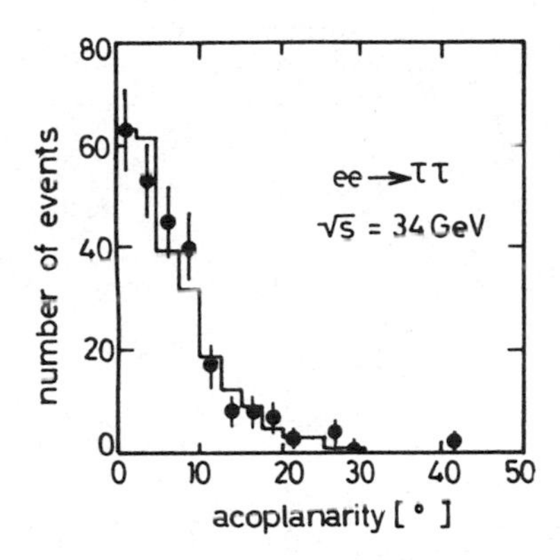

Fig. 5: The acollinearity and acoplanarity distribution for τ-pair production. The solid line is the α^3 QED and V-A charged current prediction.

The nearly complete and background-free event sample allows a precise branching ratio determination. The three charged prong decay branching ratio of 15 ± 2% has, since publication, been confirmed by other PETRA/PEP experiments (in contradiction with the old particle data book value of 28 ± 3%). Our measured τ branching ratios are given in Table 2. The ρ-signal, shown in Fig. 6, demonstrates how well we are able to identify individual decay modes[8].

Table 2: τ-Branching Ratio

Decay Mode	Branching Ratio
$\tau \to$ 1 charged prong	.840 ± .020
$\tau \to$ 3 charged prong	.150 ± .020
$\tau \to e\nu\nu$	.183 ± .024 (stat.) ± .019 (syst.)
$\tau \to \mu\nu\nu$	.176 ± .026 (stat.) ± .021 (syst.)
$\tau \to \pi\nu$	.099 ± .017 (stat.) ± .013 (syst.)
$\tau \to \rho\nu$	.228 ± .025 (stat.) ± .021 (syst.)

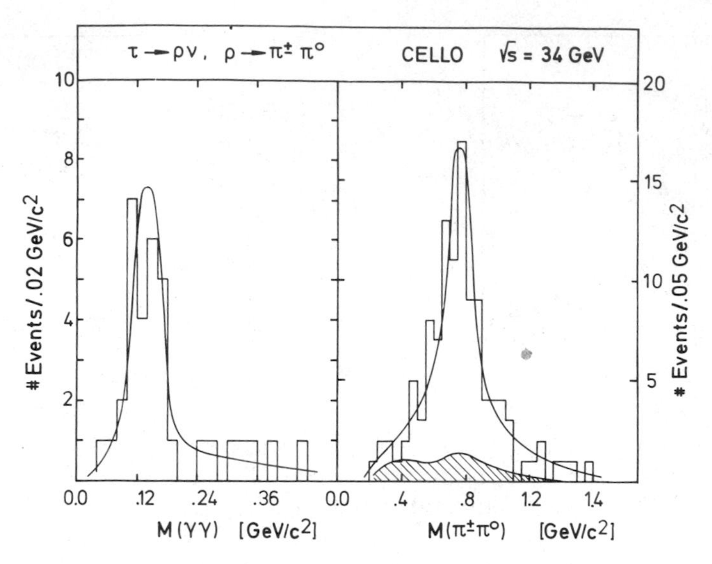

Fig. 6: Two photon invariant mass spectrum for ρ candidates with two separated showers, and the $\pi\pi^0$ invariant mass spectrum. The solid line is the MC expectation for ρ-production, the shaded area the background contribution.

I.3 Leptonic Coupling Constants

The essential non-zero weak neutral current effect, expected in the PETRA energy range, is the charge asymmetry in muon- and tau-pair production. Fig. 7 shows the measured weighted asymmetry for muon- and tau-pair production. The values for the asymmetry of $-6.4\% \pm 6.4\%$ for μ-pair production and $-10.3\% \pm 5.2\%$ for τ-pair production yield axial vector coupling values of $a_\mu = -.7 \pm .7$ and $a_\tau = -1.1 \pm .6$, in agreement with the standard model expectation $a = -1$. Since the experimental value of v_e is consistent with zero, the interference term in the total cross section yields no constraint to v_μ or v_τ.

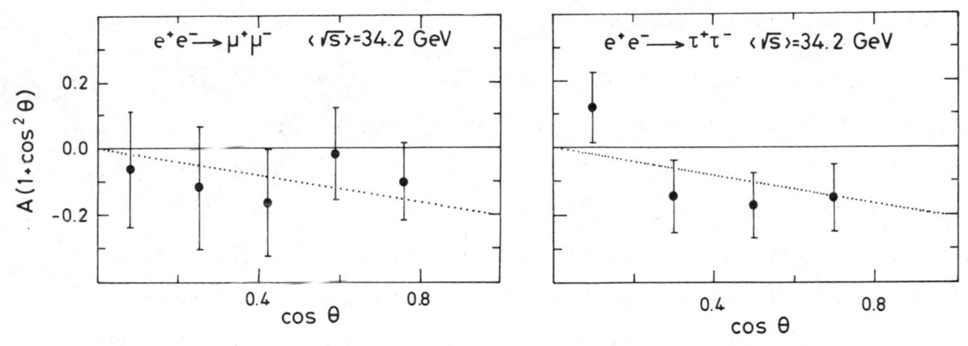

Fig. 7: The weighted asymmetry for lepton pair production. The dotted line represents the best fit for all lepton data.

Assuming lepton universality of the electroweak interaction, all the above described CELLO data are used in a simultaneous fit for v^2 and a^2 [4]. Fig. 8 shows the 95% C.L. contours obtained together with the two allowed regions from neutrino electron scattering. Our data favour the axial vector dominated solution of the neutrino data by more than four standard deviations; the central values are $a^2 = 1.22 \pm .47$ and $v^2 = -.12 \pm .33$. Fitting the single parameter $\sin^2\theta_W$ and taking into account the relation $M_Z = 37.2$ GeV/$\sin\theta_W \cos\theta_W$, we obtain $\sin^2\theta_W = .21 {}^{+.14}_{-.09}$.

With v^2 and a^2 as measured in neutrino electron scattering, we obtain a lower limit $M_Z > 57$ GeV (95% C.L.). The constraints on the vector coupling, v^2, yield limits for models with a different vector boson structure as the standard model (see talk by Schildknecht). Using the parametrization $v^2 = (4 \sin^2\theta_W - 1)^2 + 16$ C, our data give C < 0.031 (95% C.L.).

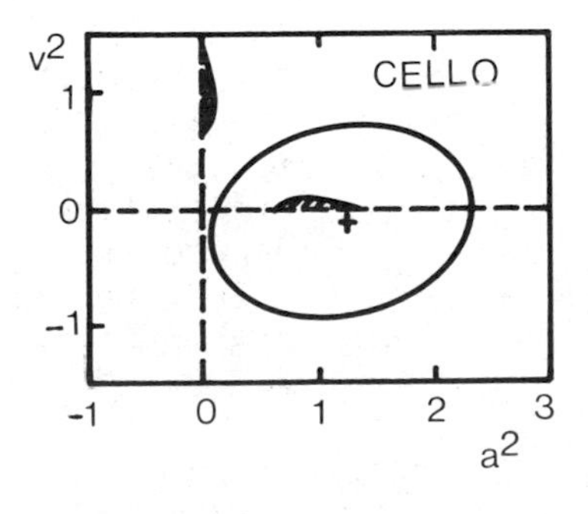

Fig. 8: The 95% C.L. contour for the coupling constants a^2 and v^2. The shaded areas are the neutrino-electron results.

I.4 Tau Final State Polarization

The charged current weak decays are a well-known polarization analyzer. Due to its short decay length, the τ particle offers an experimentally appealing possibility to observe helicity-dependent effects in e^+e^- reactions[9]. In the case of polarized τ's, the angular decay asymmetry in the rest frame leads to a distortion of the laboratory momentum spectra of the decay particles. This is easy to see in the simplest two body decay into πν, which actually yields the biggest effect.

At PETRA energies, the τ-polarization is expressed as

$$P_\tau\ (\cos\theta) = 2\ \chi\ (v_e a_\tau + 2\ v_\tau a_e\ \cos\theta/(1 + \cos^2\theta)) \qquad (2)$$

In contrast to the angular asymmetry and the total cross section measurements discussed above, here, a product of the vector and axial vector couplings is measurable, a characteristic feature for a parity violating process.

In CELLO, we made a first attempt to measure the τ-polarization by analyzing the decays into ρ, π, μ and e[8]. The results are shown in Fig. 9. The four laboratory momentum distributions are compared

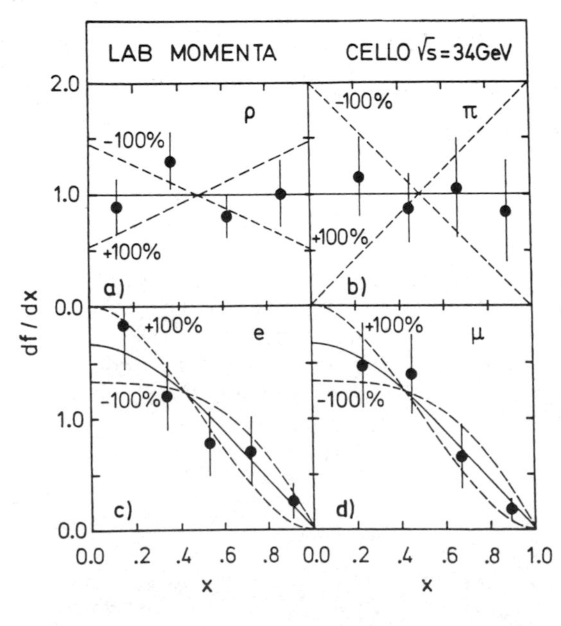

Fig. 9: The background subtracted laboratory momentum spectra for inclusive final states in τ decays. The expectation for non-polarization; ± 100% polarization is indicated.

with the expectations for zero or full polarization. The errors are dominated by the statistical errors; nevertheless, the average polarization is measured with 22% accuracy. As seen from (2), the forward-backward polarization asymmetry is sensitive to the product $v_\tau a_e$. Our measurement yields $v_\tau = -0.1 \pm 2.8$.

II. Search for New Particles

The search for new particles is a promising way to find experimental signatures leading beyond the standard model. None of the searches so far has found positive evidence. Here, we discuss the methods applied and the constraints which can be derived from the negative CELLO results.

II.1 Search for Higgses

A neutral Higgs boson required by the minimal symmetry-breaking scheme of the standard model is not expected to be produced with a reasonable rate in the e^+e^- continuum. In contrast, pair production of charged Higgses ($H^\pm$) (or technipions), predicted in other models, should be observable. The production rate is $\sigma \sim 1/4\ \beta^3\ \sigma_{\mu\mu}$. In a search for $S^\pm \rightarrow \tau^\pm \nu_\tau$[10] sensitive to large scalar masses almost up to the beam energy, we looked for two charged particles with an

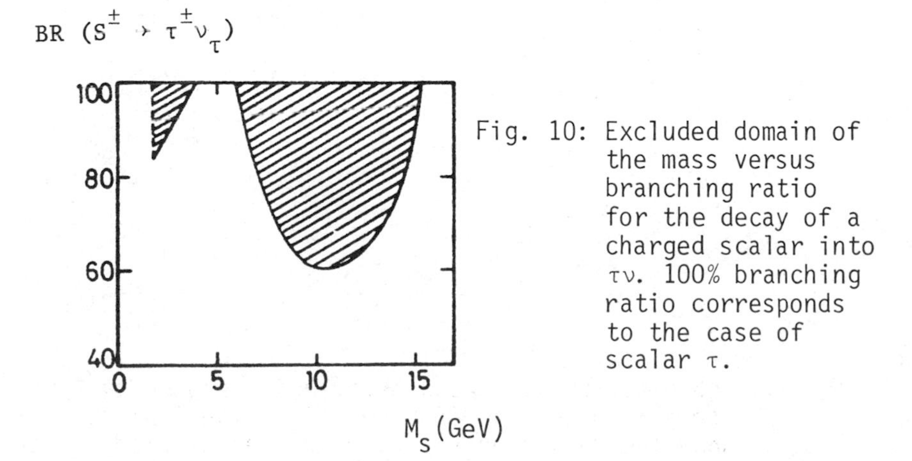

Fig. 10: Excluded domain of the mass versus branching ratio for the decay of a charged scalar into $\tau\nu$. 100% branching ratio corresponds to the case of scalar τ.

acoplanarity angle greater than 30^0, a momentum imbalance greater than 2.5 GeV and no additional photon to compensate for the missing momentum. No event candidate is observed; Fig. 10 shows the limits obtained. Event candidates from a low mass $S^{\pm}$ should show up in the τ-pair event sample. The total cross section (the limit obtained is indicated in Fig. 10), acoplanarity angle distribution (Fig. 5), and the inclusive momentum distributions (Fig. 9) are sensitive quantities; they are all well described with the reaction $ee \rightarrow \tau\tau$.

II.2 Search for Conventional or Excited Leptons

New generations or excited states of leptons could give insights into the generation problem or a composite structure of leptons.

In a search for a conventional heavy lepton, we looked for an isolated muon candidate with a momentum greater than 4 GeV recoiling against hadrons. Five event candidates are abserved, where standard quark pair production Monte Carlos predict 7.1 ± 3.3 events for a heavy lepton of 16 GeV mass. Our analysis yields a lower mass limit of 16.3 Gev with 95% C.L..

An excited heavy electron e^* [11] would modify the differential cross section for the QED reaction $ee \rightarrow \gamma\gamma$. Fig. 11 shows that our measurement is well described by the QED process[12]. From the indicated Λ^+-cut off parameter, we can set a lower mass limit of 59 GeV

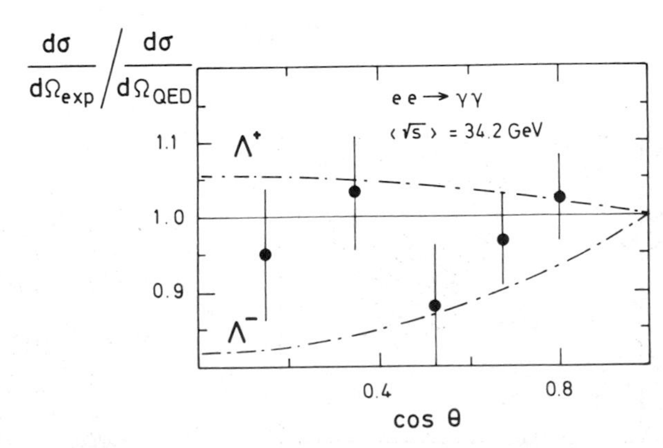

Fig. 11: The differential cross section for photon pair production normalized to the QED prediction. The indicated lines show the deviation expected from an excited electron.

for a heavy electron. Excited leptons can also be directly produced in the reaction $ee \to l^* l^*$ or $ee \to l^* l$. Since the l^* is expected to decay fast into $l\gamma$, l^* should show up in the lepton photon invariant mass combinations of the radiative lepton pair production event sample. The observed number of events in the reaction $ee \to ee\gamma$ (see Fig. 2) and $ee \to \mu\mu\gamma$ agrees with the QED expectation up to order α^3 [5].

The cross section for the reaction $ee \to \mu\mu^*$ is

$$\frac{d\sigma}{d\Omega} = \alpha^2 \lambda^2 \frac{(s - M^2_{\mu^*})}{s^3} \left| s + M^2_{\mu^*} - (s - M^2_{\mu^*}) \cos^2\theta \right| \qquad (3)$$

where λ is a parameter characterizing the $\gamma\mu\mu^*$ coupling. The result quoted above for a heavy electron assumes $\lambda = 1$. Fig. 12 shows the result obtained from the reaction $ee \to \mu^* \mu$ and $ee \to \mu^* \mu^*$.

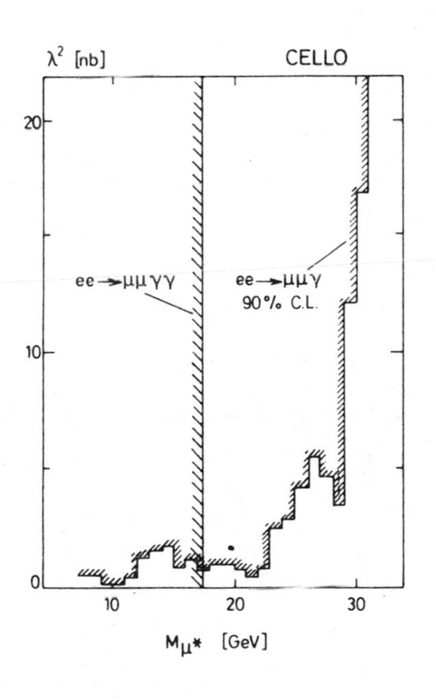

Fig. 12: The upper limits on the $\mu\mu^*\gamma$ coupling constant λ, derived from a comparison of $\mu\mu\gamma$ events with α^3-QED Monte Carlo.

II.3 Search for Supersymmetric Particles

Supersymmetry relates particles with different spins; the known particles should have supersymmetric partners, which are distinguished from ordinary particles by half a unit in spin; e.g. scalar (spin 0) leptons, partners to the ordinary leptons and a spin 1/2 photino, partner to the photon. So far, no model with definite mass predictions exists.

Scalar leptons would be pair-produced in e^+e^- physics according to[13]

$$\frac{d\sigma}{d\Omega} = \frac{\alpha^2\beta^3\sin^2\theta}{8s} \left[1 + \left(1 - \frac{4k}{1-2\beta\cos\theta+\beta^2}\right)^2\right] \tag{4}$$

where k = 0 for scalar muons (s_μ) and taus (s_τ), and k = 1 for scalar electrons (s_e).

The scalar leptons are expected to decay fast into the associated ordinary lepton and a low mass, non-interacting photino. Therefore, the experimental signature depends on the ratio of the scalar lepton mass to the beam energy. Low mass scalars should show up in the event samples for Bhabha scattering, muon- and tau-pair production. They are excluded by the cross section measurements, and the energy and momentum spectra.

High mass searches are more complicated due to the limited phase space available ($\sigma \sim \beta^3$). A clean signal is an acoplanar lepton pair with no additional photon to compensate for the missing momentum. No event candidate for a scalar electron, muon or tau is found[10], giving 95% C.L. lower mass limits of 16.8, 16.0, 15.3 GeV, respectively. Fig. 13 illustrates the result of the scalar muon search. The fall in the number of expected events is due, in the low mass region, to the applied acoplanarity cut (30°), in the high mass region to the phase space limitation.

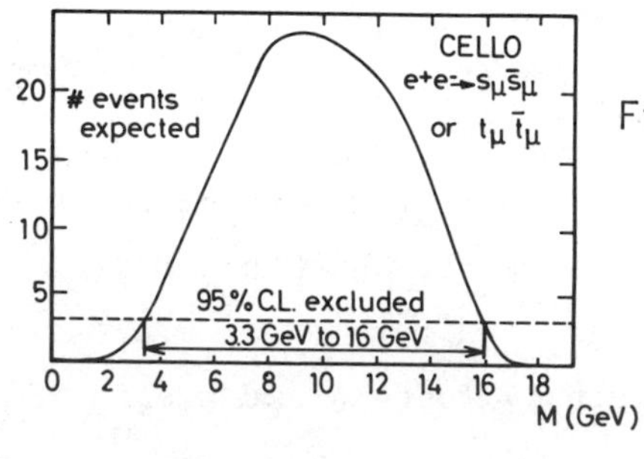

Fig. 13: The expected number of events as a function of scalar muon mass. No event candidate is observed. The corresponding mass limit is indicated.

Photinos (λ_γ) could be pair-produced by the exchange of a scalar electron. The cross section reads

$$\frac{d\sigma}{d\Omega} = \frac{\alpha^2 s}{16} \frac{2\beta^3}{m^4_{s_e}} (1 + \cos^2\theta) \quad (5)$$

The production rate is suppressed by the massive scalar electron propagator, and sizeable production rates, even for the pair production of light photinos, are expected only at high energies.

As mentioned before, the photino is expected to interact like a neutrino and, therefore, only indirect observation of photino pair production is possible.

If the photino is massive and unstable, it can be observed by its decay into a photon and a non-interacting gravitino. The lifetime of the photino depends on the symmetry-breaking parameter, d, and reads[14]

$$\tau_{\lambda_\gamma} = \frac{8\pi d^2}{m_{\lambda_\gamma}{}^5} \quad (6)$$

We searched for two photons with missing energy in the calorimeter[12]. The limits obtained on the photino mass as a function of d are shown in Fig. 14. The upper limit for the photino mass is determined by the lifetime limit, the photino must have time to decay before leaving the calorimeter; the lower limit of 13 GeV is dominated by the available phase space ($\sigma \sim \beta^3$).

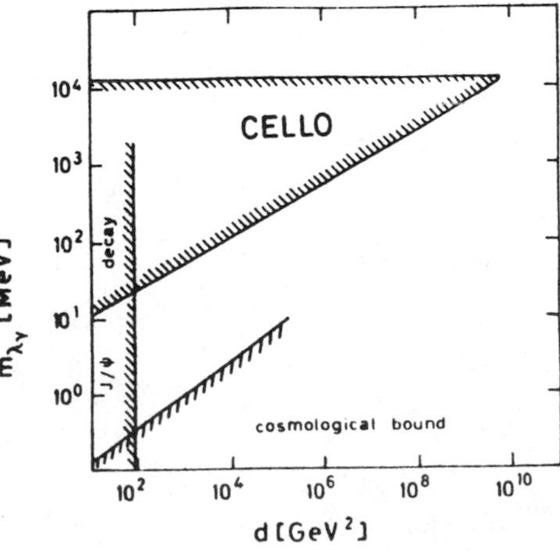

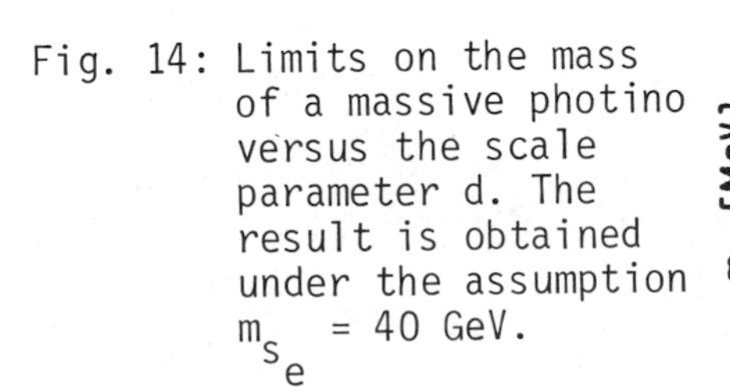
Fig. 14: Limits on the mass of a massive photino versus the scale parameter d. The result is obtained under the assumption $m_{s_e} = 40$ GeV.

Summary

We have analyzed Bhabha scattering, muon- and tau-pair production at $\sqrt{s}$ = 34 GeV. Our data can be described by the standard model using the value of $\sin^2\theta_W$ measured at low q^2, and are in agreement with universality.

An extensive search for new particles was negative. The methods applied and the limits achieved are presented.

Acknowledgement

It was an honour for me to present the work of so many people; thanks to all who contribute to the CELLO-experiment.

I wish to thank DESY for the kind hospitality and the continuous support during my stay in Hamburg.

I am grateful to the organizer of the Conference for the warm hospitality and the stimulating atmosphere at Erice.

References

|1| CELLO-Collaboration, H.-J. Behrend et al., Phys. Scripta, Vol. 23 (1981), 610

|2| G. Glashow, A. Salam, S. Weinberg Review of Modern Physics, Vol. 52/3 (1980)

For recent experimental tests see a review by M. Davier, Proceedings of the 21st Int. Conf. on High Energy Physics, Paris 1982

|3| R. Budny, Phys. Lett. 55B (1975), 227

|4| CELLO-Collaboration, H.-J. Behrend et al., Z. f. Physik C16 (1983), 301

|5| see talk of F.A. Berends this conference

|6| CELLO-Collaboration, H.-J. Behrend et al., Z. f. Physik C14 (1982), 283

|7| CELLO-Collaboration, H.-J. Behrend et al., Phys. Lett. 114B (1982), 282

|8| CELLO-Collaboration, H.-J. Behrend et al., Phys. Lett. 127B (1983), 270

|9| J.E. Augustin, Proceedings of the LEP Summer Study, CERN 79-01, 499

G. Goggi, ibid. p. 483

|10| CELLO-Collaboration, H.-J. Behrend et al., Phys. Lett. 114B (1982), 287

|11| A. Litke, Ph. D. thesis, Harvard University, 1970

|12| CELLO-Collaboration, H.-J. Behrend et al., Phys. Lett. 123B (1983), 127

|13| G.R. Farrar, P. Fayet Phys. Lett. 89B (1980), 191

|14| N. Cabbibo, G.R. Farrar and L. Miani, Phys. Lett. 105B (1981), 155

RESULTS FROM PEP-4 TPC

Presented by J.G. Layter

H. Aihara[1], M. Alston-Garnjost[1], D.H. Badtke[4], J.A. Bakken[4], A. Barbaro-Galtieri[1], A.V. Barnes[1], B.A. Barnett[4], B. Blumenfeld[4], A. Bross[1], C.D. Buchanan[2], W.C. Carithers[1], O. Chamberlain[1], C. Chen[1], J. Chiba[5], C.Y. Chien[4], A.R. Clark[1], O.I. Dahl[1], C.T. Day[1], P. Delpierre[1], K.A. Derby[1], P.H. Eberhard[1], D.L. Fancher[1], H. Fujii[5], T. Fujii[5], B. Gabioud[1], J.W. Gary[1], W. Gorn[3], W. Gu[1], N.J. Hadley[1], J.M. Hauptman[1], B. Heck[1], H-J. Hilke[1], W. Hofmann[1], J.E. Huth[1], J. Hylen[1], H. Iwasaki[5], T. Kamae[5], R.W. Kenney[1], L.T. Kerth[1], R. Koda[2], R.R. Kofler[6], K.K. Kwong[3], J.G. Layter[3], C.S. Lindsey[3], S.C. Loken[1], X-Q. Lu[4], G.R. Lynch[1], L. Madansky[4], R.J. Madaras[1], R. Majka[1], J. Mallet[1], P.S. Martin[1], K. Maruyama[5], J.N. Marx[1], J.A.J. Matthews[4], S.O. Melnikoff[3], W. Moses[1], P. Nemethy[1], D.R. Nygren[1], P.J. Oddone[1], D.A. Park[2], A. Pevsner[4], M. Pripstein[1], P.R. Robrish[1], M.T. Ronan[1], R.R. Ross[1], F.R. Rouse[1], R. Sauerwein[1], G. Shapiro[1], M.D. Shapiro[1], B.C. Shen[3], W.E. Slater[2], M.L. Stevenson[1], D.H. Stork[2], H.K. Ticho[2], N. Toge[5], U. Urban[1], R.F. vanDaalen Wetters[2], G.J. VanDalen[3], R. van Tyen[1], H. Videau[1], M. Wayne[2], W.A. Wenzel[1], M. Yamauchi[5], M.E. Zeller[6], and W-M. Zhang[4]

SUMMARY

Spatial resolution and dE/dx resolution for the PEP-4 TPC have been measured in a beam environment at PEP. Preliminary results are reported on event configuration studies, on particle fractions and inclusive cross sections, and on charge and multiplicity correlations in rapidity space.

[1]University of California, Lawrence Berkeley Laboratory
[2]University of California, Los Angeles
[3]University of California, Riverside
[4]Johns Hopkins University
[5]University of Tokyo
[6]Yale University

THE DETECTOR

The PEP-4 TPC Facility is a multifunction detector based on the Time Projection Chamber (1,2), developed at the Lawrence Berkeley Laboratory beginning in 1974. The time projection tracking concept is combined with dE/dx particle identification to form the central detector of the facility which includes drift chambers at the inner and outer radii of the TPC, primarily for triggering; electromagnetic calorimeters with projective readout, on the magnet pole tips and in six modules outside the 4Kg conventional coil; and a muon system, measuring two coordinates over 98% of the solid angle for particles that penetrate a one-meter iron filter. The TPC, the Inner Drift Chamber, and the Pole Tip Calorimeters operate at 8.5 atmospheres of Argon-Methane (80-20%) and cover an angular region down to 225 mrad in polar angle. PEP-9, the Two Gamma experiment, covers the region from 180 mrad down to 20 mrad at each end with a similar array of detectors.

All the components of the PEP-4 TPC facility were assembled in the summer of 1981, and cosmic rays were observed for the first time in the complete detector. The Facility was rolled into the PEP beam at the beginning of 1982 and has been taking data since that time. Various aspects of the performance of the system have been reported previously (3,4,5).

TPC Operation

The TPC is a two meter long, two meter diameter cylinder in each half of which an axial drift field is defined by systems of equipotential rings at the inner and outer radii, stepping the voltage down uniformly from a central membrane at -75 KV to the ends at ground. Ionization produced by charged particles traversing the TPC drifts in this field toward the end caps, each of which incorporates six identical pie-shaped MWPC modules (sectors) which detect the drifted ionization. A sector contains 183 sense wires of 20 micron gold-plated tungsten, stretched 4 mm above the cathode plane at a spacing of 4 mm along the radius bisecting the sector. Under every thirteenth wire the cathode is segmented into 7.5 mm by 7.5 mm pads which detect the charge induced by the avalanche on the wire above. Normally two to three pads produce a measurable signal for each ionization trajectory, and by determining the pulse height from each pad, one can calculate the center of gravity of the induced pulses along the pad row, thus determining an x-y position in the magnet bending plane.

The z coordinate is obtained by using the "time projection" property of the TPC. Signals from both the pads and the wires are recorded on analog shift registers or CCDs (Fairchild 321A). At any instant the CCD for a given channel contains some 20 microseconds of past history of that channel--225 "buckets" 100 nanoseconds long. For typical pad or wire hits, five to seven buckets will contain signals over threshold, so again a center of gravity can de determined, giving a precise z coordinate.

Spatial Resolution

The spatial resolution of the TPC has been analyzed(6) in terms of the processes expected to be involved and the data show a very close agreement with these expectations. The position resolution, σxy, in the bend plane can be parametrized as

$$\sigma_{xy}^2 = \sigma_N^2 + \sigma_D^2 \quad [(L/Lmax)e^{\gamma L}\sec\alpha] + \sigma_F^2 \quad [e^{\gamma L}\cos\alpha\tan^2\alpha]$$

where:

σ_N = Intrinsic spatial resolution due to electronic noise and other systematic effects

σ_D = Contribution from transverse diffusion of the ionization

σ_F = Contribution from ionization fluctuations within the collection region of each wire

L = Drift distance for a track segment

Lmax = Maximum drift distance

α = Angle of the track to the sector bisector

γ = Electron capture rate per unit length

The multiplicative factors in the second and third terms represent the effect of ionization statistics. For the data discussed here, γ was approximately 0.32 m^{-1}.

The overall spatial resolution is defined as the standard deviation of a gaussian fit to the distribution of residuals of pad hits left out of fits to cosmic ray tracks. The σ coefficients for the various contributions are

σ_N = 160 ± 2 microns

σ_D = 105 ± 6 microns

σ_F = 249 ± 7 microns

and the average resolution for this sample is about 190 microns. Improvements in the electronics and in calibration may eventually reduce the first term, σ_N, to around 120 microns. The contributions from the other two terms are very close to the calculated values and, in quadrature, contribute little to the overall resolution.

The z position is determined by fitting a parabola to the three highest buckets of the five to seven that are normally over threshold. The spatial resolution in z, determined as in the case of x-y, is 340 ± 5 microns. Since the TPC measures space points with comparable accuracy in all three dimensions, it is very insensitve to background and has very good multitrack resolution: confusion distances must be taken in real space rather than in two dimensional projected sub-spaces. In practice, less than 1% of the space points are ambiguous in hadronic events.

Momentum Resolution

The momentum resolution of the TPC is limited at the present time because of electrostatic distortions of the drift field. These give rise to spatial distortions which are appreciable for the inner three and outer two pads rows; consequently only ten pad rows are currently being used for momentum determination. By calculating the momentum of a cosmic ray separately in the two sectors it traverses and comparing the two results, one can determine the momentum resolution. This gives a result of $\Delta p/p^2 = 6.4 \pm 0.3\%$. If one constrains the two track segments to pass through the same point in the horizontal plane, this resolution improves to $\Delta p/p^2 = 2.4 \pm 0.2\%$. However, because of the distortions, the curvature difference is then no longer centered at zero.

Distortions arise primarily from positive ions feeding back into the drift volume after ionization from background processes is multiplied in the amplification region. This part of the distortion increases with beam current, but the effect can be reduced by several orders of magnitude by fitting the sectors with gating grids to form a "tetrode TPC"(7). Drifting ionization is excluded from the multiplication region by the gating fields until a pulse is applied to the grids in response to an external trigger to make them transparent to the ionization. A much smaller distortion arises from error potentials caused by inhomogeneities in the material from which the TPC is constructed. A partially conducting coating can be applied to the structure to reduce the error potentials to a negligible level. These modifications will be introduced during the 1983 summer shutdown and should improve momentum resolution for the TPC alone to better than 0.5%.

Particle Identification by dE/dx

The 183 wires per sector provide ionization information for dE/dx. The wire information for each track constitutes a high density set of points in the r-z plane, facilitating track identification, particularly at small polar angles, and making it possible to recognize secondary vertices, especially those involving undetected neutral particles. However, the primary function of the wires is particle identification.

The TPC uses the 65% truncated mean algorithm (discard the highest 35%) to define a value of dE/dx from the sample taken for each track. Results obtained from the hadronic sample using this algorithm are shown in Figure 1. Each wire signal from each track has been corrected for variations in electronic gain from channel to channel, for variations in wire gain from point to point on a given wire, from wire to wire on a sector, and from sector to sector.

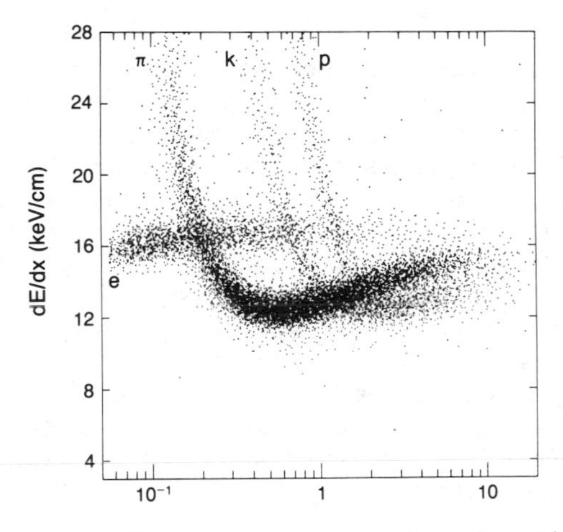

Figure 1. dE/dx energy loss versus momentum for the hadronic sample.

Samples from a track are corrected for electron capture, track length, and dip angle. Events from a given run are corrected for variations in pressure, temperature, and gas mixture. The results currently obtained for energy resolution of the hadronic sample, for tracks which have crossed 120 or more wires, is 3.6 ± 0.2%. This lead to a pion-kaon separation of 3.8 σ and a kaon-proton separation of 1.3 σ for the region above 4 GeV/c. Slightly better resolutions, approaching 3.0%, have been obtained for cosmic rays, comparing both halves of the same track, as described above, thereby minimizing systematic variations.

Recently it has been shown that some dE/dx information can be obtained from the pads. A resolution of from 7 to 10% has been obtained for tracks crossing more than 9 pad rows. This extends the unique three-dimensional character of the TPC tracking to dE/dx, so that at least some particle identification should be available for all tracks for even the most complex events.

Cosmic ray data have been taken at several gas pressures to determine the relativistic rise dependence on this variable. The results, shown in the table, indicate that, although the actual rise of dE/dx with momentum in the relativistic region is greater at lower pressure, the gain in statistics leads to a net gain in particle separation at high pressure.

Trigger

The charged particle trigger for the TPC must depend on external subsystems since the 20 microsecond maximum drift time for ionization in the chamber volume means that information will not be available for processing before the next beam crossing, every 2.4 microseconds. A charged pretrigger makes use of the drift chamber system to generate a decision with 500 ns, to give sufficient time to reset the electronics if no candidate event is found in that crossing. The inner drift chamber defines spatial roads using coincidences of alternate layers, whose sense wires are at the same azimuth; and relates the event time to the beam crossover time by forming a meantime using two adjacent layers, whose sense wires are half a cell width apart. The drift chamber provides no z coordinate however, so analog information from the TPC wires is processed by a hardwired trigger module to select among the pretriggering events those that emanate from the beam crossover point.

Table. Relativistic Rise Data

Relativistic rise is measured in terms of a difference parameter D which is defined as the kaon/pion separation at 3.5 GeV/c and is expressed in per cent and in standard deviations. D is calculated by taking the difference between the muon energy loss at 2.65 GeV/c and at 0.75 GeV/c (points at which the muon energy loss equals the kaon and pion energy losses respectively at 3.5 GeV/c) and dividing by the minimum energy loss.

p (atm)	D (%)	D (s.d.)
8.5	14.6 ± 0.3	4.8 ± 0.2
4.0	16.6 ± 0.5	3.9 ± 0.5
1.0	20.3 ± 1.1	3.3 ± 0.2

In addition to the charged particle trigger there are several neutral triggers based on information from the pole tip and hexagonal calorimeters. Redundant charged trigger modes make it possible to monitor trigger efficiency on line. Triggers from the PEP-9 experiment, suitably prescaled, can also initiate readout. Sufficient flexibility has been built into the trigger to allow it to be matched to changing beam conditions. Currently the trigger rate is between 1.5 and 2 Hz for luminosities of 2×10^{32} $cm^{-2} sec^{-1}$.

ANALYZED DATA

During the spring of 1982 PEP-4 TPC accumulated 4800 nb^{-1} of data. This was understood primarily as a shakedown run inasmuch as it was the first beam data for the detector. In addition, only two of the six calorimeter modules were installed; and the use of the inner drift chamber was lost after two months of running due to hydrocarbon polymerization on the wires. An interim pretrigger was established using the outer drift chamber as the timing element, and incorporating a "fast" TPC signal, i.e., the hits registered in the first microsecond of drift, to extend the angular coverage to smaller polar angles. This latter pretrigger element had no timing rejection, and the outer drift chamber had little spatial definition; consequently, the data were dominated by cosmic rays. Nevertheless, a sample of hadronic events was obtained at a beam energy of 14.5 GeV.

The Hadronic Sample

A sample of hadronic events has been defined using the following cuts on charged particles measured in the TPC:

Number of tracks ≥ 5 with
  Momentum > 0.1 GeV/c
  Closest approach to beam ≤ 5 cm,
  Distance from nominal interaction point in z $\leq \pm 10$ cm
Visible energy - $\Sigma|p| > 1/2$ Ebeam
Momentum balance - $\Sigma p_1/\Sigma|p| < 0.4$

Background contamination in this sample is estimated to be: < 1% beam-gas, < 2% two-photon, < 3.5% tau^+tau^-. A total of 1433 events survive the cuts.

Event Configuration Studies

From this sample the ratio R for annihilation into multihadrons is found to be $3.7 \pm 0.1 \pm 0.4$, and the average charge multiplicity per event is $\langle n_{ch} \rangle = 12.0 \pm 0.3 \pm 1.0$. These quantities have been determined mainly as a test of our understanding of the detector acceptance. The Monte Carlo program TPCLUND incorporates the Lund approach to quark fragmentation and includes radiative effects to order α^3. The second error contribution in each case reflects model dependence, uncertainties in luminosity, etc. In the same spirit the event configuration variables sphericity and thrust have be measured for the sample, and the results, $\langle S \rangle = 0.132 \pm 0.004 \pm 0.006$ and $\langle 1-T \rangle = 0.092 \pm 0.002 \pm 0.002$, are in good agreement with values obtained by the PETRA experiments.

Three-jet events have been isolated from the sample by applying standard cuts on sphericity and aplanarity: $S > 0.25$ and $A < 0.1$. The ratio F_3 of planar three-jet events thus defined to the total sample is $F_3 = 13.4 \pm 1.2\%$. Using this value for the ratio and workin with the TPCLUND fragmentation approach, the strong coupling constant α_S is found to be 0.26. Using a fragmentation model more in the spirit of Field and Feynman, the coupling constant is 0.17. A simila range of values for the coupling constant results if the calculation is based on sphericity. This dependence of the determination of α_S on the Monte Carlo treatment has been discussed extensively elsewhere(8).

Particle Fractions

The study of the fractions of various particle species as a function of momentum using the dE/dx capabilities of the TPC has been a principal goal of the analysis. Determinations of the fractions in the $1/\beta^2$ region are made easily. In the relativistic rise region, the pion fraction has been found for three momentum bins by determining the shape of the pion truncated mean distribution in the minimum ionizing region 450 MeV/c $< p <$ 740 MeV/c and using that shape to fit to the distribution in the relativistic rise region. The results are shown in Figure 2 and have been corrected for decays, nuclear interactions, and direct muons. Because of the limited statistics in the hadronic sample, it has not been possible to fit kaons and protons separately in this region. The remaining gap in the plot is the result of the kaon and proton crossover of the pion curve. It is expected that some of this gap will be filled in using likelihood fits to the detailed shape of the dE/dx distributions. Some differences between these results and similar TASSO results(9) at 34 GeV are apparent in the low momentum region, but a fuller discussion must await a larger data sample.

Using the dE/dx separation, the inclusive charged pion cross section has been determined and is shown in Figure 3, plotted with the 34 GeV TASSO points. The cross section for neutral pions has also been obtained from an analysis of hexagonal calorimeter data, where a 21.5 MeV width for the reconstructed pion has been achieved. Reconstruction efficiency is low however, since only two modules were available for this data run.

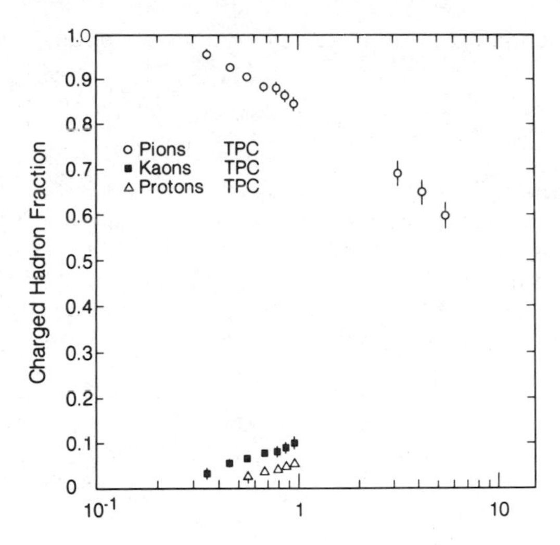

Figure 2. Fractions of particle species versus momentum.

Correlations

The track recognition capabilities of the TPC over its entire angular coverage make it well suited for the study of charge and multiplicity correlations, while the dE/dx identification of the same tracks should make it possible to investigate flavor correlations. While this latter topic has not yet been developed for lack of statistics, some preliminary results are available on the more familiar correlations. Figure 4 shows the associated charge density balance(10), calculated for a test charge in the rapidity interval between 2.0 and 5.0. The long range correlation confirms a previous such observation(11) presented in support of charged primary partons.

The average multiplicity of one ("left") jet is plotted versus the multiplicity of the other ("right") jet in Figure 5, in which two central excluded rapidity regions are shown. The striking absence of any correlation--apart from the slight rise presumably due to heavy quark production--has also been observed by other experimenters(12) and is in contrast to the strong correlations observed at the ISR and the SPS in hadronic collisions. Hadronization appears to proceed via stochastic processes and not through any sort of phase transition.

New Particles

The dE/dx energy loss of a particle in the TPC depends on the velocity of the particle and also on its charge. Consequently the TPC is ideally suited to search for fractionally charged particles. It is expected that by the summer of 1983 it will be possible to lower the cross section limit for Q=2/3 particles by nearly an order of magnitude below the present limit out to a mass of 15 GeV, both

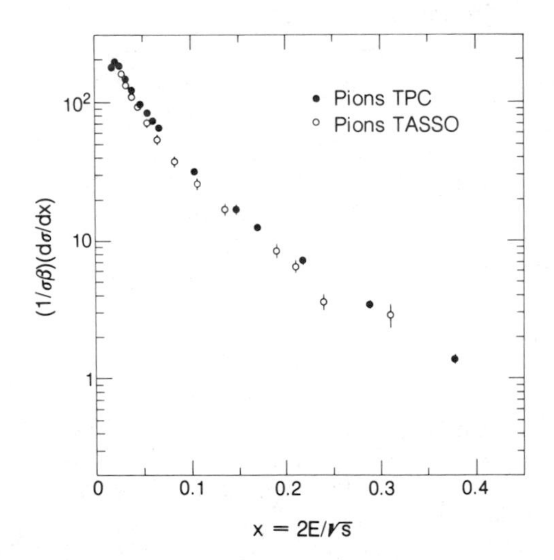

Figure 3. Charged pion inclusive cross section for TPC at 29 GeV and for TASSO at 34 GeV.

in the inclusive and exclusive modes. There exist theoretical arguments(13) for a stable diquark of charge 4/3, and a search is currently in progress for such objects. Principal backgrounds come from track overlaps and from deuterons and tritons produced in secondary interactions. Limits for Q = 4/3 particles (or Q = 4/3 particles) should be presented by the summer.

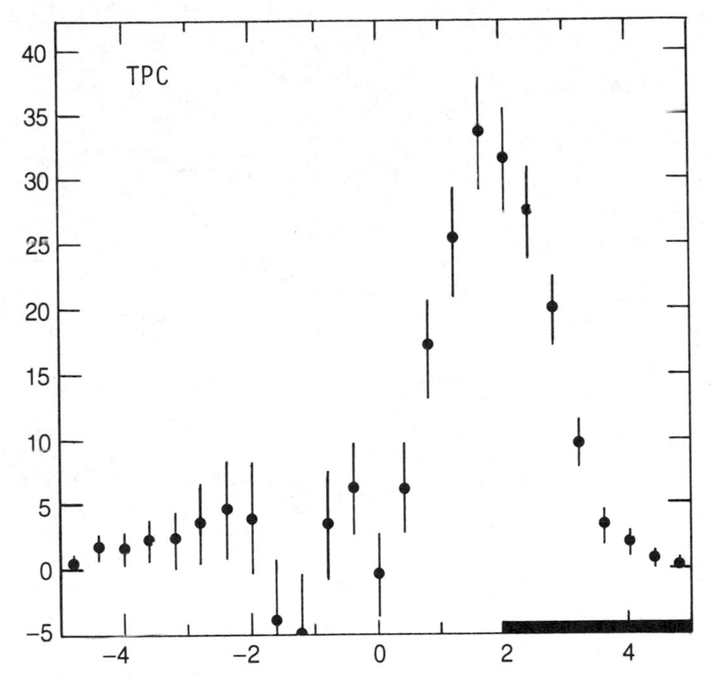

Figure 4. The associated charge density balance for a test particle in the rapidity range 2.0-5.0.

There are many measurements that can be used to exclude the existence of a t quark at PEP energies. One of these is to measure the prompt muon rate in multihadron events. The standard model predicts that $(3.21 \pm 0.14)\%$ of multihadron events should have at least one prompt muon in them with a momentum greater than 2 GeV. If a t quark exists at 10 GeV and has a branching ratio into muons of 12%, this rate increases to $(8.77 \pm 0.18)\%$. The measured prompt muon ($p > 2$ GeV) rate in multihadron events is $(3.7 \pm 1.2 \pm 0.5)\%$, in good agreement with the five quark prediction and more than three standard deviations away from the expected rate from six quarks. A similar conclusion is reached in examining the momentum distribution of the prompt muon relative to the thrust axis.

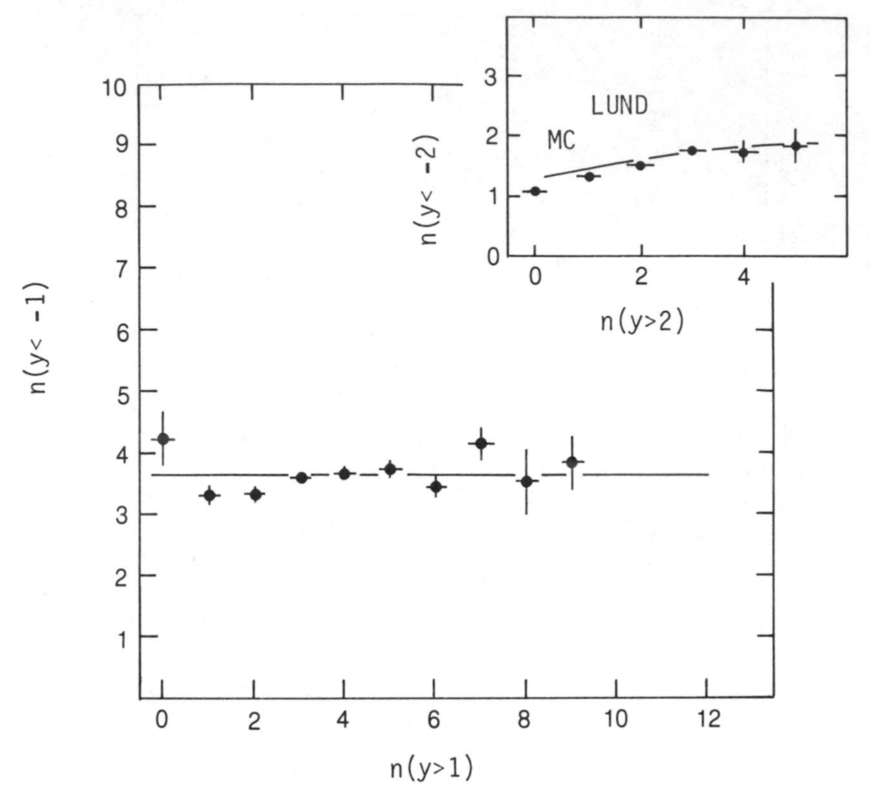

Figure 5. Average left multiplicity versus right multiplicity with the rapidity range $|y| < 1.0$ excluded, and (inset) with $|y| < 2.0$ excluded.

Prospects

In two-photon physics, f° and $\rho^\circ \rho^\circ$ signals have been observed, but not with sufficient statistics to make possible any worthwhile statement on the subject. The recently achieved interchange of event readout with the PEP-9 experiment will facilitate studies of hadron production in single- and double-tagged events and of the photon structure function with good coverage in the forward direction, down to 20 mrad.

During the summer of 1982 the full complement of hexagonal calorimeters was installed, the inner drift chamber was repaired and incorporated into the trigger, significantly reducing the cosmic ray contamination, and various noise problems in the TPC were solved.

At the same time the luminosity of PEP has increased more than three-fold with no increase in background. More than 40000 nb^{-1} of data have been accumulated in these conditions, with excellent prospects of doubling that amount before the summer of 1983, so the current statistics limitations will be largely overcome by that time. During this coming summer the superconducting coil will be installed, and the hardware corrections to the electrostatic distortions described above will be incorporated into the TPC. Completion of these modifications will entail foregoing the fall running period in 1983, but PEP-4 TPC will be at full strength and ready for data taking in the spring of 1984.

References

(1) "The Time Projection Chamber", Jay Marx and David R. Nygren, Physics Today, October, 1978.
(2) Proposal for a PEP Facility Based on the Time Projection Chamber (TPC), PEP Proposal 004 (1976).
(3) "Performance of a Time Projection Chamber", D. Fancher et al., Nucl. Inst. & Methods, 161, 383 (1979).
(4) Proceedings of the International Conference on Instrumentation for Colliding Beam Physics, SLAC-250 (1982): "Tracking with the PEP-4 TPC", A. Barbaro-Galtieri, p. 46; and "Energy Loss Measurements in the PEP TPC", G.R. Lynch and N.J. Hadley, p. 85.
(5) "Spatial Resolution of the PEP-4 Time Projection Chamber", H. Aihara et al., IEEE Trans. Nucl. Sci. NS-30-1, 76 (presented by R.J. Madaras); "Measurement of Ionization Loss in the Relativistic Rise Region with the Time Projection Chamber", Ibid. p. 76 (presented by B. Gabioud); and other papers presented by the PEP-4 TPC Collaboration at the 1982 IEEE Nuclear Science Symposium.
(6) R.J. Madaras, reference 5.
(7) "A Gated Time Projection Chamber", P. Nemethy et al., LBL-15281 (1982). To be published in Nucl. Instrum. & Methods.
(8) For example see "Jets at PETRA", H. Oberlack, Invited talk given at the 2nd International Conference on Physics in Collisions, Stockholm June 1982. MPI-PAE/EXP-El-110.
(9) M. Althoff. et al., Z. Physik C17 (1983) 5. DESY 82-070.
(10) D. Drijard et al., Nucl. Phys. B155 (1979) 269.
(11) R. Brandelik et al., Phys. Lett. 100B (1981) 357.
(12) "Correlations in Electron-Positron, Lepton-Hadron, and Hadron-Hadron Collisions', W. Koch, Invited paper presented at the XIII International Symposium on Multiparticle Dynamics, Volendam, the Netherlands, June 1982. DESY 82-072.
(13) R. Slansky, T. Goldman, and G.L. Shaw, Phys. Rev. Lett. 47, 887 (1981).

RESULTS FROM TASSO

G. Mikenberg

Weizmann Institute of Science
Rehovot, Israel

ABSTRACT

Using data obtained with the TASSO detector at PETRA, results are presented on weak-electromagnetic effects in lepton and charmed quark production. Using the semileptonic decay of the b quark, a determination of the branching ratio B.R.($B \to \ell \nu X$) is made and indications of weak-electromagnetic interference effects are presented. Inclusive $\pi, K, p, \Lambda, \rho^0$ and D^* cross-sections are presented and the consequences of these measurements for the fragmentation picture is given. Finally, upper limits are given for the production of new scalar particles.

INTRODUCTION

First observations of weak neutral current effects in e^+e^- lepton pair production can be traced to 1981.[1-3] With more than twice the data that has become available now, a more careful analysis is possible that leads to a determination of the neutral weak coupling constants with errors comparable to those obtained in νe scattering experiments. The observable charge symmetries in quark pair production are expected to be larger than in the lepton pair case, since the interference between the electromagnetic and weak interactions is inversely proportional to the charge of the pair being produced. Using the good momentum resolution qualities of the TASSO detector it has been possible to tag charmed quarks pair production via its D^{*+} decay modes and therefore measure its production charge symmetry.[4] This leads to a preliminary measurement of the weak neutral couplings of the charmed quark. A simple way to tag heavy quarks is via their semi-leptonic decay modes.

This has been done using the TASSO detector for multihadron events containing leptons in the final state. Looking at leptons produced with large $P_{\perp}$ with respect to the jet direction enhances the b quark production component and therefore measures the weak neutral coupling of the b quark.[5] The method requires, however, a very good understanding of the background component as well as knowledge of the fragmentation characteristics of the b quark. A preliminary measurement of the b quark charge asymmetry will be presented in this talk.

The TASSO detector provides full charged particle identification in a limited region of the solid angle.[6] Using these characteristics it has become possible to obtain inclusive cross-sections of various particle types. This information becomes very important to correct the various quark fragmentation parameters of the models used to understand the data. Section IV presents the various measured inclusive particle cross-sections and the consequence of those measurements for the fragmentation picture.

Various theoretical models[7-8] predict the existence of charged scalar particles. A search for these particles has been performed using the TASSO detector up to masses of 13 GeV with negative results.[10] Details of these searches are given in section V.

II. WEAK ELECTROMAGNETIC INTERFERENCE IN LEPTON PAIR PRODUCTION

Using the TASSO detector, the following lepton-production reactions have been measured:

$$e^+e^- \rightarrow e^+e^- \quad (1)$$

$$e^+e^- \rightarrow \mu^+\mu^- \quad (2)$$

$$e^+e^- \rightarrow \tau^+\tau^- \quad (3)$$

For reactions (1) and (2) events were selected having two oppositely charged tracks, each with a measured momentum larger than 20% of the beam momentum and the total measured momentum of the two tracks larger than 70% of the beam momentum. The tracks were required to come from the beam crossing point and time of flight cuts were imposed in order to reduce the cosmic background. The two tracks were required to be colinear within 10°. For reaction (2), the μ identification was done by either requiring a set of good hits in the muon detectors or a recognized minimum ionizing particle transversing the Lead-Liquid Argon electromagnetic calorimeter. A total of 73800 events satisfying the above requirements were obtained with an average c.m. energy of 34.5 GeV. A sub sample of 2673 events were found that satisfy the cuts for reaction (2). The data was corrected for the contributions of reaction (3) as well as for radiative effects. The overall systematic error is estimated to be 3.5%. Comparing the measured cross-section to the QED expectations for reaction (1) leads to an excellent agreement between the two.

Figure 1 shows the ratio between the measured cross-section for reaction (2) and the QED expectation as a function of the production angle. Corrections have been made for the geometrical acceptance and the contamination of τ pairs. A clear angular asymmetry and a strong disagreement with QED can be seen. By parametrizing the angular distribution to the form $d\sigma/d\cos\theta \propto 1+\cos^2\theta+b\cos\theta$ in order to extrapolate to $\cos\theta=\pm1$ an asymmetry of $A_\mu=-9.1\pm2.3\%$ is obtained. The origin of this asymmetry is easily explained in terms of the interference between the electromagnetic and weak interactions. The Glashow-Weinberg-Salam (G.W.S.) model predicts for this measurement a value of $A_\mu=-9.2\%$, in excellent agreement with the above result.

On the combined data of reactions (1) and (2) an overall fit of the weak neutral coupling has been made. This was done by assuming e-μ universality, and taking an overall normalization error of $\pm5\%$ and a relative normalization error between the e and μ data of 3%. Using the G.W.S. model, where only one free parameter is available, one obtains for this parameter the value

$$\sin^2\theta_W = 0.26 \pm 0.07 \qquad (4)$$

Assuming that the neutral weak boson has a mass of 90 GeV one can determine, in a model independent way, its vector and axial couplings. The values obtained are:

$$g_V^2 = -0.034 \pm 0.052 \qquad (5)$$

$$g_A^2 = 0.220 \pm 0.054 \qquad (6)$$

in excellent agreement with the G.W.S. values of $g_V^2\approx0.00$ and $g_A^2=0.25$. Although not directly measurable, a fit to the couplings g_A and g_V has been done as well, using the signs obtained in νe scattering for g_A. The 95% confidence level contours of this fit are shown in figure 2, where they are compared to the combined νe scattering data. It can be seen that the errors of the two measurements are similar. The values obtained for the weak coupling constants are:

$$g_V = 0.000 \pm 0.157 \qquad (7)$$

$$g_A = -0.481 \pm 0.054 \qquad (8)$$

Reaction (3) has also been investigated by looking for final states where τ decays into one charged particle and the second τ

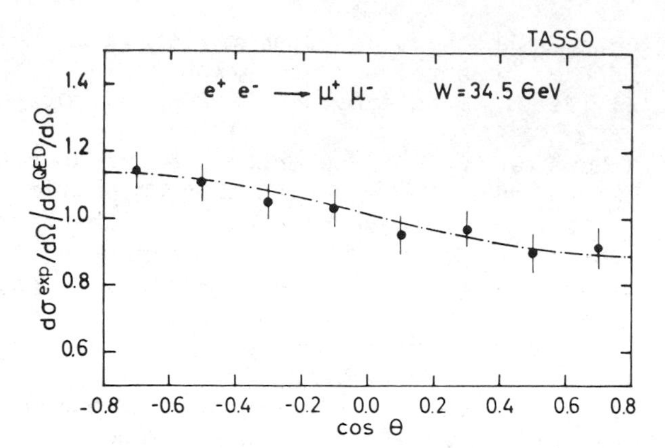

Fig. 1: Ratio between the measured angular distribution for μ pair production and the QED prediction. The curve is described in the text.

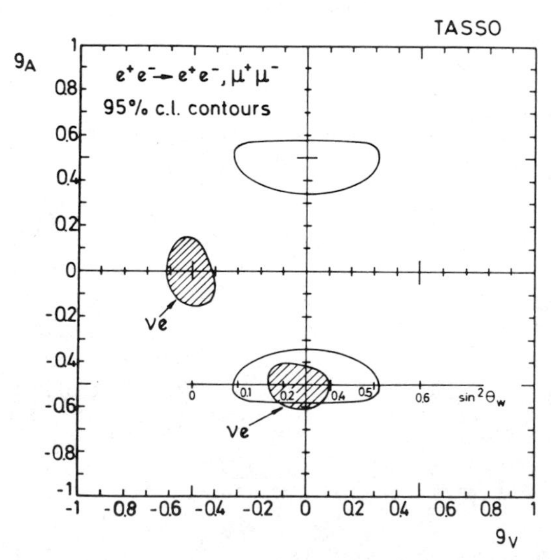

Fig. 2. 95% c.l. contours for the determination of the weak neutral coupling constants. The shaded contours are the results of the combined νe scattering data.

into three. Because of the various branching ratios of the τ decay, this method has a lower efficiency than the search for μ pairs. Presently, a total of 517 τ pair candidates satisfying the above-mentioned topological requirement have been found. The measured angular asymmetry of this preliminary sample is $A_\tau=-0.054\pm.045$ consistent with the G.W.S. prediction of $A_\tau=-0.091$.

III. DETERMINATION OF THE WEAK NEUTRAL COUPLINGS OF HEAVY QUARKS

Particles containing heavy quarks provide a very good tool for tagging the initial quark produced in e^+e^- annihilation since they are seldom produced from the vacuum during the fragmentation process. Using the good momentum resolution of the TASSO detector one has identified charmed quark production via its fragmentation into D^{*+}.[4] This has been done by making use of the small Q value for the decay $D^{*+}\to\pi^+D^0$ which provides a very narrow peak in the $M(D^*)-M(D)$ mass difference at a value of 0.145 MeV. This peak can be seen in figure 3a where the D^0 decay is identified via its $K^-\pi^+$ decay mode. Figure 3b shows the same distribution where the $K^-\pi^+$ mass distribution is cut in a control region. No such peak is observed. Figure 4 shows the $K^-\pi^+$ mass spectrum for the events contained in the peak at 0.145 GeV. A clear D^0 peak can be observed accompanied by the S^0 peak, which is a reflection of the D^0 decaying into $K^{*-}\pi^+$ or $K^-\rho^+$ with a slow unobserved π^0 produced in its decay.

Since for the observation of weak effects one is interested in the highest possible c.m. energy, for the following analysis only events with $E_{cm}\geq 34$ GeV were taken. Furthermore, in order to improve the signal-to-noise ratio, only events with high momentum tracks were taken, i.e. $P_K, P_\pi > 1.4$ GeV/c. With these cuts one obtains the shaded part in figure 4. This figure contains a total of 51 D^* decaying into D via either the $K^-\pi^+$ mode or to the S^0 reflection with an estimated background of 5 events. The angular distribution of these D^*'s was fitted to the form $d\sigma/d\cos\theta \propto 1+\cos^2\theta+b\cos\theta$, which yielded an asymmetry $A_c=-0.28\pm0.13$. The corrections due to the background events, B decays and radiative effects are small and well within the statistical errors. Using this asymmetry value one then obtains $g_A^e g_A^c=-0.49\pm0.23$. Assuming lepton universality and using an average of the results on μ pair production, $|g_A^e|^2=0.30\pm0.04$ yields $|g_A^c|=0.89\pm0.44$, consistent with the G.W.S. model prediction of $g_A^c=0.5$.

To search for weak effects in b quark production a preliminary study has been made on inclusive lepton production in hadronic final states. The leptons were required to have a $p_\perp$ with respect to the jet direction larger than 1.GeV/c. The preliminary semileptonic branching ratios obtained are: $BR(b\to\mu\nu x)=15.0\pm3.5\pm3.5\%$, $BR(b\to e\nu x)=13.6\pm4.9\pm4.9\%$ using the Liquid Argon detector and $BR(b\to e\nu x)=12.4\pm8.7\pm8.0\%$ using the shower counters in the hadron arms. The

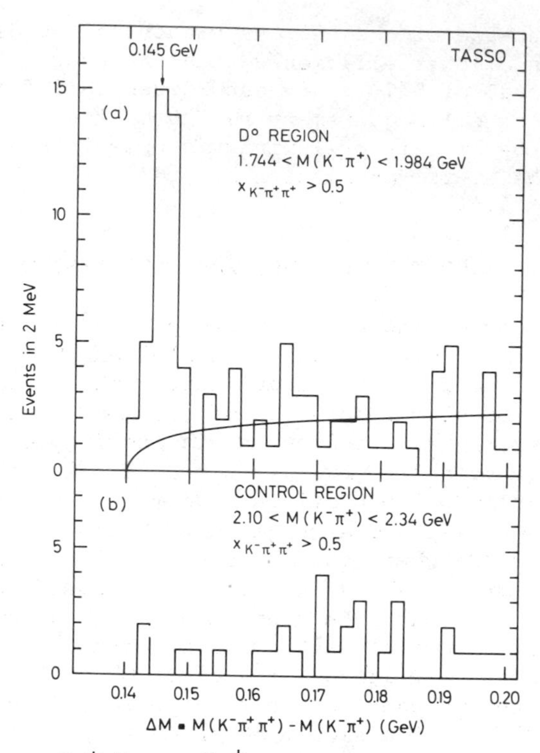

Figure 3. $\Delta M = M(K^-\pi^+\pi^-) - M(K^-\pi^+)$ mass difference for a) the $K^-\pi^+$ mass combination being in the D^o region, and b) in a control region above the D^o region.

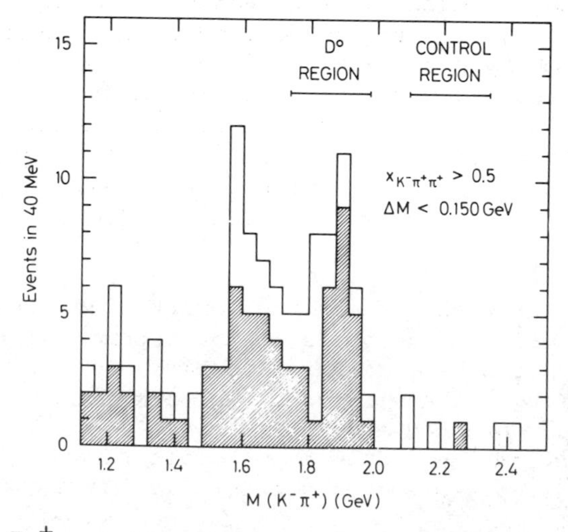

Figure 4. $M(K^-\pi^+)$ mass distribution for events with $\Delta M \leq 0.15$GeV.

preliminary values obtained for the forward-backward charge asymmetries of the leptons are $A(\mu)=-0.17\pm0.10$ and $A(e)=-0.11\pm0.17$ consistent with the G.W.S. model prediction of -0.084.

IV. INCLUSIVE HADRON CROSS-SECTIONS

In order to determine the various parameters needed for different models to describe the data, in particular the amount of strange particle as well as baryon production, one has to measure the relative production of the various particle types. The TASSO detector, with its charged particle identification capability, is ideally suited for this.[10] Fig. 5 shows the various particle multiplicities as a function of the c.m. energy. They all show an increase inconsistent with a $\ell n E_{cm}$ behaviour. Furthermore, the TASSO detector has also measured K^o and Λ^o production. The average particle production multiplicities per event obtained at 34 GeV are 10.3 ± 0.4 $\pi^{\pm}$; 2.0 ± 0.2 $K^{\pm}$; 1.4 ± 0.1 $K^o,\bar{K}^o$ (preliminary); 0.8 ± 0.1 $p,\bar{p}$ and 0.28 ± 0.04 $\Lambda,\bar{\Lambda}$, where the quoted errors are statistical only, while the systematic errors are $\sim$30%. Also ρ^o production has been measured, and an average ρ^o multiplicity of 0.73 ± 0.06 ρ^o's per event is found. From the measured ρ^o cross-section a determination of the relative production of pseudo scalar particles can be made, the result obtained is $P/(P+V)=0.42\pm0.08\pm0.15$.

Figure 6 shows the scaling cross-section $s/\beta d\sigma/dx$ for charged π, K and p as well as for Λ's. Even though their behaviour cannot be described by a simple exponential, they all decrease very rapidly as a function of x. In contrast to that, figure 7 shows the same distribution for D^{*+} production, where a clear peak at $x\sim0.7$ can be seen. This implies that the charmed quark has a hard fragmentation. Parametrizing this fragmentation by the form:[11]

$$\frac{s}{\beta}\frac{d\sigma}{dx} \propto \frac{1}{x(1-\frac{1}{x}-\frac{\varepsilon}{1-x})^2} \quad (4)$$

leads to an ε parameter of $\varepsilon=0.19\pm0.08$. Integrating the D^{*+} cross-section and assuming $D^{*+}=D^{*o}$, one obtains $R_{D^*}(x>0.3)=(\sigma_{D^*}+\sigma_{\bar{D}^*})/\sigma_{\mu\mu}=2.5\pm0.64\pm0.88$ which almost saturates the charmed pair production rate of $R_c\approx2.8$, leaving little room for pseudoscalar charm production. This last result, however, depends very heavily on the decay branching ratio used for $D^o\rightarrow K^-\pi^+$.

V. SEARCH FOR SCALAR PARTICLES

The production of scalar leptons is predicted by supersymmetric theories.[8] A search for these particles has been made with the

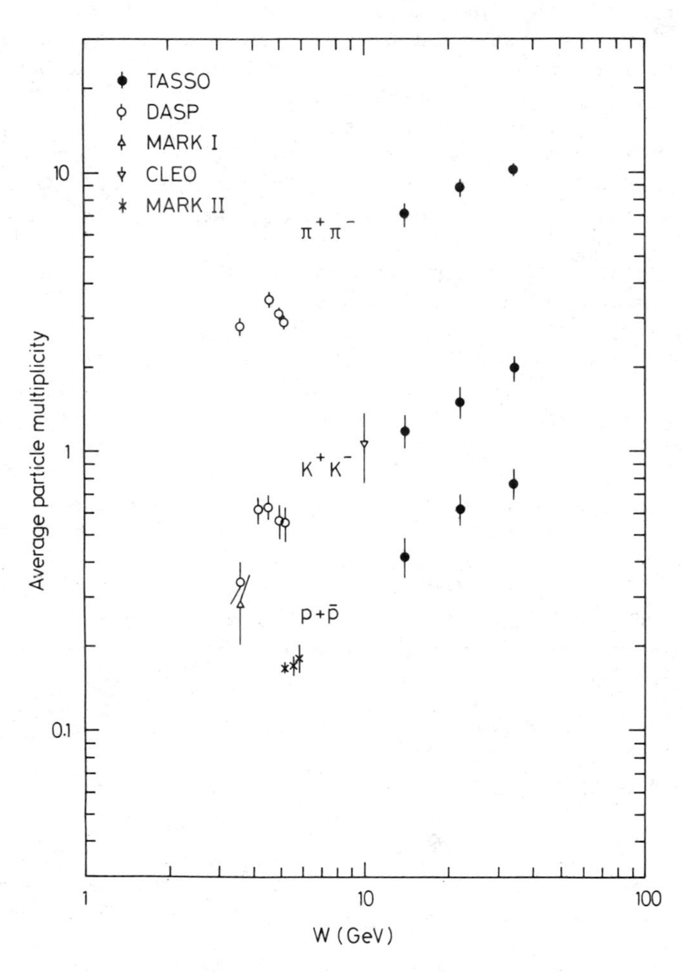

Fig. 5: Average particle multiplicity as a function of the c.m. energy.

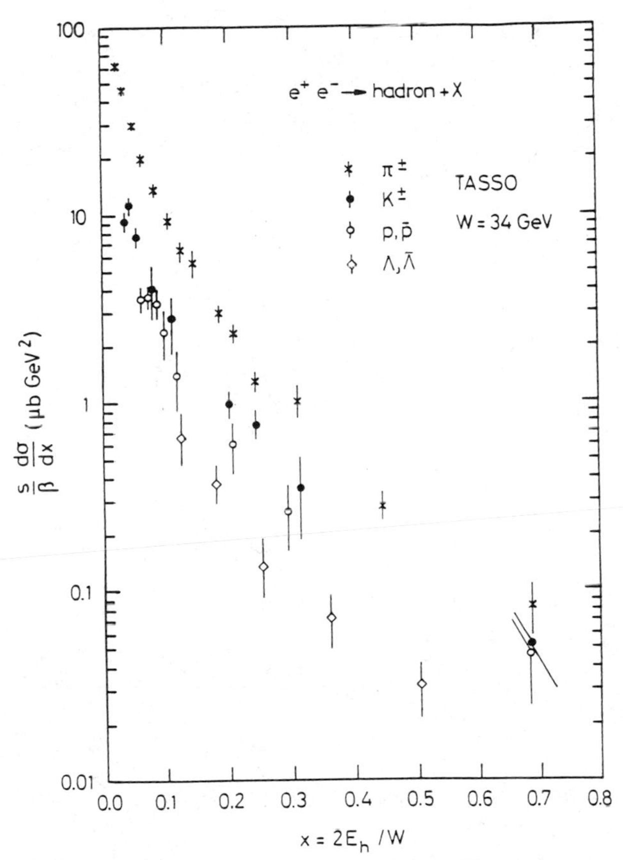

Fig. 6. Invariant particle cross-section $S/\beta d\sigma/dx$ as a function of x for charged π, K and p and for Λ.

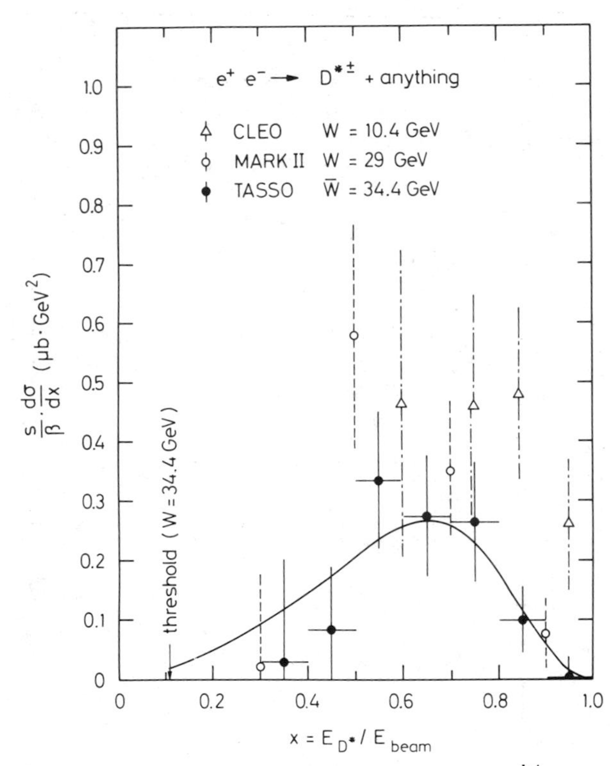

Fig. 7. Invariant particle cross-section for $D^{*\pm}$ production as a function of x.

TASSO detector by looking for final states containing two leptons and two missing photinos or goldstinos. This has been done by comparing the acoplanarity distributions for Bhabha events and muon pairs with the QED predictions, including radiative corrections. From the agreement between the two cross-sections, a 95% confidence level lower limits for the masses of scalar leptons can be established at $M_{S_e}>16.8$ GeV and $M_{S_\mu}>16.1$ GeV.

A search for Higgs and technipions produced in the process $e^+e^- \to H^+H^-$ has also been made. While for a strong decay via $H \to \tau\nu$ the mass range $4 \leq m_{H^-} \leq 13$ GeV has been excluded by other experiments,[12] a complementary search has been made for decays into $c\bar{s}$ or $c\bar{b}$ quark jets. Combining this search with the results obtained in reference 12, charged Higgs and technipions can be excluded in the mass range $5 \leq m_{H^-} \leq 13$ GeV.

ACKNOWLEDGEMENT

I would like to thank Drs. G. Wolf, Ch. Youngman, R. Cashmore and D. Lueke for their great help during the preparation of this presentation

REFERENCES

1. JADE Collaboration, W. Bartel et al., Phys. Lett. 108B:140 (1982).
2. TASSO Collaboration, R. Brandelik et al., Phys. Lett. 110B:173 (1982).
3. Mark J Collaboration, B. Adeva et al., Phys. Rev. Lett. 48:173 (1982).
4. TASSO Collaboration, M. Althoff et al., DESY publication 83-010 (1983), submitted to Phys. Lett.
5. TASSO Collaboration, D. Lueke, talk presented at the XXI Int. Conf. on High Energy Physics, Paris 1982 and DESY publication 82-073.
6. TASSO Collaboration, R. Brandelik et al., Phys. Lett. 83B:261 (1979) and Z. Phys. C4:87 (1980).
7. E. Golowich and T.C. Yang, Phys. Lett. 80B:245 (1979); G. Barbiellini et al., DESY 79-027 (1979) and references therein; S. Weinberg, Phys. Rev. D13:974 (1976); D19:1277 (1979); L. Susskind, Phys. Rev. D20:2619 (1979).
8. See for example P. Fayet and G. Farrar, Phys. Lett. 89B:191 (1980).
9. TASSO Collaboration, M. Althoff et al., Phys. Lett. 122B:95 (1983).
10. TASSO Collaboration, M. Althoff et al., Z. Phys. C17:5 (1983).
11. C. Peterson et al., Phys. Rev. D27:105 (1983).
12. JADE Collaboration, W. Bartel et al., Phys. Lett. 114B:211 (1982).

RECENT RESULTS FROM JADE ON ELECTROWEAK INTERACTIONS

Beate Naroska

DESY
Notkestr. 85
D-2000 Hamburg 52

1) INTRODUCTION

The Jade collaboration[1] has performed measurements at Petra at an average center-of-mass (cms) energy of $W \sim 35$ GeV in order to test Quantum Electrodynamics (QED) and electroweak theories. QED tests are still possible because the influence of the neutral weak current is negligible in many places, e.g. in $e^+e^- \rightarrow \gamma\gamma$, or in the total cross sections for $e^+e^- \rightarrow e^+e^-$, $\gamma\gamma$, $\mu^+\mu^-$, and $\tau^+\tau^-$. On the other hand the interference of the electromagnetic and the weak amplitudes leads to an asymmetry in the angular distributions of $\mu^+\mu^-$ and $\tau^+\tau^-$, which was established for the first time at Petra. The measurements can be compared to predictions of electroweak theories, in particular the "standard model" of Glashow, Salam, and Weinberg[2], which has successfully described neutral current phenomena in neutrino scattering[3] and polarized e-d scattering[4]. The tests at Petra were performed at a Q^2 much higher than previously available.

In these reactions the contributions from diagrams of higher orders in α, the QED coupling constant, lead to significant corrections, and it becomes important to check these corrections experimentally. Jade has performed measurements of the process $e^+e^-\gamma$, $\gamma\gamma\gamma$, and $\mu^+\mu^-\gamma$, which provide an important check of radiative corrections,i.e. available calculations to order α^3.

In the total cross section for $e^+e^- \rightarrow$ hadrons, QCD corrections play an important role in addition to the electroweak effects. A careful study of the systematic errors of R was undertaken in order to derive results on the running coupling constant α_s and the weak mixing angle $\sin^2\theta_W$.

Finally a search for new heavy leptons, which occur in some models, is reported. Limits for the mass of excited states of the electron and muon, for a new sequential heavy lepton, and for a neutral heavy lepton are given.

The Jade detector was described in ref. 5. Its central tracking device is the "jet chamber" . Hodoscopes of leadglass are used as electromagnetic shower detector and muons are detected in a segmented muon filter.

2) THE ELECTROWEAK CROSS SECTION

In lowest order the differential cross section for $e^+e^- \rightarrow f^+f^-$, where f can be a μ, τ, or quark, assuming factorisation, is[6]:

$$(1) \quad d\sigma/d\Omega = \frac{\alpha^2}{4s} \cdot (C_1 \cdot (1 + \cos^2\theta) + C_2 \cdot \cos\theta)$$

$$\text{with } C_1 = Q_f^2 - 2 \cdot Q_f \cdot v_e v_f \chi + (v_e^2 + a_e^2) \cdot (v_f^2 + a_f^2) \cdot \chi^2$$

$$\text{and } C_2 = -4 \cdot Q_f \cdot a_e a_f \chi + 8 \cdot v_e v_f \cdot a_e a_f \cdot \chi^2$$

Q_f is the electric charge of the final state fermion. The a_e, v_e, a_f, v_f, denote the vector and axial vector coupling constants of the Z_o to the electron and final state fermion currents respectively. $\chi = g \cdot s \cdot M_Z{}^2/(s - M_Z{}^2)$ and $g = G_F/(8 \cdot \pi\alpha \cdot \sqrt{2})$ where G_F is the Fermi coupling constant and M_Z the mass of the Z_o. For $e^+e^- \rightarrow e^+e^-$ the cross section is more complicated due to the presence of the t-channel.

The presence of the term $C_2 \cdot \cos\theta$ in (1) leads to an angular asymmetry: which for μ and τ pairs ($Q_f = -1$) is:

$$(2) \quad A = (F-B)/(F+B) = -1.5 \cdot a_e \cdot a_\mu \cdot \chi.$$

The small purely weak contribution $\sim \chi^2$ was neglected; F and B denote the cross sections integrated over the forward and backward regions respectively. The asymmetry depends then only on the axial coupling constants of the Z_o to the leptons and via the propagator on the mass of the Z_o. Within the standard model the coupling constants are fixed by one free parameter, the electroweak angle θ_W (see table 1). Using $\sin^2\theta_W = 0.23$ from neutrino measurements[3] the predicted asymmetry is $A = -0.094$ at $s = 1182$ GeV^2.

The total cross section is obtained by integrating (1):

$$(3) \quad \sigma_{ff}/\sigma_{QED} = C_1$$

Table 1: Weak Coupling Constants in the Standard Model

Particle	a	v	v($\sin^2\theta$=0.23)
e,μ,τ	-1	$-1+4\cdot\sin^2\theta$	-0.08
u,c,(t)	+1	$+1-8/3\cdot\sin^2\theta$	0.39
d,s,b	-1	$-1+4/3\cdot\sin^2\theta$	-0.69

where $\sigma_{QED} = 4\pi\alpha^2/3/s$ is the lowest order QED cross section. As the vector coupling constant of the electron, which is close to 0, appears in the interference term the expected deviations from QED are small.

Modifications of the expression 1-3 are expected from higher order contributions to the electroweak amplitude. Calculations to order α^3 are available for the pure QED cross sections[9]. They will be taken into account in the results quoted here. The corrections for the purely weak and electroweak interference term are partially available[7]. They are at Petra energies in general smaller than the corrections for the purely electromagnetic amplitudes and will not be corrected for. E.g. the muon asymmetry is changed from -9.4% to -(8.8±0.3)% using calculations from ref. 7c.

3) LEPTONIC REACTIONS

3.1) $e^+e^- \rightarrow e^+e^-$ and $\gamma\gamma$

For the analysis[8] of Bhabha scattering and the e^+e^- annihilation into a pair of photons an integrated luminosity of 68.9 pb^{-1} at an average cms energy of 34.6 GeV was used. The event selection was based on the barrel leadglass hodoscope and the tracking information was only used for charge separation and to distinguish Bhabha scattering from two photon production.

The principal cuts were:

i) At least two energy clusters in the leadglass of energy $E > E_{beam}/3$.

ii) The acollinearity of the clusters had to be less than 10°.

iii) Both tracks had to be within a fiducial region of $|\cos\theta| < 0.76$, where θ is the angle between the track and the incoming positron.

After the selection of events the remaining background came mainly from hadronic events, τ pair production, and $e^+e^-\gamma$, where one of the electrons in the final state did not fulfill the cuts, but the

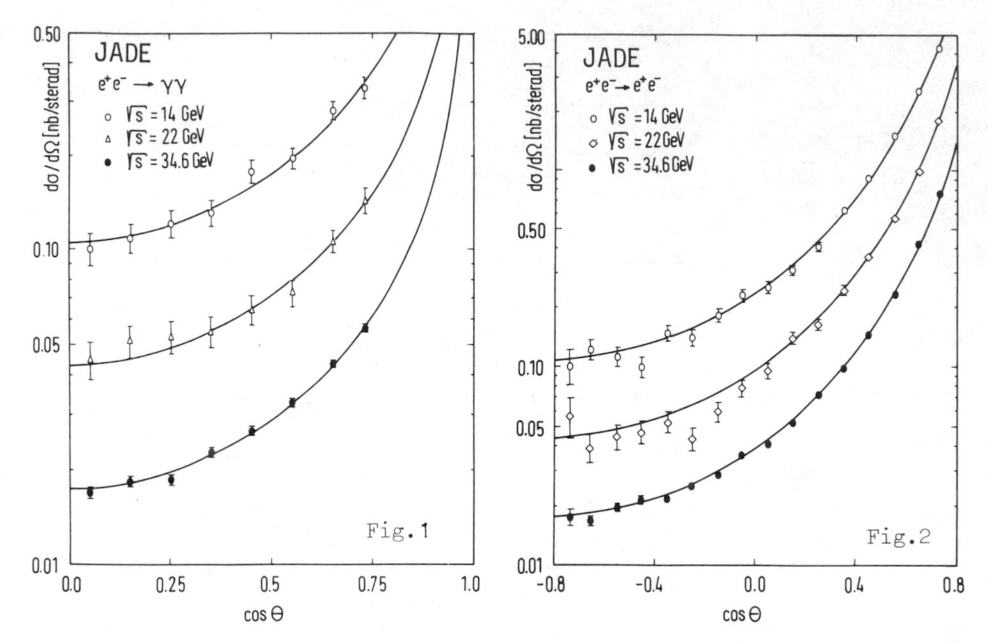

Fig.1 and 2: The differential cross section for 1) $e^+e^- \to \gamma\gamma$ and for 2) $e^+e^- \to e^+e^-$ for three different cms energies. The curves are QED predictions.

photon nearby did. The background was reduced by scanning the events. The scan was also performed for simulated events and the rejected event fractions agreed well. The remaining background was subtracted statistically. Corrections were applied for small counter gaps, doubly converted $e^+e^- \to \gamma\gamma$, and charge misassignment. Radiative corrections up to order α^3 were applied[9].

<u>3.1.1) $e^+e^- \to \gamma\gamma$.</u> In $e^+e^- \to \gamma\gamma$ weak currents do not enter in lowest order. So the reaction is good for testing QED even at high energies.

In fig.1 the differential cross section for the reaction $e^+e^- \to \gamma\gamma$ is given as a function of $\cos\theta$ for three different cms energies. The data points agree well with the prediction of lowest order QED (full lines). Cut off parameters $\Lambda_\pm$ are traditionally used to parametrize limits on the deviations from QED.

$$(d\sigma/d\Omega)_{meas.} \;/\; (d\sigma/d\Omega)_{QED} = 1 \pm s^2/2\Lambda_\pm^{\;4} \cdot (1 - \cos^2\Theta)$$

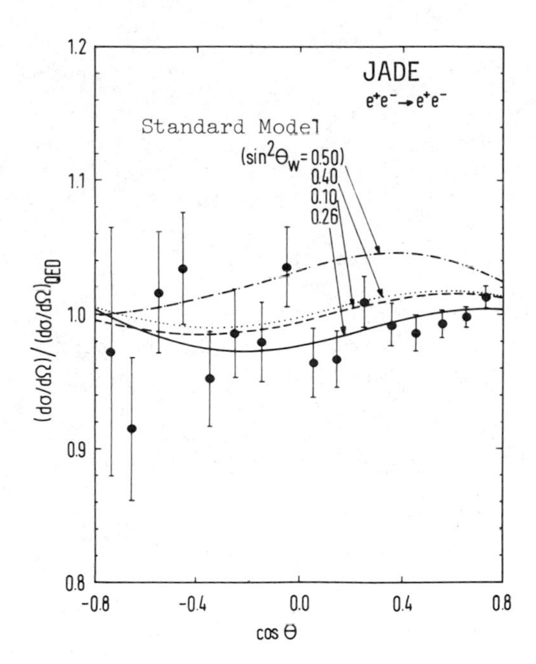

Fig.3: The ratio of the cross section for $e^+e^- \to e^+e^-$ to the QED expectation compared to predictions of the standard model for different values of $\sin^2\theta_W$.

From the angular distribution at the highest cms energy 34.6 GeV, the following limits on the cut off parameters were derived: $\Lambda_+ > 61$ GeV and $\Lambda_- > 57$ GeV. Λ_+ can be interpreted as the upper limit for M^2/λ, where M is the mass of a new heavy lepton and λ its coupling in units of the QED coupling constant e. The resulting upper limit is displayed in fig.8 (upper curve).

3.1.2) $e^+e^- \to e^+e^-$. In fig. 2 the differential cross section for Bhabha scattering is displayed as a function of $\cos\theta$ for three cms energies. Agreement with QED to lowest order (full lines) is good. The comparison of data and QED can be done more conveniently by plotting the ratio of the two as a function of $\cos\theta$ as in fig.3. The data points are compatible with 1 and therefore with QED. The effect of the electroweak interference term is displayed for different values of $\sin^2\theta_W$. The predicted deviations are small for $\sin^2\theta_W = 0.23$. The best fit to the data gives $\sin^2\theta_W = 0.26 \pm 0.13$.

3.2) $e^+e^- \to \mu^+\mu^-, \tau^+\tau^-$

Data for e^+e^- annihilation into muon pairs come from 71.2 pb^{-1} at an average cms energy of 34.4 GeV[10]; for τ pairs only 30 pb^{-1} were analysed. The τ results will be updated soon. So no details will be given here.

The selection for μ pair candidates starts by searching for pairs of tracks in a fiducial region of $|\cos\theta| < 0.80$, where θ is the angle between track and positron beam direction. The following cuts are applied to reduce background from Bhabha scattering, two photon scattering, and τ pair production:

i) The tracks must come from the interaction region of the beams. They must be back to back, with an acollinearity less than 0.2 radians.
ii) The energy deposited in the leadglass is compatible with that of a minimum ionizing particle. The particles must penetrate the muon filter.
iii) The momenta are required to be $p > E_{beam}/3$.
iv) A time-of-flight cut is made to remove cosmic ray muons.

Remaining background is removed by a scan of the events. In the final sample of 3200 muon pair candidates the background from cosmic ray muons and Bhabha scattering is negligible. Muon pairs from two photon scattering contribute 0.7% as calculated using ref.18. τ pairs, where both taus decay leptonically into a muon are calculated to contribute 2% to the final event sample.

In fig.4a and b the angular distributions for muon pairs and τ pairs are displayed. The data were corrected for QED contributions up to order α^3 using the programs by Berends and Kleiss[9]. For comparison the lowest order QED prediction (dashed line) and a fit of the parabola $1 + \cos^2\theta + 8/3 \cdot A \cdot \cos\theta$ (solid line), as predicted by the electroweak interference effect are shown in the figures. The agreement of the data with the electroweak prediction is good.

The resulting values for the forward backward asymmetry are given in table 2, firstly as calculated directly from the number of events in the forward and backward directions for $|\cos\theta| < 0.8$ (column 2). Secondly, the asymmetry from the fit mentioned above, which is extrapolated to $\cos\theta = 1$, is given in column 3 of table 1. The asymmetry expected in lowest order in the standard model using $\sin^2\theta_W = 0.23$ ($M_Z = 88.9$ GeV) is given in the last column of table 2.

The systematic error in the muon forward backward asymmetry of 1% in table 2 is mainly due to charge misassignment estimated by the number of like sign muon pairs and by performing combined fits of both tracks, if they were collinear within 1°. An error due to a possible twist of the apparatus was estimated by studying the

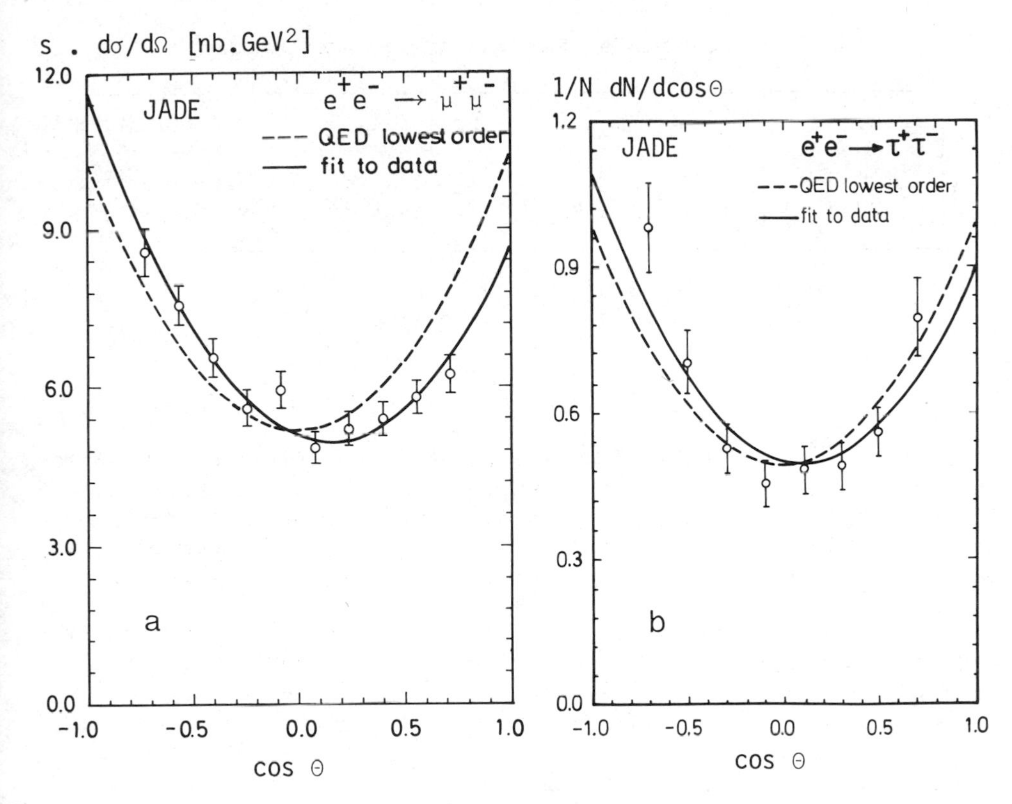

Fig.4: Angular distributions for a) $e^+e^- \to \mu^+\mu^-$ and b) $e^+e^- \to \tau^+\tau^-$, corrected for radiative effects to order α^3. The full curves are parabola fits to the data, the dashed curves are QED predictions.

correlation of the momenta of the two tracks. The error due to remaining background is small as the only sizeable background comes from τ decays, which have the same asymmetry as μ pairs. Even if more cosmic rays, two photon or Bhabha events were present they would only reduce the absolute value of the measured asymmetry. The radiative corrections up to order α^3 show a positive forward backward asymmetry of +2% for the Jade cuts. This was corrected for in the data. Including the systematic error the measured muon asymmetry is 5 standard deviations from 0 and agrees well with the expectation from the standard model.

The measurement of the forward backward asymmetry can be interpreted within the standard model. The axial coupling constants were derived using expression 2 in chapter 2. With M_Z = 88.9 GeV one gets: $a_e \cdot a_\mu = 1.24 \pm 0.21$. Using $a_e = 1.09 \pm 0.11$ from the fit to all

Table 2: Forward Backward Asymmetry for $\mu\mu$ and $\tau\tau$

Process	F-B/F+B	A from fit	A Stand.Mod.
$e^+e^- \to \mu^+\mu^-$	-10.9 ± 1.77	-11.63±1.88±1.0	-9.4
$e^+e^- \to \tau^+\tau^-$	- 6.7 ± 3.4	- 7.9 ±4.0 ±2.9	-9.2

neutrino data[3], one obtains $a_\mu = -1.14 \pm 0.26$. This result confirms the assumption of electron muon universality of the standard model. If one would use the asymmetry including higher order corrections to the weak amplitude the coupling constant would be increased by only a third of the statistical error.

Assuming that the axial couplings are as predicted by the standard model, one can derive the mass of the weak neutral boson: $M_Z = 61 \pm {}^{19}_{7}$ GeV. These limits correspond to 1 standard deviation. At the 95% confidence level $M_Z > 50$ GeV.

3.3) $e^+e^- \to e^+e^-\gamma$, $\gamma\gamma\gamma$, and $\mu^+\mu^-\gamma$

Measurements of these processes are interesting for two reasons: They provide a test of the next-to-leading order calculations of the QED processes, which have to be used to correct all e^+e^- measurements. Secondly, the existence of an excited lepton, e* or μ*, which decays into e+γ or μ+γ, can be investigated in this channel.

3.3.1) $e^+e^- \to e^+e^-\gamma$ and $\gamma\gamma\gamma^-$. A search was made for events containing at least three energy clusters in the leadglass. Two had to have an energy of $E > E_{beam}/3$, and one $E > 0.1 \cdot E_{beam}$. Background was suppressed by requiring $\theta_{ik} > 10^\circ$ and $\Sigma\theta_{ik} > 350^\circ$, where θ_{ik} were the respective angles between the clusters. 2531 (195) events were obtained for $e^+e^-\gamma(\gamma\gamma\gamma)$ compared to 2588 (178) events expected by QED. Energy and angular distributions (figs. 5 and 6) show good agreement with the curves obtained using the calculations of ref. 9.

The invariant mass distribution of eγ is shown in fig.7. No deviation from QED can be seen. The structures reflect energy and angular cuts. The 95% confidence level upper limit on the production of an e* was calculated using the expression by Terazawa et al.[11]. It is displayed in fig.8, λ is the ratio between the new e* coupling to the photon and that of the normal electron coupling.

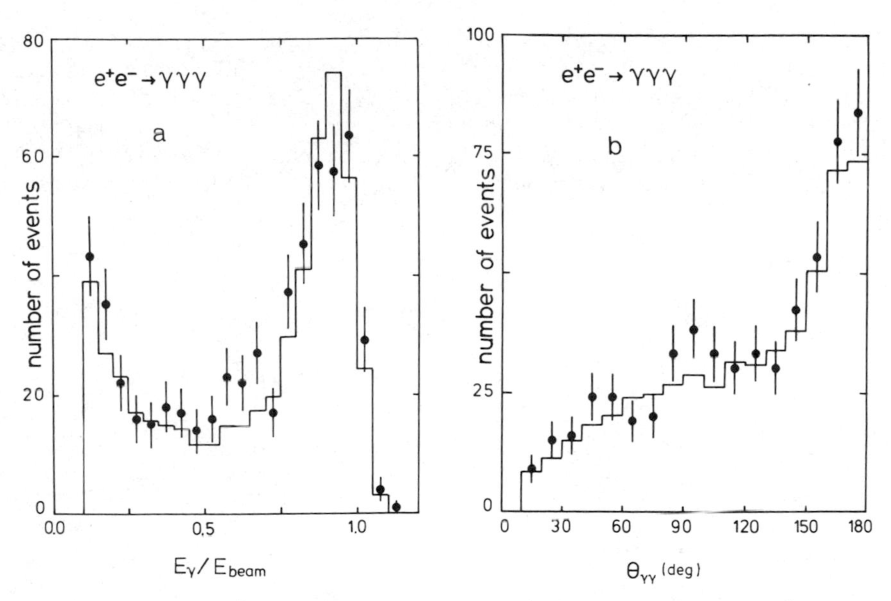

Fig.5: $e^+e^- \to \gamma\gamma\gamma$: a) Photon energy normalized to E_{beam}, b) Angle between two photons. The histograms are predictions of QED to order α^3.

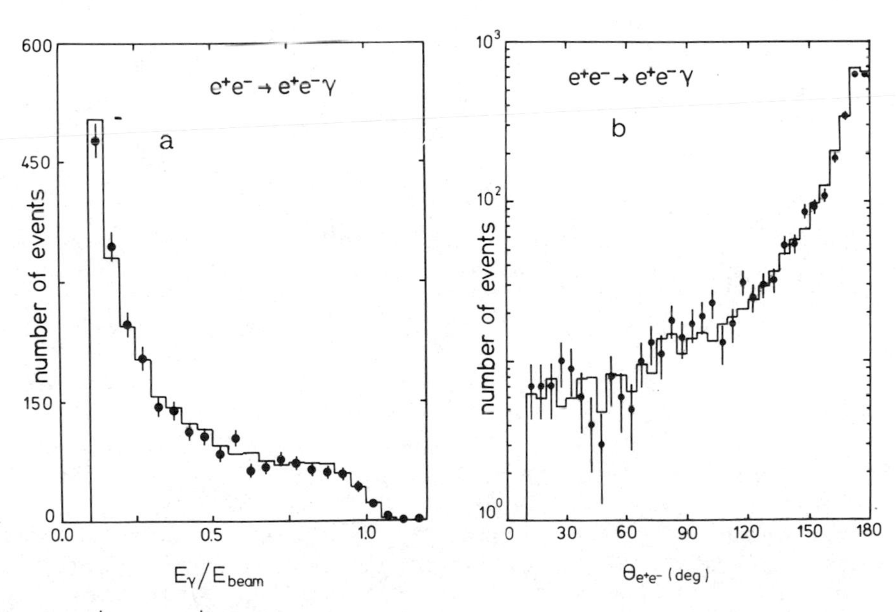

Fig. 6: $e^+e^- \to e^+e^-\gamma$: a) Photon energy normalized to E_{beam}, b) Angle between e^+e^-. The histograms are predictions of QED to order α^3.

3.3.2) $e^+e^- \to \mu^+\mu^-\gamma$. The selection for $\mu^+\mu^-\gamma$ required in addition to the 2 charged tracks a photon of $E > 1$ GeV, which had to be separated by at least 20° from the tracks. The sum of the momenta and the photon energy had to be larger than $E_{beam}/3$, the mass of the μ pair larger than 1.2 GeV, and the sum of all three angles larger than 355°. 270 events were obtained compared with 298 predicted by QED.

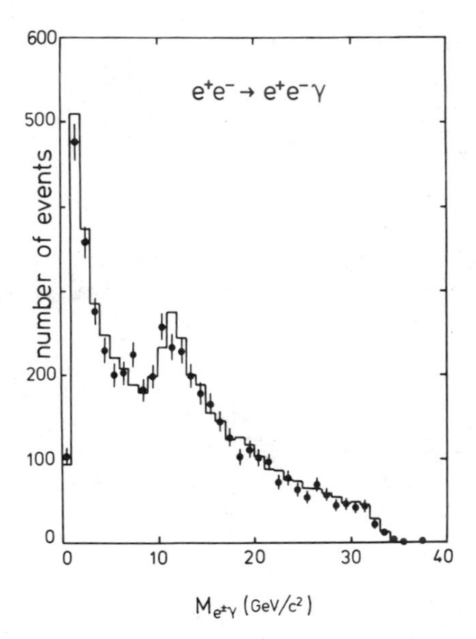

Fig.7: $e^+e^- \to e^+e^-\gamma$: Invariant mass of the system electron-photon and positron-photon. The histograms are predictions of QED up to order α^3.

The angular distribution is displayed in fig.9. It shows a strong forward backward asymmetry $A = -0.39 \pm 0.08$. This is due to the interference of the amplitudes for photon emission in the initial and final states.It is well described by the same calculations used for correcting the muon cross section[9]. The asymmetries expected from QED alone and QED + weak effect are $A_{QED} = -0.36 \pm 0.006$; $A_{QED+WEAK} = -0.42 \pm 0.006$, where the error is due to statistics of the Monte Carlo calculation.

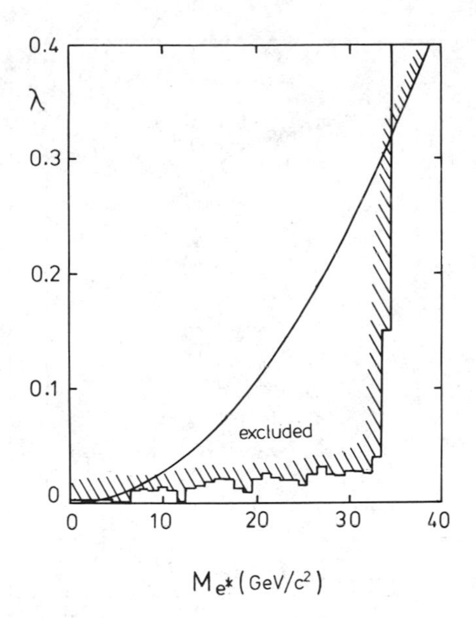

Fig.8: 95% confidence level upper limit for the production of an excited electron e*. The smooth curve comes from $e^+e^- \to \gamma\gamma$, the histogram from $e^+e^- \to e^+e^-\gamma$.

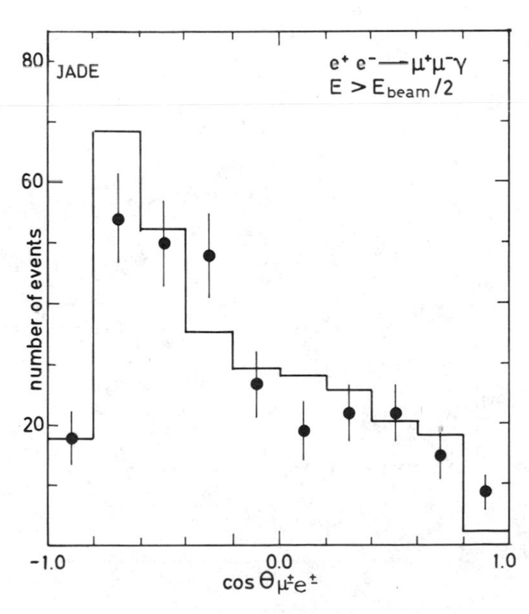

Fig.9: $e^+e^- \to \mu^+\mu^-\gamma$: Distribution of the angle between μ^+e^+ and μ^-e^-. The histograms are predictions of QED up to order α^3.

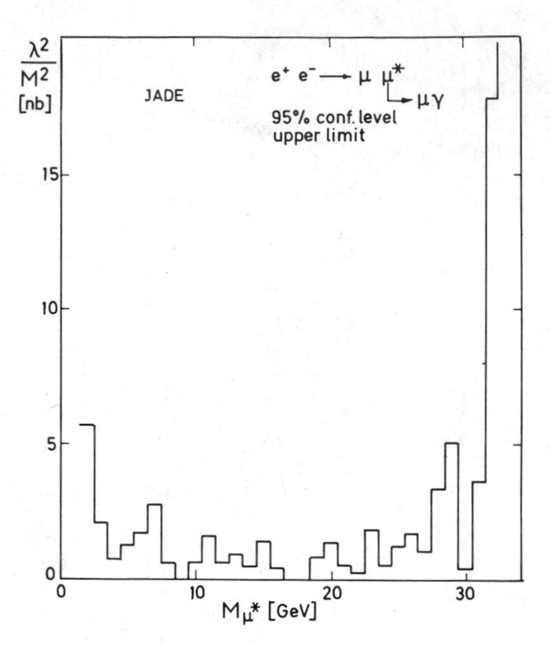

Fig.10: 95% conf. level upper limit on the production of an excited muon μ^* from $e^+e^- \to \mu^+\mu^-\gamma$.

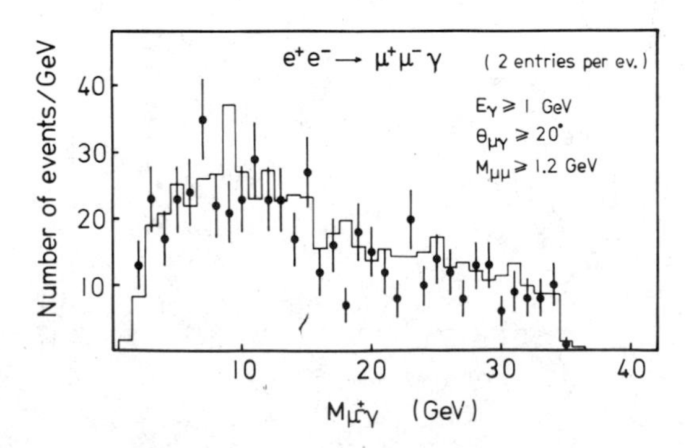

Fig.11: $e^+e^- \to \mu^+\mu^-\gamma$ invariant mass of muon and photon. The histograms are predictions of QED to order α^3.

The $\mu\gamma$ invariant mass distribution is shown in fig.11. There is no structure showing a deviation from QED. The 95% conf. level upper limit for the production of an excited muon is again obtained using the Terazawa et al.[11] expressions. It is given in fig. 10.

4) $e^+e^- \to q\bar{q} \to$ HADRONS

In the Quark Parton Model (QPM) the total cross section ratio R is given by:

$$R = \sigma(e^+e^- \to q\bar{q})/\sigma_{QED} = 3 \cdot \Sigma Q_f^2$$

where σ_{QED} is the lowest order QED cross section for muon pairs and the sum runs over all available flavours. The factor 3 takes into account the three colours of quarks. For 5 quarks the QPM prediction is R = 11/3. QCD[12] predicts a correction factor $1 + \alpha_s/\pi$ for this ratio due to gluon emission, where α_s is the running coupling constant. With $\alpha_s = 0.18$ this correction amounts to a change in R of approximately 5%, almost energy independent in the Petra range of energies. The higher order QCD corrections are negligible.

Another modification of the simple QPM formula is due to the electroweak interference, which within the framework of the standard model, using $\sin^2\theta_W = 0.23$, leads to an energy dependent modification of $\sim$0.3% at W = 14 GeV, and $\sim$1% at W = 34 GeV. Values of $\sin^2\theta_W$ that differ appreciably from 0.23 lead to a stronger energy dependence as shown in fig.13.

In order to study these effects Jade has carefully investigated the contributions of systematic error to R and tried to reduce them. The total cross section at each energy point is calculated using the following quantities:

$$R = (N_{MH} - N_{BG})/(L \cdot \varepsilon \cdot (1+\delta))/\sigma_{\mu\mu}$$

where N_{MH} is the number of multihadronic events, N_{BG} the number of remaining background events, L the luminosity, δ the radiative corrections, and ε the acceptance. In the following the determination of each of these quantities and their errors is briefly described. The number of multihadronic events N_{MH} was obtained using the following selection criteria:

1) A minimum amount of shower energy in the barrel part of the leadglass was required depending on the cms energy; at W > 24 GeV the requirement was 3 GeV.
2) At least 4 tracks should have come from the interaction region, 3 must have $p_t > 0.5$ GeV. The τ topology - namely 1 track opposite to 3 - was rejected.
3) $E_{vis} > E_{beam}$, where E_{vis} was the sum of neutral and charged energy.

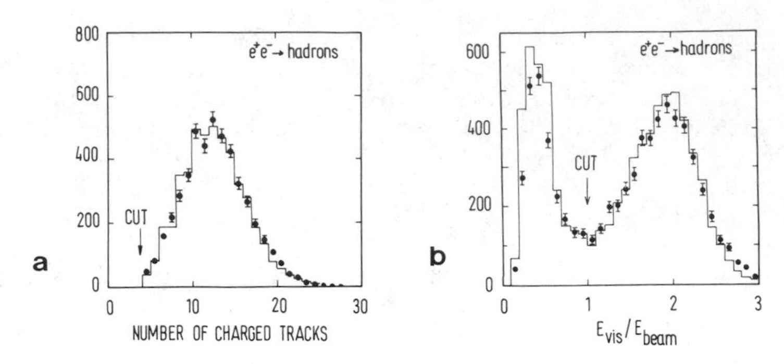

Fig.12: $e^+e^- \to$ hadrons. a) Multiplicity distribution. b) E_{vis}/E_{beam} for $|p_{bal}| < 0.4$. The histograms are Monte Carlo predictions.

4) $|p_{bal}| \leqq 0.4$; where $p_{bal} = \Sigma p_i^z/E_{vis}$ is the longitudinal energy balance, p_i^z is the projection of the momenta on the beam direction.
5) the event vertex was required to be within 150 mm from the interaction point in the beam direction.
Two examples of distribution are shown in fig.12.

All events were scanned. The main background came from hadron production via two photon exchange characterized by low visible energy (fig.12b) and τ pair production. Both contributions were calculated by Monte Carlo programs. The systematic error on the 2 photon background was estimated by varying the cut in the visible energy. Furthermore the amount of 2 photon background could be greatly reduced by cutting out events in the forward region of the detector. Within statistics no change in R was observed. The systematic error of the τ background was obtained by a variation of the decay branching ratios. The background from beam gas events, Bhabha scattering, two photon production, and cosmic rays were reduced to a negligible fraction after scanning.

The acceptance was calculated using the Lund Monte Carlo program[13], which best reproduces the experimental distributions including the multiplicity (fig.12a).

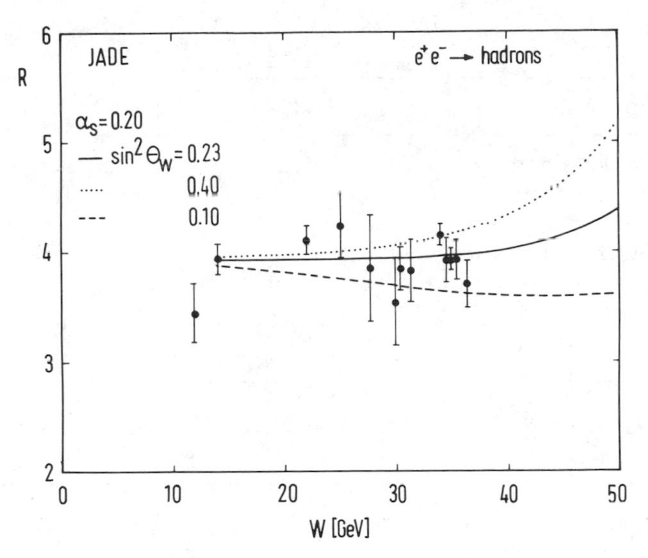

Fig.13: $e^+e^- \rightarrow$ hadrons. The total cross section normalized to $\sigma_{\mu\mu}$ as a function of cms energy. The error bars include stastistical and point-to-point systematic errors. The curves show theoretical calculations including QCD and electroweak corrections for α_s = 0.20 and different values of $\sin^2\Theta_W$.

Radiative corrections were calculated using the programs of Berends and Kleiss[9]. Mainly the uncertainty in the vacuum polarization term due to the uncertainty in the total hadronic cross section leads to the systematic errors in table 3. The error due to the radiative corrections of orders higher than α^3 was not taken into account.

The luminosity was calculated using Bhabha events in the barrel part of the detector ($|\cos\Theta|<0.76$). Uncertainties in the acceptance and radiative corrections are given as a normalization error, while errors due to small counter gaps, calibration uncertainties, and background subtraction are included in the point to point error.

The systematic errors from the various contributions are summarized in table 3 for two values of the cms energy. For W = 35 GeV they add up to 3%. The total hadronic cross section ratio is shown in fig.13 as a function of cms energy. In the error bars the statistical as well as the point to point systematic error are included. The points are compatible with being constant and $\langle R\rangle = 3.97\pm0.05$ (stat)$\pm$0.10(sys). This value is 8.3%$\pm$3.0% above prediction of the QPM for 5 quarks including colour.

A fit of the data points for W > 14 GeV to the cross section was performed including electroweak and QCD corrections[14] with α_s

Table 3: Systematic Error of R in %.

Source	$W \leqq 14$ GeV	22-37 GeV
Background	1.6	1.6
Rad. Corrections	1.1	0.8
Det. Efficiency	2.5	1.5
Lumi point-to-point	1.0	1.0
Lumi normalisation	1.5	1.5
TOTAL	3.6	3.0
Point-to-point	2.7	1.9
Normalisation	2.4	2.4

and $\sin^2\theta_W$ as free parameters. In addition to the data points the absolute normalisation with the overall systematic error was used as a measured point in calculating χ^2. The result is $\alpha_s = 0.20\pm0.08$ and $\sin^2\theta_W = 0.23\pm0.05$, where the errors include statistical as well as systematic errors. The correlation of errors is small. The best fit is shown in fig.13 as a full curve. The minimum χ^2 is 9.8 for 10 degrees of freedom. A second minimum at $\sin^2\theta_W = 0.54$ is excluded by the lepton data of the Jade experiment.

5) SEARCH FOR NEW HEAVY LEPTONS

Jade has performed searches[15] for a sequential lepton beyond the τ, which would be produced in $e^+e^- \to L^+ + L^-$, and the same time for a neutral electronlike lepton produced by W exchange $e^+e^- \to E^0\overline{\nu}_e$ or $\overline{E}^0\nu_e$. Both V-A and V+A couplings of the E^0 to the W were assumed, leading to slightly different limits[16].

The decay of these new leptons then proceeds in the standard way: $L^\pm \to \nu_L + W^\pm$ and $E^0 \to e^- W^+$. The W decays to leptons or quarks. The branching fractions depend on the mass of the heavy lepton below $\sim$14 GeV, above that they are constant, $\sim$35% to (ud) and (cs) pairs, and $\sim$10% to each pair of leptons[17].

A luminosity of 37 pb^{-1} at an average cms energy of 34.2 GeV was used. The upper limits were obtained by applying a certain set of cuts to the data and to simulated events and comparing the expected numbers to the real events.

Depending on the mass range of the lepton different event topologies were used. For L and E^0 of high masses acoplanar two jet

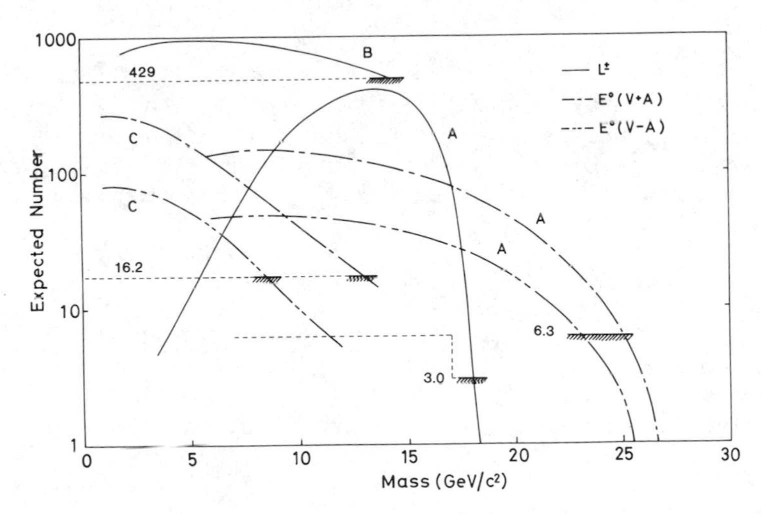

Fig.14: 95% confidence level upper limits for the production of a new heavy sequential lepton L and a new heavy electronlike lepton E^o. For an explanation of the curves see text.

events were searched for. The 95% confidence upper limits as a function of mass are given in fig.14 in the curves labelled A. A heavy lepton L of lower mass was searched for in events containing a jet opposite to an isolated track. The limit is displayed in curve B of fig.14. Finally for the low mass E^o a single jet opposite to an empty cone was searched for. The upper limits are given by curves C in fig.14.

In conclusion: A new heavy sequential lepton is excluded up to a mass of 18 GeV. A neutral heavy lepton with V+A coupling is excluded below 24.5 GeV and excluded below 22.5 GeV with V-A coupling.

SUMMARY OF RESULTS:

1) Jade has performed tests of QED and electroweak interactions using the processes: $e^+e^- \to e^+e^-$, $\gamma\gamma$, $\mu^+\mu^-$, $\tau^+\tau^-$, $e^+e^-\gamma$, $\gamma\gamma\gamma$, $\mu^+\mu^-$. For $\mu^+\mu^-$ a forward backward asymmetry of $-(11.76 \pm 1.88 \pm 1.0)\%$ was observed which agrees with the prediction of the standard model using $\sin^2\theta_W = 0.23$. The limit on the mass of the weak neutral boson is $M_Z > 50$ GeV at 95% confidence level. All other reactions agree with QED including terms up to α^3.
2) The systematic error in $e^+e^- \to$ hadrons was reduced to 3% at high energy. The measured cross section was used to determine $\alpha_s = 0.20 \pm 0.08$ and $\sin^2\theta_W = 0.23 \pm 0.05$ simultaneously. The errors include statistical as well as systematic errors.
3) Excited electrons or muons decaying into lepton and γ could be excluded for masses up to ~32 GeV.
4) A new heavy sequential lepton was excluded below 18 GeV, a new neutral electronlike heavy lepton below ~24.5 GeV if it has V + A coupling, below 22.5 GeV for V - A coupling. All limits are at 95% confidence level.

REFERENCES

1) The Jade Collaboration has members from the following institutions: DESY, Universities of Hamburg, Heidelberg, Lancaster, Manchester, Rutherford Appleton Laboratory, University of Tokyo.
2) S.L. Glashow, Nucl. Phys. 22 (1961) 579; Rev. Mod. Phys. 52 (1980) 539
A. Salam, Phys. Rev. 127 (1962) 331; Rev. Mod. Phys. 52 (1980) 525
S. Weinberg, Phys. Rev. Lett. 19 (1967) 1264;
Rev. Mod. Phys. 52 (1980) 514.
3) J.E. Kim et al., Rev. Mod. Phys. Vol. 53 (1981) 211;
F.W. Buesser, Invited Talk at the DESY Workshop on Electroweak Interactions, DESY T-82-05 (1982);
F.W. Krenz, Compilation on Neutrino Electron Scattering Data and Remarks on Lepton Coupling Constants, PITHA 82/26.
4) C. Prescott et al., Phys. Lett. 77B (1978) 347;
C. Prescott et al., Phys. Lett. 84B (1979) 524.
5) Jade Collaboration, W. Bartel et al., Phys. Lett. 88B (1979) 171
H. Drumm et al., Wire Chamber Conference 1980, Ed.W. Bartl and M. Regler, North Holl. Publ. Comp., Page 333.
6) R. Budny, Phys. Lett. 55B (1975) 227;
7) G. Passarino and M. Veltman, Nucl. Phys. B160 (1979) 151;
M. Greco et al., Nucl. Phys. B160 (1979) 208
F.A. Berends, R. Kleiss, and S. Jadach: Nucl. Phys. B202 (1982) 63
E.A. Paschos, Private Communication;
W. Wetzel, Private Communication.
8) Jade Collaboration; W. Bartel et al., Phys. Lett. 92B (1980) 206;
Jade Collaboration, W. Bartel et al., to be published.
9) F.A. Berends et al., Nucl. Phys. B57 (1973) 381;
Nucl. Phys. B61 (1973) 414; Nucl. Phys. B63 (1973) 381;
Nucl. Phys. B63 (1973 452; Nucl. Phys. B68 (1973) 541;
Nucl. Phys. B177 (1981) 237; Nucl. Phys. B178 (1981) 141;
F.A. Berends, and R. Kleiss, DESY-Report 80-66 (1980).
10) Jade Collaboration, W. Bartel et al., Phys. Lett. 108B (1982) 140.
11) Terazawa et al., Ins.-Rep.-443, Dec. 1981, University of Tokyo.
12) M. Dine et al. , Phys. Rev. Lett. 43 (1979 668;
K.G. Chetyrkin et al., Phys. Lett. 85B (1979) 277;
W. Cellmaster et al., Phys. Rev. Lett. 44 (1979) 44
13) B. Andersson et al., Phys. Lett. 94B (1980) 211
14) J. Jersak et al., Phys. Lett. 98B (1981) 363
15) Jade Collaboration, W. Bartel et al., Phys. Lett. 123B (1983) 353
16) J.D. Bjorken et al., Phys. Rev. D7 (1973) 887;
F. Bletzacker et al., Phys. Rev. D16 (1977) 2115;
K. Fujikawa, Phys. Rev. D17 (1978) 1841;
M. Gourdin et al., Nucl. Phys. B164 (1980) 387
17) Y.S. Tsai, Phys. Rev. D4 (1971) 2821; Slac-Pub-2450 (1979)
18) J.A.M. Vermaseren, Proc. of the Intern. Workshop on γγ Collisions, Amiens 1980, published by Springer Verlag: Lecture Notes in Physics 134 (1980) 269

REPORT ON THE HIGH RESOLUTION SPECTROMETER AT PEP

H. Ogren

HRS Collaboration +
Indiana University *
Bloomington, Indiana 47405

Abstract

A report on the High Resolution Spectrometer at PEP is presented. The detector is presently collecting data at PEP. The first data analysis indicates that all design criteria have been met.

The HRS detector has now been running for more than one year at PEP. The data collected during the spring run in 1982 corresponded to an integrated luminosity of 23 pb^{-1}. For the 1982-83 period that integrated luminosity has already been exceeded in Feb., 1983. We expect to obtain more than 100 pb^{-1} for this period. PEP operation has improved dramatically since the beginning of 1983. The PEP energy has remained at 14.5 GeV per beam and integrated luminosities of up to 1 pb^{-1}/day are now being recorded. The HRS operation has been very reliable. We have logged more than 90% of the delivered luminosity.

A view of the HRS detector is shown in Fig. 1. In Fig. 2 details of the internal detector can be seen: central drift chamber, barrel shower counter, outer drift chambers, Cerenkov counters, and end cap shower counters.

+ Argonne National Laboratory, Indiana University, University of Michigan, Purdue University, Lawrence Berkeley Laboratory and Stanford Linear Accelerator Center.

* Work supported by the U.S. Department of Energy.

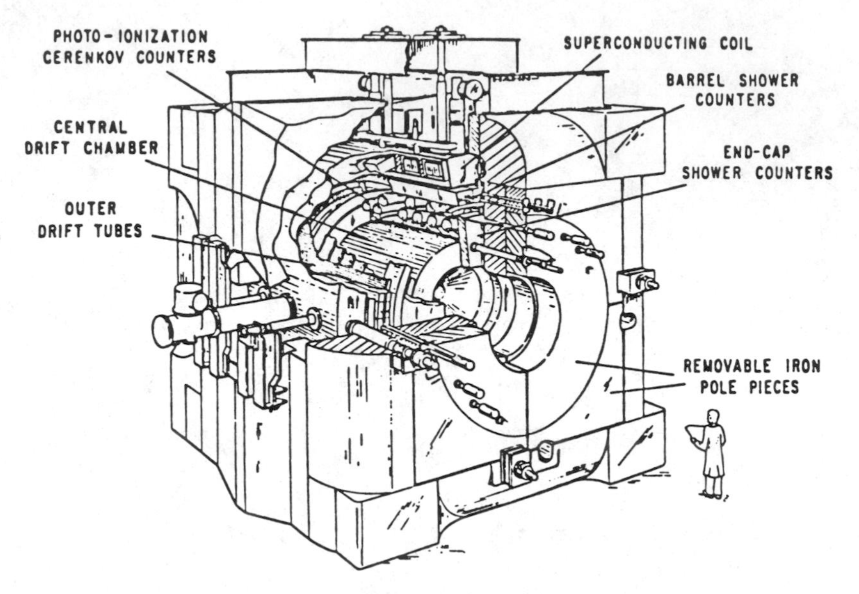

Fig. 1. View of HRS

The HRS detector is a general purpose solenoidal detector that has full neutral coverage and a large 2 meter radius tracking volume with a 16KG magnetic field. To enhance the momentum and invarient mass resolution the amount of material traversed by particles emerging from the interaction area is less than 1% X_o. The Beryllium beam pipe is 1.4 mm in thickness while the inner cylinder of the central drift chamber is 1 mm Beryllium. In fact, the principle material seen by the outgoing particles is the AR-CO_2 gas in the central drift chamber.

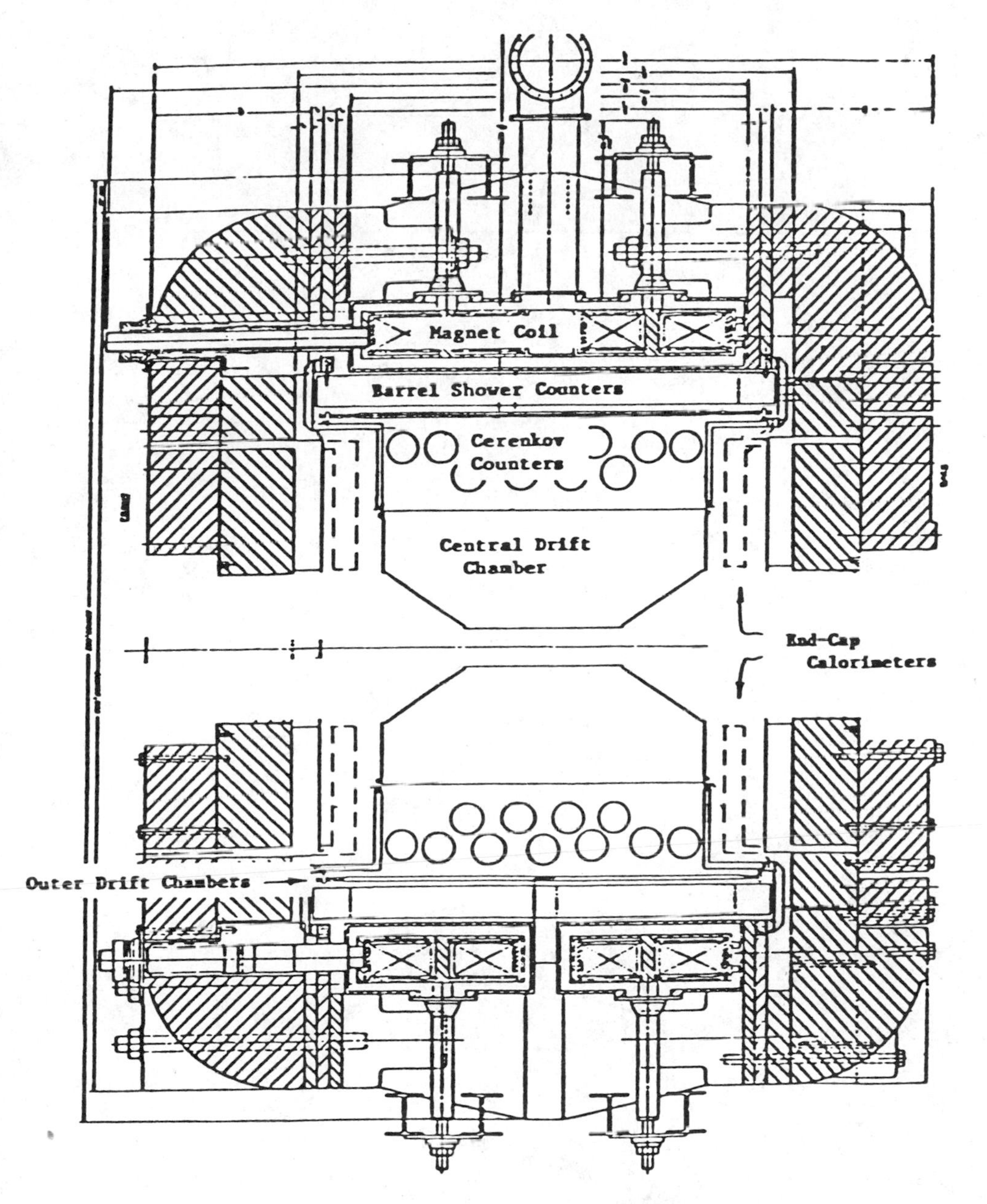

Fig. 2. Details of HRS Detector Elements

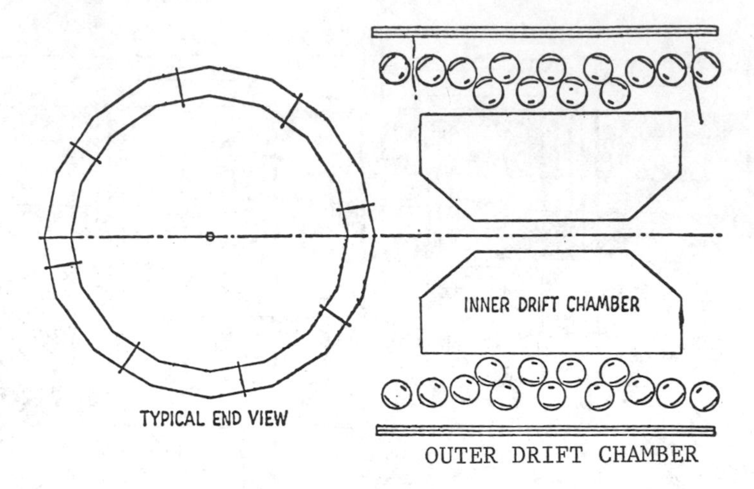

Fig. 3. Photoionization Threshold Cerenkov Counters - HRS

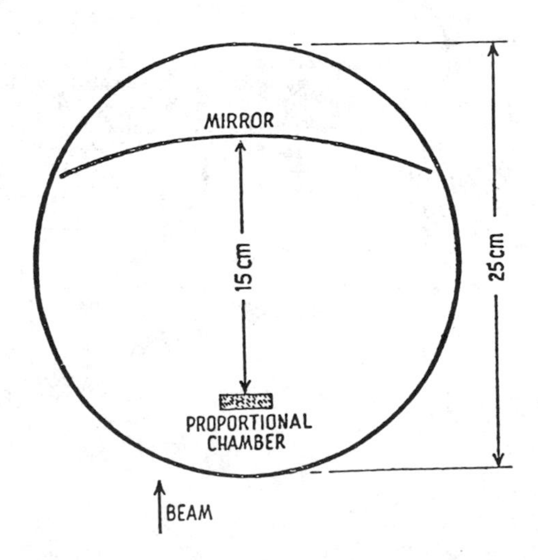

Fig. 4. Detail of photoionization Cerenkov counters

During the summer of 1982 photo ionization Cerenkov counters were added to the HRS. The Cerenkov detectors are shown in Fig. 3. There are 11 tori presently installed. Each torus contains an array of mirrors and PWCs that are configured as shown in Fig. 4. Ultraviolet Cerenkov radiation emitted by fast particles in the Ar-N_2(85,15%) radiator gas is focused on the array of PWCs. Each PWC consists of 6 cells containing 79% Ar, 20% CO_2, 1% Benzine. The ultraviolet photons photoionize the benzine and the resulting gas amplified pulse is read out at the chamber. The windows of the PWCs are MgFl.

The Cerenkov counters are presently being run at a pressure of 8 ATMS. With this pressure the pion threshold is at 1.6 GeV/c. We calculate that there will be 7-8 photoelectrons produced for a particle above threshold. We are presently bringing into operation the electronics for the Cerenkov counters. We anticipate that the Cerenkov counters will be logging useful data this spring.

With the HRS we are able to track both before and after the Cerenkov counters using the central and outer drift chambers. This will be very useful for determining the efficiency of the counters. However, multiple scattering in the Cerenkov counters will degrade the momentum resolution somewhat as shown in Fig. 5.

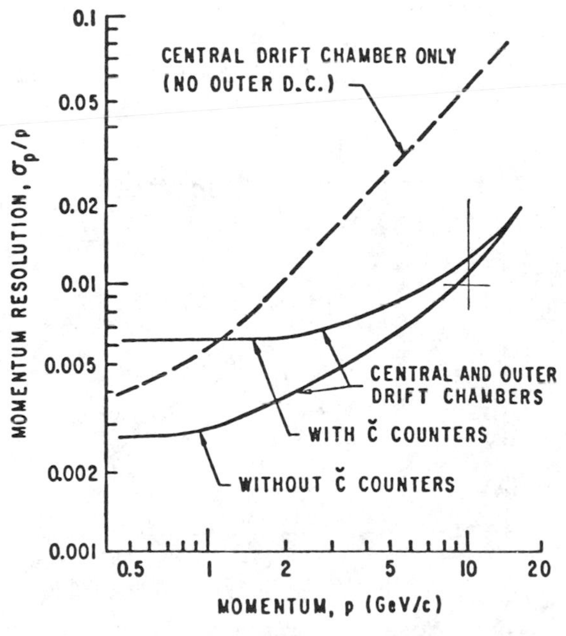

Fig. 5. Calculated Momentum Resolution - HRS

We have met the design specifications on all drift chambers. Fig. 6 shows a central drift chamber tracking residual $\sigma \sim 170\mu$. We have measured all chamber resolutions to be better than or equal to 200μ.

The momentum of Bhabha electrons that enter the outer drift chambers ($|\cos\phi|<.6$) is shown in Fig. 7. For these events we have a momentum resolution of .1%P = dp/p. Some individual chambers are better than .09%p. This is very close to our design values as indicated in Fig. 6.

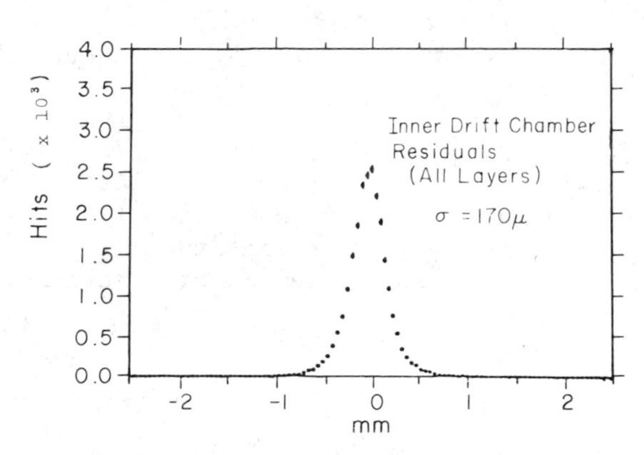

Fig. 6. Inner Drift Chamber Tracking Residual

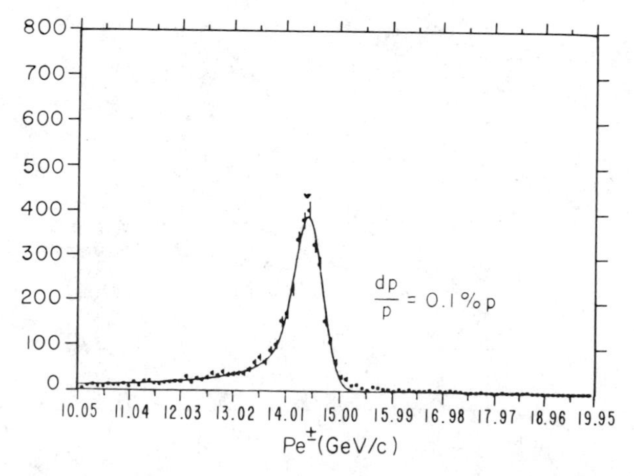

Fig. 7. Measured Momentum Resolution

The HRS shower counter system covers 90% of the solid angle. The system consists of a 'barrel' detector just inside the coil at a radius of ∿2 meters and a moveable end cap system at 1.5 meters from the interaction region. Both systems are used in the triggering system and both measure TOF. The initial $3X_0$ of the barrel system are used to measured TOF. This resolution is ∿360ps for minimum ionizing particles. For Bhabhas this resolution is ∿165ps since in this case the timing is no longer photon limited. The end cap shower counter uses BBQ as a reradiator and therefore has somewhat worse TOF resolution. These TOF's are shown in Fig. 8. The time of flight separation for a small sample of the data is shown in Fig. 9. The energy resolution of the shower counter system for Bhabha's is shown in Fig. 10 and 11. Both shower counter systems perform according to specifications.

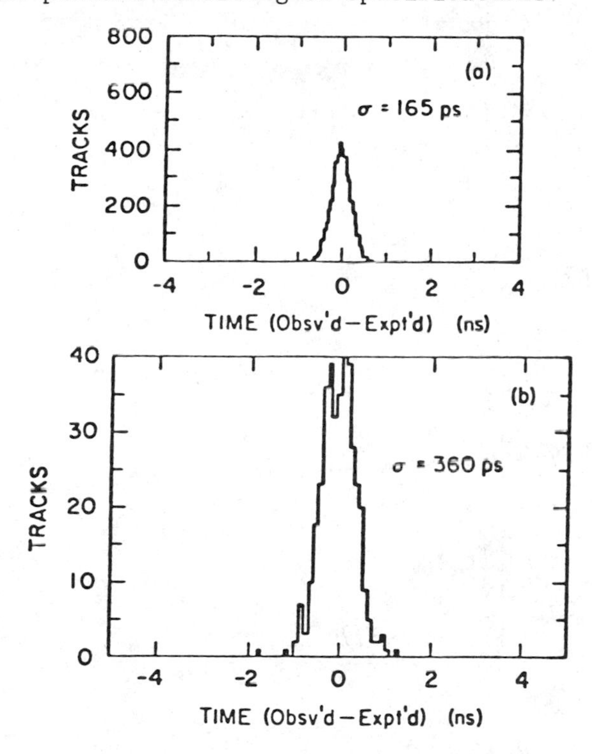

Fig. 8. a) TOF Resolution for Bhabhas

b) TOF Resolution for Minimum ionizing.

A typical hadronic event in the HRS at 16KG and 29 GeV center of mass energy is shown in Fig. 12. The tracking in the central drift chamber is satisfactory and no serious problems with curl up tracks has been observed.

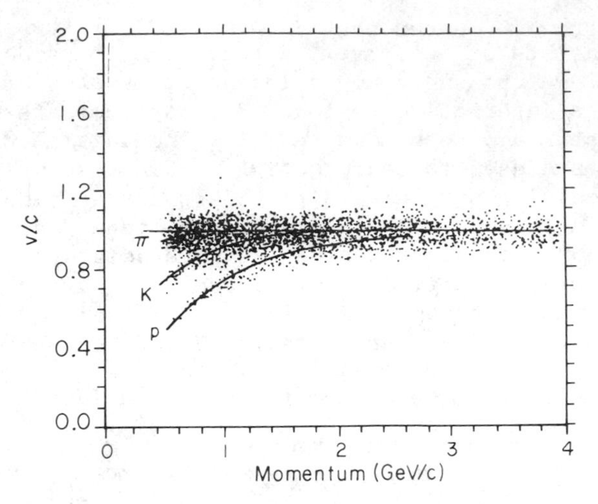

Fig. 9. TOF Particle Separation

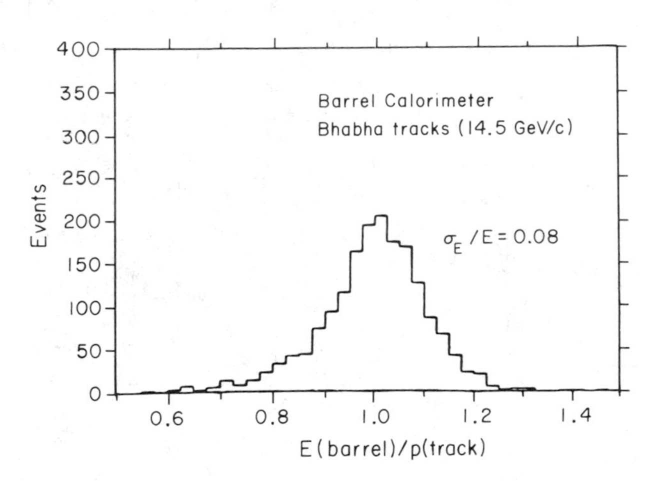

Fig. 10. Energy Resolution of Shower Counter for Bhabhas

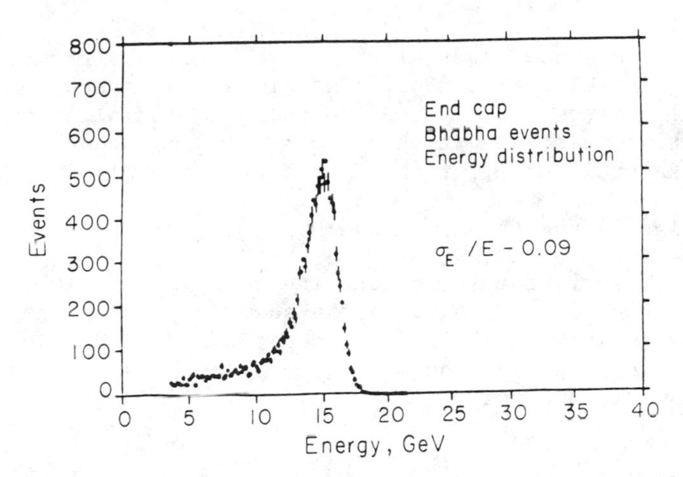

Fig. 11. Energy Resolution of End Cap Shower Counters for Bhabhas.

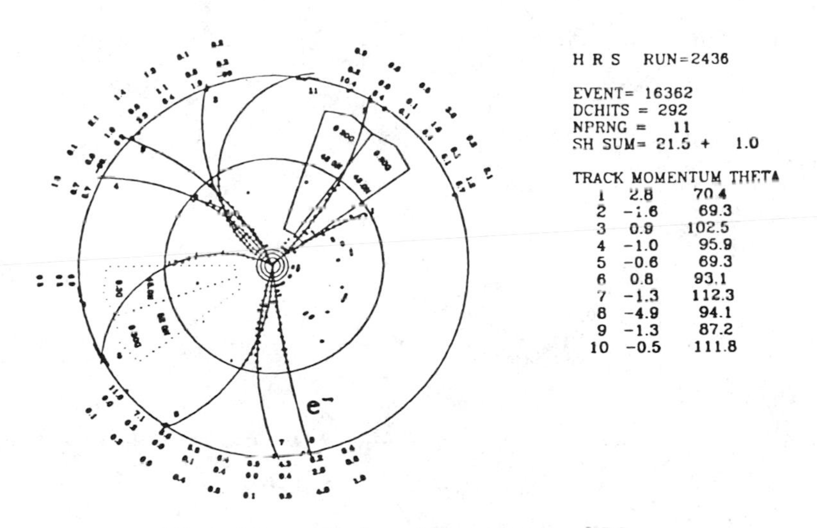

Fig. 12. Hadronic Event in HRS

We have begun a study of the hadronic events from the 23 pb^{-1} of running in the spring 1982. The plots that are shown here are still preliminary. In Fig. 13 the charged multiplicity is shown. It is quite consistent with the $\overline{N_{CH}} = 12$ as has been reported by others. An artificial cut at $NC_H < 5$ was introduced in this plot.

In Fig. 14 the thrust of the hadronic events is displayed. This shows the high thrust peaking that is expected in two jet events. Similarly the low sphericity peaking for two jet events is seen in Fig. 15. In Fig. 16 the scatter plot of sphericity vs aplanarity is shown. The regions separating 3 and 2 jet events is indicated. Analysis of the strong coupling constant, and possible correlations is underway.

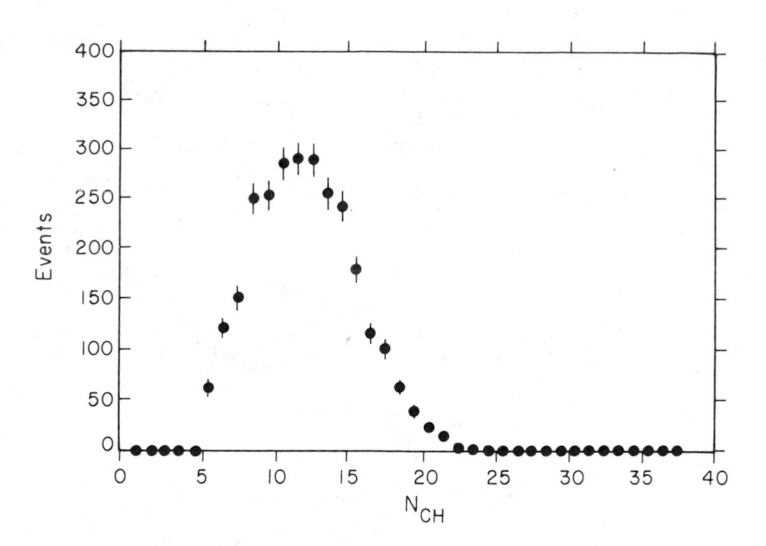

Fig. 13. Charged Multiplicity for Hadronic Events

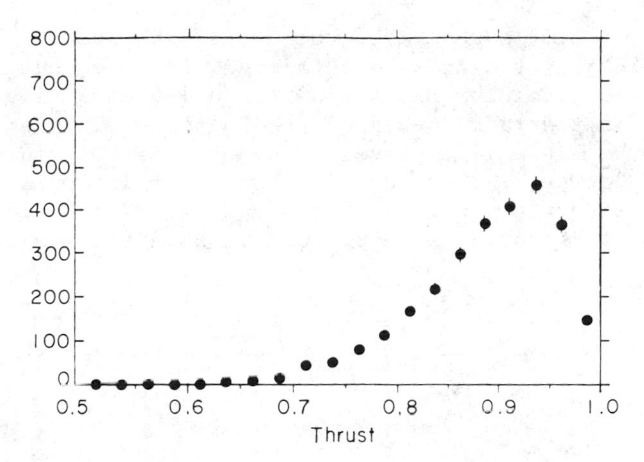

Fig. 14. Thrust Distribution for Hadronic Events

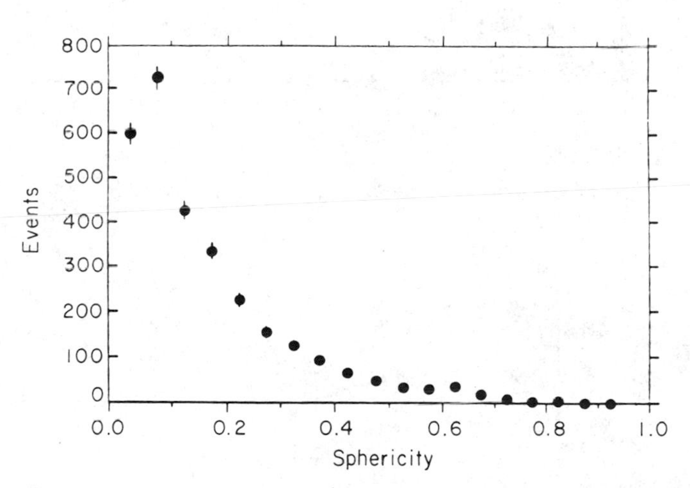

Fig. 15. Sphericity Distributions for Hadronic Events

The particular strength of the HRS is the precise invarient mass distributions that can be obtained with excellent momentum resolution, good angular resolution and a low mass detector. Fig. 17 shows a first determination of the K^o peak. The resolution is $\sigma < 4$ MeV/c. This will improve as the chamber constants are tuned up. We, of course, will study all resonances including K^*, ρ ϕ, D, D^* etc. With the completion of our first 23 pb^{-1} hadron summary tape this spring we hope to have some results on these resonances.

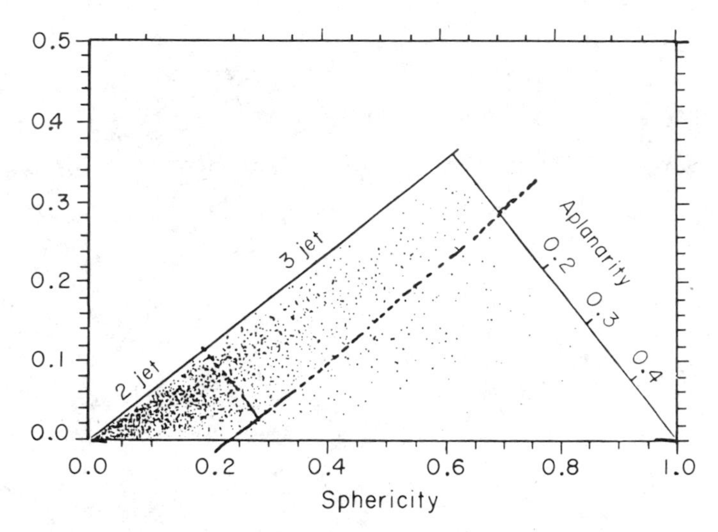

Fig. 16. Sphericity vs Aplanarity

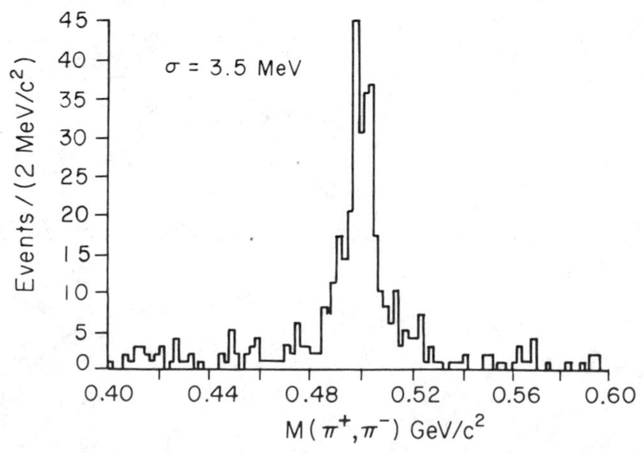

Fig. 17. Preliminary K^o Reconstruction

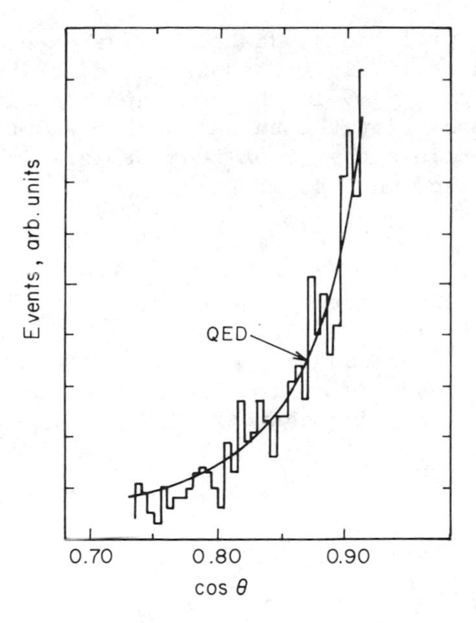

Fig. 18. Distribution of Bhabhas in Endcap

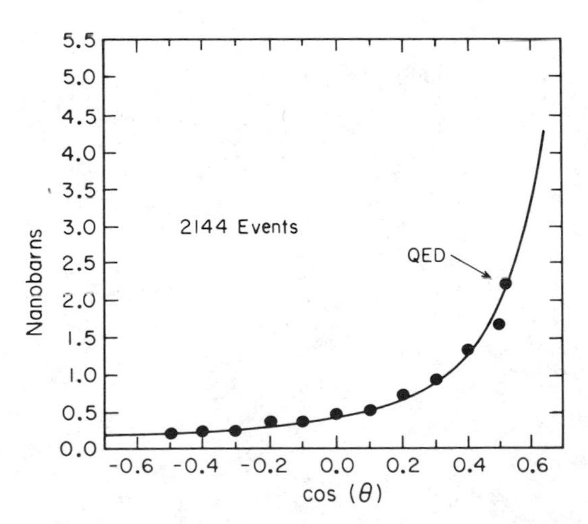

Fig. 19. Bhabha Distribution in Barrel Shower Counter

We have some preliminary distributions on Bhabhas at 29 GeV center of mass energy. Fig. 18 shows the distribution of Bhabhas measured down to $\cos\phi = .9$ in the end caps. A similar distribution is shown for the barrel shower counters in Fig. 19. Studies of any possible deviation from QED are in progress, as are determinations of muon asymmetries.

Future plans

We are now beginning to produce physics with the HRS. We anticipate that continuing improvements in PEP operations could allow $\sim$100-180 pb^{-1} for each of the next two years. This data sample size should provide a rich field of study. We also plan on incorporating a tracking chamber near the intersection point to facilitate in track finding and to veto cosmic rays. This will be installed in Fall, 1983.

STATUS AND FUTURE PLANS FOR THE MARK II DETECTOR AT SLAC*

Martin L. Perl

Stanford Linear Accelerator Center
Stanford University
Stanford, California 94305

SECTION I. INTRODUCTION AND STATUS OF PEP

In this brief talk, I want to report to you on three subjects. First, I want to tell you about the present status of PEP, where there has been a very large increase in the luminosity in the past five months. Next, I want to tell you a little about the present status of the Mark II detector, whose secondary vertex detector constitutes a very important part of the physics which our collaboration is doing at PEP. Finally, I want to review for you the design of the upgraded Mark II Detector which will be used at the Stanford Linear Collider (SLC).

In the summers of 1981 and 1982 the final focus quadrupoles at PEP were moved nearer to the interaction points; the distance of the nearest quadrupole from the interaction point being decreased from 11 meters to about 7 meters. In the fall of 1983 extensive machine physics studies and accelerator physics calculations were carried out for this new arrangement of quadrupoles. In January of 1983 a configuration was attained which has

*Work supported in part by the Department of Energy, contracts DE-AC03-76SF00515 and DE-AC03-76SF00098.

SLAC-PUB-3108
April 1983
(E)

resulted in excellent luminosity parameters for PEP. These luminosity parameters are:[1]

$$L_{max} = 3.2 \times 10^{31}\ cm^{-2}\ s^{-1}$$

$$L_{average,typical} = 1000\ nb^{-1}/day \qquad (1)$$

$$L_{average,maximum} = 1500\ nb^{-1}/day$$

These substantial accomplishments are described in detail by a recent paper of R. Helm et al.[1] In that paper the authors also describe three other factors which have contributed enormously to these improvements. These factors are: improvements in instrumentation and software; superb efforts by the PEP operators to empirically adjust many parameters of beam operation and to continually increase the productivity of PEP; and the finding of so-called 'golden orbits'.[2]

This improved luminosity has produced, of course, substantial improvements of the rate at which the detectors at PEP can acquire data. For example, our Mark II detector in the first week of April passed the 100,000 nb^{-1} mark in acquiring data. And so, if you will pardon a bit of poetics, our dream of being able to collect 100,000 nb^{-1} a year has now become a reality.

SECTION II. THE STATUS OF THE MARK II DETECTOR AT PEP

The physicists now operating and using the Mark II detector at PEP are listed in Table I. There are, of course, other physicists who have worked with the Mark II collaboration in the past and who have contributed substantially to the construction and operation of the Mark II detector at SPEAR and at PEP.

The major improvement in the Mark II detector since we moved it to PEP has been the insertion of a high spatial resolution drift chamber inside the main drift chamber. The details of the construction and use of this secondary vertex chamber have been described by J. Jaros.[3,4] Some details of the chamber construction and performance are shown in Figs. 1 and 2, and are listed in Eq. 2.

length = 120 cm

average radius of 4 inner layers = 12 cm

average radius of 3 outer layers = 30 cm (2)

measurement accuracy in actual use in Mark II = about 90 μm/layer

TABLE I: Members of the Mark II Collaboration at PEP as of April, 1983

SLAC	LBL
A. Boyarski	G. Abrams
M. Breidenbach	D. Amidei
P. Burchat	A. Baden
D. Burke	C. De La Vaissiere
J. Dorfan	G. Gidal
G. Feldman	M. Gold
G. Hanson	G. Goldhaber
C. Matteuzzi	L. Golding
R. Hollebeek	D. Herrup
L. Gladney	I. Jurici
W. Innes	J. Kadyk
J. Jaros	M. Nelson
R. Larsen	P. Rowson
B. LeClaire	H. Schellman
A. Lankford	P. Sheldon
N. Lockyer (Spokesman)	G. Trilling
V. Lüth	
R. Ong	
M. Perl	
B. Richter	**HARVARD**
M. Ross	M. Levi
J. Yelton	R. Schwitters
C. Zaiser	T. Schaad

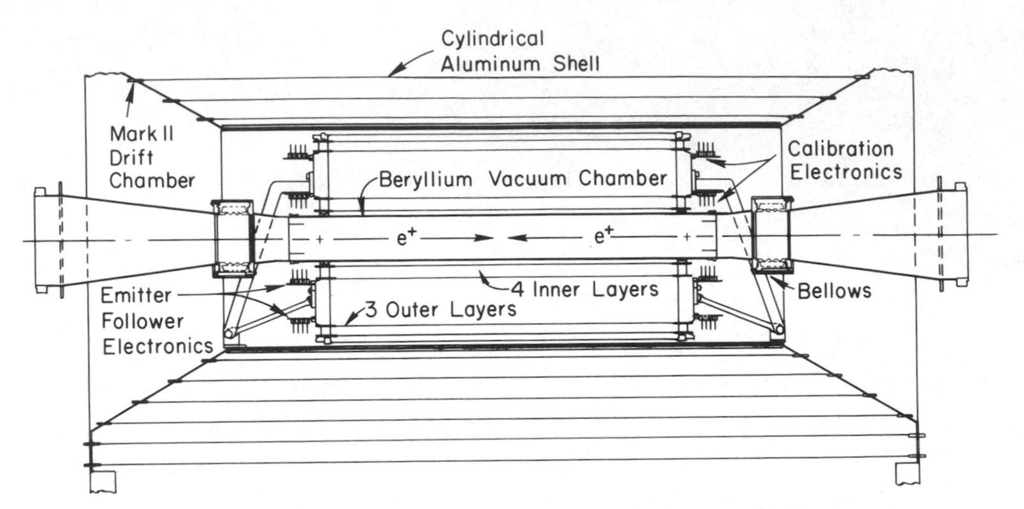

Fig. 1. Cross section of the secondary vertex chamber as installed in the Mark II Detector at PEP.

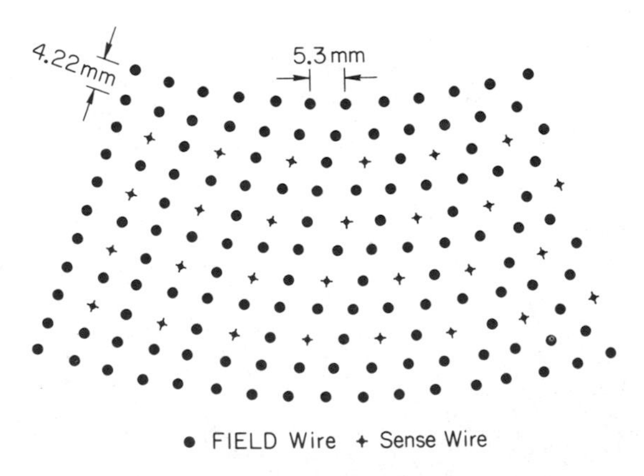

Fig. 2. Arrangement of wires in the 4 inner layers of the secondary vertex detector.

The secondary vertex detector, as its name implies, has been used to look for secondary vertices to measure the lifetimes of short-lived particles. A new measurement of the lifetime of the tau using this detector has already been presented[4] and work is progressing on further measurements of the lifetime of the tau, on a measurement of the lifetime of B mesons, and on further studies of the lifetimes of the D mesons. We are also beginning to explore how to use secondary vertices to tag those events which have short-lived particles in them. Although physicists at present and future colliding beams facilities dream of having a reliable secondary vertex, on-line, trigger; it is clear to us through our present work with this secondary vertex detector that there is a large amount of technical improvements and analytical studies which have to be done to realize that dream.

As some of you know, the Mark II detector is now operating at approximately one-half its designed field, about .24 T rather than .5 T. We have had to operate that way because in the spring of 1982 a short developed between the inner and outer layers of the magnet coil; and we continued operating by just powering the outer layer. We do not know the reason for the short, it occurred after operating the coil for almost eight years without any problems. We do know how to return the coil to full field and we are now considering the appropriate time to do so.

Our present plans are to continue to acquire data at PEP with the Mark II detector until the summer of 1984. At that time we plan to make some substantial changes in the Mark II detector to upgrade it for use at the Stanford Linear Collider. The plans for that upgrading are the subject of the next and final section of this talk.

SECTION III. THE DESIGN OF THE UPGRADE OF THE MARK II DETECTOR

A. Introduction

A collaboration, Table II, is being formed to upgrade the Mark II detector for use at the SLC. Table III lists the components of the Mark II detector which will be retained through the upgrade and those components which will be replaced. As you will note, the two major replacements are a new drift chamber and its associated electronics, and new electromagnetic shower detecting endcaps. Another replacement item is that new time-of-flight, scintillation counters will be installed. Finally, the present secondary vertex detector is not suitable for use at the SLC and will have to be replaced.

TABLE II: Physicists and Engineers (E) who are upgrading the Mark II Detector for use at the SLC as of April, 1983

<u>LBL</u>

G. Abrams, G. Gidal, G. Goldhaber, D. Herrup, J. Kadyk, P. Sheldon, G. Trilling (Spokesman), K. LEE (E), M. Nakamura (E)

<u>Caltech</u>

B. Barish, G. Fox, T. Gottschalk, C. Peck, R. Stroynowski, R. Cooper (E)

<u>U.C.Santa Cruz</u>

D. Dorfan, C. Heusch, H. Sadrozinski, T. Schalk, A. Seiden, W. Nilsson (E)

<u>University of Hawaii</u>

R. Cence, F. Harris, S. Parker, D. Yount

<u>University of Michigan</u>

C. Akerlof, J. Chapman, D. Meyer, D. Nitz, R. Thun, A. Seidl

<u>SLAC - Group A</u>

W. Atwood, H. DeStaebler, R. Pitthan, L. Rochester

<u>SLAC - Group C</u>

A. Boyarski, F. Bulos, J. Dorfan (Spokesman), R. Hollebeek, A. Lankford, R. R. Larsen, V. Lüth

<u>SLAC - Group E</u>

D. Burke, G. Feldman (Spokesman), G. Hanson, W. Innes, J. Jaros, M. Perl

<u>SLAC - Group BC</u>

J. Ballam, T. Carroll, T. Glanzman, C. Field, K. Moffeit

<u>SLAC Engineers</u>

B. Denton (E), D. Horelick (E), C. Hoard (E), D. Hutchinson (E), D. Porat (E)

Table III: Changes to be made in the Mark II Detector to upgrade it for use at the Stanford Linear Collider

Component	Changes
Secondary vertex detector	Will be changed; no design concept has been selected.
Main drift chamber	New drift chamber being built. Will use new timing electronics and new dE/dx electronics.
Time-of-flight scintillation counters	Scintillator is being replaced, the electronics is being retained.
Liquid argon, barrel, electro-magnetic shower calorimeter	Entire system is being retained, minor changes may be made in the electronics.
Magnet	The iron work is being retained; the coil is being replaced.
Muon detection system	Entire system is being retained.
Endcap electromagnetic shower calorimeters	New endcaps being built. Will use new electronics.
Data acquisition and calibration system	Being retained except for changes required by new drift chamber and new endcap.
On-line and off-line data analysis programs	Being retained except for changes required by new drift chamber, new endcaps and higher energy.

B. Schedule for the Upgrading of the Mark II Detector

Our present plans are as follows. In the summer of 1984 we plan to begin to replace the present drift chamber by a new drift chamber and its associated electronics. In the fall or early winter of 1984 we plan to resume operation of the Mark II detector at PEP. This resumption of operation of the detector at PEP has two purposes. First, we want to put the new drift chamber and its associated electronics into operation under running conditions. Second, we would like to acquire additional data at PEP. During this second period of operation of the Mark II detector at PEP we also plan to use the new electromagnetic shower endcap detectors. The moving of the Mark II detector from PEP to the SLC is now proposed to take place in the early part of 1986. The proposal is to assemble the detector off-line at the SLC and to check it out on cosmic rays. Finally, we hope that we can begin operation of the detector at the SLC in early 1987.

C. The New Drift Chamber

Figure 3 presents a schematic horizontal cross section of the upgraded detector. The new drift chamber is designed to meet the following objectives: excellent momentum resolution, good solid angle coverage, high tracking efficiency, simplicity of pattern recognition, and measurement of dE/dx to provide an independent aid to calorimetry for electron-hadron separation. The chamber is a large cell design, Fig. 4, with 6 sense wires in each cell. There are 12 concentric layers of these cells. Alternate layers have their wires parallel to the axis or at a stereo angle of $\pm 3.5^{\circ}$. Figure 5 shows the overall layout. The length of the chamber is 2.3 meters.

The chamber provides good tracking efficiency for the angular region $|\cos\theta| \leq 0.89$. The expected position accuracy per wire is $\lesssim 200$ μm; and the expected impact parameter error for track extrapolation to the beam line is 150 μm for particles which traverse most of the chamber. A precision σ_p/p^2 of 0.12% GeV^{-1} is expected with vertex constraint for $|\cos\theta| \lesssim 0.65$. This increases to 0.45% GeV^{-1} at $|\cos\theta| = 0.85$.

There are seventy-two samples of track ionization, which we calculate will provide a resolution of about 7%. Our objective is to aid the separation of pions from electrons, we calculate that this separation will be better than three standard deviations up to about 9 GeV/c.

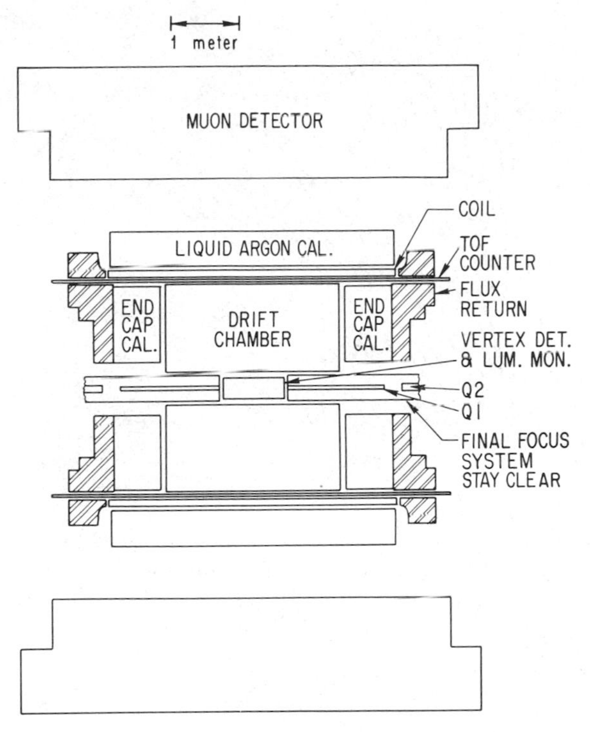

Fig. 3. Schematic, horizontal cross section of the upgraded detector.

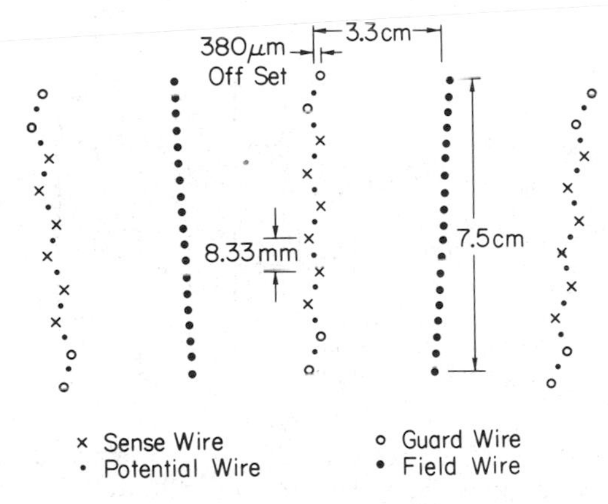

Fig. 4. Cell structure of the new drift chamber.

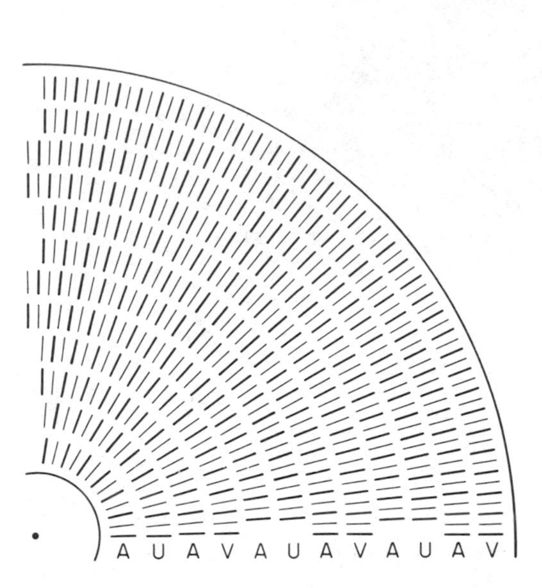

Fig. 5. The layer structure of the new drift chamber. Axial layers are indicated by A; stereo layers by U and V.

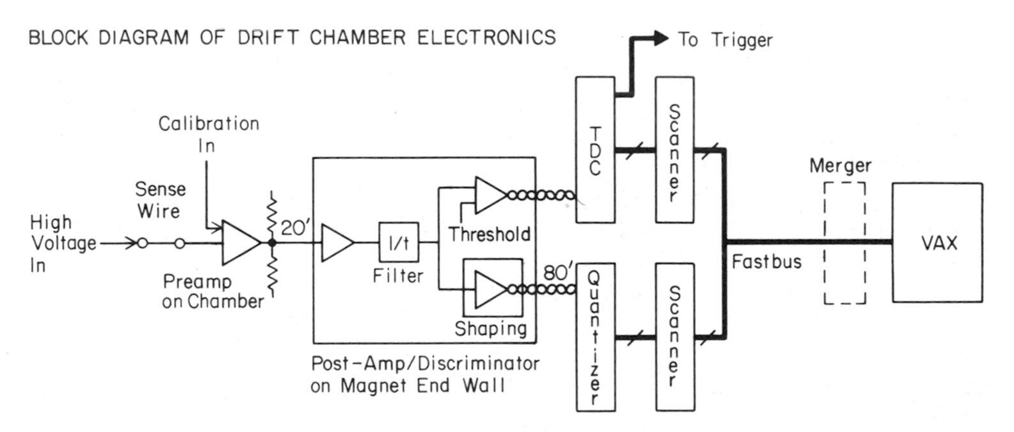

Fig. 6. Schematic of electronics for drift chamber. After the post-amp/discriminator, the upper electronics path is for processing the drift time measurement, the lower electronics path is for processing the dE/dx measurement.

Figure 6 is a schematic diagram of the new electronics to be associated with the drift chamber. Further details on the electronics and on mechanical construction can be obtained from the technical proposal[5] entitled "Proposal for the Mark II at SLC" dated April 1983.

D. The New Electromagnetic Shower Detector Endcap

The new endcaps are intended to match in performance the liquid argon, barrel, electromagnetic shower calorimeters which we are retaining. The endcap construction consists of aluminum proportional tubes sandwiched between lead sheets. Each endcap will have 36 layers of lead sheets and proportional tubes each layer corresponding to 0.5 of a radiation length. The proportional tubes are rectangular, 0.9 cm x 1.5 cm in cross section, and are arranged to lie along four different coordinate axes. We expect to obtain an energy resolution for incident photons and electrons of about $18\%/\sqrt{E}$ for energies between 1 and 50 GeV. The ability of the endcaps to separate electrons from pions is expected to be equal to that ability in the present liquid argon, electromagnetic shower detector system.

E. Time-of-flight System

The scintillation counter, time-of-flight system is being modified in order to improve the segmentation by a factor of 2. Ninety-six counters will now be used, and a resolution equal to, or better than 200 ps is expected.

F. Precision Vertex Detection

As I noted before we cannot use the present secondary vertex detector at the SLC. Three different methods for precision vertex detection are being explored. One method is an extension of our present secondary vertex detector; that is, we are exploring the use of a pressurized, very high-precision, drift chamber. The goal is 30 μm resolutions per layer with an ultimate impact parameter precision of 20 μm. The second method being explored is the use of silicon microstrip detectors. These would have a position precision of 5 μm, a double-track resolution of 30 to 50 μm, and an impact parameter resolution of about 15 μm. A preliminary design has about 30,000 strips arranged in three concentric layers at radii between 1.5 and 4 centimeters. The third method which is being studied would use solid state charge-coupled optical imagers called CCD arrays. The resolution would be of the order of 15 μm. A three-layer detector would require about 300 chips, each about 1 cm^2 in area.

G. Small-Angle Detector

It is important to be able to detect electrons at angles smaller than those encompassed by the main drift chamber. For that reason the detector will have a small-angle electron detector and luminosity monitor. Figure 3 shows one possible position for this device. An alternate position would be between the beam pipe and the endcaps.

H. New Magnet Coil

The upgraded detector will have a new room temperature, aluminum coil which will produce a field of 0.5 tesla.

I. Acknowledgment

In this very brief talk I have not been able to do justice to the immense amount of work being done on the Upgrade of the Mark II by the physicists and engineers listed in Table II. Reference 5, our proposal, gives much more detail.

REFERENCES

1. R. Helm et al., SLAC-PUB-3070 (1983).
2. To quote from Ref. 1 on 'golden orbits':
   "Once a good configuration has been established, it is always found that the orbit which works best is not the 'best' orbit as indicated by the position monitors. Evidently the inspired green-thumbing by the operators eventually compensates for effects such as position monitor errors, local phase and β errors, dispersion functions, and x-y coupling. A very useful feature of the orbit correction program is the ability to save the 'golden orbit' and subsequently correct to it rather than to the 'best' orbit."
3. J. Jaros, Proc. Int. Conf. on Instrumentation for Colliding Beam Physics (SLAC Report 250, 1982), p. 29.
4. J. Jaros, SLAC-PUB-3044 (1983).
5. Caltech, LBL, U.C.Santa Cruz, Univ. of Hawaii, Univ. of Michigan Collaboration; "Proposal for the Mark II at SLC"; April 1983; CALT-68-1015.

RECENT RESULTS FROM MARK-J AT PETRA

José Salicio

Junta de Energia Nuclear, Madrid/Spain
and
Deutsches Elektronen-Synchrotron DESY
Hamburg/Fed. Rep. of Germany

INTRODUCTION

We can test the Standard Electroweak Theory[1] at high center of mass energies and at high q^2 by studying e^+e^- interactions at the colliding beam facility PETRA at DESY. The initial state is clean since only point-like particles are involved. The dynamics of the reactions $e^+e^- \rightarrow f\bar{f}$, where f can be an e, μ, τ, or a quark, can be studied, providing the tools to check the theory.

I will first present recent results from the MARK-J Collaboration on inclusive muons concerning the measurement of semileptonic branching ratios and fragmentation functions for c and b quarks and an upper limit for the branching ratio $Br(b \rightarrow \mu^+\mu^- X)$.

Results on the measurement of the forward-backward charge asymmetry of the $e^+e^- \rightarrow \mu^+\mu^-$ and $e^+e^- \rightarrow \tau^+\tau^-$ channels at an energy of $\sqrt{s}$ = 34.6 GeV are presented in the second part.

The third and the fourth parts will be devoted to a high order QED test (α^4) at large momentum transfers (Q^2 = 100 GeV^2) and to a search for the top quark setting a limit m_t > 19 GeV.

INCLUSIVE MUONS

The standard electroweak interaction model uses the SU(2)xU(1) gauge group coupled with the 3 left handed quark doublets. In this context quarks can be mixed using the Kobayashi-Maskawa mechanism[2], thus the b quark would mix with the d and s quarks and belong together with the "top" quark to a SU(2) doublet. Quarks would decay through charged W bosons to leptons or lighter quarks. Branching ratios $c \to \mu + X$, $b \to \mu + X$ are interesting in order to get information about the Kobayashi-Maskawa[2] matrix elements.

The 6-quark standard model forbids flavor changing neutral current (FCNC) processes. Nevertheless, there are alternative models allowing this kind of process in which the b quark stays alone in a SU(2) singlet without the "top quark". Recent work of Barger and Pakvasa[3], Kane and Peskin[4], show, under the assumption of b decaying through W and Z bosons and belonging to a singlet representation of the weak symmetry SU(2), that the $b \to \ell^+\ell^- X$ decay and other FCNC processes must happen with branching ratios of the order of 1%.

Measurement of Branching Ratios and Fragmentation Functions

The inclusive μ candidate sample has been taken from a hadronic sample corresponding to an integrated luminosity of $\int L dt \sim 79\ pb^{-1}$ and an energy of $\sqrt{s} = 34.6$ GeV. The resulting sample is composed of 850 events in which we have a background of 19% from decays (e.g. $\pi^\pm \to \mu^\pm \nu$, $k^\pm \to \mu^\pm \nu$) and 13% of punch through (hadrons going through the detector reaching the outer chambers).

In Fig. 1 we show the μ transverse momentum distribution with respect to the thrust axis (P_t) for the data sample (full dots). In full line we show the Monte Carlo prediction including QCD effects. With dashed and dotted-dashed lines, the relative contribution of c and b quarks is shown. The big difference observed between the c and b quarks contributions to the P_t distribution, suggests this as good to establish a tag between the 2 flavors.

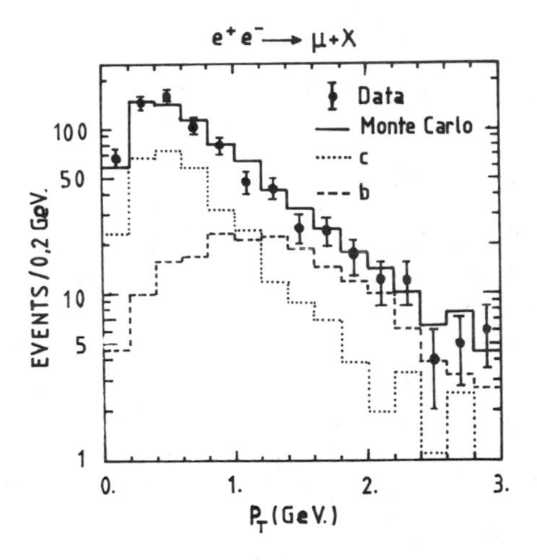

Fig. 1

Information about the energy of the primordial heavy meson[5] is contained in the μ-momentum distribution (P). We define the fragmentation variable z as the fraction of the energy of the original quark carried by the primordial heavy meson, i.e. $z = (E + P_{//})_H/(E + P_{//})_Q$. Peterson et al.[6] under the assumption that the production probability of a primordial heavy meson in the fragmentation process is inversely proportional to the square of the energy transferred from the heavy quark to the heavy meson, propose the following parametrization for the c and b quark fragmentation function:

$$F_Q(z) = \frac{1}{z\,(1 - \frac{1}{z} - \frac{\varepsilon_Q}{z})^2} \qquad (1)$$

where ε_Q is proportional to the squared ratio between light and heavy quark masses. Mixing the relative contributions of the different flavors as a function of the branching ratios Br($c \to \mu + X$), Br($b \to \mu + X$) and modifying the P distribution as a function of the parameters ε_c and ε_b, we obtain the values of the parameters which better fit the samples.

Results on Branching Ratios and Fragmentation Parameters

The Monte Carlo we use includes radiative corrections[7], a full detector simulation and the assumption that B decays via $B \to C + X$ only.

The fit between data and Monte Carlo events was made using a likelihood method in a 2-dimensional plot P vs. P_t divided into 20x20 bins in the region $0 < P < 20$ GeV and $0 < P_t < 10$ GeV.

The results coming from the fit are:

$$\mathrm{Br}(c \to \mu + X) = (8.7 \pm 0.9 \pm 2.2)\%$$
$$\mathrm{Br}(b \to \mu + X) = (10.7 \pm 1.4 \pm 1.8)\%$$

$$\varepsilon_c = 0.52 \pm 0.22 \pm 0.12$$
$$\varepsilon_b = 0.023 \pm 0.017 \pm 0.047$$

with an average z for the fragmentation distributions of $\langle z_c \rangle = 0.48$ and $\langle z_b \rangle = 0.74$.

The shape of the c and b quarks fragmentation functions os shown in Fig. 2. We observe that both of them are hard functions (especially for b), showing that heavy mesons produced in the fragmentation

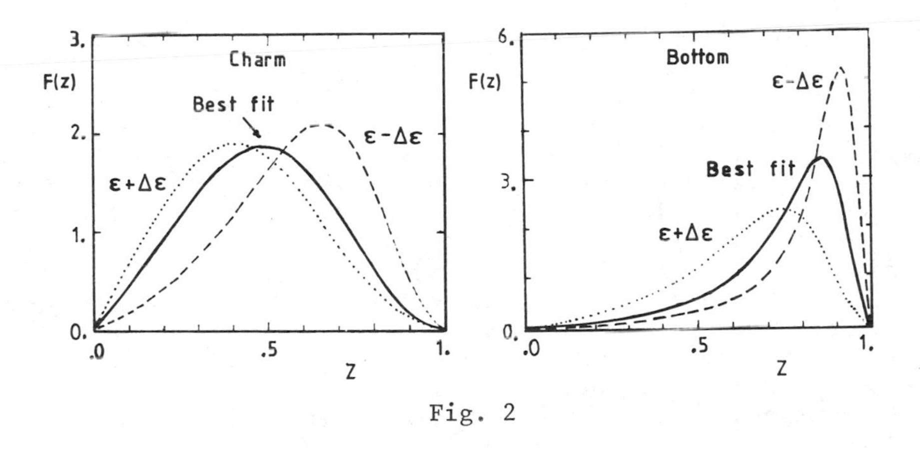

Fig. 2

process prefer carrying most of the original quark momentum. The fit stability was tested by changing the set of initial values of the parameters in a wide range. The fit sensitivity can be observed in Fig. 2 where the fragmentation functions corresponding to $\varepsilon \pm \Delta\varepsilon$ are shown. A study of systematic errors has been done by changing the distributions to fit and changing the contributions of backgrounds.

Test on Flavor Changing Neutral Currents (FCNC)

In the study of the decay $b \to \mu^+\mu^- X$ we look for both muons in the same hemisphere defined by the perpendicular plane to the thrust axis.

We accept events with an opening angle between both muons bigger than 15° and each muon momentum being bigger than 1.5 GeV. This procedure reduces significantly the background without noticable reduction of the signal. With these cuts our acceptance for the $\mu^+\mu^-$ decay of b quark is 17%.

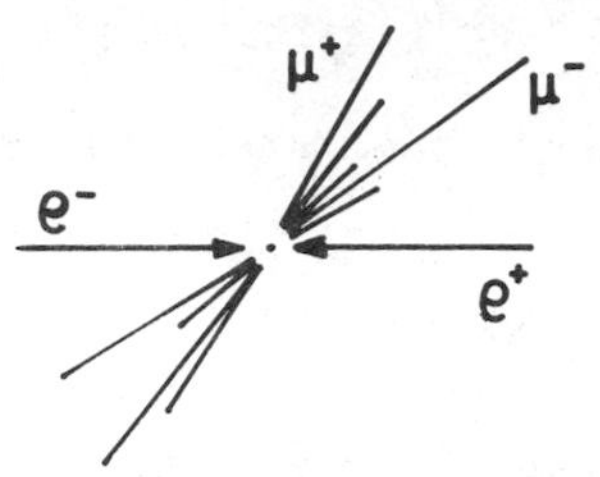

In a sample of 26,000 hadronic events with energies $\sqrt{s} > 30$ GeV we find 2 events. From conventional sources we expect 1.5 events while from FCNC the expected number of events is 18 if the branching ratio $Br(b \to \mu^+\mu^- X)$ is $\sim 1.2\%$ as predicted in some models[4].

Then we put an upper limit with the 95% C.L. to the branching ratio:

$$Br(b \to \mu^+\mu^- X) < 0.7\%$$

which rules out a large class of topless models in which the b quark is alone in a weak singlet, suggesting that the b quark is in a weak SU(2) doublet together with the "top" quark.

LEPTON PAIRS

The production of lepton pairs in e^+e^- annihilations is a clean tool to test electroweak theories. With the MARK-J detector we can measure the cross section and the forward-backward charge asymmetry with high precision. The deviation of the cross section from pure QED depends mainly on the vector coupling constant g_V, while the forward-backward charge asymmetry (which shows the interference between diagrams mediated by γ and Z^0) depends on the axial vector coupling constant g_A. Our measurements of the cross section and charge asymmetry allow us (in lowest order) to determine the coupling constants g_V and g_A and to test the universality of leptons in weak interactions up to an energy of $\sqrt{s} \sim 35$ GeV.

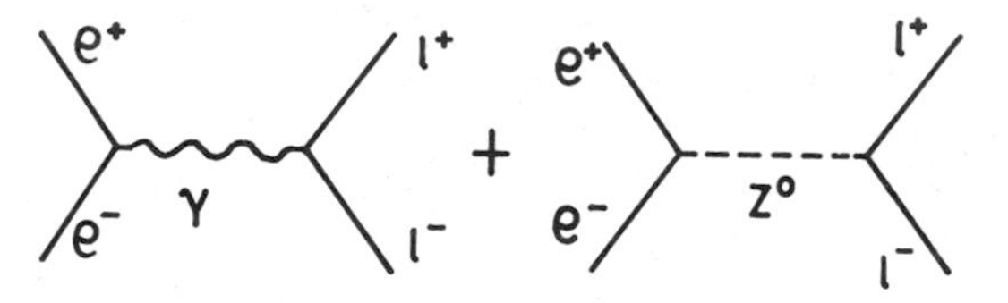

Cross Section and F-B Charge Asymmetry

The quantities we define in this paragraph are valid for both μ-pairs and τ-pairs.

We define $R_{\ell\ell}$ as the ratio between the lepton pair production cross section and the point-like cross section.

$$R_{\ell\ell} = \frac{\sigma_{\ell\ell}}{\sigma_{pt}}, \quad \sigma_{pt} = \frac{4\pi\alpha^2}{3s} \tag{2}$$

Including the weak interaction according to the standard model in lowest order, we may express $R_{\ell\ell}$ as a function of the vector and axial-vector coupling constants:

$$R_{\ell\ell} = 1 + 2g_V^2\chi + (g_V^2 g_A^2)^2\chi^2 + \ldots. \tag{3}$$

where the propagator term $\chi = (G_F/2\sqrt{2}\pi\alpha).s.\ m_Z^2/(s-m_Z^2)$, is $\chi \sim -0.26$ for $m_Z = 90$ GeV and $\sqrt{s} = 35$ GeV.

The angular distribution of lepton pairs may be expressed in lowest order in electroweak theories as:

$$\frac{d\sigma}{d\cos\theta} = \frac{\pi\alpha^2}{2s}\left[F_1\ (1+\cos^2\theta) + F_2\cos\theta\right] \tag{4}$$

where $F_1 = R_{\ell\ell} \sim 1$ and $F_2 = 4g_A^2\chi + 8g_V^2 g_A^2\chi^2$ and θ is the scattering angle defined as the angle between the e^- (beam) direction, and the direction of the negatively charged produced lepton, as shown in Fig.3.

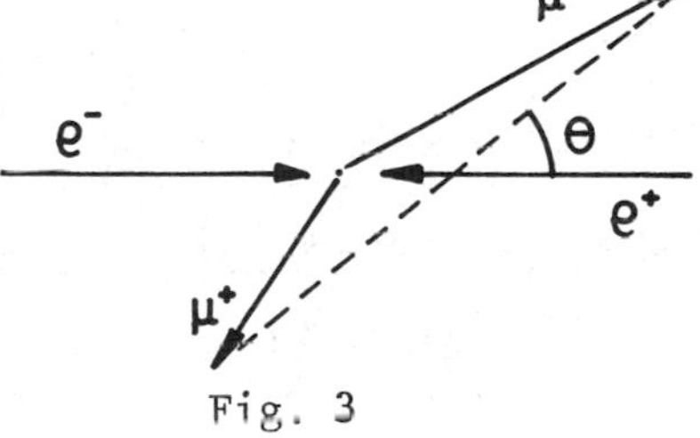

Fig. 3

The non-symmetric term "$F_2\cos\theta$" in equation (4) produces a measurable forward-backward charge asymmetry A^W. After integration over the full range of $\cos\theta$:

$$A^W = \frac{3}{8}\frac{F_2}{F_1} = \frac{2.7 \times 10^{-4}}{\mathrm{GeV}^2}\ g_A^2\ \frac{s\cdot m_Z^2}{s-m_Z^2} + \ldots + g_V^2 g_A^2 \tag{5}$$

The measured charge asymmetry as a function of θ is defined to be:

$$A_{meas} = \frac{N_{\bar{\ell}}(\theta) - N_{\bar{\ell}}(\pi-\theta)}{N_{\bar{\ell}}(\theta) + N_{\bar{\ell}}(\pi-\theta)} \tag{6}$$

What is usually quoted as charge asymmetry is the integral over the angular range: $A_{meas} = (F-B)/(F+B)$, where F and B stand for the number of ℓ^- in the forward and in the backward side of the detector.

To compare the measured asymmetry with the lowest order electroweak expectation, we correct the measured distributions subtracting the small and positive QED asymmetry A^{QED} arising from the α^3 radiative corrections[8].

$$A_{meas} = A^W_{meas} + A^{QED} \tag{7}$$

MUON PAIRS

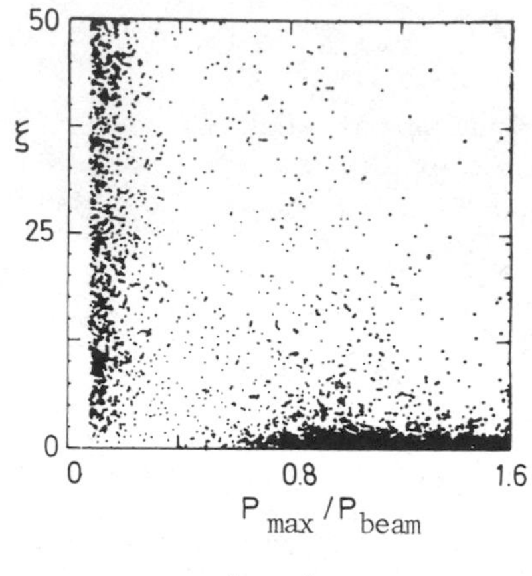

Fig. 4

The selection of μ-pairs requires 2 isolated tracks going through the detector reaching the outer chambers. The cosmic ray background is eliminated by requiring both μ-tracks to be in time in the μ-trigger counters. The final sample is obtained after applying an acollinearity cut $\xi < 20^o$ and requiring the maximum momentum of both μ's to be bigger than the 50% of P_{beam}.

With these cuts we get a clean sample of $\mu^+\mu^-$ as shown in Fig. 4 where there are two clearly separated regions. The one in the left side corresponds to 2-γ μ-pair production that we shall study later on. The other region corresponds to the events in which we are interested now.

Results on Muon-Pair Cross Section

The values of $R_{\mu\mu}$ together with the integrated luminosity and the number of μ-pairs are presented below. The systematic error in the cross section is dominated by the luminosity error of ∿ 3%. The other contributions to the error are of the order of 1%.

No deviation from QED is observed within errors. The data points lie around the QED prediction R = 1. If we assume the validity of QED we may set a limit in the vector coupling constant: $g_V^2 = 0.06\pm0.08$ which is compatible with zero.

$\sqrt{s}$ GeV	$R_{\mu\mu}=\sigma_{\mu\mu}/\sigma_{pt}$	$\int Ldt$ (pb^{-1})	$N_{\mu\mu}$
14.0	1.04±0.05	1.57	494
22.0	1.02±0.05	3.12	353
34.6	1.04±0.02	76.32	3658

Systematic Errors to the Muon Charge Asymmetry

The acceptance in the polar angle is very high, ∿ 90% and it is flat in the range $|\cos\theta| < 0.8$ (Fig. 5). That is important because we do not have to correct for acceptance.

Only 24 events out of ∿ 3600 have the same charge assigned to both of the muons which is less than 1%. Then, the double charge confusion is less than 0.01%. As a further check of this the data were processed by two completely independent teams using different event selection and momentum fitting procedures. Event by event comparisons of the two results indicate excellent agreement.

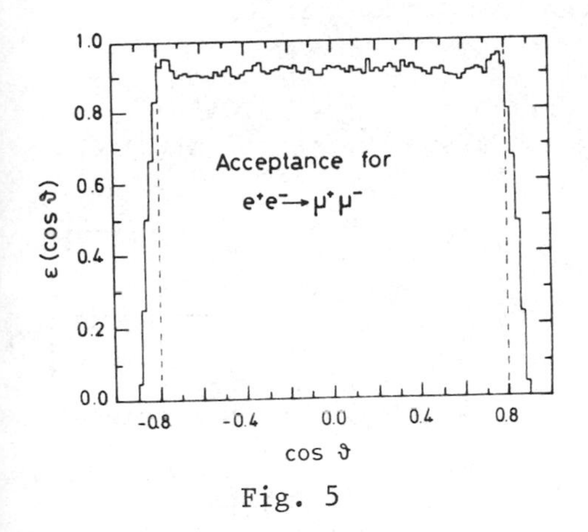

Fig. 5

The muon momentum resolution is of the order of 30% and is independent of the charge involved and of the back or front side position in the detector as shown in Fig. 6 where gaussian fits have been done to the P_{beam}/P_{μ} distributions of data. Already it is possible to see the manifestation of the charge asymmetry: there are more μ^- in the back than in the front side of the detector.

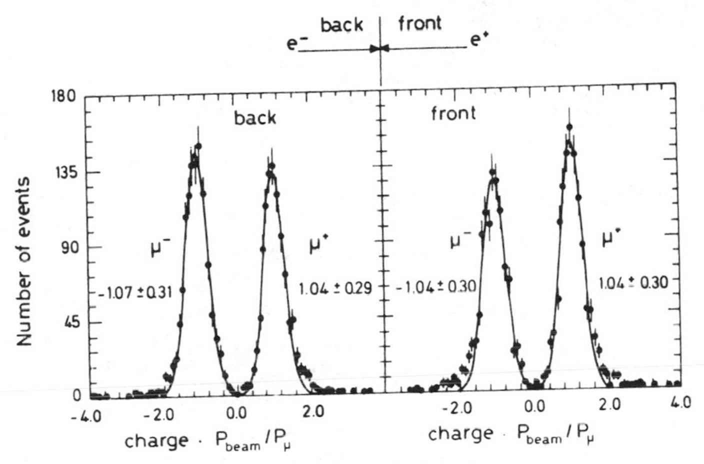

Fig. 6

The results of a study[2] of the detector asymmetry as a function of the polar angle using 20,000 cosmic ray muons is shown in Fig. 7 and gives

$$A_{det} = (-0.2\pm1.0)\%.$$

In addition this detector asymmetry is eliminated collecting equal amounts of data at both magnet polarities, which cancels all systematic errors related to the charge measurement of μ's.

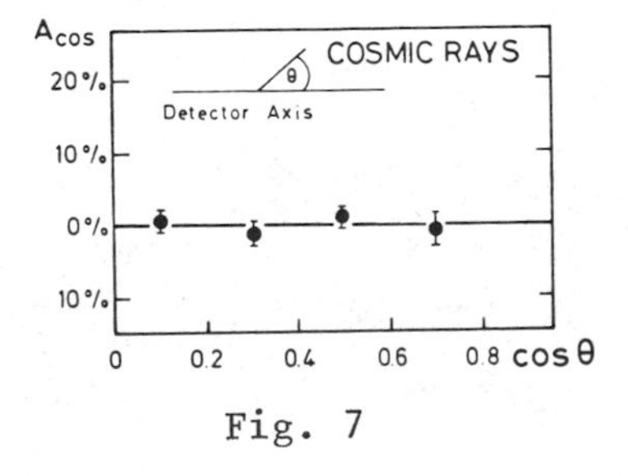

Fig. 7

The background coming from processes like τ-pair production, 2-γ μ-pairs. cosmic rays, etc., have been taken into account and is summarized in the table shown below.

From these studies we conclude that by taking equal amounts of data at both magnet polarities the total systematic error affecting the charge asymmetry is less than 1%.

Source	Measured Value(%)	
Affecting Total Muon Rate:		
Trigger Related	< 0.1	Redundant Triggers
Selection	< 0.5	2 Independent Selections
Acceptance Asymmetry:		
Double Charge Confusion	< 0.1	"like sign" Rate
Detector Asymmetry	0.2±1.0	from Cosmic Ray Analysis
Background Processes:		
Cosmic Ray Muons	< 0.1	Cosmic Ray Studies
2-Photon Muons	0.1±0.1	Monte Carlo, checked by data
Other (taus, Bhabhas, etc)	negl.	Monte Carlo, Event scan

Results on Muon Pair Asymmetry

The total asymmetry A_{mass} resulting from the simple counting of events in the forward and backward hemispheres in the angular range $|\cos\theta| < 0.8$ and $\xi < 20^o$ yields $A_{meas} = (-8.1\pm1.6)\%$. The asymmetry due to QED is $A^{QED} = (+1.4\pm0.1)\%$. Therefore, we observe a non-QED asymmetry:

$$A^W_{meas} = (-9.5 \pm 1.6)\%$$

which is more than 5σ away from zero. The GWS expectation is: $A^W_{expected} = (-7.5\pm0.2)\%$ and within errors both values agree.

If we fit equation (4) to the data as indicated by the solid curve in Fig. 8, we obtain the asymmetry value in all the angular range $-1.0 < \cos\theta < 1.0$. The resulting value is

$$A^W_{fit} = (-11.7 \pm 1.7)\%$$

and the expectation is: $A^W_{expected} = (-8.5\pm0.2)\%$.

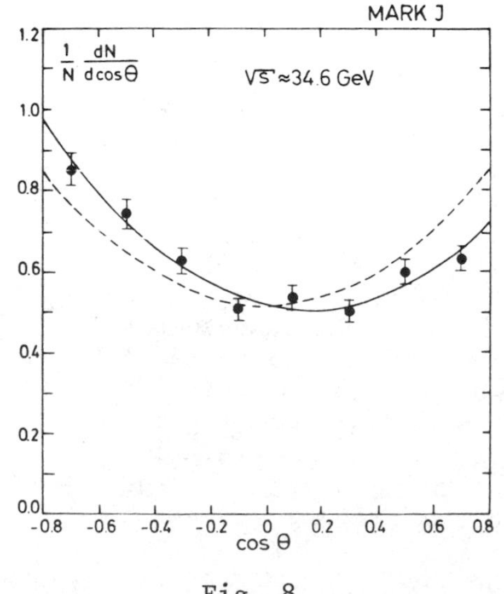

Fig. 8

From equation (5) neglecting high order terms and using the fitted value of the asymmetry and $m_Z = 90$ GeV, we determine

$$g_A^e \, g_A^\mu \quad = \quad 0.31 \pm 0.05$$

From the νe scattering experiments[10] we know $g_A^e = 0.53 \pm 0.06$ (or $g_A^e = 0$). Putting this in the expression above, we get:

$$g_A^\mu \quad = \quad -0.58 \pm 0.12$$

This means that the electron and μ axial coupling constants are equal to the third component of the left handed weak isospin current at the statistical level we are. This constitutes a test of the e-μ universality in weak interactions.

The fitted values of the asymmetry as a function of s, as well as curves corresponding to different masses of the Z^o boson assuming lepton universality are shown in Fig. 9. The highest energy point lies between curves with Z^o mass of 50 and 90 GeV. At the 2σ level we can not yet exclude the $m_Z = \infty$ limit.

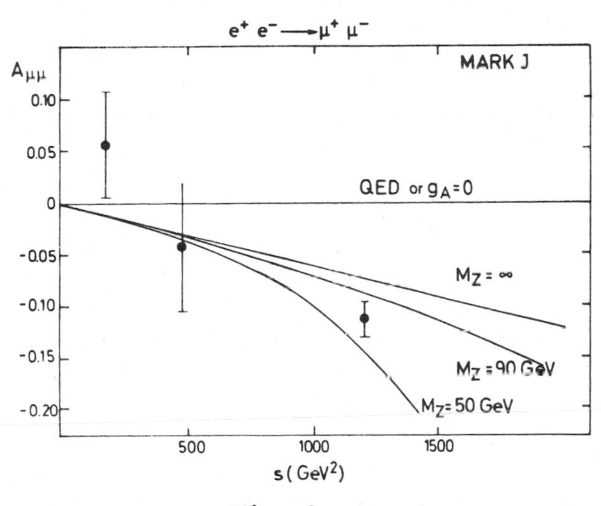

Fig. 9

TAU PAIRS

In order to have a clean τ sample we have chosen processes in which at least one of the τ decays into a μ + neutrinos and the other into e + neutrinos or hadrons + neutrino. We have found 783 events with energies above 30 GeV.

The $\cos\theta$ distribution for τ-pair data together with the $1+\cos\theta^2$ distribution and the best fitted curve are shown in Fig. 10. The value we get from the fit, extrapolated to the total angular range at an average energy $\sqrt{s} = 34.6$ GeV is

$$A^W_{fit} \quad = \quad (-7.8 \pm 4.0)\%$$

and the Glashow-Weinberg-Salam expectation is $A^W_{expected} = -8.5\%$.

Similarly to what we did with μ-pairs (using the electron axial vector coupling constant from νe scattering experiments) we can test

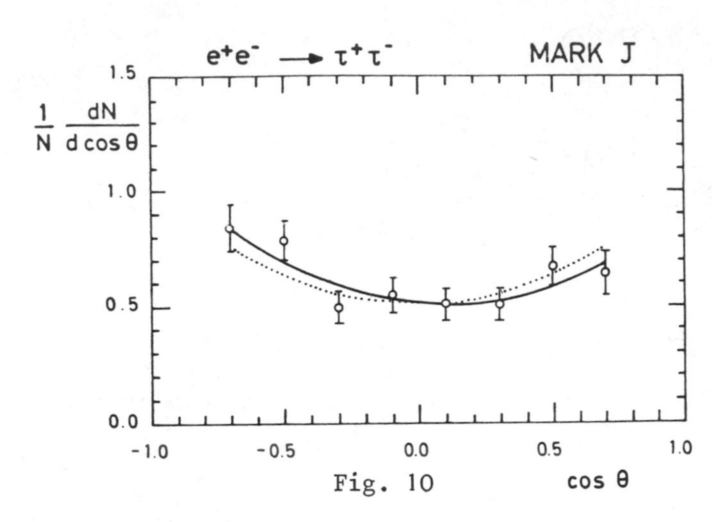

Fig. 10

the e-τ universality

$g_A^\tau = 0.39 \pm 0.25.$

Again this is compatible within errors with the expected value $g_A^\tau = I_3 = -\frac{1}{2}$.

Together with the result from μ-pairs our data confirms the e-μ-τ universality in weak interactions.

BHABHA SCATTERING

In MARK-J we use the Bhabha scattering at small θ angles for the measurement of luminosity. The region of large θ angles can be used to test electroweak theories. In Fig. 11 the quantity:
$\delta = (N_{data} - N_{QED})/N_{QED}$
is shown in each $|\cos\theta|$ bin.

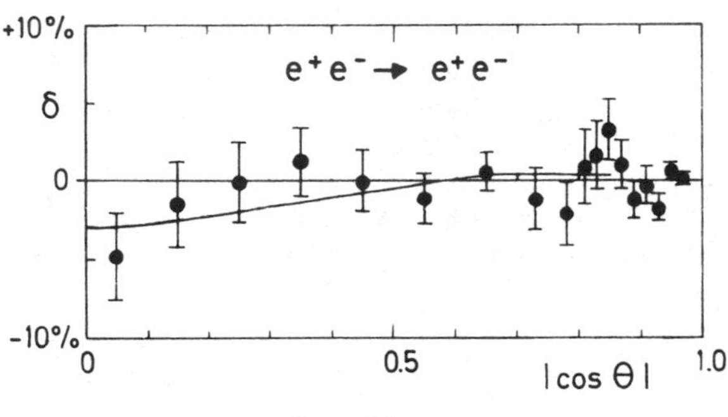

Fig. 11

The QED prediction (straight line at zero) includes up to α^3 radiative corrections according to the generator from Berends and Kleiss[8]. The GWS expectation is the full line. At our statistical level, the data agrees with the electroweak prediction, but does not disagree with the purely QED prediction.

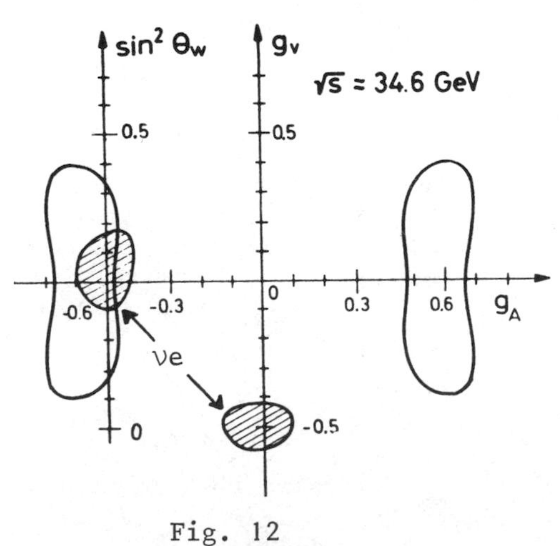

Fig. 12

DETERMINATION OF THE COUPLING CONSTANTS

Combining our results from Bhabha scattering and μ-pair production at an energy $\sqrt{s} = 34.6$ GeV we obtain the 95% C.L. contours for the variables g_A^2 and g_V^2 shown in Fig. 12. The two shaded regions are the 95% C.L. limits coming from three different νe scattering experiments[10]. As we can see the combination of both sets of contours leave only one solution for g_V^2 and g_A^2 which correspond to the values:

$$g_V^2 = 0.04 \pm 0.04$$

$$g_A^2 = 0.35 \pm 0.05$$

for a mass of the Z^0 of 90 GeV. This result is consistent with the standard model ($g_V = \frac{1}{2}(1-4\sin^2\theta_W)$, $g_A = \frac{1}{2}$) shown by the vertical line.

TWO PHOTON PROCESSES

The two photon processes provide a usefull tool to test high order QED and to study resonance production in the two photon exchange reactions. In this paragraph I present results from an experimental test of high order QED (α^4) at large momentum transfers ($q^2 \sim 100$ GeV^2).

The relevant diagrams for the process $e^+e^- \to e^+e^-\mu^+\mu^-$ are the two photon exchange diagrams and the bremsstrahlung diagrams[11] shown in Fig. 13. The biggest part of the signal comes from the observation of coplanar and acollinear μ-pairs. The momentum spectrum of these processes is peaked at small values.

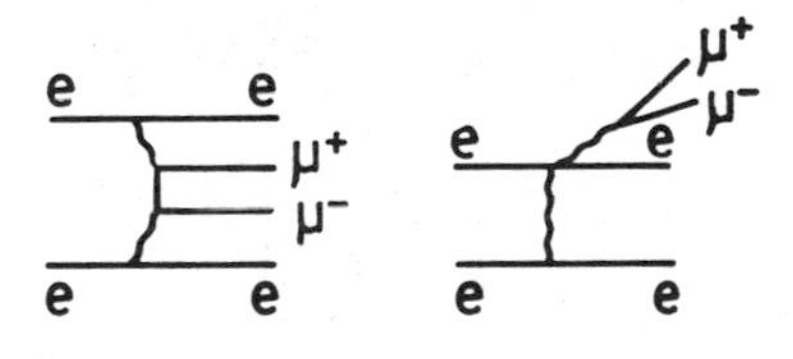

Fig. 13

We have used the Monte Carlo program from Vermaseren[12] which includes the predictions of QED for two photon processes to order α^4. Higher order radiative effects have been estimated by Defrise et al.[13] to be ∿10% and have not been included.

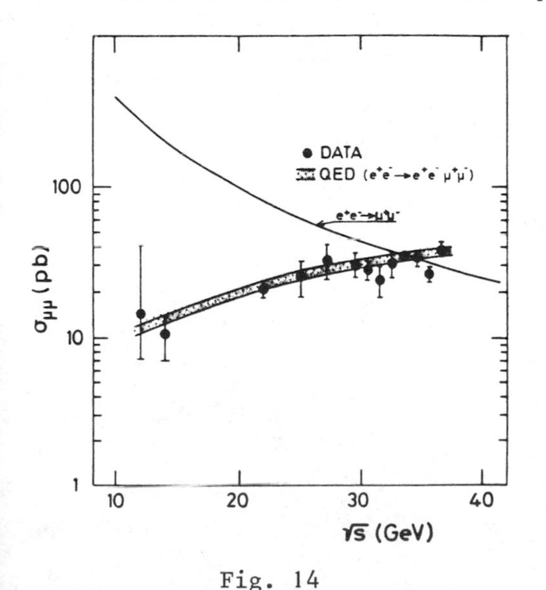

Fig. 14

As we have already seen in the study of μ-pairs, we may separate two photon μ-pair production from one photon μ-pairs (Fig. 4). The remaining background from $e^+e^- \to \mu^+\mu^-$ and $e^+e^- \to \tau^+\tau^-$ has been estimated to be ∿ 3.5%.

The observed cross section for the case in which two μ's are observed is shown in Fig.14 together with the expected (total) cross section for μ-pair production via the one photon process. The predicted (observed) cross section for two photon μ-pairs is shown with the shaded band. The agreement with the QED prediction is very good.

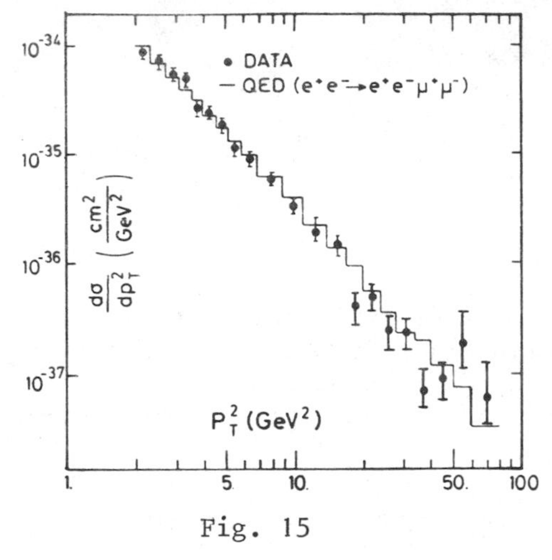

Fig. 15

The quantity $d\sigma/dP_t^2$, where P_t is the transverse momentum of the μ with respect to the beam line is shown in Fig. 15. We have fitted our measurement to the form A. P_t^{-B} in the region $P_t^2 > 3.2$ GeV^2 giving A=(8.3±1.3) nb/GeV^2 and B=4.9±0.2 with χ^2/NDF=17/16. That this result departs from the well known P_t^{-4} behavior[14] is supported by the QED Monte Carlo prediction which also disagrees with P_t^{-4} but reproduces the data quite closely (χ^2/NDF=21/18).

SEARCH FOR A NEW HEAVY QUARK

Since October 1982 the PETRA beam energy has been increased. In the period between October and December we have done an energy scan for $\sqrt{s}$ between 37.94 and 38.63 GeV in 30 MeV steps, reaching a total integrated luminosity of 1627 nb^{-1}. The MARK-J detector has taken data in this energy scan in a search for an enhancement of the cross sections in the process: $e^+e^- \to (\gamma) \to$ hadrons.

We have performed this search looking at the cross section evolution with the energy and at the thrust distribution[15] of the measured hadron events.

R-Results

The quantity R is defined as the ratio of the total hadron production cross section to the point-like QED cross section:

$$R = \frac{\sigma(e^+e^- \to \text{hadrons})}{\sigma_{pt}}$$

Fig. 16

A previous energy scan covering the $\sqrt{s}$ region from 29.9 to 31.46 GeV and from 33.0 to 36.72 GeV with an average luminosity per point of ∿ 40 nb^{-1} gives an R value of R = 3.76±0.05. The result from the recent scan in the $\sqrt{s}$ range 37.94 to 38.63 GeV with an average luminosity of ∿ 60 nb^{-1} per point, gives a mean value of R = 3.91±0.19 (Fig. 16).

Errors shown are statistical and we estimate a systematic error of 6% due to model dependence of the acceptance, event selection criteria and measurement of the luminosity. The data are consistent with the QCD prediction including only 5 quarks, namely $R \sim 3.88$. Nevertheless, if we are just on the threshold of the top production any effect will show up in the mean value of R but may be some enhancement in some region of s. An upper limit on the production cross section for a narrow resonance in the scanning data was obtained by fitting the function:

$$R = R_o + R_{RES}(B_h\Gamma_{ee}, M_{RES}) \tag{9}$$

where R_o represents the constant continuum production and $R_{RES}(B_h^o\Gamma_{ee}, M_{RES})$ describes the expected toponium resonance after corrections for machine energy resolution and radiative effects[16]. Γ_{ee} is the width for the decay to e^+e^- and B_h is the branching ratio into hadrons.

We find that the largest signal in all regions under study lies at $\sqrt{s} = 38.343$ GeV with a strength of $\int\sigma_{RES}dW = (20\pm8)$ MeV nb. Since in lowest order this cross section is $\int\sigma(e^+e^- \rightarrow V_{\bar{t}t} \rightarrow \text{hadrons})dW = 6\pi^2 B_h\Gamma_{ee}/M_{RES}^2$, we obtain the limit

$$B_h\Gamma_{ee} < 2.0 \text{ KeV} \quad (95\% \text{ C.L.})$$

On the basis of the experimental fact that Γ_{ee}/e_q^2 is approximately 10 KeV for the vector meson ground states $\rho,\omega,\phi,J,\Upsilon$ and the expectation[17] that B_h is of the order of 80%, the production of a ground state vector particle consisting of $q\bar{q}$ bound state, where the quark has charge $\frac{2}{3}e$ should give $B_h\Gamma_{ee} = 3.5$ KeV and is therefore excluded by our scanning data.

Thrust

With the thrust distribution we may check if the open "top" production threshold has been crossed. In Fig. 17 we show the data corresponding to all the energy points from the recent scan period (in full dots with error bar) and to the last four higher energy scan points (in triangles). We compare with the QCD prediction for five quarks (full line) and for six quarks (dashed line). The QCD production with 5 quarks agrees with the data showing no open top production at the energies we have reached up to now.

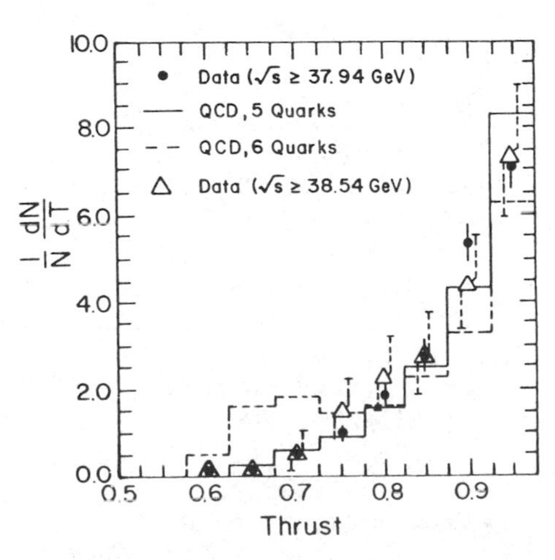

Fig. 17

CONCLUSIONS

(1) We have measured semileptonic branching ratios of c and b quarks and established the hard character of the fragmentation functions (particularly for the bottom quark).

(2) We rule out a large class of topless models allowing flavor changing neutral currents setting an upper limit on the branching ratio $Br(b \to \mu^+\mu^- X)$.

(3) We measure a significant ($> 5\sigma$) forward-backward charge asymmetry in μ-pair production $A_{\mu\mu} = (-11.7\pm1.7)\%$ with a systematic error of less than 1%. The asymmetry of τ-pair production has also been measured to be $A_{\tau\tau} = (-7.8\pm4.0)\%$.

(4) We have tested QED up to α^4 order at large momentum transfers $Q^2 \sim 100$ GeV^2 using the two photon μ-pair production.

(5) There is no evidence for "Toponium" ($e_q = 2/3$) ground state production in the range $37.94 < \sqrt{s} < 38.68$ GeV.

(6) There is no open "top" production up to an energy of $\sqrt{s} = 38.54$ GeV.

FUTURE PLANS

When PETRA reaches energies of the order of $\sqrt{s} \sim 45$ GeV we should be able, by measuring the reaction $e^+e^- \to \mu^+\mu^-$, to determine the mass of the Z^o with an accuracy of the order of 25%, under the assumption that only one neutral boson exists.

We shall keep on searching for "Toponium" bound states and for open "Top" production.

REFERENCES

1 S.L. Glashow, Nucl. Phys. 22, 579 (1968).
A. Salam and J.C. Ward, Phys. Lett. 13, 168 (1964).
S. Weinberg, Phys. Rev. Lett. 19, 1264 (1967), and Phys. Rev. D5, 1412 (1972).
2 M. Kobayashi and T. Maskawa, Prog. of Theor. Phys. 49, 652 (1973).
3 V. Barber and S. Pakvasa, Phys. Lett. 81B, 195 (1979).
4 G.L. Kane and M.C. Peskin, Nucl. Phys. B195, 29 (1982).
5 R.D. Field and R.P. Feynman, Nucl. Phys. B136, 1 (1978).
A. Ali, DESY-Report 81-016 (1981).
6 D. Bjorken, Phys. Rev. D17, 171 (1978).
C. Peterson, D. Schlatter, I. Schmitt, P.M. Zerwas, Phys. Rev. D27, 105 (1983).
7 F.A. Berends and R. Kleiss, DESY-Reports 80/66 and 80/73.
8 F.A. Berends, K.F.J. Gaemers, and R. Gastmans, Nucl. Phys. B63, 361 (1973), Nucl. Phys. B68, 541 (1974).
9 M. Pohl, MARK-J Internal Note.

10 L.W. Mo, Contr. to Neutrino 80, Erice (1980).
M. Roos and I. Liede, Phys. Lett. 82B, 89 (1979).
F.W. Buesser, Proc. of Neutrino 81, Wailea, Hawaii (1981).
11 R. Bhattacharya, J. Smith and G. Grammer, Phys. Rev. D15, 3267 (1977).
12 J. Smith, J.A.M. Vermaseren and G. Grammer, Phys. Rev. D15, 3280 (1977).
13 M. Defrise, Zeitschr. f. Phys. C, Particles and Fields, 9, 41 (1981).
M. Defrise, S.Ong, J. Silva, C. Carimalo, Phys. Rev. D23, 663 (1981).
14 S.M. Berman, J.D. Bjorken and J.B. Kogutt, Phys. Rev. D4, 3388 (1971).
15 D.P. Barber et al., Phys. Report 63, 337 (1980).
16 J.D. Jackson and D.L. Scharro, Nucl. Instr. Meth. B128, 13(1975).
17 J.P. Leveille, Univ. of Michigan Preprint, UM-HE-81-11,

CHARMED QUARK FRAGMENTATION IN e^+e^- ANNIHILATIONS AT 29 GeV*

R. Stroynowski

California Institute of Technology

Pasadena, California 91125

The production of charmed quarks has been studied now for several years. So far, the best measurements of their decays have been done at SPEAR [1]. It has been found there, that the charmed quark fragmentation results predominantly in the D and D* particles. Their branching ratios into various final states have been measured with about 15% accuracy. However, the shape of the charm fragmentation function has not been established in the SPEAR experiments. Due to the large mass of charmed mesons and low c.m. energy the kinematic range accessible there was very small, e.g., the fragmentation of charmed quark into D* was limited to $z = E_{D*}/E_{beam} \geq 0.7$. Thus the detailed study of the dynamics of the charm fragmentation process were not possible. The determination of the charm fragmentation function is much easier at higher energies, where the kinematic restrictions do not play any significant role.

I am reporting here on the measurement of the D* production at $\sqrt{s}$ = 29 GeV with the DELCO detector at PEP [2] based on the integrated luminosity of $\int Ldt = 21.5\ pb^{-1}$. DELCO is an upgraded SPEAR detector with the emphasis on good particle identification (see Fig. 1). Its main feature is the atmospheric pressure threshold gas Cerenkov counter of 36 cells covering about 70% of the solid angle. The counter is filled with isobutane allowing for the e/π separation below $\sim$ 3 GeV/c of the particle momentum and the K/π separation in the range of 2.5 to 9 GeV/c. The counter has a light yield of about 28 photoelectrons and efficiency greater than 99.97%. The charged

*Work supported in part by the U.S. Department of Energy under contract No. DE-AC03-81-ER-40050.

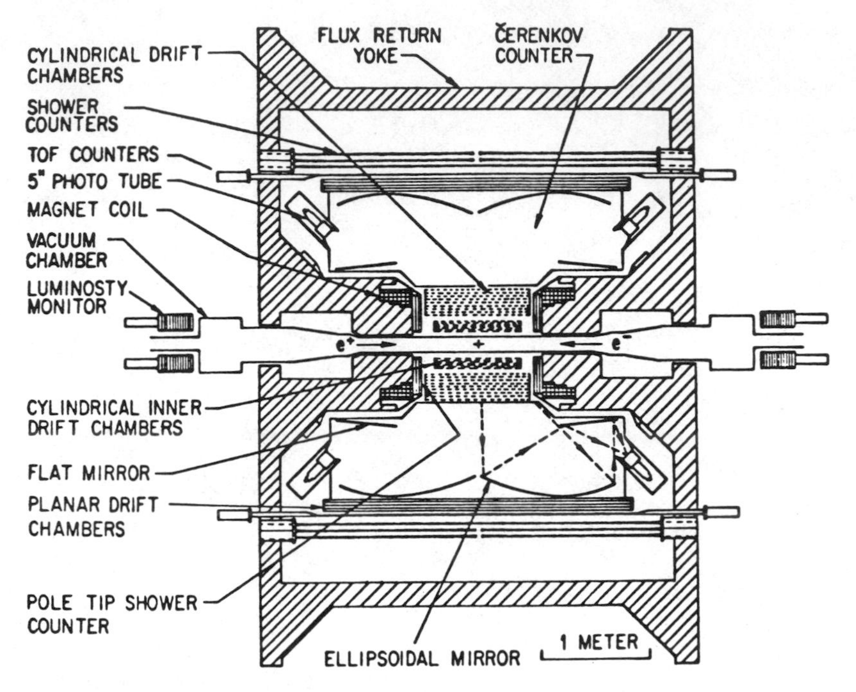

Figure 1

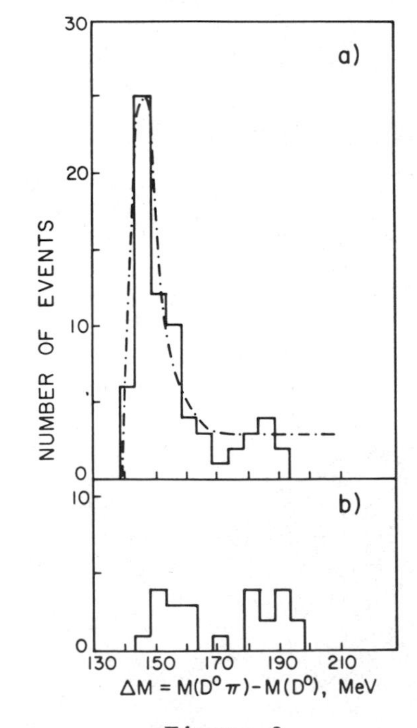

Figure 2

particle tracking is provided by 16 layers of cylindrical drift chambers situated between the beam pipe and the entrance to the Cerenkov counter and additional 6 layers of planar drift chambers at the exit of the counter. The open geometry magnet with the field integral of $\int Bdl = 1.8$ kGm and small mass in the path of the particles minimizes the Coulomb scattering and photon conversion probability. The momentum resolution achieved so far is $\Delta p \simeq 0.02p$ with additional 6% error due to the multiple scattering in the beam pipe. Additional information about the track is obtained from the system of the time-of-flight counters and the barrel shower counters placed behind the planar drift chambers. The method used to identify D*'s follows the suggestion of V.L. Fitch et al. [3], who noticed a distinct signature of the decay

$$D^* \rightarrow D^\circ \pi \ . \tag{1}$$

Due to the small mass difference of D* and D° (the pion kinetic energy is only 5.4 MeV in the D* rest frame) this decay results in a D° and a pion with almost identical velocities and directions. This in turn reduces by about two orders of magnitude the combinatorial background in the ($D^\circ\pi$ - D°) effective mass difference providing a clear enhancement due to the D* resonance.

We identify the D° candidates in the following decay mode:

$$D^\circ \rightarrow K^-\pi^+ \tag{2}$$

$$D^\circ \rightarrow K^-\pi^-\pi^+\pi^+$$

and in the charge conjugate channels corresponding to $\overline{D^\circ}$. The analysis is based on the particle identification provided by the Cerenkov counter. One track is required to have momentum above 2.7 GeV/c. This track is identified as a pion if the associated cell fired, i.e. registered a signal of more than 4 photoelectrons; otherwise the track is called a kaon. For the $K^-\pi^+$ decay mode of the D°, the second track is required to have the opposite charge to the first one and its particle identification is assigned as complementary. The minimum momentum requirement of the Cerenkov counter makes it impractical to identify also the type of the second track.

The D° candidates are then defined as the $K\pi$ mass combinations within ±300 MeV from the nominal mass of the resonance. This cut does not eliminate the background coming from the $K^{*-}\pi^+$ and the $K^-\rho^+$ decays where the final π° remained undetected. Such background is removed by additional selection on the opening angle of the $K^-\pi^+$ system correlated with its effective mass. The $K^*\pi$ and $K\rho$ decays of D° result in a smaller opening angle for a given $K\pi$ mass, thus allowing for good separation of the signal. Finally, the evidence for the D* is shown in Fig. 2a as the enhancement in the mass difference $\Delta M = M_{D^\circ\pi} - M_{D^\circ}$ below 155 MeV. The background, shown in

Fig. 2b, is estimated by plotting the mass difference using the wrong charge combinations of the "bachelor" pion, i.e. $M_{D^\circ\pi^-} - M_{D^\circ}$. For $\Delta M < 155$ MeV, the background represents about 15% of the signal. The dashed-dotted line represents a Monte Carlo prediction for the shape of the enhancement after taking into account DELCO momentum and effective mass resolution.

The analysis of the $K3\pi$ decay mode of the D° follows essentially the same prescription as outlined above. Since the method of background estimation requires the knowledge of the "right" charge combination, only events with first particle identified as a kaon were used to form the D° candidates. This, together with lower acceptance and efficiency for the five particle final state resulted in a smaller number of events as shown in Fig. 3 and summarized in Table 1. Also shown in Table 1 are the cross section values for both decay channels based on the observed number of events, acceptance of the apparatus and efficiency of the procedures. The systematic errors quoted are dominated by the errors of the D° branching ratios as given in the Particle Data Tables [4]. It is worth noting that the measured values are rather high. The total cross section for the charm and bottom quark production at $\sqrt{s}$ = 29 GeV is $\sigma_{c,b} = 0.36$ nb. Thus, the observed D* cross section (assuming $\sigma_{D^{*\circ}} = \sigma_{D^{*\pm}}$) seems to saturate the heavy quark fragmentation. Similar effects have been observed in other experiments [5] suggesting possible underestimate of the branching ratios used in the calculations.

The differential distribution $s(d\sigma/dz)$ is shown in Fig. 4. The data are peaked at large values of z and are consistent with previous measurements [4,6]. At $\sqrt{s}$ = 29 GeV, the cross section is dominated by the charm quark production, and the secondary charm contribution due to bottom fragmentation is negligible in the kinematic range of the experiment. The measured distribution reflects, therefore, the

TABLE 1

D* Decay Mode	No. of Observed Events	No. of Background Events	$D^{*\pm}$ Cross Section (nb)
$D^{*\pm} \to D^\circ\pi^\pm$			
$\to K\ \pi^\pm$	53	8	.18 ± .04 ± .06
$D^{*\pm} \to D^\circ\pi^\pm$			
$\to K\ \pi^-\pi^+\pi^\pm$	10	0	.16 ± .06 ± .03

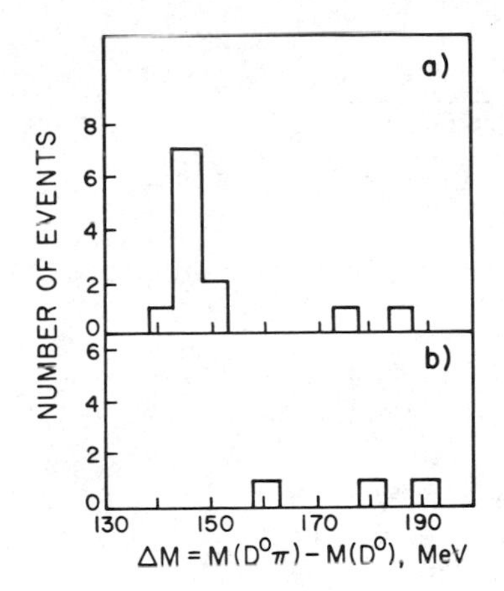

Figure 3

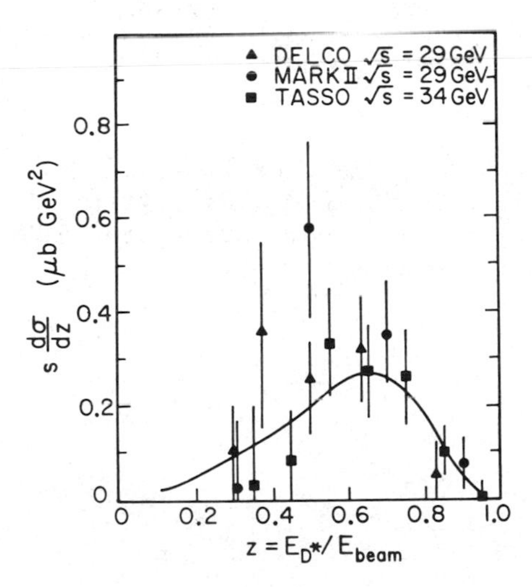

Figure 4

shape of the charm fragmentation function. The parameterization of the fragmentation function proposed by C. Peterson et al. [7] in the form

$$P(z) = \frac{1}{z\left(1 - \frac{1}{z} - \frac{\varepsilon}{(1 - z)^2}\right)^2}$$

shows good agreement with the measured points for $\varepsilon = 0.25$ (see Fig. 4).

REFERENCES

1. G. Goldhaber et al., Phys. Lett. 69B (1977) 503.
   M. Piccolo et al., Phys. Lett. 70B (1977) 260.
   G. Trilling, Phys. Reports 75 (1981) 57.
2. The members of the Caltech-SLAC-Stanford Collaboration at DELCO are: W.B. Atwood, P.H. Baillon, B.C. Barish, G.R. Bonneaud, H. DeStaebler, G.J. Donaldson, R. Dubois, M.M. Duro, S.G. Gao, Y.Z. Huang, G.M. Irwin, R.P. Johnson, H. Kichimi, J. Kirkby, D. Klem, D.E. Koop, J. Ludwig, G.B. Mills, A. Ogawa, T. Pal, D. Perret-Gallix, R. Pitthan, D.L. Pollard, C.Y. Prescott, L.Z. Rivkin, L.S. Rochester, W. Ruckstuhl, M. Sakuda, S.S. Sherman, E.J. Siskind, R. Stroynowski, R.E. Taylor, S.Q. Wang, S.G. Wogcicki, H. Yamamoto, W.G. Yan and C.C. Young.
3. V.L. Fitch et al., Phys. Rev. Lett. 46 (1981) 761.
   R.W. Kadel, Princeton University, Ph.D. Thesis C00-3072-89 (1977).
4. M. Roos et al., "Reviews of Particle Properties," Phys. Lett. 111B (1982) 1.
5. Y. Yelton et al., Phys. Rev. Lett. 49 (1982) 430.
6. M. Althoff et al., DESY83-010 (1983) preprint.
7. C. Peterson et al., Phys. Rev. D27 (1983) 105.

WEAK NEUTRAL CURRENTS IN e^+e^- EXPERIMENTS

Albrecht Böhm

ABSTRACT

Experimental results on weak neutral currents in e^+e^- experiments at PETRA and PEP are reviewed. The precise measurements of the leptonic reactions $e^+e^- \to e^+e^-$, $\mu^+\mu^-$ and $\tau^+\tau^-$ are used to determine the weak neutral current parameters. The main result is the observation of a significant forward-backward asymmetry in the angular distribution of $e^+e^- \to \mu^+\mu^-$ and $e^+e^- \to \tau^+\tau^-$, which measures directly the strength of the neutral weak interaction with respect to the electromagnetic interaction. Finally, first attempts are made to extend these studies to the reaction $e^+e^- \to q\bar{q} \to$ hadrons with the aim of measuring the weak neutral current couplings of heavy quarks.

INTRODUCTION

Since the construction of the high energy e^+e^- storage rings PETRA at DESY and PEP at SLAC, it becomes possible to study weak neutral currents in e^+e^- annihilations. The cross sections of e^+e^- reactions in the energy range of PETRA and PEP are dominated by the electromagnetic interaction and the contributions of purely weak interactions are negligibly small. However, the interference between the photon and Z^o-exchange produces measurable effects, which rise with the square of the c.m. energy. It is therefore of great importance to have data at the highest possible c.m. energy. We will see that this condition favors the experiments at PETRA, where a c.m. energy of 43 GeV has been reached in July 1983.

The study of weak neutral currents in e^+e^- annihilations is motivated by the following arguments: the electroweak theory can be tested at energies and momentum transfers, which are much higher than in other weak interaction experiments. It is of special importance that the reactions ee → ee, $\mu\mu$ and $\tau\tau$ are purely leptonic and are free from the complications appearing in lepton nucleon scattering. In addition a comparison of these reactions tests the lepton-universality of weak interactions. We also can study the weak neutral current couplings of the heavy quarks c, b which, above threshold, are copiously produced via the reaction $e^+e^- \rightarrow q\bar{q} \rightarrow$ hadrons. Last but not least, the PETRA and PEP storage rings operate at energies where for the first time we might be able to observe a deviation from the point-like four-fermion coupling. Clearly, such a deviation would either show the compositeness of leptons or, as expected, indicate the existence of the Z^o boson [1].

Finally, it should be noted that there have been extensive searches for new particles in high energy e^+e^- reactions [2]. Many results are relevant to the theory of weak interaction, which should be mentioned here. No leptons more massive than the tau, nor quarks heavier than the bottom have been found up to c.m. energies of 38.5 GeV [3]. Extensive searches for charged Higgs particle and technipions have been performed at PETRA, PEP and CESR. The result is that their existence is excluded for masses between the tau mass and 14 GeV independently of their decay branching ratio into $\tau\nu$, $c\bar{s}$ or $c\bar{b}$.

I. NEUTRAL CURRENT COUPLINGS OF LEPTONS

The reaction $e^+e^+ \rightarrow \ell^+\ell^-$ is described in a model-independent way by three parameters. Following Hung and Sakurai [4] we call them h_{VV}, h_{VA} and h_{AA}. Clearly, if lepton universality of the weak neutral currents is valid, the parameters are identical for ℓ = e, μ and τ. At high energies we have to introduce the Z^o masses as additional parameters.

In SU(2) x U(1) models these parameters are expressible in terms of the weak mixing angle $\sin^2\theta_w$ and of the weak isospin components of left-handed and right-handed leptons.

$$
\begin{aligned}
h_{VV} &= \rho(T_{3L} + T_{3R} + 2\sin^2\theta_w)^2 \\
h_{AA} &= \rho(T_{3L} - T_{3R})^2 \\
m_Z^2 &= \frac{1}{\rho}\,\frac{\pi\alpha}{\sqrt{2}\,G_F}\,\frac{1}{\sin^2\theta_w\cos^2\theta_w}
\end{aligned}
\qquad (1)
$$

The parameter ρ measures the strength of the weak neutral current interaction relative to the weak charged current interaction. It is defined by the relation

$$\rho = \frac{m_W^2}{m_Z^2 \cos^2\theta_w} , \qquad (2)$$

where m_W is the mass of the charged vector boson.

A combined fit to neutrino-electron and electron-deuteron measurements leads to [5]

$$\rho = 1.002 \pm 0.015 \quad \text{and} \quad \sin^2\theta_w = 0.234 \pm 0.013 .$$

Thus for most of the results discussed here we assume $\rho = 1$.

In the standard $SU(2)_L \times U(1)$ theory of GSW [6] the left-handed lepton fields are arranged in weak iso-doublets, the right-handed lepton fields in weak iso-singlets. Since this theory also predicts $\rho = 1$, we get

$$\begin{aligned} h_{VV} &= g_V^2 = \frac{1}{4}(1 - 4\sin^2\theta_w)^2 \\ h_{AA} &= g_A^2 = \frac{1}{4} . \end{aligned} \qquad (3)$$

Using $\sin^2\theta_w = 0.23$, we find that $g_V^2 = 0.002$, $g_A^2 = 0.25$ and $m_Z \approx 90$ GeV. Since the leptonic cross sections in e^+e^- reactions depend on the square of the vector or axial-vector coupling constant and the vector coupling vanishes, we cannot detect an effect due to the vector coupling. However, the reverse argument is also valid: if we see no effect due to g_V, the weak mixing angle $\sin^2\theta_w$ has to be very near to 0.25, in order to give $g_V \approx 0$. The parameters h_{VV} and h_{AA} are often called g_V^2 and g_A^2 in e^+e^- experiments and we follow that usage. However, be aware that the symbols g_V and g_A are usually reserved for the coupling constants of ν-e scattering.

II. MEASUREMENT OF LEPTONIC REACTIONS

The data we discuss here were obtained at PETRA and PEP until up to February 1982. While PEP has run at a fixed c.m. energy of 29 GeV, PETRA has run at various c.m. energies between 12 GeV and 38.6 GeV with an average of 34.5 GeV at high energies. It is impossible to describe the experiments and the event selection in details in this brief summary. The reader is referred to publications and to the reports of the individual experiments in these proceedings [7-12].

Leptonic reactions are easily recognized as two mainly collinear back-to-back leptons, whose momenta are approximately equal to the beam momentum. For selection one typically requires the acollinearity angle ξ of the scattered leptons to be smaller than 10° and each lepton momentum to be larger than half the beam momentum. The radiative corrections to order α^3 are calculated for these cuts by the Monte Carlo program of Berends and Kleiss [13]. All measurements presented here are corrected for QED radiative effects to order α^3 such that the data can directly be compared to the lowest order QED prediction.

III. RESULTS ON $e^+e^- \rightarrow e^+e^-$

We begin our discussion of individual leptonic reactions with Bhabha scattering, $e^+e^- \rightarrow e^+e^-$. Its measurement fulfills a twofold purpose: small angle Bhabha scattering serves as a luminosity monitor, while the angular distribution at large angles is sensitive to the weak interaction. The theoretical cross section [14] shows a rather complicated dependence on g_V^2 and g_A^2 because there are space-like and time-like exchanges of the virtual photon and the Z^o. Since the deviations from QED are small, we present the data as the ratio of the measured cross section divided by the QED prediction.

Fig. 1 displays the measurements performed at PETRA and PEP. Since the experiments MARK-J and MAC do not distinguish between electrons and positrons, their angular distribution is folded aroung 90°. The errors indicated in Fig. 1 are statistical only. In addition there are systematic errors on the overall normalization of 3 to 5% and point-to-point errors of 1 to 3% for the different experiments. We conclude from Fig. 1 that it is not yet possible to establish the effect of the weak interaction in Bhabha scattering, since the difference between the predictions of pure QED and the GSW theory with $\sin^2\theta_w = 0.23$ is too small to be resolved.

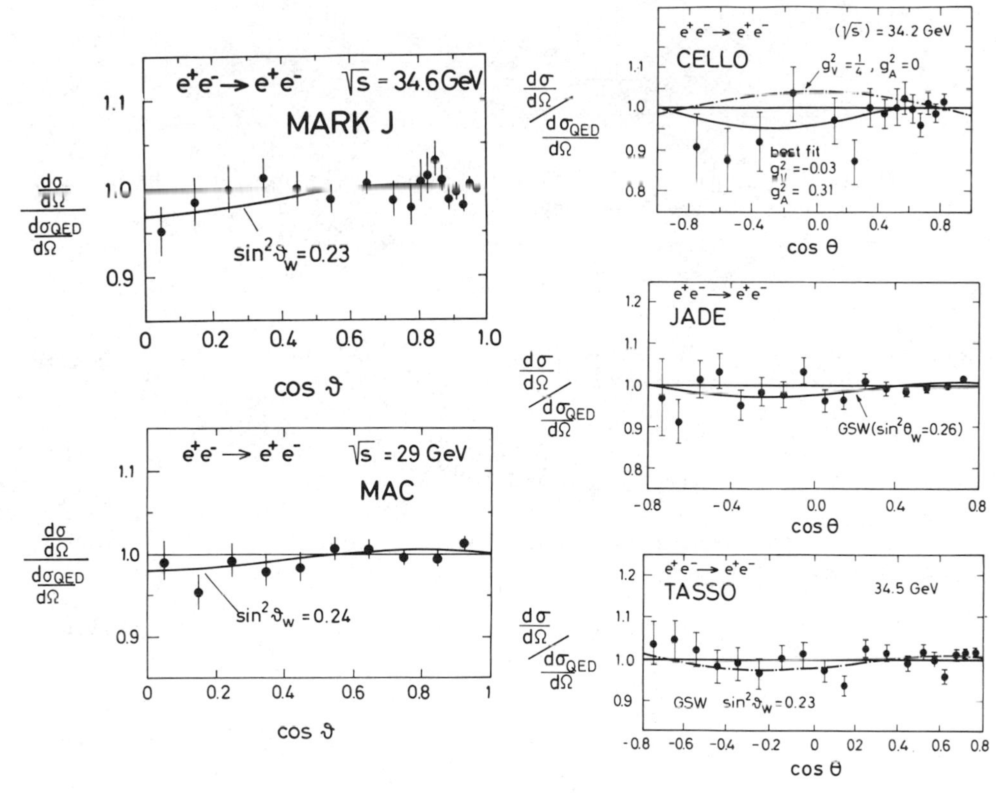

Fig. 1 Measured cross sections of Bhabha scattering divided by QED cross section and compared to the predictions of the electroweak theory.

IV. CROSS SECTION OF $e^+e^- \to \mu\mu$

The angular distribution of $e^+e^- \to \mu^+\mu^-$ and $e^+e^- \to \tau^+\tau^-$ has the form

$$\frac{d\sigma}{d\cos\theta} = \frac{\pi\alpha^2}{2s} \left[R_{\mu\mu}(1+\cos^2\theta) + B\cos\theta \right] \qquad (4)$$

where

$$R_{\mu\mu} = \sigma_{\mu\mu} \Big/ \frac{4\pi\alpha^2}{3s} = 1 - 2g_V^2\chi + (g_V^2+g_A^2)^2\,\chi^2 \qquad (5)$$

and

$$\chi = \frac{G_F}{2\sqrt{2}\pi\alpha} \frac{s \cdot m_Z^2}{m_Z^2 - s} . \tag{6}$$

The scattering angle θ is defined as the angle between the μ^- and the outgoing e^- beam. The factor

$$B = -4g_A^2\chi + 8g_V^2 g_A^2 \chi^2 \tag{7}$$

depends mainly on the axial vector coupling and it determines the forward-backward asymmetry which is given by

$$A_{\mu\mu} = \frac{N(\theta<90^o) - N(\theta>90^o)}{N(\theta<90^o) + N(\theta>90^o)} = \frac{3}{8} \frac{B}{R_{\mu\mu}} . \tag{8}$$

$N(\theta<90^o)$ is the number of events, where the μ^- is in the forward hemisphere, defined by the outgoing e^- beam.

This asymmetry depends on g_A^2 and is practically independent of the value of g_V^2

$$A_{\mu\mu} = \frac{3}{8} \frac{B}{R_{\mu\mu}} = \frac{-3g_A^2\chi(1-2g_V^2\chi)}{2(1 - 2g_V^2\chi + ...)} \approx -\frac{3}{2} g_A^2\chi . \tag{9}$$

Lepton universality is assumed in expressions (5) to (9), otherwise should replace g_A^2 by $g_A^e \cdot g_A^\ell$ and g_V^2 by $g_V^e \cdot g_V^\ell$, where ℓ is the lepton produced in the final state. The predicted asymmetry is negative and its absolute value increases with s. At $\sqrt{s}$ = 34.5 GeV we expect a value of $A_{\mu\mu}$ = -9.5% for $g_A^2 = 1/4$ and m_Z = 90 GeV.

We first discuss the measurements of $R_{\mu\mu}$, the cross section normalized to the point-like cross section (Fig. 2). The systematic error on $R_{\mu\mu}$ is dominated by an overall normalization error due to luminosity and acceptance, which ranges between 3% and 5% for different experiments. The statistical error at the highest energies of $\sqrt{s} \approx$ 34.5 GeV is about 2% as typically 3000 muon pairs are observed.

The measurements of Fig. 2 agree with the pure QED prediction within the statistical and systematical errors. The effect of weak interactions on $R_{\mu\mu}$ arises mainly from the vector coupling as shown in equation (5). We observe no effect and conclude that $g_V^e \cdot g_V^\mu$ is close to zero, as we expect it in the GSW theory for

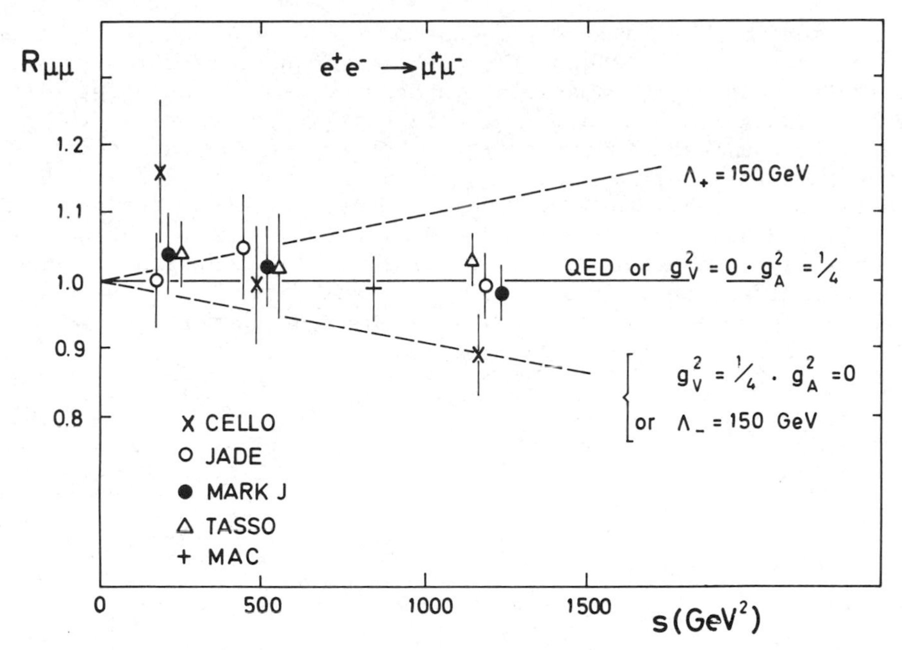

Fig. 2 Results on $R_{\mu\mu}$ compared to different predictions from QED and electroweak theory.

$\sin^2\theta_W = 0.23$. On the other hand we can neglect the weak interaction, which would in any case be unmeasurable for $\sin^2\theta_W = 0.23$ and test QED. This is usually done in terms of a cut-off parameter, for which the experiments obtain a lower limit of about 200 GeV with 95% C.L. This value indicates, that muons (and electrons) behave as point-like particles down to a distance of 10^{-16} cm.

V. ASYMMETRY OF $e^+e^- \rightarrow \mu^+\mu^-$

The measurement of the forward-backward asymmetry is one of the central experimental tasks of high energy e^+e^- experiments. The asymmetry arises from the interference between the electromagnetic and weak interaction. Thus it measures directly the strength of the neutral weak interaction with respect to the electromagnetic interaction. The GSW-model predicts a negative asymmetry, which means that there are more $\mu^+\mu^-$ events with the μ^- in the backward hemisphere than events with the μ^- in the forward hemisphere.

To measure the asymmetry one needs to determine the direction and the charge of the muons. The asymmetry can be measured very precisely, because it is a relative measurement, independent of the luminosity measurement. It is insensitive to errors in the acceptance and reconstruction efficiency as long as the acceptance is the same for positive and negative muons. For this reason, the experiments are able to limit the systematic error of the asymmetry to $\lesssim 1\%$. In addition, MARK-J and MAC took equal amounts of data with positive and negative magnet polarity. The addition of both data sets effectively cancels all systematic errors, which relate to the charge measurement, and leaves no doubt that the systematic error is much smaller than 1%.

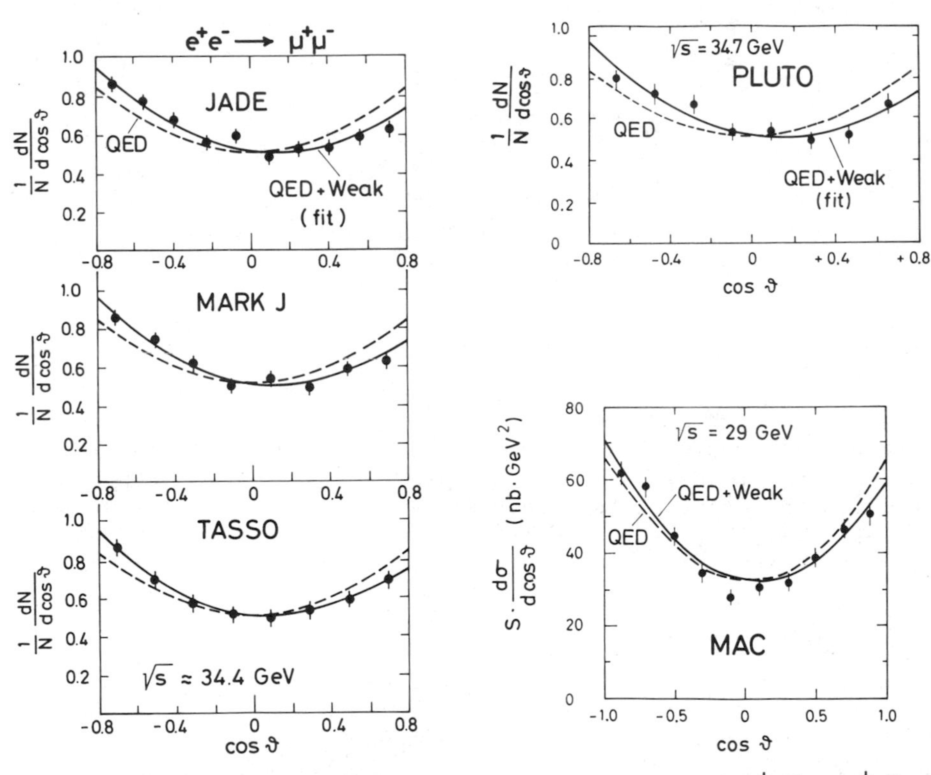

Fig. 3 Measurements of the angular distribution of $e^+e^- \to \mu^+\mu^-$ compared to the prediction of QED (dashed line) and to a fit including the weak interaction (solid line). All measurements are corrected for QED radiative effects to order α^3, except the data from MAC.

Fig. 3 shows the measurements of four PETRA experiments, JADE, MARK-J, PLUTO and TASSO, at $\sqrt{s} \approx 34.5$ GeV and of the MAC experiment at PEP at $\sqrt{s} = 29$ GeV.

All measurements deviate clearly from the prediction of pure QED and show a negative asymmetry as predicted by the electroweak theory. Table I summarizes the high energy measurements of the asymmetry. The systematic errors are estimated to be smaller than 1% and are included in the combined results. The asymmetries in Table I are obtained from a fit of (4) to the measured angular distribution. The values expected by the GSW theory are listed in the last column. They differ from the values obtained with (9) by about -0.7% due to electroweak radiative effects for which the measurements were not corrected (see next chapter). The measurements agree within errors with the prediction of the GSW theory

TABLE I

RESULTS FROM PEP AND PETRA ON THE ASYMMETRY OF $e^+e^- \to \mu^+\mu^-$

Experiment	$\sqrt{s}$ (GeV)	Measured $A_{\mu\mu}$ in %	Expected in GSW for $\sin^2\theta_W = 0.23$ in %
MAC	29	$-7.6\pm1.8\pm0.3$	-6.0
MARK II	29	-9.6 ± 4.5	-6.0
PEP Results combined		-7.9 ± 1.7	-6.0
CELLO	34.2	-6.4 ± 6.4	-8.7
JADE	34.4	$-11.1\pm1.9\pm(<1)$	-8.7
MARK-J	34.6	$-11.7\pm1.7\pm(<1)$	-8.6
PLUTO	34.7	$-12.0\pm3.2\pm(<1)$	-9.0
TASSO	34.5	$-9.1\pm2.3\pm0.5$	-8.8
PETRA Results combined	34.5	-10.9 ± 1.2	-8.7 for $m_Z = 90$ GeV -7.4 for $m_Z = \infty$

for $\sin^2\theta_W \approx 0.23$ which gives $g_A^\mu = g_A^e = -1/2$, $g_V \approx 0$ and $m_Z \approx 90$ GeV. The combined PETRA asymmetry is higher than the predicted value by nearly two standard deviations. Although the difference is within normal statistical fluctuations, its existence will keep interest in future measurements above 40 GeV at PETRA.

We will now turn to the interpretation of the measured asymmetry. We rewrite (9) in a somewhat different form to show explicitly the parameters, which determine the asymmetry: the axial vector coupling g_A^e and g_A^μ, the parameter ρ and the mass of the Z^o boson:

$$A_{\mu\mu} = -\frac{3}{4}\rho \frac{G_F}{\sqrt{2}\,\pi\alpha} \frac{s\cdot m_Z^2}{m_Z^2 - s} g_A^e g_A^\mu \qquad (10)$$

We now make different assumptions about these parameters and see what kind of conclusions we can draw from the measurements at PETRA and PEP as listed in Table I.

(1) If we set $\rho = 1$ and $m_Z = 90$ GeV

$$g_A^e \cdot g_A^\mu = 0.32 \pm 0.04 \quad .$$

If we insert $g_A^e = -0.51 \pm 0.06$ from νe-scattering |33|, we determine the axial-vector coupling of the muon

$$g_A^\mu = T_{3L} - T_{3R} = -0.63 \pm 0.15 \quad .$$

The value is in good agreement with the assumption that the left-handed muon is the lower member of a weak iso-doublet and that the right-handed muon is in a weak iso-singlet. Comparing g_A^e with g_A^μ we get an impressive confirmation of the μ-e universality of the weak neutral current interactions.

(2) Alternatively, again with the assumption of $\rho = 1$, we can determine the Z^o mass, $m_Z = 61^{+10}_{-6}$ GeV, if we assume e-μ universality and $g_A^2 = 1/4$ and $g_V^2 \approx 0$. The one standard deviation error is somewhat misleading, because the errors are non parabolic. It is better to state that the Z^o mass ranges between 52 and 97 GeV with 95% C.L. The measured asymmetry differs by three standard deviation from the value of -7.4% which would be expected for a point-like four-fermion coupling i.e. $m_Z = \infty$. It therefore indicates the effect of a propagator term and the existence of a Z^o boson with finite mass [1].

(3) The asymmetry measures in a very direct way the strength of the weak neutral current interaction with respect to the electromagnetic interaction. Since both are related via (10) and the Fermi coupling G_F to the weak charged current, we measure directly the

parameter ρ. If we take the weak mixing angle $\sin^2\theta_W \approx 0.23$ from low q^2 data, we find

$$\rho = 1.25 \pm 0.14$$

Clearly the error is large, but the measurement is obtained from purely leptonic reactions. It can be compared to the measurement of neutrino and antineutrino scattering on electrons, where the CHARM collaboration obtains $\rho = 1.12 \pm 0.12 \pm 0.11$ [15].

Fig. 4 summarizes all existing measurements on the muon asymmetry. The low energy points are from PETRA and from SPEAR [16]. The measurements show a dependence on the c.m. energy s as predicted for an electroweak interference and prefer a value of Z^0 mass between 50 and 90 GeV. Since PETRA runs now at energies above 40 GeV i.e. $s \geq 1600\ GeV^2$, we expect a strong improvement of the mass limits and other electroweak parameters within about a year.

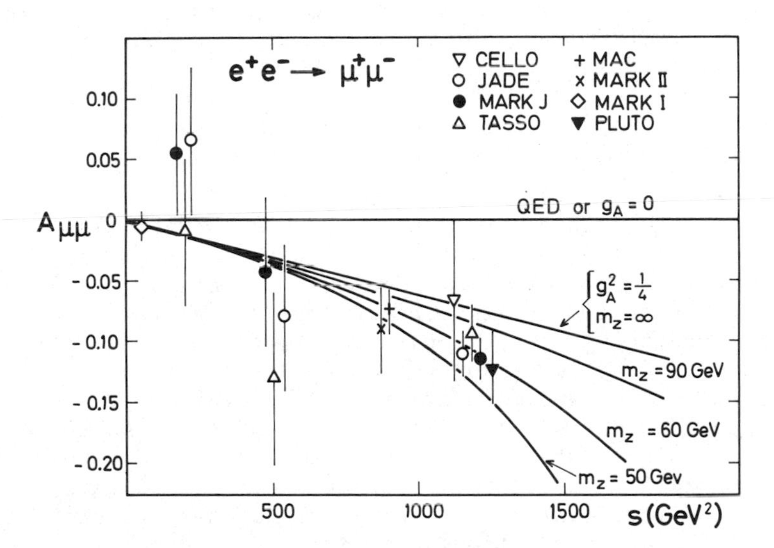

Fig. 4 Measurements of the forward-backward asymmetry of the reaction $e^+e^- \to \mu^+\mu^-$ as a function of the c.m. energy squared. The data are compared to the predictions of an electroweak interference with $g_A^2 = 1/4$ and different masses of the Z^0-boson. Since the measurements are corrected for radiative effects, pure QED or an electroweak theory with $g_A = 0$ predicts no asymmetry.

VI. RADIATIVE CORRECTIONS ON THE ASYMMETRY

Pure QED also produces a forward-backward asymmetry mainly due to the interference between the one and two photon exchange and between the initial and final state bremsstrahlung

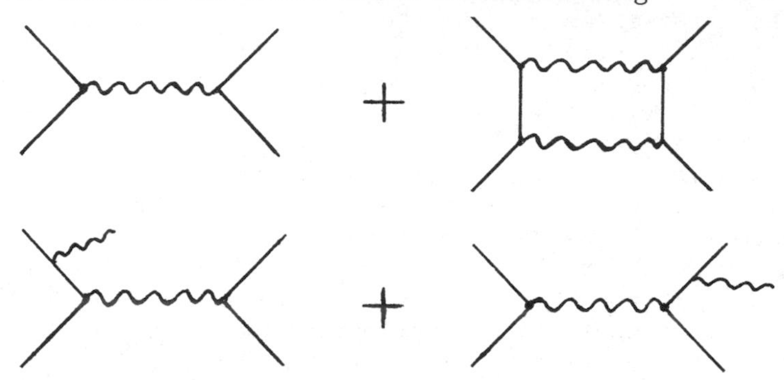

The QED asymmetry is much smaller than the electroweak asymmetry and it is positive. A Monte Carlo program developed by Berends and Kleiss |13| gives A_{QED} = +1.5% for $|\cos\theta| \leq 0.8$ and an acollinearity cut of 20°. Since we determine the electroweak asymmetry from a fit to the angular distribution, we determine the radiative correction δ at each angle by comparing the α^3 QED cross section to the lowest order QED cross section

$$\frac{d\sigma_{QED}}{d\cos\theta} = \frac{d\sigma_o}{d\cos\theta}(1+\delta) = \frac{\pi\alpha^2}{2s}(1+\cos^2\theta)(1+\delta). \quad (11)$$

The measured angular distribution is then divided by $(1+\delta)$ and fitted by equation (4) to obtain the electroweak asymmetry. This procedure corrects only for QED radiative effects to the lowest order QED cross section. It does not include the full one loop electroweak radiative correction to the lowest order electroweak cross section. A Monte Carlo program of Berends et al. [17] including all QED corrections of order α^3 to the photon <u>and</u> Z^o exchange leads to an additional correction of +0.7% to the asymmetry for the same cuts as stated above. This is mainly due to a QED radiative correction to the electroweak interference term, which has been neglected previously. We are making further investigations to find out if there are additional corrections. Theoretical calculations indicate that these corrections are small [18].

The MARK-J group has studied the radiative corrections by calculating the asymmetry for different bins of the acollinearity angle. The QED predicts a positive correction to collinear events and a negative correction to acollinear events originating from hard photon bremsstrahlung (Fig. 5). The GSW theory predicts a similar shape but shifted downwards by about -9% due to the electroweak asymmetry [17].

Since the MARK-J detector has a very high angular resolution for muons one can see this effect in the data (Fig. 6). The measured asymmetries as a function of the acollinearity angle follow well the prediction of the electroweak theory (GSW). This shows for the first time the presence of electroweak radiative corrections, or more precisely expressed, the QED corrections to the electroweak interference. Thus it constitutes an important check of our understanding of radiative corrections.

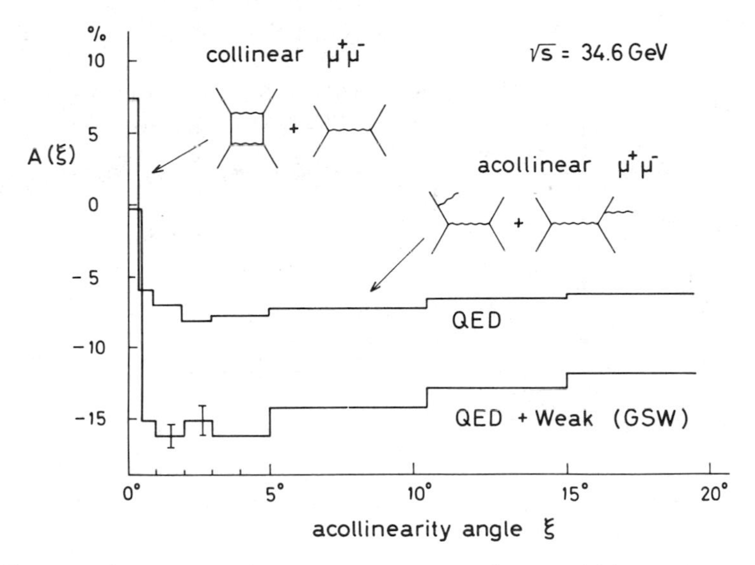

Fig. 5 Dependence of the asymmetry on the acollinearity angle ξ.

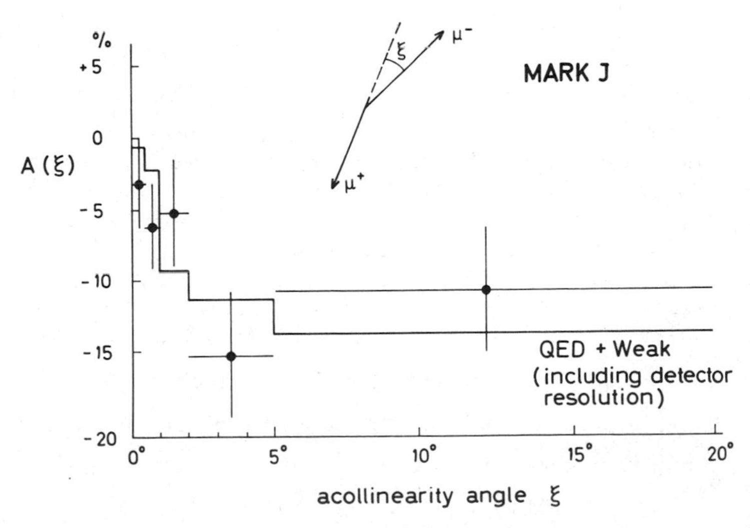

Fig. 6 Measurements of the forward-backward asymmetry as a function of the acollinearity angle ξ.

VII. COMBINED ANALYSIS OF $e^+e^- \to e^+e^-$ AND $e^+e^- \to \mu^+\mu^-$

The strongest constraints on the vector and axial-vector couplings of leptons can be obtained by combining the measurements of the cross sections and angular distributions of ee → ee and ee → μμ. Fig. 7 displays the 95% C.L. contour in the (g_V, g_A) plane, determined by data from three PETRA experiments, JADE, MARK-J and TASSO.

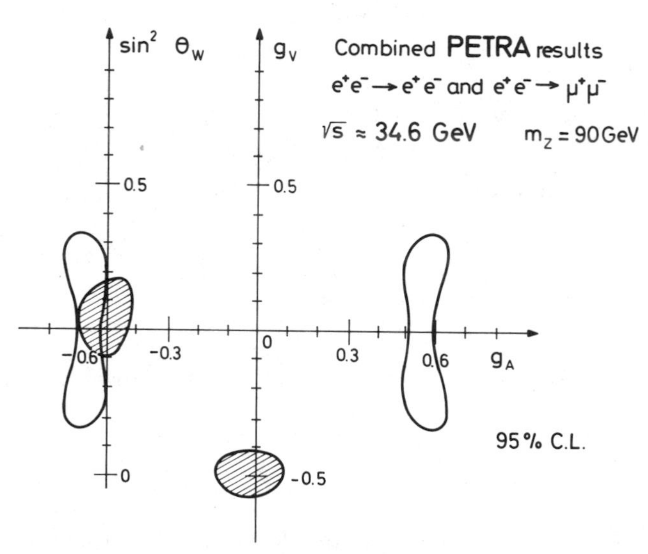

Fig. 7 Allowed regions (95% C.L.) for the vector and axial vector coupling of leptons determined by neutrino electron scattering (shaded area) and by PETRA experiments (unshaded area).

The allowed region for g_V and g_A has a fourfold symmetry because e^+e^- experiments measure the squares of the coupling constants. Neutrino scattering alone limits the values to two regions, a vector-like and a axial-vector-like solution. To resolve this ambiguity one previously had to consider lepton-hadron scattering with the inherent complications of hadronic targets. Now a unique solution can be determined from purely leptonic reactions [19]. To reach this conclusion we have assumed the relations $h_{VV} = g_V^2$ and $h_{AA} = g_A^2$. Sakurai has proposed that one compares the ratios h_{VV}/h_{AA} and g_V^2/g_A^2 to exclude one solution [20]. In this way we do not make an assumption about the coupling of the Z^o to neutrinos in the process $\nu + e \to \nu + e$. From the results shown in Fig. 7 we find

$$\frac{g_V^2}{g_A^2} = \frac{h_{VV}}{h_{AA}} < 0.27 \text{ with 95\% confidence.}$$

This limit excludes the vector-like solution of the neutrino electron scattering data on more general grounds, namely in the framework of models with a single Z^o.

VIII. $e^+e^- \rightarrow \tau^+\tau^-$: CROSS SECTION AND SEMILEPTONIC BRANCHING RATIO

The measurements of tau pair production $e^+e^- \rightarrow \tau^+\tau^-$ is more complicated because the tau decays after a flight path of less than a few millimeters. However, tau leptons are well recognized by their characteristic decays, either into lepton and neutrinos or into hadrons (low multiplicity) and a neutrino. The statistical and systematic errors of the cross section measurements are generally larger than for $\sigma_{\mu\mu}$ and depend on the fraction of decays which are used to select taus. We refer to the original publications for a detailed description of the selection and the measurement of tau events [7-12,21,22].

The measurements of $R_{\tau\tau}$, the cross section scaled by the point-like cross section, are summarized in Fig. 8. The measurements agree well with pure QED and also with the prediction of GSW for $\sin^2\theta_W = 0.23$. The effect of weak interaction depends on $g_V^e \cdot g_V^\tau$

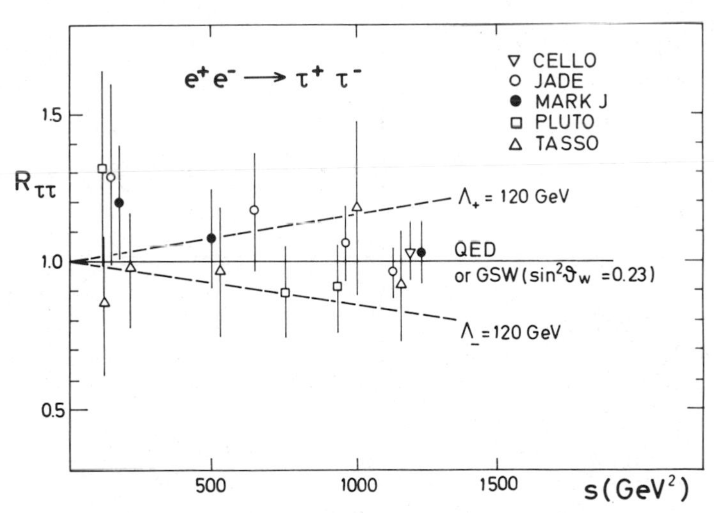

Fig. 8 Measurements of $R_{\tau\tau}$ at different c.m. energies squared. The solid line shows the prediction of pure QED which coincides practically with the prediction of the GSW-theory for $\sin^2\theta_W = 0.23$. Also indicated are the deviations expected if the cut-off parameter $\Lambda_\pm$ would be 120 GeV.

and since the vector coupling of the electron is close to zero, there is practically no sensitivity to the vector coupling of the tau. We therefore draw the following conclusions from Fig. 8:

1. The agreement of the cross section with the QED prediction or with the GSW theory with $\sin^2\theta_w \sim 0.23$ allows us to set an upper limit of the cut-off parameter, which is $\Lambda \approx 150$ GeV with 95% confidence level. We conclude that the tau does not show a structure up to an energy scale of 150 GeV and interacts as if pointlike, down to a distance of about 10^{-16} cm. This is especially remarkable in view of the fact that the tau lepton has a mass which is about twice the proton mass.

2. Alternatively, we assume the validity of QED and determine the branching ratios of tau decay modes which were used for the selection of $ee \to \tau\tau$. Table II shows as an example, the measurements of the leptonic branching ratio for the decay $\tau \to \mu\nu\nu$. We see that the recent measurements have improved the precision of this branching ratio by a factor two.

TABLE II: RESULTS ON THE BRANCHING RATIO FOR $\tau \to \mu\nu\nu$

Experiment	$B\ (\tau \to \mu\nu\nu)$ in %
CELLO [21]	17.6 ± 2.6 ± 2.1
MAC [9]	17.6 ± 1.5 ± 1.0
MARK-J [10]	17.8 ± 1.6
MARK II (SPEAR) [22]	17.1 ± 0.6 ± 1.0
AVERAGE	17.4 ± 0.8
previous world average	17.5 ± 1.7

IX. ASYMMETRY OF $e^+e^- \to \tau^+\tau^-$

The angular distribution of tau leptons should show a forward-backward asymmetry proportional to $g_A^e \cdot g_A^\tau$, which is completely analogous to that of the reaction $ee \to \mu\mu$. If $g_A^e = g_A^\tau = -1/2$ we expect an asymmetry of -9.4% at 34.5 GeV. This means that negative taus go more frequently in the direction of the positron beam and positive taus tend to follow the electron beam. The charge of the tau lepton is easily determined from the sum of the charges of its decay products. Most of the experiments do not detect all decay modes and have therefore a smaller number of events than in the reaction ee → μμ, with attendant larger statistical errors.

Fig. 9 displays the measurements of CELLO, JADE, MARK-J and TASSO at PETRA.

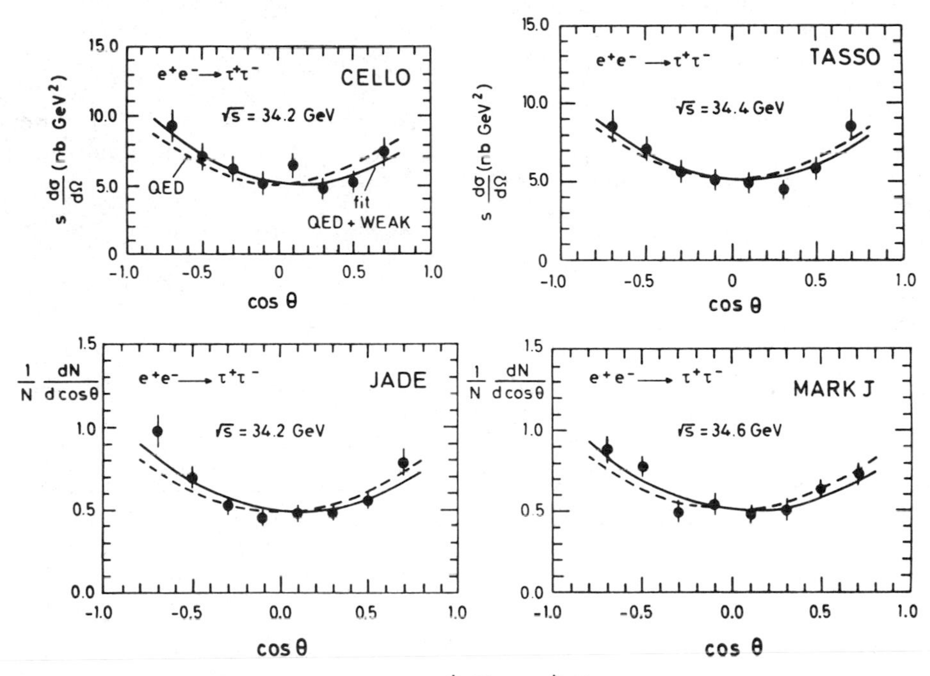

Fig. 9 Angular distribution of $e^+e^- \to \tau^+\tau^-$ measured by CELLO, TASSO, JADE and MARK-J at PETRA. The dashed line is the lowest order QED prediction of the form $(1 + \cos^2\theta)$. The solid line is a fit to the data, which includes weak interaction in the form of (4).

The angular distributions are corrected for radiative effects of order α^3 and must therefore be compared with the lowest order QED or electroweak prediction. All four measurements depart from the symmetric $(1 + \cos^2\theta)$ distribution predicted by pure QED and prefer a small negative asymmetry. The results on the tau asymmetries from PEP and PETRA are listed in Table III. The measurements have a systematic error of 1% to 2% in addition to the statistical error. The combined values include the systematic errors. The experiments find a negative asymmetry with a typical significance of about two standard deviations. The combined value from PETRA is (-7.4 ± 2.3)%. It is the first significant observation of electroweak interference associated with the tau. The measured values agree within their errors with the prediction of the GSW theory for $\sin^2\theta \approx 0.23$. Since the asymmetry measures the product

$g_A^e \cdot g_A^\tau$, we find by combining the PETRA and PEP results (for $m_Z = 90$ GeV)

$$g_A^e \cdot g_A^\tau = 0.18 \pm 0.06$$

If we use $g_A^e = -1/2$ we obtain for the axial-vector coupling constant of the tau

$$g_A^\tau = T_{3L} - T_{3R} = -0.36 \pm 0.12 \quad .$$

This value is consistent with the weak isospin assignment of leptons in $SU(2)_L \times U(1)$, i.e. $T_{3L} = -1/2$ and $T_{3R} = 0$. Thus we observe e-μ-τ universality in weak neutral currents.

TABLE III: RESULTS FROM PEP AND PETRA ON THE ASYMMETRY IN THE REACTION $e^+e^- \to \tau^+\tau^-$

Experiment	$\sqrt{s}$ GeV	$A_{\tau\tau}$ in %	Expected in GSW for $\sin^2\theta_W = 0.23$ [23]
MAC	29	-1.3 ± 2.9	-6.0
MARK II	29	-3.2 ± 5.0	-5.0 ($\lvert\cos\theta\rvert < 0.7$)
PEP combined	29	-1.7 ± 2.5	-5.8
CELLO	34.2	-10.3 ± 5.2	-8.6
JADE	34.6	-6.9 ± 3.4	-8.6
MARK-J	34.6	$-7.8 \pm 4.0 \pm 2.0$	-8.5
TASSO	34.4	-5.4 ± 4.5	-8.7
PETRA combined	34.5	-7.4 ± 2.3	-8.6

X. AXIAL-VECTOR COUPLING CONSTANT OF CHARM AND BOTTOM [24]

In analogy to leptons we expect a forward-backward asymmetry of the reaction $e^+e^- \to q\bar{q}$, which depends on the axial-vector coupling of the quark of flavor f and its charge Q_f

$$A = \frac{3}{2} \times \frac{1}{Q_f} g_A^e g_A^f \quad . \tag{13}$$

The factor $\frac{1}{Q_f}$ has the consequence that the asymmetries of quarks are much larger than for leptons. We expect an asymmetry of -14% for charm quarks and -25% for bottom quarks at 34.5 GeV. A measurement of these asymmetries requires a determination of the quark flavor and the quark charge, i.e. particle-antiparticle separation. Two methods have been employed so far.

In the first method [25] one studies the reaction $e^+e^- \to q\bar{q} \to \mu$ (or e) + hadrons selecting prompt leptons, which appear to come from the interaction vertex. These events are explained by the semileptonic decay of a heavy quark, more precisely by the decay of a shortlived hadron, which is built up by that quark in the first step of the fragmentation. The sign of the prompt lepton charge indicates the sign of the charge of the parent quark. Examples are $b \to \mu^- X$ and $c \to \mu^- X$. The cascade decay $b \to c \to \mu^+ X$ creates some charge confusion but the lepton spectrum is softer and so a cut on the lepton momentum eliminates a large number of these events. Thus we find that positive muons select c-quarks and b-antiquarks; negative muons select c-antiquarks and b-quarks. If we are not able to separate the quark flavors b and c, their asymmetry will partially cancel, because we add particle and antiparticle asymmetries. Therefore a good flavor separation is of major importance for this measurement.

Quark flavors can be selected with variables, which are sensitive to the quark mass. The experimenters use thrust, which measures the width of a jet and the transverse momentum of a prompt lepton with respect to the thrust (jet) axis [26]. The observed asymmetries are substantially reduced, because the separation of b and c quarks is incomplete and the event sample contains background from π- and K-decay and punch through.

The second method of flavor tagging selects charged D^* mesons by reconstructing the decay modes $D^{*+} \to D^o\pi^+ \to K^-\pi^+\pi^+$ and $D^{*-} \to K^+\pi^-\pi^-$. After a cut on the mass difference between D^* and D^o one observes a D^o peak and uses these events to calculate the asymmetry. This method has a low background, but also a very low number of D^* events, typically about 50 events [27]. The hope is that with better mass resolution and with higher acceptance the statistical significance of this measurement will improve in the future.

Table IV summarizes the preliminary results for charm and bottom [27,28]. The measurements are corrected for acceptance and radiative effects and can be compared to the lowest order electroweak prediction. Although the results, especially for the bottom quark, are not very significant, I made the attempt to combine them. The axial-vector coupling constant of the charm quark is $g_A^c = +0.77 \pm 0.23$ where we expect $g_A^c = T_{3L} - T_{3R} = +1/2$ in the $SU(2)_L \times U(1)$ theory with $\binom{c}{s'}_L$. For the bottom quark we find $g_A^b = -0.45 \pm 0.25$, which agrees astonishingly well with $g_A^b = -1/2$ as expected for $\binom{t}{b'}_L$. Is this an indication that the top quark exists? The answer might be affirmative, since we have further evidence for the existence of the top quark coming from a search for flavor changing neutral currents. If the top quark does not exist or if the b-quark is in a weak SU(2) singlet, one expects the branching ratio for $b \to \mu^+\mu^- X$ to be at least 1% [29]. CESR and PETRA experiments study the reaction $e^+e^- \to b\bar{b}$ and search for a $\mu^+\mu^-$-pair in one of the bottom jets. They find the following upper limits with 95% confidence [30]:

CLEO (CESR)	$B(b \to \mu^+\mu^- X) < 0.4\%$
JADE and MARK-J	$B(b \to \mu^+\mu^- X) < 0.7\%$.

Thus we conclude that the t-quark is a member of a weak isodoublet and we have still some hope to find it at PETRA in autumn 1983 when we scan up to 46 GeV.

TABLE IV: ASYMMETRY AND AXIAL-VECTOR COUPLING OF CHARM AND BOTTOM

Experiment	A_b in %	expected (%)	g_A^c	method
JADE	-27 ± 14	-14	$+1.0 \pm 0.5$	D^*
MARK-J	-17 ± 9	-14	$+0.6 \pm 0.5$	$c \to \mu X$
TASSO	-28 ± 13	-14	$+1.0 \pm 0.5$	D^*
Experiment	**A_b in %**	**expected (%)**	**g_A^b**	**method**
MAC	-7.4 ± 9.2	-12.2	-0.3 ± 0.4	$b \to \mu X$
MARK-J	-15 ± 22	-25	-0.3 ± 0.4	$b \to \mu X$
TASSO	-17 ± 10	-8	-1.1 ± 0.6	$b \to \mu X$

XI. ALTERNATIVE MODELS

Alternative models to the standard electroweak theory [31] seem to be strongly restricted since the discovery of the $W^{\pm}$ and the Z^{o}. Many composite models or models with a larger gauge group SU(2) x U(1) x G or general electroweak mixing schemes have an effective neutral current lagrangian which in the low q^2 limit becomes [20,32]

$$L_{eff}^{NC} = - \frac{4G_F}{\sqrt{2}} (J_\mu^3 - \sin^2\theta_w J_\mu^{em})^2 + C(J_\mu^{em})^2 \qquad (14)$$

It differs from SU(2) x U(1) by a term proportional to the square of the electromagnetic current which is parity conserving and which is therefore invisible in the neutrino experiments and in polarized electron-deuteron scattering. However, it modifies the vector coupling and h_{VV}, previously given by (3), becomes

$$h_{VV} = \frac{1}{4} (1 - 4\sin^2\theta_w)^2 + 4C \quad . \qquad (15)$$

To reproduce the low energy neutrino and electron-deuteron results we use $\sin^2\theta_w = 0.23$ and determine upper limits of the parameter C from leptonic e^+e^--reactions (Table V). The resulting limits impose strong restrictions to the mass spectrum and couplings of Z^o's in these models.

TABLE V: UPPER LIMITS TO THE PARAMETER C WITH 95% CONFIDENCE

CELLO [7]	C < 0.031	MARK-J [10]	C < 0.025
JADE [8]	C < 0.031	TASSO [11]	C < 0.011

CONCLUSIONS

1. e^+e^- experiments at PETRA and PEP have been testing the electroweak theories up to $s \approx 1200$ GeV^2 and $q^2 \approx 1000$ GeV^2 and find good agreement with the GSW model.
2. No new lepton, nor quark, nor fundamental scalar or pseudoscalar particles have been observed.
3. Leptons, including the tau, and quarks interact as if pointlike down to a distance of 10^{-16} cm.
4. The asymmetries due to the axial-vector coupling of the muon and tau have been unambiguously observed.

5. The neutral current couplings of leptons agree within errors with the standard model and we observed e-μ-τ universality of the weak neutral current interaction.
6. The study of neutral current couplings of heavy quarks in $e^+e^- \rightarrow$ hadrons is more difficult. First progress has been made indicating $T_{3L}(c) = +1/2$ and $T_{3L}(b) = -1/2$.
7. Furthermore, since we do not observe flavor changing decays $b \rightarrow \mu^+\mu^-X$, we are confident that the top quark exists. So lets find it!

ACKNOWLEDGEMENT

I wish to thank the members of the PETRA and PEP collaborations and my colleagues from the MARK-J group for providing me with unpublished data. In particular I am very grateful to G. Bella, P. Grosse-Wiesmann, G. Herten, H.-U. Martyn, B. Naroska, F.P. Poschmann and L. Sehgal for discussions and helpful comments. I also wish to thank Mrs. I. Gojdie and G. Zomerland for typing of the manuscript. Finally, it is a pleasure to thank the organizers, especially Harvey and Linda Newman, for this very interesting and enjoyable conference.

REFERENCES

1. At the time of writing this review the Z^o boson has been discovered: UA1 Coll.: G. Arnison et al., Phys. Lett. 126B, 398 (1983) and Phys. Lett. 129B, 273 (1983);
   UA2 Coll.: P. Bagnaia et al., Phys. Lett. 129B, 130 (1983).
2. For reviews see:
   A. Böhm, Proc. of the XVII Rencontre de Moriond, ed. J. Tran Tanh Van, Les Arcs 1982, p.159, and DESY Report, DESY 82-027 (1982);
   S. Yamada, Proc. of the 1983 Int. Symp. on Lepton and Photon Interactions at High Energies, Ithaca 1983, to be published; and references therein.
3. The energy of PETRA has recently been extended to 43 GeV but no new quark or leptons have been found.
4. P.Q. Hung and J.J. Sakurai, Phys. Lett. 69B, 323 (1977).
5. Y.E. Kim et al., Rev. Mod. Phys. 53, 211 (1981).
6. S.L. Glashow, Nucl. Phys. 22, 579 (1961);
   S. Weinberg, Phys. Rev. Lett. 19, 1264 (1967) and Phys. Rev. D5, 1412 (1972);
   A. Salam, in Elementary Particle Theory, ed. N. Svartholm, Stockholm 1968, p.361;
   S.L. Glashow, J. Iliopoulos and L. Maiani, Phys. Rev. D2, 1285 (1970).

7. CELLO Coll.: H.J. Behrend et al., Phys. Lett. 114B, 282 (1982); Z. Phys. C14, 283 (1982) and Z. Phys. C16, 301 (1983) and P. Grosse-Wiesmann, these proceedings.
8. JADE Coll.: W. Bartel et al., Phys. Lett. 99B, 281 (1981); Phys. Lett. 108B, 140 (1982); Z. Phys. C19, 197 (1983) and B. Naroska, these proceedings.
9. MAC Coll.: D.M. Ritson, Proc.of the XXI Int. Conf. on High Energy Physics, ed. P. Petiau and M. Porneuf, Paris 1982, p.52; E. Fernandez et al., Phys. Rev. Lett. 50, 1238 (1983) and G.B. Chadwick, these proceedings.
10. MARK-J Coll.: D.P. Barber et al., Phys. Reports 63, 337 (1980); Phys. Rev. Lett. 46, 1663 (1981); B. Adeva et al., Phys. Rev. Lett. 48, 1701 (1982) and J. Salicio, these proceedings.
11. PLUTO Coll.: Ch. Berger et al., Z. Phys. C7, 289 (1981); Phys. Lett. 99B, 489 (1981) and DESY preprint DESY 83-084 (1983) and G. Bella, these proceedings.
12. TASSO Coll.: R. Brandelik et al., Phys. Lett. 92B, 199 (1980); Phys. Lett. 110B, 173 (1982); Phys. Lett. 117B, 365 (1982) and G. Mikenberg, these proceedings.
13. F.A. Berends and R. Kleiss, Nucl. Phys. B177, 237 (1981).
14. R. Budny, Phys. Lett. 55B, 227 (1975).
15. F. Bergsma et al., Phys. Lett. 117B, 272 (1982).
16. T. Himel et al., Phys. Rev. Lett. 41, 449 (1978).
17. F.A. Berends, R. Kleiss and S. Jadach, Nucl. Phys. B202, 63 (1983).
18. G. Passarino and M. Veltman, Nucl Phys. B160, 151 (1979); G. Passarino, Nucl. Phys. B204, 237 (1982); M. Böhm and W. Hollik, Nucl. Phys. B204, 45 (1982) and DESY Report 83-060 (1983); M. Greco, G. Pancheri-Srivastava and Y. Srivastava, Nucl. Phys. B171, 118 (1980); W. Wetzel, Heidelberg preprint HD-THEP-82-18 and preprint May 1983.
19. MARK-J Coll.: D.P. Barber et al., Phys. Lett. 95B, 149 (1980).
20. J.J. Sakurai, Proc. of the XVII Rencontre de Moriond, ed. Y. Tran Tanh Van, Les Arcs 1982, p.241.
21. CELLO Coll.: H.J. Behrend et al., Phys. Lett. 127B, 270 (1983).
22. MARK II Coll.: C.A. Blocker et al., Phys. Lett. 109B, 119 (1982).
23. The expected values include a correction (of about -0.7% at PETRA energies) due to radiative corrections on the interference terms, which have not been applied to the measurements.
24. Since the results on this subject have improved substantially in the months following this conference, I have included the new results in this article.
25. A. Ali, Z. Phys. C1, 25 (1979); M.J. Puhala et al., Phys. Rev. Lett. D25, 95 (1982) and Phys. Rev. Lett. D25, 695 (1982).

26. MARK II Coll.: M.E. Nelson et al., Phys. Rev. Lett. 50, 1542 (1983);
MAC Coll.: E. Fernandez et al., Phys. Rev. Lett 50, 2054 (1983);
MARK-J Coll.: B. Adeva et al., Phys. Rev. Lett. 51, 443 (1983);
CELLO Coll.: H.-J. Behrend et al., Z. Phys. C19, 291 (1983).
27. TASSO Coll.: M. Althoff et al., Phys. Lett. 126B, 493 (1983).
28. MARK-J Coll.: G. Herten, Proc. of the Int. Europhysics Conf. on High Energy Physics, Brighton 1983.
JADE Coll.: B. Naroska, private communication;
MAC Coll.: G.B. Chadwick, private communication.
29. G.L. Kane and M.E. Peskin, Nucl. Phys. B195, 29 (1982).
30. CLEO Coll.: H. Kagan et al., Proc. of the XVIII Rencontre de Moriond, 1983, to be published;
JADE Coll.: W. Bartel et al., DESY preprint DESY 83-049 (1983);
MARK-J Coll.: B. Adeva et al., Phys. Rev. Lett. 50, 799 (1983).
31. For a review see D. Schildknecht, these proceedings.
32. M. Kuroda and D. Schildknecht, Phys. Rev. Lett. D26, 3167 (1982) and references therein.
33. W. Krenz, Aachen Report PITHA 82/26 (1982) and references therein.

BOUND STATE EFFECTS IN HEAVY FLAVOUR DECAY

Guido Altarelli

Dipartimento di Fisica, Università "La Sapienza" di Roma
INFN - Sezione di Roma

1 - INTRODUCTION

The study of weak decays of heavy flavoured particles is important as a source of information on many interesting aspects of the $SU(3)\otimes SU(2)\otimes U(1)$ standard theory. Strong and electroweak effects all contribute to the determination of the decay properties. New light can be shed on new and old important problems like the understanding of parton dynamics, the riddle of non leptonic weak decays of strange particles and the determination of fundamental parameters in the theory as the K-M mixing angles. The inclusive (semileptonic and total) decay rates are especially important in that they are simplest and allow the least of model dependency. Many interesting problems are already found at this level, some of them not yet clarified, so that we shall concentrate on inclusive rates in this lecture. Due to limits of space in these proceedings I shall only cover some recent results [1,2] on bound state effects in the inclusive rates. These results refer to the charged lepton spectrum and the inclusive rate for semileptonic decay and to a quantitative study of the interference with the spectator in D^+ non leptonic decays.

2 - SEMILEPTONIC WIDTH OF A HEAVY QUARK

The semileptonic width of a heavy flavoured particle is the simplest inclusive rate. When the momentum of the lepton pair is sufficiently large with respect to the hadron binding energy a parton description suggests that the rate for semileptonic decay is reduced to the corresponding quark decay: $Q \rightarrow q+l+\nu_e$. Deviations from the parton model associated with interference with spectators and annihilation are absent in this case or suppressed so that quark decay is appropriate in most cases. We then concentrate on the decay of a heavy free quark. The semileptonic width is given by:

$$\Gamma_{SL}(Q) = \sum_q |U_{Qq}|^2 I\left(\frac{m_q}{M_Q}, \frac{m_e}{M_Q}, 0\right)\left(\frac{M_Q}{m_\mu}\right)^5 \Gamma_\mu \cdot \cdot \left\{1 - \frac{2}{3}\frac{\alpha_s(M_Q^2)}{\pi}\left(\pi^2 - \frac{25}{4}\right) f\left(\frac{m_q}{M_Q}\right) + \cdots\right\} \quad (1)$$

where U_{Qq} is the entry $Q \rightarrow q$ of the quark mixing matrix [3], $I(x,y,z)$ is the three body phase space factor [4] and the last bracket describes the QCD leading correction for real and virtual gluon emission [5]. The function $f(x)$ decreases monotonically from $f(o)=1$ to $f(1) \simeq 0.41$. The QCD correction for $\Lambda \simeq 250$ MeV is $\sim 25\%$ for $c \rightarrow s$ and $\sim 15\%$ for $b \rightarrow c$.

The main theoretical uncertainties on $\Gamma_{SL}(Q)$ arise from a) the value of M_Q (as $\Gamma_{SL} \sim M_Q^5$) and b) the incidence of bound state effects. Information on both can be obtained from a study of the lepton spectrum through a bound state model [1]. According to fig.1, Γ_{SL} for Q bound in a meson is given by a convolution of the decay width in flight of Q (taken as off shell, with virtual mass W and momentum $\vec{q}$) with the probability of finding Q with momentum q in the meson:

$$\Gamma_{SL} = \int d^3q \, |\psi(q)|^2 \, \Gamma_{SL}(\vec{q}, W) \quad (2)$$

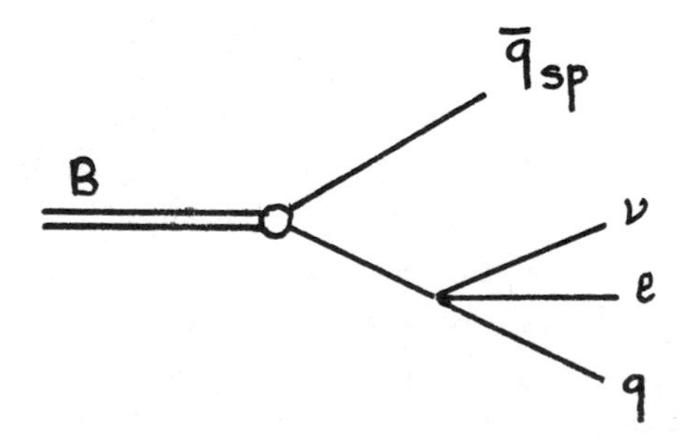

Fig. 1

Treating the spectator as a on shell particle one immediately obtains:

$$W^2 = M_B^2 + m_{sp}^2 - 2M_B\sqrt{\vec{q}^2 + m_{sp}^2} \tag{3}$$

where M_B and m_{sp} are the meson and spectator masses. Similarly for the lepton spectrum:

$$\frac{d\Gamma_{SL}}{dE} = \int d^3q \, |\Psi(q)|^2 \, \frac{d\Gamma_{SL}^Q}{dE}(\vec{q}, W, E) \tag{4}$$

The QCD correction of order α_s to $d\Gamma_{SL}/dE$ as computed in refs.6.7 is included. The main virtue of this bound state model is that it provides us with a parametrization of the Fermi motion of the heavy quark inside the meson in a way that fully respects the kinematics. We take a gaussian form for the wave function:

$$\Psi(q) = (\sqrt{\pi}\, p_F)^{-3/2} \exp\left(\frac{-\vec{q}^2}{2p_F^2}\right) \tag{5}$$

p_F fixes at the same time the shape of the spectrum near the end point and the effective value of the Q mass: $M_Q \simeq \langle W \rangle$.

2.1 Charm decay

For charm [8] the electron spectrum in D decay has been measured by DELCO. The results [1] on P_F can be read from figs. 2-3. The spectator and strange quark masses were taken in the ranges $m_{sp} \simeq (0-150)$ MeV , $m_s \simeq (300-500)$ MeV, which were chosen considering that the final state of D semileptonic decays is mainly made up of a kaon plus pions. The value of α_s is only important for the total width and not for the spectrum shape. Small values of P_F are indicated by the spectrum, in the range:

$$P_F \lesssim 300 \text{ MeV} \tag{6}$$

with preference for $P_F \simeq (0-200)$ MeV. The information obtained from the spectrum can be used to evaluate Γ_{SL} for D mesons. With $\theta_c \sim 0$, $m_{sp}=(100-150)$ MeV, $m_s \simeq (300-400)$ MeV, $P_F = (0-200)$ MeV, $\alpha_s = 0.20-0.36$ (corresponding to $\Lambda = 100-400$ MeV) one obtains

$$\Gamma_{SL}^D = \cos^2\gamma\,(1.6\div3.4)10^{11}\,\text{sec}^{-1} \qquad (P_F \lesssim 200 \text{ MeV}) \tag{7}$$

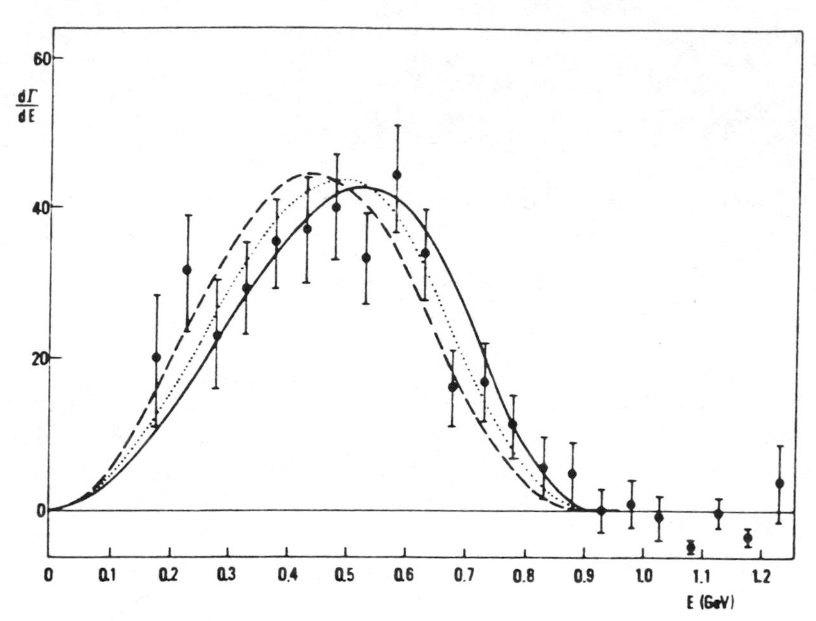

Fig.2 - Charged lepton spectrum in D decay for M_D=1.866 GeV, m=0.3 GeV, m_{sp}=0.15 GeV, α_s=0.38, P_{com}=0.26 GeV and P_F=0 (solid), P_F=0.15 GeV (dotted), P_F=0.3 GeV (dashed). The normalization is fixed to the number of events.

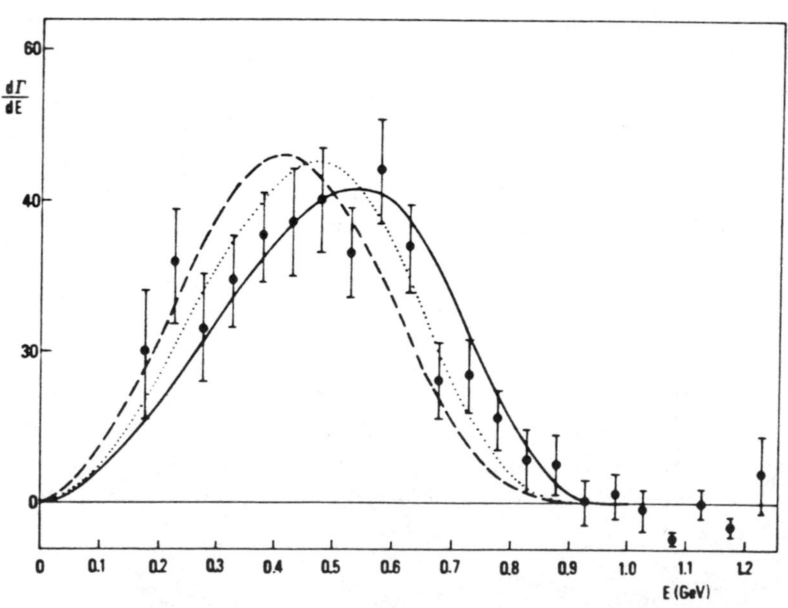

Fig.3 - Charged lepton spectrum in D decay for M_D=1.866 GeV,m=0.5 GeV, m_{sp}=0, α_s=0.38, P_{com}=0.26 GeV and P_F=0 (solid), P_F=0.15 GeV (dotted, P_F=0.3 GeV (dashed). The normalization is fixed to the number of events.

We follow the Maiani definition of mixing angles [9] :

$$
\begin{aligned}
u &\rightarrow \cos\beta\, d_c + \sin\beta\, b \\
c &\rightarrow \cos\gamma\, e^{i\delta} s_c + \sin\gamma(-\sin\beta\, d_c + \cos\beta\, b) \\
t &\rightarrow -\sin\gamma\, e^{i\delta} s_c + \cos\gamma(-\sin\beta\, d_c + \cos\beta\, b)
\end{aligned}
\tag{8}
$$

The existing limit on $\cos^2\gamma$ is derived by the CDHS [10] collaboration from the production of dimuons in ν and $\bar{\nu}$ deep inelastic scattering and is given by $\cos^2\gamma \gtrsim 0.4$.

If one knows the semileptonic branching ratio B_{SL} of D mesons then the estimate in eq.12 of the semileptonic width leads to a predicted range for the total lifetime. The average B_{SL} of 56% D^o and 44% D^+ is known with good accuracy from experiments at the ψ'' resonance:

$$\bar{B}_{SL}(D^{+,o}) = (8.2 \pm 1.2)\ \% \tag{9}$$

As the dominant transition $c \rightarrow s$ is isoscalar, Γ_{SL} must be the same for D^o and D^+ when isospin violations are neglected and $\theta_c \approx 0$ Then from eqs.7,9 one obtains:

$$\cos^2\gamma\ \bar{\tau}(D) = (2 \div 6) 10^{-13}\ \text{sec.} \quad (P_F \lesssim 200\ \text{MeV}) \tag{10}$$

Note that if the range of allowed values of P_F in eq.7 is extended up to 300 MeV the upper value 6.10^{-13} s in eq.10 is increased up to 9.10^{-13} s The present world averages of $D^{+,o}$, F^+ lifetimes are given by [11, 12]

$$
\begin{aligned}
\tau_{EXP}(D^o) &= (4.0 \pm^{1.2}_{0.9}) 10^{-13}\ \text{s} \\
\tau_{EXP}(D^+) &= (9.3 \pm^{2.7}_{1.8}) 10^{-13}\ \text{s} \\
\tau_{EXP}(F^+) &= (2.9 \pm^{1.8}_{0.9}) 10^{-13}\ \text{s}
\end{aligned}
\tag{11}
$$

which leads to

$$\bar{\tau}_{EXP}(D) = (6.65 \pm^{2}_{1.4}) 10^{-13}\ \text{s} \tag{12}$$

Similarly for $B(D^+) \simeq$ 15% (a value favoured by theory) one obtains

$$\cos^2\gamma\, \tau(D^+) \simeq (4\div 9)10^{-13}\,\mathrm{s} \qquad (P_F \lesssim 200\ \mathrm{MeV}) \tag{13}$$

We also mention that from $B_{SL}(\Lambda_c) \simeq 5\%$ as known from experiment, if one takes, for purposes of illustration, $\Gamma_{SL}(\Lambda_c) \simeq \Gamma_{SL}(D)$ (as would be the case for free heavy quark decay) then one obtains:

$$\cos^2\gamma\, \tau(\Lambda_c) \simeq (1.5\div 3)10^{-13}\,\mathrm{s} \quad (P_F \lesssim 200\ \mathrm{MeV}) \tag{14}$$

which is to be compared with the measured value [11]

$$\tau_{EXP}(\Lambda_c) \simeq (2.2^{+0.9}_{-0.5})10^{-13}\,\mathrm{s} \tag{15}$$

In conclusion the theoretical extimates of Γ_{SL} for charm lead to a satisfactory agreement with the measured lifetime values, once the values of B_{SL} are taken from experiment.

2.2 Beauty decay

The electron spectrum of beauty decay is particularly interesting in that it is sensitive to the ratio of the $(b \to c) \sim \sin\gamma\cos\beta \sim \sin\gamma$ and the $(b \to u) \sim \sin\beta$ transitions:

$$\frac{d\Gamma^b_{SL}}{dE} = \sin^2\gamma \frac{d\Gamma^c}{dE} + \sin^2\beta \frac{d\Gamma^c}{dE} \tag{16}$$

The possibility of separating the two components depends on the theoretical control of the hard tail of the charged lepton spectrum. This is sensitive to QCD corrections and to bound state effects. In particular the case of a final massless quark is of special interest and is relevant for the $b \to u + e + \nu$ channel. The electron spectrum is given by

$$\frac{d\Gamma}{dx} = \frac{d\Gamma^0}{dx}\left[1 - \frac{2\alpha_s}{3\pi} G(x,o)\right] \tag{17}$$

where $x = 2E/M_Q$ is the electron energy fraction in the rest frame of the heavy quark of mass M_Q, $d\Gamma^0/dx$ is the uncorrected spectrum while $G(x,o)$ is a complicated function which describes the QCD correction [6,7]. Both $d\Gamma^0/dx$ and $G(x,o)$ are different for the $c \to s$ and the $b \to u$ transitions. In fact, as a consequence of the V-A structure of the current the ν spectrum in the $c \rightarrow s$ transition is the same as the charged lepton spectrum in the $b \rightarrow u$ transition and conversely. In particular:

$$\frac{d\Gamma^0}{dx} = \begin{cases} \sim x^2(1-x) & \text{for} \quad c \rightarrow s \\ \sim x^2(2-x) & \text{for} \quad b \rightarrow u \end{cases} \tag{18}$$

In both cases near $x \rightarrow 1$ [1]

$$G(x,o) \underset{x \rightarrow 1}{\longrightarrow} \ln^2(1-x) \tag{19}$$

While in the $c \rightarrow s$ case the singular behaviour of $G(x,o)$ at $x \rightarrow 1$ is damped by the corresponding vanishing of $d\Gamma^0/dx$, it cannot be ignored in the $b \rightarrow u$ case where the bare spectrum is non vanishing at $x=1$. In the latter case first order perturbation theory becomes inadequate near $x \simeq 1$, at values of $x \cong x_c$ such that:

$$\frac{2\alpha_s}{3\pi} \ln^2(1-x_c) \simeq 1 \Rightarrow M_{QCD} = M(1-x_c)^{1/2} \simeq 0.55 \text{ GeV} \tag{20}$$

(for $M \sim 5$ GeV). M_{QCD} is the minimum invariant mass of the recoiling hydronic system at which first order perturbation theory can still be applied. The double logs are of infrared origin and can be resummed [1] at all orders leading to a Sudakov exponential

$$\frac{d\Gamma}{dx} = \frac{d\Gamma^0}{dx} \exp\left(-\frac{2\alpha_s}{3\pi} \ln^2(1-x)\right) \left[1 - \frac{2}{3}\frac{\alpha_s}{\pi} \tilde{G}(x,o)\right] \tag{21}$$

where

$$\tilde{G}(x,o) = G(x,o) - \ln^2(1-x) \tag{22}$$

The resulting spectrum now vanishes for $x \rightarrow 1$ and one cannot distinguish a massless final quark from a quark with mass M_{QCD}. Since, however, the charm quark mass is much larger than M_{QCD} the possibility of distinguishing the $b \rightarrow c$ and the $b \rightarrow u$ transitions is not spoiled.

Bound state effects can be taken into account by the same model as already described [1]. The same value for P_F as found from the analysis of D decays can be tentatively assumed. This is justified in a crude non relativistic picture of c and b flavoured mesons where the wave function is determined by the reduced mass which is insensitive to the heavy quark mass provided it is sufficiently large.

The results obtained by the CUSB and CLEO collaborations at Cornell can be summarized as follows. The $b \rightarrow c$ transition is found to be dominant with respect to the $b \rightarrow u$ transition and the limit

$$\frac{\sin^2\beta}{\sin^2\gamma} < 0.04 \tag{23}$$

is derived. The total lifetime of a b flavoured meson can be written down in the form:

$$\tau(B) = \left(\frac{B_{SL}}{0.131}\right) \frac{10^{-14}\,s}{\sin^2\gamma\left[Z_c + \frac{\sin^2\beta}{\sin^2\gamma} Z_u\right]} \tag{24}$$

For $P_F \lesssim 300\,MeV$ and $\Lambda = 250\,MeV$ one finds [1] :

$$Z_c = 2.5 \div 3.5 \qquad\qquad Z_u = 5.6 \div 7.2 \tag{25}$$

From the experimental value of B_{SL} measured at the Υ''' , which refers to a weighted average of 60% B^+ and 40% B^o :

$$\bar{B}^{B}_{SL} = 0.131 \pm 0.012 \tag{26}$$

and the limit in eq.23 one obtains

$$\bar{\tau}(B) = \frac{1}{\sin^2\gamma}\,(2.4 \div 4.4)\,10^{-15}\,s \tag{27}$$

The experimental value of $\tau(B)$ is not yet known. The limit

$$\tau(B) < 1.4\;10^{-12}\,s \qquad (JADE) \tag{28}$$

implies $|\sin\gamma| \gtrsim 0.04 \sim \theta_c^2$

3 - NON LEPTONIC RATES AND THE INTERFERENCE WITH SPECTATORS

In the non leptonic sector the theory of the inclusive rate [13] can be summarized as follows:

a) The effective hamiltonian for non leptonic weak amplitudes appears to be well understood. It has the form of a superposition of 4-fermion operators with coefficients which deviate from the free field limit by QCD effects which have been computed up to too-loop accuracy. In addition there may be penguin diagram contributions which are irrelevant for charm decay and presumably for beauty decay.

b) The evaluation of matrix elements of the effective hamiltonian is a more complicated problem. The "annihilation" contribution (with gluon emission) [14] is the main effect to be added to the conventional free heavy quark decay mechanism in order to explain the observed differences in the lifetimes of charmed particles. For charm decay a contribution of the annihilation term of roughly the same magnitude as that of quark decay appears both theoretically understandable and sufficient to explain the data [12] .

Here we address the question of possible spectator effects. An interesting point was put forward in ref.15. At the Cabibbo allowed level a c quark decays according to $c \rightarrow su\bar{d}$. Among mesons the D^+ is unique in that the spectator $\bar{d}$ finds an identical antiquark in the final state of c decay. Note parenthetically that also for Λ_c there would be a u spectator identical with a u quark from c decay. Assuming for a moment that the D^+ wave function was a δ-function at the origin, then the two $\bar{d}$'s would be emitted in the same space-time point. The final state would be the same as produced from the vacuum by an interaction where $\bar{c}$ is replaced by $\bar{d}$: for example $(\bar{c}s)_L\,(\bar{d}u)_L \rightarrow (\bar{d}s)_L\,(\bar{d}u)_L$,because the D^+ is color singlet and spin singlet, which means that color and helicity are just right for the replacement of an incoming c with an outgoing $\bar{d}$. Thus the analogue of O_- would vanish and the non-leptonic rate of D^+ would be reduced by a factor

$$\frac{4c_+^2}{2c_+^2+c_-^2} \sim \frac{1}{2}$$

Now, of course this is an upper bound becuase the wave function is not a δ at the origin and the overlap between spectator and active antiquark must be far smaller. This matter can be quantitatively studied [2] by going back to the bound state model developed in the analysis of semileptonic decays. We write the non leptonic width of D^+ as a sum of a direct plus an interference term arising from the diagrams in fig. 4a,b respectively:

$$\Gamma_{NL}(D^+) = \Gamma_Q + \Gamma_{INT} \tag{29}$$

Omitting mixing angles for simplicity we have:

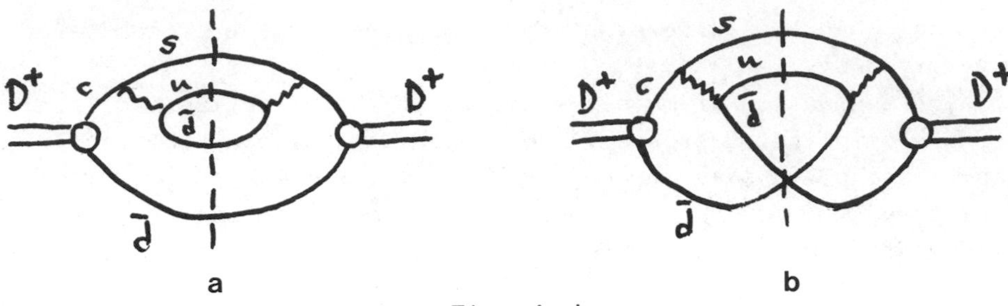

Fig. 4a,b

$$\Gamma_Q = (2c_+^2 + c_-^2) J_{2L} \int d^3q\, |\psi(q)|^2 \left(\frac{W}{E_W}\right) \frac{G_F^2 W^5}{192\pi^3} I\left(\frac{m_s}{W}, \frac{m_{sp}}{W}, \frac{m_{sp}}{W}\right) \tag{30}$$

where $(2c_+^2+c_-^2)$ arises from QCD leading logs, J_{2L} is the two loop QCD correction [16] , W is defined in eq.3, W/E_W is the Lorentz factor for the decay in flight of a c quark of mass W and I is the phase space factor ($m_u \sim m_d \approx m_{sp}$). Correspondingly:

$$\Gamma_{INT} = -(c_-^2 - 2c_+^2) \frac{G_F^2}{\pi} \int \frac{d^3q_1 d^3q_2}{(2\pi)^3} \psi(q_1) \psi^*(q_2) \left(\frac{W_1 W_2}{E_{W_1} E_{W_2}}\right)^{1/2} f(\vec{q}_1, \vec{q}_2) \tag{31}$$

where $f(\vec{q}_1,\vec{q}_2)$ is a kernel arising from matrix elements and phase space which is given in ref.2. The limits on P_F obtained from the lepton spectrum lead to rather strong upper bounds for the ratio $R = -\Gamma_{INT}/\Gamma_Q$ which are shown in fig.5. One realizes that the interference mechanism, although working in the right direction, can only play a marginal role in explaining a lifetime difference between D^+ and D^0 by a factor 2-3 as observed. The theoretical expectation for $B_{SL}(D^+)$ is plotted in fig.6 as a function of P_F and α_s . The resulting prediction turns out to be:

$$B_{SL}(D^+) = (15 \pm {}^{2}_{3})\ \% \tag{32}$$

This prediction is particularly important because it is not affected by the uncertainties connected with the "annihilation" contribution which for D^+ are only present at the Cabibbo suppressed level. Thus it is an important test of the QCD short distance effect in the non leptonic effective hamiltonian.

Finally for beauty decay the interference mechanism is completely negligible. Actually, since all non parton effects are going rapidly down with increasing heavy quark mass, one expects that

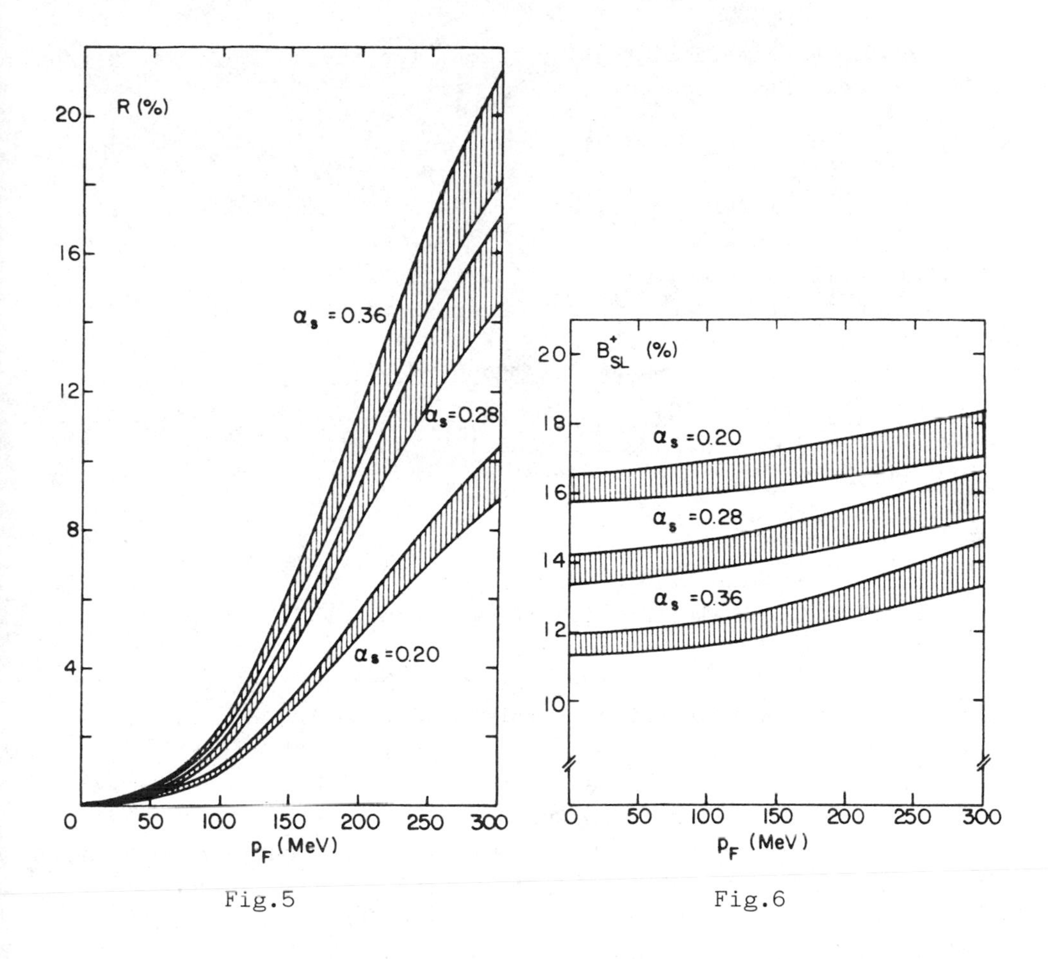

Fig.5

Fig.6

Fig.5 - The ratio $R=-\Gamma_I/\Gamma_Q$ of the interference and the direct term in D^+ non leptonic decay, as a function of p_F for different values of $\alpha_s(M_D)$ and of final quark masses. The upper curve in each shaded band is for m_S=300 MeV, m_{sp}=150 MeV, while the lower curve corresponds to m_S=400 MeV, m_{sp}=100 MeV.

Fig.6 - The semi-leptonic branching ratio B^+_{SL} of D^+ as a function of p_F for different values of $\alpha_s(M_D)$ and of final quark masses. The upper curve in each shaded band is for m_S=300 MeV, m_{sp}=150 MeV, while the lower curve corresponds to m_S=400 MeV, m_{sp}=100 MeV.

b decays are a formidable testing ground for the QCD improved parton model of heavy flavor decay.

References

1. G.Altarelli, N.Cabibbo, G.Corbò, L.Maiani, G.Martinelli; Nucl. Phys. B208 (1982) 365.
2. G.Altarelli, L.Maiani; Phys. Lett. 118B (1982) 414.
3. M.Kobayashi, T.Maskawa; Progr. Theor. Phys. 49 (1973) 652.
4. J.L.Cortes, X.Y.Pham, A.Tounsi; Phys.Rev. D25 (1982) 188.
5. N.Cabibbo, L.Maiani; Phys. Lett. 79B (1978) 109; M.Suzuki; Nucl. Phys. B145 (1978) 420.
6. N.Cabibbo, G.Corbò, L.Maiani; Nucl. Phys. B155 (1979) 93.
7. G.Corbò; Phys. Lett. 116B (1982) 298; Nucl. Phys. B212 (1983) 99.
8. For a review of charm decay, see for example, G.Trilling; Phys. Rep. 75 (1982) 57.
9. L.Maiani; Phys. Lett. 62B (1976) 183.
10. H.Abraimowicz et al.; Zeitschrift f. Phys. C15 (1982) 19.
11. G.Kalmus, Proceedings of the Paris Conference, 1982.
12. L.Maiani, Proceedings of the Paris Conference, 1982.
13. For a short review see for example: G.Altarelli; Phys.Reports 81 (1982) 1.
14. M.Bander, D.Silverman and A.Soni; Phys. Rev. Lett. 44 (1980) 7; E 44 (1980) 962. H.Fritzsch and P.Minkowsky; Phys. Lett. 90B (1980) 455; Nucl. Phys. B171 (1980) 413. W.Bernreuther, O.Nachtmann and B.Stech; Z.Phys. C4 (1980) 257.
15. B.Guberina, S.Nussinov, R.D.Peccei and R.Rückl; Phys. Lett. 98B (1979) 111; see also R.D.Peccei, R.Rückl ; MPI-PAE/PTH 75/81/ 1981, T.Kobayashi, N.Yamazaki; Progr. Theor. Phys. 65 (1981) 775.
16. G.Altarelli, G.Curci, G.Martinelli, S.Petrarca; Phys. Lett. 99B (1981) 141; Nucl. Phys. B187 (1981) 461; see also G.Altarelli; Phys. Reports 81 (1982) 1.

WEAK MIXING AND THE STRUCTURE OF CHARGED CURRENTS

Ling-Lie Chau

Physics Department
Brookhaven National Laboratory
Upton, N.Y.

ABSTRACT

The knowledge of the quark mixing matrix and experimental implications from the Kobayashi-Maskawa model are presented. Nonleptonic decays, neutral particle-antiparticle mixing and CP violations in heavy quark systems are discussed.

INTRODUCTION

The standard model [1] is in great shape: The $W^{\pm}$ have been found [2]! The result of six events of UA1 was reported by C. Rubbia at the American Physics Society Meeting on Jan. 26, '83, and at Brookhaven National Laboratory on Jan. 27, '83. They have a mass of $M_W = (81 \pm 5)$ GeV/c^2, and a production cross section of few $\times 10^{-33}$ cm^2, as predicted by theoretical calculations [3,4,5]. The next exciting thing is to see if the Z^0 is where it should be, the t quark exists and the predicted $e^{\pm}$ asymmetry [3] is true (i.e. e^+ from W^+ tends to move in the $\bar{p}$ direction which e^- from W^- tends to move in the p direction). The observation of the $e^{\pm}$-asymmetry will be a direct confirmation of the quark-parton Brownianion in hadronic interactions! It reminds me of proving the corpuscular structure of matter via Einstein's theory of Browning motion.

Now while we are waiting for these exciting developments, I want to report on the status of some other aspects of the standard model. The outline of the talk is as follows:

I. Quark mixing in weak interaction, and the determination of the Kobayashi-Maskawa mixing matrix.
II. Neutral particle-antiparticle mixing $P^0 \leftrightarrow \bar{P}^0$ and CP violation in heavy-quark systems D^0, B^0, T^0.

III. The physics of ε'.
IV. CP violation in partial decays.
V. Comparison of CP violation effects from various sources.
VI. Rare decays.
VII. Concluding remarks.
For more detailed discussion, see Ref. (6).

I. Quark mixing in weak interactions and the determination of the Kobayashi-Maskawa mixing matrix

This is well known history now. Demanding that CP violation effects come from the complexity of the W bosons' coupling to quark-antiquark Wqq, in 1973 Kobayashi and Maskawa [7] proposed the extension of the quark qork of Cabibbo, Glashow-Maiani-Iliopoulos [8], to three pairs of left-handed doublets and three right-handed singlets

$$\binom{u}{d'}_L, \binom{c}{s'}_L, \binom{t}{b}_L; \ u_R, \ c_R, \ t_R, \ d_R, \ s_R, \ b_R. \tag{1.1}$$

Now the quark mixing matrix is a 3 × 3 unitary matrix where

$$V = \begin{pmatrix} V_{ud} & V_{us} & V_{ub} \\ V_{cd} & V_{cs} & V_{cb} \\ V_{td} & V_{ts} & V_{tb} \end{pmatrix}$$

$$= \begin{pmatrix} c_1 & s_1c_3 & s_1s_3 \\ -s_1c_2 & c_1c_2c_3-s_2s_3e^{i\delta} & c_1c_2s_3+s_2c_3e^{i\delta} \\ -s_1s_2 & c_1s_2c_3+c_2s_3e^{i\delta} & c_1s_2s_3-c_2c_3e^{i\delta} \end{pmatrix} . \tag{1.2}$$

Note that there are three mixing angles and one phase δ, which is the source of CP violation in this model.

V_{ud} can be determined [9-10] from

$0^+ \to 0^+$ nuclear β-decay of ^{14}O, $^{26m}A\ell$

$$V_{ud} = 0.9737 \pm 0.0025 \quad . \tag{1.3}$$

And V_{us} can be determined from K_{e3} and hyperon decays

$$V_{us} = 0.219 \pm 0.002, \tag{1.4a}$$

$$s_1 = 0.227 \pm 0.010, \tag{1.4b}$$

From unitarity one obtains

$$s_3 = 0.28 {}^{+0.21}_{-0.28} . \tag{1.5}$$

Recently there have been two new developments in the hyperon decay fit. It was noticed in Ref. [10] that the inclusion of the $\Delta s = 0$ hyperon decay $\Sigma \to \Lambda e\nu$ significantly worsens the χ^2 in the fit then from pure $\Delta s = 1$ hyperon decays. Recently the measurements have been improved and $\Sigma \to \Lambda e\nu$ data on rate [11] can not be consistently fitted. Another new result is that from the asymmetry of the electron distribution in $\Sigma \to ne\nu$, a positive value of $g_1/f_1 = 0.44 \pm 0.03$ has been measured [12], while the theoretical calculation gives a negative number. Theorists need to work further on this question. Such important experimental results ought to be checked by other independent measurements.

Since the model is designed to provide CP violation, some of the parameters must be determined from the CP violation. The K_L, K_S system is still the only experimentally established system having CP violation since its observation in 1964. To constrain the other two parameters V_{cs}, V_{cd}, two sets of experimental information can be used, i.e. the K_L, K_S mass difference $\Delta_m = m_L - m_S$ and the CP violation parameter $\varepsilon \approx A(K_L \to \pi\pi)/A(K_S \to \pi\pi)$. The mass difference comes about the from $K^0 \leftrightarrow \bar{K}^0$ transition, see Fig. (1.1), and the CP violation effect ε comes from the complexity of the mixing matrix element in the Wqq coupling in Fig. (1.1)

Given s_1, ($s_3 < 0.5$ from Eq. (1.5)), we can [13, 14] determine s_2, s_3 in terms of δ. Because of the redundancy in the contribution of the mixing angles to Δm and ε, one can pick the convention that θ_1, θ_2, θ_3 are all in the first quadrant, and let δ vary in all four quadrants. The results are that Δm and ε do not allow δ to be in the fourth quadrant, $-\pi/2 < \delta < 0$, but δ is allowed in the whole region of $0 < \delta < \pi$, especially including $\delta = 90°$. Here we list some examples of the mixing matrix obtained, given $s_1 = 0.23$:

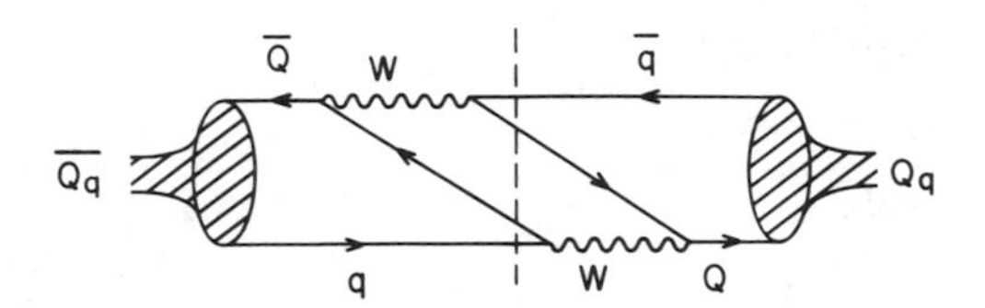

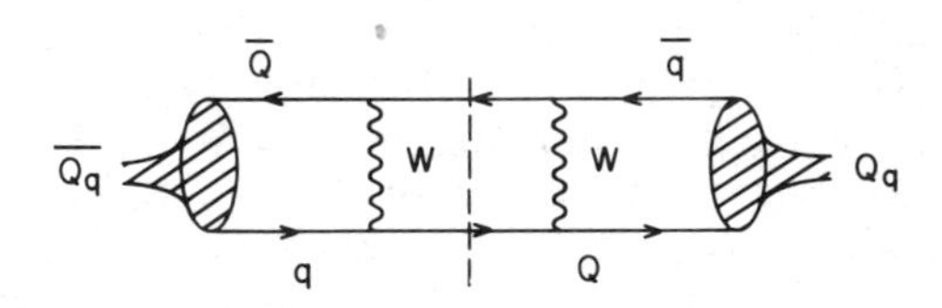

Fig. 1.1 The box diagram for $Q_q^0 \leftrightarrow \overline{Q}_q^0$ transition. For the $K^0 \leftrightarrow \overline{K}^0$, Q = s, q = d. The dashed lines are cut for the absorptive part.

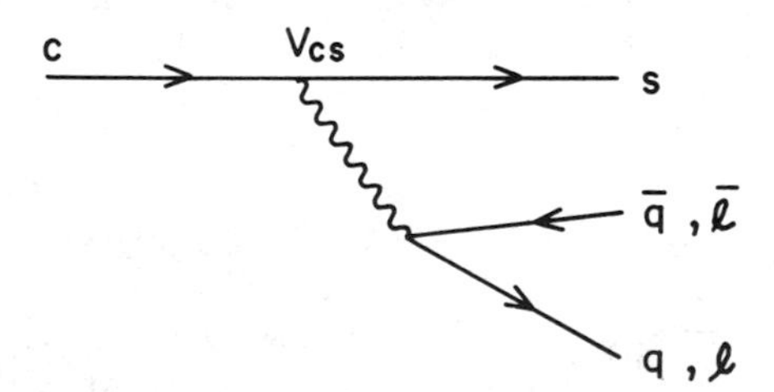

Fig. 1.2 The W-emission diagram for decay.

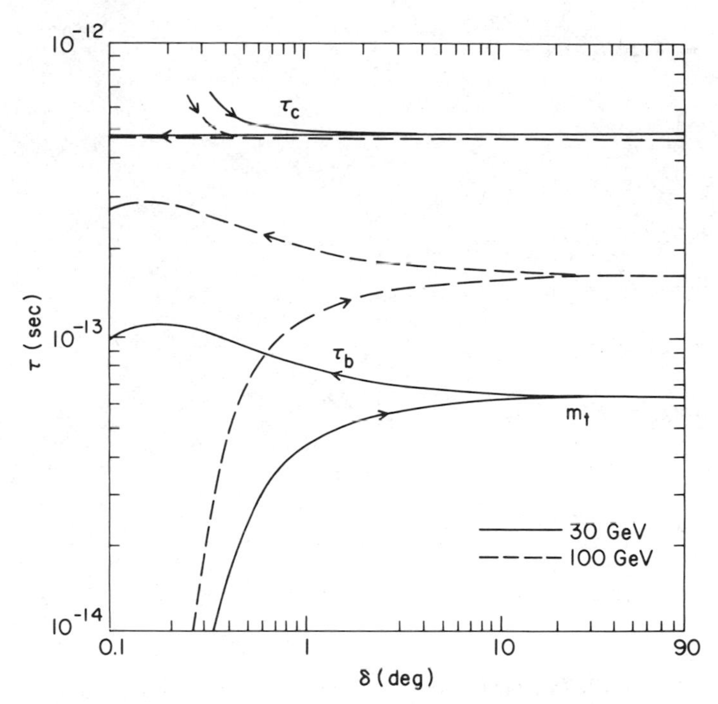

Fig. 1.3 Lifetimes τ_c, τ_b for charmed and b-flavored particles.

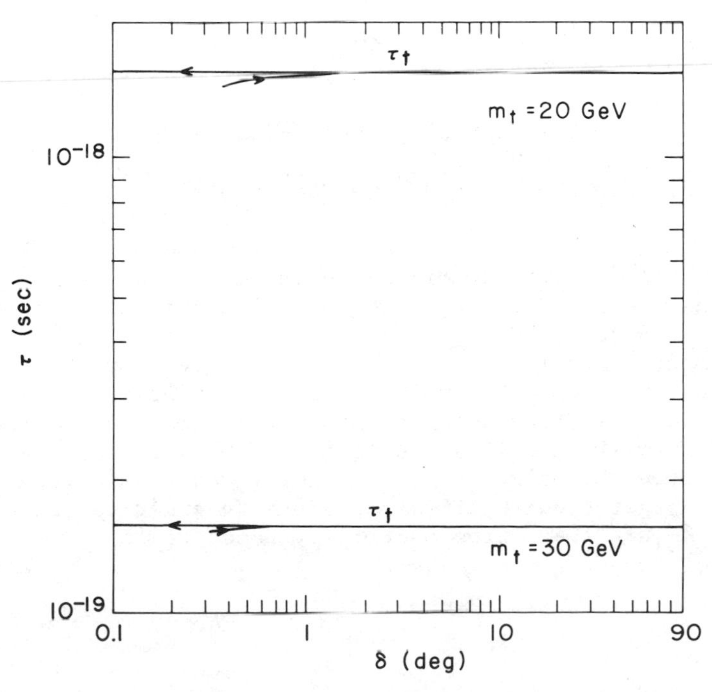

Fig. 1.4 Lifetimes τ_t for the t particles.

$\delta_{minimum} = 0.34°$, $s_3 = 0.5$,

$$V_{B=0.4} = \tag{1.6}$$

$$\begin{array}{c} \\ \\ \\ \end{array}\begin{matrix} d & s & b \end{matrix}$$

$$\begin{pmatrix} 0.9737 & 0.197 & 0.114 \\ -0.227 & 0.789-i0.29\times 10^{-3} & 0.571+i0.51\times 10^{-3} \\ -0.0228 & 0.582+i0.29\times 10^{-2} & -0.813-i0.50\times 10^{-2} \end{pmatrix} \begin{matrix} u \\ c \\ t \end{matrix}$$

$\delta = 90°$, $s_3 = 0.00135$,

$$V_{B=0.4} = \tag{1.7}$$

$$\begin{pmatrix} 0.9737 & 0.228 & 0.308 \times 10^{-3} \\ -0.222 & 0.947-i0.31\times 10^{-3} & 0.12\times 10^{-2}+i0.233 \\ -0.053 & 0.227+i0.13\times 10^{-2} & 0.61\times 10^{-3}-i0.972 \end{pmatrix}$$

$\delta_{maximum} = 179.93°$, $s_3 = 0.5$

$$V_{B=0.4} = \tag{1.8}$$

$$\begin{pmatrix} 0.9737 & 0.197 & 0.114 \\ -0.183 & 0.975-i0.36\times 10^{-3} & -0.122+i0.62\times 10^{-3} \\ -0.135 & 0.0979+i0.49\times 10^{-3} & 0.986-i0.85\times 10^{-3} \end{pmatrix}$$

With these results of mixing angle, and using the naive W-emission diagram Fig. (1.2), we can calculate the lifetimes of the D,B,T decays, Figs. (1.3, 1.4). The charm life was quite correctly predicted (with the modification of other effects that result in the possible D^+, D^0 lifetime difference). Now it is interesting to note that the b life predicted from the model varies from 2×10^{-13} to 10^{-14} sec. depending upon the values of δ. Larger t quark mass tends to result in longer b quark lifetime, since in order to fit ε and Δm of K_S, K_L the mixing matrix V_{cb} needs to be smaller for larger t quark mass.

Some other interesting general results have also emerged from this analysis. The diagonal matrix elements of the mixing matrix are always bigger than the off diagonal elements. The magnitudes of matrix elements decrease as they move away from the diagonal. This says that quarks like to keep their original generation identity though there is some quark mixing among different generations of quark doublets, i. e. a phenomenon of generation gap. In physical terms, quarks decay in a cascade fashion, e.g the b particles will prominently decay into charm particles, then charm to strange. This is now supported by experiment from CESR [15]. This analysis also predicts that the t particles will decay mainly into b particles. However this phenomenon of generation gap also prevents us from knowing whether there are more doublets, just like the Cabibbo theory fit strange particle very well though the third doublet can exist.

Besides the region of $0.35° < \delta < 179.92°$, there is a small region in the third quadrant of δ that is allowed from this K-system analysis, $180.1° < \delta < 180.25°$, $0.14 < s_2 < 0.38$, $0.45 < s_3 \lesssim 0.5$. The distinct feature of the solution in this region is $|V_{ub}|/|V_{cb}| \geq 0.3$. **However this region has now been firmly ruled out by the latest CESR data [15] of b decay in the T(4s) region, $V_{ub}/V_{cb} \lesssim 0.2$.**

A word of caution is in order on the way to determine the quark mixing matrix using Fig. (1.1). It has a matrix element

$$\langle \bar{K}^0 | [\bar{s}\, \gamma_\mu (1-\gamma_5) d][\bar{s}\, \gamma^\mu (1-\gamma_5) d] | K^0 \rangle ,$$

which contains the strong interaction contributions of the making of K^0 from $\bar{s}d$ quarks. To calculate this quantity accurately requires our fundamental understanding of the non-perturbative part of the strong interaction or the long distance effect, in the current QCD terminology. We certainly do not have this knowledge. Conventionally it was estimated in two ways, the vacuum insertion calculation and the MIT-bag calculation. Recently the extreme uncertainty in the latter calculation has been noted and new methods are used to estimate this matrix element [16]. Actually it will be interesting to calculate this matrix element using the lattice gauge theory method [17].

The absolute values of the mixing matrix elements can also be determined from many other reactions involving charm and b-flavor particle productions and decays [14,18,19], which we list in Table I, as taken from Ref. [14]. To summarize we list the range of the absolute values of the matrix elements determined:

$$V_{abs} = \left(\left| V_{ij} \right| \right) =$$

$$\begin{array}{ccc} d & s & b \\ \end{array}$$

$$\begin{pmatrix} 0.9712-0.9762 & 0.217-0.221 & 0.098-0.0 \\ 0.17-0.23 & 0.66-1.0 & 0.73-0.0 \\ 0.17-0.0 & 0.72-0.0 & 0.67-1.0 \end{pmatrix} \begin{array}{c} u \\ c \\ t \end{array} \qquad (1.9)$$

Table I Summary of $|V_{ij}|$ Determination from Sources Other than Box-graph Fittings of Δm, ε of the $K^0\bar{K}^0$ Systems

(1)	$V_{ud} = c_1 = 0.9737 \pm 0.0025$	$0^+ \rightarrow 0^+$ nuclear β decay
(2)	$V_{us} = s_1c_3 = 0.219 \pm 0.002$	K_{e3} and semileptonic decay of
(2a)	$s_3 = 0.28 \begin{smallmatrix} +0.21 \\ -0.28 \end{smallmatrix}$	hyperons
(3)	$0.64 < \lvert V_{cs}\rvert < 1$	$c \rightarrow s\, e\, \nu_e$
(3a)	$\lvert V_{cs}\rvert = 0.8 \pm 0.2$	$D^+ \rightarrow K^0 e^+ \nu_e$
(4)	$\lvert V_{cs}\rvert^2 + \lvert V_{cd}\rvert^2 = 0.49 - 1.04$	from $\bar{\nu}\,(\bar{d}+\bar{s}) \rightarrow \mu^+ c\, X \rightarrow \mu^+ \ell^- X$
(4a)	$\lvert V_{cb}\rvert^2 = 0.0 - 0.51$	from $\lvert V_{cs}\rvert^2 + \lvert V_{cd}\rvert^2 + \lvert V_{cb}\rvert^2 = 1$
(5)	$\lvert V_{cd}\rvert = 0.17 - 0.23$	from the difference of $\nu_\mu(d+s) \rightarrow \mu^- c\, X' \rightarrow \mu^- \ell^+ X$ and $\bar{\nu}_\mu(\bar{d}+\bar{s}) \rightarrow \mu^+ c\, X' \rightarrow \mu^+ \ell^- X$
(5a)	$\lvert V_{cd}\rvert = \lvert s_1c_2\rvert$, $s_2 < 0.6$	using (5)
(6)	$\lvert V_{cs}\rvert = 1.0 - 0.66$	from (4) and (5)
(7)	$\lvert V_{cd}\rvert / \lvert V_{cs}\rvert = 0.17 - 0.35$	from (5) and (6)
(8)	$\lvert V_{cd}\rvert / \lvert V_{cs}\rvert < 0.4$	from (5a), $s_3 < 0.5$, $s_2 < 0.5$
(9)	$\lvert V_{cd}\rvert / \lvert V_{cs}\rvert < 0.77$	from $\Gamma(D^+ \rightarrow \pi^0\pi^+)/\Gamma(D^+ \rightarrow \bar{K}^0\pi^+) < 0.3$ and $\Gamma(D^+ \rightarrow \pi^0\pi^+)/\Gamma(D^+ \rightarrow \bar{K}^0\pi^+) = 1/2\, \lvert V_{cd}/V_{cs}\rvert$
(10)	$\lvert V_{cd}\rvert / \lvert V_{cs}\rvert < 0.30 \pm 0.06$	from $\Gamma(D^0 \rightarrow K^+K^-)/\Gamma(D^0 \rightarrow K^-\pi^+)$, and $\Gamma(D^0 \rightarrow \pi^+\pi^-)/\Gamma(D^0 \rightarrow K^-\pi^+)$
(11)	$\lvert V_{ti}\rvert$ in Eq. (1.9)	from (1), (2), (3), (4), (5) and the unitarity of the V Matrix
(12)	$0.06 < \lvert V_{cb}\rvert$	from $\tau_b < 1.4 \times 10^{-12}$ sec
(13)	$\lvert V_{ub}\rvert = s_1s_3 < 0.112$	from $s_3 < 0.5$
(14)	$\lvert V_{ub}\rvert / \lvert V_{cb}\rvert < 0.2$	$b \rightarrow c\, \ell^- \bar{\nu}_\ell$

From the values of the matrix elements $|V_{ij}|$ we can ask what constraints they put on the mixing angles.

In Fig. (1.5a,b,c) the constant contours of $|V_{ij}|$ in s_2-s_3 plane from various experimental limits are given. $s_3 < 0.5$ from values $|V_{ud}|$, $|V_{us}|$ obtained from nuclear β decay and strange particle decay; $s_2 < 0.5$ from $|V_{cd}|$ obtained from $\bar{\nu},\nu N \to \mu^+\mu^- X$; $|V_{cd}/V_{cs}| < 0.3$ from $\Gamma(D^0 \to \pi^+\pi^-)/\Gamma(D^0 \to K^-\pi^+)$ value; $|V_{ub}/V_{cb}| < 0.2$ from $B \to \ell^- X$ measurement; $|V_{cb}| > 0.06$ from $\tau_b < 1.4\times10^{-12}$ sec. $|V_{cb}| = 0.19$ is a calculation for $\tau_b = 1.4\times10^{-13}$ sec. The circles in the figures are values of s_2, s_3 determined from ε and Δm; the dashed circles are from B = 1, the vacuum insertion calculation, and the solid curves are from the B = 0.4 calculation.

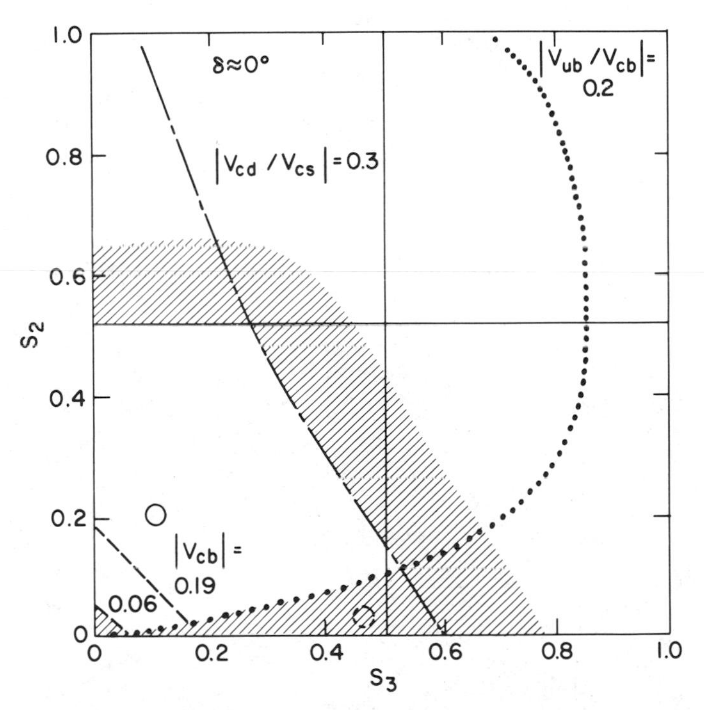

Fig. 1.5a Limits on the region of s_2, s_3 for $\delta \approx 0$ from various experiments: See text.

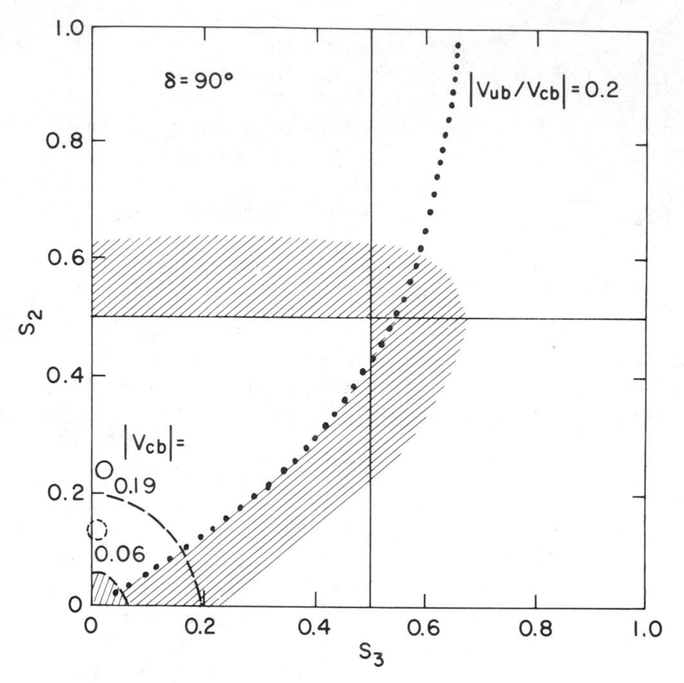

Fig. 1.5b Same as Fig. 1.5a for $\delta = 90°$.

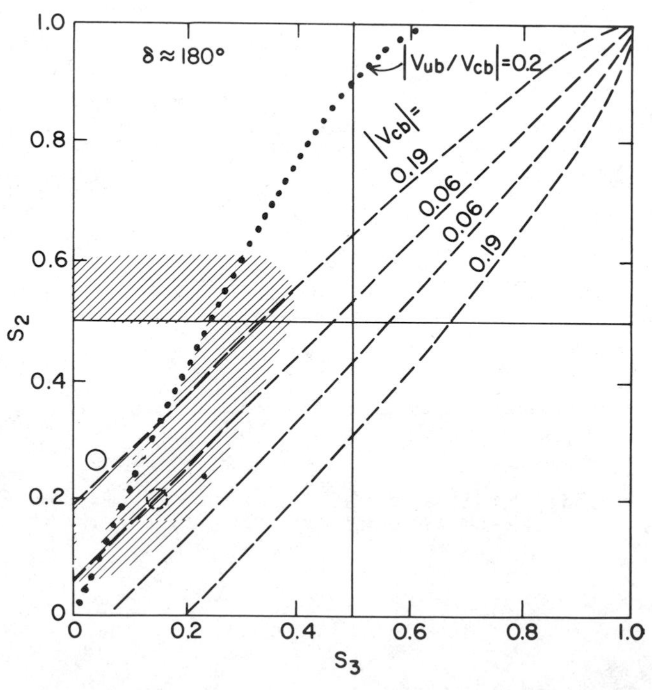

Fig. 1.5c Same as Fig. 1.5a for $\delta = 180°$.

We see that in some regions of δ, the B = 1 calculation is already ruled out. So we see that future experimental improvement on the information for $|V_{ij}|$, not only will narrow down the region of the mixing angles and the phase, but will also shed light on the weak decay dynamics.

To summarize for this section:

(1) Approximate ranges of $|V_{ij}|$ are known,

(2) we need more data of charm and b-flavor particles, decays and production.

(3) The further determination on the value of δ depends on future information of CP violation.

II. Neutral particle-antiparticle mixing and CP violation in heavy quark systems

Due to the neutral particle-antiparticle mixing $p^0 \leftrightarrow \bar{p}^0$, the physical states become

$$|P^0_\pm> = (1 + \varepsilon)|P^0> \pm (1 - \varepsilon)|\bar{P}^0> \quad , \tag{2.1}$$

and their time evolutions are

$$|P^0_\pm(t)> = e^{im_\pm t - 1/2\, \Gamma_\pm t}|P^0_\pm(0)> \quad . \tag{2.2}$$

For a state being p^0 at $t = 0$, $|\phi(t=0)> = |p^0>$, and later at time t

$$|\phi(t)> = a_p(t)|p^0> + a_{\bar{p}}(t)|\bar{p}^0> \quad , \tag{2.3}$$

where,

$$a_p(t) = [e^{im_+t - 1/2\, \Gamma_+t} + e^{im_-t - 1/2\, \Gamma_-t}],$$

$$a_{\bar{p}}(t) = [e^{im_+t - 1/2\, \Gamma_+t} - e^{im_-t - 1/2\, \Gamma_-t}] .$$

For heavier particles like $D^0\bar{D}^0$, $B^0\bar{B}^0$, and $T^0\bar{T}^0$ due to their short lifetimes the neutral particle-antiparticle mixing and CP violation must be studied via the time integrated results:

$$r(p) \equiv \frac{"p^0 \to \bar{p}^0"}{"p^0 \to p^0"} = \frac{\int |a_{\bar{p}}(t)|^2\, dt}{\int |a_p(t)|^2 dt} = \eta^{-2}\Delta \quad , \tag{2.4}$$

where "$p^0 \to \bar{p}^0$" denote the t-integration probability of a particle p^0 at $t = 0$ to develop into $\bar{p}^0$, and

$$\eta = \left|(1 - \varepsilon)/(1 + \varepsilon)\right| \quad , \tag{2.5}$$

$$\Delta = \frac{(\delta m/\bar{\Gamma})^2 + (1/2\ \delta\Gamma/\bar{\Gamma})^2}{2 + (\delta m/\bar{\Gamma})^2 - (1/2\ \delta\Gamma/\bar{\Gamma})^2} \quad , \tag{2.6}$$

with $\delta m = m_+ - m_-$, $\delta\Gamma = \Gamma_+ - \Gamma_-$, and $\bar{\Gamma} = 1/2\ (\Gamma_+ + \Gamma_-)$.
Similar discussions can be done for a $|\bar{p}^0\rangle$ state at $t = 0$, and define

$$\bar{r}(p) \equiv \frac{"\bar{p}^0 \to p^0"}{"\bar{p}^0 \to \bar{p}^0"} = \eta^{-2}\Delta \quad . \tag{2.7}$$

For small CP violation, $\eta = 1$,

$$r(p) \approx \bar{r}(p) \approx \Delta \quad . \tag{2.8}$$

There are two cases of maximal mixing, i.e.

$\bar{r}(p),\ r(p) \approx \Delta \approx 1$,
(1) $|1/2\ \delta\Gamma/\Gamma| \approx 1 \gg (\delta m/\Gamma)$, i.e. $\Gamma_+ \gg \Gamma_-$, or $\Gamma_- \gg \Gamma_+$, this is the K^0_S, K^0_L case. In this situation, either with the K^0 or $\bar{K}^0$ to begin with, it will quickly end up as $K^0{}_L$, which is an almost equal mixture of K^0 and $\bar{K}^0$.
(2) $|\delta m/\Gamma| \gg 1$. In this situation, before decaying, the system oscillates very quickly between p^0 and $\bar{p}^0$ and appears as an equal mixture of p^0 and $\bar{p}^0$. This turns out to be the case for the $\bar{B}^0 B^0$ system. When mixing is appreciable, we can study CP violation by measuring $a(p)$,

$$a(p) = \frac{("p^0\bar{p}^0 \to \bar{p}^0\bar{p}^0") - ("p^0\bar{p}^0 \to p^0p^0")}{("p^0\bar{p}^0 \to \bar{p}^0\bar{p}^0") + ("p^0\bar{p}^0 \to p^0p^0")} \approx -4\left|\varepsilon_p\right| \tag{2.9}$$

To calculate Γ, we use the W-emission diagram Fig.(1.2). To calculate δm, we use the box graphs shown in Fig. (1.1), and $\delta\Gamma$ the imaginary part of Fig. (1.1). It turns out that the neutral particle-antiparticle mixing $r(p)$, $\bar{r}(p)$ and the CP violation $a(p)$ for D and T particles are extremely small. This is due to the simple fact that the decay width Γ is quark mixing-matrix V_{ij} nonsuppressed for the D and T particles, yet $\delta\Gamma$, δm are always quark mixing-matrix V_{ij} suppressed. Therefore $\delta m/\Gamma$, $\delta\Gamma/\Gamma$ are always very small in Eq. (2.6), see Fig. (2.1), that gives $\bar{r}(p)$, $r(p) \approx \Delta \ll 1$. Once the mixing is small, there is little chance to see CP violation effect $a(p)$, see Fig. (2.2).

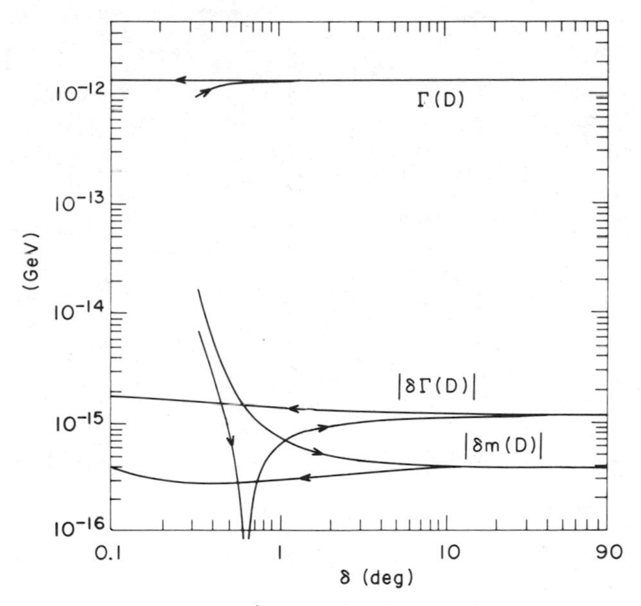

Fig. 2.1 Calculated Γ(D), |δΓ(D)|, |δm(D)| as a function of δ. For details see Refs. [6] and [14].

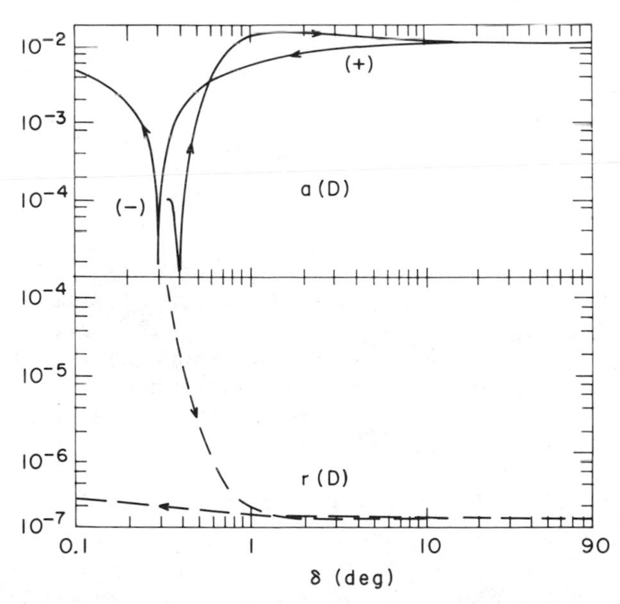

Fig. 2.2 Calculated a(D), r(D) as function of δ; for definition of a(D), r(D) see Eqs. (2.4-9). Since it is a logplot, only the absolute values of |a(D)| are given, the sign is indicated by (+) or (-)

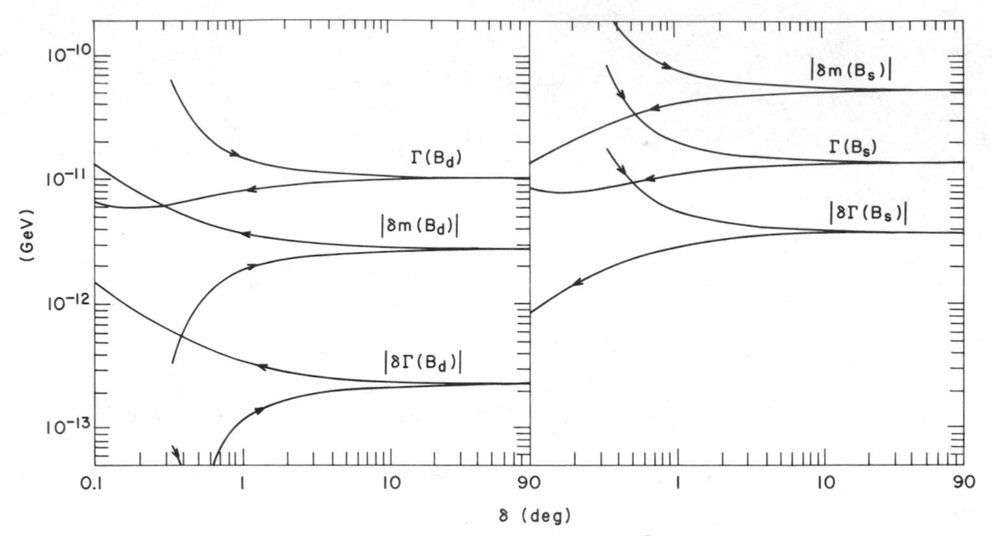

Fig. 2.3 Same as Fig. (2.1) for B_d^0, B_s^0 Systems.

However for the B system, the situation is different. Γ_B is also quark mixing-matrix suppressed so that there are chances $\delta m_B/\Gamma_B$, $\delta\Gamma_B/\Gamma_B$ are appreciable. It turns out to be the case for some reasonable choice of the parameters in the calculation, Figs. (2.3) (2.4). However the CP effect is small as shown in Fig. (2.4b) for $\gamma \cdot a$.

To look for such $B^0\bar{B}^0$ mixing, some characteristic decay B^0, $\bar{B}^0$ can be used, e.g.:

$$B^0 \to \ell^- X,\ \bar{K}\ X,\ c\ X \quad , \tag{2.10a}$$

$$\bar{B}^0 \to \ell^+ X,\ K\ X,\ \bar{c}\ X \quad . \tag{2.10b}$$

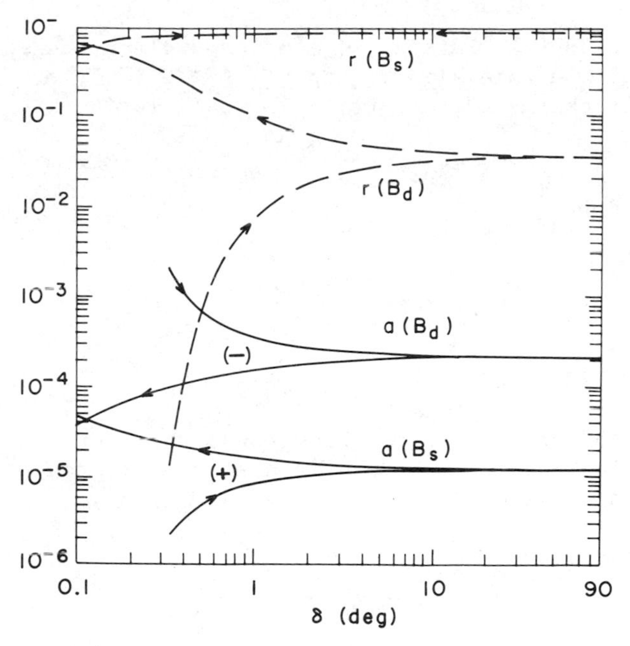

Fig. 2.4a Same as Fig.2.2 for B_d, B_s systems.

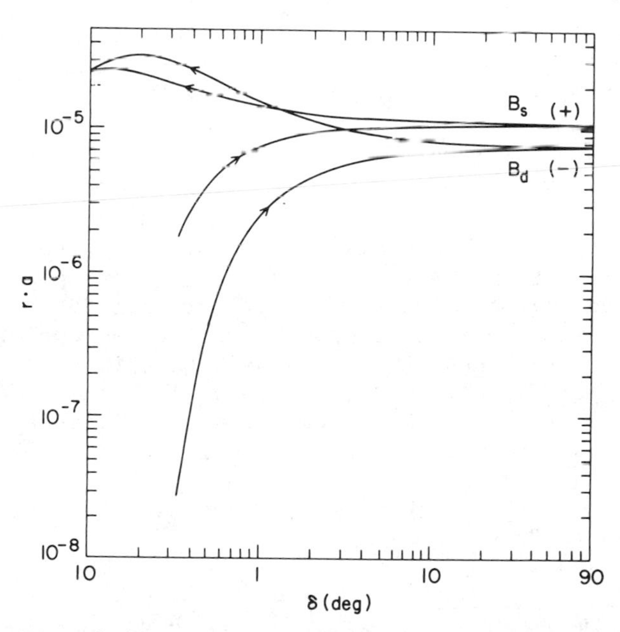

Fig. 2.4b a.γ, a measure of CP violation, for the b particle system.

The signature for $B^0\bar{B}^0$ mixing in e^+e^- or nucleon nucleon reactions will be same sign di-leptons $\ell^+\ell^+$, $\ell^-\ell^-$; or di-kaons KK, $\bar{K}\bar{K}$; or di-charms cc, $\bar{c}\bar{c}$ from $B^0\bar{B}^0$ oscillation

$$e^+e^- \to B^0\bar{B}^0 \nearrow \bar{B}^0\bar{B}^0 \to \ell^+\ell^+ X \ , \qquad (2.11a)$$

$$\searrow B^0B^0 \to \ell^-\ell^- X \ ; \qquad (2.11b)$$

$$NN \nearrow B^0\bar{B}\ X \to B^0\bar{B}^0 X \to \ell^+\ell^+ X \ , \qquad (2.12a)$$

$$\searrow B\bar{B}^0\ X \to B^0 B X \to \ell^-\ell^- X \ ; \qquad (2.12b)$$

Experimentally, of course, the background subtraction is very important. The estimate of such same sign di-lepton production can be done as follows.

$$\frac{\sigma(e^+e^- \to \ell_b^+\ell_b^+ X)}{\sigma(e^+e^- \to B\bar{B})} = \frac{\sigma(e^+e^- \to B_s^0\bar{B}_s^0) + (\sigma(e^+e^- \to B_d^0\bar{B}_d^0)}{\sigma(e^+e^- \to B\bar{B})}$$

$$\times \left[Br(B^0 \to \ell X)\right]^2 \times \frac{1}{2} \times \frac{1}{2}$$

$$\simeq \left[(\frac{1}{3} + \frac{1}{3}) \text{ or } (\frac{1}{5} + \frac{2}{5})\right] \times (10^{-1})^2 \times (\frac{1}{2})^2$$

$$\approx (1.7 \text{ or } 1.5) \times 10^{-3} \ , \qquad (2.13)$$

where (1/3 + 1/3) are the probability of creating a $s\bar{s}$, $d\bar{d}$, pair respectively for equal probability of $d\bar{d}$, $s\bar{s}$, $u\bar{u}$ creation; (1/5 + 2/5) for a factor 2 suppressed creation of $s\bar{s}$ pair comparing to $d\bar{d}$ and $u\bar{u}$. The factors 1/2 are for "$B^0 \to \bar{B}^0$" and "$\bar{B}^0 \to B^0$" when maximal mixing between $B^0\bar{B}^0$. For an optimistic 50,000 $B\bar{B}$/yr at $\Upsilon(4s)$ at CESR, there will be only 85-75 $\ell^+\ell^+$, or $\ell^-\ell^-$ events. For KK, $\bar{K}\bar{K}$, or cc, $\bar{c}\bar{c}$ production, the event number are increased by $(10)^2$ because $Br(B \to c \to s) \sim 1$ rather than the $Br(B \to \ell X) \sim 10^{-1}$. However the difficulty is to identify them.

For nucleon-nucleon scattering, the same sign dilepton production can be from

$$NN \to B^0\bar{B}\ X \xrightarrow[B^0 \to \bar{B}^0 \text{ from mixing}]{} \bar{B}^0\bar{B}\ X \to \ell_b^+\ell_b^+ X \ , \qquad (2.14)$$

where $\bar{B}$ can be $\bar{B}^0$ or B^+.

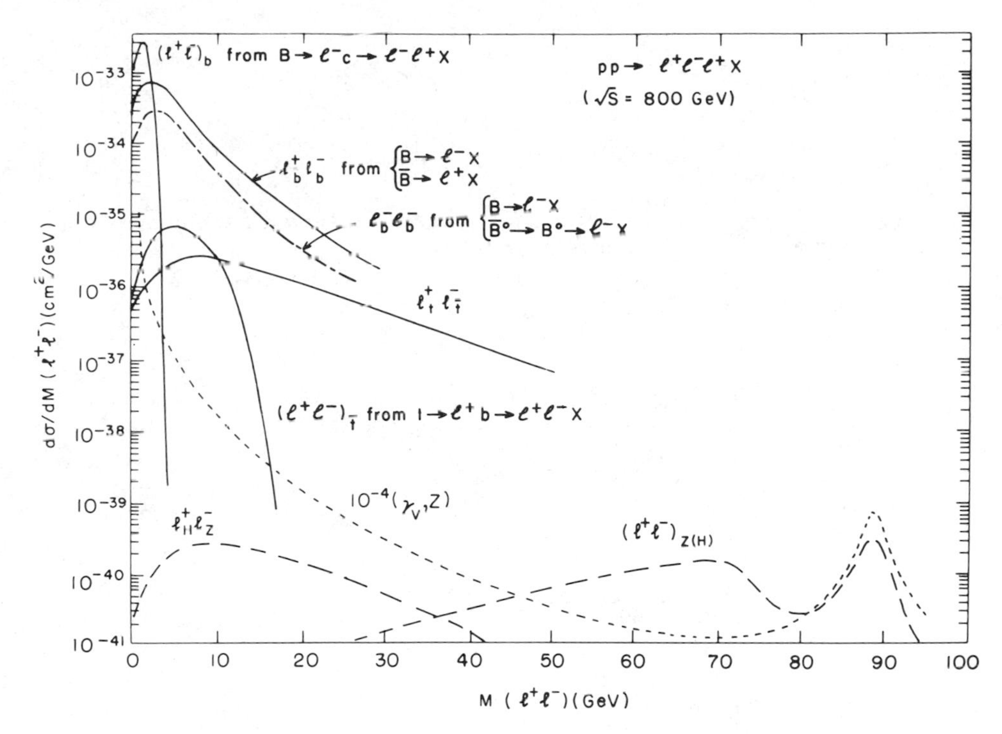

Fig. 2.5 A panorama of lepton distributions. The dash-dot curve is for $pp \to B^0\bar{B}X \to \bar{B}^0\bar{B}X \to (\ell^+\ell^+)\ \bar{c}X \to (\ell^+\ell^+)\ \ell^-X$. For other details see Ref. [20].

$$\frac{\sigma(NN \to \ell_b^+\ell_b^+\ X)}{\sigma(NN \to B\bar{B}\ X)} = \frac{\sigma(NN \to B_s^0\bar{B}) + \sigma(NN \to B_d^0\bar{B})}{\sigma(NN \to B\bar{B})}$$

$$\times \left[Br(B \to \ell\ X)\right]^2$$

$$= \left(\frac{2}{3} \text{ or } \frac{3}{5}\right)\ (10^{-1})^2 \quad ; \tag{2.15}$$

where 2/3 is for equal probability of creating a $s\bar{s}$, $d\bar{d}$ pair of quark; while 3/5 if the $s\bar{s}$ pair creation is only half of the $u\bar{u}$, $d\bar{d}$ creation.

From gluon-gluon interaction in $N\bar{N}$ scattering to produce $B\bar{B}$,

$$\sigma(NN \to B\bar{B}\ X) \approx 10^{-31}\ cm^2, \quad \text{at } \sqrt{s} = 800\ GeV/c, \tag{2.16}$$

$$= 10^{-33}\ cm^2, \quad \text{at } \sqrt{s} = 62\ GeV/c, \tag{2.17}$$

$$\sigma(NN \to \ell_b^+ \ell_b^+ X) \approx 10^{-33}\ \mathrm{cm}^2, \quad \text{for } \sqrt{s} = 800\ \mathrm{GeV}, \tag{2.18}$$

$$\approx 10^{-35}\ \mathrm{cm}^2, \quad \text{for } \sqrt{s} = 62\ \mathrm{GeV}\ . \tag{2.19}$$

So there are many such events in nucleon-nucleon scattering, if same sign di-lepton from other sources can be subtracted. Fig. (2.5) shows a panorama of di-lepton distribution in nucleon-nucleon scattering.

To summarize:

(1) CP violation effects from zero particle antiparticle mixing, i.e. "$p^0 p^0 - \bar{p}^0 \bar{p}^0$" are very small from neutral D, B, and T particles.

(2) We need to look for CP violation in other ways, see next Section.

(3) $B^0 \bar{B}^0$ mixing can be maximal. Its signals are same sign di-lepton production; or KK; $\bar{K}\bar{K}$; cc; $\bar{c}\bar{c}$ productions.

III. The Physics of ε'

The CP violation parameter ε', which is related to the $\Delta I = 2$ part of the K decay, is given by [21]:

$$\eta_{00} \equiv \frac{A(K_L \to \pi^0\pi^0)}{A(K_S \to \pi^0\pi^0)} = \varepsilon - 2\varepsilon' \ , \tag{3.1}$$

$$\eta_{+-} \equiv \frac{A(K_L \to \pi^+\pi^-)}{A(K_S \to \pi^+\pi^-)} = \varepsilon + \varepsilon' \ , \tag{3.2}$$

$$\varepsilon' = (\eta_{+-} - \eta_{00})/3$$

$$= \frac{i}{\sqrt{2}} \, \mathrm{Im}\left(\frac{a_2}{a_0}\right) e^{i(\delta_2 - \delta_0)}$$

$$\approx \frac{i}{\sqrt{2}} \left|\frac{\mathrm{Re}\ a_2}{\mathrm{Re}\ a_0}\right| (t_2 - t_0)\, e^{i(\delta_2 - \delta_0)}, \quad t_i \equiv \frac{\mathrm{Im}\ a_i}{\mathrm{Re}\ a_i} \tag{3.3}$$

Our present experimental knowledge is [22]:

$$\delta_2 - \delta_0 = \frac{\pi}{4} \Longrightarrow \varepsilon,\ \varepsilon' \text{ are } \pm \text{ parallel;} \tag{3.4}$$

$$\left|\frac{a_2}{a_0}\right| \approx \frac{1}{20}, \text{ from } \frac{\Gamma(K^+ \to \pi^+\pi^0)}{\Gamma(K_S \to \pi\pi)} \approx \frac{1}{670} \quad ; \tag{3.5}$$

$$\frac{\varepsilon'}{\varepsilon} = -\,0.003 \pm 0.015, \text{ from } \frac{\eta_{00}}{\eta_{+-}} = 1.008 \pm 0.04. \tag{3.6}$$

It was proposed in Ref. (23), the K decay is dominated by the "Penguin" diagram, Fig. (3.1). In that scenario of explaining $\Delta I = 1/2$ dominance with the K-M phase convention, a_0 is complex and has a rather small imaginary part, but a_2 is real. This phase difference between a_0 and a_2 gives $\varepsilon' \neq 0$. This was the point made by Gilman and Wise [24]:

Assuming dominance from "Penguin" diagrams alone, elaborate calculations have been made including various considerations of perturbative QCD effects [25]. The value of Im a_0/Re a_0, thus ε'/ε, can vary a great deal

$$2 \times 10^{-4} < \left|\text{Im } a_0/\text{Re } a_0\right| < 8 \times 10^{-4} \quad , \tag{3.7}$$

which implies

$$1/600 < \left|\varepsilon'/\varepsilon\right| < 1/80 \quad , \tag{3.8}$$

depending on the approximation scheme, and the QCD scale 0.1 GeV $< \Lambda < 0.7$ GeV. The smaller values of Λ tend to give smaller values of ε'/ε. Combining this calculation of ε' with that of ε in the K-M model, as given in Section I, it was noted by Hagelin [25] that ε' and ε are parallel as $s_\delta > 0$, and antiparallel as $s_\delta < 0$. Now that the $s_\delta < 0$ solution is ruled out, so the prediction from the model is ε' and ε are parallel.

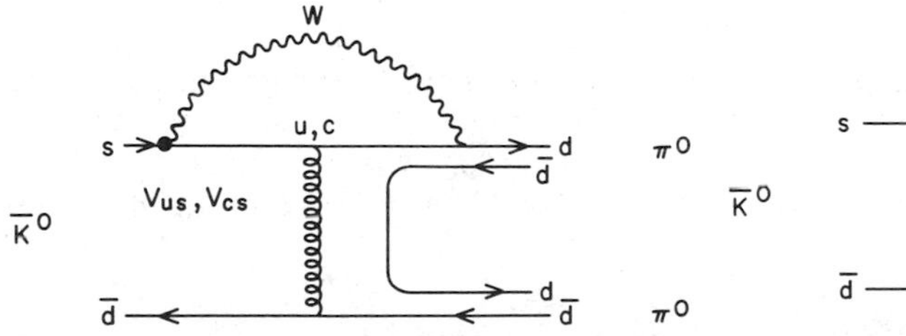

Fig. 3.1 The "Penguin" diagram for $\bar{K}^0 \to \pi^0\pi^0$; for $\bar{K}^0 \to \pi^+\pi^-$ $\bar{K}^0 \to \pi^+\pi^-$ we simply replace dd-quark line by uu.

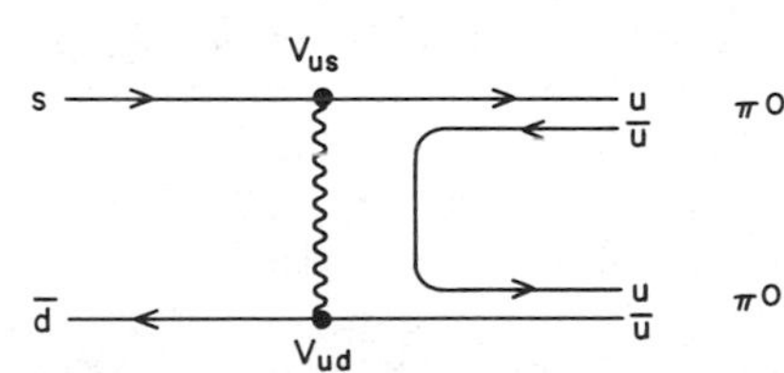

Fig. 3.2 The W-exchange diagram for $\bar{K}^0 \to \pi^\circ\pi^\circ$; for $\bar{K}^0 \to \pi^+\pi^-$ we simply replace uu quark by dd.

Caution [26]: If the W-exchange diagram, Fig. (3.2) is important in the K non-leptonic decay, it contributes the same way as the "Penguin" diagram but non-complex. It can change the "Penguin"-diagram calculation.

To conclude: measurements of ε'/ε should provide a better understanding of the dynamics of non-leptonic weak decay and help distinguish different sources of CP violation. There are two experiments measuring ε': Chicago, Stanford, Saclay, FNAL Exp. #617; Yale-Brookhaven, BNL Exp. #749. Both had sensitivity of $|\varepsilon'/\varepsilon|$ to 0.2%. The CERN experiment claims to have a sensitivity of $|\varepsilon'/\varepsilon|$ to 0.01%.

IV. CP Violation in Partial Decays

Besides contributing CP violation effects in the mass matrix, the complexity in the mixing matrix can also give rise to CP violation effects in the partial decay rates due to interference between the weak interaction amplitudes and the strong interaction amplitudes. It was discussed in generality quite some time ago by the authors of Ref. [27] that, though CPT predicts equal total decay rate for particle and antiparticle, the partial decay rates of particle and antiparticle into CP conjugated final particles can be different if CP is not invariant. The complexity of V_{ij} in the six-quark model of K-M provides an explicit mechanism to such difference in partial decay rates [6, 28, 29].

First we discuss the case of K^0, $\bar{K}^0 \to \pi^+\pi^-$. The decay amplitudes can be decomposed into amplitudes of I = 0, I = 1 or the $\pi^+\pi^-$ system

$$A(K^0 \to \pi^+\pi^-) = -(1/\sqrt{3})a_2 e^{i\delta_2} + (\sqrt{2}/\sqrt{3})a_0 e^{i\delta_0} , \qquad (4.1a)$$

$$A(\bar{K}^0 \to \pi^+\pi^-) = -(1\sqrt{3})a_2^* e^{i\delta_2} + \sqrt{2}/\sqrt{3}\ a_0^* e^{i\delta_0} . \qquad (4.1b)$$

Note that the phases δ_0, δ_2 from strong interaction says the same for K^0 and $\bar{K}^0$, but the weak phases for K^0 and $\bar{K}^0$ are of opposite sign as demonstrated in Fig. (4.1). (Changing the $d\bar{d}$ creation in Fig. (3.1) to $u\bar{u}$ creation we obtain the $K^0 \to \pi^+\pi^-$ amplitude.) Now we can calculate

$$|A(K^0 \to \pi^+\pi^-)|^2 - |A(\bar{K}^0 \to \pi^+\pi^-)|^2$$
$$= (\sqrt{2/3})(\mathrm{Re}\ a_2)(\mathrm{Re}\ a_0)(t_2 - t_0)\sin(\delta_2-\delta_0) , \qquad (4.2a)$$

and

$$|A(K^0 \to \pi^+\pi^-)|^2 + |A(\bar{K}^0 \to \pi^+\pi^-)|^2 = \frac{4}{3}(\mathrm{Re}\ a_0)^2 . \qquad (4.2b)$$

$$\Delta_{K^0\to\pi^+\pi^-} \equiv \left|A(K^0 \to \pi^+\pi^-) - \bar{A}(\bar{K}^0 \to \pi^+\pi^-)\right| / (|A|^2 + |\bar{A}|^2)$$

$$= \frac{i}{\sqrt{2}} \left(\frac{\text{Re } a_2}{\text{Re } a_0}\right) (t_2 - t_0) \sin(\delta_2 - \delta_0) \tag{4.3}$$

Comparing to Eq. (3.3), we obtain

$$\Delta_{K^0\to\pi^+\pi^-} = -2 \text{ Re}(\varepsilon') \quad . \tag{4.4}$$

Another example is $\Lambda\to\pi^- P$, $\bar{\Lambda}\to\pi^+\bar{P}$ to demonstrate the point.Using the quark diagram description [6, 30], the amplitudes of such decays are of the form:

$$A(\Lambda \to \pi^- P) = V_{us}V_{ud}A_1 + V_{cs}V_{cd}A_2 \quad , \tag{4.5}$$

$$\bar{A}(\bar{\Lambda} \to \pi^+\bar{P}) = V^*_{us}V^*_{ud}A_1 + V^*_{cs}V^*_{cd}A_2 \quad . \tag{4.6}$$

where $A_1 = a' + b' + c' + e'$, the sum of all four graphs in Fig. (4.1a), and $A_2 = e'$, the amplitudes a', b', c', e' are the external W-emission, internal W-emission, the W-exchange and the "Penguin" diagram respectively, as shown in Figs. (4.1a,b). We obtain the difference in partial decay rates,

$$\Delta_\Lambda = \frac{Br(\Lambda \to \pi^- P) - Br(\bar{\Lambda} \to \pi^+\bar{P})}{Br(\Lambda \to \pi^+ P) + Br(\bar{\Lambda} \to \pi^+\bar{P})}$$

$$= \frac{4(s_2 s_3 s_\delta c_1 c_2 c_3)\ \text{Im}(A_1 A_2^*)}{(|A|^2 + |\bar{A}|^2) s_1^{-2}} \quad . \tag{4.7}$$

As in ε, here appears the factor $s_2 s_3 s_\delta \sim 10^{-3} - 10^{-4}$. It is difficult to estimate the amplitude factor. One roughly expects it to be $\text{Im}(a_{1/2} a_{3/2})/|a_{1/2}|^2$, which is not large, from $\Delta I = 1/2$ dominance.

The difference of $A(\Lambda \to \pi^- p)$, $A(\bar{\Lambda} \to \pi^+\bar{p})$, as given in Eqs. (4.5,4.6) can also give CP violation effects in the pion momentum distribution against the polarization of Λ in its rest frame. It is known that the π^- distribution is very much peaked in the direction of the Λ spin, i.e.,

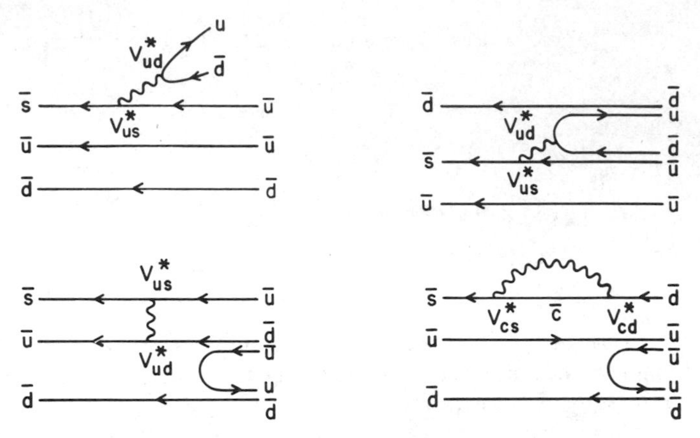

Fig. 4.1a Quark diagram amplitudes
a', b', c', e', $\Lambda \rightarrow \pi^- P$

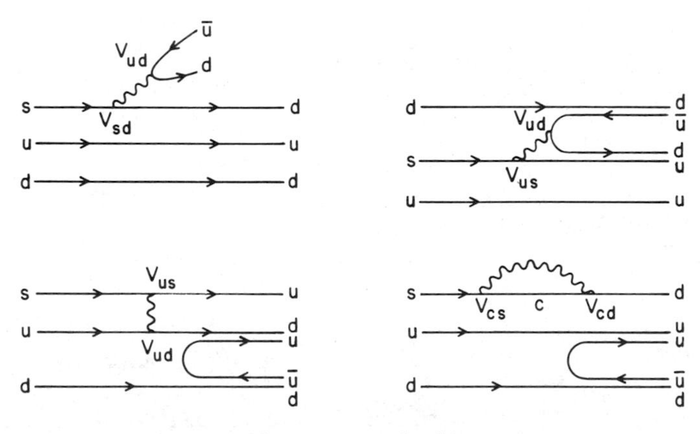

Fig. 4.1b Quark-diagram
amplitudes for $\bar{\Lambda} \rightarrow \pi^+ \bar{p}$

$$\frac{d\sigma}{d\Omega} = (1 + \alpha \cos\chi) \quad , \tag{4.8}$$

where χ is the angle between the π^- momentum and the Λ spin in the rest frame of Λ; for $\bar{\Lambda}$

$$\frac{d\bar{\sigma}}{d\Omega} = (1 - \bar{\alpha} \cos\chi) \quad . \tag{4.9}$$

Current experimental measurements gives:

$$\begin{aligned} \alpha = \bar{\alpha} &\approx 0.647, \text{ for } \Lambda \to \pi^- p \quad , \\ &\approx -0.979 \text{ for } \Sigma^+ \to p\pi^0 \quad . \end{aligned} \tag{4.10}$$

This parameter α was calculated by Lee and Yang in Ref. (27) and was given by the interference between the s-wave and the p-wave amplitudes in the Λ decay, i.e.,

$$\alpha = \frac{2\, \mathrm{Re}(A_s A_p^*)}{|A_s|^2 + |A_p|^2} \quad , \tag{4.11}$$

and similarly for $\bar{\Lambda} \to \pi^+ p$

$$\bar{\alpha} = \frac{2\, \mathrm{Re}(\bar{A}_s \bar{A}_p^*)}{|\bar{A}_s|^2 + |\bar{A}_p|^2} \quad . \tag{4.12}$$

Using Eqs. (3.1) and (3.2), we find

$$\begin{aligned} \Delta\alpha = \frac{\alpha - \bar{\alpha}}{\alpha + \bar{\alpha}} &= -\,2(s_2 s_3 s_\delta c_1 c_2 c_3) \times \frac{[\mathrm{Im}(A_{1,s} A_{2,p}^*) + \mathrm{Im}(A_{2,s} A_{1,p}^*)]}{(|A_s^2| + |A_p^2|)\, s_1^{-2}} \\ &\approx (10^{-3} \sim 10^{-4}) \sin(\delta_s - \delta_p) \times |A_s||A_p| / (|A_s|^2 + |A_p|^2) \end{aligned} \tag{4.13}$$

where $\delta_s - \delta_p = (7.0 \pm 1.0)°$, unfortunately not very large.

For the K-M scheme it is natural to have these CP violation effects in Λ and Σ decays. It is worthwhile to embark on a systematic search for such effects [31].

In charm decay, such difference in partial decay rates have been estimated by Bernabéu and Jarlskog [28] for $B \to D^{*}\bar{D}$ and $\bar{B} \to \bar{D}^{*} D$, as shown in Fig. (4.2).

To summarize: The K-M mechanism provides sources of CP violation effects in many reactions. For a general discussion see Ref. (6). It is important to look for CP violation effects in the corresponding partial decay rates of particles and antiparticles, e.g.

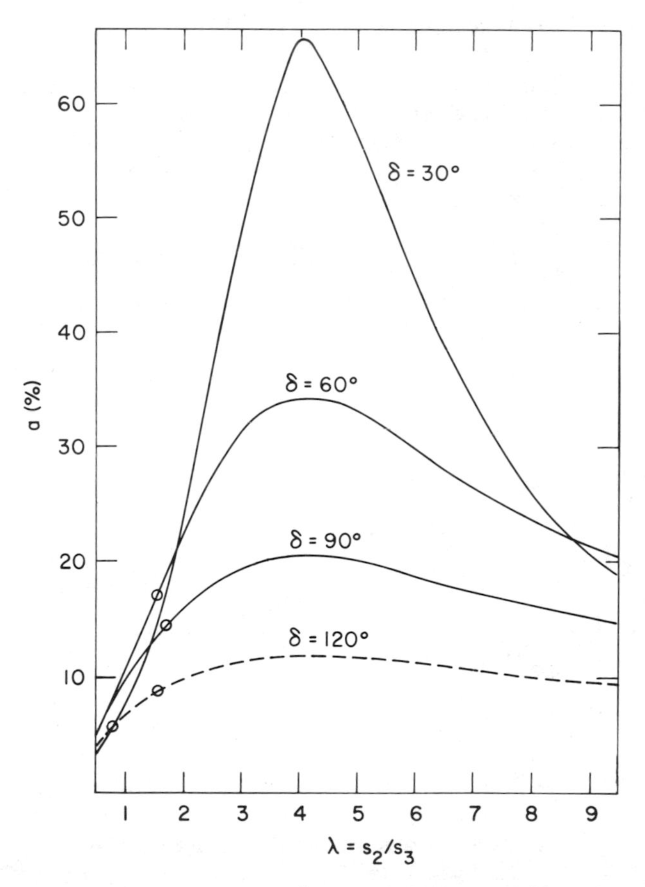

Fig. 4.2 $\Delta_{B\to D^*D}$ function in s_2/s_3, from Bernabéu and Jarlskog [28]. The circles are the points relating δ to s_2/s_3, given by the fit to the Δm, ε in the K^0 system from Ref. (14).

$$\frac{Br(\Lambda \to \pi^- p) - Br(\bar{\Lambda} \to \pi^+ \bar{p})}{Br(\Lambda \to \pi^- p) + Br(\bar{\Lambda} \to \pi^+ \bar{p}} = ? \quad ,$$

$$\frac{Br(B \to D^* D) - Br(\bar{B} \to \bar{D}^* \bar{D})}{Br(B \to D^* D) + Br(\bar{B} \to \bar{D}^* \bar{D})} = ? \quad .$$

V. Comparison of CP Violation Effects from Various Sources

In Table V we compare predictions of CP violation effects from various sources other than the K-M model.

TABLE V Summary of CP Violation Effects from Various Models

models	ε	ε'/ε	r (p)	a(p)	Δ_p	d_n(e cm)
superweak	fitted	0	non-zero only in the K^0 system	non-zero only in the K^0 system	zero	0
the K-M model	fitted	1/600-1/80	fitted in the K^0 system; mixing can be large in the $B_s^0\bar{B}_s^0$ system and in the $B_d^0\bar{B}_d^0$ systems for $\delta > 177°$; small for the $D^0\bar{D}^0$ & $T^0\bar{T}^0$ systems	appreciable ($\approx 10^{-2}$) only in the $D^0\bar{D}^0$ system for $\delta \approx 0°$ system	zero for all semileptonic decays; non-zero for many hadronic channels	$< 10^{-30}$
complex Higgs (a la Weinberg) no flavor-changing neutral coupling	fitted	> 1/50 close to exceed experi-mental bounds	appreciable only for the $T^0\bar{T}^0$ system; small for the $D^0\bar{D}^0$, $B^0\bar{B}^0$ systems	appreciable only for the $T^0\bar{T}^0$ system; small for the $D^0\bar{D}^0$, $B^0\bar{B}^0$	non-zero for many channels including exclusive semileptonic decays	10^{-25} close to experi-mental bounds
strong CP violation	0	0	0	0	0	accommo-dated

r(p) = parameter for neutral particle-antiparticle mixing, Eq. (2.4)

a(p) = asymmetry between $p^0 p^0$ and $\bar{p}^0 \bar{p}^0$ due to $p^0 \bar{p}^0$ mixing, Eq. (2.9)

Δ_p = partial decay rate difference between particle and its antiparticle decaying into their corresponding CP conjugated final states, see Eqs. (4.7 and 4.13).

VI. Rare Decays

It is important to learn from rare decays. They test our theoretical models in a "nut shell". The $K_L \to \mu^+\mu^-$ serves as a prime example of the importance of knowing the precise value of the rare decays.

The question to what degree the neutral b-flavor changing is suppressed is still extremely important to know. Since if our theory is composed of only lefthanded doublets, the generalized GIM mechanism will suppress very much the inclusive opposite charge dilepton $\ell^+\ell^-$ decays of the B, just like for $K^0 \to \mu^+\mu^-$ and $K \to \pi e^+e^-$, [32]. Some estimates show [33]

$$Br(B^0 \to \tau^+\tau^-) \approx 10^{-8} \quad , \tag{6.1}$$

$$Br(B \to X_s\ell^+\ell^-) \approx 10^{-5} \quad . \tag{6.2}$$

In some horizontal gauge formulations to connect quark generations, flavor changing neutral coupling can be present. They are mediated by neutral Higgs bosons or horizontal gauge bosons, and can give much larger branching ratios $Br(b \to \ell^+\ell^-X)$ than those given by the K-M model, Eqs. (6.1,6.2). Therefore, the oservation of neutral b-flavor changing coupling at a level above that predicted by the K-M model has the interesting implication that either the t quark is absent, or there are neutral flavor changing Higgs fields or horizontal gauge bosons. All these possibilities are extremely interesting.

VII. Concluding Remarks

1) Up to now the standard electroweak unified theory is in good shape in comparing with experiments. Direct observation of $W^\pm$, Z^0 is the strongest recent evidence.

2) Approximate ranges of the quark-mixing matrix elements V_{ij} are known, the angles and the phase are poorly known. A wide range of δ is allowed, $0° < \delta < 180°$. To determine it better, we need future observation of new CP-violation effects.

3) CP violation from neutral particle-antiparticle transitions are very small in the heavy quark systems. It is important to look for the CP violations in the differences of partial decay rates, e.g.

$$\frac{Br(\Lambda \to \pi^-p) - Br(\bar{\Lambda} \to \pi^+\bar{p})}{Br(\Lambda \to \pi^-p) + Br(\bar{\Lambda} \to \pi^+\bar{p})} = ? \quad ,$$

$$\frac{Br(B \to D^* \bar{D}) - Br(\bar{B} \to \bar{D}^* D)}{Br(B \to D^* \bar{D}) + Br(\bar{B} \to \bar{D}^* D)} = ? \quad .$$

The K-M model provides a natural mechanism for such CP violation effects.

4) It is important to measure ε', d_n^e, and the CP violation effects in exclusive semileptonic decays to distinguish different CP-violation sources.

5) The K-M model can give maximal $B^0\bar{B}^0$ mixing. It is interesting to study its effects in $N\bar{N}$, e^+e^- interaction (signatures $\ell^+\ell^+$, $\ell^-\ell^-$, $\bar{K}\bar{K}$, KK, $\bar{c}\bar{c}$, cc).

6) There is still a lack of understanding in the origin of quark mixing in weak interactions.

7) Left-handed models without the t quark are ruled out by CESR experiments.

8) It is important to search for rare decays,e.g., the K-M model gives very small

$$Br(B^0 \to \tau^+\tau^-) \approx 10^{-8} \quad ,$$

$$Br(B \to X_s\ell^+\ell^-) \approx 10^{-5} \quad .$$

Observation of these decays above these levels implies no t quark, or the existence of horizontal gauging boson and Higgs boson.

9) The searching questions: where is the t quark; why do the quark generations repeat; and why do they mix in a certain way in weak interactions?

Aknowledgments

I would like to thank Dr. W.-Y. Keung for his help in preparing material for this talk.

This manuscript has been authored under Contract No. DE-AC02-76CH00016 with the U.S. Department of Energy.

References

1. S.L. Glashow, Nucl. Phys. 22 (1961) 579;
   A. Salam and J.C. Ward, Phys. Lett 13 (1964) 168;
   S. Weinberg, Phys. Rev. Lett. 19 (1967) 1264;
2. UA1 Collaboration at CERN, G. Arnison et al., Phys. Lett. 122B (1983) 103.
3. R.F. Peierls, T. Trueman and L.L. Wang, Phys. Rev. D16 (1977) 1397.
4. L.B. Okun and M.B. Voloshin, Nucl. Phys. B120 (1977) 459;
   C. Quigg, Rev. Mod. Phys. 94 (1977) 297;
   J. Kogut and J. Shigemitsu, Nucl. Phys. B129 (1977) 461;
   R. Horgan and M. Jacob, Proc. CERN School of Physics, Malente (FRG), 1980 (CERN-81-04), p. 65.
5. F.E. Paige, Proc. Topical Conf. on the Production of New Particles at Super-High Energies, Univ. Wisconsin, Madison, 1979.
6. Ling-Lie Chau, "Quark Mixing in Weak Interactions", BNL preprint BNL-31859, Aug. 1982 to be published as a Physics Report.
7. M. Kobayashi and T. Maskawa, Prog. Theor. Phys. 49 (1973) 652.
8 N. Cabibbo, Phys. Rev. Lett. 10 (1963) 531.
   S. Glashow, J. Iliopoulos and L. Maiani, Phys. Rev. D2 (1970) 1285.
9. M. Roos, Nucl. Phys. B77 (1974) 420M.
10. R. Shrock and L.-L. Wang, Phys. Rev. Lett. 41 (1978) 1692.
11. J.A. Thompson et al., Phys. Rev. D21 (1980) 25;
   J. Wise et al., Phys. Lett. 91B (1980) 165; ibid. 98B (1981) 423;
   D.A. Jesen et al., Proc. of the XX Int'l. Conf. High Energy Physics, Wisconsin 1980, p. 364. CERN WA2 Collaboration, M Bourquin et al., "Measurement of $\Sigma \to \Lambda e\nu$, $\Lambda \to pe\nu$, $\Xi \to \Lambda e\nu$, $\Xi \to \Sigma e\nu$ and $\Sigma \to ne\nu$, Branching Ratios and Form Factors", Contributed paper to Int'l. Conf. on High Energy Physics, Lisbon, Portugal, 9-15 July 1981.
   For more recent analysis, see
   J.F. Donoghue and B.R. Holstein, Phys. Rev. D25 (1982) 2015;
12. Ohio-Chicago-Argonne Collaboration, P. Keller et al., Phys. Rev. Lett. 48 (1982) 971; and see some attempts to fit, A. Garcia and P. Kielanowski, "Symmetry Breaking and Higher Representation in the Cabibbo Theory", del Institute Politéchnico National preprint, 1982, Mexico; and A. Bohm, "Fits of Hyperon Data", U. of Texas at Austin preprint 1982, the fit though does not answer the fundamental question of the source of the breaking, demonstrated the SU(3) breaking of mass equality can accomodate the data.

13. For the original detailed analysis of mixing matrix from CP violation see, R.E. Shrock, S.B. Treiman and L.-L. Chau Wang, Phys. Rev. Lett. 42 (1979) 1589;
V. Barger, W.F. Long and S. Pakvasa, Phys. Rev. Lett. 42 (1979) 1585.
14. For a recent detailed quark mixing matrix, see L.-L. Chau, W.-Y. Keung and M.D. Tran, Phys. Rev. 27D (1983) 2145 and Ref. [6].
15. For reviews see A. Silverman, Proc. of the 1982 Int'l. Symp. on Lepton and Photon Interactions at High Energies, Bonn, Aug. 24-29, 1982;
P. Franzini and J. Lee-Franzini, "Upsilon Physics at CESR", Phys. Report, 81 (1982) 239-291.
See the talk by Dr. D. Kreistick in this conference.
16. For the original bag calculation of the B, see R. Shrock and S.B. Treiman, Phys. Rev. D19 (1979) 2148;
B. McWilliams and O. Schanker, Phys. Rev. D22 (1980) 2853.
For a recent discussion on bag calculation see P. Colic, D. Tadic and J. Trampetic, "K_L-K_S Mass Difference and Quark Models", Max-Planck Institute, etc., preprint, MPI-PAE/PTh, 1982.
(For some parameters of the bag model B can be negative. However even for negative B there are still solutions in the allowed ranges of s_2, s_3, though very limited, as noted by the authors of Ref. (15)); and
J.F. Donoghue, E. Golowich, B.R. Holstein, Phys. Lett. 119B (1982) 412.
17. K. Wilson, Phys. Rev. D10 (1974) 2445; A.M. Polyakov, Phys. Lett. 59B (1975) 82; G. 't Hooft, Phys. Rev. D14 (1976) 3432. For recent work, see M. Creutz, Phys. Rev. Lett. 43 (1979);
M. Creutz, L. Jacobs, C. Rebbi, Phys. Rev. D20 (1979) 1915;
I would like to thank Prof. N. Cabibbo for an enlightening discussion on the topic.
18. S. Pakvasa, S.F. Tuan and J.J. Sakurai, Phys. Rev. D23 (1981) 2799.
19. E.A. Paschos, U. Türke, "Charged Current Coupling in the Six Quark Model", Univ. Dortmund preprint, DO-TH 82/07, 1982. (Their results of V_{cs} and V_{cd} from ν, $\bar{\nu}$ reactions are different from those of Ref. [14] due to some difference in the analysis; see comments in Ref. [14].
CERN, Dortmund, Heidelberg, CEN Saclay, and Beijing Collaboration, H. Abramowicz et al., CERN preprint CERN-EP-82/77.
20. L.-L. Chau, W.-Y. Keung and S.C.C. Ting, Phys. Rev. D24 (1981) 2861.
21. See Sections I.3 and III.3 of Ref. [6].
22. Aachen, CERN, Torino collaboration, M. Holder et al., Phys. Lett. 40B (1972) 141;

Princeton, BNL experiment, M. Banner et al., Phys. Rev. 28 (1972) 1997;
K. Kleinknecht, Proc. of the XVII Int'l. Conf. on High Energy Physics, London, 1974, ed. J.R. Smith, (Rutherford Lab., Chilton, Didcot, Berkshire, England (1974) p. III-23; and Ann. Rev. Nucl. Sci., 26 (1976) 26.

23. A.I. Vainshtein, V.I. Zakharov and M.A. Shifman, Pisma Zh. Eksp. Teor. Fiz. 22 (1975) 123 [JETP Lett. 22 (1975) 55], Nucl. Phys. B120 (1977) 316, Zh. Eksp. Teor. Fix. 72 (1977) 1275 [Sov. Phys. JETP 45 (1977) 670].
24. F.J. Gilman and M.B. Wise, Phys. Lett. 83B (1979) 83.
25. F.J. Gilman and M.B. Wise, Phys. Rev. D20 (1979) 2392;
M.B. Wise and E. Witten, Phys. Rev. D30 (1979) 1216;
L. Wolfenstein, Nucl. Phys. B150 (1979) 501;
B. Guberina and R.D. Peccei, Nucl. Phys. B163 (1980) 289;
C.T. Hill, Phys. Lett. 97B (1980) 275;
V.V. Prokhorov, Yad. Fiz. 31 (1979) 1019, [Sov. J. Nucl. Phys. 31 (1980) 527].
J.S. Hagelin, Phys. Rev. D23 (1981) 119;
26. The relevance of the W-exchange diagram in K decays and its implication on ε'/ε was pointed out in, L.L. Chau Wang, "Quark Mixing and Decay", Proc. of the Int'l. Workshop on High Energy Physics, Serpukov, U.S.S.R, Sept. 1980.
27. T.D. Lee and C.N. Yang, Phys. Rev. 108 (1967) 395;
A. Pais and S.B. Treiman, Phys. Rev. D12 (1975) 2744;
L.B. Okun, V.I. Zakharov and B.M. Pontecorvo, Lett. al Nuovo Cim. 13 (1975) 218.
28. M. Bander, D. Silverman and A. Soni, Phys. Rev. Lett. 43 (1979) 242.
B. Carter and A.I. Sanda, Phys. Rev. Lett. 45 (1980) 953; Phys. Rev. D23 (1981) 1567; I.I. Bigi and A.I. Sanda, Nucl. Phys. B193 (1981) 85.
J. Bernabéu and C. Jarlskog, Z. Phys. C Particle and Fields 8, (1981) 233.
29. L.L. Chau Wang, "Phenomenology of CP Violation from the Kobayashi-Maskawa Model", AIP Conf. Proc. No. 72, Particle and Fields, Subseries No. 23, "Weak Interactions as Probes of Unification", Virginia Polytechnic Inst. 1980, eds. G.B. Collins, L.N. Chang and J.R. Ficenec.
30. The development of the complete six quark diagram for meson decays was reported in L.-L. Chau Wang, "Flavor Mixing and Charm Decay", Proc. of 1980 Guangzhou (Canton) Conf. on Theoretical Particle Physics, Jan. 5-14, 1980. The importance of the W-exchange diagram for the D^0 decay was also discussed.
31. I would like to thank Drs. K. Kilian, T.K. Kalogeropoulos, P. Pavlopoulos and R.R. Rau for enlightening discussions on the experimental possibility of such studies.

32. J. Ellis, M.K. Gaillard and D.V. Nanopoulos, Nucl. Phys. B109, (1976) 213;
B.W. Lee, Phys. Rev. D10 (1974) 897, D15 (1977) 3394;
A.I. Vainshtein, V.I. Zakharov, L.B. Okun, M.A. Shifman, Yad. Fiz. 24 (1976) 820 [Sov. J. Nucl. Phys. 24 (1976) 427];
F.J. Gilman and M.B. Wise, Phys. Rev. D21 (1980) 3150;
R. Decker and E.A. Paschos, Phys. Lett. 106B (1981) 211.
33. T.G. Rizzo, Phys. Rev. D21 (1980) 2692;
R. Decker and E.A. Paschos, Phys. Lett. 106B (1981) 211;
B.A. Campbell and P.J. O'Donnel, Phys. Rev. 25D (1982) 1989.

WEAK DECAYS OF b-FLAVORED MESONS

D. L. Kreinick

Wilson Synchrotron Laboratory
Cornell University
Ithaca, New York 14853 USA

INTRODUCTION

This paper has two goals: to review the experimental study of b quark decays in e^+e^- collisions and to present new CLEO data reporting the observation of the B meson.

EVIDENCE FOR THE b QUARK

The first experimental evidence for the existence of the fifth quark, the "b", was the discovery of the first Upsilon states at FNAL.[1] These apparently narrow resonances were immediately interpreted as bound quark states. The first observations of the first two resonances at the e^+e^- storage ring DORIS[2] confirmed the very small width and established precise masses. When CESR began taking data in the fall of 1979[3], the picture of Fig. 1 began to emerge. The first three states are narrower than the beam energy spread (about 10 MeV FWHM), but the fourth is visibly broad (22 MeV). This suggests that the first three states are below threshold for the production of mesons with b-flavor, but the fourth is slightly above threshold and decays strongly to b-flavored mesons, B mesons. The fourth Upsilon is still very narrow for a meson of its mass because of the small phase space available for its decay.

If this picture is right, we should be able to observe enhanced production of high energy leptons from weak decays of the daughter B mesons. Fig. 2 shows that at the $\Upsilon(4S)$ the inclusive electron cross section jumps by over a factor of two where the hadronic cross section is enhanced by only 40% (Fig. 1). In fact, about a third of the resonant events at the $\Upsilon(4S)$ contain a lepton of momentum above 1 GeV.[4]

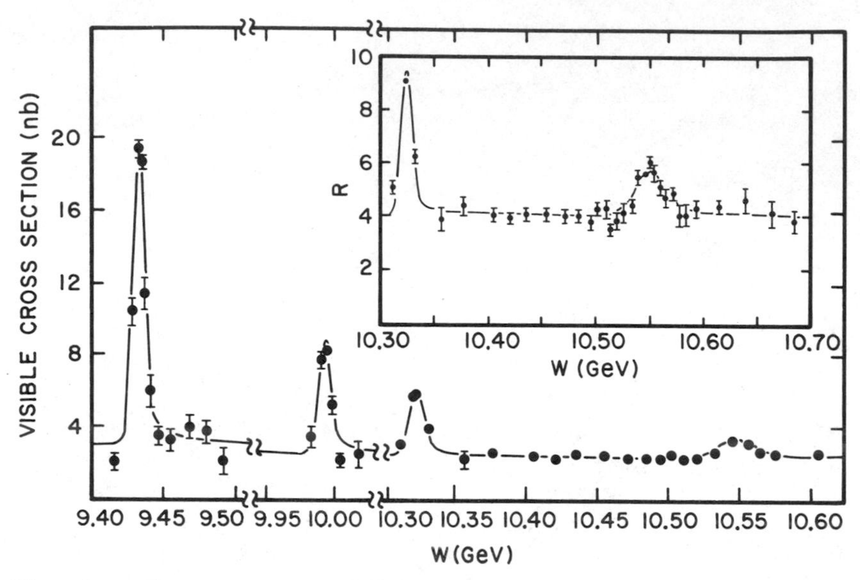

Fig. 1. The spectrum of Υ mesons measured at CESR by CLEO.

Further evidence supporting the existence of the b quark comes from the ratio R of the hadronic to the dimuon cross section. R is proportional to the sum of the square of charges of all particles which can couple to the virtual photon produced

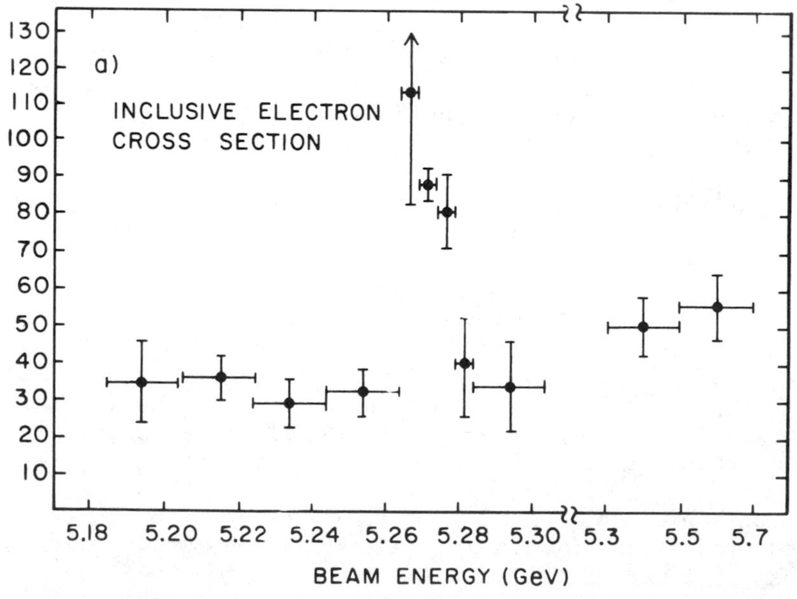

Fig. 2. Inclusive electron cross section at and near the Υ(4S).

in the e^+e^- annihilation. If a new charge 1/3 quark threshold is passed, R should increase by 1/3 after threshold behavior has smoothed out. This appears to be the case.[5]

WEAK b-DECAY MODELS

I will list classes of theories of b-quark decay and give their experimental status. If for simplicity you assume that b quarks decay to fewer than five leptons and quarks and furthermore assume that spin and charge are conserved, there are only a few allowed b decay modes:

$b \to c\, q_1 \bar{q}_2,\ u\, q_1 \bar{q}_2$ "hadronic"

$b \to c\, \ell\, \bar{\nu}\ ,\ u\, \ell\, \bar{\nu}_\ell$ "semileptonic"

$b \to s\, \ell\, \bar{\ell},\ d\, \ell\, \bar{\ell}$ "FCNC"

$b \to q\, \ell\, \ell,\ q\, \bar{\ell}\, \bar{\ell},\ q\ \ell_i\, \ell_j \quad i{=}j$ "exotic with dileptons"

$b \to \bar{q}\, \bar{q}\, \ell,\ q\, q\, \bar{\ell}$ "exotic with baryons"

An additional possible decay of b is to a new kind of particle, for example to a light charged Higgs:

$b \to c\, a^-,\ u\, a^-$ "light charged Higgs"

A number of higher symmetry models[6] predict semileptonic and dilepton-exotic or baryon-exotic b decays. These models and the Higgs decay can be excluded using data from CLEO because they fail to simultaneously account for the observed average amount of charged energy per $B\bar{B}$ event, the electronic and muonic branching ratios and the branching rates to protons and lambdas. Take the exclusion of the Higgs, for example: If you cook up Higgs decay modes to maximize the number of hadrons, you can just match the observed fraction of charged energy per event, but you are left with almost no leptons. To match the observed lepton production rates, you could postulate a large branching rate of the Higgs to τ, but then you lose too much energy to the inevitable neutrinos to match the observed charged energy fraction. All possible combinations are excluded at the 99.5% confidence level or more. If a Higgs of mass less than 5 GeV existed it would dominate b decay modes, having only one weak coupling factor. Therefore, we conclude that it does not exist. Similar analyses exclude both types of exotic decays from accounting for all b decays.

In the standard model 6 quarks are arranged in 3 generations of left-handed weak doublets. The b quark is supposed to be paired with the t or "top" quark. A large fraction of the experimental effort at PETRA and PEP has been spent looking for evidence of the top quark, but no top has been seen.[7] The simplest explanation for this is that these machines do not run at high enough energy to produce top, but it is also possible that the standard model is wrong and that b is a weak singlet. Were this the case, the GIM mechanism would break down and there might well be flavor changing neutral currents.[8] Mark J[9] has sought the decay $b \to \mu^+\mu^- X$ in e^+e^- collisions at c.m. energies above 30 GeV. The muons would have a distinctively large opening angle. Mark J sees only two candidate events, which limits the branching fraction to less than 0.7% at 95% confidence level. CLEO produces its b's nearly at rest and must therefore rely on a subtraction to separate $b \to \ell^+\ell^- X$ from the case where both b and $\bar{b}$ decay to leptons. This disadvantage is more than compensated by vastly greater b counting rate; CLEO's upper limit is 0.4% at 95% confidence level. This excludes these models, which demand a rate of over 1% for $b \to \ell^+\ell^- X$.

While these results exclude the most plausible topless theories, it is still possible to invent models without tops. For example, the b quark could be coupled to the c quark in a RIGHT-handed doublet, suppressing FCNC by orthogonality.[10] The lepton momentum spectra from the V+A decay are hard to distinguish from the standard model's V-A. There is some hope to exclude this model using correlations between the c-quark and lepton momenta, but quite possibly the top will be found before enough data have been accumulated to do this.

For the remainder of this paper I will assume the standard model. This model postulates three generations of quarks in left-handed weak doublets:

$$\begin{pmatrix} u \\ d \end{pmatrix} \qquad \begin{pmatrix} s \\ c \end{pmatrix} \qquad \begin{pmatrix} t \\ b \end{pmatrix}$$

Transitions between the charge 2/3 member of the pair and the -1/3 charge partner are mediated by a charged W. If the weak eigenstates were the same as the physical states, there would be no transitions between generations, but there is mixing, which allows decays at a Cabibbo-suppressed rate. The mixing is described by the Kobayashi-Maskawa matrix, which can be parametrized with three angles and one phase. If the mixing is relatively small, c and t decay to s and b with amplitudes proportional to cosines of small angles, so the most sensitive way to evaluate the angles is by studying s and b decays.

In the spectator model the b quark decays to a u or c quark by emitting a virtual W^- boson while its antiquark partner in the B meson stands by idly. The W^- then decays to lepton-neutrino or quark-antiquark as indicated in Fig. 3. The decay coupling constant of the W, combined with phase space, predicts the semileptonic decay rate of the B to be 15% to 17%, depending on

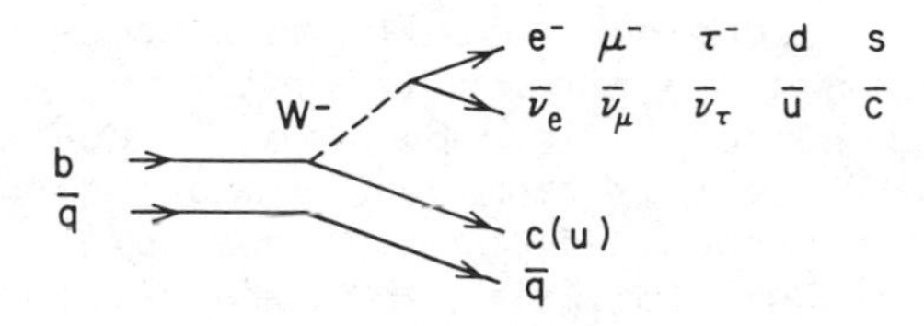

Fig. 3. Spectator model of B decay.

whether the b decays predominantly to c or to u. Corrections for QCD and non-spectator effects[11] bring the estimate down to 11% - 13%. Experimental results are listed in Table 1. Note the two groups of experimental results measure quantities which are in principle different. The CLEO and CUSB groups, operating at the $\Upsilon(4S)$, have only enough energy to produce B^0 and $B^\pm$, whereas the other groups are also seeing leptons from strange B, B baryons, etc. In any case, the semileptonic branching ratios agree with predictions of the standard model. While the spectator model may not be completely correct, the predicted semileptonic branching fractions are close, so I will use it as an intuitive guide.

CONSTRAINTS ON K-M MATRIX ELEMENTS

The B lifetime measures one combination of the $b \to u$ and $b \to c$ K-M matrix elements--the sum of their squares. By analogy with D decay the B lifetime is expected to be on the order of 10^{-13} sec., clearly a very difficult measurement. JADE has a 95% confidence level upper limit on the B lifetime of 1.4×10^{-12} sec, and MAC a limit of 3.7×10^{-12} sec.[13]

Table 1. Semileptonic Branching Ratios[12]

Who	Electron		Muon
Spectator model		.15 to .17	
Corrected		.11 to .13	
TASSO	.136±.05 ±.02		.15 ±.035±.035
MARK II	.11 ±.03 ±.02		
MARK J			.107±.014±.018
CLEO	.127±.017±.013		.122±.007±.015
CUSB	.137±.009±.020		

Another useful and experimentally more accessible handle on the K-M matrix elements is the measurement of how often b decays to u compared with how often it decays to a c quark. Limits on the angles in the K-M matrix from decays among the four lightest quarks suggest that b should decay to c more often than to u. The most straightforward way to verify this and thereby provide much more stringent limits is to search for s or c-containing mesons from the ϒ(4S). At CLEO we have used $K^{\pm}$, K^0_S, D^0, $D^{*\pm}$ and ψ mesons. The best of these straightforward evaluations is the ratio of the number of kaons observed per event on the ϒ(4S) to the number per event on the continuum, measured to be 1.58±0.15. A Monte Carlo calculation containing ocean production of kaon pairs predicts 1.8±.1 if b always decays to c and 0.95±.1 if b always decays to u. Thus one calculates b decays to c (74±25±9)% of the time. This measurement is indecisive because the uncertainty in the number of kaons produced by $s\bar{s}$ pairs popped from the ocean. Other straightforward estimates of the fraction of b decays to u using D's or ψ's are limited at present by statistics.

A somewhat less direct measurement exploiting the difference in mass between the u and c quarks provides more precision. In the semileptonic decay $B \to Xe\bar{\nu}$, the smallest mass that X can have if it contains a charm quark is the D mass. If b decays to u, X will typically have masses below 1 GeV. Thus the lepton spectrum will be stiffer for u production than for c. The electron spectra as seen by CUSB[14] is shown in Fig. 4.

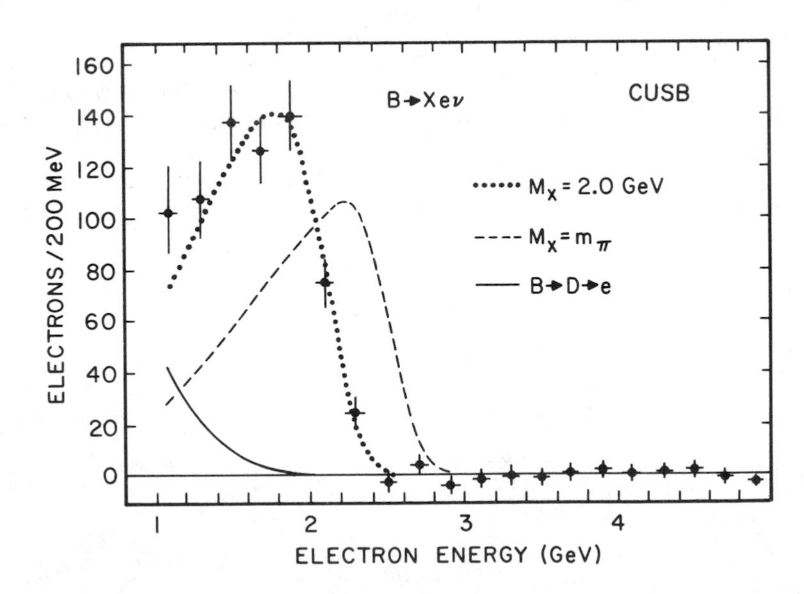

Fig. 4. Electron momentum spectrum from $B \to e\bar{\nu}X$. Curves indicate the background from decays of daughter D's and the expected spectra for recoil masses of m_{π} and 2.0 GeV.

CUSB uses the model of Altarelli et al.[15] to account for Fermi motion of the quarks bound in the B and other kinematical effects. They obtain a limit on $BR(b \to u)/BR(b \to c)$ less than 9% if $M_X < 0.5$ GeV. CLEO[16] finds an upper limit of 5% for $M_X = 1.0$ GeV which becomes somewhat more stringent for lower M_X.

RECONSTRUCTION OF B DECAYS

Recently the CLEO group has succeeded in reconstructing 18 low multiplicity B decays. The important features of the CLEO detector[17] for this work are outlined in Fig. 5 and Table 2.

As you might guess from the fact that it took over three years of data taking and 80,000 B decays to see our first 18 reconstructed B mesons, there are severe experimental difficulties:

1.) B and $\overline{B}$ from $\Upsilon(4S)$ decay are nearly motionless. Thus

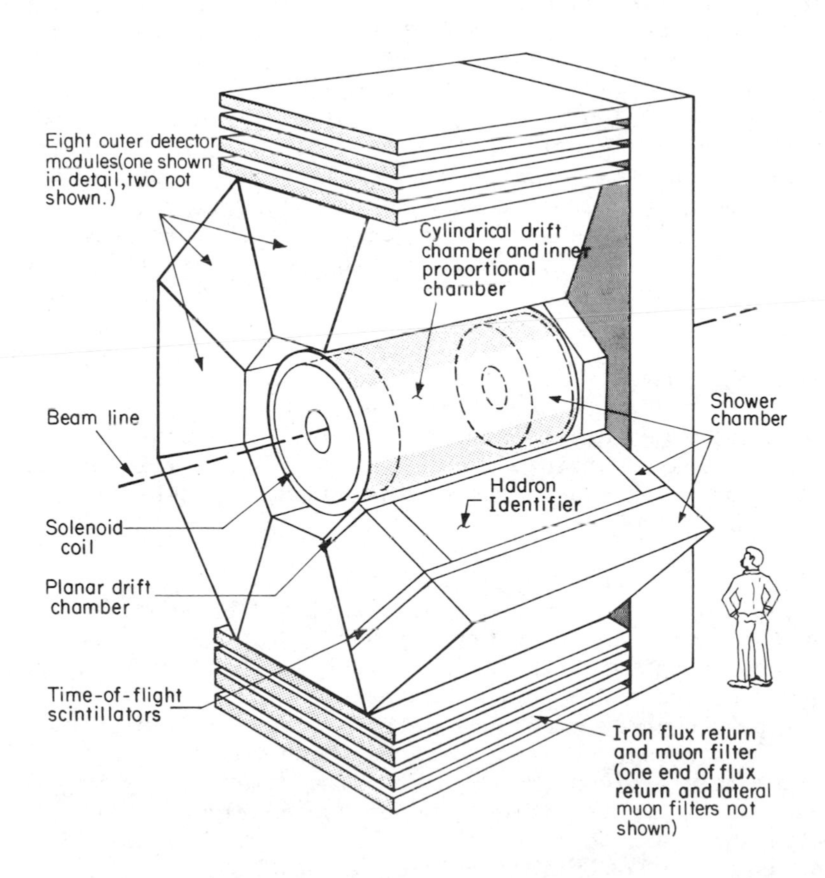

Fig. 5. The CLEO Detector.

Table 2. Properties of CLEO Relevant to B Reconstruction

Detector	% Solid Angle	Resolution	Comments
Drift chamber	90%	$\Delta p/p = .012p$ at 1.0 Tesla	
dE/dx chamber	46%		K: 0.45 to 0.95 GeV/c p: 0.65 to 1.45 GeV/c
Time of flight	50%	400 psec 2 m path	K: 0.50 to 1.00 GeV/c p: 0.85 to 1.45 GeV/c
Shower (octant only)	50%	$\Delta E/E = .17/\sqrt{E}$	π:e = 1:150 in conjunction with dE/dx
Muon chamber	78%		minimum p 0.9 to 1.6 GeV/c $\pi:\mu$ = 1:150

decay products from the B will, in general, be confused with those from the $\overline{B}$.

2.) From the observed charged multiplicity of Υ(4S) decays[18] (obtained by subtracting the continuum) and from the multiplicity of events containing an identified high momentum lepton from the Υ(4S), one can unfold the semileptonic B decay charged multiplicity (4.1±0.4±0.2) and the hadronic B decay charged multiplicity (6.3±0.2±0.2). Since a typical mixture of D and D^* contains 2.5 charged particles[19], a typical hadronic decay makes a $D(D^*)$ plus about four charged pions. This very large number of tracks has a huge number of combinatorial possibilities.

3.) Because CLEO measures charged particles much more accurately than π^o it is advantageous to demand that the B decays only to charged particles.

It is hopeless to try to see high multiplicity B decays by calculating the mass of all combinations of many tracks. CLEO had to hope that despite the large average B decay multiplicities there would be a measurable number of low multiplicity decays. Kinematical tricks, particle decay sequences and particle identification were used to reduce the number of combinations that had to be tried.

One approach, first suggested by Fritzsch[20], is to seek B decays to ψ-K or ψ-K-π. ψ's can be identified by their leptonic decays. For the ee analysis CLEO demands positive identification of one electron and scans the second track for all track pairs forming a mass above 2.5 GeV. The second track is discarded if it points to the dE/dx or shower detectors and is identified as a pion. Fig. 6a shows the mass spectrum which results. A 2 to 3σ enhancement is visible at the ψ mass. The dimuon spectrum, in which both muons are required to penetrate nearly a meter of iron, also shows a 2 to 3σ enhancement (Fig. 6b). This corresponds to a 90% confidence level upper limit of 1.7% on $B \to \psi$ + anything, or, if the signal is assumed real, a branching fraction of (1.0±0.4)% (continuum is not subtracted).

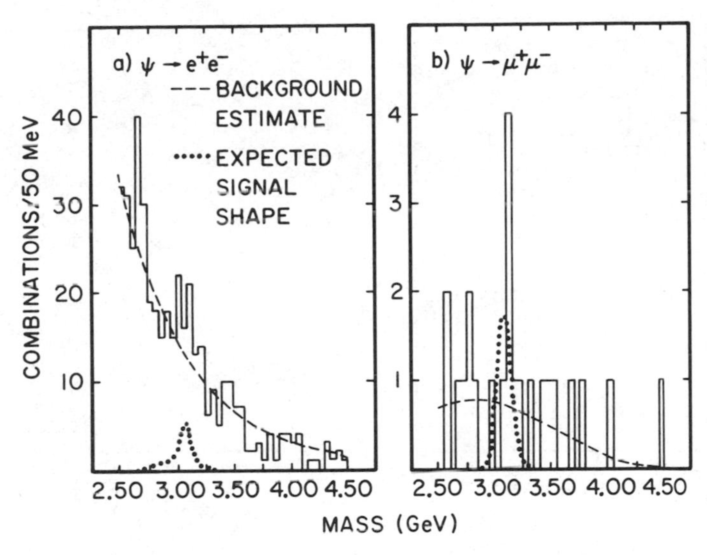

Fig. 6. Mass spectra for a) dielectrons and b) dimuons showing possible ψ production.

Unfortunately, when a further search for low multiplicity decays is made no enhancement in the ψ-K or ψ-K-π invariant mass can be discerned above background. So this approach fails.

We do better when attempting to reconstruct[21]

$B^- \to D^0\pi^- \to K^-\pi^+\pi^-$ (1)

$\overline{B}^0 \to D^0\pi^+\pi^- \to K^-\pi^+\pi^+\pi^-$ (2)

$\overline{B}^0 \to D^{*+}\pi^- \to K^-\pi^+\pi^+\pi^-$ (3)

$B^- \to D^{*+}\pi^-\pi^- \to K^-\pi^+\pi^+\pi^-\pi^-$ (4)

(I have listed only the reactions with a b quark. The $\overline{b}$ quark reactions are charge conjugates).

The first step in reconstructing events is to pick out the D^0 or D^*. D^0 is seen using $D^0 \to K^-\pi^+$. If no particle identification is used, a D^0 signal is visible, but the background is high. The background is much reduced if the kaon is identified by time of flight and dE/dx (Fig. 7a). In selecting D^0 candidates for B reconstruction CLEO demanded that the K-π mass be within 40 MeV of the known D^0 mass 1863 MeV.

D^* are observed in $D^{*+} \to D^0\pi^+$ using a kinematic trick. The Q-value of this decay is so small that the π^+ has little momentum and is therefore very well measured. By demanding that the K-π-π - K-π mass difference be within 3 MeV of 154.5 MeV so much discrimination is obtained that it is not necessary to identify the K in the D^0 decay (see Fig. 7b). In selecting D^* for B reconstruction CLEO demanded that the mass of the K-π inside the D^* be within 80 MeV of 1863 MeV.

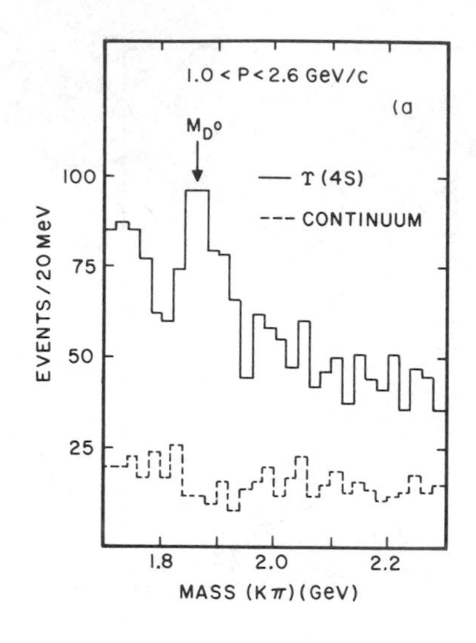

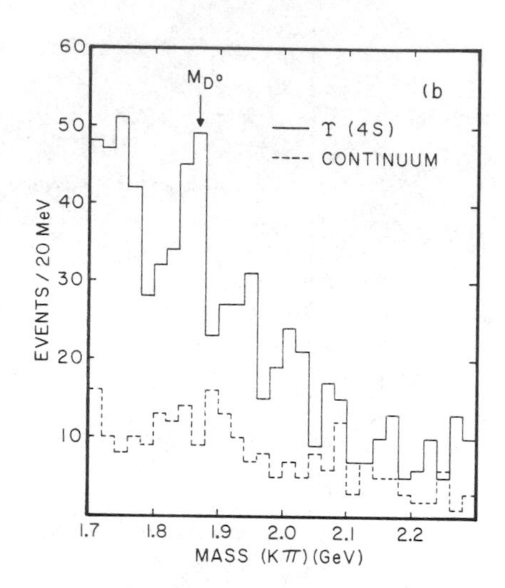

Fig. 7. K-π mass spectra from a) identified K with second track assumed to be pion; b) D^* candidates with no particle identifications.

Step 2 in the reconstruction of B events is kinematic fitting to the reactions (1) to (4). The B mass is essentially given by $M_B^2 = E_{beam}^2 - (\Sigma p)^2$, so it is measured relative to the beam energy. The fit has two constraints: 1) the B energy must equal the beam energy and 2) the K-π or K-π-π mass must equal the D^o or D^{*-} mass. A good fit must have $\chi^2 < 14$. Also, since the $B\bar{B}$ threshold lies between the Υ(3S) and Υ(4S), good events must have $5.18 < M_B < 5.28$ GeV. If an event has two acceptable fits the one with the lower χ^2 is used. Finally, candidate events are scanned to eliminate those using drift chamber tracks misfit by the trackfinding program.

Since b → c → s, certain sign combinations cannot come from B decay and are not considered. For example, in reaction (1) the direct π and the K must have the same sign because otherwise the c or s would be changed to $\bar{c}$ or $\bar{s}$. This not only reduces the number of combinations to try, but provides a way ("wrong sign") to estimate backgrounds.

The resulting mass spectrum is shown in Fig. 8a. The CESR energy scale is shifted to agree with the VEPP4 measurement of the Υ(1S) mass.[22] Of the 34 events in the 21 mass bins of the plot, 18 lie in the four bins near 5275 MeV. The width of the peak is consistent with the resolution expected from Monte Carlo studies.

CLEO uses several ways to estimate the background. The simplest is to extrapolate the number of events outside the peak into the peak region. A Monte Carlo shows that the background remains essentially flat. A second method estimates the effect of fake D's by attempting to reconstruct B's containing a "D" of mass 1663 or 2063 MeV, using otherwise the same cuts as the real analysis ("sideband" analysis, Fig. 8b). Another method uses the "wrong sign" events which either violate $b \to c \to s$ or lead to doubly charged B's (Fig. 8c). Finally, the D analysis was run for 19.6 pb^{-1} of continuum data taken between the $\Upsilon(3S)$ and $\Upsilon(4S)$ (Fig. 8d). This figure also includes wrong sign and sideband off-4S events to improve statistics. A Monte Carlo shows that

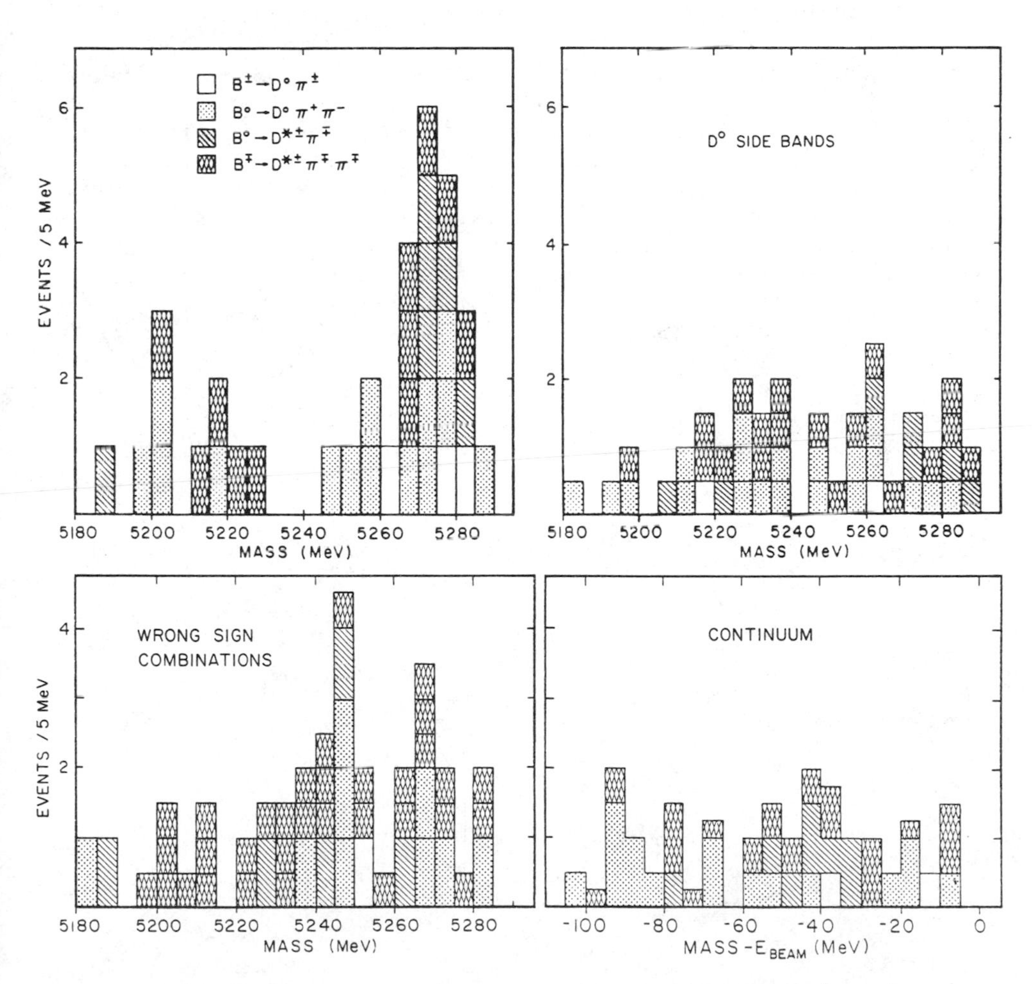

Fig. 8. B mass spectra for signal and three background estimates: a) signal, b) sidebands, c) wrong charge, and d) continuum.

there is negligible feed down to reactions (1) - (4) from events with higher multiplicity.

The event cuts used above were chosen to maximize the number of B's found. The signal to noise ratio can be maximized by tightening the cuts on χ^2 from 14 to 10, the D^o mass from 40 to 30 MeV and the mass of the D^o in the D^* from 80 to 60 MeV. The remaining mass plot (Fig. 9) has 15 events in the B peak region and only 6 outside.

If a B decays to D^*-π or D^*-π-π and we lose the soft photon (from a neutral D^*) or soft pion (from a $D^{*\pm}$), we will still reconstruct a B, but will obtain a slightly shifted B mass and (in the latter case) the wrong B charge. For this reason only those 11 events which contain a D^* can be used to determine the charged and neutral B masses. CLEO finds the neutral B mass to be 5274.2±1.9±2.0 MeV and the charged B mass 5270.8±2.3±2.0 MeV, where the first error is statistical and the second systematic. The difference of these masses agrees within large error limits with the 4.4 MeV that Eichten predicted.[23] However, the Q value of the Υ(4S) decay to $B\bar{B}$, measured to be 32.4±3.0±4.0 MeV, is not well described by Eichten, who gets too large an estimate, or by Gronau et al., and Bigi and Ono[24], who have too small a result. Using the theoretical 4.4 MeV mass difference and the measured Q-value of the Υ(4S) decay, one calculates that the Υ(4S) decays 60±2% of the time to B^+B^- and 40±2% to $B^o\bar{B}^o$.

The branching ratios of B's into the decay modes (1) - (4) are estimated to be a few percent each. This is similar to D branching ratios into analogous channels, a little surprising

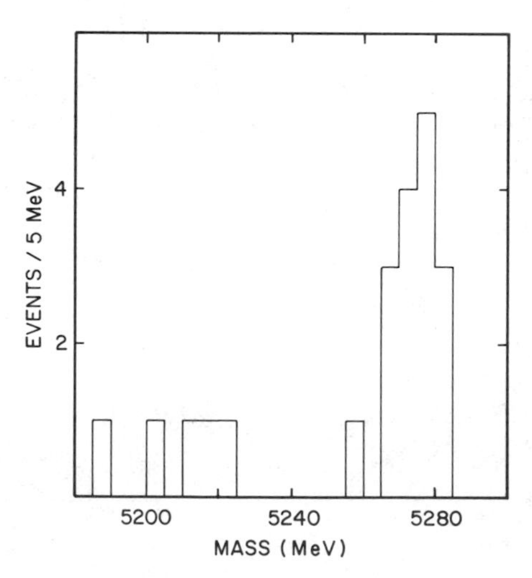

Fig. 9. B mass spectrum with stringent quality requirements.

considering the high average B multiplicity and the additional energy available to the B for its decay.

CONCLUSIONS

The standard model describes b decay well. Most competitive exotic decay models have been ruled out. Strong limits of FCNC rule out most topless models.

The semileptonic b quark decay rate has been measured in seven independent ways by five different experiments. The result, about 12.5%, is in good agreement with the standard model.

Only upper limits can be obtained as yet for K-M angles for b quark transitions. The B lifetime has not been measured precisely enough yet to improve much on constraints from unitarity. The branching ratio for b decay to u is less than 5% of b decay to c and is consistent with b always decaying to c.

CLEO has seen B mesons in low multiplicity decay modes. The B masses are $M_{\bar{B}} = 5270.8 \pm 2.3$ MeV and $M_{B}^{0} = 5274.2 \pm 1.9$ MeV.

ACKNOWLEDGEMENTS

I would like to thank my CLEO colleagues, especially Sheldon Stone and Bernie Gittelman, for helpful discussions.

REFERENCES

1. S. W. Herb et al., Phys. Rev. Lett. 39, 252 (1977).
2. C. Berger et al., Phys. Lett. 76B, 243 (1978); C. W. Darden et al., Phys. Lett. 76B, 246 (1978), and 78B, 364 (1978); J. Bienlein et al., Phys. Lett. 78B, 360 (1978).
3. D. Andrews et al., Phys. Rev. Lett. 44, 1108 (1980); T. Bohringer et al., Phys. Rev. Lett. 44, 222 (1980); D. Andrews et al., Phys. Rev. Lett. 45, 219 (1980); G. Finocchiaro et al., Phys. Rev. Lett. 45, 222 (1980).
4. C. Bebek et al., Phys. Rev. Lett. 46, 84 (1981); K. Chadwick et al., Phys. Rev. Lett. 46, 88 (1981); L. J. Spencer et al., Phys. Rev. Lett. 47, 771 (1981).
5. E. Rice et al., Phys. Rev. Lett. 48, 906 (1982); R. K. Plunkett, Ph.D. thesis, Cornell Univ., 1982 (unpublished).
6. E. Derman, Phys. Rev. D19, 317 (1979); R. N. Mohpatra, Phys. Lett. 82B, 101 (1979); H. Georgi and M. Machacek, Phys. Rev. Lett. 43, 1639 (1979); H. Georgi and S. L. Glashow, Nucl. Phys. B167, 173 (1980).

7. A. Boehm, "Recent Results from MARK J", invited talk at this conference.
8. G. L. Kane and M. E. Peskin, Nucl. Phys. B195, 29 (1982).
9. B. Adeva et al., "Search for Top Quark and a Test of Models Without Top Quark at the Highest PETRA Energies", MIT Tech. Report 128 (1982).
10. M. E. Peskin and S.-H. H. Tye, Proceedings of the Cornell Z^0 Theory Workshop, Cornell Laboratory of Nuclear Studies, CLNS 81-485 (1981); V. Barger, W. Y. Keung, R. J. N. Phillips, Phys. Rev. D24, 1328 (1981).
11. J. P. Leveille, University of Michigan UMHE 81-18 (1981).
12. Data are taken either from the Proceedings of the 21st International Conference on High Energy Physics (Paris, 1982) or from data presented at this conference.
13. W. Bartel et al., DESY preprint 82-014 (1982); D. M. Ritson, Proceedings of the 21st International Conference on High Energy Physics (Paris, 1982).
14. L. J. Spencer et al., Phys. Rev. Lett. 47, 771 (1981).
15. G. Altarelli et al., Nucl. Phys. B208, 365 (1982).
16. A. Brody et al., Phys. Rev. Lett. 48, 1070 (1982); D. Andrews et al., Cornell preprint CLNS 82/547, 14 (1982).
17. D. Andrews et al., "The CLEO Detector", Cornell preprint CLNS 82/538, to be published in Nucl. Instrum. Methods 211, 47 (1983).
18. M. S. Alam et al., Phys. Rev. Lett. 49, 357 (1982).
19. R. H. Schindler et al., Phys. Rev. D24, 78 (1981).
20. H. Fritasch, Phys. Lett. 86B, 343 (1979).
21. Since this conference, B reconstruction results have been published in S. Behrends et al., Phys. Rev. Lett. 50, 881 (1983).
22. A. S. Artamonov et al., Phys. Lett. 118B, 225 (1982).
23. E. Eichten, Phys. Rev. D22, 1819 (1980).
24. M. Gronau et al., Phys. Rev. D25, 3100 (1982); I. I. Bigi and S. Ono, Nucl. Phys. B189, 229 (1981).

EXPERIMENTAL CONSTRAINTS ON WEAK MIXING ANGLES IN THE SIX QUARK SCHEME

K. Kleinknecht

Institut für Physik der Universität Dortmund

Dortmund, Federal Republic of Germany

ABSTRACT

Experimental results on charm production by neutrinos and on B meson decays are used to derive bounds on the quark mixing angles θ_2 and θ_3 in the six quark model of Kobayashi and Maskawa.

1. INTRODUCTION

The quark mixing scheme in the six quark model proposed by Kobayashi and Maskawa (KM) [1] has been accepted as a useful parametrization of the connection betwen generations of quarks in terms of three angles, θ_1, θ_2, θ_3, and one phase δ, - apart from the fact that the sixth quark, t, has not been found yet. Over the last years, experimental data have been obtained which allow us to put rather stringent constraints not only on the first angle θ_1, but also to give restricted allowed domains in the plane of sin θ_2 versus sin θ_3. In this paper, we summarize the present picture emerging from this data. We first go through the constraints on the coupling parameters, i.e. the elements of the KM matrix, and then derive bounds on the angles.

2. KOBAYASHI-MASKAWA MATRIX

In this generalization of the Glashow-Iliopoulos-Maiani four quark mixing [2], the 3x3 mixing matrix U for six quarks can be expressed by the three angles $\theta_i (i = 1,2,3)$ and the phase δ in the way shown in fig.1. If CP violation is due to quark mixing, then the phase δ is related to the CP violation parameter ε in the neutral K meson system describing the admixture of wrong CP parity in the long- and shortlived K states, measured to be $|\varepsilon| = (2.28 \pm 0.05)10^{-3}$ [3]. An

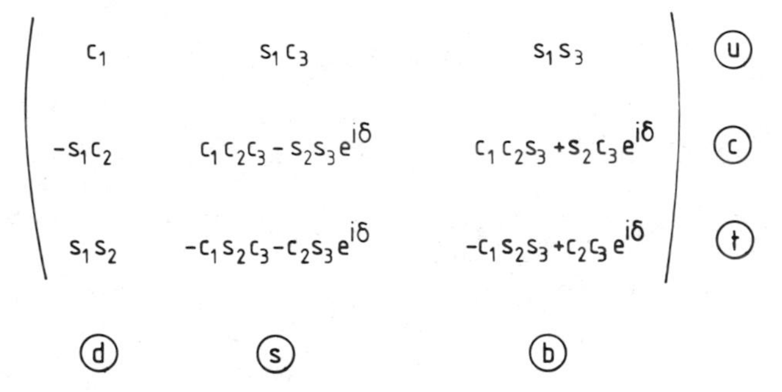

Fig.1: Kobayashi-Maskawa quark mixing matrix (ref.[1])

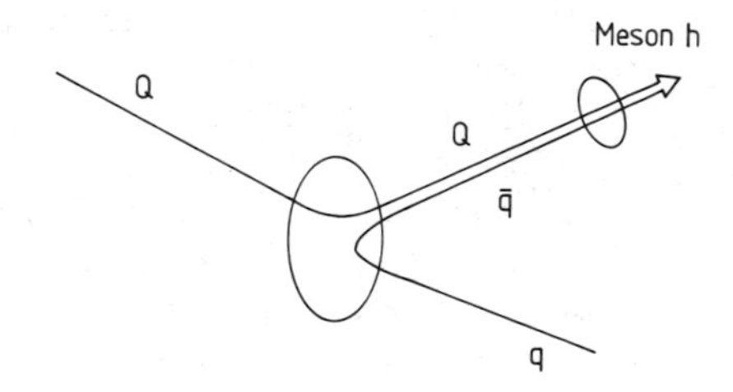

Fig.2: Formation of a heavy meson $h = Q\bar{q}$ containing heavy quark Q

approximate relation derived by Pakvasa and Sugawara [4] is

$$|\varepsilon| \sim |(m_t - m_c)/m_c| \sin 2\theta_2 \tan\theta_3 \sin\delta/(2\sqrt{2}\cos\theta_1)$$
$$\tilde{\sim} \sin 2\theta_2 \tan\theta_3 \sin\delta \qquad (1)$$

where m_t and m_c are the top and charm quark masses.

3. LIGHT QUARK COUPLINGS

3.1 U_{ud}

This coupling parameter can be determined from a comparison of measured rates of nuclear beta decays, in particular $0^+ \to 0^+$ transitions, with the muon decay rate. Two different evaluations of this quantity have been done [5,6] and their results are $U_{ud}=0.937\pm0.0025$ [5] and $U_{ud}=0.9730\pm0.0024$ [6], respectively. Combining these two, one obtains $U_{ud}=0.9733\pm0.0024$.

3.2 U_{us}

Strange particle decays, in particular kaon and hyperon semileptonic decays, are used to extract the values $U_{us}=0.219\pm0.002\pm0.011$ [5] and $0.227\pm0.003\pm0.013$ [6], where the first error is the statistical one and the second gives the estimate of systematic uncertainties. When combining the two, the error remains the same: $U_{us}=0.223\pm0.013$.

4. CHARM QUARK COUPLINGS

4.1 Production of single charmed quarks in neutrino and antineutrino reactions

In the six quark model [1], charmed quarks can be produced in the following reactions

$\nu + d \to \mu^- + c$ with coupling U_{cd}

$\nu + s \to \mu^- + c$ with coupling U_{cs}

$\bar{\nu} + \bar{d} \to \mu^+ + \bar{c}$ with coupling U_{cd}

$\bar{\nu} + \bar{s} \to \mu^+ + \bar{c}$ with coupling U_{cs}.

In the quark-parton picture, this leads to the differential cross-sections for charm production on isoscalar targets:

$$\frac{d\sigma^\nu}{dxdy} = \frac{G^2ME_\nu x}{\pi}\left[U_{cd}^2\ (u(x) + d(x)) + |U_{cs}|^2\ 2s(x)\right] \qquad (2)$$

$$\frac{d\sigma^{\bar{\nu}}}{dxdy} = \frac{G^2ME_\nu x}{\pi}\left[U_{cd}^2\ (\bar{u}(x) + \bar{d}(x)) + |U_{cs}|^2\ 2\bar{s}(x)\right] \qquad (3)$$

where u(x), d(x), and s(x) are the quark density distributions in the proton. G is the Fermi coupling constant, M the nucleon mass, E_ν the neutrino laboratory energy, and x and y the Bjorken scaling variables. Experimentally, the observation of charm production has been done mainly by three methods: 1. direct observation of the short-lived decay of charmed hadrons in emulsions, 2. observation of semileptonic charm decay $c \to s + \mu^+ + \nu_\mu$ in counter experiments, 3. observation of semileptonic charm decay $c \to s + e^+ + \nu_e$ in bubble chambers.

4.2 Fragmentation function of charmed quarks

In experiments, the observed particle is always a charmed hadron, not the charmed quark produced in the hard scattering. The process is usually described in terms of the fragmentation variable z, which, in the case of lepton-hadron scattering is defined as $z = P \cdot Q_h / P \cdot Q$ with P, Q_h and Q being the four-momenta of nucleon, outgoing hadron and struck quark, respectively, and in the case of electron-positron annihilation given by $z = E_h/E_{beam}$, where E_h is the outgoing hadron energy, and E_{beam} the energy of the colliding beam particles. The underlying picture for the process of heavy meson formation is shown in fig.2. Here the struck quark Q picks up a light antiquark $\bar{q}$ to form a meson h.

Results on charm fragmentation have been obtained so far by experiments on neutrino-nucleon scattering and on electron-positron annihilation.

First results on this fragmentation function were obtained by the CDHS-collaboration [7]. The charmed quark is tagged here by its semileptonic decay $c \to s\mu\nu$ or $d\mu\nu$, and the events used for the analysis are 10381 neutrino induced dimuon events of the type $\nu_\mu + Fe \to \mu^- + \mu^+ + X$. The variable used is the fragmentation variable $z_L = p_{\mu 2}/(p_{\mu 2} + E_{sho})$, where $p_{\mu 2}$ is the momentum of the μ^+ from semileptonic charm decay, and E_{sho} is the measured energy of the shower of hadrons labeled X. Therefore $p_{\mu 2} + E_{sho}$ is the total hadronic energy in the reaction, apart from the neutrino from charm decay escaping detection, and is, in good approximation, the energy E_c of the charmed quark produced in the interaction. The true fragmentation variable is $z = E_D/E_c$. The distribution of events in z_L is incompatible with computed curves based on $D(z) \propto (1-z)^2$, and by an unfolding procedure the best fit for D(z) in 5 bins of z is obtained, as given in fig.3. In this procedure, the kinematical threshold for D meson production $w > (m_p + m_D + m_\pi)$ was taken into account, where w is the invariant mass of the hadronic system. This threshold appears in the z distribution as a threshold at $z_{min} = m_D/w_{max} = m_D/\sqrt{s}_{max}$; with $\sqrt{s}_{max}$ being the maximum center-of-mass energy accessible in the experiment.

The unfolding procedure assumes D production to be the only source of dimuons. This is justified by the identification of charmed particles produced in the neutrino emulsion experiment [8] in a nearly identical beam. Applying the kinematical cuts of the CDHS experiment, in particular requiring a visible energy of more than 30 GeV, the composition of the charmed hadrons with C=+1 is D^0(45 ±13%), D^+(40 ±16%), Λ_c^+(10 ±6%), F^+(<11%) and neutral charmed baryons (4^{+9}_{-3}%), all corrected for detection efficiency.

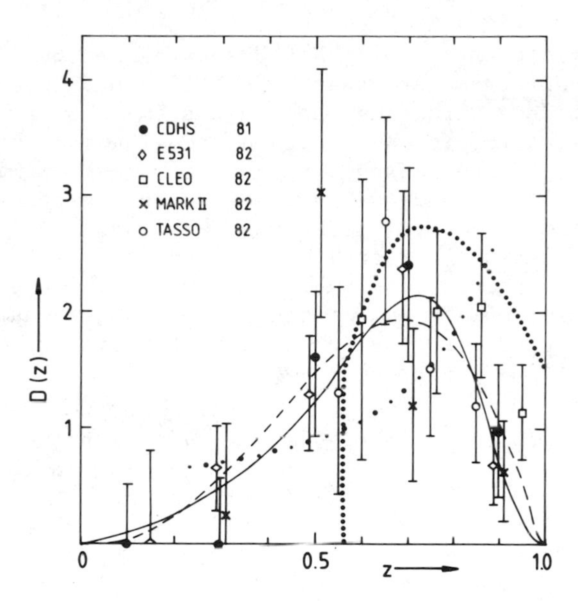

Fig.3: Experimental results on the fragmentation function of a charmed quark into D or D^* mesons. Data from CDHS [7], E 531 [8], Mark II [9], TASSO [10] and CLEO [11]. Models from Ref. [13] (dotted line), Ref. [15] for $Q^2 = 25$ (GeV/c^2) (small dots). Fits according to functions $D_1(z)$ (dashed line) and $D_2(z)$ full line), see text.

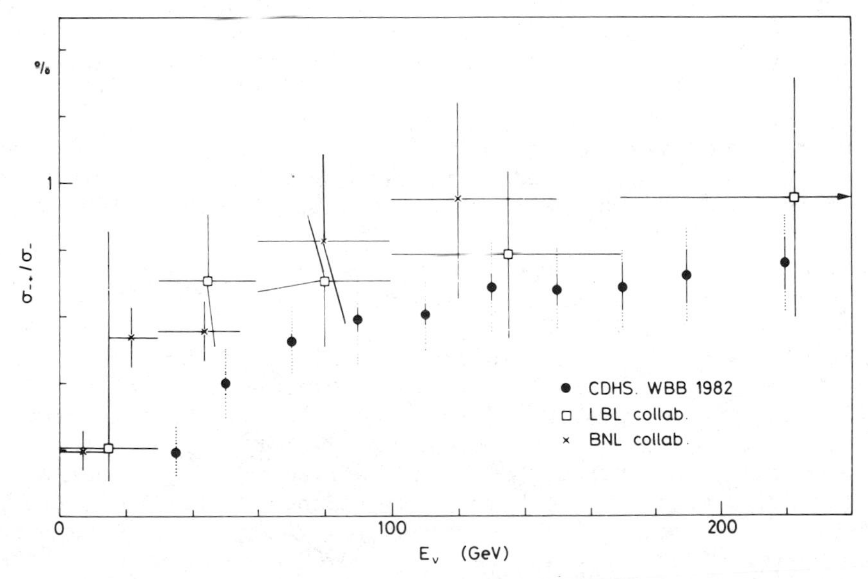

Fig.4: σ^{ν}(dilepton)/σ^{ν}(1μ) vs. visible neutrino energy E_{ν} from CDHS [7], LBL coll. [19] and BNL-coll. [20].

Another measurement of D(z) has been obtained by the Fermilab hybrid emulsion-spectrometer experiment E 531 [8]. In this case, all momenta of charm decay products seen in the emulsion are measured in the downstream spectrometer and added up to give the momentum and energy E_h of the charmed hadron. The charmed quark energy E_c is again given by the total hadronic energy observed in the reaction, E_H. The resulting distribution in $z = E_h/E_H$ based exclusively on the charmed particles identified as D mesons, normalized to $\int_{z_{min}}^{1} D(z)dx = 1$ is given in figure 3. Here again, $z_{min} = m_D/\sqrt{s_{max}}$ is the kinematic limit for the reaction. It is, of course, not known whether these D mesons are produced directly or whether they stem from a D^* produced in the neutrino reaction.

Independent measurements of fragmentation functions have been reported by e^+e^- annihilation experiments [9-11]. In this process, a charm-anticharm pair is produced, $e^+e^- \rightarrow c\bar{c}$, such that only half of the total c.m. energy $\sqrt{s} = 2E_{beam}$ is available for one of the charmed quarks, and the fragmentation variable is $z = 2E_h/\sqrt{s}$ with E_h being the observed energy of the charmed hadron. The minimum value of z is then $z_{min} = 2m_h/\sqrt{s}$ if m_h is the mass of the charmed meson produced. Results are given by the experiments in terms of cross-sections $d\sigma/dz$, and since the kinematical factor in the electromagnetic cross-section for the production of a spin 1/2 pair is $2\pi\alpha^2(1-\beta^2/3)\beta/s \propto \beta/s$ where β is the c.m. velocity of the charmed hadron, the conversion into a fragmentation function requires to consider the quantity $d\sigma/dz \cdot s/\beta$ and to normalize it by the condition $\int_{z_{min}}^{1} D(z)dx = 1$. In some cases, where data do not extend over the full range from z_{min} to 1, we have normalized D(z) by assuming a linear rise from the zero value at z_{min} to the first measured point, and a linear fall-off from the last measured point up to the point at z=1 with D(1)=0. The resulting correction in the normalization is substantial only for the experiment [11], and the error on this correction, assumed to be equal to the correction, increases only moderately the error on D(z). Details about the experiments are contained in the table 1 and the fragmentation functions D(z) extracted from them are plotted in fig.3.

The data from different experiments can be compared because the difference in the fragmentation function between D produced directly and a D coming from $D^* \rightarrow D\pi$ decay is small: the average z of the D from D^* is calculated to be lower by 7 ± 2%. With the present measurement errors it is therefore negligible which fraction of D mesons originated in D^* decays. The data agree very well wherever their z ranges overlap. The general behaviour is that of a hard fragmentation function, in contrast to those observed for light quarks, and in reasonable agreement with theoretical models [12 - 15], except perhaps the one of ref. [15]. This is in line with the first moments given by two of the experiments, $\langle z \rangle = 0.68 \pm 0.08$ at $\langle Q^2 \rangle = 20 (GeV/c)^2$ [7] and $\langle z \rangle = 0.61 \pm 0.04 (stat.) \pm 0.12 (syst.)$ [8,16] at $\langle Q^2 \rangle = 15 (GeV/c)^2$.

In order to parametrize D(z), two different functions can be used. One of them is $D_1(z) = z^{\alpha}(1-z)^{\beta}/B(\alpha+1,\beta+1)$ with B being the beta

Table 1

Collaboration	Ref.	Reaction	Beam	Events	z_{min}	$\langle z \rangle_{exp}$	$\langle z \rangle_{fit}$	$\langle Q^2 \rangle / (GeV/c)^2$
Mark II	[9]	$e^+e^- \to D^{*+}X$	E_{beam} = 14.5 GeV	16	0.14	--	0.54 ± 0.04	
TASSO	[10]	$e^+e^- \to D^{*+}X$	E_{beam} = 16 GeV	54	0.12	--	$0.60^{+0.10}_{-0.18}$	
CLEO	[11]	$e^+e^- \to D^{*+}X$	E_{beam} = 5.2 GeV	37	0.39	--	$0.68^{+0.07}_{-0.10}$	
CDHS	[7]	$\nu N \to \mu^- \mu^+ X$	350 GeV WBB $\langle E_\nu \rangle$ = 93 GeV	10381	$\sim$0.09	0.68 ± 0.08	$0.66^{+0.07}_{-0.11}$	20
E 531	[8]	$\nu N \to \mu^- DX$	350 GeV WBB $\langle E_\nu \rangle$ = 78 GeV	28	$\sim$0.12	0.61 ± 0.04(stat) ± 0.12(syst)	0.62 ± 0.08	15

function, which is suggested by the zeros of D(z) at z=0 and z=1. A satisfactory fit with $\chi^2/DF = 12.4/15$ is obtained [17] with $\alpha = 2.8 \pm 0.8$ and $\beta = 1.3 \pm 0.4$ as given in fig.3.

The other kinematical form of D(z) was suggested by Peterson et al. [18] using kinematical considerations:

$$D_2(z) = A/(z(1 - \frac{1}{z} - \frac{\varepsilon}{1-z})^2) \qquad (4)$$

where the parameter ε should be approximately the ratio $\varepsilon \sim m_q^2/m_Q^2$ with m_q the effective light quark mass and m_Q the effective heavy quark (charm) mass, and A fixes the normalization. This function also gives a good fit with $\chi^2/DF = 15.3/15$ to the data of fig.3, with the result $\varepsilon = 0.11 \pm 0.03$ and $A = 0.96 \pm 0.03$ [17]. In the model considered above this would mean $m_q \sim 1/3\ m_c$.

Using the distribution functions $D_1(z)$ fitted to each data set separately, we determine the corresponding first moments of $D_1(z)$, $\langle z \rangle_{fit}$. They are given in table 1 and agree with each other. Combining all data, the first moment comes out to be $\langle z \rangle_{fit} = 0.62 \pm 0.04$.

4.3 Neutrino and antineutrino charm production cross-sections and coupling parameter U_{cd}

A summary of recent experiments is given in table 2. The advantage of emulsion experiments [8] is the fact that the semileptonic branching ratio does not enter, bubble chamber experiments [19-22] can observe electrons down to a momentum of 0.3 GeV/c, while counter experiments [23-26], in particular the experiment of the CDHS collaboration [7] have by far the largest event sample.

In order to extract the charm production cross-section, several corrections have to be applied to the observed ratio of charm events to the total number of neutrino events. For the emulsion experiment, these are corrections for geometrical and kinematical acceptance and a correction for the kinematical suppression of charm production due to the charmed quark mass, the "slow rescaling" correction [27]. For the dilepton experiments, in addition there is the neutrino from charm semileptonic decay which escapes detection, and therefore the total energy visible in a dilepton event is incomplete. This missing energy has to be corrected for before rates of dilepton and single muon events are compared in bins of energy. The percentage of missing energy in dimuon events has been measured in the CDHS experiment [7] to be $(12 \pm 1)\%$, in good agreement with the expectation, $(12.3 \pm 2)\%$ from a simulation assuming only D mesons to be produced and decaying to $K^*\mu\nu$, $K\mu\nu$ and $\pi\mu\nu$ with branching ratios of 0.37/0.65/0.06 [28].

In table 2 is indicated which of these corrections have been applied in the experiments.

The resulting ratio on neutrino induced charm production and single muon cross-section from dilepton experiments is shown in fig.4, not corrected for slow rescaling. μ-e universality requires

Table 2 Recent experiments on single charm production by neutrinos

Collab. [Ref.]	Year	Target	Beam	Techn.	Events ν ind.	Events $\bar{\nu}$ ind.	p_{lept} cut	Corrections applied: geom. acc.	kin. acc.	miss. energ.	slow resc	Fragm. function used
					$\mu^-\mu^+$	$\mu^+\mu^-$						
CFR [24]	77	Fe	NBB var.	Ctr.	67	28	range >1.8m	x	-	-	-	
CDHS [25]	77	Fe	NBB 200GeV	Ctr.	257	58	3.5GeV/c	-	-	-	-	flat
FHOPRW [23]	78	Fe/Sci	QT, SSBT	Ctr.	199	49	5 GeV/c	x	-	-	-	e^{-3z}
CHARM [26]	81	C	WBB	Ctr.	495±32	285±29	4 GeV/c	-	-	-	-	?
					μ^-e^+	μ^+e^-						
Col-BNL [20]	77/81	Ne	WBB 400GeV	BC	81(249)	--	0.3GeV/c	x	x	x	x	
BFHSW [19]	81	Ne/H_2	QT 400 GeV	15'FNAL	49	14	0.3GeV/c	x	x	-	-	
					charmed hadron							
E 531 [8]	82	Em.	WBB 350GeV	Em.	41	--	--	x	x	-	-	$\langle z\rangle=0.61 \pm 0.12$
					$\mu^-\mu^+$	$\mu^+\mu^-$						
CDHS [7]	82	Fe	WBB 350/400	Ctr.	10381	3513	4.7GeV/c	x	x	x	x	$\langle z\rangle=0.68 \pm 0.08$

equal branching ratios for $c \to e\nu$ and $c \to \mu\nu$, such that the $\mu^- e^+$ and $\mu^-\mu^+$ cross-sections should agree, which is reasonably well fulfilled [29]. A comparison of dilepton and emulsion cross-sections requires a knowledge of the semileptonic branching ratio B of that mixture of charmed particles which is produced in the neutrino reactions. This number can be obtained [7] from the branching ratio B' of the mixutre produced in e^+e^- reactions [28,30,31], 44% D^+:56% D^0, the information on the composition of charmed particles produced above 30 GeV neutrino energy in a 350 GeV wide band beam as given by the emulsion experiment [8], and the (coarse) measurement [31] of semileptonic branching ratios of D^0 and D^+separately. The result is $B = (7.1 \pm 1.3)$ %.

Using this number, we can obtain total charm prodution cross-sections from the CDHS experiment and compare them to the emulsion results. Fig.5 shows that above 100 GeV neutrino energy, this cross-section converges to about 10 % of the total cross-section, and that counter and emulsion experiment agree.

In order to obtain the coupling parameter U_{cd}, the contribution of charm produced from the strange sea s and $\bar{s}$ quarks has to be eliminated. According to the cross-sections given in sect. 4.1, this can be done by using the weighted difference of neutrino and antineutrino cross-sections:

$$BU_{cd}^2 = \frac{(\sigma^{\nu}_{\mu^-\mu^+}/\sigma^{\nu}_{\mu^-}) - (R\sigma^{\bar{\nu}}_{\mu^+\mu^-}/\sigma^{\bar{\nu}}_{\mu^+})}{1 - R}\,\frac{2}{3} \qquad (5)$$

where R is the ratio of antineutrino to neutrino total cross-sections, $R = \sigma^{\bar{\nu}}_{\mu^+}/\sigma^{\nu}_{\mu^-} = 0.48 \pm 0.02$ [32]. The dimuon to single-muon cross-section ratio obtained by the CDHS collaboration, corrected for acceptance and slow rescaling, are shown in figs. 6 and 7. The average for $U_{cd} \cdot B$ in the energy region $80 < E_\nu < 160$ GeV is $U_{cd}^2 B = (0.41 \pm 0.07) 10^{-2}$. With the value for B quoted above, we obtain $|U_{cd}| = 0.24 \pm 0.03$. In the GIM model U_{cd} is just the sine of the Cabibbo angle, and this experimental result is in good agreement with the accepted value $\sin\theta_c = 0.230 \pm 0.003$ [33]. In the KM model, $U_{cd} = \sin\theta_1 \cos\theta_2$, where $\sin\theta_1 = 0.228 \pm 0.011$ [5]. This therefore implies $\cos\theta_2 = 1.05 \pm 0.14$. Upper limits are $\sin\theta_2 < 0.50$ and $\theta_2 < 33^\circ$, at the 90 % confidence level.

4.4 Strange sea structure function and the coupling U_{cs}

This coupling appears in charged-current reactions always together with the strange sea structure function xs(x) or its integral $S = \int xs(x)dx$. The quantity measured is $|U_{cs}|^2 \cdot 2S$, as shown in eqs. (2,3). In the absence of an independent determination of S, only the upper limit of an SU(3) symmetric sea $2S \leq \bar{U} + \bar{D}$ and a corresponding lower limit on $|U_{cs}|$ can be obtained. In principle, an independent measurement of S can be done by studying the y distribution of

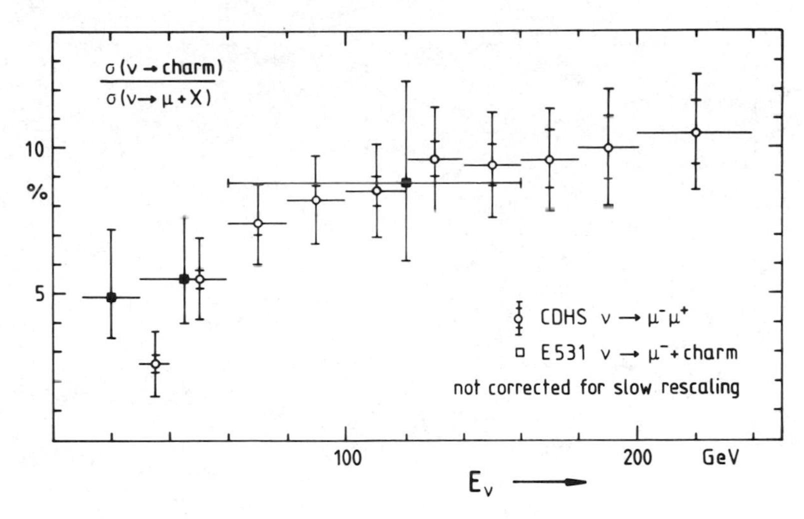

Fig.5: Cross-section for single charm production from E 531 (ref. [8]) and CDHS (ref. [7]) using semileptonic branching ratio of 7.1 ± 1.3 % [7].

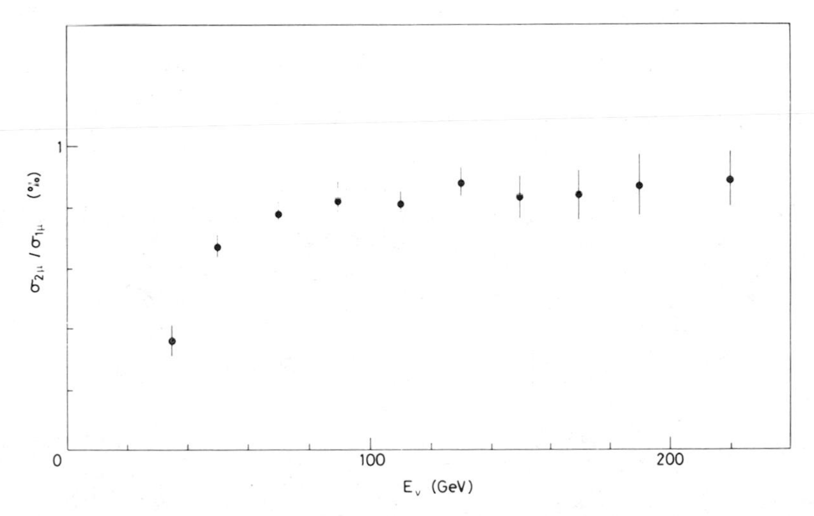

Fig.6: Cross-section ratio $\sigma(2\mu)/\sigma(1\mu)$ corrected for slow rescaling for neutrinos (ref. [7]).

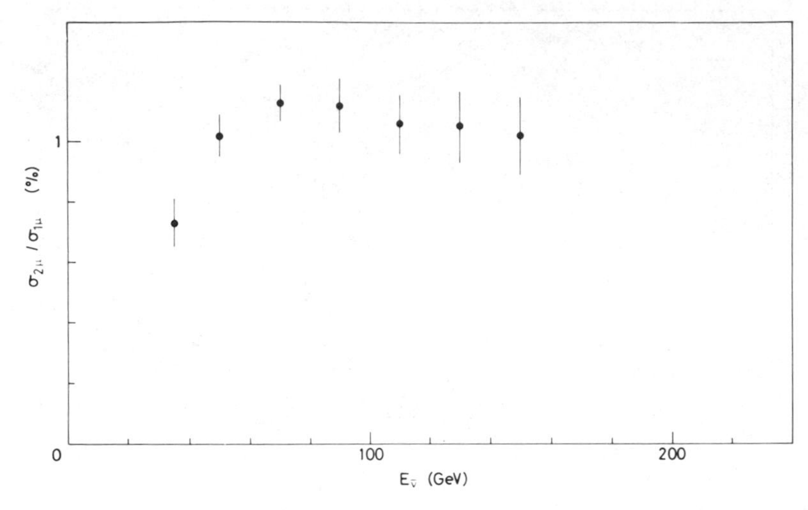

Fig.7: Cross-section ratio $\sigma(2\mu)/\sigma(1\mu)$ corrected for slow rescaling for antineutrinos (ref. [7]).

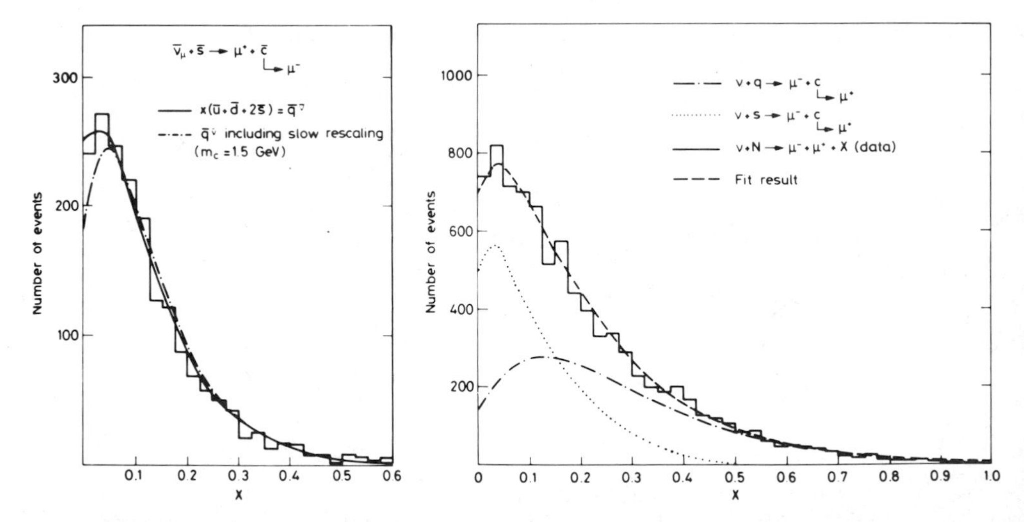

Fig.8: x_{vis} distribution for dimuon events. The histograms represent the data.a) Antineutrino; the solid curve is the sea distribution obtained from single-muon data (ref.[32]), the dashed-dotted curve demonstrates the effect of slow rescaling. b) Neutrino; the curves show the decomposition into 48% strange-sea contribution taken from the data of fig.8a (dotted curve) and 52% quark contribution (dashed-dotted curve). The dashed curve is the sum of both.

neutral current neutrino interactions, but the experimental data [34] do not allow to set a more stringent limit on S than the one given above.

The product $|U_{cs}|^2 \cdot 2S$ can be extracted in three ways from the neutrino and antineutrino dimuon production data. The first method [7] to determine this quantity uses the shape of the x distribution of neutrino and antineutrino induced dimuons, as shown in fig.8. According to eqs.(2,3), the antineutrino dimuons are mainly produced by scattering from the strange sea at low x, while the neutrino dimuons are produced from s and d quarks. One can therefore fit the x distribution of neutrino induced dimuons with a mixture of xs(x) and x(u(x) + d(x)). This has been done by the CDHS collaboration [7] with the result:

$$\frac{|U_{cs}|^2}{U_{cd}^2} \cdot \frac{2S}{U+D} = 1.19 \pm 0.09 \tag{6}$$

where U and D are the integrated momentum fractions of up and down quarks, e.g. $U = \int xu(x)dx$.

If we define the relative momentum fraction of the strange sea as compared to the non-strange sea by $\alpha = 2S/(\bar{U}+\bar{D})$, then $0 < \alpha \leq 1$, and the upper limit corresponds to SU(3) symmetry of the sea. Using the integrals of structure functions from ν-Fe inclusive scattering [32], $\bar{U}+\bar{D}+2S = (7.00\pm0.56)\%$ and $\int F_2 dx = 0.438 \pm 0.022$, one obtains

$$(\bar{U} + \bar{D})/(U + D) = (0.19 \pm 0.02)/(1 + \alpha) \tag{7}$$

depending on the strange sea strength α. This yields

$$|U_{cs}|^2/U_{cd}^2 = (6.26 \pm 0.79)(1 + \alpha)/\alpha \tag{8}$$

and

$$|U_{cs}|^2 = (0.36 \pm 0.10)(1 + \alpha)/\alpha. \tag{9}$$

This minimum value of the r.h.s. is reached for $\alpha = 1$, and the corresponding lower limit is

$$|U_{cs}| > 0.68 \quad (90\ \%\ \text{C.L.}) \tag{10}$$

In the second method [35], the product $|U_{cs}|^2 \cdot 2S$ can be extracted from the two cross-sections eqs.(2,3) in sect.4.1 by eliminating the U_{cd} terms with the result

$$\frac{|U_{cs}|^2}{U_{cd}^2} \cdot \frac{2S}{U+D} = \frac{R\,R^{\bar{\nu}} - R^{\nu}(\bar{U} + \bar{D})/(U + D)}{R^{\nu} - R\,R^{\bar{\nu}}} \tag{11}$$

where $R^{\nu} = \sigma^{\nu}_{\mu^-\mu^+}/\sigma^{\nu}_{\mu^-}$ and $R^{\bar{\nu}} = \sigma^{\bar{\nu}}_{\mu^+\mu^-}/\sigma^{\bar{\nu}}_{\mu^+}$ are the cross-section ratios for dimuon production relative to single muon production. With the experimental values from [7], one obtains

$$|U_{cs}|^2 = (0.42 \pm 0.13)(1 + \alpha)/\alpha \tag{12}$$

and the lower limit for $\alpha = 1$

$$|U_{cs}| > 0.70 \quad (90\ \%\ \text{C.L.}) \tag{13}$$

While these two methods use only ratios of cross-sections and do not depend on the semileptonic branching ratio B of charmed particles, the third method [6] is based on cross-sections and the value of B:

$$U_{cs}^2 \frac{3S}{U+D} = \frac{R^{\bar{\nu}} R - R^{\nu}(\bar{U} + \bar{D})/(U + D)}{B\ (1 - R)} \tag{14}$$

Using also eq.(7) for the non-strange sea, this leads to

$$U_{cs}^2 = (0.41 \pm 0.09)/(1 + \alpha)/\alpha \tag{15}$$

and to a lower limit

$$|U_{cs}| > 0.77 \quad (90\ \%\ \text{C.L.}) \tag{16}$$

Of these three methods, the first one (using the x distribution of neutrino-induced events) and the third one (using the absolute cross-section ratio and the semileptonic branching ratio) are independent, while the second one is related to the third one. We can therefore use eqs.(8) and (15) as independent constraints. The results obtained with the three methods agree well.

It is also worth noticing that from these data, using $|U_{cs}| \leq 1$, a lower limit on the strange sea fraction, $\alpha = 2S/(\bar{U}+\bar{D}) > 0.41$ is obtained at 90% confidence.

4.5 U_{cb}

From unitarity within the KM matrix follows $|U_{cb}|^2 = 1 - |U_{cd}|^2 - |U_{cs}|^2$ and, with the limits above, this results in $|U_{cb}| < 0.6$ with 90 % confidence.

5. BOTTOM QUARK COUPLINGS

5.1 Limit on $b \to u$ coupling from B decays into K^o or $K^{\pm}$

At the CESR storage ring, the reaction $e^+e^- \to \Upsilon(4S) \to B\bar{B}$ can be used as a B meson factory. In the CLEO detector [36], amongst 21000 hadronic events on the $\Upsilon(4S)$, $K^{\pm}$ mesons were identified by time-of-flight and K^o mesons by their $\pi^+\pi^-$ decay. The visible cross-sections for hadronic final states with a K^o or with a $K^{\pm}$ show clear enhancements around the $\Upsilon(4S)$. This signal amounts to a total of 2.82 ± 0.25 K mesons per $\Upsilon(4S)$ decay assuming the $\Upsilon(4S)$ to decay into $B\bar{B}$. In order to obtain a limit on $b \to u$ decays, the double ratio R = (Kaons/Υ(4S)/(Kaons/continuum event) is considered. CLEO obtains $R_{exp} = 1.58 \pm 0.15$, while the theoretical expectation is

$$R_{theo} = \begin{cases} 1.80 \pm 0.10 \text{ for } b \to c \\ 0.95 \pm 0.10 \text{ for } b \to u. \end{cases} \tag{17}$$

Denoting the fraction of $b \to c$ decay by f, one obtains $f = \Gamma(b\to c)/\Gamma(b\to all) = 0.74 \pm 0.18(stat) \pm 0.09(syst)$ and a 90 % C.L. upper limit $\Gamma(b\to u)/\Gamma(b\to c) < 0.79$.

Taking into account the phase space factors, the total b decay rate is [37]

$$\Gamma^b_{total} = \Gamma^b_o \{2.75\,|U_{cb}|^2 + 7.7\,|U_{ub}|^2\} \tag{18}$$

where $1/\Gamma^b_o = 0.93 \cdot 10^{-14}$ sec. The limit above is therefore transformed into a limit on the coupling parameters

$$\frac{|U_{ub}|^2}{|U_{cb}|^2} = \frac{2.75}{7.7}\,\frac{\Gamma(b\to u)}{\Gamma(b\to c)} < 0.28 \qquad (90\ \%\ \text{C.L.}) \tag{19}$$

or $|U_{ub}|\ /\ |U_{cb}| < 0.53$.

5.2 Lepton momentum spectrum from B decays

Recently, much more sensitive limits have been derived from the lepton momentum spectrum in semileptonic B decays. Both the CLEO [38] and the CUSB [39] collaborations working at Cornell have reported on measurements of the electron momentum spectrum as shown in figs. 9 and 10, and the CLEO [38] group has in addition given the muon momentum spectrum. In principle, the two possible decays, $B \to e\nu X_u$ and $B \to e\nu DX$ can be distinguished by measuring the end point of the β spectrum. In practice, the analysis is model-dependent because of the theoretical uncertainty about which X_u state with which mass is populated in the decay. Martinelli et al. [40] have calculated the expected lepton momentum spectra, and based on this model, both groups report a limit of

$$|\Gamma(b\to u)/\ \Gamma(b\to c)\ | < 0.1 \text{ at } 90\ \%\ \text{C.L.} \tag{20}$$

which corresponds to

$$|U_{ub}/U_{cb}| < 0.2 \qquad \text{at } 90\ \%\ \text{C.L.} \tag{21}$$

Fig.11 shows the limit of the CLEO collaboration eq.(20) as a function of the mass of the X_u state.

5.3 Limit on B lifetime

From the JADE group [41], an upper limit has been obtained: $\tau_b < 1.4\ \ 10^{-12}$ sec at 95 % C.L. Using the relation for the total decay rate eq.(11), this implies the following limit:

$$|U_{cb}|^2 + 2.8\,|U_{ub}|^2 > 3.1 \cdot 10^{-3} \quad (90\ \%\ \text{C.L.}) \tag{22}$$

which means that U_{cb} and U_{ub} cannot vanish at the same time.

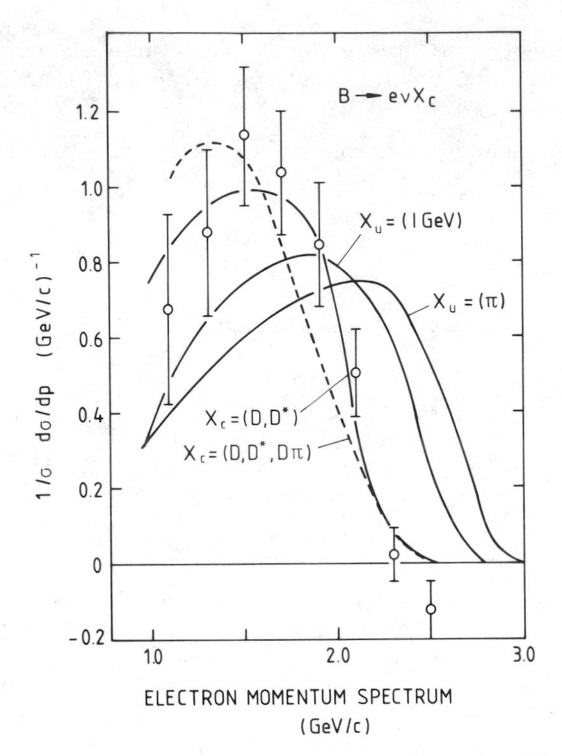

Fig. 9: Electron momentum spectrum from semileptonic B decays from CLEO Collaboration [38].

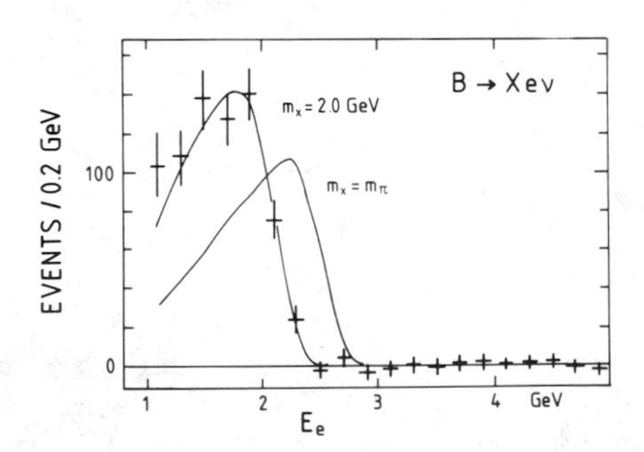

Fig.10: Electron momentum spectrum from semileptonic B decays from CUSB Collaboration [39]

5.4 Like-sign dimuon production in antineutrino reactions; limits on u-b-coupling

The prompt $\mu^+\mu^+$ rate in $\bar{\nu}$ reactions can be used to set a limit on b production from u quarks by antineutrinos: $\bar{\nu} + u \to \mu^+ + b$, because this reaction could be detected through the decay chain $b \to c$ and $c \to \mu^+\nu$.

The cross-section for $\mu^+\mu^+$ production from an isoscalar target would then be

$$\frac{d\sigma}{dy}(\bar{\nu} \to \mu^+\mu^+) = \frac{G^2ME}{\pi} U_{ub}^2 \, (U + D) \, T_b \, B(b \to c) \, B(c \to \mu\nu) \cdot (1-y)^2 \quad (23)$$

where U_{ub} is the u-b-coupling, U and D are the integrated quark momentum fractions of up and down quarks, T_b the threshold factor caused by the quark mass, $B(b \to c)$ the fraction of $b \to c$ decays, and $B(c \to s\mu\nu)$ the semileptonic charm branching ratio.

This can be compared to opposite-sign dimuon production from d quarks only:

$$\sigma(\nu \to \mu^-\mu^+)_{(d)} = \frac{G^2ME}{\pi} U_{cd}^2 \, (U + D) \, T_c \, B(c \to \mu\nu).$$

In the ratio of these two cross-sections, the semileptonic branching ratio and other factors cancel, such that we have, for an interval $0 \le y \le \alpha$

$$\frac{\sigma(\bar{\nu} \to \mu^+\mu^+)}{\sigma(\nu \to \mu^-\mu^+)_d} = \frac{U_{ub}^2}{U_{cd}^2} \, B(b \to c) \, \frac{T_b}{T_c} \, \frac{\int_0^\alpha (1-y)^2 dy}{\alpha} \quad (24)$$

In order to deduce an upper limit on U_{ub} from this, one can use the prompt experimental [42] like-sign rate in $\bar{\nu}$ reactions in the interval $0 \le y \le 0.4$ as an upper limit. Using for normalization ref. [7] and taking the value for $B(b \to c)$ from section 5.2, one obtains $U_{ub}^2 < 0.62 \cdot 10^{-2} T_c/T_b$ with 90 % confidence. This limit can be improved by assuming that the origin of the neutrino induced like-sign dimuons (where, in the 6 quark scheme with left-handed currents, the contribution of $\nu\bar{u} \to \bar{b}\mu^-$ is negligibly small) is also contributing to the $\bar{\nu}$ induced signal in proportion to the total cross-section. If this contribution is subtracted, the limit becomes $U_{ub}^2 < 0.38 \cdot 10^{-2} T_c/T_b$. For the masses $m_c = 1.5$ GeV and $m_b = 5$ GeV, $T_c/T_b \sim 8$ for 350 GeV wide band beams. Due to the severe kinematical suppression of b production at present energies, the limit becomes $U_{ub} < 0.18$. This is a direct determination of this coupling.

A more sensitive method to obtain a limit on U_{ub} within the KM scheme [5,43] uses the known coupling U_{ud} from a comparison of neutron and nuclear beta decay to muon decay and U_{us} from strange particle decays, and the unitarity relation $|U_{ub}|^2 = 1 - |U_{ud}|^2 \, |U_{us}|^2 = (0.3 \pm 0.7)10^{-2}$, which gives $|U_{ub}| < 0.11$ at 90 % C.L.

6. CONSTRAINTS ON WEAK QUARK MIXING ANGLES IN THE KOBAYASHI-MASKAWA SCHEME

The angle θ_1 is determined very precisely from weak decays of nonstrange and strange particles, with the result $s_1 = \sin\theta_1 = 0.228 \pm 0.011$. We can then, for a fixed phase δ, use the constraints on the elements of the KM-matrix obtained in the preceeding sections in order to obtain coupled constraints on the angles θ_2 and θ_3 [43,44]. Choosing for graphical representation the plane $\sin\theta_2$ vs. $\sin\theta_3$, we indicate the forbidden regions.

There are two contraints where only one of the angles (and no phase δ) enters:

i) the limit from dimuons on the cd coupling (sect.4.3)
   $0.24 \pm 0.03 = U_{cd} = \sin\theta_1 \cdot \cos\theta_2$
   gives directly $s_2 = \sin\theta_2 < 0.5$ at 90% C.L.

ii) the limit on $U_{ub} = \sin\theta_1 \sin\theta_3$ from the unitarity argument
   yields (sect.54.) $s_3 = \sin\theta_3 < 0.44$ at 90% C.L.

For other constraints, the two angles and the phase δ are coupled.

iii) the limit on c-s coupling from dimuons (sect.4.4) excludes, via the relation
   $U_{cs} = c_1c_2c_3 - s_2s_3e^{i\delta}$
   for $\delta = 0$: a large band in the s_2s_3 plane
   for $\delta = \pi$: the corners near $s_3=1$ and $s_2=0$ and near $s_3=0$ and $s_2=1$.

iv) the limit on the b quark couplings reads (sect.5.2)
   $|U_{ub}|/|U_{cb}| = |s_1s_3|/|(c_1c_2s_3 + c_3s_2e^{i\delta})| < 0.2$.
   Again, for $\delta = \pi$ a large region around the diagonal in the (s_2,s_3) plane is excluded.

v) the b lifetime limit (sect.5.3) $|c_1c_2s_3 + c_3s_2e^{i\delta}|^2 + 2.8|s_1s_3|^2 > 3.1 \cdot 10^{-3}$ excludes a small corner around the origin of the (s_2,s_3) plane.

Superimposing the excluded regions, one obtains the remaining white regions for s_2 and s_3, for the three cases $\delta = 0$, $\pi/2$ and π (figs. 12,13 and 14). For the case $\delta = \pi/2$, a theoretical limit from eq.(1) can be drawn if CP violation in the mass matrix is due to quark mixing. All the experimental constraints can be used for a combined fit of the angles θ_2 and θ_3.

We obtain for a minimum $\chi^2 = 1.3/4$ D.F. the values $\sin\theta_1 = 0.228 \pm 0.011$, $\alpha = 2S/(\bar{U}+\bar{D}) = 0.47 \pm 0.06$ and best values of $\sin\theta_2 \approx 0.12$ and $\sin\theta_3 \approx 0$ with the error contours in the $(\sin\theta_2, \sin\theta_3)$ plane given in fig.15. These contours vary slightly with the value of the phase δ. It is interesting to see that the phenomenological analysis [45] using the K_L-K_S mass difference Δm and the CP violation parameter ε yields values of $\sin\theta_2$ between 0.11 and 0.24 for a top quark mass $m_t = 30$ GeV and between 0.07 and 0.14 for $m_t = 100$ GeV, if $\sin\theta_3 = 0$. For a wide range of the phase δ, $1^o \lessapprox \delta \lessapprox 179^o$, the values of $\sin\theta_2$ and $\sin\theta_3$ from this phenomenological analysis as given by the curves in fig.15 lie within the error contours of our experimental evaluation.

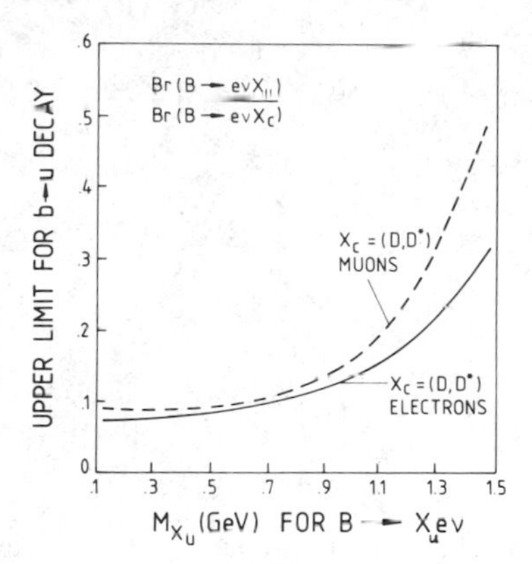

Fig. 11: Upper limit for the ratio $Br(B\to e\nu X_u)\,/Br(B\to e\nu X_c)$ vs. mass of the X_u state, from CLEO [38].

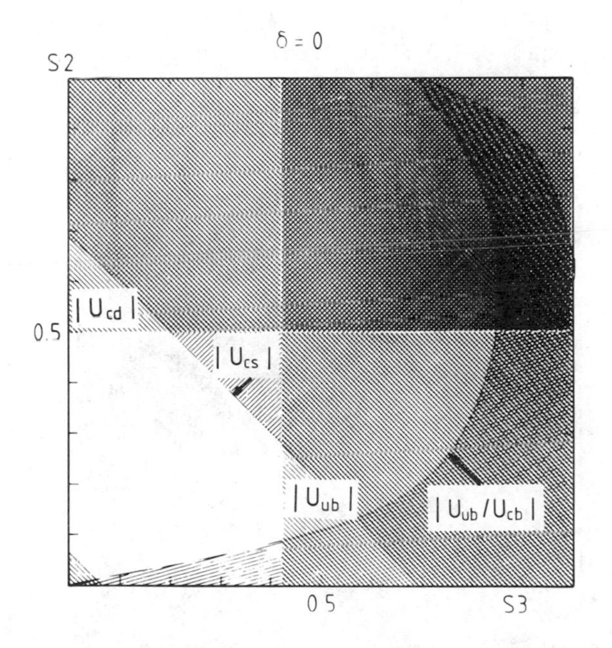

Fig.12: Excluded regions in the ($\sin\theta_2$, $\sin\theta_3$)-plane by experimental limits on KM matrix elements for a phase angle $\delta = 0$. All contours are 90 % C.L. limits.

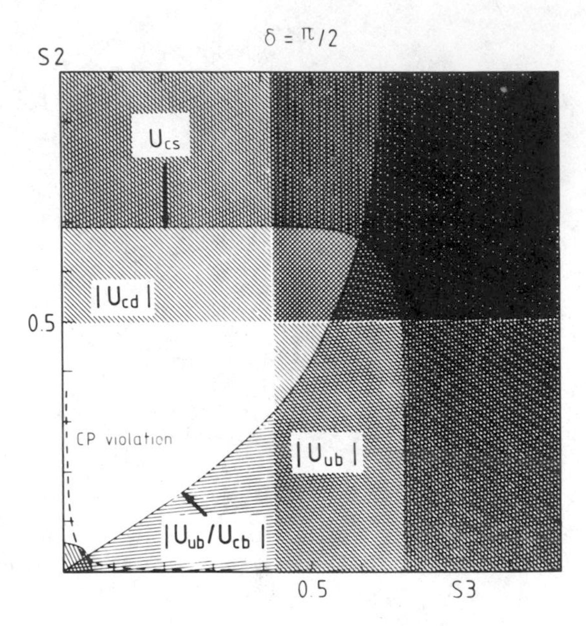

Fig.13: Same as in fig.12 for $\delta = \pi/2$. The dotted line is a theoretical bound if CP violation is due to quark mixing [4]. Contours are 90% C.L. limits.

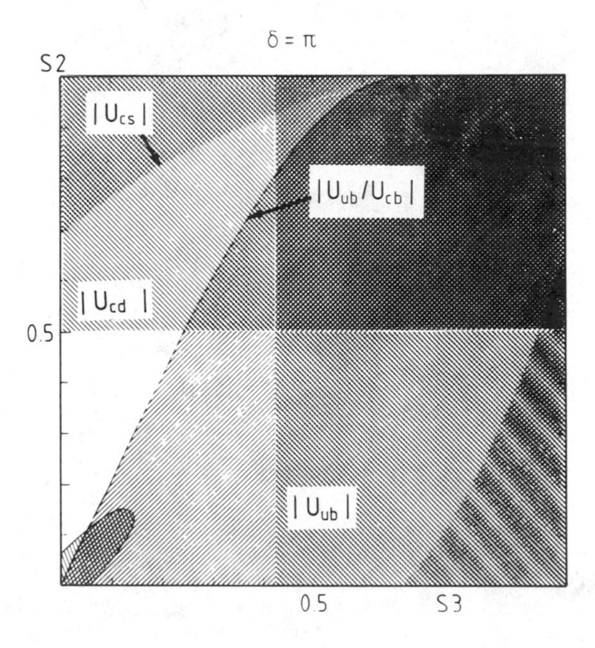

Fig.14: Same as in fig.12 for $\delta = \pi$. Contours are 90% C.L. limits.

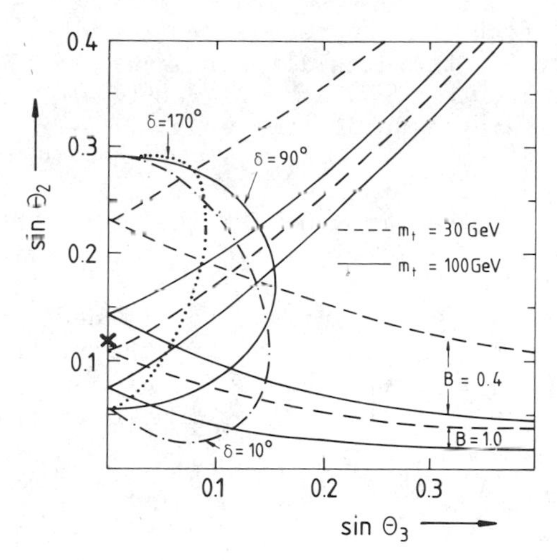

Fig.15: Best fit value (marked by cross x) and 1σ error contours in the ($\sin\theta_2$, $\sin\theta_3$)-plane of mixing angles for three values of the phase δ(10°, 90°, 170°). Also shown are four curves with two branches each derived from a theoretical analysis [45] of the K_L-K_S mass difference and the CP violation parameter ε for two values of the box-graph parameter B and of the top quark mass m_t. These curves give correlations between $\sin\theta_2$ and $\sin\theta_3$, where along the lower branch the phase angle δ varies from 0° at $\sin\theta_3 = 1$ to 90° at $\sin\theta_3 = 0$, while on the upper branch δ varies from 90° at $\sin\theta_3 = 0$ to 180° at $\sin\theta_3 = 1$.

But still there is a lot of work to do before we can really pin down the values of the angle θ_2 and θ_3 responsible for the mixing between different generations of quarks in our present picture.

7. FUTURE PROGRESS

Experimental progress in this field could happen in the following directions:

7.1 A very precise measurement of the $\pi^+ \to \pi^o e^+ \nu$ decay rate could give U_{ud} without a knowledge of nuclear corrections. The matrix element is known, but the branching ratio for this π decay is only 10^{-8}. An experiment of a Bern-SIN-group [46] aims at 10^5 decays and an error in U_{ub} of $\Delta U_{ub} = \pm 0.0025$.

7.2 Observation of the KM-suppressed decay mode $b \to u$ would be a unique possibility of obtaining a lower limit on U_{ub}. One channel where this coupling enters would be $B^+ \to \tau^+ \nu$ [47], and this could be detected in an associated production, e.g.: $e^+e^- \to B^-(\to D\pi(\pi), D^*\pi(\pi)) + B^+(\to \tau^+ \nu)$

7.3 A measurement of the B meson lifetime would become possible with precise vertex detectors. It will give a lower limit on $\sin \theta_2$.

7.4 Observation of single top production, by neutrinos, $\nu + d \to \mu^- + t$, with semileptonic top decay $t \to \mu^+ b$ may become feasible at the Tevatron if the mass $m_t \lesssim 20$ GeV. The ratio r of opposite dimuon production via charm or top production is estimated to be

$$r \sim \frac{U_{td}^2}{U_{cd}^2} \cdot \frac{T_t}{T_c} \cdot \frac{B(t \to \mu\nu b)}{B(c \to \mu\nu s)} \sim 2 \times 10^{-5}$$

where T_t and T_c are the kinematical threshold factors, and $U_{td} = \sin \theta_1 \sin \theta_2 \lesssim 0.228 \cdot 0.05 \approx 0.01$.

7.5 The search for a second kind of CP violation in the weak decay Hamiltonian (amplitude ε') may shed some light into the question whether the 6 quark model gives a correct picture of CP violation. Three experiments are planned or starting now to measure the ratio of rates

$$\left|\frac{\eta_{oo}}{\eta_{+-}}\right|^2 = 1 - 6 \left|\frac{\varepsilon'}{\varepsilon}\right| \tag{25}$$

where $\eta_{oo} = a(K_L \to \pi^o\pi^o)/a(K_S \to \pi^o\pi^o)$ and $\eta_{+-} = a(K_L \to \pi^+\pi^-)/a(K_S \to \pi^+\pi^-)$ are amplitude ratios of CP violating and CP conserving K^o decays, and $|\varepsilon| = (2.28 \pm 0.05)\ 10^{-3}$ is the amplitude admixture of wrong CP parity in the long-lived K state. One experiment (Chicago-Stanford-Saclay) has already registered 3500 $K_L \to \pi^o\pi^o$ events, which gives a statistical error on $|\varepsilon'/\varepsilon|$ of 0.3% [48]. A Yale-BNL experiment aims at 0.2% and a CERN-Dortmund-Siegen-Pisa-experiment at 0.1% in this quantity.

Acknowledgements

It is a pleasure to than B. Renk for fruitful discussions.

References

[1] M.Kobayashi, K.Maskawa, Progr.Theor.Phys.49, 652 (1973)
[2] S.L.Glashow, J.Iliopoulos, L.Maiani, Phys.Rev.D 2, 1285 (1970)
[3] See review by K.Kleinknecht, Ann.Rev.Nucl.Sci.26, 1 (1976)
[4] S.Pakvasa and H.Sugawara, Phys.Rev.D 14, 305 (1976); L.Maiani, Phys.Lett. 62 B, 183 (1976); J.Ellis et al., Nucl.Phys.B 109, 213 (1976); J.S.Hagelin,Nucl.Phys.B 193, 123 (1981)
[5] R.E.Shrock, L.L.Wang, Phys.Rev.Lett.41, 1692 (1978) and 42, 1589 (1979)
[6] E.A.Paschos and U.Türke, Phys.Lett. 116 B, 360 (1982)
[7] J.Knobloch et al., (CDHS Collaboration), Proc. of the 9th. Int. Conf. on Neutrino Physics and Astrophysics, Maui, Hawaii 1981, Vol.I, p.421; H.Abramowicz et al., Z.Physik C 15, 19 (1982)
[8] N.Ushida et al., Characteristics of Charmed Hadrons Produced by Neutrino Interactions, Canada-Japan-Korea-USA Emulsion Collaboration E 531, contribution to the 10th Int.Conf. on Neutrino Physics and Astrophysics, Balatonfüred, June 13-20, 1982, quoted in Ref [16], Phys.Lett. 121 B, 287 and 292 (1983)
[9] J.N.Yelton et al., Mark II Collaboration, Phys.Rev.Lett 49, 430 (1982)
[10] TASSO Collaboration, talk by G.Wolf at 21st Conf. on High Energy Physics, Paris, July 26-31, 1982 and priv. comm. by G.Wolf
[11] CLEO Collaboration, C.Bebek et al., Inclusive Charged D^* Production in e^+e^- Annihilations at W = 10.4 GeV, CLEO 82-01, contribution to 21st Conf. on High Energy Physics, Paris, July 26-31, 1982
[12] J.D.Bjorken, Phys.Rev.D 17, 171 (1978)
[13] M.Suzuki, Phys,Lett 71 B, 139 (1977)
[14] H.Georgi and D.Politzer, Nucl.Phys. B 136, 445 (1978)
[15] L.M.Jones and R.Migneron, preprint DESY 82-038 (June 1982)
[16] K.Kleinknecht, Proc. of the 10th Int.Conf. on Neutrino Physics and Astrophysics, Balatonfüred, June 13-20, 1982, Vol.I, p.115
[17] K.Kleinknecht and B.Renk, Z.Phys.C (1983) in press
[18] C.Peterson et al., SLAC-PUB-2912 (April 1982)
[19] H.C.Ballagh et al., Phys.Rev.D 24, 7 (1981)
[20] C.Baltay, Recent Results from Neutrino Experiment in Heavy Neon Bubble Chambers, in Proc. 1979 JINR-CERN School of Physics, September 1979, p.72, Budapest, Hungarian Academy of Sciences, 1980; C.Baltay et al., Phys.Rev.Lett 39, 62 (1977)
[21] J.Blietschau et al., Phys.Lett.58 B, 361 (1975)
J.Blietschau et al., Phys.Lett.60 B, 207 (1976)
B.C.Bosetti et al., Phys.Lett 73 B, 380 (1978)
[22] J. von Krogh et al., Phys.Rev.Lett.36, 710 (1976)
B.C.Bosetti et al., Phys.Rev.Lett.38, 1248 (1977)

[23] A.Benvenuti et al., Phys.Rev.Lett.34, 419 (1975)
A.Benvenuti et al., Phys.Rev.Lett.35, 1199 (1975)
A.Benvenuti et al., Phys.Rev.Lett.41, 1204 (1978)
[24] B.C.Barish et al., Phys.Rev.Lett. 36, 939 (1976)
B.C.Barish et al., Phys.Rev.Lett. 39, 981 (1977)
[25] M.Holder et al., Phys.Lett 69 B, 377 (1977)
[26] M.Jonker et al., Phys.Lett. 107 B, 241 (1981)
[27] R.Brock, Phys.Rev.Lett. 44, 1027 (1980)
[28] W.Bacino et al., Phys.Rev.Lett. 43, 1073 (1979)
[29] This agreement is at variance with a recent theoretical suggestion than an e/μ asymmetry in the semileptonic branching ratios of charmed particles (caused by a Higgs boson exchange) could explain a possible ν_e/ν_μ asymmetry in the CERN beam dump experiments, V.Barger, F.Halzen, S.Pakvasa and R.J.N.Philips, Preprint UH-511-465-82 (April 1982)
[30] J.M.Feller et al., Phys.Rev.Lett. 40, 274 (1978)
[31] R.H.Schindler, Ph.D.Thesis, SLAC report 219 (1979)
[32] J.G.H.de Groot et al., Z.Phys. C 1, 143 (1979)
[33] M.Roos, as quoted by K.Kleinknecht, in: Proc. 17th Int. Conf.on Hich Energy Physics, London, England, 1974, p.III, 23.Chilton, Didcot, Berks.: Rutherford High Energy Laboratory, 1975, and by M.Nagel et al., Nucl.Phys. B 109, 1 (1976). See also M.Roos, Nucl.Phys. B 77, 420 (1974)
[34] M.Jonker et al., Phys.Lett. 102 B, 67 (1981)
[35] K.Kleinknecht and B.Renk, Z.Phys. C 16, 7 (1982)
[36] A.Brody et al., Phys.Rev.Lett.48, 1070 (1982), D.Kreinick, contr. to this conference
[37] M.K.Gaillard and L.Maiani, Cargèse 1979, Plenum Press, p.433 (1980)
[38] B.Gittelmann, CLEO Collaboration, paper given at 21st. Int.Conf. on High Energy Physics, Paris, July 26-31, 1982
[39] P.Franzini, CUSB Collaboration, paper given at 21st. Int.Conf. on High Energy Physics, Paris, July 26-31, 1982
[40] G.Martinelli, paper given at 21st. Int.Conf. on High Energy Physics, Paris, July 26-31, 1982
[41] W.Bartel et al., Phys.Lett. 114 B, 71 (1982)
[42] J.G.H.de Groot et al., Phys.Lett. 86 B, 103 (1979)
[43] J.J.Sakurai, Proc.Int.Conf. on Neutrino Physics and Astrophysics, Maui, Hawaii 1981, Vol,II, p.457
[44] S.Pakvasa et al., Phys.Rev. D 23, 2799 (1981)
[45] L.L.Chau (review), preprint BNL-31859-Rev. (Aug.1982), to be published in Phys.Reports
[46] P.Minkowski, priv. comm. (Feb. 1983)
[47] Discussions with L.L.Chau at this conference
[48] J.W.Cronin, priv. comm. (Feb. 1983)

PROGRESS IN MEASUREMENT AND UNDERSTANDING OF BEAM POLARIZATION IN ELECTRON POSITRON STORAGE RINGS

Presented by D.P. Barber*

The DESY Polarization Group: D.P. Barber, H.D. Bremer, J. Kewisch, H.C. Lewin, T. Limberg, H. Mais, G. Ripken, R. Rossmanith, R. Schmidt

Deutsches Elektronen-Synchrotron DESY
2000 Hamburg 52, Federal Republic of Germany

ABSTRACT

A report is presented on the status of attempts to obtain and measure spin polarization in electron-positron storage rings. Experimental results are presented and their relationship to predictions of calculations discussed. Examples of methods for decoupling orbital and spin motion and thus improving polarization are discussed.

INTRODUCTION

Since the realisation in the 1960s[1] that the emission of synchrotron radiation in electron positron storage rings can lead to a build up of spin polarization along the direction of the guide field, there have been many theoretical and experimental studies of the phenomenon. However, although polarization has been observed at several centres, it has remained something of a curiosity. It has indeed been used in investigations of quark angular distributions at SPEAR[2], and to obtain accurate calibration of beam energies. However it becomes most powerful at high energies, above a few tens of GeV, where longitudinal polarization provides a tool for investigation of the couplings in weak interactions[3] and where, unfortunately, the generation of polarization becomes more difficult.

*On leave of absence from Laboratory for Nuclear Science Massachusetts Institute of Technology.

In this article I report on the experimental study of the machine physics aspects of polarization at SLAC and Novosibirsk and on the status of the successful attempts to obtain polarization at the highest energies available, namely at PETRA.

I restrict discussion to vertical polarization which is the necessary precursor to longitudinal polarization.

Before proceeding, I remind the reader of some basic properties of the polarization mechanism in storage rings[4].

a) Synchrotron photon emission can lead to a build up of vertical polarization due to the difference between the spin-up to spin-down transition rate and the spin-down to spin-up transition rate[1,4]. The maximum achievable polarization due this mechanism is 92.38 %.

b) The polarization vector points along the so called $\hat{n}$ axis, a periodic unit vector for the spin direction which is dependent on the shape of periodic (closed) orbit of the beam and on the energy. An arbitrary spin vector precesses around $\hat{n}$ by $a\gamma$ times per revolution where γ is the Lorentz factor, $a = (g-2)/2$ where g is the electron g factor and $a\gamma$ is called the spin tune.

c) Synchrotron radiation both creates polarization and causes its destruction: sudden energy loss by photon emission causes a particle suddenly to follow a new orbit in the focussing system with the result that the correlation between orbital and spin motion is lost and depolarization can occur.

d) In travelling around the ring, particles experience horizontal and vertical betatron oscillations and longitudinal or energy oscillations known as synchrotron oscillations.

e) The depolarizing effects are strongest at the so called depolarizing resonances where:

$$a\gamma = k \qquad \text{(imperfection resonances)} \qquad (1a)$$

$$a\gamma = k \pm Q_I,\ I = x,y,s \qquad \text{(linear intrinsic resonances)} \qquad (1b)$$

$$a\gamma = k \pm n_x Q_x \pm n_y Q_y \pm n_s Q_s \quad \text{(non-linear intrinsic resonances)} \quad (1c)$$

where k and n_x, n_y, n_s are integers and Q_x, Q_y, Q_s are the orbital tunes.

In short, electron beam behaviour is dominated by focussing, damping and stochastic excitation effects and these have a profound effect on polarization.

MEASUREMENT OF POLARIZATION

In order to investigate the behaviour of polarization and to use it to specify the running conditions of the physics experiments we clearly need polarimeters which provide reliable and reproducible measurements which are independent of the High Energy Physics effort. The polarimeters should be fast so that feed back can be applied to the machine conditions and so that measurements can be statistically accurate. It should be possible to make measurements on single beams.

The best way to satisfy these requirements is to backscatter circularly polarized laser light off the electrons as shown in Fig. 1 which illustrates the layout of the PETRA polarimeter. Electrons entering a straight section from the left collide with photons of 2.42 eV from an argon ion laser. The photons are Compton scattered through 180° into a shower counter 50 meters down stream behind a vertically steerable slit. The scattering angle θ and the photon energy are correlated by kinematics and θ is of order $1/\gamma$. For example for 16 GeV electron energy, θ is typically 30 microrad and the scattered photon energy $\leq$ 5 GeV. If the electrons are vertically polarized, the angular distribution of scattered photons has the following form[5,6,7] :

$$\frac{d\sigma}{d\theta\ d\varphi} \propto [f(\theta) \pm P \cos\varphi\ g(\theta)] \qquad (2)$$

and where P is the electron polarization and the choice of + or - sign depends on the laser photon helicity. Eqn. 2 shows that there is an up/down asymmetry and an asymmetry under reversal of photon helicity. The asymmetries are proportional to the degree of vertical polarization of the beam which can then be measured by recording the rate in the shower counter for each photon helicity and as a function of the vertical position of the slit. In practice we can reduce systematic errors by combining these rates into on overall asymmetry:

$$\hat{A}(y) = \frac{1}{2}\,(A(+y)-A(-y))$$

where $\quad A(y) = (N_{+}(y)-N_{-}(y)) \,/\, (N_{+}(y)+N_{-}(y))$

and where $N_{+/-}(y)$ is the photon rate for positive (negative) helicity at position y. The maximum of $\hat{A}(y)$, $\hat{A}_m$ occurs at $\theta \approx 1/\gamma$ i.e. at $|y| \approx 1.5$ mm at PETRA.

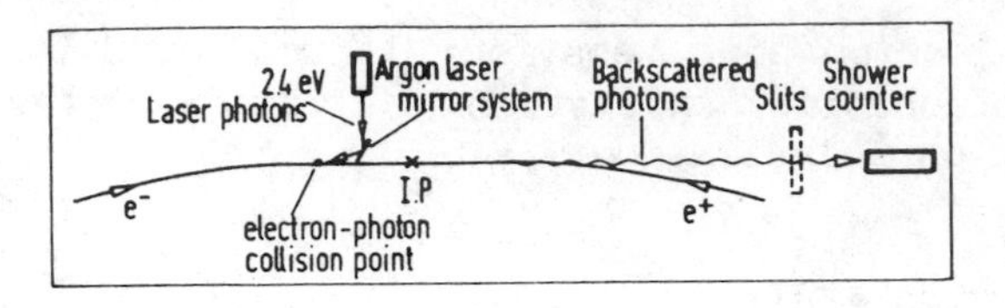

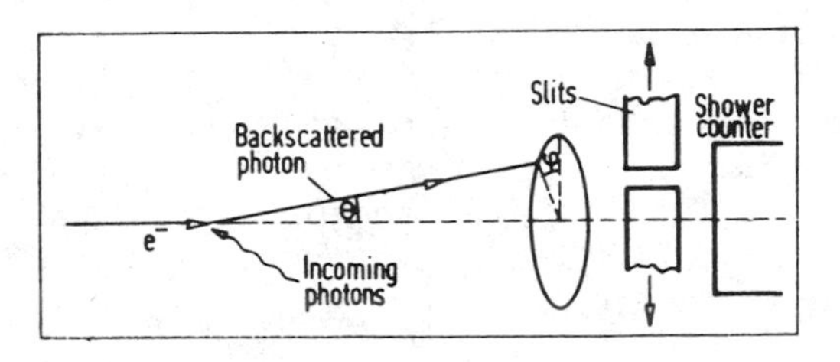

Fig. 1. Layout of PETRA polarimeter

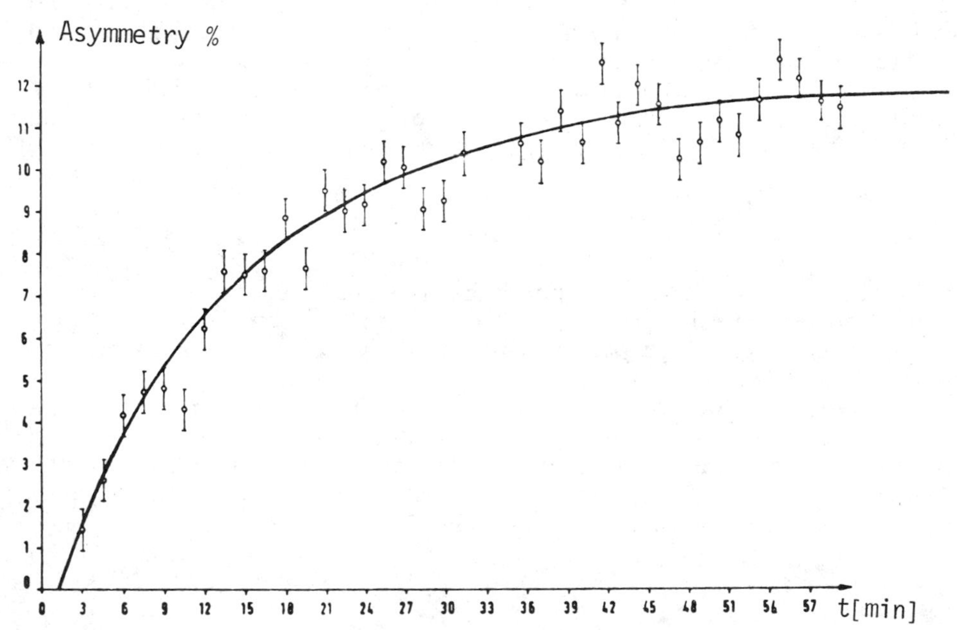

Fig. 2. Exponential build up in time of the polarimeter asymmetry at PETRA at 16.5 GeV

Since from Eqn. 2, $\sigma_p/P \approx \sigma_A/A$ and since $\sigma_A \approx 1/\sqrt{N}$, the error on polarization is minimized if the rate is high and A is large. This implies use of high laser power and high electron-photon collision luminosity. The high rate also reduces contamination of the signal by backgrounds. At PETRA, where $\hat{A}_m$ is ≈ 10 % it is routinely possible to achieve $\sigma_p/P \approx 2$ % with count rates of order 20 kHz.

If we begin with an initially unpolarized beam, the asymmetry increases in time as:

$$\hat{A}_m(t) \propto P(t) = 0.92 \frac{\tau_e}{\tau_p} \, [1\text{-}\exp\text{-}t/\tau_e] \qquad (3)$$

where τ_p is the polarization build up time for a perfect machine, and is proportional to $(\text{Energy})^{-5}$, $\tau_e = \tau_p \tau_d/(\tau_p + \tau_d)$ and τ_d describes depolarising effects in a real machine. Fig. 2 shows a typical measurement of the rise of $\hat{A}_m$ at PETRA at 16.5 GeV. The smooth line is the fit to the exponential shape of Eqn. 3 and corresponds to $\tau_e = 13.6 \pm 0.8$ min. $\tau_p = 17.9$ min. and from Eq.3 we deduce that the asymptotic polarization, P (∞) was 70 ± 4 % in agreement with the degree of polarization deduced from the asymptotic value of $\hat{A}_m$ and knowledge of the polarimeter acceptance.

We can see clearly from this example that use of laser polarimeters is a well established technique. Nevertheless improvements in count rate are welcome and a high power Nd-Yag laser is now being installed at PETRA.

CALCULATION OF POLARIZATION

In addition to our being able to make measurements it is obviously essential that we are able to calculate and predict polarizations as an aid in optimising machine design. Unfortunately, since the mechanism of depolarization is so complex, it is extremely difficult to obtain complete meaningful analytic predictions. Until recently, the only practical method which incorporates all aspects of the lattice of a particular machine has been to use the program SLIM written by A. Chao[4,8]. This program is based on a linear perturbation theory which extends the 6x6 transport matrix formalism of conventional machine optics to a linear, coupled spin-orbit formalism using 8x8 matrices. Thus the formalism only predicts linear intrinsic resonances but it does superimpose all resonance effects simultaneously.

Lack of space prohibits further discussion but further information may be found in Ref. 9. SLIM is the program that everyone uses and I will comment on comparison of predictions with experimental results below.

A more complete description of depolarization including non-linear effects can be obtained by a tracking simulation in which a bunch of electrons is followed around the ring, emitting photons. This approach of course consumes large amounts of computer time but it has recently borne fruit[10] with the prediction of non-linear resonance effects in PETRA.

CONTROL OF DEPOLARISATION EFFECTS

In a perfectly aligned (flat) storage ring particles travelling on the closed orbit feel vertical magnetic fields and the $\hat{n}$ axis is vertical. In reality, storage rings are never flat and the closed orbit is "wavy". Since electrons on the closed orbit now feel the radial fields of the quadrupoles, the $\hat{n}$ axis is no longer exactly vertical. Within the framework of the first order perturbation theory of SLIM it may then be shown that the spin vectors of individual particles which are executing horizontal betatron oscillations experience varying degrees of precession around the $\hat{n}$ axis depending on their amplitude and phase. The ensemble of spin vectors thus becomes smeared or depolarized. Also, since the beam now has vertical thickness, similar smearing results from precession of the (almost) vertical spins around the radial fields of the quadrupoles during vertical betatron motion. These effects are strongest near energies satisfying the resonance conditions of Eqn. 1 and they increase with energy.

Clearly , the effect of horizontal motion can be suppressed if the $\hat{n}$ axis can be returned to vertical with the aid of correction coils which control the vertical closed orbit. This would then be an example of how to decouple spin and horizontal motion. It may be shown that the deviation of the $\hat{n}$ axis from vertical is proportional to[5] :

$$[\int B_r(s) \cos \varphi_s \, ds]^2 + [\int B_r(s) \sin \varphi_s \, ds]^2 \tag{4}$$

where φ_s is the spin precession phase and $B_r(s)$ is the radial field on the closed orbit. The chief contributions to these integals come from Fourier harmonics in the periodic $B_r(s)$ which are closest to the spin tune and which may comprise only a small part of B_r. If these harmonics can be empirically suppressed by selective excitation of vertical correction coils in a way that does not cause undue additional distortion of the orbit by excitation of other harmonics, then $\hat{n}$ may again be brought close to vertical.

This scheme, has been successfully applied at PETRA and experimental results are discussed below.

EXPERIMENTAL STUDIES

a) Single beams at low energy

As mentioned above, depolarizing effects increase with energy and it is expected that it is easier to obtain polarization at low energies than at high energies. With this in mind, Fig. 3a shows an amalgamation of single beam measurements for SPEAR [11] obtained over many different machine runs spanning 3.52-3.76 GeV. We see regions of high polarization partitioned by linear and non-linear resonances. The smooth curve is hand drawn to guide the eye. By selecting suitable energies in this range there seems to be no particular difficulty in obtaining good single beam polarizations. Fig. 3b shows the measured dependence of τ_e on energy at SPEAR away from resonances. The straight line is the absolute theoretical prediction and is in good agreement with the measurements. Our earlier comments on SLIM are illustrated in Fig. 3c [4] which shows a comparison between measurement and SLIM prediction near 3.6 GeV. SLIM indeed predicts the linear resonances $a\gamma - Q_x = 3$ and $a\gamma - Q_y = 3$ but as expected fails to predict the non-linear resonance at $a\gamma - Q_x + Q_s = 3$.

b) Single beams at high energy

At the higher energies of PETRA, [6] success was initially more elusive, with measured polarizations rarely above 30 % and not reproducible. However the situation improved radically with the introduction of the closed orbit correction scheme described above. The measurements were made near 16.52 GeV. ($a\gamma = 37.5$) so that the dominant closed orbit harmonics were the 37th and 38th. Figs. 4a and 4b show the dependence of the asymmetry on the 38th sine and cosine harmonics and Fig. 4c shows the energy dependence of the asymmetry after optimisation of all four harmonics at 16.52 GeV. There is then a range of about 150 MeV where the asymmetry is high. This behaviour is reproducible and after calibration with a rise time measurement is seen to correspond to maximum polarizations above 70 %. This represents a major success in attempts to understand and control depolarization effects at high energy [6].

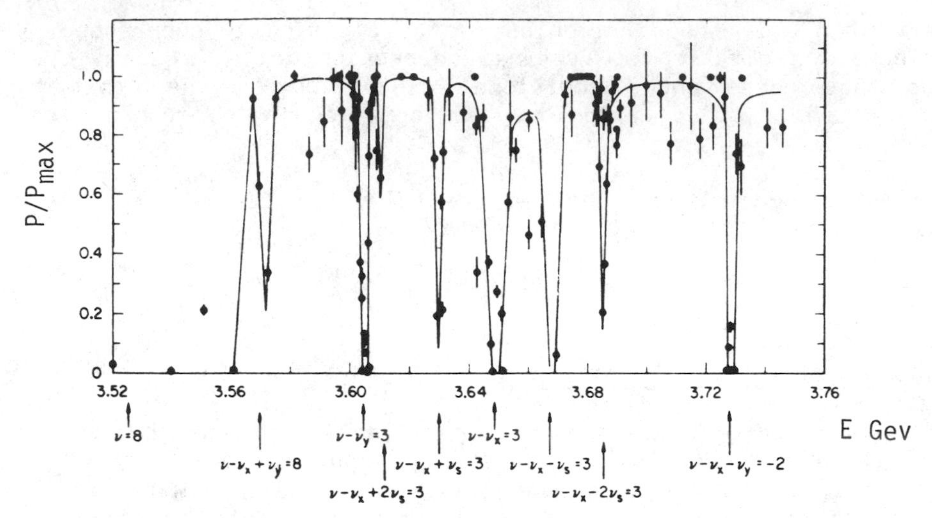

Fig. 3a. Polarization as a function of energy measured at SPEAR. (From Ref. 11)

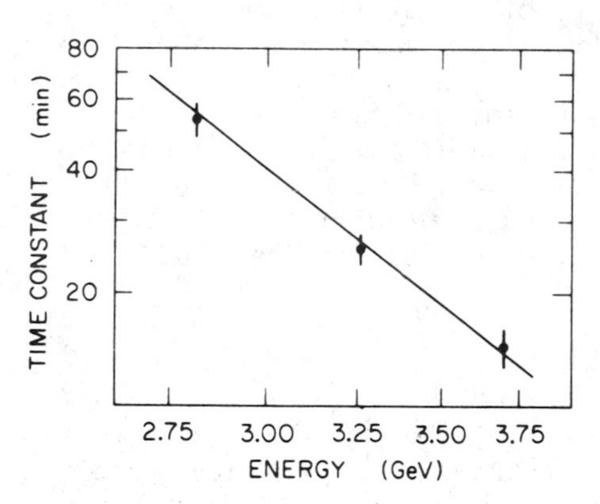

Fig. 3b. Polarization build up time as a function of energy measured at SPEAR.(From Ref. 11)

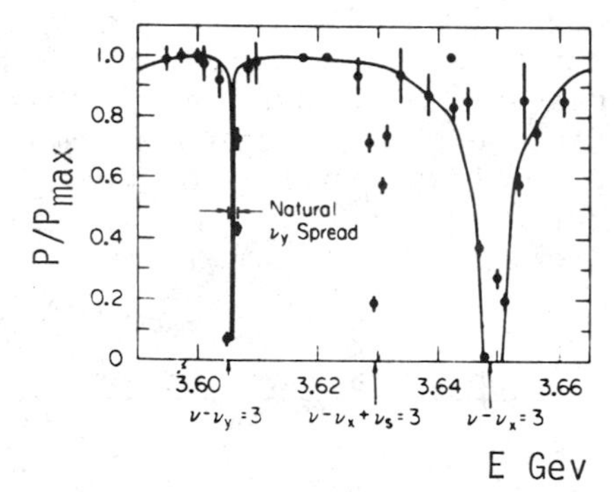

Fig.3c. Detail of SPEAR polarization curve showing non-linear resonance: $a\gamma - Q_x + Q_s = 3$ (From Ref. 11)

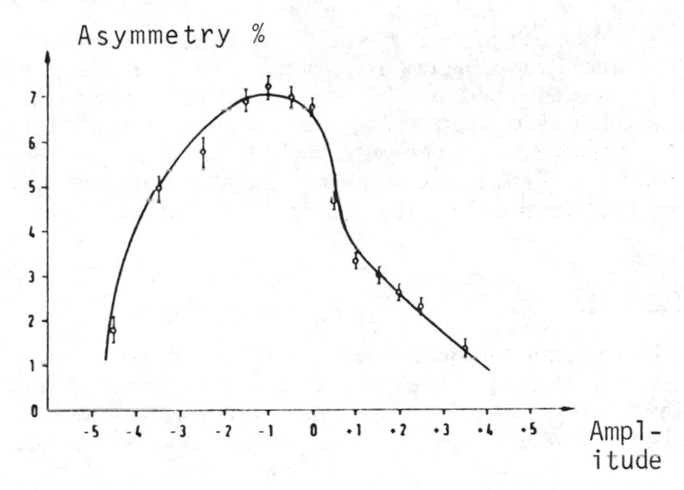

Fig. 4a. Polarimeter asymmetry at PETRA vs. the strength of the 38th cosine harmonic in the closed orbit (arbitrary units).

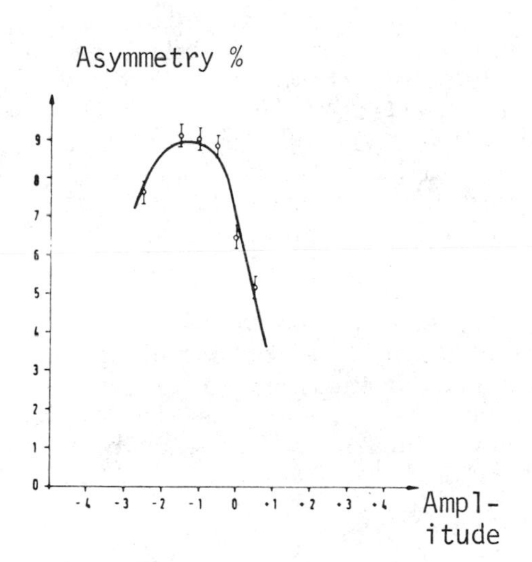

Fig. 4b. As in Fig.4a but for the 38th sine harmonic.

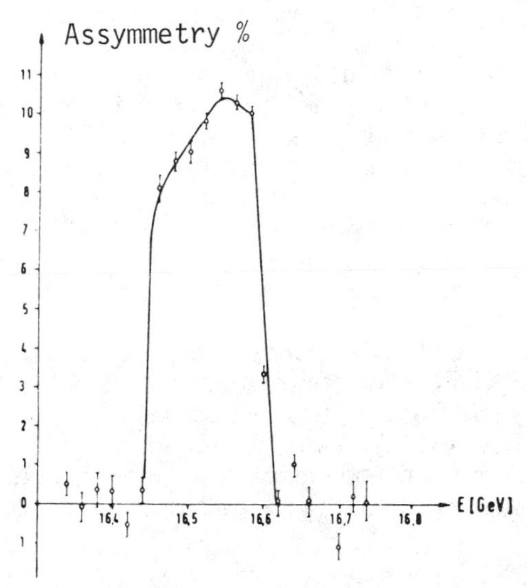

Fig. 4c. Polarimeter asymmetry at PETRA as a function of energy after closed orbit optimisation.

At these high energies, where the depolarizing effect of closed orbit distortions becomes strong, quantitative predictions of SLIM must be treated with some caution since the orbit distortions which would need to be fed into SLIM are difficult to measure with precision. However, SLIM still provides a very useful qualitative framework for simulating correction schemes and the gross features of depolarizing effects.

c) Beam-beam effects

The results presented above were all obtained with a single beam circulating. With both electron and positrons circulating, the particles are subject to strong nonlinear electromagnetic forces as the bunches pass through each other. There is a tendency for the beam diameters to increase and for the luminosity to suffer. The beam-beam force is also a source of depolarization.

The beam-beam effect has also been investigated at PETRA[6]. After optimising the closed orbit as described and establishing a single beam polarization of more than 70 %, measurements were made with a range of equal electron and positron currents. The ratio of single to double beam asymmetries is shown in Fig. 5 and it is seen that the polarization is unaffected until currents of $\sim$8 ma are reached. Then, not only the polarization but also the luminosity decreases and the vertical beam height increases. At PETRA the luminosity and polarization are thus limited at similar currents and if beam blow up can be avoided it looks as if polarization will probably also remain high.

Colliding beam polarization has also been observed and used[11,12] at SPEAR. Broadening of resonances by non-linear effects has been seen and high polarizations have been obtained at normal luminosity but in general they find that their results are somewhat unpredictable. Colliding beam polarization has also been observed at VEPP4 but not at full luminosity[12].

DEPOLARIZERS

Surprisingly, it is also very useful to be able to depolarize the beam in a controlled manner with the aid of a weak oscillating radial magnetic field[13]. Depolarization occurs if the resonance condition:

$$w_d = w_c\,(a\gamma - k) \quad \text{or} \quad w_c\,(k' - a\gamma) \qquad (5)$$

is satisfied where w_c is the circulation frequency, w_d is the depolarizer drive frequency and $a\gamma - k$ or $k' - a\gamma$ are the fractional

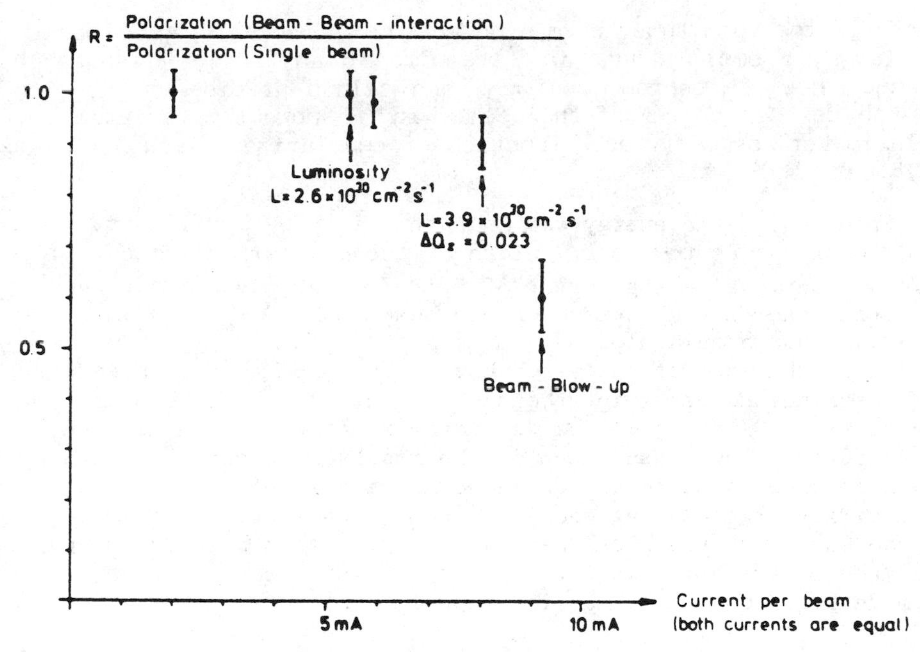

Fig. 5. Ratio of colliding beam polarization to single beam polarization vs. current at PETRA.

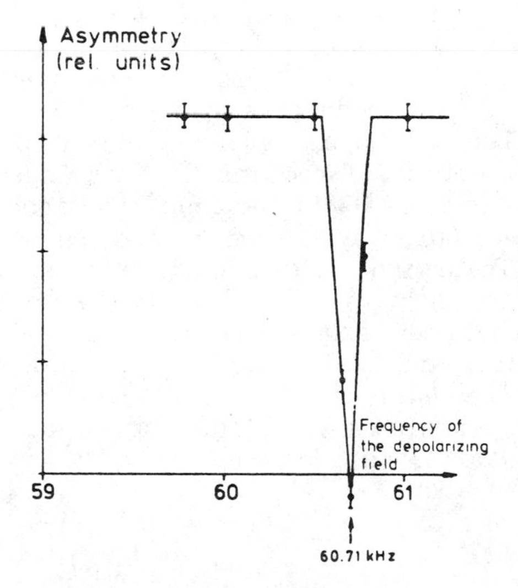

Fig. 6. Polarimeter asymmetry vs. depolarizer frequency at PETRA in the vicinity of 16.5 GeV.

parts of the spin tune. w_d can be measured precisely and since w_c is large, a small change in γ results in an easily measureable change in w_d. Measurement of w_d then allows precise calibration of the central energy of the beam. Fig. 6 shows the variation of polarimeter asymmetry as a function of depolarizer frequency near 16.5 GeV at PETRA[5].

From eqn. 5 the energy was found to be 16509 ± 0.13 Mev. The error corresponds to the 1/2 width of the depolarization curve. Thus a fractional energy error of $\sim 10^{-5}$ is achievable although the fractional energy spread of the beam is $\sim 10^{-3}$. This technique has been exploited at VEPPA[12 14], for example, to determine the mass of the γ to be 9459.7 ∓ 0.6 MeV and in earlier VEPP machines to determine the mass of the ρ, K^{+-o}, $J/\psi,\psi'$ to one part in 10^4 or 10^5. The depolarizer technique can also be extended to allow measurement of the momentum compaction factor and incoherent synchrotron frequency thus opening up the possibility of measuring machine dynamics parameters otherwise not so easily accessible. Depolarizers can also be used to ensure that the polarization is zero or for selectively depolarizing one beam during High Energy Physics experiments.

SOLENOIDS

Finally, I report on some recent studies of the depolarizing effects of detector solenoids . With vertically polarized beams, uncompensated solenoids cause depolarization by tilting the $\hat{n}$ axis and because spin is disturbed by the betatron motion in the solenoid field. For example, Fig. 7 shows how the PETRA polarization falls as two uncompensated solenoids are run up. In the case of longitudinal polarization the $\hat{n}$ axis is unaffected but the betatron effect still remains. The preferred remedy is to compensate each solenoid by adjacent opposite strength "antisolenoids" but normally there is no space owing to the proximity of the focussing quadrupoles. It has now been found that a solution to this second difficulty is to place the antisolenoids further out among the quadrupoles and to use an optics in which the particle direction at the solenoid is the same as that at the antisolenoid[15,16]. This is then a second example of how the machine conditions can be adjusted so as to decouple spin and orbital motion.

SUMMARY

In summary it is clear that there is now no particular difficulty in obtaining good vertical polarization in single beams at low energies. The generation of polarization at high energies in the PETRA range is much more difficult but there has

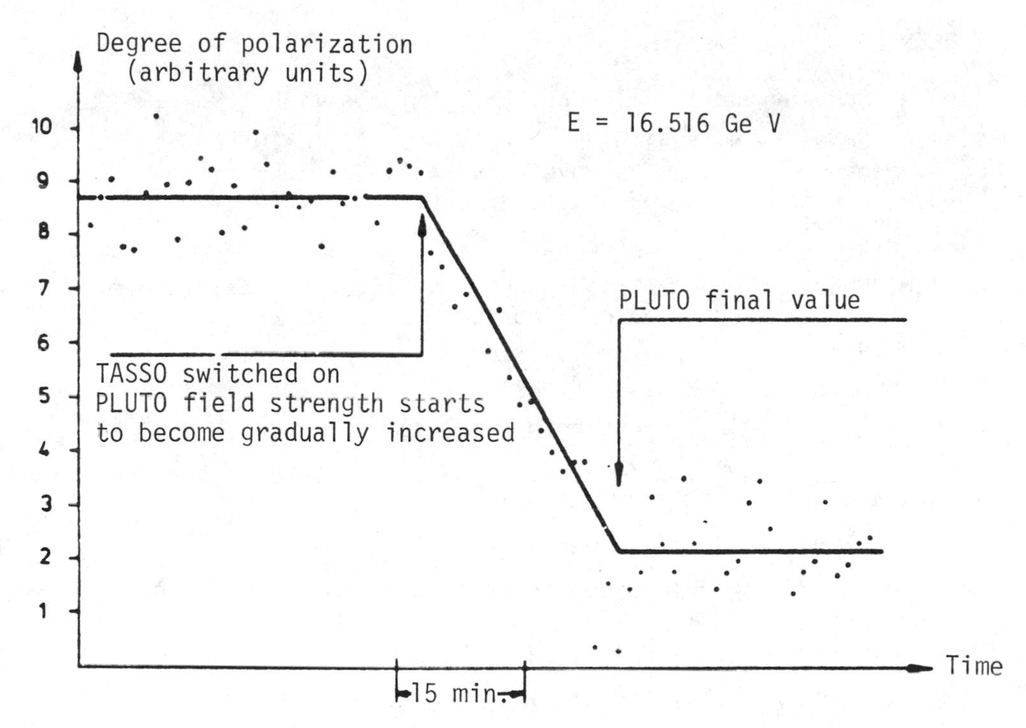

Fig. 7. Polarization vs. time as two experimental solenoids are run up at PETRA.

been a big increase in confidence and understanding following the success of the PETRA closed orbit correction scheme. However much more remains to be done, the region of high polarization at PETRA must be widened and polarization must be obtained at the new maximum PETRA energy of about 22 GeV and finally, realistic schemes for obtaining longitudinal polarization must be found.

ACKNOWLEDGEMENTS

We wish to thank Prof. G.-A. Voss for continued encouragement and support.

REFERENCES

1) A.A. Sokolov and I.M. Ternov, Sov. Phys.Dokl. 8:1203 (1964)

2) R.F. Schwitters et.al., Phys. Rev.Lett. 35: 1320 (1975)
G. Hanson et.al., Phys. Rev.Lett. 35: 1609 (1975)
J.G. Learned et.al., Phys. Rev.Lett. 35: 1688 (1975)

3) C. Prescott, These proceedings

4) A.W. Chao, in Physics of High Energy Particle Accelerators, R.A. Carrigan ed. American Institute of Physics No.87 New York 1982

5) R. Schmidt, DESY M-82-22 (1982)

6) H.D. Bremer et.al., DESY 82-026 (1982)

7) F. W. Lipps and H.A. Tolhoek, Physica XX: 85-98, 395-405 (1954)

8) A. W. Chao, Nucl. Inst. Meth. 180: 29, (1981)

9) A. W. Chao in Polarized Electron Acceleration and Storage, DESY M-82-09 (1982)

10) J. Kewisch, DESY 83-032 (1983)

11) J. R. Johnson et.al., Nucl. Inst. Meth. 204: 261 (1983)

12) A. S. Artamonov et.al., Institute of Nuclear Physics Novosibirsk, Preprint 82-94 (1982)

13) R. Neumann and R. Rossmanith, Nucl. Inst. Meth. 204: 29 (1983)

14) Ju. Shatunov in Polarized Electron Acceleration and Storage, DESY M-82-09 (1982)
L. M. Barkov et.al., Nucl. Phys. 148B: 53 (1978)
A. D. Bukin et.al., Yad. Fiz. 27: 976 (1978)
A. A. Zholentz et.al., Phys. Lett. 96B: 214 (1980)

15) D. P. Barber et.al., DESY 82-076 (1982)

16) K. Steffen, DESY M-82-25 (1982) and DESY HERA 82-11 (1982)

TECHNIQUES FOR LONGITUDINAL POLARIZATION IN ELECTRON STORAGE RINGS

J. Buon

Laboratoire de l'Accélérateur Linéaire
Université de Paris-Sud, 91405 ORSAY - FRANCE

ABSTRACT

Progress in studies for designing polarized beams in very high energy electron storage rings is reviewed. Main emphasis is concentrated on devices delivering special spin manipulations for obtaining highly and longitudinally polarized electrons of both helicities at interaction points in collision mode with opposite beams of positrons (or protons). These devices are : 90° spin rotators, 180° spin rotators (Siberian snakes), asymmetric wigglers, spin flippers and finally spin matching procedures for suppressing beam-beam depolarization. Recent theoretical progress allows now experimental developments at future high energy electron rings like HERA, TRISTAN and LEP.

1. INTRODUCTION

The interest in physics with polarized electron and positron beams increases with energy, due to the more and more important role of electroweak interactions. Polarized electron and positron beams are foreseen in both ep and e^+e^- new facilities.

At low energies (below 10 GeV) a natural transverse polarization has been generally observed in electron storage rings, as expected from the Sokolov-Ternov polarization mechanism[1]. A high degree of polarization is obtained apart at energies corresponding to strong depolarization resonances which occur when the spin tune $\nu = E(\text{GeV})/.44$ (number of spin precessions per revolution) is an integral and linear combination of betatron tunes $Q_{x,z}$ and synchrotron tune Q_S : $\nu = k + k_X Q_X + k_Z Q_Z + k_S Q_S$.

In particular at SPEAR the observed (Fig. 1) polarization has been used in one e^+e^- experiment[2], which has checked the spin 1/2 nature of two-jet events in e^+e^- annihilation. In this experiment the polarization level was about 70 % at 7.4 GeV centre of mass energy, and the polarization time was 15 min. Depolarizing effects, due to field errors in the ring, are small. They are responsible for the difference between the observed polarization and the maximum value (92.4 %) expectable from the Sokolov-Ternov mechanism.

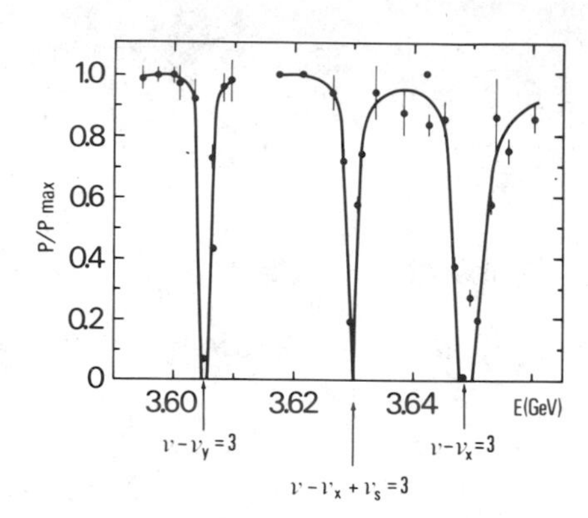

Fig. 1 - SPEAR polarization normalized to P_{max} = 92.4% versus beam energy.

However, at high energies (above 10 GeV) depolarizing effects are expected to become stronger and stronger as energy increases. Effectively all magnetic fields, including field errors, are ramped in energy. Spin tilts, due to field errors, scale linearly with energy. Depolarization, proportional to the square of spin tilts, scales quadratically. Extrapolating the natural polarization level observed at PETRA, one expects no more than 20 % and 10 % natural polarization at HERA (30 GeV) and LEP (50 GeV) respectively.

Now at high energies the beam energy spread (50 MeV at 50 GeV for LEP) becomes not negligible compared to the depolarization resonance spacing (440 MeV). Consequently, depolarization is greatly enhanced together with the appearance of many resonance satellites. For LEP at 50 GeV the depolarization enhancement[3] is never less than about a factor 3. (Fig. 2), and this factor increases very quickly with energy.

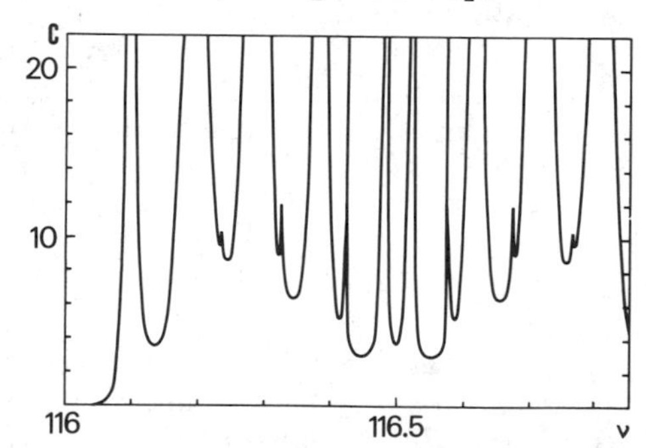

Fig. 2 - Depolarization enhancement factor due to energy spread versus spin tune ν = 116,117 resonances, assumed equally strong).

A third difficulty arises from beam-beam interaction. Spin motion is perturbed by the space charge field of opposite beams as well as the orbital motion is. It results a depolarization even at lower current than the beam-beam limit[4-6]. Consequently polarization would be obtained only at reduced luminosity, limiting practical interest of polarized beams.

A last difficulty in the case of LEP is a too long polarization time in its low energy range (Fig. 3) : 3h at 50 GeV. It is due to the weak bending field at these ener-

At last one must add the requirement of obtaining longitudinal polarization at interaction points for experiments. The natural transverse polarization has to be rotated by 90° which can only be obtained at high energies by special arrangements of vertical bends. Moreover it is desirable to obtain both electron helicities. That would need to reverse the spin 90° rotation.

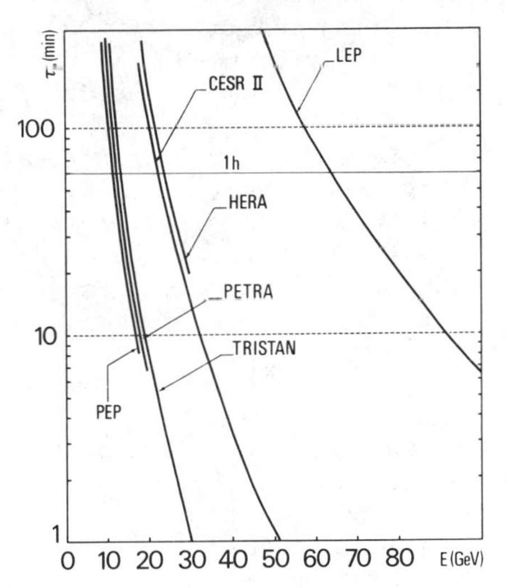

Fig. 3 - Polarization time τ_p vs. beam energy E for several electron rings.

Certainly it would be almost hopeless to obtain significant longitudinal polarization (at least 50 %) at high energies due to all these difficulties without special cures.

At first depolarization due to field errors can be reduced by correction procedures similar to those used for reducing orbit displacements. The polarization at PETRA was thus increased from 20 % - 30 % up to 80 % by adjusting small currents in some correcting coils[6]. No fundamental limitation lies in corre-tion procedures. They need only to be more clever at higher energies.

Special devices can also be found which bring about proper spin manipulation in order to overcome the other difficulties for obtaining longitudinally polarized electron (or positron) beams of both helicities at interaction points. The following list gives the presently studied devices together with their effect on polarization :

Effect		Devices
longitudinal polarization	:	90° spin rotators
avoid depolarization resonances	:	180° spin rotators (Siberian snakes)
short polarization time	:	asymmetric wigglers
helicity flip	:	spin flippers
suppress beam-beam depolarization	:	spin matching procedures

Recent progress in studies of these spin manipulation devices is reviewed here.

2. 90° SPIN ROTATORS

At high energies solenoïdal magnetic fields are quite inefficient for rotating the spin. A 180° rotation would need a solenoïd delivering 524 Tm at 50 GeV. Only transverse magnetic bends can practically be used. Their rotation power is independent of energy. A 180° spin rotation needs only a 4.6 Tm integrated field.

Schemes of 90° spin rotators using vertical bends are easily derived. However, if they allow to obtain longitudinal polarization at interaction points, they also strongly depolarize the beam. Effectively vertical bends of the rotators act on polarization like large radial field errors. This large depolarisation must be prevented by a proper design, known[7] as "spin matching".

The idea[8-9] of spin matching is to achieve a complete decoupling of spin motion from orbital motion. It is possible as spin motion is a new degree of freedom of particle kinematics which has no influence on orbital motion. Moreover spin rotation is not simply proportional to the amplitude of orbital motion since spin rotations in quadrupoles are mixed with normal spin precession in bending magnets. Therefore the ring optics can be designed to achieve this decoupling without losing the necessary optical properties of the ring. Then the ring is said to be completely spin transparent, i.e. whatever is the orbital motion the overall spin rotation does not deviate from normal spin precession in the standard bending magnets. Depolarization is so completely suppressed.

Spin matching can be illustrated by the simple example[8] of a ring with an interaction region equipped with a pair of 90° spin rotators (Fig. 4). Spin is longitudinal between these two rotators and vertical in the rest of the ring. When an electron experiences a horizontal betatron oscillation between the rotators, its spin is rotated in each quadrupole by an angle $\delta\alpha$ proportional to its horizontal displacement. Spin matching requires that the overall spin rotation in this interaction region cancels out exactly (there is no spin rotation due to horizontal oscillations in the rest of the ring since the spin is there vertical). The ring is thus spin transparent for horizontal betatron oscillations. Similar spin matching conditions for vertical and synchrotron oscillations have been derived[9]. They involve also the optics in the rest of the ring.

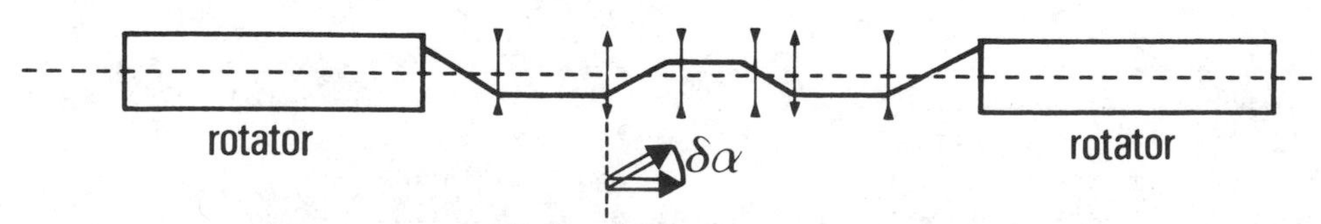

Fig. 4 - Schematic top view of an interaction region equipped with two 90° spin rotators. Solid line is a trajectory with a magnified horizontal betatron oscillation. The angle $\delta\alpha$ is the radial kick on spin, occurring in one quadrupole.

Spin matching is normally designed at one energy. At different energies the spin matching conditions are not the same as spin precession varies with the energy E at B/E constant. It is the main limitation of spin matching, which is not too serious for an ep facility as HERA or TRISTAN, since there is apparently no great interest of changing electron energy which is much smaller than proton energy.

The four most commonly studied 90° rotators are :

i) Vertical S-bend rotator[10,11] (Fig. 5) : simple vertical bends provide a vertical beam slope at interaction point which is adjusted to $90° \times \frac{.44}{E_{GeV}}$ for obtaining 90° spin rotation. This rotator can be ramped in energy in a ± 30 % energy range, around the designed energy, with less than 20 % decrease in the longitudinal component of polarization. However, this rotator has two great disadvantages : it does not allow to invert helicity and vertical bends quite near the interaction point produce high synchrotron radiation background.

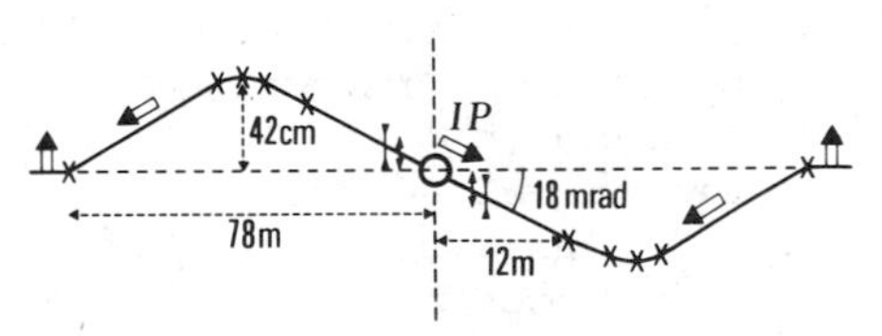

Fig. 5 - Schematic side view of an interaction region equipped with a vertical S-bend rotator for CESR II (X : vert. bend, ↑ spin direction

ii) H-V mixed bend rotator[12] (Fig. 6) : a pair of 90° rotators is symmetrically installed at the ends of the ring arcs on each side of the long straight section. Each rotator consists of a local vertical beam bump combining vertical bends and a horizontal bend. At the designed energy each vertical bend rotates the spin by 45°. The horizontal bend, which is part of the normal radial bending in the ring, rotates the spin by 180°. At the end of each rotator the orbit is again in the plane of the ring.

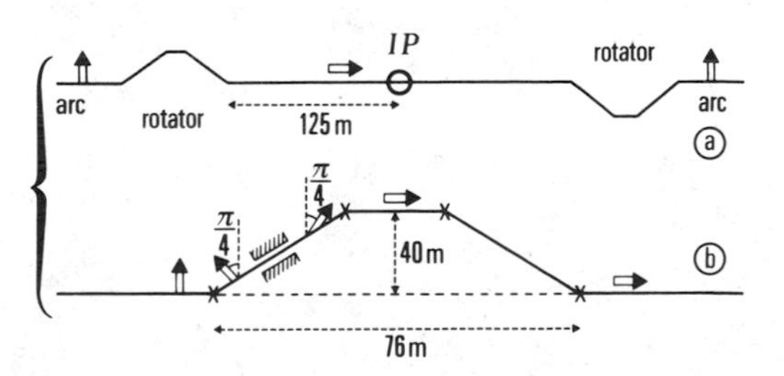

Fig. 6 - a) Schematic side view of an interaction region equipped with a pair of H-V mixed bend rotators

b) Magnification of one H-V mixed bend rotator for HERA (X : vert. bend, ▨ horiz. bend, ⇨ spin direction)

The corresponding vertical bends in the two rotators of an interaction region are opposite. Therefore each pair of rotators is antisymmetric in the vertical plane with respect to the interaction point. However, the corresponding horizontal bends are identical as they both participate to the normal radial bending. Each pair of rotators is symmetric in the horizontal plane.

The magnetic field in these horizontally bending magnets must be ramped in energy as normal magnets in the arcs. The 180° spin rotation is only obtained at the designed energy (27.5 GeV for HERA). What is more, the vertical spin direction in the arcs of the ring is only restored at this designed energy. The tilt of spin direction at an off-energy set must be corrected for avoiding depolarization, but correction can only be done[13] in a small energy range (± a few percents).

This type of rotator could allow to invert helicity at the interaction point by reversing the fields in vertical bends. However, some of these bends must also be displaced and this operation cannot be easily repeated very often. Synchrotron radiation background would also be less severe than for the S-bend rotator since the involved beam bumps are far away (at least 100 m for HERA) from the interaction region. Moreover these bumps can be used for shielding the interaction region from the arcs. The main limitation of this rotator type is its limited energy range, although it may not be a great disadvantage for an ep facility like HERA.

iii) Hybrid rotator[14] (Fig. 7) : it is a combination of the two first rotators, and is obtained from the H-V mixed bend rotator by omitting the beam translation that restores the horizontal beam line around the interaction point. The beam line has now a vertical slope as for the S-bend rotator, but much smaller (3.6 mrad for HERA). Synchrotron radiation background will be less severe as for the H-V mixed bend rotator, but electron helicity cannot be repeatedly inverted.

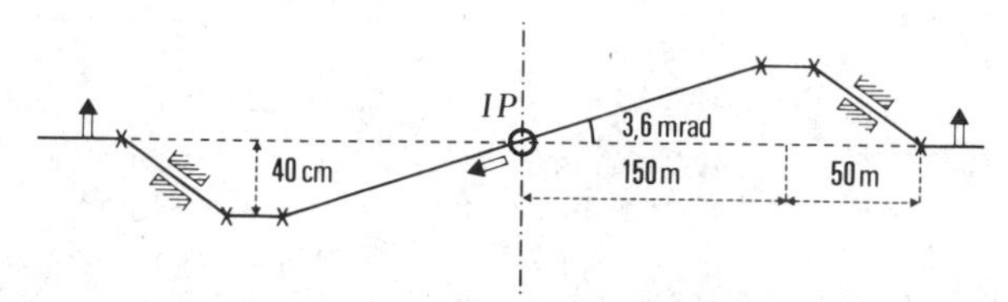

Fig. 7 - Schematic side view of an interaction region equipped with an hybrid spin rotator for HERA (X : vert. bend, [hatched symbol] horiz. bend, [arrow symbol] spin direction)

iv) Antisymmetric H-V mixed bend rotator[15] (Fig. 8) : the previous H-V mixed bend rotator is modified for reducing vertical displacement and for achieving complete antisymmetry in horizontal and vertical planes. Reduction of vertical displacement is obtained by weaker vertical bends (rotating the spin by 22°5 only), but incorporating a second horizontal bend. Antisymmetry is obtained by alternating the signs of the horizontal bends. At the ends of the rotator two additional weak horizontal bends restore the horizontal beam line direction. Full antisymmetry allows to widen the possible energy range in comparison with the H-V mixed bend rotator.

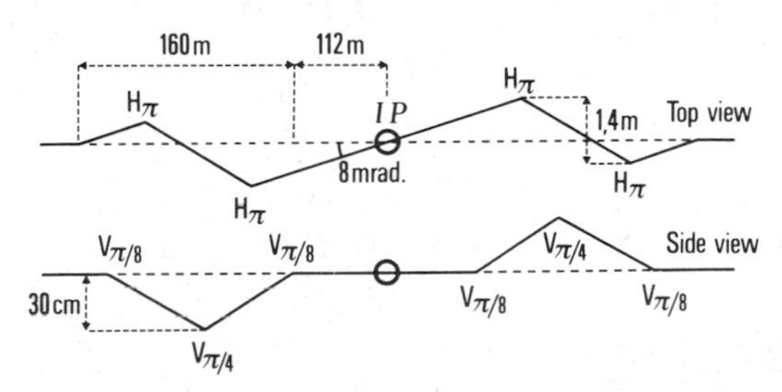

Fig. 8 - Schematic side and top views of an interaction region equipped with an antisymmetric H-V mixed bend rotator for LEP (H_α, V_β horiz. and vert. bends rotating spin by α, β angles resp.)

The former rotators have been studied in more detail than the last one. They have been incorporated in the design procedure of ring optics. Spin matching has also been achieved[14] for HERA with the hybrid rotator at 27.5 GeV (spin tune is 62.4). Fig. 9 shows the polarization estimated by the SLIM program around this energy and shows the large improvement obtained at the design energy by spin matching. If not, polarization level would only be about 20 % without taking account field errors. When the energy differs from 27.5 GeV polarization is reduced as the rotator has not yet been corrected for off-energy operation and as spin matching conditions are no more completely fulfilled.

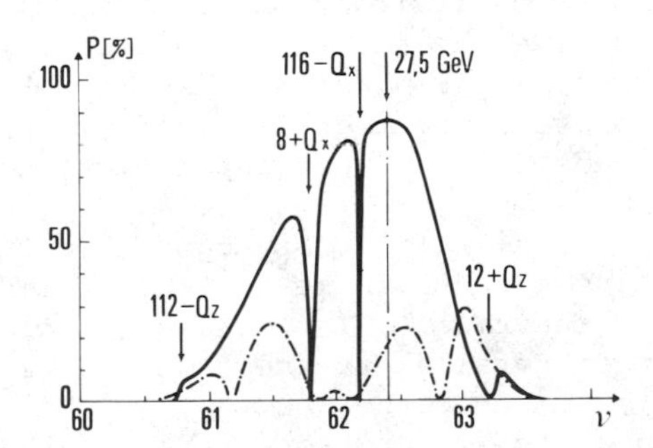

Fig. 9 - Estimated polarization vs. spin tune ν, in HERA with hybrid rotators (designed at 27.5 GeV)
———: with spin matching
·—·—: without spin matching

It is worth noting that there is no universally good rotator. Each type has advantages and also some disadvantages. Choice of a rotator type is always a compromise. Spin matching is an essential tool for avoiding large depolarization due to vertical bends. From recent studies it is believed that practical designs of 90° spin rotators are now available.

3. 180° SPIN ROTATORS ("Siberian Snakes")

Ya. Derbenev et al.[19] have proposed to avoid depolarization resonances by implementing 180° spin rotators, named after "Siberian Snakes". In principle such a device allows for rotating the spin by 180° around an axis which is either longitudinal (1st kind) or radial (2nd kind).

In a ring equipped with a 180° spin rotator the spin tune becomes exactly half-integer independent of energy E. Depolarization resonances, occuring when the energy is such that spin tune ν is equal to some linear combination of betatron and synchrotron tunes, are avoided since the fractional part of the latter tunes is always less than .5.

Therefore the spread in spin tune cancels out independently of energy spread $\sigma(E)/E$. The depolarization enhancement due to energy spread is suppressed.

These two properties of 180° spin rotators are quite attractive for high energy rings like LEP, since they will get rid of the greatest difficulties at high energy.

However, in an electron ring equipped with a single 180° rotator of either 1st kind or 2nd kind, the synchrotron radiation becomes strongly depolarizing. Effectively in that case the stable and closed direction of polarization lies in the plane of the ring outside the rotator (Fig. 10a). There is no more an asymmetry in the spin-flip component of synchrotron radiation. The Sokolov-Ternov polarization mechanism has disappeared and the spin-flip component leads only to fast depolarization[20].

This difficulty can be overcome by using two 180° spin rotators (double snake system) of different kinds, on opposite sides of the ring[21]. The stable and closed direction of polarization is then parallel to the magnetic field in one half-arc and antiparallel in the other half (Fig. 10b). The system is neutral with respect to synchrotron radiation. Neither polarization or depolarization occurs. A wiggler, or other means, is then required for polarizing the electron beams.

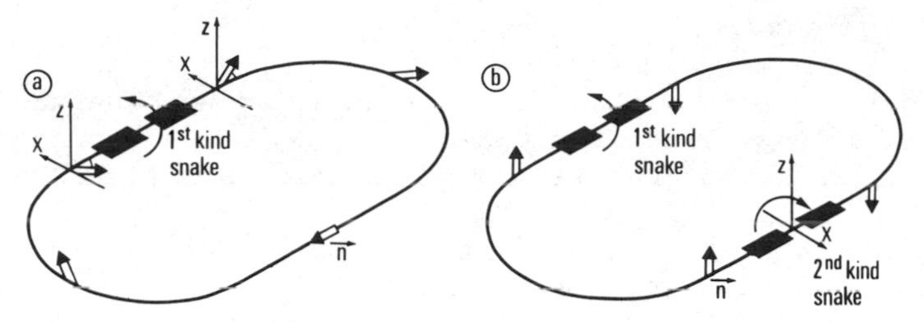

Fig. 10 - Schematic view of a ring equpped with
a) a 1st kind "Siberian Snake"
b) a double Snake system

Fig. 11 shows a double snake system which has been studied[11] for CESR II. It is more complicated, more difficult to fit with the ring optics and more delicate to operate than a simple 90° spin rotator. Its properties are maintained when energy varies only if spin rotation stays equal to 180°. The two 180° spin rotators must then be operated at fixed fields, implying variable geometry of beam line within the rotators. Siberian Snakes appear more complicated for electron rings than for proton rings. Search for finding a better design is highly desirable.

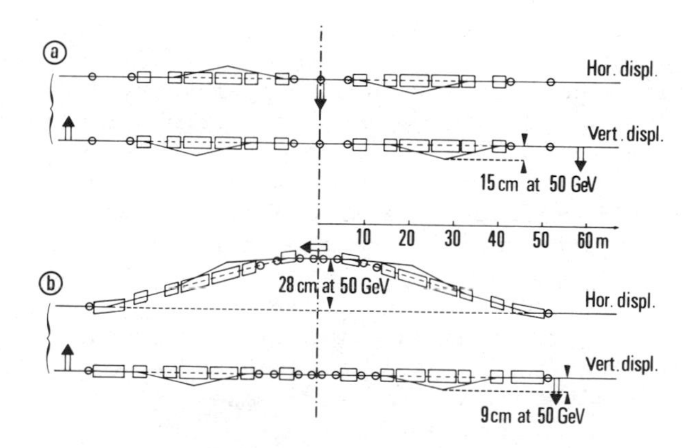

Fig. 11 - Double snake system for CESR II
a) 1st kind "Siberian Snake"
b) 2nd kink "Siberian Snake"
Solid lines : horizontal or
vertical beam displacement

4. ASYMMETRIC WIGGLERS

A wiggler consists of a magnet with a strong bending field that is symmetrically preceded and followed by oppositely bending magnets such that there is no overall beam deflection nor displacement.

When the field of the oppositely bending magnets is much smaller than the central strong bending field, the wiggler is said to be asymmetric (Fig. 12).

The polarization rate due to synchrotron radiation emitted in a magnet is proportional to the cube of the bending field. The strong field of a wiggler will then accelerate the polarization build-up, i.e. will reduce polarization time[22] as required for LEP at low energy for instance (Fig. 3). The field of the oppositely bending magnets tends to produce an opposite polarization and reduces the average level of polarization obtained, but only a little in the case of an asymmetric wiggler.

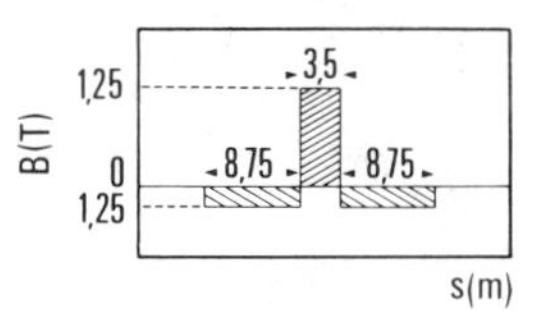

Fig. 12 -Magnetic configuration of a moderate asymmetric wiggler for LEP

Moreover the direction of polarization is determined by the direction of the wiggler strong field which is normally much stronger than the bending field in ring magnets. Therefore the direction of polarization can be easily changed by inverting the wiggler fields. That would allow to get both electron helicities at interaction points equipped with 90° rotators.

An asymmetric wiggler will also polarise the electron beams in a ring equipped with a double snake system, which otherwise would stay unpolarized.

Nevertheless, wigglers have two important drawbacks which unfortunately limit their interest. They are a very intense source of synchrotron radiation and they increase the beam energy spread.

For example (Fig. 12) a moderate wiggler for LEP (1.25T, 3.5m central magnet and .25T, 8.75m side magnets) will reduce the polarization time from 3h to 1h. It will generate locally 60 KW of additional synchrotron radiation power, and it will increase by 80 % the beam energy spread. Consequently the depolarization enhancement due to energy spread is increased by about one order of magnitude.

5. SPIN FLIPPERS

Inverting spin rotation in a 90° spin rotator involves not only reversing of fields in vertical bends but also displacement of some magnets and quadrupoles which is not an operation easy to repeat. The flip of electron helicity at interaction points cannot thus be done very often.

In principle spin flip in an electron ring could be achieved in a similar way to spin flip of protons in a synchrotron. It has been obtained at ZGS[23] and more recently at SATURNE II[24], without significant loss of polarization at the latter synchrotron.

Let one recall the principle of spin flip in a proton synchrotron : it happens during acceleration by adiabatic crossing of an integer depolarization resonance. In the vinicity of such a resonance with an integer number K, proton spin tune ν varies linearly with time t :

$$\nu = K + \lambda t$$

With respect to a frame rotating with rotation frequency number K, the proton spin precesses (Fig. 13) around a vector $\vec{\Omega}$ with vertical component ν-K and radial (or longitudinal) component ε which is the result of the perturbing field component driving this resonance.

During acceleration the direction of the precession vector $\vec{\Omega}$ varies. It was antiparallel to Oz below resonance. It becomes parallel to Ox just on resonance and will be parallel to Oz above resonance. During that time spin continues to precess around $\vec{\Omega}$. This precession evolves adiabatically if the $\vec{\Omega}$ vector variation is sufficiently slow : $\lambda \ll \varepsilon^2$. Therefore spin direction is ultimately inverted as the vector $\vec{\Omega}$ is.

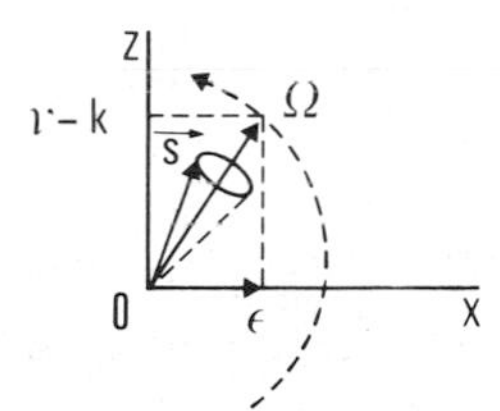

Fig. 13 -Adiabatic crossing of an integer resonance

A similar process can be designed for electrons in a storage ring, as initially proposed by the Novosibirsk group[25]. Spin flip is achieved again by adiabatic crossing of a depolarization resonance. Variation of spin tune with energy due to acceleration in a synchrotron is replaced by variation of frequency f of a perturbing field in an electron storage ring :

$$f = \nu + \lambda t$$

Resonance occurs when the frequency f coincides with spin tune.

With respect to a frame rotating with frequency f, the electron spin (Fig. 14) precesses around a vector $\vec{\Omega}$ with vertical component ν - f and radial component ε proportional to the strength of the perturbing field. The features of the resonance crossing are then quite similar to those of resonance corssing in a proton synchrotron.

However, in the case of electrons one has to consider quantum fluctuations of synchrotron radiation[26]. As a result electron spins will lose some of their coherence during the resonance crossing. The time spent on the resonance must be sufficiently short in order to avoid some depolarization. It needs a strong perturbing field in order that the crossing stays adiabatic, especially at high energy since depolarization effects of quantum fluctuations increase with energy.

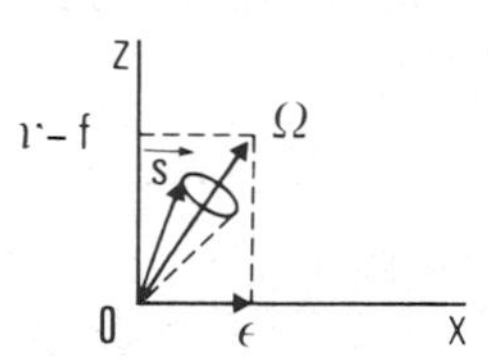

Fig. 14 - Adiabatic crossing of a resonance due to an A.C. perturbing field with modulated frequency f

Such a spin flipper has been worked out at VEPP2-M[25]. Spin flip has been observed at .65 GeV without apparently significant depolarization. This positive result encourages further investigation[26] on spin flippers, although at high energies a limitation comes from the required perturbing field strength which may lie in the range of several kGm. Moreover, when polarization direction is opposite to the natural one due to the Sokolov-Ternov mechanism, the latter tends to decrease the polarization in order to restore the natural direction. Therefore running with opposite direction of polarization obtained with a spin flipper will lead to some loss of polarization in average during running time.

A spin flipper offers also the possibility of flipping selectively the spins of one bunch of a beam (electron or positron). One can thus obtain e^+e^- collisions of same or opposite helicities, maximizing polarization asymmetries in electroweak interactions.

6. SPIN MATCHING FOR SUPPRESSING BEAM-BEAM DEPOLARIZATION

Perturbation of spin motion due to beam-beam interaction can be suppressed by spin matching as well as depolarization due to rotators. Spin motion will thus be decoupled from beam-beam kicks occuring at each interaction point, as initially suggested by K. Steffen[27].

Following a recent study[28], lets one consider only a planar electron ring with vertically polarized beams. Horizontal kicks do not affect vertical polarization. Only vertical kicks must be considered.

A particle crossing an opposite beam at one interaction point, with some vertical displacement, experiences a vertical kick due to the beam-beam force. Its vertical betatron oscillation is modified by the vertical kick. Assuming a linear optics between interaction points, the new oscillation is the sum of the preceding oscillation plus an additional oscillation initiated by the kick (Fig. 15).

At the following interaction point another vertical kick, and another additional oscillation, occur. The vertical motion is thus a superimposition of successive additional oscillations.

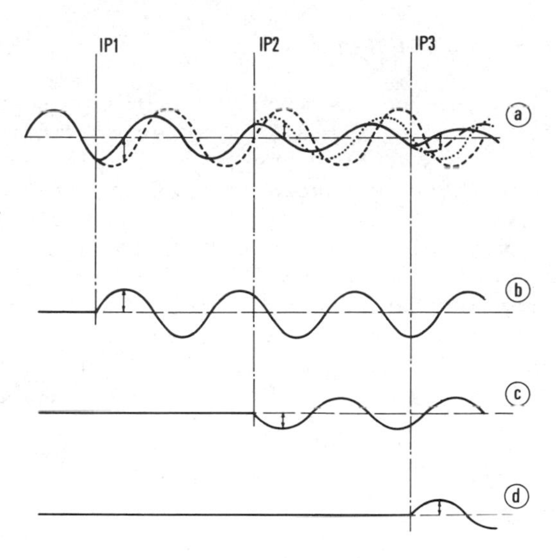

Fig. 15 - a) Schematic vert. betatron oscillation with kicks at interaction points IP1, IP2, IP3. Dashed lines: continuation of oscillation after IP1, IP2 and IP3 respectively, ignoring the corresponding kicks.
b), c) and d) : additional betatron oscillation initiated by the kicks at IP1, IP2 and IP3 respectively

Spin matching consists of imposing that spin rotation, produced by one beam-beam kick, just cancels out the spin rotation produced by the subsequent additional oscillation.

Both spin rotations are proportional to the considered beam-beam kick. Therefore the spin matching conditions are independent of the kick strength. Whatever is the kick strength : non linear in vertical displacement, variable with beam intensity and dimensions, the spin matching conditions remain the same. Beam-beam depolarization is thus exactly suppressed.

This spin matching for beam-beam interaction can easily be extended to the case of a non-planar ring, in particular to a ring equipped with 90° spin rotators. One has just to consider also horizontal kicks. The number of spin matching conditions is doubled.

7. CONCLUSION

In the last few years great progress has been accomplished for reducing the difficulties encountered in the design of polarized electrons in high energy storage rings. This report gives the present status of these theoretical studies. At the same time experimental studies at PETRA and other rings were very encouraging. One must recall again that 80 % transverse polarization at 15 GeV has been obtained at PETRA with a new correction procedure[6]. The main step is now to obtain longitudinally polarized electrons with 90° spin rotators. It can really be foreseen with great optimism for HERA and TRISTAN around 30 GeV due to the recent progress. The next step would be for LEP at 50 GeV and above.

REFERENCES

1. I.M. Ternov, Yu.M. Loskutov and L.I. Korovina, Sov. Phys. - JETP. 14, 921 (1962)
   A.A. Sokolov and I.M. Ternov, Sov. Phys. - Dokl 8, 1203 (1964)

2. G. Hanson et al., PRL 35, 1609 (1975) and R.F. Schwitters, SLAC-PUB 2258 and IIIth Int. Symp. on H.E.P. with Polarized Beams and Polarized Targets, Argonne (1976)

3. C. Biscari, J. Buon and B. Montague, CERN/LEP-TH/83-8
   J. Buon, Orsay report, LAL-RT/83-02 (1983).
   (The minimum enhancement factor is only 3. instead of 5.4 as quoted by error, when two nearby integer resonances are considered)

4. A.S. Artamonov et al., Novosibirsk preprint INF 82-94 (1982)

5. J.R. Johnson et al., NIM 204 (1983) 261

6. H.D. Bremer et al., DESY Report 82-026 (1982)

7. K. Steffen, Cornell report CBN 81-29 (1981)
   K. Steffen, DESY HERA report 82-02 (1982)

8. J. Buon, Orsay Internal Report NI/07-81 (1981)

9. A.W. Chao and K. Yokoya, KEK TRISTAN report 81-7 (1981)

10. R. Schwitters and B. Richter, SPEAR-175 and PEP note 87 (1974)

11. K. Steffen, Cornell report CNB 81-30 (1981)

12. K. Steffen, DESY HERA report 80-08 (1980)

13. J. Buon, Orsay Internal Report NI/35-81 (1981)

14. K. Steffen, Brookhaven Spin Physics Conference (1982)

15. B.W. Montague, unpublished

16. A. Skuya, Polarized Electron Acceleration and Storage Workshop, Hamburg (1982), DESY M-82/09, p.T10

17. K. Steffen, Cornell reports CBN 82-07 and CBN 82-08 (1982)

18. HERA proposal, DESY HERA report 81/10 (1981) and K. Steffen, DESY HERA report 81/05 (1981)

19. Ya.S. Derbenev and A.M. Kondratenko, Novosibirsk preprint INF 76-84 (1976)

20. R. Schwitters, Proc. of the LEP Summer Study, Les Houches (1978), CERN 79-01, vol. 1 p. 239 and ECFA/LEP note 40 (1978)

21. Ya.S. Derbenev and A.M. Kondratenko, Proc. of Int. Symp. on Polarized Beams and Targets, Argonne (78) and Novosibirsk preprint INF 78-74 (1978)

22. A.M. Kondratenko and B.W. Montague, CERN-ISR-TH/80-38 and LEP note 255 (1980)

23. Y. Cho et al. in "H.E.P. with Polarized Beams and Targets", AIP Conf. Proc. n°35 (Argonne 76) p. 396

24. E. Grorud, Polarized Electron Acceleration and Storage Workshop, Hamburg (1982) DESY M-82/09, p. N1

25. Ju. Shatunov, Polarized Electron Acceleration and Storage Workshop, DESY M-82/09, p. 01

26. K. Yokoya, KEK TRISTAN preprint 82-32 (1983)

27. K. Steffen, Cornell Report CBN 81-29 (1981) and K. Steffen, DESY HERA Report 82-02 (1982)

28. J. Buon, Orsay Report LAL/RT/83-04

POLARIZED BEAMS IN HIGH-ENERGY e^+e^- STORAGE RINGS

Bryan W. Montague

CERN, Geneva, Switzerland

INTRODUCTION

The polarization of electron/positron beams in storage rings is fundamentally governed by the quantized nature of synchrotron radiation. On the one hand, beams can become polarized by the Sokolov-Ternov spin-flip asymmetry of the photon emission, whereas on the other hand, the abrupt energy loss excites orbit oscillations which perturb the spin motion and give rise to depolarizing effects. It is the balance between these two aspects of the same phenomenon that determines the degree of beam polarization that can be built up and maintained in a machine.

In this paper we shall first recall the main features of the radiative polarization mechanism, followed by some important properties of solutions of the Thomas-BMT equation. The motion of the spin-precession axis in the presence of orbit oscillations will then be discussed, leading to the concept of the spin-orbit coupling vector. The influence of radiation damping and the cumulative effects of many photon emissions leads to spin diffusion and thus to depolarization in the vicinity of spin resonances. The origin of the driving terms for these resonances will be examined, in particular for high-energy electron storage rings, and means of reducing their depolarizing influence will be outlined. Finally, expectations for polarized beams in LEP will be presented.

Before proceeding on these topics it is instructive to consider the time scales of the various physical processes involved; these are shown against a logarithmic scale in Fig. 1 for a typical (but non-existent) electron storage ring of about 25 GeV energy. It is seen that characteristic times span more than fourteen orders of

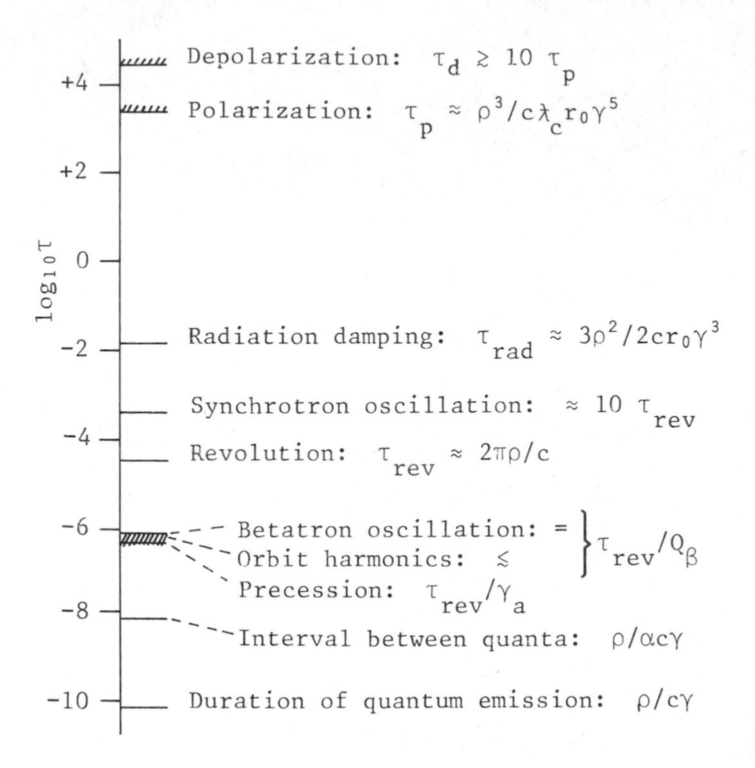

Fig. 1 Characteristic times of processes at 25 GeV energy; γ = Lorentz factor, ρ = bending radius, r_0 = classical radius, Q_β = betatron tune, α = fine structure constant, λ_c = Compton wavelength, a = gyromagnetic anomaly.

magnitude, ranging from less than 10^{-10}s for the duration of the quantum emission process to over 10^4s for a depolarizing time desirably exceeding the polarization time by a factor of ten.

An important feature of Fig. 1 is the clustering together of betatron-oscillation, orbit-harmonic and spin-precession periods, a factor which is of some significance in the spectrum of depolarizing resonances. By contrast, the large separations between times for polarization, radiation damping and oscillation modes helps to simplify calculations by the use of averaging methods. The separation of time scales between the oscillation modes, the interval between quanta emitted by a single electron and the duration of the quantum emission process enable the latter to be considered as an abrupt random process with no correlation between successive photons.

RADIATIVE POLARIZATION

Spontaneous polarization of an electron or positron moving in a transverse magnetic field $\vec{B}$ results from an asymmetry in the probability of photon emission according to the initial spin state, parallel (+) or anti-parallel (−), to B. This process, predicted by Ternov, Loskutov and Korovina,[1] and calculated in detail by Sokolov and Ternov[2] for a uniform field, gives the probability with spin-flip as:

$$w\uparrow\downarrow(\pm) = \frac{1}{2\tau_p}\left(1 \pm \frac{8}{5\sqrt{3}}\right) = \frac{\xi^2}{6}\, w_0\left(1 \pm \frac{8}{5\sqrt{3}}\right) \tag{1}$$

where

$$\frac{1}{\tau_p} = \frac{5\sqrt{3}}{8}\,\frac{c r_0 \lambda_c \gamma^5}{\rho^3}$$

is the polarization rate

$$w_0 = \frac{5}{2\sqrt{3}}\,\frac{c r_0 \gamma}{\lambda_c \rho}$$

is the photon emission rate and

$$\xi = \frac{3}{2}\,\frac{\lambda_c \gamma^2}{\rho} = \frac{\hbar\omega_{cr}}{mc^2\gamma}$$

is the ratio of the critical photon energy and the particle energy, and is typically of the order of 10^{-6}. The probability of photon emission with spin flip is evidently very small, which accounts for the long polarization time, but the asymmetry with respect to initial spin state is large, which leads to a high degree of final polarization in a beam.

There is also an asymmetry in the probability of photon emission without spin flip, given by

$$w\uparrow\uparrow(\pm) = w_0\left[1 - \xi\left(\frac{16}{15\sqrt{3}} \pm 1\right)\right]. \tag{2}$$

Here the asymmetry is small but its probability relatively large, being proportional to ξ rather than ξ^2. Under some conditions this could be used to enhance the polarization, as mentioned below.

Generalization of the spin-flip probability to the case of arbitrary magnetic fields, by Baier and Katkov,[3] gives

$$w\uparrow\downarrow(\vec{\zeta}) = \frac{1}{2\tau_p}\left[1 - \frac{2}{9}(\vec{\zeta}\cdot\vec{v}) + \frac{8}{5\sqrt{3}}\,\vec{\zeta}\cdot(\vec{v}\times\dot{\vec{v}})\right], \tag{3}$$

where $\vec{\zeta}$ is the initial spin vector and $\vec{v}$, $\dot{\vec{v}}$ the velocity and acceleration vectors, respectively, of the particle.

POLARIZATION WITH MACHINE IMPERFECTIONS

Nearly all depolarizing phenomena may be expressed in terms of the spin-orbit coupling vector $\vec{\Gamma} = \gamma\partial\vec{n}/\partial\gamma$, sometimes designated $\vec{d}$. This quantity describes the energy dependence of the spin precession axis, and a description of its physical origin will be discussed in a later section.

It has been shown by Derbenov and Kondratenko[4] that the polarization time τ_p and the asymptotic level of polarization $P(\infty)$ in a storage ring with alignment and field errors are given quite generally by

$$\frac{1}{\tau_p} = \frac{5\sqrt{3}}{8} c r_0 \lambda_c \gamma^5 \left\langle \frac{1}{|\rho^3|} \left[1 - \frac{2}{9}(\vec{n}\cdot\vec{v})^2 + \frac{11}{18}|\vec{\Gamma}|^2\right] \right\rangle \tag{4}$$

and

$$P(\infty) = -\frac{8}{5\sqrt{3}} \frac{\left\langle \frac{1}{|\rho^3|} \hat{b}\cdot(\vec{n} - \vec{\Gamma}) \right\rangle}{\left\langle \frac{1}{|\rho^3|}\left[1 - \frac{2}{9}(\vec{n}\cdot\vec{v})^2 + \frac{11}{18}|\vec{\Gamma}|^2\right]\right\rangle}, \tag{5}$$

where $\hat{b}$ is the unit vector parallel to the magnetic field $\vec{B}$, $\vec{n}$ is the unperturbed precession axis, and the averages are taken over the machine circumference and the particle distribution. For a perfect planar machine both $\vec{\Gamma}$ and $(\vec{n}\cdot\vec{v})$ vanish, leading to an asymptotic polarization of $8/5\sqrt{3} = 92.4\%$. A value of $|\vec{\Gamma}|^2$ around unity or more leads to a substantial depolarization.

The term $\vec{\Gamma}$ in the numerator of Eq. (5) together with the non-spin-flip asymmetry in Eq. (2) can in principle be used to increase the level of polarization to about 95%,[4] but the practical application has not yet been sufficiently studied so we shall not discuss this in detail.

THOMAS-BMT EQUATION

The relativistic spin-precession equation, first derived by Thomas[5] and later reformulated in covariant four-vector notation by Bargmann, Michel and Telegdi,[6] may be written in the form

$$\frac{dp}{dt} = \vec{\Omega} \times \vec{P} ,$$

with

$$\vec{\Omega} = -\frac{e}{m\gamma}\left[(\gamma a)\vec{B} - (\gamma-1)a\frac{(\vec{v}\cdot\vec{B})\vec{v}}{v^2} + \left(\gamma a - \frac{\gamma}{\gamma^2-1}\right)\frac{\vec{E}\times\vec{v}}{c^2}\right], \qquad (6)$$

where $a = (g-2)/2$ is the gyromagnetic anomaly, about 1.16×10^{-3} for electrons and positrons, $\vec{B}$ and $\vec{E}$ are the magnetic and electric field vectors, and $\vec{v}$ is the particle velocity vector. In eq. (6), $\vec{\Omega}$ is referred to a coordinate system $(\hat{x},\hat{y},\hat{z})$ rotating with the particle trajectory, $\hat{y}$ being the tangent vector to the orbit and $\hat{z}$ the normal to the orbit plane. In the case of a perfect planar orbit $\vec{B} = \vec{B}_z$, and with $\vec{E} = 0$ the precession axis $\vec{\Omega}$ reduces to

$$\vec{\Omega} = -\frac{e\vec{B}_z}{m\gamma}(\gamma a) .$$

The precession wave number or spin tune $\nu = \gamma a$ is the number of spin precession turns per orbital period; it has a value of about 104 at 46.5 GeV.

Solutions of Eq. (6) correspond, in general, to rotations about some axis $\vec{n}$. An important property of any two such solutions $\vec{P}_1$, $\vec{P}_2$ for a given $\vec{\Omega}$ is

$$\frac{d}{dt}\left(\vec{P}_1\cdot\vec{P}_2\right) = 0 , \qquad (7)$$

which implies a constant angle between $\vec{P}_1$ and $\vec{P}_2$ and a constant length $|\vec{P}|$, by putting $\vec{P}_1 = \vec{P}_2$. The precession axis $\vec{n}$ is evidently also a solution and, locally, $\vec{n} = \vec{\Omega}$. The properties of $\vec{n}$ under general conditions determine the polarization state of the beam.

PERIODIC ORBIT

We first consider the spin motion of a particle moving on a periodic (closed) orbit at constant energy and without oscillations. It is then convenient to replace the independent variable t by the azimuthal variable $\theta = ct/R$, where R is the average radius of the machine, and the precession equation becomes:

$$\frac{d\vec{P}}{d\theta} = \vec{\Omega}(\theta) \times \vec{P} \qquad (8)$$

with the property

$$\vec{\Omega}(\theta) = \vec{\Omega}(\theta + 2\pi)$$

on a periodic orbit. Now since all solutions of Eq. (8) correspond to rotations about some axis $\vec{n}_0$, this axis is also a solution with the particular periodic property

$$\vec{n}_0(\theta) = \vec{n}_0(\theta + 2\pi) .$$

It is also evident that, the orbit being periodic in θ, $\vec{n}_0(\theta)$ repeats itself from one revolution to the next indefinitely.

More generally, close orbits exist for arbitrary (constant) energies deviating from a nominal energy γ_0 (within practical aperture limits). Except for a perfect planar orbit, the periodic spin solution is then energy dependent and we write, to first order:

$$n(\theta+2\pi,\ \gamma_0+\delta\gamma) = \vec{n}(\theta,\gamma_0+\delta\gamma) = \vec{n}_0(\theta) + \gamma\frac{\partial\vec{n}}{\partial\gamma}(\theta)\ \frac{\delta\gamma}{\gamma_0} \qquad (9)$$

for an energy deviation $\delta\gamma$.

Particles in general execute betatron and synchrotron (energy) oscillations around a closed orbit; their trajectories cannot be periodic since this would lead to instability of the orbital motion. We are therefore led to seek a formulation in which the precession axis $\vec{n}$ can conveniently be defined in the presence of aperiodic orbit oscillations.

ORBIT OSCILLATIONS

Recalling from Fig. 1 the large separation of time scales between radiation damping and quantum processes on the one hand, and oscillation periods on the other, we examine the transformation properties of the orbital motion and of the spin motion over one revolution of the machine. During one revolution period the effects of radiation damping and quantum diffusion have a very small influence and we can for the moment neglect them in discussing the orbital and spin motions.

We let $X = (x,x',y,y',z,z')$ represent an orbit coordinate vector, where x, z are, respectively, the radial and vertical positions, y the position along the direction of motion, and the primes denote the corresponding conjugate variables; all are taken with respect to a reference particle moving on the ideal closed orbit with azimuthal variable θ. Then from Eq. (6), the statement

$$\vec{\Omega}(X,\theta) = \vec{\Omega}(X,\theta+2\pi)$$

simply specifies that the structure is periodic. We may define a general mapping of the orbit X and the spin $\vec{P}$ between two arbitrary positions θ and θ' by

$$M(\theta,\theta') : (X,\vec{P}) \rightarrow (MX,\vec{P}') ,$$

which implies that the spin motion depends on the orbital motion but that the orbital motion does not depend on the spin motion. (The magnetic moment of a particle has a negligible influence on its trajectory.) In general $\vec{P}' \neq \vec{P}$; also, the mapping need not necessarily be linear.

We now consider a one-turn mapping from θ to $\theta + 2\pi$ which we designate M_θ. Then for all X, θ there exists $\vec{n}(X,\theta)$ such that

$$M_\theta\{X, \vec{n}(X,\theta)\} = \{M_\theta X, \vec{n}(X,\theta)\} .$$

This corresponds to the "hairy ball" theorem which states that a vector field on the surface of a sphere vanishes at at least one point. Physically the meaning is that whilst the orbit vector X transforms to $M_\theta X$ ($\neq X$ in general), arbitrary solutions $\vec{P}$ of the Thomas-BMT equation corresponding to the orbit X do not transform into themselves; however, there is one solution $\vec{n}(X,\theta)$ which does, and this is the precession axis of the other spin solutions.

It should be noted that $\vec{n}$ is thus defined for one turn, starting at θ, and must be redefined for successive turns, consequently

$$\vec{n}(X,\theta) \neq \vec{n}(M_\theta X, \theta+2\pi) .$$

In the original formulation of Derbenev and Kondratenko[7] the orbit vector X was expressed in action-angle variables I_j, ψ_j ($j = x,y,z$), with the equality:

$$\vec{\Omega}(I_j,\psi_j,\theta+2\pi) = \vec{\Omega}(I_j,\psi_j,\theta) = \vec{\Omega}(I_j,\psi_j+2\pi,\theta) .$$

The first part of the equality expresses the periodicity of the structure and the second part the periodicity with oscillation phase. The precession axis $\vec{n}$ then satisfies the same periodicity conditions

$$\vec{n}(I_j,\psi_j,\theta+2\pi) = \vec{n}(I_j,\psi_j,\theta) = \vec{n}(I_j,\psi_j+2\pi,\theta) .$$

In this formulation is is important to note that the I_j, ψ_j represent points in the phase space and not the coordinates of a particular orbit. However, both representations are equivalent and in the limit of vanishing oscillation amplitudes $X \to 0$ or $I_j \to 0$, the precession axis $\vec{n}(\theta) \to \vec{n}_0(\theta)$, the unperturbed axis, as would be expected from continuity.

Now, since $\vec{n}(\theta)$ is the axis around which all other spin solutions precess with spin tune ν, a Fourier analysis of $\vec{n}$ does not contain the frequency ν but only a spectrum of orbit-oscillation frequencies. This is a useful property, because it enables us to calculate most depolarization processes in terms of the perturbation of $\vec{n}$, without considering individual spins, as long as the spin tune is not too close to a resonance.

SYNCHROTRON OSCILLATIONS

To illustrate the behaviour of the precession axis in the presence of a perturbation we consider a simple example involving only synchrotron (energy) oscillations excited by an energy jump $\delta\gamma$ (< 0) resulting from the emission of a single photon. We neglect betatron oscillations, a good approximation away from the betatron spin resonances, and for the moment we ignore the radiation damping, which is small in one turn.

Before the photon emission the spin $\vec{P}$ of a particle is precessing around the axis $\vec{n}_0(\gamma_0)$ with no orbit oscillation, as in Fig. 2a. After photon emission the energy loss $\delta\gamma$ results in a new precession axis $n(\gamma_0+\delta\gamma)$ around which the spin $\vec{P}$ starts to precess, with $\vec{n}(\gamma_0+\delta\gamma) = \vec{n}_0(\gamma_0) + \vec{\Gamma}(\delta\gamma/\gamma_0)$ to first order. The spin-orbit coupling vector $\vec{\Gamma} = \gamma\partial\vec{n}/\partial\gamma$ characterizes the sensitivity of $\vec{n}$ to energy fluctuations and oscillations.

In Fig. 2b we follow the evolution of $\vec{n}$ projected on a plane tangent to the unit sphere at $\vec{n}_0$, but considering successive one-turn transformations from θ to $\theta + 2\pi$, $\theta + 4\pi$, and so on. The corresponding precession axes are denoted $\vec{n}_1$, $\vec{n}_2$, etc. Now since we are neglecting radiation damping and considering only a single

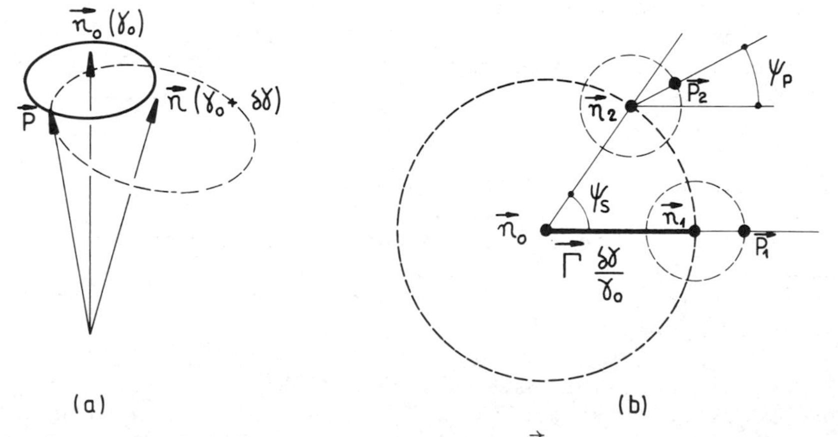

Fig. 2 a) Perturbation of precession axis $\vec{n}$ from emission of a photon.
b) Evolution of one-turn precession axis with synchrotron oscillation.

quantum emission, the perturbation $\vec{\Gamma}(\delta\gamma/\gamma_0)$ has a constant magnitude and the locus of $\vec{n}_1$, $\vec{n}_2$, etc., must lie on a circle centered on $\vec{n}_0$. Furthermore, the phase advance on this circle between successive one-turn axes $\vec{n}_1$, $\vec{n}_2$, ... is evidently the synchrotron-oscillation phase advance per turn $\psi_s = 2\pi Q_s$, where Q_s is the synchrotron tune which has typically a value of around 0.1 in high-energy e^+e^- storage rings.

Meanwhile, the spin is precessing around $\vec{n}_1$, $\vec{n}_2$, ... with a phase advance of $\psi_p = 2\pi\nu$ per turn, also shown in Fig. 2b. A resonant situation occurs if $\psi_p = 2\pi k_0 + k_s\psi_s$, where k_0, k_s are integers, i.e. if:

$$\nu = k_0 + k_s Q_s \ .$$

Such resonances are excited by errors in the closed orbit with a non-vanishing Fourier component k_0 near to the spin tune ν; the synchrotron oscillations modulate the perturbation, giving rise to spin satellite resonances of order k_s. These can cause depolarization even if the spin tune is not very close to the integer k_0, especially in high-energy e^+e^- rings which have a relatively large Q_s and large-amplitude synchrotron oscillations.

NUMERICAL INTERLUDE

The scale of the diagrams in Fig. 2 has been exaggerated for clarity and it is instructive to examine typical numerical values. The emission of a single photon of critical energy $\hbar\omega_c$ causes a relative energy loss $\delta\gamma/\gamma_0$ of about 10^{-6}; this is the parameter ξ of Eq. (1). We also saw in Eq. (5) that a magnitude $|\vec{\Gamma}| \approx 1$ gives a substantial degree of depolarization, and corresponds to an angular perturbation $\vec{n} - \vec{n}_0 = \vec{\Gamma}\delta\gamma/\gamma_0$ of 10^{-6} radian. Thus, first-order perturbation theory is quite adequate if $|\vec{\Gamma}| \not\gg 1$, i.e. not too close to a resonance.

We can compare this perturbation of 10^{-6} from a single photon with a simple classical representation of a beam 92.4% polarized, in which all the spins are imagined to lie on the surface of a cone of half angle arc cos (0.924) = 0.39 radian. This number, which characterises the average spread in spin orientation of an almost fully polarized beam, is large compared with the perturbation from a single photon emission. However, many photons are emitted randomly, about $5\pi\alpha\gamma/\sqrt{3}$ per turn, causing a random walk of $\vec{n}$ reaching 0.39 radian in a time which can be shown to be approximately equal to the polarization time. We thus have a simple semi-quantitative physical model of the spin-diffusion mechanism arising from quantum fluctuations of the precession axis, showing that $|\vec{\Gamma}| \sim 1$ corresponds to approximately equal polarization and depolarization times.

RADIATION DAMPING

In the presence of radiation damping of orbit oscillations the locus of $\vec{n}$ is no longer a circle as in Fig. 2b but a spiral converging on $\vec{n}_0$. For $|\vec{\Gamma}| << 1$ this spiral is very fine, deviating only slightly from a circle over one oscillation period. The spin vector $\vec{P}$ can then follow $\vec{n}$ adiabatically during the damping process; furthermore, the quantum fluctuations of $\vec{n}$ are also small so there is a high degree of coherence in the spin motion and therefore little depolarization.

In contrast near a resonance $|\vec{\Gamma}| \gtrsim 1$, and a loss of coherence arises both from the enhanced effects of the quantum fluctuations and from the coarser nature of the spiral, the latter inducing a non-adiabatic component of spin precession around $\vec{n}$.

It is important to draw a distinction between the effects of quantized energy loss on the orbital motion and on the spin motion. In the case of orbital motion, the average energy loss is recovered from the RF acceleration system, a classical process which leads to damping of oscillations and an equilibrium with the quantum excitation. Spin motion has, in the normal sense, no kinetic damping mechanism associated with it. Fortunately, spin-flip polarization fulfils an equivalent function but it can hardly be considered as analogous to orbit damping in view of the much longer time constant and the subtle quantum-mechanical asymmetry involved.

BETATRON OSCILLATIONS

The energy loss from photon emission gives rise to excitation of betatron oscillations because of the non-vanishing energy dispersion in a storage ring. As in the case of synchrotron oscillations a one-turn precession axis can be defined whose locus is a spiral together with a random walk. The spin resonance condition is then

$$\nu = k_0 + k_z Q_z$$

for vertical betatron oscillations, which are usually the more important. For $k_0 = 0$, pure betatron resonances are driven only by the normal oscillation amplitude; since $Q_z >> 1$ these resonances are widely spaced. The presence of orbit-error harmonics introduces driving terms for $k_0 \neq 0$, which are modulated by the betatron oscillations, producing families of spin satellites analogous to those of synchrotron oscillations but with wider spacing in general.

WHAT MAKES $\vec{\Gamma}$ LARGE?

In a perfect storage ring without errors the orbit lies everywhere in the median plane and particles experience essentially only a vertical component B_z of the magnetic field, since the vertical oscillation amplitudes are exceedingly small. The precession axis $\vec{n}_0$ of the spin motion is then everywhere vertical, independent of energy and of horizontal betatron amplitude, and $\vec{\Gamma} = 0$ as a consequence.

Practical storage rings have errors due to magnet misalignments and field imperfections, causing the orbit to deviate from the ideal median plane. The focusing forces from quadrupoles and the end fields of bending magnets then subject particles both to radial field components B_x and to longitudinal fields B_y, which cause the precession axis $\vec{n}$ to vary with azimuth, energy and oscillation amplitudes. In high-energy e^+e^- storage rings, longitudinal field components from magnet ends are relatively weak and their effects on spin motion are further weakened by the factor (γa) in the Thomas-BMT equation. However, detector solenoids are normally strong enough that their influence on the spin motion must be compensated locally, either by "anti-solenoids" or by special orbit bumps.

The major contribution to large values of $\vec{\Gamma}$ remains the radial field components B_x associated with orbit deviations and energy dispersion in the vertical plane, though vertical betatron oscillations enhanced by coupling from the horizontal motion also play an important role. In addition, a storage ring equipped with spin rotators, to obtain longitudinal polarization at the interaction point, necessarily has some intentional vertical bending; the resulting $\vec{\Gamma}$ induced locally must be compensated nearby to prevent it propagating around the machine and generating a large contribution to $|\vec{\Gamma}|^2$ in Eq. (5).

VERTICAL ORBIT-ERROR HARMONIC

Vertical orbit deviations are of particular importance since they can combine with betatron and synchrotron oscillations to excite a wide spectrum of resonances given by the general form

$$\nu = \gamma a = k_0 + k_x Q_s + k_z Q_z + k_s Q_s \ .$$

The integer k_0 identifies the Fourier harmonics of orbit errors which can drive integer spin resonances; these are spaced throughout the spectrum at intervals of $mc^2/a \approx 440$ MeV. Considering these integer resonances alone we can expand the orbit deviation in a Fourier series:

$$z(\theta) = \sum_{k=1}^{\infty} z_k \cos(k\theta+\phi_k)$$

and take only the harmonic $k \sim \nu$ nearest to the spin tune which makes the dominant contribution. It can then be shown[8] that the corresponding component of the spin-orbit coupling vector is given by

$$|\vec{\Gamma}|_k = \frac{2\nu^2 k^2}{(\nu^2-k^2)} \frac{z_k}{R}, \tag{10}$$

where R is the average radius of the machine.

The resonance denominator demonstrates how $\vec{\Gamma}$ can become large if ν is close to an integer k for which there is an appreciable amplitude z_k of orbit harmonic. Equation (10) also shows how the sensitivity to orbit errors increases at high energies, being proportional to γ^3 for $k \approx \nu = \gamma a$.

More generally one must take account of the spectrum of perturbations including also those arising from betatron and synchrotron oscillations, and sum the perturbations over all significant terms.

Synchrotron-oscillation satellite resonances

A further complication arises at high energies because of the large energy spread of the beam, given by

$$\sigma_E = \left[\frac{55}{64\sqrt{3}} \hbar c\, mc^2\right]^{\frac{1}{2}} \frac{\gamma^2}{\sqrt{\rho}},$$

leading to a correspondingly large spread in spin tune $\sigma_\nu = a\sigma_E$. When this spread occupies a large fraction of the space between integer resonances (1 in spin tune, 440 MeV in energy), particles in the Gaussian tails of the beam are unavoidably close to the integer resonances, as shown schematically in Fig. 3a. The situation is aggravated in the presence of synchrotron-oscillation spin satellites, which further reduce the space available for the spread in spin tune, as in Fig. 3b.

These spin satellites of the integer resonances are of considerable importance in LEP since they impose tight limits on the permissible strengths of nearby orbit harmonics[9]. The "error-sensitivity enhancement factor" C is the ratio by which the relevant orbit harmonics must be reduced in the presence of synchrotron-oscillation satellites compared with the case of no satellites; the example shown in Fig. 4 indicates that values of C approaching 10 can occur even if the spin tune ν is mid-way between two satellites. The apparent poles at the satellites are unrealistic, since the first-order perturbation theory used in Ref. 9 is inaccurate for large C.

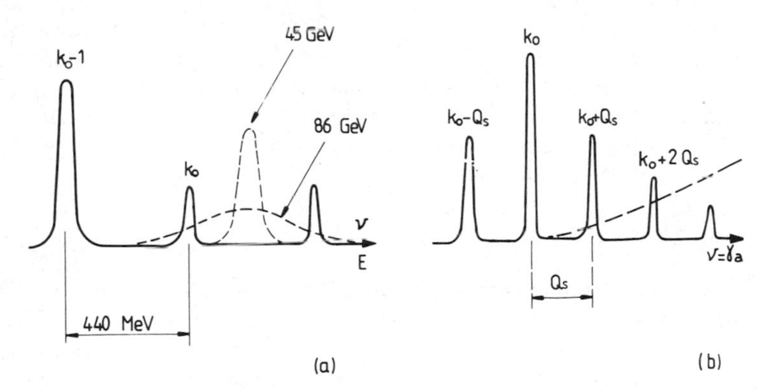

Fig. 3 a) Integer resonances $\nu = k_0$. b) Synchrotron-oscillation spin satellites overlapping energy distribution function.

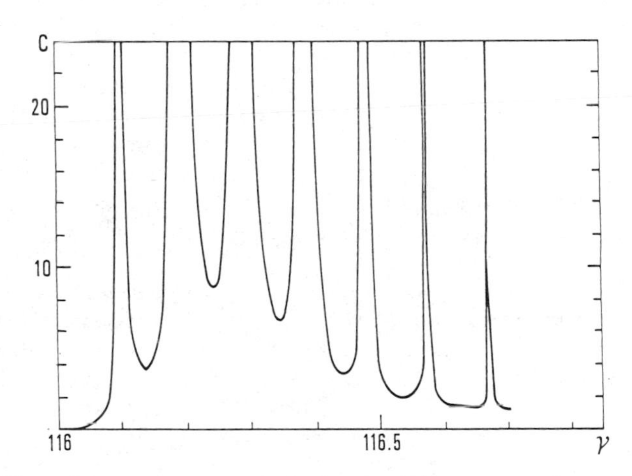

Fig. 4 Error-sensitivity enhancement factor C.

Preliminary results of a second-order theory (C. Biscari, paper in preparation) indicate that the resonances are indeed narrower but that the minima of C between the resonances are not reduced by taking account of second-order terms.

The nominal parameters of LEP in the 50 GeV energy region are such as to make the value of C rather critically dependent on the energy spread and on the synchrotron tune Q_s. This is unfortunate for predictive purposes but leaves open the possibility of using small changes in some machine parameters to minimize experimentally the depolarizing effects of the satellites.

POSSIBLE CURES

We have seen that the major cause of depolarizing resonances is the ubiquitous spectrum of vertical closed-orbit harmonics, which not only drive the integer spin resonances and their synchrotron-oscillation satellites but also extend the otherwise sparse spectrum of betatron resonances. The first priority is therefore to reduce the more important harmonics to an acceptably low level, which can be done in principle by arrangements of dipole orbit correctors excited in a suitable patten.

The main problem arises in measuring the amplitude and phase of orbit harmonics in order to know what corrections to apply. Normal methods of correction aim to reduce the maximum orbit excursions, which are associated mainly with harmonics near to the betatron tune, and will even tend to excite higher harmonics. Now high-energy storage rings have spin tunes typically a factor of two greater than the betatron tunes and the corresponding orbit harmonics contribute very little to the orbit displacement. It is therefore very difficult, perhaps impossible, to measure and analyse the orbit to the precision required for these higher harmonics.

Failing direct measurement of the errors, one can measure the polarization of the beam and apply systematic, predictable compensation of the appropriate harmonics by a searching procedure in order to increase the polarization. This requires a fast, high-resolution polarimeter, which is no great problem, but also some initial measurable polarization level, around 5% or more.

The effects of machine errors on the spin motion may be reduced by a procedure known as "spin matching", proposed by Chao et Yokoya[10] and extended by Steffen[11]. This involves choosing the orbit parameters in such a way that ten integrals can be made to vanish, or at least become small. The spin matching conditions are energy dependent and, even at selected energies, it may not be possible to safisfy them all because of machine design constraints. Nevertheless this method looks promising for reducing the sensitivity of the spin motion to orbit perturbations. Spin matching could also be used to reduce depolarization from pure betatron resonances[12] and from the non-linear beam-beam effect[13].

Depolarization from synchrotron-oscillation spin satellites is due to the extended tails of the normal Gaussian energy distribution. Non-linear wiggler magnets could be used to produce stronger damping for large-amplitude synchrotron oscillations[9] and modify the energy distribution function to reduce the density of the tails overlapping the satellites. The same non-linear wigglers which may be necessary at low energies in LEP to reduce collective effects[14] could be operated in a different mode at higher energies to reduce the density of the energy tails.

POLARIZATION RATE AT THE LOWER ENERGIES

The natural polarization rate of a beam from the Sokolov-Ternov effect is a very steep function of energy, varying as γ^5 in a given machine as seen in Eq. (1). Although the nominal design energy of LEP is about 85 GeV, it will operate for a considerable time in the 50 GeV range where the natural polarization time is about 3 hours. This can be reduced by the installation of asymmetric polarizing wigglers[15] consisting of a sequence of three dipole magnets bending the orbit in the horizontal plane. The centre magnet is short and strong with a field B_+ and the outer magnets are long and weak, with field B_-, such that the total field integral over the three magnets vanishes leaving the orbit outside the wiggler unperturbed. Since the rate of polarization varies as B^3, it is strongly enhanced by the centre magnet and only weakly reduced by the outer ones, leading to an overall increase in polarization rate by a factor

$$\left(\frac{\tau_p^W}{\tau_p}\right)^{-1} = 1 + \frac{B_+^3\eta_+ + 2|B_-|^3\eta_-}{\langle B_z\rangle\, B_z^2} ,$$

where $\eta_\pm = \ell_\pm/2\pi R$ are the normalised lengths of the bending magnets, B_z is the normal bending field in the machine arcs and $\langle B_z\rangle$ is this field averaged over the whole circumference.

The price to pay for the enhanced polarization rate is a reduction in the asymptotic polarization level to

$$P(\infty) = \frac{8}{5\sqrt{3}}\,\frac{B_+^2 - B_-^2}{B_+^2 + B_-^2}$$

and an increase in both the synchrotron radiation power and the beam energy spread. A large ratio $|B_+/B_-|$ minimizes these undesirable effects.

For the LEP machine a first step in speeding up the polarization will be made by a small modification to the emittance-control wrigglers,[16] which are required in any case to control the beam size for optimum luminosity. Making these somewhat asymmetric detracts in no way from their primary function, costs very little and decreases the polarization time by a factor of 2 or 3, to give about 90 minutes at 50 GeV with a polarization level of around 75%.

POLARIZED BEAMS IN LEP

There are encouraging prospects that transversely-polarized beams could be observed in LEP fairly early in Phase 1 at around 50 GeV. To make this possible certain modest provisions must be made during the construction period of the machine.

The most important requirement is a laser back-scattering polarimeter which would be necessary even if polarized beams were not required, in order to avoid confusion of physics results by unsuspected partially-polarized beams. It was concluded at the DESY Workshop[17] that an argon-laser polarimeter, similar to those used at SPEAR and PETRA, would be quite suitable for LEP. However, a faster response and somewhat better resolution would be desirable, and a Nd-YAG laser, as recently tested at PETRA, might be preferred.

The normal LEP orbit-correction magnets will be largely adequate for compensating harmonics near the spin tune, though a few extra may have to be installed at strategic points for the best results. Special correction algorithms will be needed for minimizing these higher harmonics.

Simulation of machine errors and their influence on spin resonances will be used to determine the most favourable energies and beam-optics parameters for obtaining a measurable degree of polarization initially, and for working out correction strategies. The possibility of adjusting the optics parameters to satisfy the spin-matching constraints will also be examined.

In anticipation of success in obtaining transverse polarization, some effort will be devoted to studies of spin rotators for longitudinal polarization at the interaction points, and dedicated asymmetric wigglers should be studied for increasing further the rate and level of polarization.

During early operation in Phase 1, initial polarization studies will be made under the most favourable conditions possible. These include a low beam current to avoid instabilities, detuned low-β sections to reduce orbit errors and a suitable choice of energy as suggested by theory and computer simulations. In addition one would make full use of possibilities for varying critical machine parameters such as Q_s.

With these precautions we would expect to find a measurable level of polarization which could then be improved by application of the orbit-correction schemes, until finally we would hope to achieve a useful degree of polarization under normal operational conditions for physics.

The apparent optimism on which the above scenario is based stems from a number of causes. The earlier theoretical work of the Novosibirsk group has become more widely disseminated and better understood in recent years, and its extension to the higher range of energies has been considerably clarified. Experimental results from PETRA have demonstrated the feasibility of spin-resonance compensation[17] and also shown that beam-beam depolarization is much less serious than had previously been believed[17]. The increasing interest in spin motion that has been developing in the last few years, particularly for electron storage rings, will stimulate new ideas and increase the chances of overcoming the extra difficulties at the higher energies. Finally, the physics motivation for polarized e^+ and e^- beams is now much stronger than had been the case for the smaller storage rings at lower energies[18].

ACKNOWLEDGEMENTS

It is a pleasure to acknowledge many stimulating discussions with C. Biscari, J. Buon, J.M. Jowett and R.D. Ruth.

REFERENCES

1. I. M. Ternov, Yu. M. Loskutov and K. L. Korovina, Sov. Phys. JETP 14:921 (1962).
2. A. A. Sokolov and I. M. Ternov, Sov. Phys. Doklady 8:1203 (1964).
3. V. N. Baier and V. M. Katkov, Sov. Phys. JEPT 25:944 (1967).
4. Ya. S. Derbenev and A. M. Kondratenko, Sov. Phys. JETP 37:968 (1973).
5. L. H. Thomas, Phil. Mag. 3:1 (1927).
6. V. Bargmann, L. Michel and V. L. Telegdi, Phys. Rev. Lett. 2:435 (1959).
7. Ya. S. Derbenev and A. M. Kondratenko, Sov. Phys. JETP 35:230 (1972).
8. J. Buon, Rpt LAL/RT/80-08 (1983).
9. C. Biscari, J. Buon and B. W. Montague, Rpt CERN/LEP-TH/83-8 (1983).
10. A. W. Chao and K. Yokoya, Rpt KEK 81-7 (1981).

11. K. Steffen, Internal Rpt CBN 81-29, Newman Lab. Cornell (1981).
12. K. Yokoya, Rpt KEK 81-19 (1982).
13. J. Buon, Rpt LAL/RT/83-04 (1983).
14. A. Hofmann, J. M. Jowett and S. Myers, IEEE Trans. Nucl. Sci. NS-28, 2392 (1981).
15. A. M. Kondratenko and B. W. Montague, Rpt CERN/ISR-TH/80-38 (1980).
16. J. M. Jowett and T. M. Taylor, Rpt CERN/LEP-MA-TH/83-10 (1983), Presented at the Particle Accelerator Conf. Santa Fe, 1983.
17. Workshop on Polarized Electron Acceleration and Storage, Hamburg, (1982), Rpt. DESY M-82/09 (1982).
18. Ch. Prescott, H. E. Spin Physics-1982 (Brookhaven), AIP Conf. Proc. No. 95 p. 28 and this conference.

PHYSICS WITH POLARIZED BEAMS IN e^+e^- COLLIDERS*

Charles Y. Prescott

Stanford Linear Accelerator Center

Stanford University, Stanford, California 94305

ABSTRACT

The spin structure of the standard model of electroweak interactions is described, with emphasis on the relevance to polarized beam phenomena. The polarization dependence of the cross section, charge asymmetries, and examples of experimental measurements using longitudinal polarization are given. Longitudinal beam polarization is discussed in some detail, including electroweak radiative corrections. Applications to testing the standard model and looking beyond for additional gauge bosons are considered.

INTRODUCTION

The inherent left-handedness of the weak part of the electroweak interactions makes polarization phenomena at high energies an important experimental tool for studying the basic phenomenology. The use of polarized beams for studying electroweak phenomenology is described in this talk. It is emphasized that polarization phenomena exist even in the absence of beam polarization and that there exist interesting experiments which study these effects. This talk focusses however on the much broader class of polarization phenomena resulting from beam polarization. This talk is structured in the following way: (i) a brief discussion of the spin structure of the standard model with emphasis on the couplings and the polarization dependences; (ii) consequences of polarized beams and opportunities for experiments; (iii) status of radiative corrections; (iv) looking beyond the standard model.

* Work supported by the Department of Energy, contract number DE-AC03-76SF00515.

The experimental and technical problems in providing polarized beams for experiments have been studied at considerable length. In circular colliders depolarizing effects compete with the natural buildup of polarization. Equilibrium values of polarization depend on details of the machine. An active program to understand and control the depolarization in PETRA is underway, and has been discussed in this conference.[1] In linear colliders, production of polarization and depolarizing effects come from entirely different processes. These too have been studied in the SLC workshops and are rather well understood. Polarized beams are expected to exist eventually, if not in the early operations, at both LEP and the SLC at SLAC. This talk concerns some of the interesting phenomena associated with polarization.

THE SPIN STRUCTURE OF THE STANDARD MODEL

The electroweak part of the standard model, relevant to polarization, can be summarized in the well-known weak isospin structure, whereby left-handed fermions are placed in doublets and right-handed fermions in singlets:

$$\begin{pmatrix}\nu_e \\ e^-_L\end{pmatrix}, \begin{pmatrix}\nu_\mu \\ \mu^-_L\end{pmatrix}, \begin{pmatrix}\nu_\tau \\ \tau^-_L\end{pmatrix}, \begin{pmatrix}u_L \\ d_L\end{pmatrix}, \begin{pmatrix}c_L \\ s_L\end{pmatrix}, \begin{pmatrix}t_L \\ b_L\end{pmatrix} \quad \text{(doublets)}$$

$$e^-_R,\ \mu^-_R,\ \tau^-_R,\ u_R,\ d_R,\ c_R,\ s_R,\ t_R,\ d_R \quad \text{(singlets)}$$

Such structure is fundamental to minimal SU(2) × U(1). Right-handed currents are experimentally ruled out within this minimal structure, but may exist at higher energies if more gauge bosons exist. The couplings of a Z^o boson to a fermion pair $f\bar{f}$ are given by

$$g^f_L = \frac{e}{\sin\theta_W \cos\theta_W}\left[T^f_{3L} - q^f \sin^2\theta_W\right]$$
$$g^f_R = \frac{e}{\sin\theta_W \cos\theta_W}\left[T^f_{3R} - q^f \sin^2\theta_W\right] \qquad (1)$$

where T_3 refers to the weak isospin with $T_{3L} = \pm 1/2$ and $T_{3R} = 0$. The electric charge of the fermion is q^f, and $\sin^2\theta_W$ is the weak mixing parameter. Note that $g^f_L \neq g^f_R$ because of the values of T_{3L} and T_{3R} being different. Parity violation in the neutral current interactions is the consequence of these couplings being unequal. Coupling strengths for left-handed fermions (i.e., longitudinal polarization with spin projection negative relative to momentum) to the Z^o are proportional to $(g^f_L)^2$, while for right-handed fermions, $(g^f_R)^2$. In the decay of an unpolarized Z^o, for example, the ratio of left-handed to right-handed polarized fermions is g^2_L/g^2_R. Table I shows the values for this ratio for four different

TABLE I. Neutral Current Couplings ($\sin^2\theta_W$ = .23)

fermion type	g_L^2/g_R^2
u	5.1
d	30.5
ν_e	∞
e^-	1.38

fermions and for the mixing parameter $\sin^2\theta_W$ = .23. In this table we see the preference of the Z^o for the left-handed couplings. For neutrinos, neutral current couplings are purely left-handed. If right-handed neutrinos exist, which must be so if they are massive, they do not couple through the neutral currents in the standard model.

Universality of neutral current couplings is contained in Eq. (1). The coupling strengths are the same for e, μ, and τ as are those of u,c,t and d,s,b. Test of universality are important to checks of the standard model. These tests are underway at PETRA and PEP for the leptons e, μ and τ with reasonable accuracy, and for the quarks with less sensitivity. Precision tests of these relations await the production and decay of the Z^o. The relations of Eq. (1) imply large spin dependent effects at the Z^o occur because $|g_R - g_L| \sim O(g)$. Experimentally, this implies that polarization phenomena should lead to accurate studies of these parameters.

It is customary to define vector and axial-vector couplings to Z^o by

$$g_V^f = \frac{1}{2}\left(g_R^f + g_L^f\right)$$

and (2)

$$g_A^f = \frac{1}{2}\left(g_R^f - g_L^f\right)$$

Although relations (1) and (2) are equivalent, use of vector and axial-vector couplings has been more common. Notation and normalization of these couplings vary considerably in the literature. The values of these couplings depends on the value used for $\sin^2\theta_W$. The success of the standard model comes from the unification of widely differing interactions with one value of $\sin^2\theta_W$. The current experimental determination of this parameter is .230 ± .010.[2] For the future, electroweak studies at the Z^o promise to improve the accuracies in these couplings by a order-of-magnitude. One of the best determinations will be possible using polarized beams at the Z^o pole.

What can polarized beams do? Measurements with polarized beams can enhance the physics relative to that for unpolarized beams. Examples of physics phenomena enhanced by polarization are total cross sections, charge asymmetries, and final state lepton and quark polarizations. There are also measurements which are unique to polarization. Longitudinal asymmetries, an example of this, lead to sensitive tests for extra gauge bosons and tests of electroweak radiative corrections.

EXPERIMENTAL CONSEQUENCES OF POLARIZED ELECTRON BEAMS

The cross section, in lowest order, for $e^+e^- \to f\bar{f}$ is given by[3]

$$\begin{aligned}\frac{4s}{\alpha^2}\frac{d\sigma}{d\Omega} = &\left(1 - P_L^+P_L^-\right)\left[\sigma_u^{\gamma} + \sigma_u^{\gamma z} + \sigma_u^{z}\right] \\ &\left(P_L^+ - P_L^-\right)\left[\sigma_L^{\gamma} + \sigma_L^{\gamma z} + \sigma_L^{z}\right] \\ &\left(P_T^+P_T^- \sin^2\theta \cos\psi\right)\left[\sigma_T^{\gamma} + \sigma_T^{\gamma z} + \sigma_T^{z}\right] \\ &\left(P_T^+P_T^- \sin^2\theta \sin\psi\right)\left[\tilde{\sigma}_T^{\gamma} + \tilde{\sigma}_T^{\gamma z} + \tilde{\sigma}_T^{z}\right]\end{aligned} \tag{3}$$

where P_L^+ and P_L^- refer to longitudinal polarization of e^+ and e^- incident beams, P_T^+ and P_T^- refer to transverse components of the polarization, θ is the polar angle of the outgoing fermion, and $\psi = 2\phi - \varphi^+ - \varphi^-$, with ϕ the azimuthal angle of the outgoing fermion, and φ^+ and φ^- refer to the azimuthal angle of the transverse component of the spin of the e^+ and e^- beams. The superscripts γ, γz, and z refer to cross section terms from pure photon exchange, interference between photon and Z^o, and pure Z^o exchange, respectively.

Polarization - driving mechanisms occurring in circular colliders lead to transverse polarization of both e^+ and e^- beams. It is rather natural in circular rings for $P_T^+P_T^-$ terms to appear, and indeed these effects have been experimentally observed. Some discussion of the consequences of $P_T^+P_T^-$ terms are in literature.[3] For linear colliders, providing polarized electron beams is rather easy, but not easy for positron beams. Longitudinal polarization effects for electrons polarized and for positrons unpolarized lead to significant effects at high energies. Those effects are the ones I wish to discuss here. They are important to polarized beams in the SLC. So for this talk, I will assume $P^+ = 0$. Equation (3) then reduces to

$$\frac{4s}{\alpha^2}\frac{d\sigma}{d\Omega} = \sigma_u^{\gamma} - P_L^-\left[\sigma_L^{\gamma z} + \sigma_L^{z}\right] \quad . \tag{4}$$

Note that the term σ_L^{γ} is 0 and that $\sigma_L^{\gamma z} \ll \sigma_L^{z}$, if the energy chosen is that of the Z^o-pole, $\sqrt{s} = M_Z$. For the discussion, neglect the contribution from $\sigma_L^{\gamma z}$. In the figures, however, this term remains. Figure 1 shows the total cross section for unpolarized

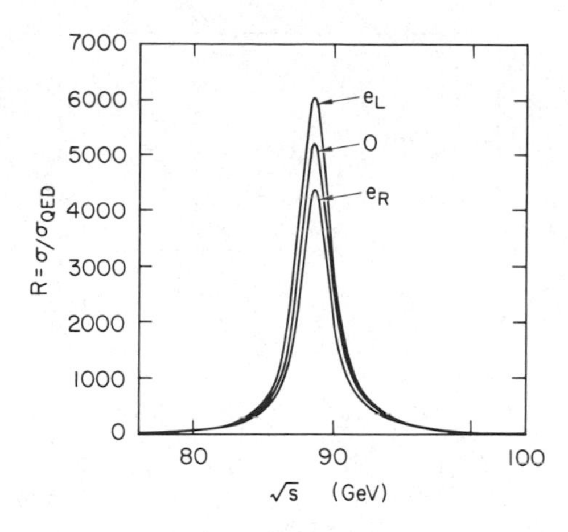

Fig. 1. The cross section ratio R in the vicinity of the Z^o, in lowest order, for three e^- beam polarization states. ($\sin^2\theta_W = .23$)

beams, in lowest order. For polarized beams, at the Z^o-pole, the rate of production of Z^o's is proportional to $N_R g_R^2 + N_L g_L^2$, where N_R, N_L are the number of right-handed and left-handed incident electrons, and g_R^2, g_L^2 are the electron-Z^o couplings squared, respectively. The beam polarization is defined $P_e = (N_R - N_L)/(N_R + N_L)$. Using the definitions of Eq. (2), gives the rate of Z^o production proportional to $(g_V^2 + g_A^2 + 2P_e g_V g_A)$, where g_V and g_A are the vector and axial-vector electron couplings. The longitudinal asymmetry is defined

$$A_L = (\sigma_R - \sigma_L)/(\sigma_R + \sigma_L)$$

which becomes

$$A_L = \frac{g_R^2 - g_L^2}{g_R^2 + g_L^2} = \frac{2g_V/g_A}{(1 + g_V^2/g_A^2)} \quad . \tag{5}$$

A measurement of A_L through (5) provides a determination of g_V/g_A. For the mixing parameter value $\sin^2\theta_W = .23$, the standard model predicts $A_L = -.16$. Figure 2 shows the polarization dependence of the cross section in the region $\sqrt{s} = M_Z$. Present experimental information lead to an accuracy on $\sin^2\theta_W = \pm.01$ at low energies. Longitudinal asymmetry measurements promise to provide accuracies an order-of-magnitude better for $\sin^2\theta_W$, and also similar improvements in g_V and g_A.

The Z^o's are polarized in polarized e^+e^- collisions. For a right-handed e^- beam, the Z^o spin aligns in the direction of the e^- beam. For left-handed e^- beams, the Z^o spin points in the opposite direction. That is, $P_Z = +1$ and -1 for these two cases.

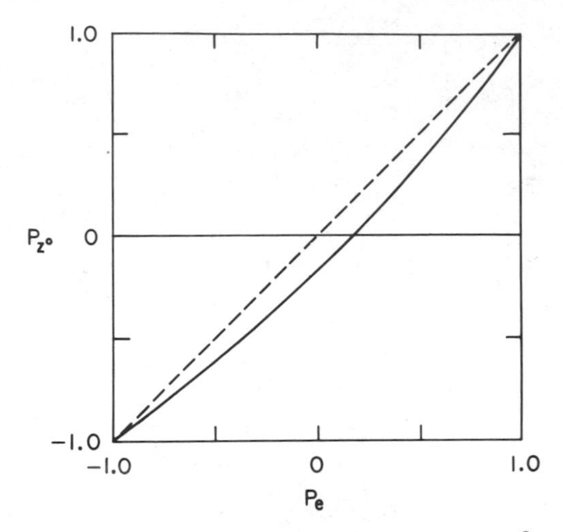

Fig. 2. The polarization of the Z^o in the e^- beam direction versus P_e for $\sin^2\theta_W = .23$. The dashed curve is for $\sin^2\theta_W = 1/4$.

However for unpolarized beams, the Z^o is still polarized. Why is this so? Because the Z^o prefers left-handed couplings, as shown in Table I. Figure 2 shows the polarization of the Z^o for P_e from −1 to +1. The form can be written as

$$P(Z^o) = (P_0(Z^o) + P_e)/(1 + P_e P_0(Z^o)) \qquad (\text{at } \sqrt{s} = M_Z) \qquad (6)$$

where $P_0(Z^o)$ is the value for unpolarized beams. Figure 2 shows that Z^o's are highly polarized if the beam is highly polarized, and are polarized in the direction of the e^- beam even if $P_e = 0$.

Polarized Z^o's lead to nonzero charge asymmetries. Charge asymmetries for $e^+e^- \to f\bar{f}$ are defined as

$$A_{CH} = (N(f) - N(\bar{f}))/(N(f) + N(\bar{f}))$$

where $N(f)$, $N(\bar{f})$ refer to the observed number of primary fermions (such as $\mu^-\mu^+$, $\tau^-\tau^+$, ...) in a detector. Figure 3 shows the case of a Z^o aligned in one case parallel to the e^- beam, and in the other case antiparallel. The forward going μ^- has its spin aligned along its motion. The rate for this spin orientation is proportional to g_L^2, while for the forward going μ^+, the spin of the μ^- is opposite its motion, giving a rate proportional to g_R^2. So at 0°, the charge asymmetry for $Z^o \to \mu^-\mu^+$ is

$$A_{CH} = P_Z \frac{2g_V^\mu/g_A^\mu}{\left(1 + (g_V^\mu/g_A^\mu)^2\right)} = \begin{Bmatrix} -.16 & \text{for } P_e = +1 \\ +.026 & \text{for } P_e = 0 \\ +.16 & \text{for } P_e = -1 \end{Bmatrix}$$

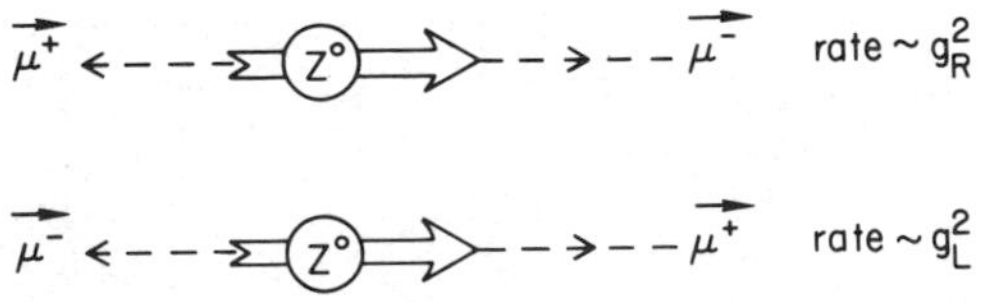

$\vec{\mu}^+ \leftarrow - - (Z^o) \Rightarrow - - \rightarrow - - \vec{\mu}^-$ rate $\sim g_R^2$

$\vec{\mu}^- \leftarrow - - (Z^o) \Rightarrow - - \rightarrow - - \vec{\mu}^+$ rate $\sim g_L^2$

Fig. 3. Polarized Z^o's decaying into forward and backward μ's. The forward going μ^- has necessarily its spin aligned forward with a rate proportional to $g_R^{\mu 2}$. The backward going μ^- has a corresponding rate proportional to $g_L^{\mu 2}$.

for $\sqrt{s} = M_Z$ and $\sin^2\theta_W = .23$. Figure 4 shows the complete calculation in lowest order, including γ-exchange and average over $\cos\theta$ in a 4π detector. The angular dependence of A_{CH} is

$$A_{CH} = f(s)\,\frac{2\cos\theta}{1 + \cos^2\theta} \quad .$$

and is shown in Fig. 5. The forward and backward portions of the solid angle are most important for charge asymmetries. Numerical estimates show that 1/2 of the sensitivity comes from the forward cone inside 40° (approximately), relatively independent of fermion type. Charge asymmetry measurements are best done by detectors which cover forward and backward regions of solid angle.

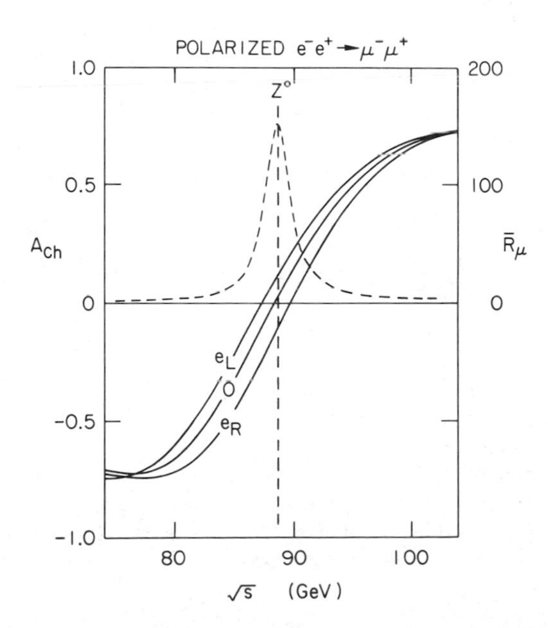

Fig. 4. A_{CH} for μ-pairs in the vicinity of the Z^o, for three polarizations of the e^- beam.

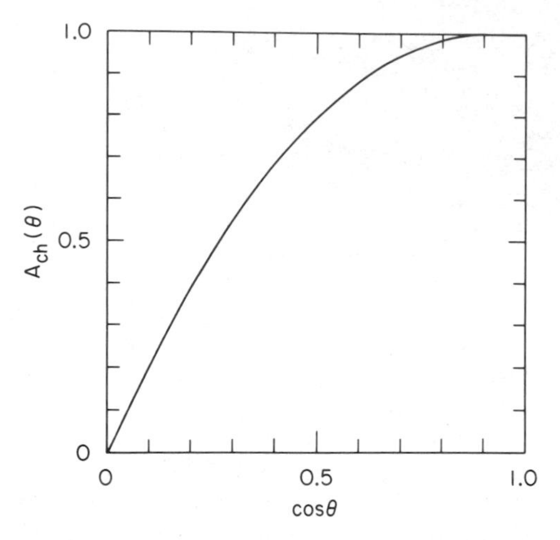

Fig. 5. The angular dependence for A_{CH}. Forward/backward directions are important to charge asymmetry measurements.

Figures 6(a) and 6(b) show charge asymmetries for hadron jets, separated into two cases; (i) primary quarks are $u\bar{u}$, $c\bar{c}$ and $t\bar{t}$, and (ii) $d\bar{d}$, $s\bar{s}$, and $b\bar{b}$. Charge asymmetries for jets are expected to be large, but sensitivity to values of $\sin^2\theta_W$ are not as large as for the leptons. Additional experimental problems of flavor identification, determination of the parent quark, and effects of gluon radiation make charge asymmetry measurements for the quarks harder than for the leptons. Semi-leptonic decays of the heavy quarks will be one area of work that will contribute to charge asymmetries.

Polarization of the final state fermions has the form

$$P_f = \frac{2g_V^f/g_A^f + P_Z\left(1 + (g_V^f/g_A^f)^2\right)}{\left(1 + (g_V^f/g_A^f)^2\right) + 2P_Z(g_V^f/g_A^f)} \qquad \begin{array}{l} \text{at } \sqrt{s} = M_Z \\ \theta = 0^o \\ \text{no } \gamma\text{-exchange} \end{array} \tag{9}$$

and the values of 1, -.31, -1. for P_e = 1,0,-1, respectively, for $\sin^2\theta_W$ = .23, and $f\bar{f} = \mu^-\mu^+$ or $\tau^-\tau^+$. Figure 7 shows the polarization of τ's for energies near the Z^o-pole, and averaging over 4π solid angle. Polarizations of the τ^- are large, even for unpolarized beams. Beam polarization permits a measure of control of the final state polarization. Studies of weak decays of heavy quarks should be substantially enhanced by this control.

Longitudinal asymmetries, defined in Eq. (5), are the quantities often called spin-flip asymmetries. The cross section in Eq. (5) may refer to total cross sections or to specific final states.

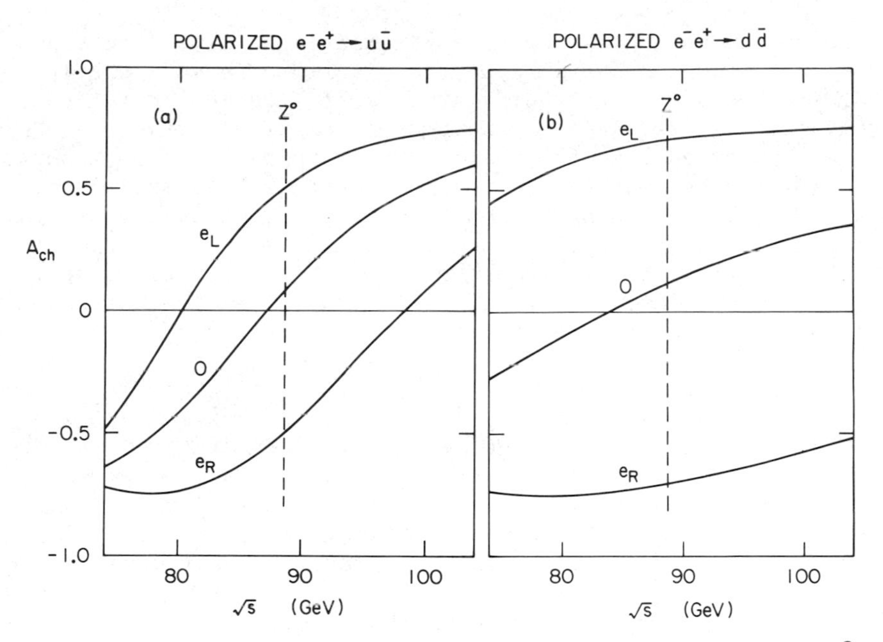

Fig. 6. A_{CH} for primary quarks in the vicinity of the Z^0, for three polarizations of the e^- beam.

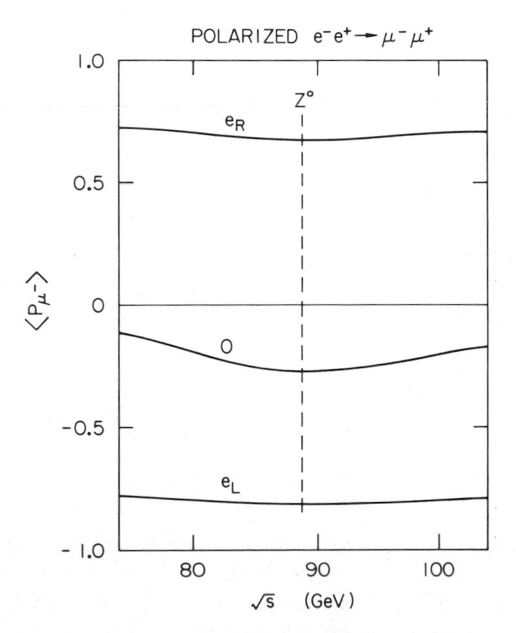

Fig. 7. Polarizations in the final state for μ-pairs or τ-pairs, near the Z^0, for the three beam polarizations indicated.

This parameter is easiest to measure experimentally. For example, A_L does not require charge identification (possibly difficult for forward-backward μ-pairs), or flavor identification in the case of hadron jets. A_L is extremely sensitive to effects from heavy Z^o's (such as contained in $SU(2) \times U(1)$, for example) and most importantly, this asymmetry is little affected by electroweak radiative corrections. This latter point is most significant, because distinguishing between radiative corrections and influences of extra gauge bosons may be difficult. The influence of radiative corrections on charge asymmetries is relatively large, making these measurements much more sensitive to such higher order effects.

RADIATIVE CORRECTIONS

Radiative corrections have been treated at different levels by a number of authors.[4] For polarized beams, the best work currently is that of M. Bohm and W. Hollik.[3] Their calculations include bubbles, vertex corrections, box diagrams, and soft bremsstrahlung of photons. Diagrams containing more than one heavy gauge boson are excluded, but are expected to be small. Figure 8 shows examples of electroweak processes which contribute to the radiative corrections. Comparison of radiative corrections for A_L and A_{CH} are given in Fig. 9. The conclusion, important to future experiments, is that

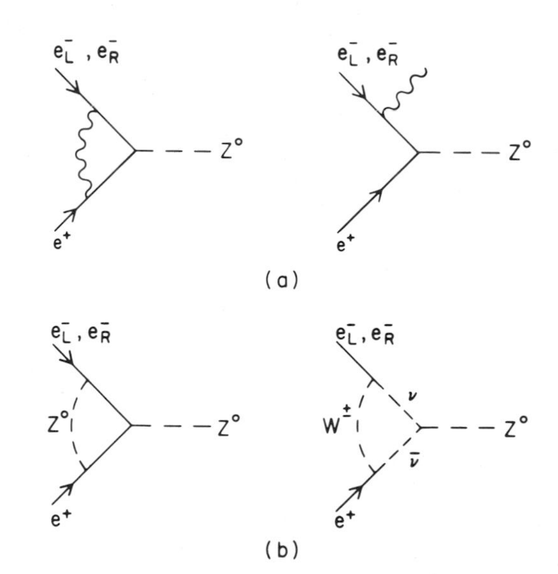

Fig. 8. Processes important to radiative corrections. The first two diagrams are examples which do not contribute to A_L, while the latter two are examples which do. Existing calculations of radiative corrections to A_L do not include the latter diagrams, but they are expected to be quite small.

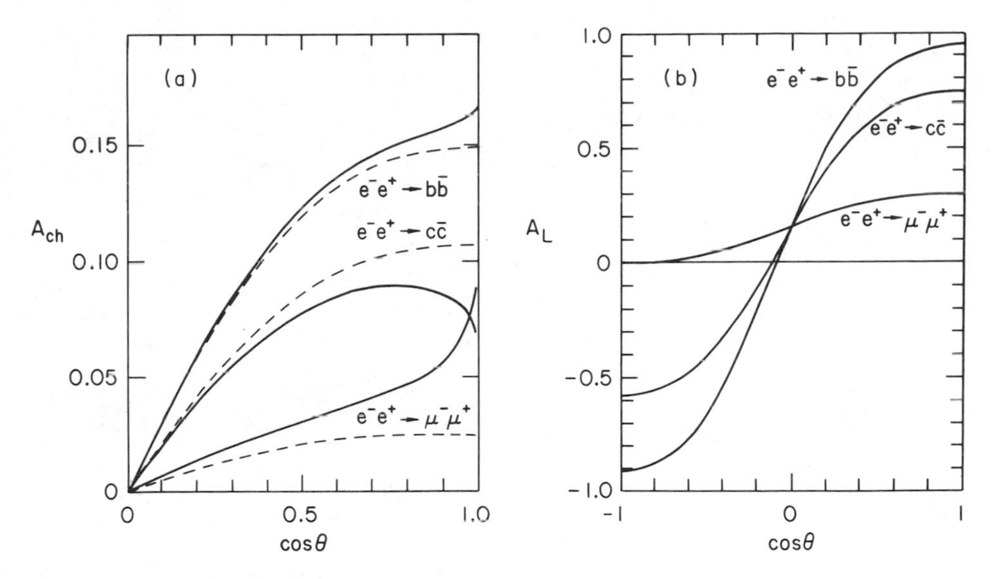

Fig. 9. (a) The on-resonance ($\sqrt{s} = M_Z$) unpolarized charge asymmetries for $e^-e^+ \to \mu^-\mu^+$, $c\bar{c}$, $b\bar{b}$ in lowest order (dashed curve) and corrected, with $\Delta E/E = 0.01$ (solid curve) (from Ref. 3). (b) The on-resonance ($\sqrt{s} = M_Z$) longitudinal polarization asymmetries A_L for $e^-e^+ \to \mu^-\mu^+$, $c\bar{c}$, $b\bar{b}$. Lowest order and corrected curves with $\Delta E/E = 0.01$ are practically identical (from Ref. 3).

radiative corrections to A_L are expected to be small. The measurement of A_L should provide an accurate measure of $\sin^2\theta_W$, insensitive to theoretical uncertainties. Combining two independent measurements, A_L and M_Z, should provide a precise check of electroweak radiative corrections.

LOOKING BEYOND THE STANDARD MODEL

Longitudinal asymmetries are sensitive to the presence of heavy gauge bosons lying well above the standard model Z^0. The effects arise through the interference between the Z^0 and a heavier $Z^{0\prime}$, in analogy to the effects of γ-Z^0 interference at low energies. Precision measurements of A_L may be possible. Studies of the sensitivity to heavy Z^0's have been carried out at the SLC Workshop[5] and elsewhere.[6] Figure 10 is the result of the SLC Workshop, showing that polarization measurements are sensitive to Z^0's with masses up to 400 GeV. The conclusions depend on details of gauge models, of course, but in fact occur in similar fashion in several models studied.

Let me then conclude. Polarization phenomena play a significant role in high energy processes of the future because the weak

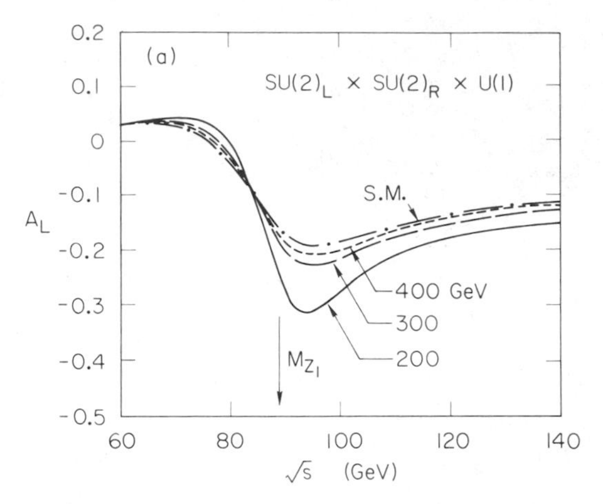

Fig. 10. Longitudinal asymmetry calculated in the standard model (S.M.) and in the $SU(2)_L \times SU(2)_R \times U(1)$ model. The numbers 200, 300 and 400 GeV refer to the mass of the second Z^o.

interactions become dominant and have an intrinsic spin structure that leads to large spin effects. Longitudinal polarization enhances many measurements such as total cross sections, charge asymmetries, and final state polarizations. Longitudinal asymmetries add additional possibilities for experiments not available to experiments without polarized beams. We can expect to see interesting work resulting from use of polarized beams at high energies.

REFERENCES

1. D. P. Barber, this conference. See also H. D. Bremer et al., DESY Report DESY-82/026 (1982).
2. M. Davier, Rapporteur's talk, International Conference on High Energy Physics, Paris, France (1982); . Kim et al., Rev. Mod. Phys. 53:211 (1981).
3. M. Bohm and W. Hollik, Nucl. Phys. B204:45 (1982).
4. G. Passarino and M. Veltman, Nucl. Phys. B160:151 (1979); M. Consoli and M. Greco, Frascati Report LNF-82/21-P (1982); M. Greco et al., Nucl. Phys. B171:118 (1980); F. Berends et al., Nucl. Phys. B202:63 (1982).
5. Proceedings of the SLC Workshop Report, SLAC Report SLAC-246 (1982).
6. W. Hollik, Z. Phys. C8:149 (1981).

ELECTROWEAK EFFECTS AT E-P COLLIDERS

K.-H.Mess

Deutsches Elektronen-Synchrotron DESY

Hamburg, Germany

INTRODUCTION

Apart from very recent discoveries at the $p\bar{p}$ collider[1] much of the experimental foundation for our present picture of nature are results from deep inelastic neutrino-, muon- and electron-hadron scattering. Still a large number of questions, repeatedly addressed during this conference, remains open. To deepen the understanding, new experimental data are necessary - in particular information about the forces around 100 GeV and above. This of course requires new devices to measure lepton-lepton, quark-quark and deep inelastic scattering (lepton-quark scattering) at higher energies.

THE E-P COLLIDER PROJECTS

Several e-p colliders have been proposed. In table 1 a few of the relevant parameters[2] are summarized. Two of the five projects are dedicated for e-p collisions and hence higher in total costs. All add-on proposals have one interaction region only and a fairly loose timeschedule. Operation in the e-p mode is likely to be delayed. The initial part of the Japanese project TRISTAN, an e^+e^- collider, is approved. The final e-p version of TRISTAN will contain an electron ring of 25 GeV and a proton ring of 300 GeV yeilding a center-of-mass energy of $\sqrt{s}$ = 173 GeV.

DESY has proposed a dedicated electron-proton colliding beam facility, HERA, on a site adjoining the present one. HERA is designed to collide 820 GeV protons with 30 GeV electrons. This corresponds to $\sqrt{s}$ = 314 GeV. For both proposed machines a luminosity above 10^{31} cm^{-2} s^{-1} is expected.

Table 1.

	TRISTAN	HERA	FNAL	LEP	LEP	CBA
e Energy (GeV)	25	30	10	51.5	51.5	20
p Energy (GeV)	300	820	1000	300	300	400
s (GeV^2)	30000	98000	40000	60000	60000	32000
Luminosity $10^{31} cm^{-2} s^{-1}$	1.8	6	4	0.1	1.3	2
Interaction Regions	3	4	1	1	1	1
Tune shift (electron)	0.03	0.014	0.03	0.022	0.006	
Tune shift (proton)	0.005	0.003	0.0009	0.003	0.0005	
Earliest Turnon	1989	1989	postponed x+4 years	?	?	?
Remarks:	dedicated		no fixed target physics	fixed E_p		

It is clear that e-p machines open a new kinematical region well outside that available at present or planned fixed target machines and even beyond Q^2 values corresponding to the characteristic interaction mass squared.

The final state topology in deep inelastic electron-proton interactions is striking and easy to recognize. As indicated in Fig. 1b and 1c, the scattered lepton appears at a large angle with respect to the beam axis and the corresponding transverse momentum is balanced by the struck quark which fragments into a jet of hadrons appearing at large angles on the opposite side of the beam axis. The remainder of the proton give rise to a forward jet of hadrons focused along the proton beam axis with no net transverse momentum. Both, the lepton and the quark jet are in one plane which also contains the proton direction.

On the basis of topology it seems unlikely to confuse a deep inelastic electron-proton event with a background event, such that the accessible Q^2 range appears to be limited by rate and not by background. Note that the particles from the lepton vertex and from the quark vertex are kinematically well separated. In the standard model only single neutrinos or electrons are allowed at the lepton vertex. The observation of jets emerging from the lepton vertex is a unique signature of new physics.

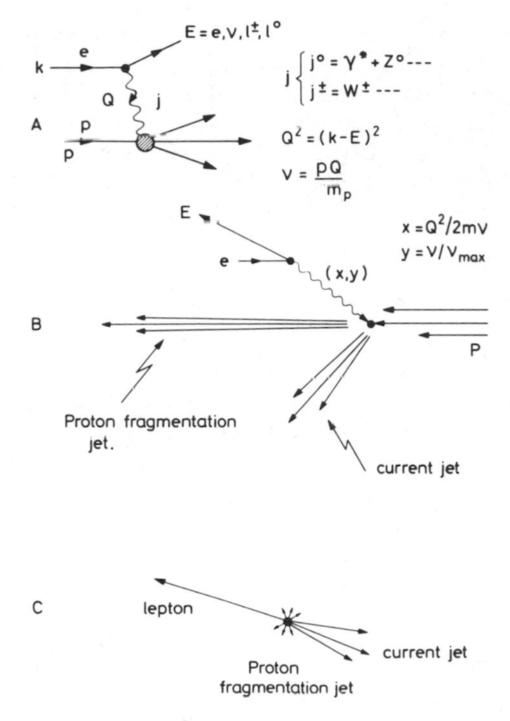

Fig. 1 - Topology of e-p interactions.

EXPLORING THE CURRENTS

Charged Currents

All present data are consistent with a left handed current which is mediated by a single charged vector boson with a mass around 80 GeV. However, the weak interactions have been probed only at Q values which are small compared to this mass scale. The apparent simplicity of the charged current might well reflect the fact that only the static limit has been studied so far. A rich structure with many vector bosons, some perhaps giving rise to right handed currents, might appear at high energies.

Compared to present fixed target experiments, e-p colliders have a number of unique features: a high center of mass energy (equivalent to a monoenergetic neutrino beam with an energy of up to 52 TeV in the case of HERA), the choice of helicity, the massless target and the favourable kinematics.

The x, y distribution of charged current events in bins of $dxdy = (0.2)^2$ expected after 500 h of data taking with an 30 GeV beam of lefthanded electrons colliding with a proton beam of 820 GeV is shown in Fig.2a. The rates were estimated using the standard model[3] with M_W = 80 GeV and formfactors parametrized according to Glück, Hoffmann and Reya[4], and assuming a luminosity of 10^{32} cm^{-2} s^{-1}. Fig.2b shows the same x, y distribution under the same assumptions for an electron energy of 25 GeV and proton energy of 300 GeV (TRISTAN). Note the different Q^2-scale.

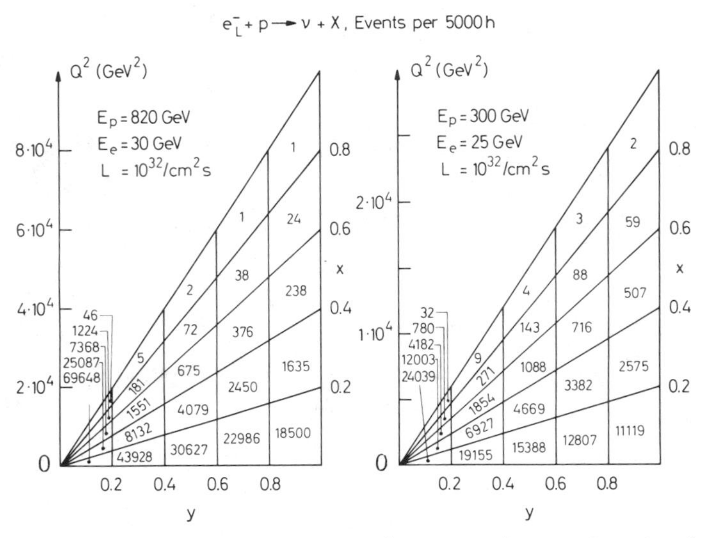

Fig. 2 - Number of charged current interactions of polarized electrons collected in 5000 h in bins of x and y.

A second pair of W would enhance the cross section. The enhancement over the standard case assuming equal couplings and a mass of the normal W of 78 GeV is shown in Fig. 3 as a function of Q^2 for various masses of the second W.

Note that for HERA about 2700 events per 5000 h at $L = 10^{32}$ cm^{-2} s^{-1} are expected in the interval $20000 \leq Q^2 \leq 30000$.

Righthanded currents do not occur in the standard model. Sensitive searches for these currents may be carried out using righthanded electrons and lefthanded positrons. Such measurements will reveal the existence of righthanded currents even if the mass of the propagator is 600 GeV provided the longitudinal beam polarization is at least 80% and known to an accuracy of 1%. This mass limit is valid even if the electron is partnered with a massive neutrino in a righthanded doublet.

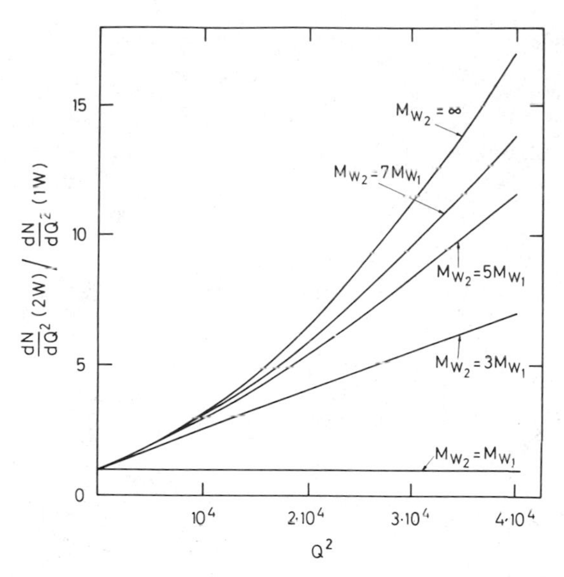

Fig. 3 - The ratio of the cross section for charged current reactions for a model with two W bosons as compared to the standard model.

Neutral Currents

One-photon exchange and Z^o exchange contribute coherently to $e + p \rightarrow e' + x$. Both contributions are of similar strength at high Q^2. Therefore measurements of this process can help to decide whether the electromagnetic and weak interactions are manifestations of a single force and this unification occurs as conjectured in the standard model of more complicated mechanisms involving several neutral vector bosons are realized in nature. The number of neutral current events produced by lefthanded electrons per nominal year in a bin dxdy = $(0.2)^2$ using the same assumptions as for Fig. 2 are plotted in Fig. 4.

The presence of the weak current gives rise to parity violation, apparent C-violation and a $(1-(1-y)^2)$ term.

The two-photon exchange, which also gives rise to a charge asymmetry, has a different Q^2 behaviour and is much smaller at large Q^2. The $1-(1-y)^2$ term cannot come from photon exchange.

The size of these effects in the standard model is shown in Fig. 5 where the ratio $\sigma(\gamma + Z^o)/\sigma(\gamma)$ evaluated for left- and righthanded electrons and positrons is plotted as a solid line versus y for x = 0.25.

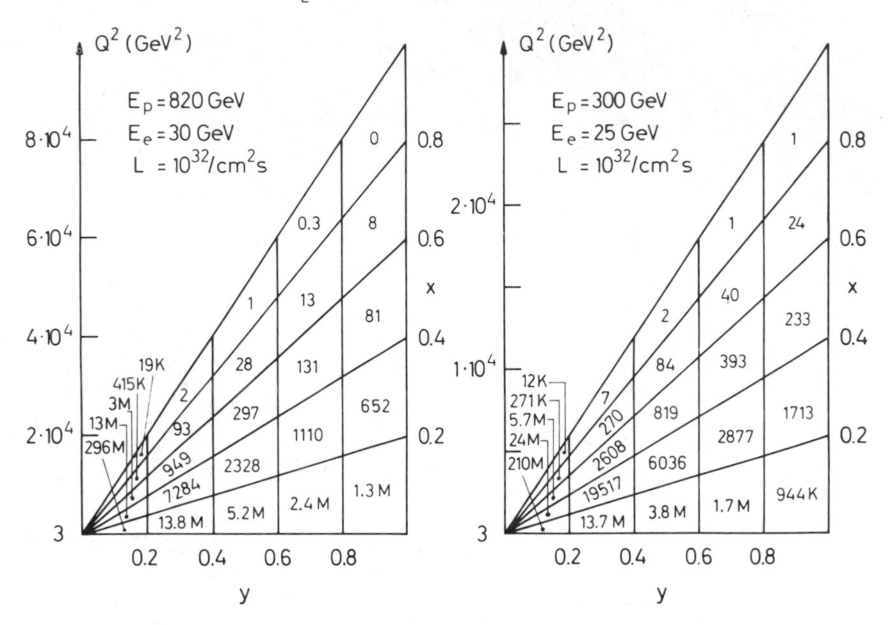

Fig. 4 - Number of neutral current interactions of lefthanded electrons collected in 5000 h in bins of x and y.

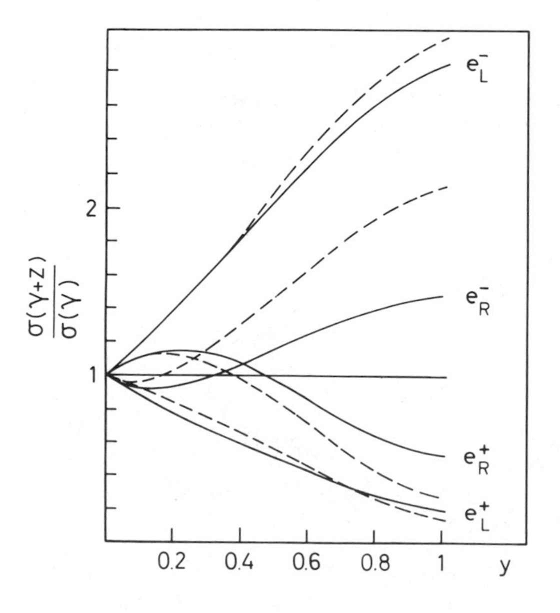

Fig. 5

The ratio $d^2\sigma/dxdy(\gamma+Z^o)/d^2\sigma/dxdy(\gamma)$ at x = 0.25 and s = 9.6 x 10^4 GeV for different weak interaction models

(Buras-Gaemers QCD parametrization).

The dashed curves serve as an example for a model with both righthanded and lefthanded currents. A measurement of the cross section for e_L^-, e_R^-, e_L^+ and e_R^+ allows to determine the Z mass to ± 5 GeV and $\sin^2\theta_W$ to ± 0.01. Note that these measurements are complementary to measurements at SLC or LEP. An additional Z^o could be sensed up to masses of 400 GeV.

The appearance of falvour changing neutral currents like $e^- + d \rightarrow \tau^- + b$ would be another deviation from the standard scenario. In this case the decay products of the τ would be measured instead of single electrons; such spectacular events would be easy to observe.

EXPLORING THE PROTON

The combination of high luminosity and the large kinematical region makes e-p machines particularly well suited to explore the anticipated phenomena and to search for new ones.

Form factors

The resolving power of a spacelike electroweak current increases with Q^2 such that the proton may be probed at shorter and shorter distances with increasing values of Q^2. The Q^2 evolution of the form factors (and maybe at some time even the moments of the form factors themselves) can be computed in Quantum Chromodynamics[5]. Using neutral and charged currents and polarized electrons and positrons the contents of the proton can be determined unambigously.

There might be deviations from the slow log Q^2 behaviour of the structure functions. As an example, quarks and leptons may have finitie size or these particles may be composite[6], made of new building blocks. With HERA the fermion structure can be probed down to 10^{-19} m. If the leptons have a size, the cross section would be modified by a form factor, $F(Q^2) = 1/(1+Q^2/M^2)$, giving rise to scaling violations which are very different from that expected in QCD. A 10% measurement at 4×10^4 GeV^2 would be sensitive to a mass of the order of 1 TeV.

An excited lepton could decay into $e + \gamma$, $e + Z^o$ or $e + W$ leading to peaks in the invariant mass spectrum. Note that the topology of such a final state with several particles emerging from the lepton vertex makes it easy to find.

If the quark has a structure, the cross section would be modified in a similar manner - i.e. also the quark structure is probed down to distances of $(1\ TeV)^{-1}$. In this case the form factors may increase or decrease depending on the charges and the weak coupling constants of the new constituents.

New Fermions

Electron-proton collisions are ideally suited to produce electronlike charged or neutral leptons and new heavy quarks which couple to the u or d quarks in the proton. Such couplings are known to be rather weak in the standard model: however, new currents may exist. Indeed, if the basic fermions are not pointlike, they must have excited states which couple to the ground state. The rate for producing a heavy quark from a light quark is plotted in Fig. 6 with the mass of the outgoing lepton as a parameter. The rates were evaluated with the additional assumption that the new current couples with the same strength as the old one.

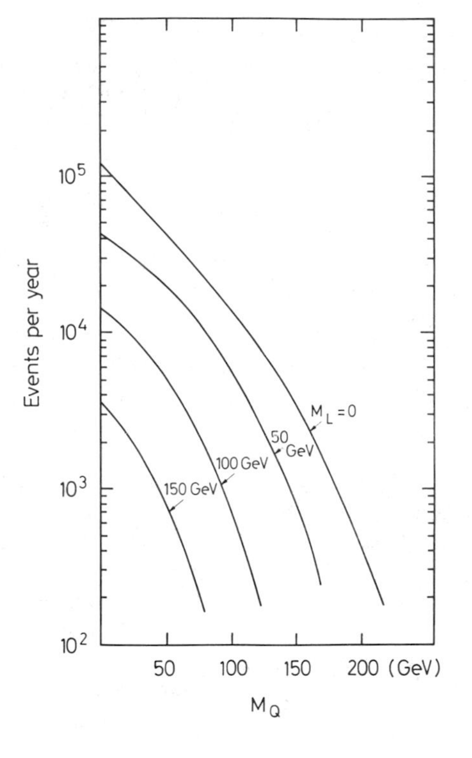

Fig. 6

Rate of heavy quarks as a function of the quark and lepton mass (5000 h).

In this way leptons and quarks with masses as high as 150 - 200 GeV can be found. The decays of these particles lead to rather spectacular signatures on the lepton side: $L^o \rightarrow e^- Q q'$ - i.e. the single lepton emerging from the lepton side in the standard model will be replaced by a high multiplicity jet containing leptons.

Leptoquarks

It is generally accepted that the gauge symmetry must be spontaneously broken to give mass to the intermediate vector bosons and to make the theory renormalizable. As an alternative to the standard Higgs mechanism it has been proposed that the symmetry breaking arises dynamically from the gauge interactions themselves. In these models a new set of unbroken non Abelian gauge interactions with a mass scale as low as 1 TeV is introduced. This "technicolour" interaction[7] gives rise to a complicated spectrum of "technicolourless" bound states with masses starting around 1 TeV. In addition, the technicolour interaction will result in leptoquarks, fundamental particles with combined lepton and baryon numbers and a mass predicted around 160 GeV.

The cross section resulting from the Feynman graphs in Fig.7a, is plotted[8] in Fig.7b versus the mass of the leptoquark. Roughly 200 events per year are expected for a leptoquark mass of 160 GeV.

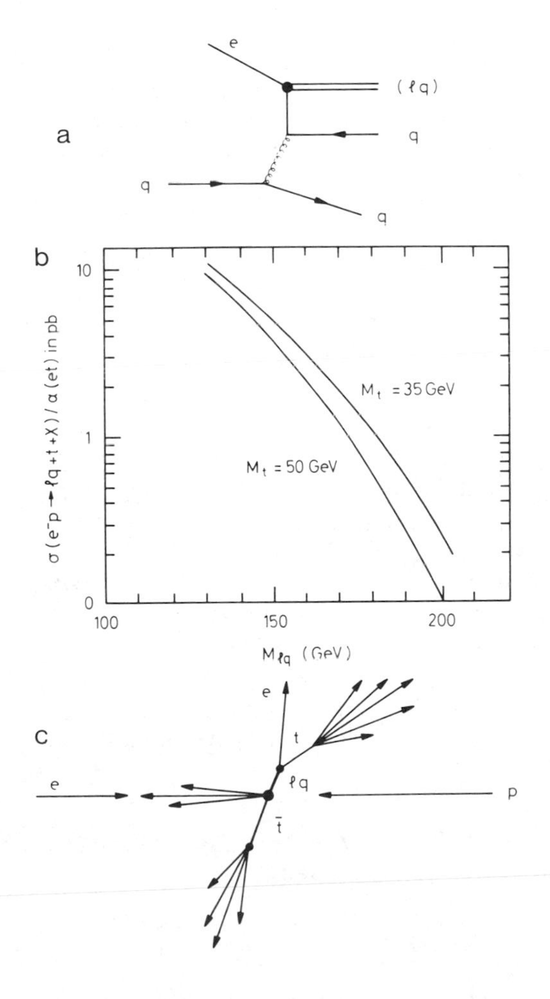

Fig. 7

a) The Feynman graph for the leptoquark production.

b) The cross section for $e\ p \rightarrow \ell_q + t + x$.

c) The final state topology.

Leptoproduction of Supersymmetric Particles

Supersymmetric theories have been addressed by several speakers at this conference. Supersymmetry can generate fermions from bosons and vice versa. Thus, for every particle with spin J there will be in principle two new particles with spin $\pm$ 1/2. The details depend on the treatment of the Higgs sector and whether gravitation is included (supergravity)[9].

Present limits allow the mass of the (scalar) quarkinos and leptinos to be as low as 20 GeV.

Electron-proton collisions at high energies are well suited to search for supersymmetric particles, because scalar quark and scalar leptons can be produced[10] directly

$$e + p \rightarrow \tilde{E} + \tilde{Q}$$

$$e + p \rightarrow \tilde{N} + \tilde{Q}$$

The observable mass range for supersymmetric particles depends on the signature. Cross sections down to 10^{-38} cm^{-2}, corresponding to 20 events a year, may be observable if the scalar leptons decay into a jet of particles. For dominant decay modes of the type

$$\tilde{L} \rightarrow \ell \, \tilde{G} \quad \text{and} \quad Q \rightarrow q \, \tilde{G}$$

(where $\tilde{G}$ is the undetectable Goldstino) the background will be more severe. The background is probably prohibitive for charged current events, in spite of the rather large cross section as high as 10% of the ordinary charged current cross section (assuming masses of 40 GeV).

The neutral current events, with at least two measurable jets, are more promising despite the very severe background.

Fig. 8 shows the neutral current cross section[10] for the production of 40 GeV electrinos and quarkinos as a function of s. The contribution from Goldstino exchange is neglected. The structure functions are taken from Reference 4. Note that 10 pb corresponds to roughly 20 000 events per "year". For a centre-of-mass energy squared of 10^5 GeV^2 a sizeable effect would be expected.

Both supersymmetric scalar particles, the $\tilde{E}$ and the $\tilde{Q}$, will presumably decay isotropically into a (light) ordinary quark and a (light?) Goldstino or Higgsino $\tilde{H}$. The latter would probably be invisible. These events have therefore three jets in the final state which are uncorrelated.

To estimate the signal to background ratio for neutral current reactions, a Monte Carlo simulation is necessary.

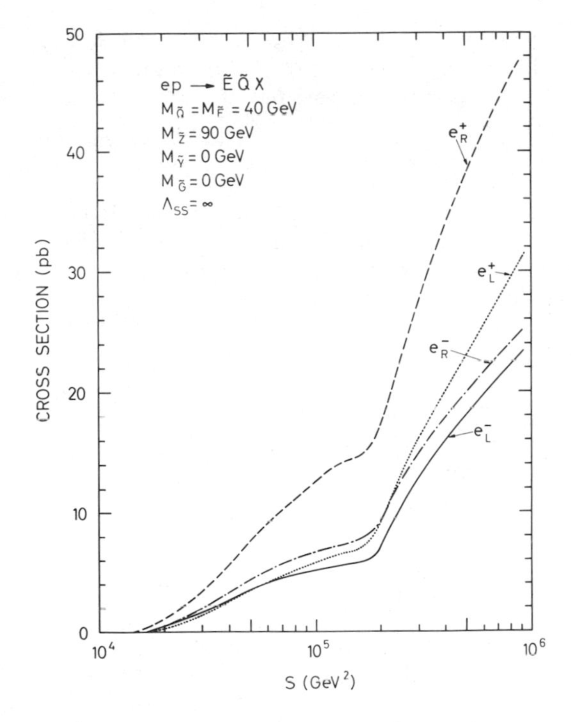

Fig. 8

Neutral current cross section for the production of 40 GeV electrinos and quarkinos as a function of s.

Supersymmetric and ordinary neutral current events were simulated according to the cross sections (with $M_{\tilde{E}} = M_{\tilde{Q}} = 40$ GeV, $M_{\tilde{Z}} = M_Z = 90$ GeV, Goldstino contribution neglected.) The supersymmetric particles decay into two light particles. The detector was assumed to be a calorimeter of the CHARM[11] type with the corresponding resolution. The distance from the interaction point to the first layer of the calorimeter was taken as 1 m. Clearly, better detectors could be invisaged.

Ordinary neutral current events, which are many orders of magnitude more abundant, can be analyzed by reconstructing the kinematics from the lepton and the quark jets separately. The corresponding x, y, Q^2 and ν agree within the experimental uncertainties. Likewise, the angle between the lepton-proton plane and the quark-jet-proton plane is approximately zero. SUSY events, if reconstructed from the lepton, assuming massless quarks and leptons, show an enhancement at high visible y (and small x) or equivalently at high visible ν. On the other hand, normal NC events are mainly due to one-photon-exchange at low ν and therefore a cut for $\nu_{vis} > 10\,000$ GeV ($y_{vis} > 0.2$) reduces the background.

The reconstruction of y from the quark yields an lower limit on the true y. Hence, the difference $y_{quark} - y_{lepton}$ will mostly be negative for SUSY events while it is centered around zero for norman events. Furthermore, the azimuthal engle ϕ between final state lepton and the quark jet is centered around π for normal events and widely spread for SUSY events.

Fig. 9 shows the SUSY events in a scatter plot as a function of the difference in y and the azimuthal angle. The size of the dots is proportional to the cross section. In the corresponding plot for normal events no events come near to $\Delta y < -0.3$ and $2.9 < \phi < 3.4$. Cutting out this rectangular area - as done in Fig. 9 - removes all normal events but leaves 88% of all SUSY events.

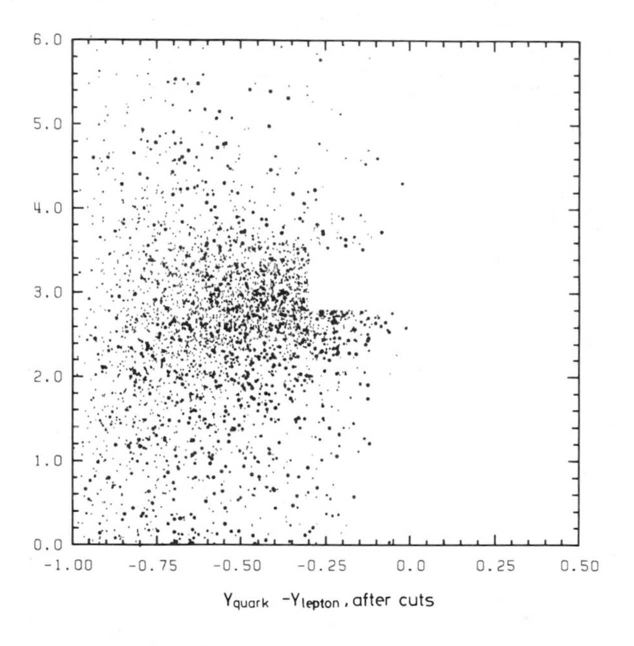

Fig. 9

Scatter plot of supersymmetric events as a function of the difference in y and ϕ (in radian) from the lepton and quark.

THE PROJECT HERA

General Description

The layout of HERA is shown in Fig.10. The machine has a fourfold symmetry; four 360 m long straight sections are joined by four arcs with a geometric radius of 779.2 m yielding a total circumference of 6336 m. HERA consists of two rings, one for electrons (positrons), the other for protons. The rings cross in the middle of the long straight sections. The rings will be buried some 10 - 20 m below the surface to avoid any disturbance of the suburban surroundings. The tunnel traverses largely land belonging either to the Federal Government or to the City of Hamburg, in particular a recreational area. It intersects the PETRA ring some

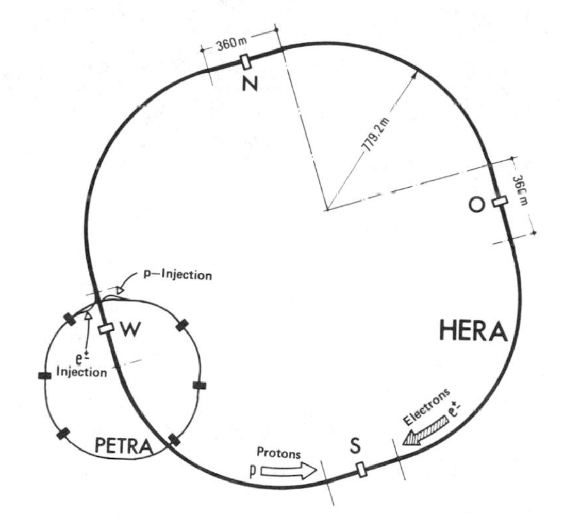

Fig. 10

The layout of HERA.

20 m below the surface. The physical plant can thus be located on the DESY site and only short injection paths are needed to connect PETRA and HERA.

A total floorspace of 875 m^2 is available for each experiment compared to the 650m^2 available in a typical PETRA hall. The beam traverses the halls 4.5 m above the floor level. Only a small access shaft and a loading ramp will be above the surface to keep the suburban and recreational areas as little disturbed as possible.

All legal questions related to the site have been settled with the passing of a law by the Parliament in Hamburg and a construction permit can be issued.

The Parameters and Performance

The general parameters of HERA are listed in Table 2. The energy of the electron beam can be varied between 10 and 35 GeV where the upper limit is determined by the available r.f. power of 13.2 MW and the lower limit by the damping time. At the nominal energy of 30 GeV the transverse polarisation builds up in 19.5 min compared to an expected lifetime of several hours.

The maximum induction of the proton ring magnets is chosen to be 4.53 Tesla. This yields a maximum proton energy of 820 GeV. The lower limit on the proton energy for long term storage is determined by the effect of persistent currents in the superconducting coils. The relative importance of these currents decreases with energy as they cause constant higher multipole fields disturbing the dipole field. An estimate of these effects based on the FNAL magnets, shows that it is possible to inject protons as 40 GeV

Table 2 - Basic parameters of HERA

	p-ring		e-ring	units
Nominal energy	820		30	GeV
$s = Q^2_{max}$		98400		GeV^2
Luminosity		$0.6x10^{32}$		$cm^{-2}s^{-1}$
Polarization time			20	min
Number of interaction points		4		
Length of straight sections		360		m
Free space for experiments		15		m
Circumference		6336		m
Bending radius	603.8		540.9	m
Magnetic field	4.53		0.1849	T
Total number of particles	$6.3x10^{13}$		$0.76x10^{13}$	
Circulating current	480		58	mA
Energy range	200→820		10→35	GeV
Emittance $(\varepsilon_x/\varepsilon_z)$	0.47/0.24		1.6/0.16	$10^{-8}m$
Beta function β^*_x/β^*_z	3.0.3		3/0.15	m
Dispersion function D^*_x/D^*_z	0/0		0/0	m
Beam-beam tune shift $\Delta Q_x/\Delta Q_z$	0.0006/0,0009		0.008/0.014	
Beam size at crossing σ^*_x	$0.12(0.91)^{**}$		0.22	mm
Beam size at crossing σ^*_z	0.027		0.013	mm
Number of bunches		210		
Bunch length	9.5		0.93	cm
RF frequency	208.189		499.667	MHz
Maximum circumferential voltage	100^{***}		290	MV
Total RF power	4-6		13.2	MW
Filling time	20		15	min
Injection energy	40.0		14.0	GeV
Energy loss / turn	$1.4x10^{-10}$		142.3	MeV
Critical energy	10^{-6}		111	keV

---

* At the interaction point
** Including the bunch length
*** 25 MV is foreseen initially corresponding to 1 - 1.5 MW.

and store them down to energies of about 250 GeV.

The proton r.f. system is designed to produce a circumferential voltage of 25 MV at a frequency of 208.2 MHz. The HERA r.f. system could be similar to the system being built for the SPS in its rôle as an injector for LEP.

For the electron r.f. system the frequency of 500 MHz, adopted for the other DESY machines, has been chosen. The system employs 192 cavities and a total r.f. power of 13.2 MW, sufficient to reach an electron energy of up to 35 GeV.

The Injection

The injection system is capable of filling the electron ring of HERA with 210 bunches of 14 GeV electrons (positrons) with a maximum intensity of 1.3×10^{11} particles per bunch in 15 min (25 min). The proposed injection system is based on existing accelerators.

The proton injection scheme is, to a large extent, based on existing accelerators. Protons from a new 50 MeV linear accelerator are injected into the DESY synchrotron, accelerated to 7.5 GeV and transferred to PETRA where they are accelerated to 40 GeV, the maximum possible energy, and injected into HERA.

The Interaction Regions

The interaction region in an electron-proton colliding ring is rather complex. Firstly, it must bring the two beams with rather different properties into collision. Secondly, as described during this conference[12], the arcs must be matched into the long straight sections such that the dispersion is suppressed. Furthermore the spin of the electron must be turned by $\pm \pi/2$ to become parallel or antiparallel to the beam direction in the interaction point and be restored to the transverse direction upon reentering the arcs. Minimizing the depolarization caused by the spin rotator introduces further constraints.

The beams cross in the horizontal plane of the electron ring at an angle of ± 10 mrad in the middle of the long straight sections. The large crossing angle chosen makes it possible to design the machines without common elements such that the energies can be varied independently. A free distance of ± 7.5 m around the interaction point is available for experiments. In this design the spin is turned into the longitudinal direction by an 80 m long rotator installed at the end of the arc and restored to the transverse direction by a similar rotator positioned at the entrance to the next arc. The "HERA-LEP" spin rotator has been described in detail in the talk of Prof. Buon during this conference[12].

As reported at this conference[12] transversely polarized electrons and positrons have been observed at PETRA in the colliding beam mode at beam tune shifts similar to those foreseen for HERA.

The rest of the electron ring of HERA is similar to the PETRA machine except that the circumference is larger by a factor of 2.75. Although the PETRA components in principle could be used directly, some changes, based on PETRA experience, are made to simplify the design and to reduce the cost.

The proton ring of HERA will be constructed using superconducting magnets. The dipole magnets developed at DESY are rather similar to the FNAL magnets and have a warm return yoke iron and a 75 mm cold bore. Three 1 m long superconducting magnets of a similar design but with a 100 mm warm bore have been built and tested at DESY during the past year. The magnets reached inductions in excess of 5.2 Tesla with little training. The field quality was within the tolerances and reproducible. Similar results were recently obtained with two 1 m long magnets with a 75 mm bore.

During 1983 several 1 m long coils and four full sized magnets will be built. These will be followed by a series of 12 prototype magnets in 1984.

The quadrupole magnets are being built at Saclay. Two prototypes will be delivered in 1983. The first has already been measured and shows the expected behaviour.

The correction quadrupole and sextupole magnets are wound directly on the dipole beam pipe. The dipole steering magnets are incorporated into the quadrupole cryostat. The correction system is developed in collaboration with NIKHEF and industry in The Netherlands.

REFERENCES

1) G.Salvini, this conference.
2) G.A.Voss, DESY HERA 81/18 C1
3) S.L.Glashow, Nuclear Phys. 22, 579 (1961)
   S.Weinberg, Phys.Rev.Lett. 19, 1264 (1967)
   A.Salam, Elementary Particle Theory, ed.N.Svartholm, Almquist and Wiksell, Stockholm, 1968 - p. 367.
4) M.Glück et al., Zeit-Phys. C13 (2) 119 (1982)
5) H.Fritzch et al., Phys.Lett. 47B, 365 (1973)
   D.J.Gross, F.Wilczek, Phys.Rev.Lett.30, 1343 (1973)
   H.D.Politzer, Phys.Rev.Lett. 30, 1346 (1973)
   S.Weinberg, Phys.Rev.Lett. 31, 31 (1973)
6) See for example:
   M.E.Peskin, 1981 International Symposium on Lepton and Photon Interactions at High Energies, Bonn, 1981

7) J.Schwinger, Phys.Rev. 125, 397, 2286 (1973)
R. Jackiw, K.Johnson, Phys.Rev. D8, 2286 (1973)
J.M.Cornwall, R.E.Norton, Phys.Rev. D8, 3338 (1973)
M.A.Beg, A.Sirlin, Ann.Rev. Nucl.Sci.24, 379 (1974)
S.Weinberg, Phys.Rev. D13, 979 (1976) and
Phys.Rev. D19, 1277 (1979)
L.Susskind, Phys.Rev. D20, 2619 (1979)

8) S.Rudaz, J.Vermaseren, CERN Preprint TH-2961 (1981)

9) Y.A.Gal'fand, E.P.Likthman - JETP Letters B, 323 (1971)
P.V. Volkov, V.P.Akalov, Phys.Lett. 46B, 109 (1973)
J.Wess, B.Zumino, Nucl.Phys. B70, 39 (1974)

10) S.K.Jones, C.H.Llewellyn-Smith, Oxford Preprint 73/82

11) M.Jonker et al., NIM 200, 183 (1982)

12) B.Montague, contribution to this conference
D.Barber, contribution to this conference
J.Buon, contribution to this conference.

RADIATIVE CORRECTIONS IN ELECTRON-POSITRON COLLISIONS AT PRESENT AND FUTURE ENERGIES

F.A.Berends

Instituut-Lorentz
University of Leiden
Leiden, The Netherlands

ABSTRACT

The various ingredients, which make up a radiative correction calculation are discussed. Known results for electron-positron collisions at PETRA/PEP energies are reviewed. From this an extrapolation to LEP/SLC energies is made, for which results on mupair production are presented.

INTRODUCTION

This lecture reviews what is known about radiative corrections in e^+e^- collisions of PETRA/PEP energies and extrapolates from there to LEP/SLC energies, for which not yet all the required calculations have been done.

Once one knows how to do the radiative corrections for a specific experiment, it usually becomes just one of the many steps in the analysis of the data and therefore often disappears in the final presentation of the physics results of the experiment. This does not mean that the problem can be discarded. On one hand, a part of the theory is involved:higher order diagrams. As long as this theory is QED one is inclined to take its validity for granted. For the electroweak theory the correctness of higher order corrections still has to be established, which is in fact one of the aims of future precision experiments. On the other hand, from the practical point of view radiative corrections cannot be neglected in accurate experiments where one wants to detect certain small effects like the asymmetry in mupair production at current energies.

Before starting a detailed account let me give two examples [1] of radiative effects. The first one is a measure of the maximum first order radiative correction which may occur in mupair production. The total lowest order cross section is given by

$$\sigma^o = 4\alpha^2\pi/3s \tag{1.1}$$

and we write for the α^3 total cross section for mupair and radiative mupair production

$$\sigma'(e^+e^- \to \mu^+\mu^-, \mu^+\mu^-\gamma) = \sigma^o(1 + \delta_T). \tag{1.2}$$

The "correction" δ_T is very large e.g. for a beam energy E=10GeV it is around 0.5, whereas for E=100GeV it is about 1. This large radiative correction can in practice easily be decreased by imposing certain experimental selection criteria on the events which effectively decrease the phase space for bremsstrahlung. Nevertheless in cases where it is hard to impose stringent selection criteria like in the measurement of the total hadronic cross section the correction remains large.

The second example is concerned with a QCD prediction. One way to establish gluon emission is to measure the thrust distribution [2] in the reaction e^+e^- into hadrons. On the parton level the $q\bar{q}$ final state does not have a thrust distribution, but a $q\bar{q}g$ final state does have one. The latter thrust distribution is represented by the dashed line in fig 1. If a photon is emitted instead of a gluon the quarks

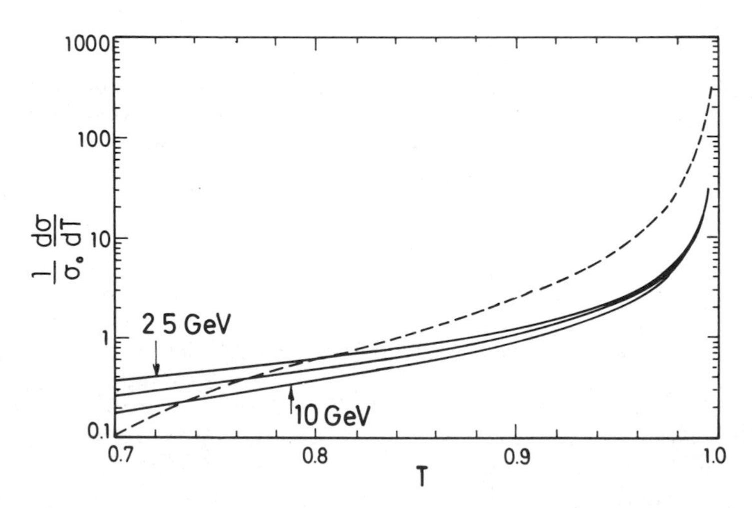

Fig 1. QCD and QED-faked thrust distributions.

get an apparent thrust distribution. Its size depends on the phase space one allows for the hard bremsstrahlung. Suppose one requires a $q\bar{q}$ pair to have at least an invariant mass of 7GeV then depending on the beam energy (10,15 or 25GeV) thrust distributions are faked (fig 1). The actual situation is more complex since both distributions will be smeared by the fragmentation of the partons. The radiative corrections should be build in directly in the hadron production Monte Carlo program.

This review is organized as follows. In order to illustrate how various ingredients make up a radiative correction calculation we first consider the simplest process possible, that of mupair production in QED. Then the actual relative importance of the various contributions to a total radiative correction is discussed and moreover the status of the calculations for the most relevant reactions will be summarized. Finally we go back to the example of mupair production and consider the situation at LEP/SLC energies.

INGREDIENTS FOR A RADIATIVE CORRECTION CALCULATION

The radiative correction calculation will be illustrated for the reaction

$$e^+(p_+) + e^-(p_-) \rightarrow \mu^+(q_+) + \mu^-(q_-) \quad . \tag{2.1}$$

When this reaction is measured one obtains at the same time data of the reaction

$$e^+(p_+) + e^-(p_-) \rightarrow \mu^+(q_+) + \mu^-(q_-) + \gamma(k) \quad , \tag{2.2}$$

of which some will fake reaction (2.1). The amount of radiative events entering in the pure mupair sample depends on the experimental selection criteria. From this it is clear that the radiative correction will always depend on the specifics of the experiment. Nevertheless for every calculation the same ingredients are required: a bremsstrahlung cross section for reaction (2.2), virtual corrections to the lowest order diagram for reaction (2.1) and a technique to calculate numerically the bremsstrahlung cross section over the phase space allowed by the selection criteria. These points will be discussed successively.

Bremsstrahlung Cross Section

Restricting ourselves for the moment to pure QED, there are 4 diagrams which contribute to reaction (2.2). (cf fig 2). The cross sections for the respective reactions are given by (taking the ultra-relativistic limit m_e/E, $m_\mu/E \rightarrow 0$ where possible):

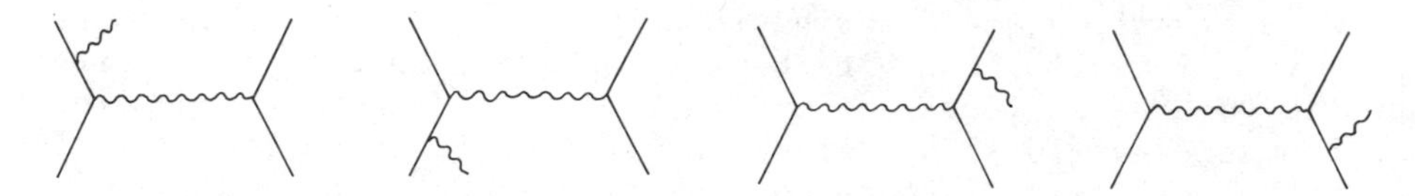

Fig 2. Bremsstrahlung diagrams for γ exchange.

$$\frac{d\sigma^o}{d\Omega_{\mu^+}} = \frac{1}{4s}\frac{1}{(4\pi)^2}\frac{1}{4}\Sigma|M^o|^2 = \frac{1}{4s}\frac{e^4}{(4\pi)^2}\,2\,\frac{(uu'+tt')}{s^2} \tag{2.3}$$

$$= \frac{\alpha^2}{4s}(1+c^2) \quad , \quad c = \cos\theta \quad , \tag{2.4}$$

$$\frac{d\sigma^B}{d\Omega_{\mu^+}d\Omega_\gamma dk} = \frac{|\vec{q}_+|k}{2p_{+o}-k+k\cos\theta_\gamma}\frac{1}{8\pi^2 s}\frac{1}{(4\pi)^3}\frac{1}{4}\Sigma|M'|^2 \quad . \tag{2.5}$$

In eqs (2.3) and (2.5) the solid angles of the μ^+ and γ are introduced. Also the photon energy k is introduced and two angles θ and θ_γ being respectively the angles between $\vec{q}_+$ and $\vec{p}$ or $\vec{q}_+$ and $\vec{k}$. Furthermore we use the quantities

$$s = (p_+ + p_-)^2 \quad , \quad t = (p_+ - q_+)^2 \quad , \quad u = (p_+ - q_-)^2 ,$$
$$s' = (q_+ + q_-)^2 \quad , \quad t' = (p_- - q_-)^2 \quad , \quad u' = (p_- - q_+)^2 . \tag{2.6}$$

In the soft photon limit (k << E) the expression for $\Sigma|M'|^2$ is well known: it consists of the lowest order matrix element squared times an infrared factor:

$$\frac{1}{4}\Sigma|M'|^2 = -e^2\left(\frac{p_+}{p_+k} - \frac{p_-}{p_-k} + \frac{q_-}{q_-k} - \frac{q_+}{q_+k}\right)^2\frac{1}{4}\Sigma|M^o|^2 , \tag{2.7}$$

leading to a factorization of the bremsstrahlung cross section in the soft photon limit

$$\frac{d\sigma^S}{d\Omega_{\mu^+}d\Omega_\gamma dk} = \frac{-\alpha k}{4\pi^2}\left(\frac{p_+}{p_+k} - \frac{p_-}{p_-k} + \frac{q_-}{q_-k} - \frac{q_+}{q_+k}\right)^2\frac{d\sigma^o}{d\Omega} \quad . \tag{2.8}$$

It now turns out that the full bremsstrahlung matrix element squared in the ultrarelativistic limit is a generalization of (2.7):

$$\frac{1}{4}\sum_{spins}|M'|^2 = -2e^6\left\{\frac{m_e^2}{s'^2}\left[\frac{t^2+u^2}{(p_-k)^2} + \frac{t'^2+u'^2}{(p_+k)^2}\right]\right.$$

$$+\frac{m_\mu^2}{s^2}\left[\frac{t^2+u'^2}{(q_-k)^2}+\frac{t'^2+u^2}{(q_+k)^2}\right]\Bigg\}+e^4\,S\,\frac{t^2+t'^2+u^2+u'^2}{ss'}\;, \tag{2.9}$$

where

$$S = e^2\left[\frac{s}{(p_+k)(p_-k)}+\frac{s'}{(q_+k)(q_-k)}-\frac{t}{(p_+k)(q_+k)}-\frac{t'}{(p_-k)(q_-k)}\right.$$

$$\left.+\frac{u}{(p_+k)(q_-k)}+\frac{u'}{(p_-k)(q_+k)}\right]\;. \tag{2.10}$$

In the pure massless case we would only have the last term in (2.9) which is again the infrared factor (but now with $s \neq s'$ etc) times a generalized lowest order matrix element squared. (cf eq(2.3)). This phenomenon has been established [3] for single bremsstrahlung QED and QCD reactions. These formulae are simpler than previously calculated expressions and are very transparent as to their peaking structure. Selecting the various propagators one easily divides the expression (2.9) into three terms, initial, final state radiation and their interference.

Integrating (2.5) analytically over a part of the phase space will in general be very involved. In special cases it is possible e.g. for soft isotropic photon emission:

$$\int_0^{k_m}\frac{d\sigma^s}{d\Omega_{\mu^+}d\Omega_\gamma\,dk}\,d\Omega_\gamma\,dk = \delta_s\,\frac{d\sigma^o}{d\Omega_{\mu^+}}\;, \tag{2.11}$$

where

$$\delta_s = \frac{2\alpha}{\pi}\Bigg\{\left[\frac{-\pi^2}{6}+\frac{1}{2}\ln\frac{s}{m_e^2}+\frac{1}{4}\ln^2\frac{s}{m_c^2}-\left(1-\ln\frac{s}{m_e^2}\right)\ln\frac{2k_m}{\lambda}\right]$$

$$+\,(m_e \leftrightarrow m_\mu) + 4\ln\left(\tan\frac{\theta}{2}\right)\ln\frac{2k_m}{\lambda}+2\ln^2\left(\sin\frac{\theta}{2}\right)$$

$$-\,2\ln^2\left(\cos\frac{\theta}{2}\right)-Li_2\left(\sin^2\frac{\theta}{2}\right)+Li_2\left(\cos^2\frac{\theta}{2}\right)\Bigg\}\;. \tag{2.12}$$

In this formula Li_2 denotes a dilogarithm and λ is a small fictitious photon mass introduced to regularize the infrared divergent integral (2.11).

Virtual Corrections

These arise from the diagrams in fig 3 which give in combination with the lowest order diagram the cross section

$$\frac{d\sigma^v}{d\Omega}=\frac{d\sigma^o}{d\Omega}\left(1+\delta_v\right)\;. \tag{2.13}$$

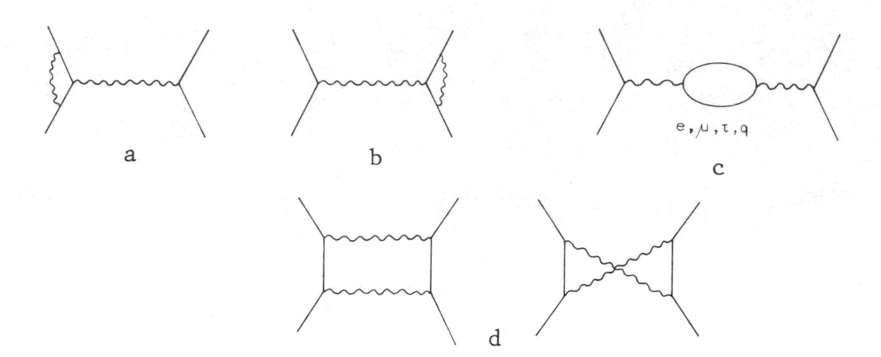

Fig 3. Virtual corrections for γ exchange.

Thus δ_v consists of contributions from the vertex correction (fig 3a,b) the vacuum polarization (fig 3c) and the box diagrams (fig 3d). Explicitly we have

$$\delta_v = \delta^e_{vc} + \delta^\mu_{vc} + \delta^e_{vp} + \delta^\mu_{vp} + \delta^\tau_{vp} + \delta^h_{vp} + \delta_{\gamma\gamma} \tag{2.14}$$

where

$$\delta_{vc} = \frac{2\alpha}{\pi} \left[(\ln \frac{s}{m^2} - 1) \ln \frac{\lambda}{m} - \frac{1}{4} \ln^2 \frac{s}{m^2} + \frac{3}{4} \ln \frac{s}{m^2} + \frac{\pi^2}{3} - 1 \right] , \tag{2.15}$$

$$\delta_{vp} = - 2\mathrm{Re}\Pi(s,m^2) = - \frac{2\alpha}{\pi} (- \frac{1}{3} \ln \frac{s}{m^2} + \frac{5}{9}) , \tag{2.16}$$

$$\delta^h_{vp} = - 2\mathrm{Re}\Pi_h(s) , \tag{2.17}$$

$$\delta_{\gamma\gamma} = \frac{2\alpha}{\pi} \Big\{ - 4\ln(\tan \theta/2) \ln \frac{2E}{\lambda} - \frac{2}{1+c^2} \Big[c(\ln^2 \sin \frac{\theta}{2} + \ln^2 \cos \frac{\theta}{2}) + \sin^2 \frac{\theta}{2} \ln(\cos \frac{\theta}{2}) - \cos^2 \frac{\theta}{2} \ln(\sin \frac{\theta}{2}) \Big] \Big\} . \tag{2.18}$$

The hadronic vacuum polarization is due to quark loops. It can be calculated by means of a dispersion integral, using the known cross section $\sigma(s)$ for $e^+e^- \to$ hadrons and extrapolating it to infinity with the present known value of R (for a numerical evaluation see ref 4):

$$\mathrm{Re}\Pi(s) = \frac{s}{4\pi^2\alpha} P \int_{4m_\pi^2}^{\infty} \frac{\sigma(s')}{s'-s} ds' . \tag{2.19}$$

The Total Correction For "Standard" Cuts

For two body reactions like the one under consideration it has become standard to apply selection of the data on the basis of two criteria

i) The muons should be back to back within a certain maximum acollinearity angle ζ (typically 10°-20°)

ii) The muons should have a sizeable fraction of the beam energy i.e. $q_{\pm o} > E_{th}$, where the threshold energy is typically 0.5E

These two constraints specify the phase space for the bremsstrahlung reaction (2.2). A numerical evaluation of the cross section (2.5) has to be performed over this phase space and to this the expression for δ_V is added. Special care has to be taken with the cancellation of the infrared singularities and the accurate numerical intergration over the huge peaking structures in the cross section. Standard programs have been around for many years (cf ref 5). The result is expressed in terms of a total correction δ_T:

$$\frac{d\sigma}{d\Omega} = \frac{d\sigma^{o}}{d\Omega}(1+\delta_{V}) + \int \frac{d\sigma^{B}}{d\Omega_{\mu+}d\Omega_{\gamma}dk} d\Omega_{\gamma}dk = \frac{d\sigma^{o}}{d\Omega}\left[1+\delta_{T}(E,\theta,\zeta,E_{th})\right] \tag{2.20}$$

Although these numerical calculations are useful, they are obtained for the standard cuts. A more flexible approach will be discussed next.

Arbitrary Experimental Cuts: A Monte Carlo Approach

In the Monte Carlo approach one simulates reactions (2.2) and (2.1) i.e. one constructs a large set of momenta

$$\{q_{+}, q_{-}, k\} \tag{2.21}$$

and when the photon energy is below some prescribed value k_o (e.g. 0.01E) one produces for all practical purposes in fact a set of two momenta

$$\{q_{+}, q_{-}\} \tag{2.22}$$

The sets should be constructed in such a way that the ratio between the numbers in set (2.21) and (2.22) is the same as the ratio between the total soft cross section and the total hard bremsstrahlung cross section. Moreover the momenta distribution of the set (2.21) should be that of eq.(2.5) and that of set (2.22) should follow the distribution

$$\frac{d\sigma}{d\Omega} = \frac{d\sigma^{o}}{d\Omega}\left[1+\delta_{V}+\delta_{S}(k_{o})\right] \quad . \tag{2.23}$$

One also knows the size of the total cross section σ_T corresponding to the generated set of N events. Once a set of Monte Carlo generated events is available it can be used for various purposes:

i) By imposing the cuts of the experiment under consideration one is left with N_s selected events which correspond to a cross section $N_s\sigma/N$ and therefore the radiative correction is known. One does not depend anymore on specific selection criteria as in (2.20).

ii) Since the cuts are arbitrary one can also specifically select hard bremsstrahlung events. This is relevant for certain background calculations e.g. in the search for a μ^*

$$e^+e^- \rightarrow \mu^*\mu \rightarrow \mu^+\mu^-\gamma \quad . \tag{2.24}$$

iii) It turns out that one can use a simple algorithm to polarize the event sample. For instance, one may consider transverse beam polarization. The generated data set for unpolarized beams has an overall isotropic azimuthal distribution. In a reweighting procedure based on the theoretical expressions for the virtual corrections and bremsstrahlung cross sections for transversely polarized beams the istropic ϕ distribution of the sample can be transformed into the ϕ-anisotropic one, leaving the momenta otherwise the same. The method outlined in the previous subsection would in contrast require the numerical integration for every ϕ.

iv) The procedure is also useful to generate quark pair events, which subsequently in a second Monte Carlo step fragment into hadrons. The procedure is then a first step in some of the known hadron Monte Carlo programs [e.g. the Lund one, ref 6].

The detailed procedure for the Monte Carlo simulation of events depends on the process, since an efficient method should be based on the specific structure of the bremsstrahlung cross section. For mu-pairs it is discussed in refs [7,8,9 and 1].

RESULTS FOR SOME RADIATIVE CORRECTIONS

Mupair Production

First we discuss the total cross section (1.2)[1]. The integration over θ removes the asymmetric terms in the differential cross section i.e. the interference terms in the bremsstrahlung cross section and the $\delta_{\gamma\gamma}$ term in the virtual corrections. The correction δ_T (table 1) is made up of several vacuum polarization (δ_{vp}) contributions, an electron vertex contribution (δ_{vc}) which we combine with initial state bremsstrahlung (δ_{Bi}) and a muon vertex contribution combined with final state radiation (δ_{Bf}). Of the latter two corrections the initial state radiation is by far the dominant one. This is caused by the 1/s' propagator effect: by losing energy from the initial state the propagator decreases the cross section to a lesser extent.

Table 1. The correction δ_T(%) to the total mupair cross section and its various contributions, for two beam energies. The last column gives δ_T for taupair production.

E	δ^e_{vp}	δ^μ_{vp}	δ^τ_{vp}	δ^h_{vp}	$\delta^e_{vc}+\delta_{Bi}$	$\delta^\mu_{vc}+\delta_{Bf}$	δ_T μ's	δ_T τ's
10	3.0	1.4	0.5	4.3	44.3	0.2	53.7	27.2
50	3.5	1.9	1.0	6.6	68.7	0.2	81.9	51.3

For completeness δ_T is also given for τ-pair production. The difference is caused entirely by initial state radiation, s' cannot become as small as in mupair production.

The correction [10,5] to the angular distribution has the effect of causing a forward backward asymmetry. The quantity δ (cf 2.20) is shown in fig 4 for specific experimental selection criteria [1,5] ($E=15$GeV, $\zeta=10^o$, $E_{th}=0.5E$). Of course this correction should be accurately known when one wants to establish the effect of Z_o exchange, which causes an opposite asymmetry. Whereas final state radiation can be neglected for the total cross section, it contributes through its interference with initial state radiation to the asymmetry.

The Reaction $e^+e^- \rightarrow \gamma\gamma$

Again one can study the total cross section correction δ_T and corrections δ to the angular distributions. For δ_T one finds e.g. 19% and 23% for beam energies of 15 and 50GeV. Although the correction increases with energy its size is not as large as for mupair production

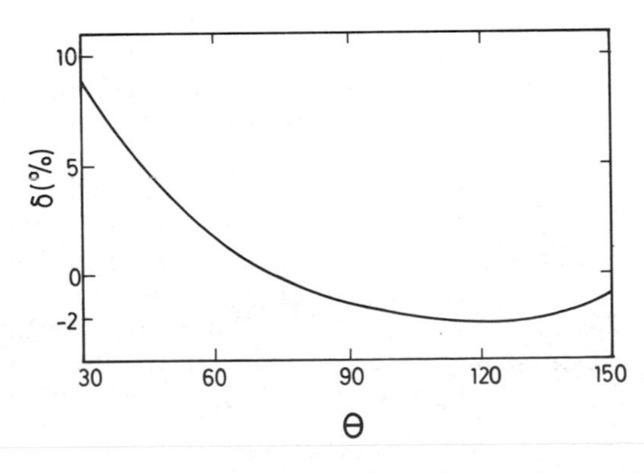

Fig 4. The total radiative correction δ to the differential QED mupair cross section.

Table 2. The correction $\delta(\%)$ to the differential cross section for $e^+e^- \to \gamma\gamma$ [cf eq.(2.20)].

θ	20	40	60	90
δ	-0.9	-6.0	-8.3	-9.4

since no propagator effect is present. Table 2 gives some typical values for the differential cross section ($E=15GeV$, $E_{th}=3GeV$, $\zeta=10^o$). For this the numerical integration method can be used [12] or the Monte Carlo method [11]. It is to be expected that this reaction remains at high energies a pure QED reaction, since only in higher order electroweak effects occur [13]. Therefore its study could be a useful luminosity check.

Bhabha Scattering

Here again, the two approaches for obtaining a numerical answer for the radiative corrections for the differential cross sections have been used [14,15]. The reaction serves two purposes. Its small angle scattering is commonly used to determine the luminosity of the machine. Its full angular dependence can be used to study QED. The electroweak effects at PETRA/LEP energies are not as marked as in mupair production. It is interesting to note that whereas in mupair production the hadronic vacuum polarization only contributes to the normalization of the cross section, here its contribution becomes angle dependent [4]. Omitting it would upset the present good agreement with QED. The sizes of radiative corrections are typically of the same order as the ones for mupair and the two photon annihilation reaction.

Hadronic Final States

For these reactions one is interested in the ratio R of total cross section to the mupair cross section and in specific jet structures. For the measurement of R the discussion of the total mupair cross section can be taken over. The initial state corrections, together with the vacuum polarization will give a good description of the radiative corrections. Typically the applied cuts to the data, which reduce the bremsstrahlung, still lead to radiative corrections of the order of 15-20%. This is large in view of the 5% effect one likes to measure from the QCD α_s/π radiative correction. (This is in fact the QCD analogy of the neglected QED final state correction $\delta^{\mu}_{VC}+\delta_{BF}$.)

Also for the jet studies it seems appropriate to use initial state corrections only. They have amongst others an effect on the determinated value of α_s.

Other Reactions

Of the other reactions, those connected with two photon physics are most important. Some calculations for the reaction $e^+e^- \to e^+e^-\mu^+\mu^-$ have been done [16]. More detailed studies are under way. [17].

EXTENSION TO LEP/SLC ENERGIES

In this section the situation in the Z_o region will be discussed. Two new aspects show up. One is the rapid change of the cross section due to Z_o exchange. The other one is the occurrence of weak virtual corrections. We review the status of the radiative corrections by discussing the various ingredients.

The bremsstrahlung cross sections needed for the calculation now also incorporate diagrams with Z_o exchange. These have been calculated in a reasonably compact form [18]. They are a generalization of eq.(2.9) and exist for mupair, quark pair production and Bhabha scattering.

The virtual corrections for the full electroweak theory consists of many contributions which makes the calculation very complicated. In an analytic form they do not yet exist, which prevents a straightforward Monte Carlo approach for the complete correction. For mupair production there exists a virtual correction calculation [19], where the Z_o width is neglected and which is therefore not applicable in the Z_o region. Recently also a calculation keeping the Z_o width has been reported [20]. Whether this calculation reproduces the results of ref 19 in those instances, where they can be compared is not known to me. For Bhabha scattering again a calculation neglecting the Z_o width exists[21].

The third ingredient, a numerical calculation incorporating bremsstrahlung and virtual corrections exists [7,8] for mupair production, only in the sense that the virtual corrections are approximated by a subset of diagrams. One can take the point of view [22] that one considers those virtual corrections which are infrared divergent and some vacuum polarization diagrams (figs 3 and 5). Combined with bremsstrahlung corrections (figs 2 and 6) this then gives an estimate of the total correction, which is expected to be determined to a large extent by hard bremsstrahlung effects in such a resonance region. In ref 7 this approximate virtual correction is compared to the results of ref 19 and found to be similar in size. (Therefore this approach is in particular believed to be reasonable at present PETRA/PEP energies). The advantage of taking this approximate virtual correction is that it is known analytically and can therefore be easily incorporated in a Monte Carlo simulation program. In a later stage this part of the program should be modified to incorporate the full virtual

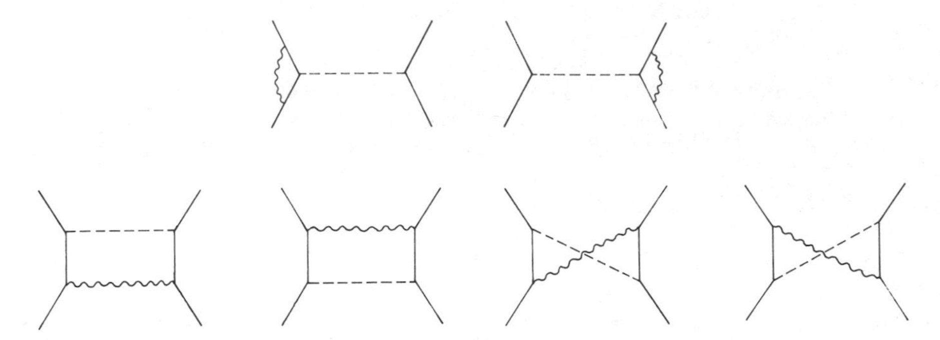

Fig 5. QED virtual corrections for Z exchange.

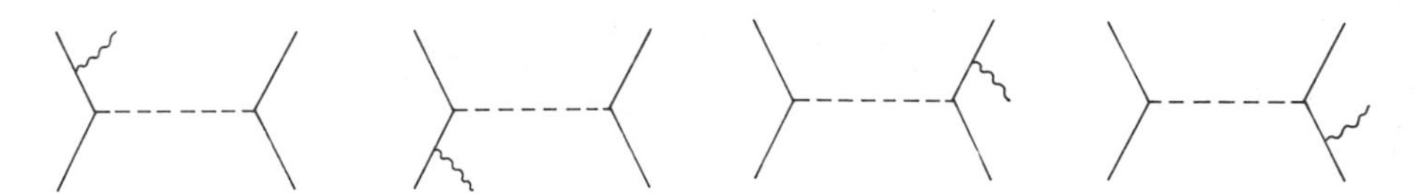

Fig 6. Bremsstrahlung diagrams for Z exchange.

corrections (i.e. the generation of set (2.2) will change). Some results of ref 7 will be shown below.

For completeness we note that for Bhabha scattering the situation is less complete. Although also here a subset of the virtual corrections have been calculated [23], they have not yet been combined into a full radiative correction calculation, including hard bremsstrahlung. In principle the techniques of the present Monte Carlo approach [15] could be used.

We now discuss some results of the total radiative corrections for mupair (or quarkpair) production using the approximate virtual correction calculation. We show the Z_0 excitation curve in fig 7 for the lowest order cross section (dashed curve) and for the first order corrected case with no cut on the photon energy (solid line). In fig 8 the integrated asymmetry is shown for the lowest order cross section (dashed curve), and the corrected ones with bremsstrahlung up to its maximum (solid line) and up to 0.2E (dotted-dashed line). The integrat asymmetry is defined as the ratio of the difference of the cross section in the forward and backward hemisphere and the total cross section. Other distributions can be easily obtained in the Monte Carlo approach and are shown in ref 7.

In conclusion, the Monte Carlo approach at present used at PETRA/ PEP energies can again be exploited at higher energies. Studies which

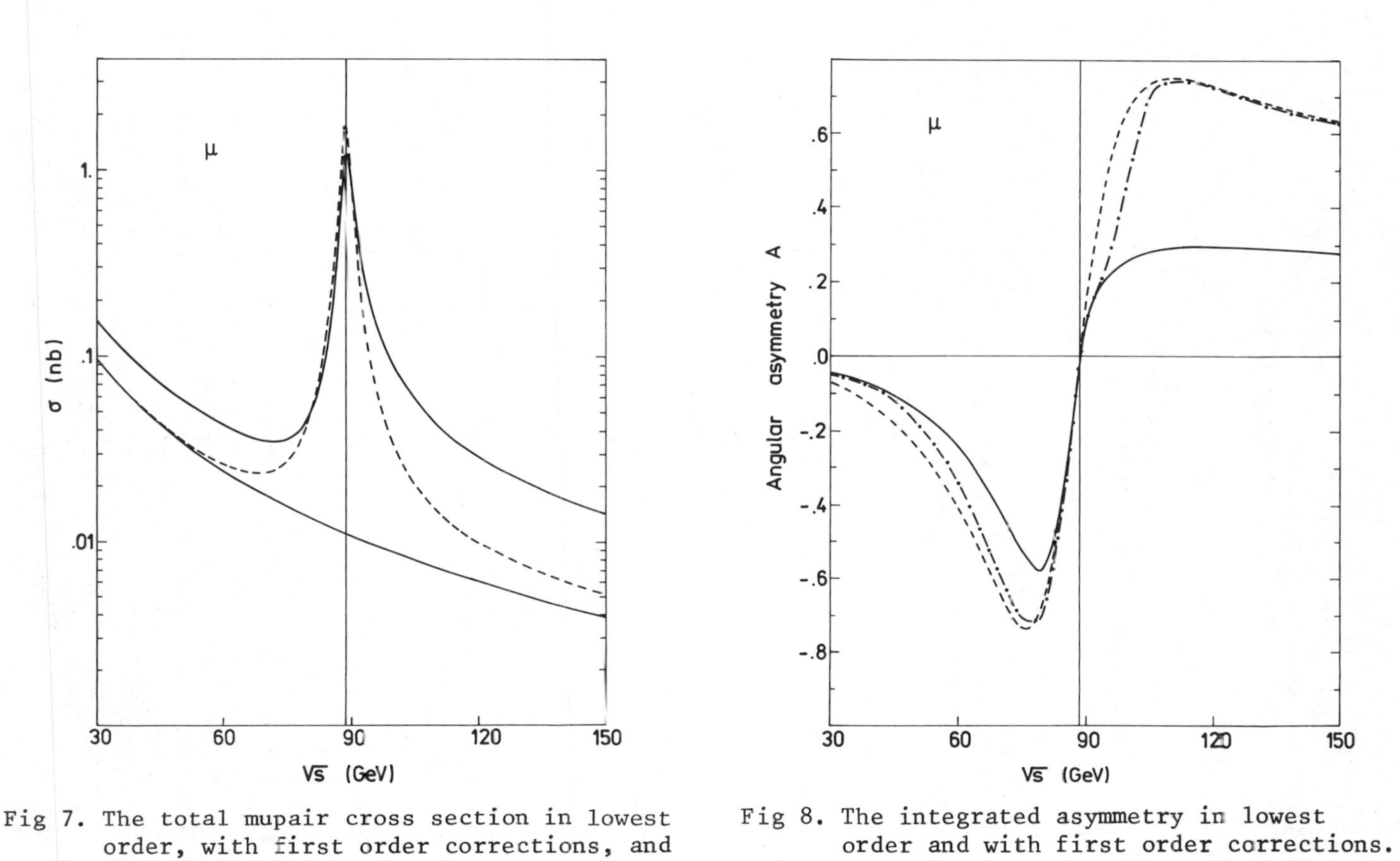

Fig 7. The total mupair cross section in lowest order, with first order corrections, and without Z exchange.

Fig 8. The integrated asymmetry in lowest order and with first order corrections.

aim at the gross features of the radiative effects in the Z_0 region could at present be carried out, using for the moment approximate analytical virtual corrections. For high precision tests the virtual corrections, including all the diagrams required by the electroweak theory, have still to be calculated for the Z_0 region.

ACKNOWLEDGEMENT

It is a pleasure to thank Professor Harvey Newman for the kind invitation to a very stimulating conference.

REFERENCES

1. F.A.Berends and R.Kleiss, Nucl.Phys. B177, 237 (1981).
2. A.de Rújula, J.Ellis, E.G.Floratos and M.K.Gaillard, Nucl.Phys. B138, 387 (1978).
3. F.A.Berends, R.Kleiss, P.de Causmaecker, R.Gastmans and T.T.Wu, Phys.Lett. B103, 124 (1981). See also the contribution of R.Gastmans to this volume.
4. F.A.Berends and G.J.Komen, Phys.Lett. 63B, 432 (1976).
5. F.A.Berends,K.J.F.Gaemers and R.Gastmans, Nucl.Phys. B57, 381 (1973), Nucl.Phys. B63, 381 (1973).
6. T.Sjöstrand, Lund preprint LU TP 82-7, june 1982.
7. F.A.Berends, R.Kleiss and S.Jadach, Nucl.Phys. B202, 63 (1982).
8. F.A.Berends, R.Kleiss and S.Jadach, Computer Physics Communications, in press.
9. R.Kleiss, thesis Leiden University (1982).
10. I.B.Khriplovich, Sov.J.Nucl.Phys. 17, 298 (1973); D.A.Dicus, Phys.Rev. D8, 890 (1973); R.W.Brown, V.K.Cung, K.O.Mikaelian and E.A.Paschos, Phys.Lett. 43B, 403 (1973).
11. F.A.Berends and R.Kleiss, Nucl.Phys. B186, 22 (1981).
12. F.A.Berends and R.Gastmans, Nucl.Phys. B61, 414 (1973).
13. M.Capdequi-Peyranere, G.Grunberg, F.M.Renard and M.Talon, Nucl.Phys. B149, 243 (1979).
14. F.A.Berends, K.J.F.Gaemers and R.Gastmans, Nucl.Phys. B68, 541 (1974).
15. F.A.Berends and R.Kleiss, to be published.
16. M.Defrine, S.Ong, J.Silva and C.Carimalo, Phys.Rev. D23, 663 (1981); M.Defrise, Zeitsch.f.Phys. C9, 41 (1981).
17. Contributions of P.H.W.M.Daverveldt et al. and of J.A.M.Vermaseren et al. to the 5th International Workshop on Photon Photon Collisions, Aachen 1983.
18. F.A.Berends, R.Kleiss, P.de Causmaecker, R.Gastmans, W.Troost and T.T.Wu, Nucl.Phys. B206, 61 (1982).
19. G.Passarino and M.Veltman, Nucl.Phys. B160, 151 (1979).
20. W.Wetzel, University of Heidelberg preprint HD-THEP-82-18.
21. M.Consoli, Nucl.Phys. B161, 208 (1979).
22. M.Greco, G.Pancheri-Srivastava and Y.Srivastava, Nucl.Phys. B171, 118 (1980). (E:B197, 543(1982)).
23. M.Consoli, S.Lopresti and M.Greco, Phys.Lett. 113B, 415 (1982).

RADIATIVE CORRECTIONS TO GAUGE BOSON MASSES

K.J.F. Gaemers

Institute for Theoretical Physics
University of Amsterdam

INTRODUCTION

There is by now abundant experimental support for the standard $SU(2) \otimes U(1)$ theory of electroweak interactions. The discovery of the Z-boson since this conference only strengthens this evidence. One of the reasons for the popularity of the standard model is the fact that it is a renormalisable theory, so that it is possible to calculate higher order effects in a consistent manner. It means on the other hand that there are parameters which have to be given as inputs. In practice already a few parameters are sufficient to be able to give predictions. If one works in lowest order, the fine-structure constant α, the Fermi-constant G_F and the Weinberg-angle $\sin^2\theta_W$ are sufficient inputs to predict (again to lowest order) the intermediate bosons through:

$$M_W^{(0)} = \frac{1}{\sin\theta_W}\left(\frac{\Pi\alpha}{G_F\sqrt{2}}\right)^{\frac{1}{2}}, \qquad M_Z^{(0)} = \frac{1}{\sin\theta_W\cos\theta_W}\left(\frac{\Pi\alpha}{G_F\sqrt{2}}\right)^{\frac{1}{2}} \tag{1}$$

One would like to test these relations as precisely as possible. The main difficulties here are the fact that $\sin^2\theta_W$ is at present not known very accurately and of course that the Z and W are not very well known [1].

Theoretically the relations in (1) are lowest order results. They will be changed by higher order effects. A discussion of these effects within the framework of the standardmodel will be the main subject of this talk.

CORRECTIONS (one-loop)

Higher-order corrections to the mass formulae (1) have been considered by several people[2)-6)]. As is argued in refs. [2)-3)] the main correction can be traced to self energy insertions into the gauge-boson propagators. The formulae in (1) express the masses of the W and Z in terms of measurable quantities. The finestructure constant α for instance can be "measured" from $e\mu$ elastic scattering at zero momentum transfer. In lowest order this reaction is just described by one photon exchange. Up to one loop we have to include a charge renormalisation due to the presence of fermion and boson loops. In the same way the Fermi constant G_F is measured in muon decay. The lowest order diagram consists of W exchange which also gets renormalized through loops of fermions and bosons. Finally the Weinberg angle can be measured from the ratio $\sigma(\nu_\mu e^-)/\sigma(\bar{\nu}_\mu e^-)$. In lowest order these processes are described by single Z-exchange. Fermion loops now introduce both a change in the Z-propagator as well as additional photon-Z mixing.

These three effects change the relations given in (1). Finally there is an effect because the loop insertions in the Z and W propagator shift the positions of the poles which define the Z and W masses.

If we write the vacuum polarisation tensor as:

$$\Pi^{\mu\nu}_{ij}(q) = \Pi_{1ij}(q^2)g^{\mu\nu} + \Pi_{2ij}(q^2)q^\mu q^\nu$$

$$i,j = W,Z,\gamma \tag{2}$$

The following expression is found for the mass-shifts of the W and Z.[2)-3)]:

$$\frac{\delta M_W^2}{M_W^2} = \left[\frac{\Pi_{1WW}(q^2)}{M_W^2} + \frac{\cos\theta_W}{\sin\theta_W}\frac{\Pi_{1\gamma Z}(q^2)}{q^2} + \frac{\Pi_{1\gamma\gamma}(q^2)}{q^2}\right]_{q^2=0} - \frac{\mathrm{Re}\,\Pi_{1WW}(M_W^2)}{M_W^2}$$

$$\tag{3}$$

$$\frac{\delta M_W^2}{M_2^2} = \left[\frac{\Pi_{1WW}(q^2)}{M_W^2} - \frac{(1-2\cos^2\theta_W)}{\cos\theta_W\sin\theta_W}\frac{\Pi_{1\gamma Z}(q^2)}{q^2} + \frac{\Pi_{1\gamma\gamma}(q^2)}{q^2}\right]_{q^2=0} - \frac{\mathrm{Re}\,\Pi_{1ZZ}(M_Z^2)}{M_Z^2}$$

If we consider only one-loop effects there are contributions from fermions (quarks and leptons) and from bosons (W,Z and Higgs) to $\Pi_1(q^2)$. It was found [3)] that the effects due to boson loops are such that $\delta M \simeq +100$ MeV. It was further found that this result depends only logarithmically on the mass of the Higgs-boson.

The mass-shift due to fermion loops turns out to be much larger. A calculation with charged leptons and neutrino's can be performed without complications. A problem arises however if one wants to calculate quark-loops. It can be seen from (3) that one has to calculate $\Pi_1(q^2)$ both at $q^2=M_W^2$, M_Z^2 and at $q^2=0$. Now for large q^2 one can use the fact that quarks are described by QCD so that lowest order perturbation theory will give a reasonable result. At $q^2=0$ however this argument cannot be used. A way out of this problem is to note that $\Pi_1(q^2)$ obeys a dispersion relation. The imaginary part of $\Pi_{1\gamma\gamma}(q^2)$ for instance is related to $\sigma(e^+e^- \to \text{hadrons})$. If one calculates this cross-section and hence $\Pi_{1\gamma\gamma}$ with free quarks, one can choose a quark mass-value such that Im $\Pi_{1\gamma\gamma}$ describes on average the experimental data. Quark masses of the order of 250 MeV have to be chosen for u and d in order that we get this average description. We have to assume that this argument is valid not only for $\Pi_{1\gamma\gamma}$ but also for the other Π_1 occurring in (2) and (3).

It should be noted here, that the full hadronic contribution to Im $\Pi_1(q^2)$ shows resonance behavior and that it is possible to use experimental data for a calculation of Π_1.[7]

With this in mind we can calculate the contributions to δM_W and δM_Z for each of the fermion doublets. The results are given in table 1.

These values have been calcutated with the following inputs: $M_u=M_d=.25$ GeV, $M_s=0.3$ GeV, $M_c=1.5$ GeV, $M_b=5$ GeV and $M_t=20$ GeV. For the Weinberg angle we have taken: $\sin^2\theta_W=.238$. With these values we find: $M_W^{(0)}=76.429$ GeV and $M_Z^{(0)}=87.555$ GeV. The total shifts are: $\delta M_W=2.998$ GeV and $M_Z=3.204$ GeV.[8] As was already mentioned in the introduction, there is an uncertainty in the values for $M_W^{(0)}$ and $M_Z^{(0)}$ due to $\sin^2\theta_W$ which is at present not known very accurately.

Table 1.
Mass shifts due to fermions (one loop)

	δM_W (GeV)	δM_Z (GeV)
e,ν	.689	.787
μ,ν	.358	.413
τ,ν	.213	.242
d,u	.912	.916
s,c	.677	.679
b,t	.150	.167

QCD-CORRECTIONS

It is clear from table 1 that the presence of quarks is responsible for a large mass-shift. One may wonder therefore whether QCD corrections will have an extra effect. It should be noted in this respect that the expressions in (3) are correct to all orders in the strong interaction.

The order α_s contributions have been calculated in [8)-9)]. It turned out that a calculation of the $\Pi_1(q^2)$ by means of the usual Feynman parameter techniques was not practical. Instead dispersion relations for Π_1 were used. The imaginary parts could be calculated using the method given in [10)]. The resulting expressions are very long and may be found in [8)-9)].

For the same quark mass values that were used in the one loop calculation, the order α_s mass-shifts are: δM_W=.166 GeV and δM_Z=.153 GeV. These should be added to the one-loop result.

CONCLUSIONS

It has been shown that higher order corrections shift the masses to the W and Z upward. A detailed comparison between these theoretical results and experimental masses will provide a strong test for the standard model. For these tests to be meaningfull it is necessary to have accurate mass determinations and a determination of $\sin^2\theta_W$ with as small an error as possible.

I like to thank T.H. Chang and W.L. v. Neerven for many discussions on the subject treated here. I also like to thank the organisers of this conference for having provided such a stimulating atmosphere.

REFERENCES

[1] At the time of the conference the Z-boson had not yet been discovered. There is evidence now. See e.g.: G. Arnison et al, CERN-EP/83-73.
[2] F. Antonelli, M. Consoli and G. Corbo, Phys. Lett. 91B (1980) 90.
[3] M. Veltman, Phys. Lett. 91B (1980) 95.
[4] A. Sirlin, Phys. Rev. D22 (1980) 971.
[5] W.J. Marciano, A. Sirlin, Phys. Rev. D22 (1980) 2695.
[6] F.M. Renard, Erice Workshop, "Physics at LEP", PM/80/3.
[7] W. Wetzel, Z. Phys. C11 (1981) 117.
[8] T.H. Chang, K.J.F. Gaemers and W.L. v. Neerven, Nucl. Physics B202 (1982) 407.
[9] T.H. Chang, K.J.F. Gaemers and W.L. v. Neerven, Phys. Lett. 108B (1982) 222.
[10] K. Schilder, M.D. Tran and N.F. Nasrallah, Nucl. Physics B181 (1981) 91 (E: B187 (1981) 594).

ELECTROWEAK INTERACTIONS ON TOPONIUM

L.M. Sehgal

III. Physikalisches Institut, Technische Hochschule

Aachen

ABSTRACT

A heavy $Q\bar{Q}$ resonance such as toponium can be a fascinating laboratory for the study of electroweak interactions. We describe the branching pattern of toponium decay, and discuss some special features connected with γ-Z interference, γ-Z-W interference, neutrino counting and single-quark decay.

1. INTRODUCTION

There is general anticipation that the sequence of vector mesons ρ, ω, ϕ, J/ψ, Υ ... will be augmented by at least one further narrow resonance, $V(t\bar{t})$ or toponium. Additional quarkonium states associated with further sequential flavours might also exist. One might even speculate about narrow resonances formed by unconventional quarks (e.g. colour non-triplets, quarks with non-standard weak couplings, or supersymmetric scalars.) The search for such systems will obviously continue to be an important preoccupation of e^+e^- physics at high energies.

Experiments at PETRA have set a lower limit of 38 GeV on the mass of toponium.[1] It is possible that a clue to the existence and mass of the t-quark will come from the ongoing $\bar{p}p$ collider experiment at CERN. Interest in toponium obviously stems from the fact that it would be a new arena for testing QCD; in particular the spectroscopy of this system would yield information about the shape of the $Q\bar{Q}$ potential at a separation $r \lesssim 0.1$ fm.[2] However, what gives this system an added dimension of interest is the fact that for a sufficiently heavy quarkonium state ($M(Q\bar{Q}) \gtrsim 50$ GeV) the decay properties are strongly influenced by the weak interactions. A heavy toponium resonance would be a veritable laboratory for the study of electroweak interactions. It is this aspect that we wish to focus on in this paper.

2. PROFILE OF QUARKONIUM DECAY

In the absence of weak interactions, all decays of a quarkonium resonance $V(Q\bar{Q})$ are supposed to occur through the annihilation of the $Q\bar{Q}$ pair, and are accordingly proportional to $|\psi(0)|^2$, the square of the wave-function at the origin. The dominant decay modes, and the associated widths are as follows:

a) $V(Q\bar{Q}) \to l^+l^-$ $(l=e,\mu,\tau)$

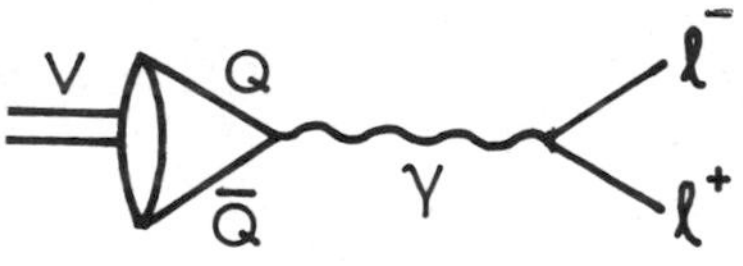

$$\Gamma(l\bar{l}) = 16\pi\alpha^2 \frac{|\psi(0)|^2}{M_V^2} e_Q^2$$

$$\equiv \Gamma_o e_Q^2 \qquad (1)$$

b) $V(Q\bar{Q}) \to q\bar{q}$ (q = light quark)

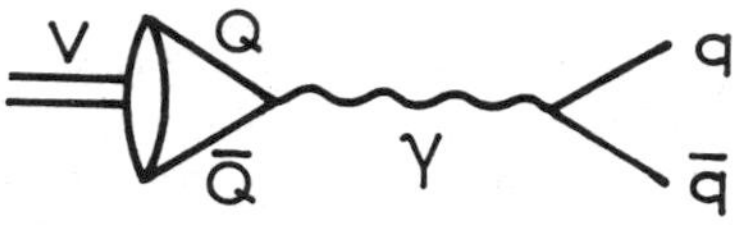

$$\Gamma(q\bar{q}) = 3e_q^2 e_Q^2 \Gamma_o \qquad (2)$$

c) $V(Q\bar{Q}) \to 3g$ (g = gluon)

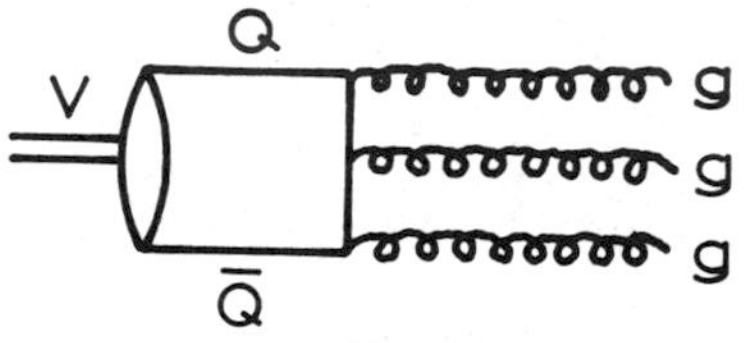

$$\Gamma(3g) = \frac{10(\pi^2-9)}{81\pi} \frac{\alpha_s^3}{\alpha^2} \Gamma_o \qquad (3)$$

d) $V(Q\bar{Q}) \to gg\gamma$

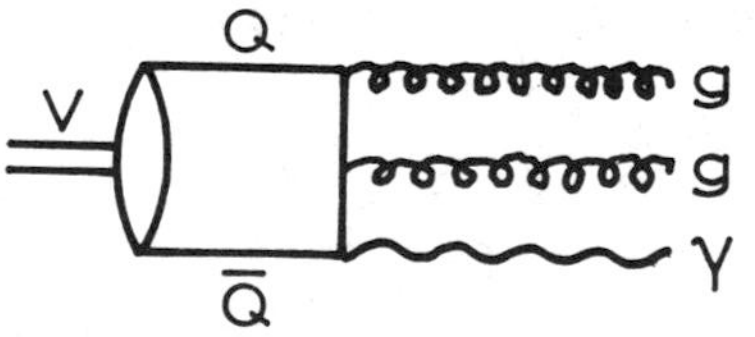

$$\Gamma(gg\gamma) = \frac{36}{5} e_Q^2 \frac{\alpha}{\alpha_s} \Gamma_{3g} \qquad (4)$$

As an illustration, assuming $\alpha_s = 0.12$, a value that may be appropriate to a toponium state with mass $M_V \simeq 50$ GeV, the above decay modes have a relative branching ratio

$$\Sigma l\bar{l} : \Sigma q\bar{q} : 3g : 2g\gamma = 1 : 1.2 : 0.8 : 0.15 \qquad (5)$$

The total width is governed by the scale Γ_o. It is a remarkable feature of the known quarkonium systems[3] that the width into $l\bar{l}$ appears to scale as e_Q^2, the proportionality constant Γ_0 having a universal value of about 11 keV. Assuming this scaling behaviour to extend to heavier $Q\bar{Q}$ states, one obtains for a 50 GeV toponium state a width of about 50 keV.

The inclusion of the weak interactions has a number of consequences.

a. Effect on branching ratios. The decays $V \to l\bar{l}$ and $V \to q\bar{q}$ receive corrections from Z-exchange which become increasingly important as M_V approaches M_Z. If the coupling of fermions to the Z-boson is parametrised as

$$\mathcal{L}_Z = \left(\frac{G_F M_Z^2}{2\sqrt{2}}\right)^{1/2} Z_\alpha \bar{f}\gamma^\alpha (v_f - a_f \gamma_5) f \qquad (6)$$

the amplitude for $V(Q\bar{Q}) \to f\bar{f}$, including γ- and Z-exchange is[4]

$$\mathcal{M} = \frac{M_V^2}{f_V} \varepsilon_\mu \bar{u}(p)\gamma^\mu (\lambda_f - \lambda_f' \gamma_5) v(\bar{p}) \qquad (7)$$

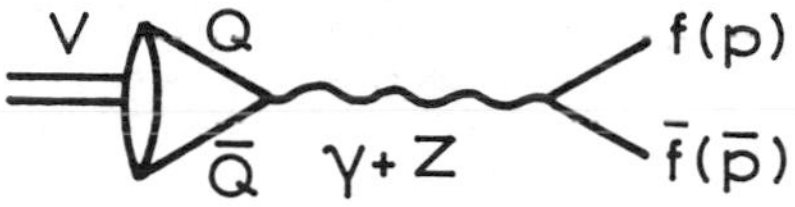

with

$$\lambda_f = \frac{e^2 e_f e_Q}{M_V^2} + \frac{GM_Z^2}{2\sqrt{2}} \frac{v_f v_Q}{M_V^2 - M_Z^2 + iM_Z\Gamma_Z}$$

$$\lambda_f' = \frac{GM_Z^2}{2\sqrt{2}} \cdot \frac{a_f v_Q}{M_V^2 - M_Z^2 + iM_Z\Gamma_Z} \qquad (8)$$

The coupling constants, in the standard model, are

$$v_f = \begin{bmatrix} 1 \\ -1 + 4x \\ 1 - \frac{8}{3}x \\ -1 + \frac{4}{3}x \end{bmatrix}, \quad a_f = \begin{bmatrix} 1 \\ -1 \\ 1 \\ -1 \end{bmatrix}, \quad e_f = \begin{bmatrix} 0 \\ -1 \\ 2/3 \\ -1/3 \end{bmatrix}, \quad \text{for } f = \begin{bmatrix} \nu_e \\ e^- \\ u \\ d \end{bmatrix} \tag{9}$$

where $x = \sin^2\Theta_W$. Thus the width for $V(Q\overline{Q}) \to l\overline{l}$ is

$$\Gamma(l\overline{l})\Big|_{\gamma+Z} = \Gamma_o \left[\left|-e_Q + Kv_e v_Q\right|^2 + \left|Ka_e v_Q\right|^2 \right] \tag{10}$$

while that into a quark-pair is

$$\Gamma(q\overline{q})\Big|_{\gamma+Z} = 3\Gamma_o \left[\left|e_Q e_q + Kv_q v_Q\right|^2 + \left|Ka_q v_Q\right|^2 \right] \tag{11}$$

where

$$K = \frac{GM_V^2}{2\sqrt{2}e^2} \cdot \frac{M_Z^2}{M_V^2 - M_Z^2 + iM_Z\Gamma_Z} \quad . \tag{12}$$

(The case $V(t\overline{t}) \to b+\overline{b}$ is special and is discussed in Sec. 4 below.) As shown in Fig. 1 (which is taken from Ref. 5) the effect of the weak corrections is to greatly enhance the branching ratio into $f\overline{f}$ pairs ($f\overline{f} = l\overline{l}$ or $q\overline{q}$) as compared to the 3-gluon mode, as M_V increases. (The width into 3 gluons may actually decline slightly, because of the asymptotic decrease of α_s). Thus a heavy toponium state will have a high probability of decaying into a lepton pair or a pair of quark jets.

b. New decay modes. The weak couplings of heavy quarks give rise to a number of new decay channels for a $V(Q\overline{Q})$ resonance. The neutral current coupling to the Z^o permits the decay mode

$$V(Q\overline{Q}) \to \nu_i\overline{\nu}_i \quad (i = e,\mu,\tau)$$

with a width

$$\Gamma(\nu_i\overline{\nu}_i) = 2|K|^2 v_Q^2 \Gamma_o \quad . \tag{13}$$

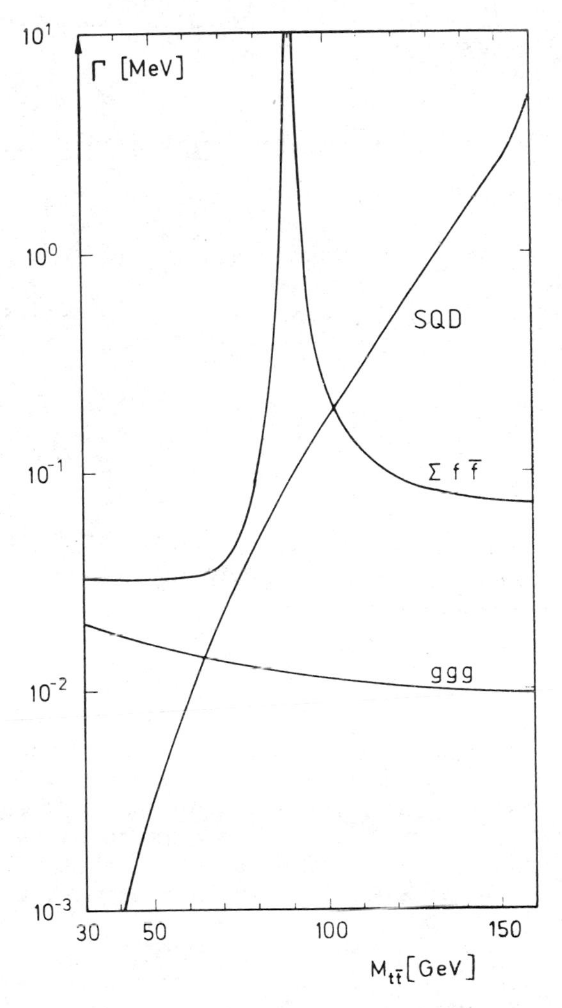

Fig. 1: Partial decay widths of toponium.

The Higgs coupling of the heavy quark Q makes possible the decay[6)]

$V(Q\bar{Q}) \to H^o\gamma$

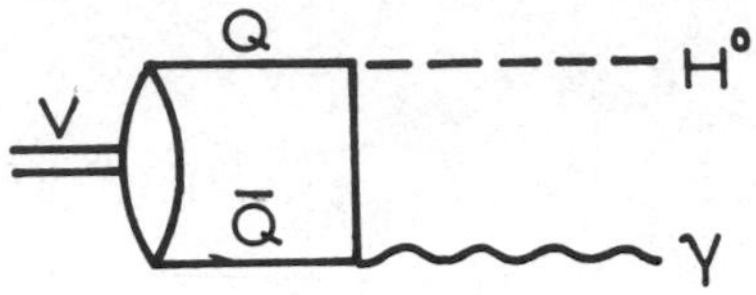

with a width

$$\Gamma(H^o\gamma) = \frac{G_F}{\sqrt{2}e^2} e_Q^2 (M_V^2 - M_H^2)\, \Gamma_o \quad . \tag{14}$$

Finally, in the case of toponium, the charged current coupling

$$\mathcal{L}_W = \left(\frac{G_F M_W^2}{\sqrt{2}}\right)^{1/2} U_{tb}\, \bar{t}\, \gamma_\mu (1-\gamma_5)\, b\, W^\mu \tag{15}$$

opens up a new class of decays not involving $Q\bar{Q}$ annihilation. These are based on the single-quark transitions

$t \to b l^+ \nu_l \qquad (l = e, \mu, \tau)$

or $\quad t \to b\, u\, \bar{d}, \quad t \to b\, c\, \bar{s}$

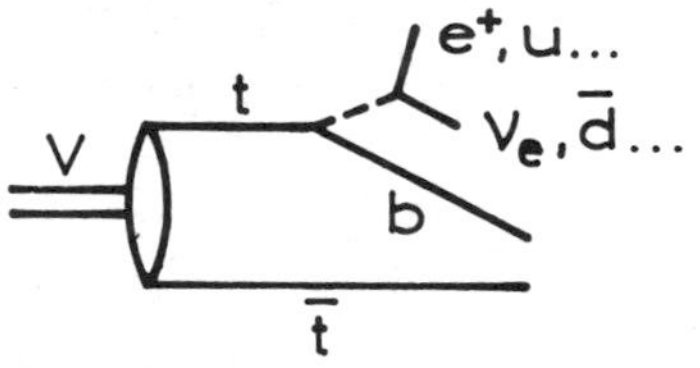

for which the total decay width is[5)]

$$\Gamma(SQD) = 18\, |U_{tb}|^2 \frac{G_F m_t^5}{192\,\pi^3} f\left(\frac{m_t^2}{M_W^2}, \frac{m_b^2}{m_t^2}\right) \tag{16}$$

$f(x,y)$ being a phase space factor, equal to unity when x and y are zero . Unlike the annihilation modes, which are inhibited by the factor $|\psi(0)|^2$, single-quark decays increase as m_t^5, and at large masses become comparable to decays into fermion-pairs or 3 gluons (Fig. 1). For a toponium state of mass $\gtrsim$ 100 GeV, single-quark decays dominate all other modes. We shall come back to these decays in Sec. 6.

c. Asymmetries. A final manifestation of weak effects in quarkonium is the appearance of asymmetries in the reactions $e^+e^- \to \mu^+\mu^-$ and $e^+e^- \to q\bar{q}$ on and around the resonance. These asymmetries arise from an interplay of weak and electromagnetic interactions, and are sensitive to the flavour (and colour) quantum numbers of the heavy quark. We study these asymmetries in the following two sections.

3. γ-Z INTERFERENCE

Let us examine the behaviour of the reaction $e^+e^- \to f\bar{f}$ in the neighbourhood of a $Q\bar{Q}$ resonance. The amplitude may be written as[4])

$$\mathcal{M} = \mathcal{M}_\gamma + \mathcal{M}_Z + \mathcal{M}_V \tag{17}$$

where $\mathcal{M}_\gamma + \mathcal{M}_Z$ is the amplitude off-resonance, and $\mathcal{M}_V$ the pole-contribution of the $Q\bar{Q}$ state. On top of the resonance, the piece $\mathcal{M}_V$ obviously dominates, and has the structure

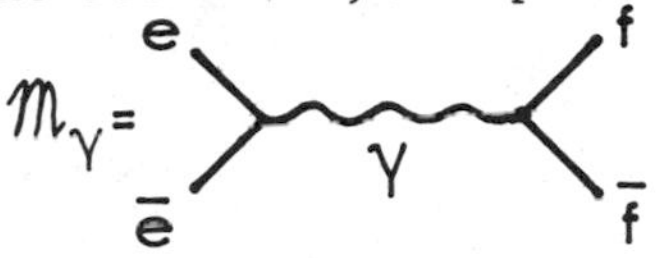

$$\mathcal{M}_V \propto \bar{u}(p)\gamma_\mu\ (\lambda_f - \lambda_f'\gamma_5)\ v(\bar{p}) \cdot \bar{v}(\bar{k})\gamma_\mu\ (\lambda_e - \lambda_e'\gamma_5)\ u(k) \tag{18}$$

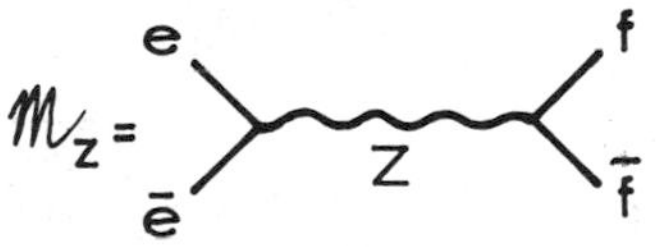

where the momenta are defined as

$$e^-(k) + e^+(\bar{k}) \to f(p) + \bar{f}(\bar{p})\ .$$

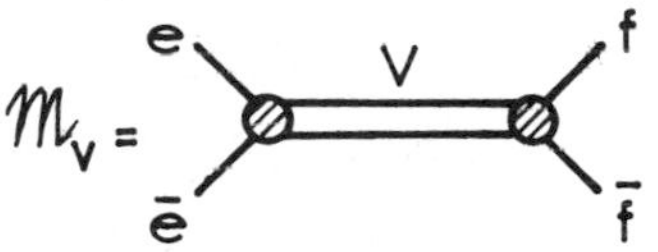

From the general expression for λ, λ' (Eq. (8)) it is clear that the behaviour on resonance depends on the quantum numbers e_Q and v_Q of the heavy quark. In this sense, this reaction serves as a probe of the flavour quantum numbers of Q.

From the expression (18) it follows that for any final state $f\bar{f}$, the cross section on resonance is different for left and right handed electrons, the asymmetry being

$$\alpha_{RL} = \frac{\sigma_R - \sigma_L}{\sigma_R + \sigma_L} = -\frac{(\lambda_e - \lambda_e')^2 - (\lambda_e + \lambda_e')^2}{(\lambda_e - \lambda_e')^2 + (\lambda_e + \lambda_e')^2}\ . \tag{19}$$

This is plotted in Fig. 2 as a function of M_V for both types of quarkonia, i.e. $e_Q = 2/3$ and $e_Q = -1/3$. Also shown is the right-left asymmetry to be expected in the continuum, for the reaction $e^+e^- \to \mu^+\mu^-$, arising from the amplitude $\mathcal{M}_\gamma + \mathcal{M}_Z$. One observes that the asymmetry on resonance is very different for $e_Q = 2/3$ and $e_Q = -1/3$, and also considerably different from that in the continuum.

A further effect resulting from weak-electromagnetic interference is a forward-backward asymmetry in the angular distribution of the reaction $e^+e^- \to f\bar{f}$. If this distribution is written as

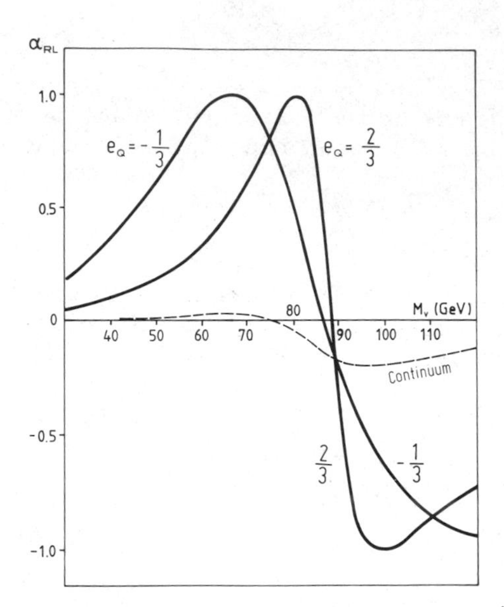

Fig. 2: Right-left asymmetry for the reaction $e^+e^- \rightarrow \mu^+\mu^-$ on top of a $Q\overline{Q}$ resonance and in the continuum.

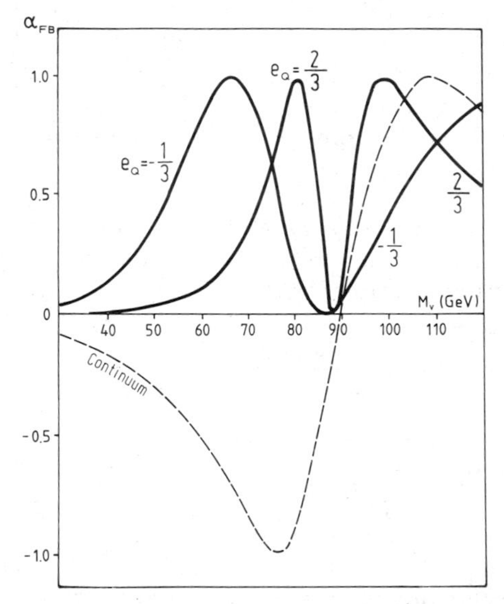

Fig. 3: Forward-backward asymmetry in $e^+e^- \rightarrow \mu^+\mu^-$ on top of a $Q\overline{Q}$ resonance and in the continuum.

$$\frac{d\sigma}{d\cos\Theta} = \text{constant}\left[(1+\cos^2\Theta) + \alpha_{FB} 2\cos\Theta\right] \tag{20}$$

the asymmetry parameter α_{FB} is

$$\alpha_{FB} = \frac{(\lambda_e+\lambda_e')^2(\lambda_f+\lambda_f')^2 + (\lambda_e-\lambda_e')^2(\lambda_f-\lambda_f')^2 - (\lambda_e+\lambda_e')(\lambda_f-\lambda_f') - (\lambda_e-\lambda_e')(\lambda_f+\lambda_f')}{[\quad '' \quad + \quad '' \quad + \quad '' \quad + \quad '' \quad]} \tag{21}$$

For the special case $e^+e^- \to \mu^+\mu^-$,

$$\alpha_{FB}^{\mu\bar{\mu}} = \frac{(\lambda_e+\lambda_e')^4 + (\lambda_e-\lambda_e')^4 - 2(\lambda_e+\lambda_e')^2(\lambda_e-\lambda_e')^2}{[\quad '' \quad + \quad '' \quad + \quad '' \quad]} \quad . \tag{22}$$

This asymmetry is plotted in Fig. 3 for $e_Q = 2/3$ und $e_Q = -1/3$, and contrasted with the asymmetry in the continuum outside the resonance.

It should be remarked that the width of a heavy $Q\bar{Q}$ resonance (typically 50 KeV) is very much smaller than the energy resolution of the e^+e^- beams (typically tens of MeV) that are likely to produce it. Thus the actual asymmetry observed at the peak of the resonance will be a weighted average of the asymmetries on- and off-resonance, i.e.

$$\alpha(\text{observed}) = \frac{\alpha(\text{off}) + \eta\alpha(\text{on})}{1+\eta} \tag{23}$$

where η is the signal-to-background ratio

$$\eta = \frac{R(\text{peak})}{R(\text{continuum})} \quad . \tag{24}$$

This ratio depends not only on the quantum numbers of the quark, but also on the shape and width of the electron beams, which in turn are functions of the beam energy. Some illustrative estimates of η are given in Refs. 7 and 8.

4. γ-Z-W INTERFERENCE

γ-Z interference effects of the type discussed above occur both in l^+l^- and $q\bar{q}$ final states. A special situation exists in the case of

$$e^+e^- \to V(t\bar{t}) \to b\bar{b} \tag{25}$$

which we describe now. By virtue of the charged current coupling (Eq. (15)), the transition $V(t\bar{t}) \to b\bar{b}$ receives an additional contribution from W-exchange in the t-channel, which interferes with the usual γ and Z exchanges.

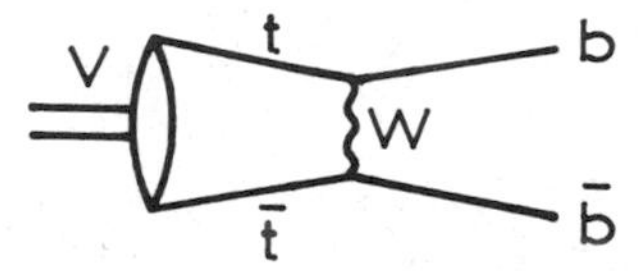

This phenomenon of "γ-Z-W interference"[4] has important repercussions. If the matrix element of the reaction (25) is written in the general form (18) (with f = b), the coefficients λ_b, λ_b' are[4]

$$\lambda_b = \frac{e^2 e_t e_b}{M_V^2} + \frac{GM_Z^2}{2\sqrt{2}} \frac{v_b v_t}{M_V^2 - M_Z^2 + iM_Z\Gamma_Z} - \frac{G}{\sqrt{2}} |U_{tb}|^2 \eta_W$$

$$\lambda_b' = \frac{GM_Z^2}{2\sqrt{2}} \frac{a_b v_t}{M_V^2 - M_Z^2 + iM_Z\Gamma_Z} - \frac{G}{\sqrt{2}} |U_{tb}|^2 \eta_W \tag{26}$$

where

$$\eta_W = \frac{1}{3} \left(1 + \frac{1}{8}\frac{M_V^2}{M_W^2}\right) \Bigg/ \left(1 + \frac{1}{4}\frac{M_V^2}{M_W^2}\right) \tag{27}$$

and $|U_{tb}|$ is, presumably, close to unity. The factor 1/3 in η_W reflects the existence of the colour quantum number: whereas Z- and γ-exchanges connect each component of the colour-singlet toponium wave-function $[|t_1\bar{t}_1\rangle + |t_2\bar{t}_2\rangle + |t_3\bar{t}_3\rangle]/\sqrt{3}$ to each component of the final $b\bar{b}$ state, the W-exchange graph connects only the components with matching colours. It follows that the weak-electromagnetic interference effects in this reaction not only involve simultaneously the electromagnetic, neutral current and charged current couplings of t and b quarks, but also probe the existence of the colour quantum number in a sensitive way. As an illustration of this, we show in Fig. 4 the branching ratio of $V(t\bar{t}) \to b + \bar{b}$ as a fraction of all two-jet events.

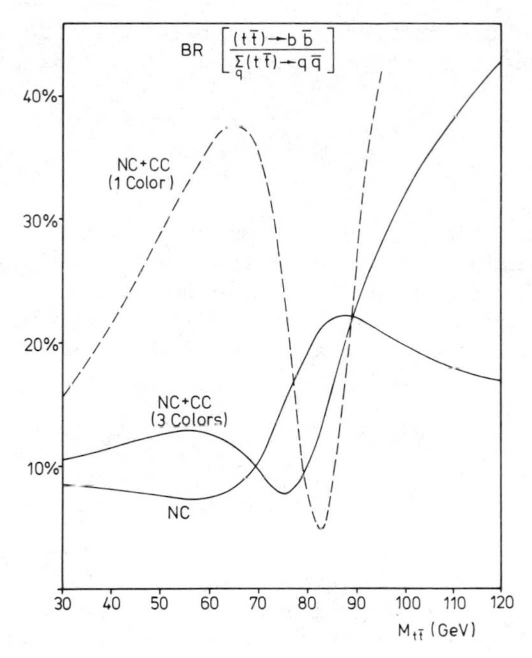

Fig. 4: Influence of γ-Z-W interference on the branching ratio of $[tt] \to b + \bar{b}$.

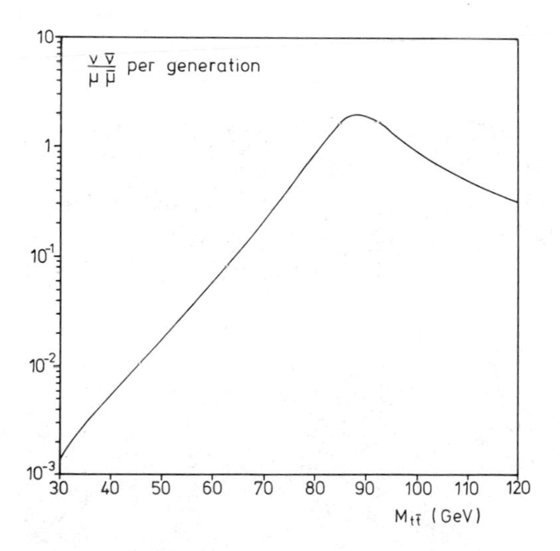

Fig. 5: Ratio of $\nu\bar{\nu}$ and $\mu\bar{\mu}$ decays as a function of toponium mass.

The importance of the charged current term, and the large difference between the three-colour and one-colour situations, is evident.

5. NEUTRINO COUNTING

The fact that a heavy quarkonium state such as $V(t\bar{t})$ can decay into the various $\nu\bar{\nu}$ states opens up the possibility of neutrino counting. The branching ratio per neutrino species, compared to that into $\mu\bar{\mu}$, is shown in Fig. 5, and varies from $\sim 2\%$ at $M_V = 50$ GeV to $\sim 100\%$ at $M_V = 80$ GeV. One possibility of measuring the ratio

$$\frac{\sum_{\nu} \Gamma\ (V(t\bar{t}) \rightarrow \nu\nu)}{\Gamma\ (V(t\bar{t}) \rightarrow \mu\bar{\mu})} \tag{28}$$

is to look for events of the type $e^+e^- \rightarrow \gamma + (\nu\nu \text{ or } \mu\bar{\mu})$ where the invariant mass recoiling against the photon is that of toponium. For a hard photon carrying away a fraction Z_γ of the beam energy at an angle Θ_γ, the cross section is[4]

$$\frac{d\sigma(e^+e^- \rightarrow \gamma + 1\bar{1})}{dZ\quad d\cos\Theta_\gamma} = \frac{\alpha}{2\pi}\ \frac{Z_\gamma(1+\cos^2\Theta_\gamma) + 4(1 - Z_\gamma)/Z_\gamma}{1 - \cos^2\Theta_\gamma} \cdot \sigma_{M^2}\ (e^+e^- \rightarrow 1\bar{1}) \tag{29}$$

where $M^2 = s(1-Z_\gamma)$. The photon produced in association with the final toponium state is monochromatic, with $Z_\gamma = 1 - M_V^2/s$. As an alternative, one can excite the 2S state of toponium, and look for reactions $e^+e^- \rightarrow V(t\bar{t})_{2S} \rightarrow \pi^+\pi^- + (\nu\bar{\nu} \text{ or } \mu\bar{\mu})$, where the mass recoiling against the $\pi^+\pi^-$ pair is the 1S toponium mass.

6. SINGLE-QUARK DECAY

As shown in Sec. 2 and Fig. 1, the decay of a toponium state through the disintegration of the t or $\bar{t}$ quark (without $t\bar{t}$ annihilation) is a significant decay mode at high $[t\bar{t}]$ masses. In one-third of such decays, the final state will contain a lepton-neutrino pair, i.e.

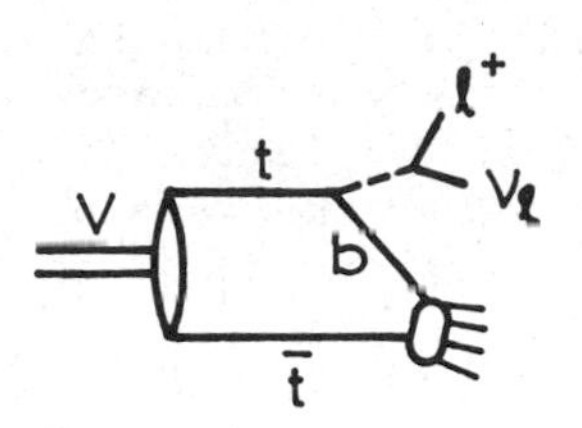

$$[t\bar{t}] \to b + \bar{t} + l^+\nu_l$$
$$\text{or} \quad [t\bar{t}] \to \bar{b} + t + l^-\bar{\nu}_l \tag{30}$$

and so will be characterised by a prompt lepton and missing energy. The l^+ spectrum, for instance, will be that of a t-quark decaying essentially at rest in the laboratory. If the toponium state is produced by e^+e^- annihilation, the t-quark has a non-zero polarization in the e^- direction,

$$P = \frac{N(S_Z = 1/2) - N(S_Z = -1/2)}{N(S_Z = +1/2) + N(S_Z = -1/2)} = \alpha_{RL} \tag{31}$$

with α_{RL} given by Eq. (19) and plotted in Fig. 2. The resulting l^+ spectrum is[9,10)]

$$\Gamma \frac{d\Gamma}{d\varepsilon_e d\cos\Theta_e} = 48\, \varepsilon_e^2\, (1-2\varepsilon_e)\, (1+\alpha_{RL}\cos\Theta_e) \tag{32}$$

where $\varepsilon_e = 2E_e/m_t$. This spectrum is relatively hard and has a forward-backward asymmetry proportional to α_{RL}. Such leptons may be separated from the background of non-prompt or secondary leptons by means of an energy cut. Notice that because of the factorization of the ε_e and $\cos\Theta_e$ dependence in (32), such a cut does not affect the angular asymmetry. (I am indebted to J. Kühn for this observation.) The spectrum of neutrinos may also be predicted[10)].

$$\frac{1}{\Gamma} \frac{d\Gamma}{d\varepsilon_\nu d\cos\Theta_\nu} = 16\, \varepsilon_\nu^2 \left[(\frac{3}{2} - 2\varepsilon_\nu) + \alpha_{RL}\, (\frac{1}{2} - 2\varepsilon_\nu) \cos\Theta_\nu \right] . \tag{33}$$

The hadronic system recoiling against the $l\nu$ pair consists initially of a rapidly moving b quark, and a spectator $\bar{t}$ quark practically at rest. Upon hadronization, this system will contain, besides some low energy fragments, a pair of heavy mesons, e.g.

$$\{T_{u,d} \text{ or } T^*_{u,d}\} \text{ plus } \{B_{u,d} \text{ or } B^*_{u,d}\} \quad . \tag{34}$$

In some instances, the $\overline{t}$ and b quarks could fuse to form a T_b or T_b^* meson. What is remarkable is that T and T* mesons would be produced singly well-below the conventional threshold for $T\overline{T}$ production. Topologically, the hadronic state would have a four-jet structure, the jets being associated with the b-quark and with the decay products of $\overline{t} \to \overline{b}\ \overline{u}\ d$ or $\overline{t} \to \overline{b}\ \overline{c}\ s$.

An alternative source of prompt leptons would be the situation in which one of the t-quarks decays non-leptonically, and the prompt lepton emerges from the decay of the T or T* meson that is formed around the spectator quark. (Recall that for a heavy t-quark, the mass difference between T and T*, which scales as $1/m_t$, can become very small, thus prohibiting the hadronic decay of T* and severely suppressing the radiative decay $T^* \to T\gamma$ as well. Thus both the T and T* mesons decay via weak interactions[7].) If the lepton originates from the spin-zero T, the angular distribution in the T-rest frame is isotropic. If, however, it is the decay product of the T*, some anisotropy is possible, since the polarization of the parent t-quark can be partially transmitted to the T* [5]. Assuming that a spectator t-quark, with $S_Z = +1/2$, converts into $T^*(S_Z = 1)$, $T^*(S_Z = 0)$ and $T(S = 0)$ with relative weight 2 : 1 : 1, the angular distribution of the lepton emerging from the spectator system will have the form (32), but with α_{RL} replaced by $\alpha_{RL}/2$. Altogether, therefore, the spectrum of prompt leptons resulting from single-quark decay of toponium, would have the distribution given by Eq. (32), with α_{RL} replaced by $3/4\ \alpha_{RL}$.

In the special case that the single-quark decay of toponium produces a T_b^* meson among the hadronic debris, one can envisage a two-body decay

$$T_b^* \to l\nu$$

which can have a branching ratio of several per cent.[7] The resulting high energy lepton would be a spectacular (if rare!) signature indeed.

Acknowledgments

I have profited from discussions with J. Kühn, P. Zerwas and I. Bigi.

Compliments and thanks to Harvey Newman for organising a fine conference.

References

1. J. Salicio, B. Naroska, and G. Mikenberg, these Proceedings.

2. W. Buchmüller and S.-H.H. Tye, Phys. Rev. D24, 132 (1981), and references therein.

3. J.J. Sakurai, Physica 96A, 300 (1979).

4. L.M. Sehgal and P.M. Zerwas, Nucl. Phys. B183, 417 (1981).

5. J.H. Kühn, lectures at the XX. Cracow School of Theoretical Physics (1980), Acta Physica Polonica B12, 347 (1981); and Schladming lectures (1982), Acta Physica Austriaca, Suppl. XXIV, 203 (1982).
   K. Fujikawa, Prog. Theor. Phys. 61, 1186 (1979).

6. F. Wilczek, Phys. Rev. Lett. 39, 1304 (1977).

7. I. Bigi and H. Krasemann, Z. Physik C7, 127 (1981).

8. G. Goggi and G. Penso, Nucl. Phys. B165, 429 (1980).

9. J.H. Kühn and K.H. Streng, Nucl. Phys. B198, 71 (1982).

10. D. Rein, L.M. Sehgal and P.M. Zerwas, Nucl. Phys. B138, 85 (1978).

BASICS OF THE $W^{\pm}$ AND Z PHYSICS IN $p\overset{(-)}{p}$ COLLISIONS

Paul Sorba

L.A.P.P., BP 909

74019 Annecy-Le-Vieux, France

INTRODUCTION

The first observations[1,2] of W events at the CERN Super Proton Synchrotron (SPS) have been known last January. Today, while completing these notes, more electron neutrino pairs are interpreted as leptonic decays of the W boson, and four electron-positron pairs and one muon pair appear with the signature of the expected Z boson[3]. The first results for the W-mass $m_W=(81\pm2)\text{GeV}/c^2$ and for the Z-mass: $m_Z \sim 95$ GeV, agree remarkably with the predictions of the minimal (i.e. with one Higgs doublet) electroweak $SU(2)\times U(1)$ gauge model[4].

Due to lack of space, we cannot reproduce here the content of our two seminars in Erice on the physics of the weak vector bosons (the interested reader is invited to consult refs 5-6-7-8) and we restrict in these pages on the W and Z production in pp and $p\bar{p}$ collisions.

As a first remark, we note that the total $p\bar{p}$ cross-section[9] at the SPS energy $\sqrt{s} = 540$ GeV, appears to be of the order 6×10^7 times the W and Z production cross-section[10].

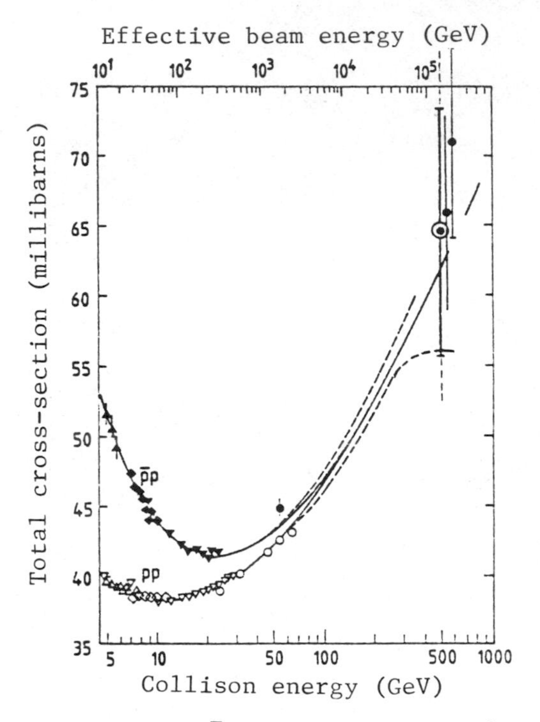

FIG. 1 Total p$\bar{p}$ and pp cross-sections

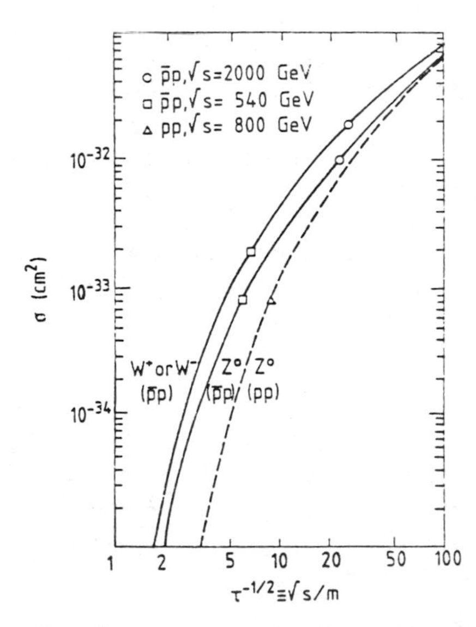

FIG. 2 Predicted production cross-sections for W and Z^{o} in pp and p$\bar{p}$ collisions. The calculated cross-sections include scale-breaking effects.

A. Cross-sections of Production

The production of W and Z and more generally any hadron-hadron collision at high energy are studied through their constituents. Already in 1970, Drell and Yan[11] gave an explanation for the production process of direct massive $\mu^+\mu^-$ pairs in terms of the quark parton model: it is the same type of diagram we will use here, but now the exchange boson will not only be the photon, but also a Z^0 or a charged W.

relevant subreactions[12] are:

$u\bar{u}$, $d\bar{d}$, $s\bar{s}$,... $\rightarrow$ Z^0

$u\bar{d}$, $u\bar{s}$, ... $\rightarrow$ W^+

$\bar{u}d$, $\bar{u}s$, ... $\rightarrow$ W^-.

Before studying the cross-section production, let us introduce some useful variables. If p_A and p_B are the four momenta of colliding hadrons, we will denote:

$$p_a = x_a P_A \qquad p_b = x_b P_B$$

the four momenta of colliding constituents: x_a and x_b will be the fractions of the parent particle momenta carried by the fusing quarks. If $\hat{s}$ and s are the square of the energy at the constituent and hadron level respectively, we have:

$$M^2 = \hat{s} = (p_a + p_b)^2 \simeq 2 p_a \cdot p_b \; ; \; s = (p_A + p_B)^2 \simeq 2 P_A \cdot P_B$$

that is: $\hat{s} = x_a x_b s$

In the c.m. system of the reaction $\vec{p}_A + \vec{p}_B = 0$ and assuming $\vec{p}_A$ and $\vec{p}_B$ along the z-axis, we get for the boson momentum:

$$P_{BOSON} = \left(\frac{x_a + x_b}{2} \sqrt{s} \; ; \; P_L = \frac{x_a - x_b}{2} \sqrt{s} \; ; \; P_T = 0 \right)$$

Two scale variables appear naturally:

$$\tau = \frac{\hat{s}}{s} = x_a x_b \quad \text{and} \quad x_F = \frac{2 P_L}{\sqrt{s}} = x_a - x_b$$

A third important scale variable is the rapidity: $y = 1/2 \ln E+p_L/E-p_L$.
We note: $y_{Boson} = 1/2 \ln x_a/x_b$ or: $x_a - x_b = 2\sqrt{\tau}$ Sh y.
It follows that at a first approximation, one will expect $y_{aver.} \sim 0$ in $p\bar{p}$, but a maximum of production away from y=0 in pp collisions since the antiquark in the proton will come from the sea and therefore will carry less momentum than the valence quark in the second proton: W is then expected to move in the direction of the valence quarks.

The cross-section calculation for W or Z production has to be made in two steps. First one must calculate the subreaction cross-section $\hat{\sigma}(a+b \to \gamma^*)$, with γ^* = W or Z, which is easily calculable, as far as we neglect strong interactions - or gluon - effects, i.e.:

$$\hat{\sigma}(a+b \to \gamma^*) = \sqrt{2}\, G_F \tau s\, \delta(\hat{s} - M_{\gamma^*}^2)$$

then we "dress" the quarks with their structure functions $G_{q/A}(x)$, which is the probability of finding in the hadron A a quark q carrying the momentum $p_a = x\, p_A$, to obtain the total cross-section of production:

$$\sigma(A+B \to \gamma^* + X) = \int_0^1 dx_a \int_0^1 dx_b \sum_{\substack{\text{all} \\ \text{flavors}}} \Big[G_{q/A}(x_a) G_{\bar{q}/B}(x_b) + G_{\bar{q}/A}(x_a) G_{q/B}(x_b) \Big] \hat{\sigma}(q\bar{q} \to \gamma^*)\, \delta\big[M^2 - (p_a + p_b)^2\big]\, dM^2$$

which can be rewritten as:

$$\sigma(A+B \to \gamma^* + X) = \pi \sqrt{2}\, G_F \tau \int_\tau^1 \frac{dx}{x} \mathcal{L}^{\gamma^*}(x, \tau/x)$$

we note that σ depends only on $\tau = M^2_{\gamma^*}/s$ before defining:

$$\mathcal{L}^{W^+}(x, \tau/x) = \frac{1}{3} \Big\{ \big[u_A(x) \bar{d}_B(\tau/x) + \bar{d}_A(x) u_B(\tau/x) \big] \cos^2\theta_C + \big[u_A(x) \bar{s}_B(\tau/x) + \bar{s}_A(x) u_B(\tau/x) \big] \sin^2\theta_C$$

$$\mathcal{L}^{Z^0}(x,\tau/x) = \frac{1}{3}\Big\{[u_A(x)\bar{u}_B(\tau/x) + \bar{u}_A(x)u_B(\tau/x)]\left(\frac{1}{4} - \frac{2}{3}\sin^2\theta_W + \frac{8}{9}\sin^4\theta_W\right)$$
$$+ [d_A(x)\bar{d}_B(\tau/x) + \bar{d}_A(x)d_B(\tau/x) + s_A(x)\bar{s}_B(\tau/x) + \bar{s}_A(x)s_B(\tau/x)]$$
$$\times\left(\frac{1}{4} - \frac{\sin^2\theta_W}{3} + \frac{2}{9}\sin^4\theta_W\right)\Big\}$$

where the structure function of a quark is denoted by the quark symbol. $\mathcal{L}^{W^-}(x,\tau/x)$ is immediately obtain from $\mathcal{L}^{W^+}(x,\tau/x)$ by interchanging q and $\bar{q}$ densities; the front factor 1/3 is the color factor.

The structure functions are deduced from the study of deep inelastic lepton-hadron scattering which at large square transfer momentum $Q^2 = -q^2$ probes short distance behaviour of the proton. Charge invariance implies:

$$G_{u/p}(x) = G_{\bar{u}/\bar{p}}(x) \equiv u(x) \;;\; G_{d/p}(x) = G_{\bar{d}/\bar{p}}(x) \equiv d(x)$$

The proton appearing as made of three valence quarks and a neutral sea of gluons and $q\bar{q}$ pairs, it is usual to define V(x) and S(x) by:

$$x\,u(x) = V(x) + S(x)$$
$$x\,d(x) = \frac{1}{2}V(x) + S(x)$$
$$x\,\bar{u}(x) = x\,\bar{d}(x) = S(x)$$

Actually $G_{q/p}(x)$ depends of $Q^2 = \mathcal{O}(\hat{s})$. With increasing Q^2, quarks will radiate more gluons, which in turn will decay in $q\bar{q}$ pairs: the number of partons will therefore increase and thus their energy decrease; it will follow a softening of the quark distributions, the low x region becoming more populated while the large x region depressed (see Fig. 3). It is interesting to remark that at energy collider $\sqrt{s} \sim$ 500-1000 GeV (CERN SPS) scaling violations are still modest (the main value of x <x> being of the

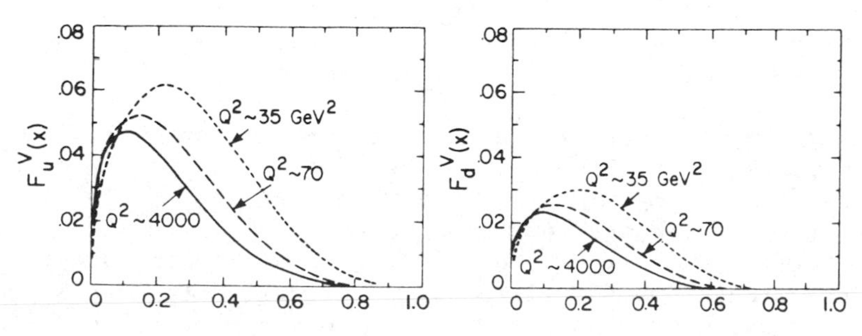

FIG. 3 Evolution of the valence quark (u,d) distributions with Q^2.

order $\sqrt{\tau} = M_W/\sqrt{s} \sim 0.1-0.2$) while at Fermilab Tevatron energy $\sqrt{s} = 2000$ GeV, they appear important (then $\langle x \rangle \sim \sqrt{\tau} \sim 0.05$). The W and Z production cross-sections[10] are presented in Fig. 3 with scaling violations effects.

First order QCD contributions[13] are of two sorts:

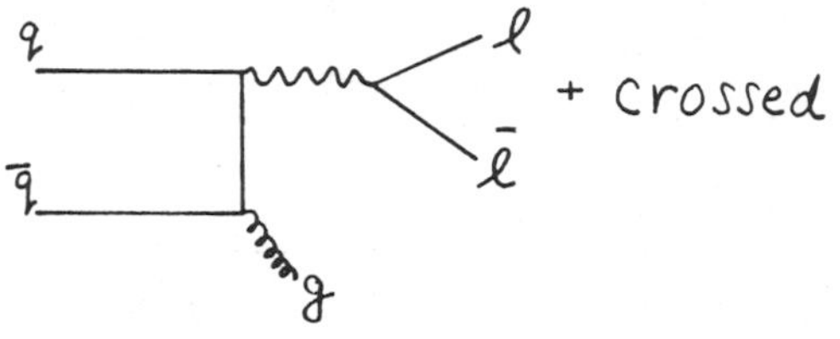

$q\bar{q}$ annihilation 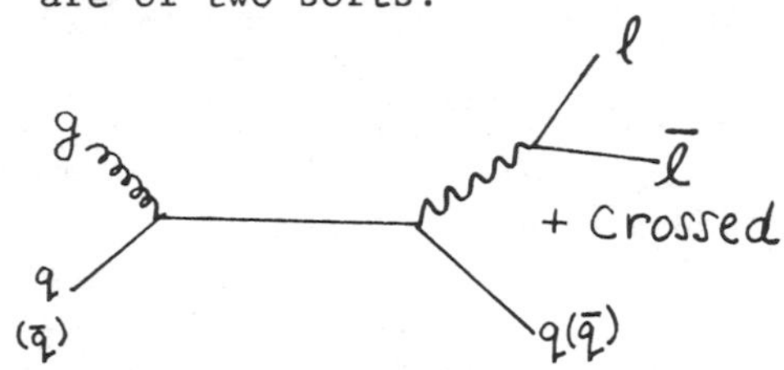 $\qquad\qquad$ $\overset{(-)}{q}g$ "Compton" scattering

Two interesting points have to be mentioned. First, gluon emission gives to the gauge boson a transverse kick which is not negligible. Secondly, because of virtual gluon exchange the cross-section is multiplied by a factor 2 [14,15] (as an example[16] calculation of $\sigma(p\bar{p} \xrightarrow{Z^o} \mu^+\mu^- + \text{jet } (p_T > 3 \text{ GeV}))$ gives 34 pb which means for the total first order QCD Z^o production the same order of magnitude as the Drell-Yan one). Due to the smallness of the cross-sections, this factor 2 might be very important (for the measure of the asymmetry for example).

The production of pairs of weak bosons seems more problematic, the corresponding cross-sections being at the order of the picobarn[17,18], while the production of pairs $W^\pm\gamma$ [18,19] is found to be higher by a factor 10: although still very small, it might be possible to study the sensitivity of $\sigma(p\bar{p} \to W^\pm\gamma+X)$ to the magnetic moment of the W, which is found to be K=1 when derived from the Yang-Mills Lagrangian.

B. Leptonic Final States; The Jacobian Peak

Leptonic decays modes are the easiest to select for a final state analysis. We recall that the branching ratio of the Z^o into charged leptons $B(Z^o \to e^+e^-, \mu^+\mu^-) \sim 3\%$ (in the case $N_G = 3$ generations) is smaller that the corresponding $W^\pm$ branching ratios $B(W^\pm \to \ell^\pm \overset{(-)}{\nu}_\ell) \sim 8\%$. If we remember (Fig.2) that the production cross-section of the Z^o is also - slightly - smaller than that of the W's, we may understand why the W has been seen before the Z. In Fig. 4 where the production $Z^o \to \ell\bar{\ell}$ is described, the background due to competing mechanisms (Drell-Yan with virtual photon and leptonic decays of heavy quarks) appear much below the peak at the Z^o mass in the lepton pair mass spectrum[20].

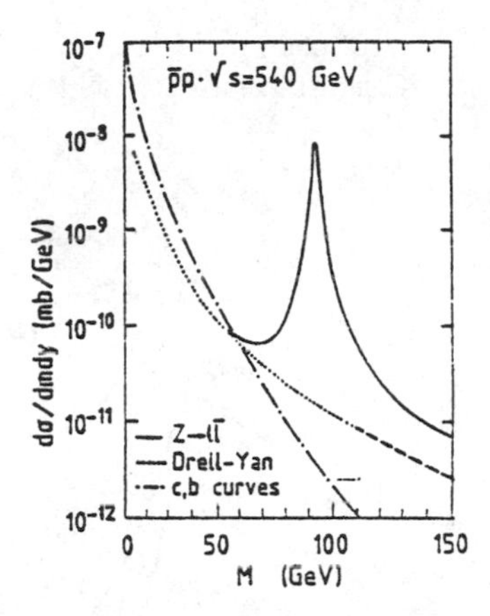

FIG. 4 Lepton-pair mass spectrum in $p\bar{p}$ collisions at $\sqrt{s}$ = 540 GeV

The situation is a priori different for the W leptonic modes, since then the second lepton is a (anti)neutrino which cannot be detected. This experimental problem will be solved by the kinematics. Indeed, considering the following distribution:

$$\ell^0 \frac{d^3\sigma}{d^3\vec{\ell}} = \sum_{a,b} \int dx_a dx_b \mathcal{L}(x_a, x_b) \ell^0 \frac{d^3\sigma}{d^3\vec{\ell}}(a+b \to W \ell \bar{\nu}_\ell) \delta(x_a x_b s - \hat{s}) d\hat{s}$$

where ℓ is the momentum of the charged lepton, it is possible[12] using suitable changes of variables to rewrite it as:

$$\ell^0 \frac{d^3\sigma}{d^3\vec{\ell}} = \sum_{a,b} \int \frac{d\hat{s}}{s} \mathcal{L}(\ldots,\ldots) \frac{|m|^2}{32\pi^2 \hat{s}^{3/2}} \frac{1}{\sqrt{\hat{s} - 4\ell_T^2}}$$

where $|\ \ |^2$ is the squared amplitude matrix element and $\vec{\ell}_T$ the momentum part orthogonal to the incident beam. In the NWA

$$\frac{1}{(\hat{s}-M_W^2)^2 + (M_W \Gamma_W^{tot})^2} \longrightarrow \frac{\pi}{M_W \Gamma_W^{tot}} \delta(\hat{s} - M_W^2)$$

we can write:

$$\frac{|m|^2}{\sqrt{\hat{s} - 4\ell_T^2}} \sim A \frac{\delta(\hat{s} - M_W^2)}{\sqrt{\hat{s} - 4\ell_T^2}}$$

Note that if $\Gamma=0$, we get a cut-off at $\ell_T = M_W/2$ and the above distribution goes to infinity. But Γ although small is not zero. Moreover the intrinsic transverse momentum of the constituents and the QCD corrections[20,21] will induce a smearing of the peak with events at $\ell_T > M_W/2$ (see Fig. 5). This peak at $\ell_T = M_W/2$ is known as the Jacobian peak. It should emerge from the above mentioned background. However a precise measurement of M_W will be difficult, due to the smearing.

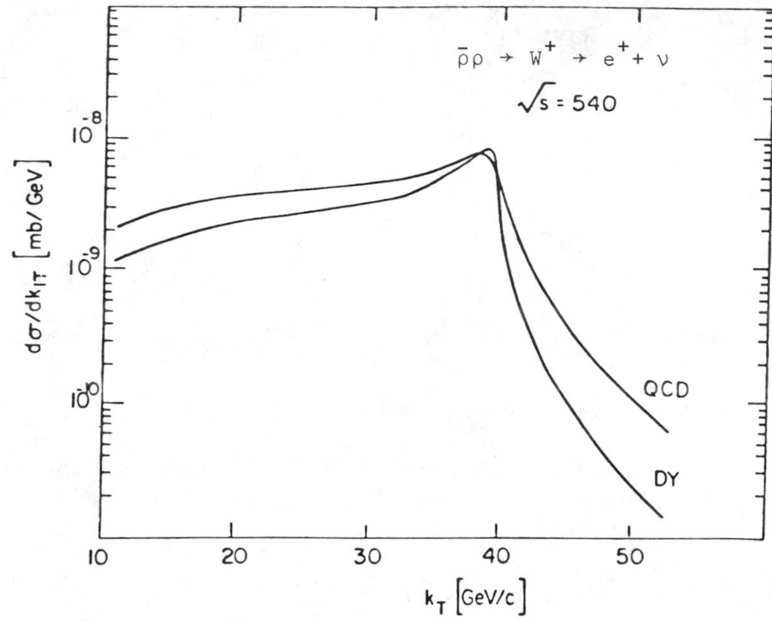

FIG. 5 The Jacobian peak of the single-lepton spectrum from W^+ production and decay. The calculation is indicated by "DY", while "QCD" contains the QCD corrections.

C. Front-Back Asymmetry of the Lepton-Spectrum

W and Z will manifest themselves as bumps in cross-sections. But a typical weak interaction effect will be given by the front-back asymmetry of the outgoing lepton.

Let us consider once more the diagram:

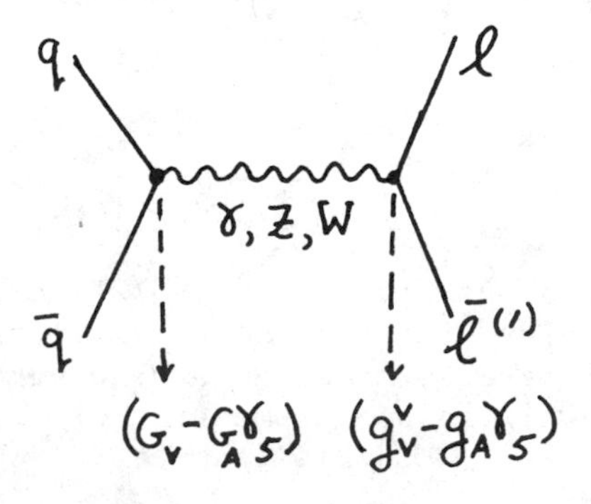

The calculation of the amplitude gives:

$$|\mathcal{M}^2| = \frac{4\hat{s}^2}{[\text{Propagator}]^2}\left\{(G_A^2+G_V^2)(g_A^2+g_V^2)(1+\cos\hat{\theta}) + 8G_AG_Vg_Ag_V\cos\hat{\theta}\right\}$$

where $\hat{\theta}$ is the scattering angle between the outgoing lepton (ℓ) and the incoming particle (q) in the c.m. of the subreaction: $\vec{p}_q + \vec{p}_{\bar{q}} = 0$

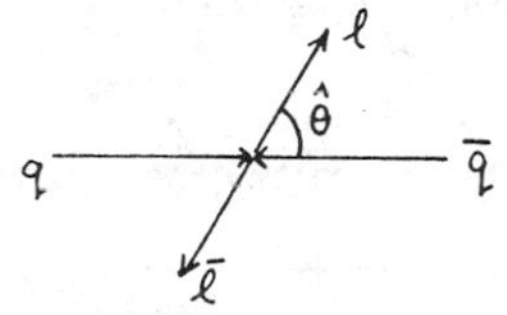

Note that in the case where interferences are present, we will get:

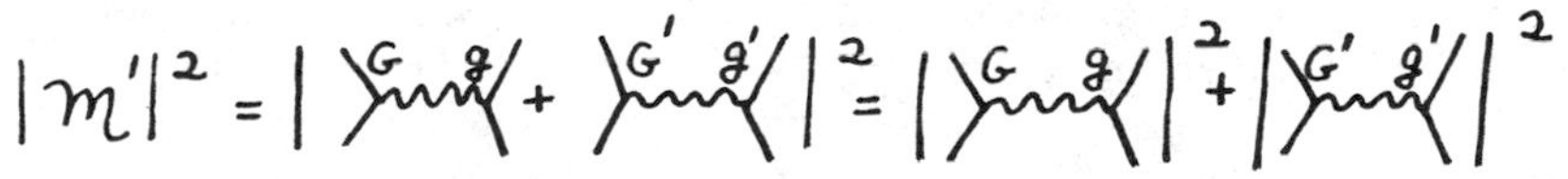

$$+\ 2\,\frac{4\hat{s}}{[\text{Prop}][\text{Prop}]'}\left\{(G_AG_A'+G_VG_V')(g_Ag_A'+g_Vg_V')(1+\cos^2\hat{\theta}) + 2(G_VG_A'+G_AG_V')(g_Vg_A'+g_Ag_V')\cos\hat{\theta}\right\}$$

When calculating the total cross-section, one will integrate with respect to $\cos\hat{\theta}$, or more precisely to $\cos\theta$, θ being scattering angle between the outgoing lepton (ℓ) and the incoming proton, obtained from $\hat{\theta}$ by a boost: it follows that only the even part in $\cos\theta$ will contribute, and therefore the cross-section will give information on terms $(G_A^2+G_V^2)(g_A^2+g_V^2)$,... But considering $d\sigma/d\cos\theta$ and more precisely:

$$A_{F.B}(s,\cos\theta) = \frac{\frac{d\sigma}{d(\cos\theta)}(\cos\theta) - \frac{d\sigma}{d(\cos\theta)}(-\cos\theta)}{\frac{d\sigma}{d(\cos\theta)}(\cos\theta) + \frac{d\sigma}{d(\cos\theta)}(-\cos\theta)}$$

we will get information on products of coupling constants $G_AG_Vg_Ag_V$,..

Note that in pp collisions where all incident particles are the same, the asymmetry will be exactly 0.

Qualitatively $A_{F.B}$ is expected to be large in the $W^{\pm}$ case due to the pure V-A coupling (see Table section 1), and small for the Z^o case because of the vector coupling of Z^o to ℓ ($\sim 1/4 - \sin^2\theta_W$).

Helicity considerations inform us that the asymmetry of W^+ (respectively W^-) must be negative (respectively positive). Indeed, considering for example W^+: since only left-handed quarks are involved for the production of the weak boson (see section 1), one will have:

$p \quad \bar{p}$

$u_L \quad \bar{d}_L$

and therefore the helicity (that is the spin direction along the propagator axis) of u_L and that of $\bar{d}_L$ will be in the direction of the $\bar{p}$ beam. The outgoing neutrino being also left-handed, there is only one possibility for the helicity of ℓ^+, i.e. the direction of the $\bar{p}$ beam:

$u_L \quad \bar{d}_L \qquad W^+ \qquad \nu_L \quad \ell^+$

Calculations confirm these qualitative results[22]. The asymmetry of Z^o is small: integrating over $\cos\theta$, one finds about 7%; a rather large statistics will be necessary to measure it. The situation is different for $W^{\pm}$: see Fig. 6. We finally note that QCD contributions do not modify consequently the results[16].

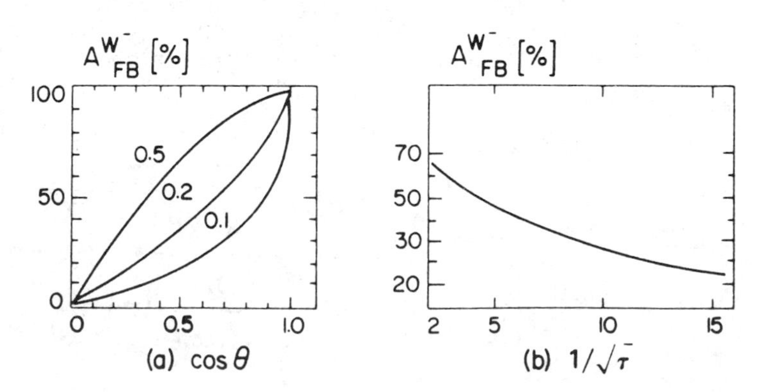

FIG. 6 The front-back W^- asymmetry (a) versus $\cos\theta$ for various values of $\sqrt{\tau}$, (b) integrated version as a function of $1/\sqrt{\tau}$.

D. Hadronic Final States

Weak bosons are supposed to decay mainly into hadronic modes. Hadronic jets with a weak origin are expected to be present in $p^{(\bar{p})}$ collisions, but overwhelmed in the terrible strong interaction background. A rapid and simple exercise shows that at the lowest order in QCD several diagrams will produce hadronic jets while only one diagram will give a pair of quarks via a weak boson. Moreover one can compare the strong coupling constant $\alpha_s(M_W^2) \sim 0.1-0.2$ with the weak one $\alpha_W = g^2/4\pi \sim 0.03$.

The following curves[23] illustrate the situation: one can see the position and the shape of the different curves corresponding to the weak interactions, to the weak and strong interactions with and without gluons in the colliding nucleons:

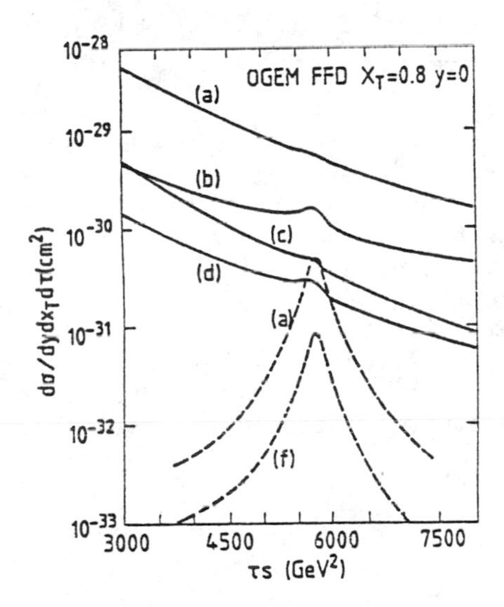

FIG. 7 Differential cross-sections at y=0 and x_T=0.8 with respect to the effective energy τs for $p\bar{p}$ jets.

	$\sqrt{s}$ = 350 GeV	$\sqrt{s}$ = 200 GeV
weak inter.	(e)	(f)
weak + strong inter. without gluons	(c)	(d)
weak + strong inter. with gluons	(a)	(b).

A suggestion has been made[24] to select heavy quark jets with a weak origin from those produced by strong interactions. It consists in considering the semi-leptonic decays of heavy quarks:

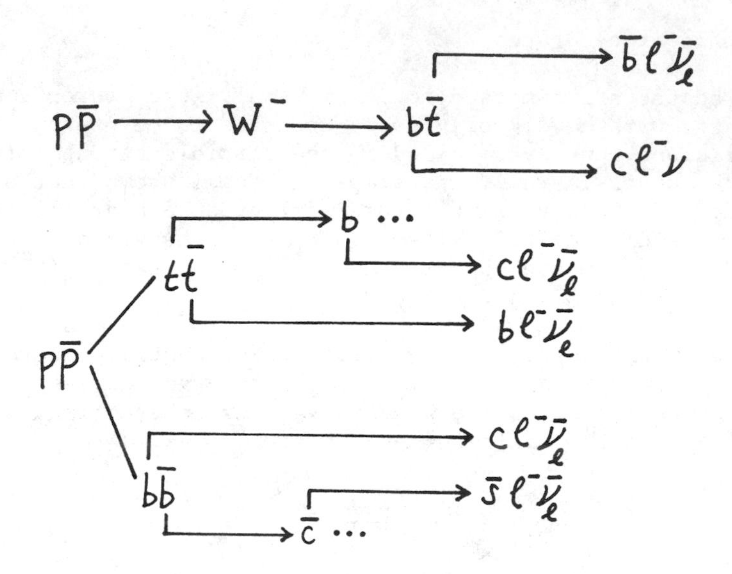

From the above decay chains, one can check that for $b\bar{b}$ and $t\bar{t}$ final states coming by strong interactions, there must be cascade decay if one look for two leptons of the same sign, while weak interaction will produce $b\bar{t}$ and $t\bar{b}$ pairs which in turn will give $\ell^-\ell^-$ or $\ell^+\ell^+$ pairs by direct decay. Since a lepton produced in cascade decay will be in general less energetic than a lepton coming from a direct decay, demanding two very energetic leptons of the same sign might be a way to select weak interaction jets from the background. The difficulties of this kind of analysis are of course numerous. One has first to be able to distinguish a heavy quark jet from a light one; secondly a careful study of the final state is also necessary.

ACKNOWLEDGEMENTS

Special thanks are due to G. Girardi and P. Mery for numerous discussions.

REFERENCES

(1) G. Arnison et al., Phys. Lett. 122B (1983) 103.
(2) G. Banner et al., Phys. Lett. 122B (1983) 476.
(3) G. Arnison et al., Preprint CERN-EP/83-73.
(4) S. Weinberg, Phys. Rev. Lett. 19 (1967) 1264.
A. Salam, Proc. 8th Nobel Symposium, Aspenäsgarden, 1968 (Almqvist and Wiksell, Stockholm (1968) p. 367-77).
S.L. Glashow, Nucl. Phys. 22 (1961) 579.
(5) J. Ellis, M.K. Gaillard, G. Girardi and P. Sorba, "Physics of Intermediate Vector Bosons", Ann. Rev. of Nucl. Part. Sci. (1982) 32, 443.
(6) P. Sorba, "Basics of the W and Z physics", LAPP-TH-84.

(7) R. Horgan and M. Jacob, Proc. of the CERN Summer School 1980.
(8) For results on W and Z in e^+e^- collisions, see for example, Proc. LEP Summer Study, Les Houches and CERN, Sept. 1978, CERN Report N° 79-01, vols 1 and 2.
For results on W and Z in ep collisions, see for example, Proc. ep Facility for Eur. DESY 79/48.
(9) G. Arnison et al., Preprint CERN-EP/83-70.
R. Battiston et al., Phys. Lett. 117B (1982) 126.
(10) F. Paige, Proc. Topical Workshop Product New Part. Super High Energy Collisions, ed. Barger and F. Halzen, Univ. Wis. Madison.
(11) S.M. Drell and T.M. Yan, Ann. Phys. 66 (1971) 578.
T.M. Yan, Ann. Rev. Nucl. Sci. 26 (1976) 199.
(12) R.F. Peierls, T.L. Trueman, L.L. Wang, Phys. Rev. D16 (1977) 1397.
C. Quigg, Rev. Mod. Phys. 49 (1977) 297.
(13) E. Reya, Phys. Rep. 69 (1981) 195.
(14) G. Altarelli, R.K. Ellis, G. Martinelli, Nucl. Phys. B147 (1979) 461.
(15) P. Minkowsky, CERN Preprint CERN-TH/3519.
(16) J. Finjord, G. Girardi, M. Perrottet and P. Sorba, Nucl. Phys. B182 (1981) 427.
(17) R.W. Brown and K.O. Mikaelian, Phys. Rev. D19 (1979) 922.
(18) R.W. Brown, D. Sahdev and K.O. Mikaelian, Phys. Rev. D20 (1979) 1164.
(19) K.O. Mikaelian, Phys. Rev. D17 (1978) 750.
K.O. Mikaelian, M.A. Samuel and D. Sahdev, Phys. Rev. Lett. 43 (1979) 746; T.R. Grose and K.O. Mikaelian, Phys. Rev. D23 (1981) 123.
(20) S. Pakvasa, M. Dechantsreiter, F. Halzen and D.M. Scott, Phys. Rev. D20 (1979) 2862.
(21) P. Aurenche and J. Lindfors, Nucl. Phys. B185 (1981) 274; ibid. p. 301.
(22) M. Perrottet, Ann. Phys. 115 (1978) 107.
(23) M. Abud, R. Gatto and C.A. Savoy, Ann. Phys. 121 (1979).
(24) M. Abud, R. Gatto and C.A. Savoy, Phys. Lett. 79B (1978) 435; Phys. Rev. D20 (1979) 1164.

B-L GAUGE MODELS*

R. E. Marshak

Physics Department
Virginia Polytechnic Institute and State University
Blacksburg, Virginia 24061

The standard low energy electroweak and strong interaction gauge group $SU(3)_C \times SU(2)_L \times U(1)_Y$, broken spontaneously by a Higgs doublet, has been highly successful in the low energy region ($\lesssim$ 100 GeV) [1]. On the other hand, the SU(5) gauge group, the lowest rank simple group containing $SU(3)_C \times SU(2)_L \times U(1)_Y$, appears to be in serious trouble [2]. The failure to observe proton decay at the level predicted by SU(5) GUT throws doubt on the existence of the "great desert" between 10^2GeV and 10^{15}GeV and leaves room for intermediate mass scales of various types. Within this context, and for other reasons (see below), I wish to point out in this talk some attractive features of those groups that have B-L as a generator (and which I call B-L gauge groups).

I first note that neither the low energy group $SU(3)_C \times SU(2)_L \times U(1)_Y$ nor SU(5) GUT contains B-L among its generators. This means that B-L can not be a local symmetry. However, both the standard electroweak group and SU(5) give intimations of the importance of the combination (B-L). It has been shown [3] that global B and L non-conservation are separately induced by instanton effects in the case of $SU(2)_L \times U(1)_Y$ but that global B-L conservation remains intact. At the SU(5) level, B-L is an accidental global symmetry that governs the predicted proton decay modes. In addition, SU(5) GUT assumes maximal parity violation up to the highest energy (the grand unification energy), considers all neutrinos to be massless [4], surrenders the full quark-lepton correspondence (since there

*Some new results have been included since the talk was given at Erice.

is no right-handed neutrino) and claims no explanation of the repetition of quark and lepton families. (I leave the gauge hierarchy and fine tuning problems to others.)

We shall show how the enlargement of the rank 2 standard electroweak group $SU(2)_L \times U(1)_Y$ into a B-L gauge group of rank 3 (e.g. $SU(2)_L \times SU(2)_R \times U(1)_{B-L}$) and the subsequent enlargement (after adjoining $SU(3)_C$) to a B-L gauge group of rank 5 (e.g. the partial unification gauge group $SU(4)_C \times SU(2)_L \times SU(2)_R$ [5]) and finally to the rank 5 B-L grand unification group SO(10), can deal with these questions and predict some interesting new physics to boot.

Perhaps it would not be amiss to mention a bit of history in these informal surroundings. When Mohapatra and I reexamined the structure of the standard $SU(2)_L \times U(1)_Y$ group in late 1979, we were struck by the asymmetrical character of the weak hypercharge current associated with the group $U(1)_Y$, as seen from its explicit form (L is left-handed, R right-handed):

$$J_\lambda^Y = \frac{1}{3} (\overline{q_i^+}, \overline{q_i^-})_L \gamma_\lambda \begin{pmatrix} q_i^+ \\ q_i^- \end{pmatrix}_L - (\overline{\ell_i^+}, \overline{\ell_i^-}) \gamma_\lambda \begin{pmatrix} \ell_i^+ \\ \ell_i^- \end{pmatrix} \qquad (1)$$

$$+ \frac{4}{3} \overline{q^+_{iR}} \gamma_\lambda q^+_{iR} - \frac{2}{3} \overline{q^-_{iR}} \gamma_\lambda q^-_{iR} - 2\overline{\ell^-_{iR}} \gamma_\lambda \ell^-_{iR}$$

In Eq. (1), q_i^+, q_i^- are the up and down quarks respectively of the ith generation ($i = 1, 2, 3$) and ℓ_i^+, ℓ_i^- are the up and down leptons of the ith generation. By using the "weak" Gell-Mann-Nishijima relation $Q = I_{3L} + Y/2$, the same values of Y_L and Y_R are obtained for the quarks and leptons of each generation, as shown in Table 1 (F is the weak flavor - see below). The weak hypercharge Y_L is

Table 1. Quantum Numbers of Quarks and Leptons

	B-L	Q	I_L	I_{3L}	Y_L	F_L	I_R	Y_R	F_R
q_i^+	1/3	2/3	1/2	1/2	1/3	0	0	4/3	1
q_i^-	1/3	-1/3	1/2	-1/2	1/3	0	0	-2/3	-1
ℓ_i^+	-1	0	1/2	1/2	-1	0	0	-	1
ℓ_i^-	-1	-1	1/2	-1/2	-1	0	0	-2	-1

evidently B-L for the left-handed weak isodoublet states (of the quarks and leptons) but takes on completely different values for the right-handed weak isosinglet states. Nevertheless, anomaly cancellation occurs because the condition $\Sigma\ (Y_L{}^3 - Y_R{}^3) = 0$ holds for each generation. If we write $Y = B\text{-}L + F$, it is obvious that F_L is 0 for quarks and leptons of both helicities while $F_R = +1$ for the up quarks and up leptons and $F_R = -1$ for the down quarks and down leptons. In other words, we can write the weak Gell-Mann-Nishijima relation as $Q = I_3 + (B\text{-}L\text{+}F)/2$. But this is precisely the relation that was postulated on the basis of baryon-lepton symmetry by Gamba, Marshak and Okubo in 1959 [6]. These authors argued at that time for a symmetry between the basic constituents (p, n, Λ) of the Sakata model [7] and the three known leptons (ν, e^-, μ^-). The extension of this symmetry principle to quark-lepton symmetry is obvious.

Another possible role for B-L was noted by Mohapatra and myself [8] in looking at the masses of the three generations of quarks and leptons as seen in Table 2. If one studies the masses in Table 2, one notices that the mass differences between the up and down quarks and the up and down leptons of the first generation have the same sign whereas the mass spectra of the second and third generation quarks and leptons have an inverted pattern. Since the small mass difference between the u and d quarks can easily be turned around due to the mixing effects of the heavier quarks, we focus on the second and third generations of leptons and quarks. The clearcut mass inversion effect could then be explained qualitatively by a dependence of the mass operator on B-L [8].

These observations led Mohapatra and myself [8] to argue that B-L should be treated as a local symmetry (that is necessarily broken [9]) in enlarging the standard electroweak group. The weak Gell-Mann-Nishijima relation $Q = I_3 + (B\text{-}L\text{+}F)/2$ suggests two possibilities for the enlarged electroweak group: $SU(2)_L \times U(1)_{B-L} \times U(1)_R$ or $SU(2)_L \times SU(2)_R \times U(1)_{B-L}$. In the first case, $U(1)_R$ gives the correct weak flavor quantum numbers and has been considered by Davidson who predicts some interesting deviations from the standard

Table 2. Mass Spectra of Quarks and Leptons (in MeV)

q_i^+	$m(u) \sim 5$	$m(c) \sim 1200\text{-}1500$	$m(t) > 18000$
q_i^-	$m(d) \sim 10$	$m(s) \sim 200\text{-}300$	$m(b) \sim 5000$
ℓ_i^+	$m(\nu_e) \lesssim 10^{-5}$	$m(\nu_\mu) < 0.5$	$m(\nu_\tau) < 250$
ℓ_i^-	$m(e) = 0.51$	$m(\mu) = 106$	$m(\tau) = 1782$

electroweak group [10]. For reasons that should become apparent, I find more attractive the second enlarged B-L electroweak group, namely $SU(2)_L \times SU(2)_R \times U(1)_{B-L}$ (to be called the left-right symmetric (LRS) group.)

The LRS group implies a weak Gell-Mann-Nishijima relation of the form [8]:

$$Q = I_{3L} + I_{3R} + \frac{B-L}{2} \tag{3}$$

where I_{3R} has replaced $F_R/2$ in Eq. (2). It follows from Eq. (3) that the weak hypercharge Y = B-L for both helicities of quarks and leptons (so that $Y_L = Y_R$); this is not surprising since the hypercharge current associated with the LRS group takes on the very symmetric form:

$$J_\lambda^Y = \frac{1}{3}\,(\overline{q_i^+},\ \overline{q_i^-})_L\,\gamma_\lambda \begin{pmatrix} q_i^+ \\ q_i^- \end{pmatrix}_L - (\overline{\ell_i^+},\ \overline{\ell_i^-})_L\,\gamma_\lambda \begin{pmatrix} \ell_i^+ \\ \ell_i^- \end{pmatrix}_L + L \rightarrow R \tag{4}$$

The structure of the LRS group also implies that parity is restored at higher energies; indeed, above 100 GeV (where $SU(2)_L$ is no longer broken), Eq. (3) implies that:

$$-\Delta I_{3R} = \frac{\Delta(B-L)}{2} \tag{5}$$

which relates the (spontaneous) breaking of parity to the (spontaneous) breaking of B-L local symmetry. We shall see presently how Eq. (5) leads to Majorana neutrinos ($\Delta L = 2$) and neutron oscillations ($\Delta B = 2$) [11]. Implicit in the LRS group is the acceptance of a finite mass neutrino so that the right-handed neutrino states, together with the right-handed charged lepton states, constitute doublets of $SU(2)_R$; it will turn out that the LRS model predicts two Majorana neutrinos (one light and one heavy) per generation rather than one Dirac neutrino with mass [12]. In other words, the postulation of B-L as the weak hypercharge for both helicities automatically recaptures the full quark-lepton correspondence for all three generations. Finally, it should be noted that equality $Y_L = Y_R$ gives a natural explanation of anomaly cancellation and instanton effect cancellation.

Let me now give you the gauge boson and Higgs content of the LRS model [13] and then summarize some novel physical consequences of this enlarged B-L electroweak group. There are three additional gauge bosons compared to the standard electroweak group, namely $W_R^\pm$, $Z^{0\prime}$; the masses of these three bosons must be substantially larger than $W_L^\pm$, Z^0 in accordance with the breaking down of $SU(2)_L \times SU(2)_R \times U(1)_{B-L}$ to $SU(2)_L \times U(1)_Y$. If we insist that the Higgs scalars be bound states of existing fermions - so that the results do not depend on the existence or non-existence of physical Higgs but simply on the pattern of symmetry breaking - then the relevant Higgs multiplets are:

$$\phi\ (2, 2, 0);\ \Delta_L\ (3, 1, 2);\ \Delta_R\ (1, 3, 2) \tag{6}$$

where the quantum numbers refer to $SU(2)_L \times SU(2)_R \times U(1)_{B-L}$. Assigning appropriate vev's to the Higgs [13], one can spell out some of the consequences. One immediate consequence is the implied existence of three new gauge bosons $W_R^{\pm}$ and $Z^{0'}$. The parameter $\sin^2\theta_W$ and the masses $M(W_R)$ and $M(Z^{0'})$ can be reconciled with the charged and neutral weak current scattering data by using values in the ranges [14]:

$$\begin{aligned} &\sin^2\theta_W = 0.25 \pm .02 \\ &M(W_R) \sim 200 - 300\ \text{GeV} \\ &M(Z^{0'}) \sim 300 - 400\ \text{GeV} \end{aligned} \tag{7}$$

The addition of the three gauge bosons $W_R^{\pm}$, $Z^{0'}$ can not only lead to a modification of the parameter $\sin^2\theta_W$ but also to changes in the masses of $W_L^{\pm}$, Z^0. The combined effect is that the LRS model predicts deviations from the standard model, particularly at the higer energies. For example, the forward-backward asymmetry parameter $A_{\mu\mu}$ in the reaction $e^+e^- \to \mu^+\mu^-$ is a fairly sensitive measure of the effects of the enlarged LRS electroweak group, as can be seen from Fig. 1* [15] where the standard model prediction of $A_{\mu\mu}$ is compared with that of the LRS model for two sets of parameters fitting the neutral and charged current data [14]. Since the additional gauge bosons in the LRS model are coupled to right-handed weak currents, experiments with polarized beams of electrons, say, can provide a probe of the onset of right-handed weak currents; in Fig. 2*, the asymmetry parameter $B = \frac{\sigma_L - \sigma_R}{\sigma_L + \sigma_R}$, where $\sigma_L(\sigma_R)$ is the cross section for the inelastic scattering of left (right) helicity electrons on protons (for x, y → 1 [16] is plotted as a function of energy for the same two sets of parameters in the LRS model as in Fig. 1. Fig. 1 should be of interest in planning for LEP experiments and Fig. 2 for HERA. Calculations have been made of other parameters in e^+e^- and e^-p collisions for both unpolarized and polarized beams (νp and pp collisions are not as useful) with results similar to those shown in Figs. 1 and 2; the general feature of deviations from the standard model only becoming appreciable at energies substantially above 100 GeV is maintained and arises from the fact that the g_V associated with the neutral electron current is approximately zero because $\sin^2\theta_W \sim 0.25$. A cynic may argue that the LRS model will only be tested when energies are sufficient to produce W_R's (and $Z^{0'}$'s), and

*Z in figures is $M(Z^0)$ in GeV for standard model in text, $Z_1 \equiv M(Z^0)$, $Z_2 \equiv M(Z^{0'})$ for LRS Model.

(prepared by P. Wang)

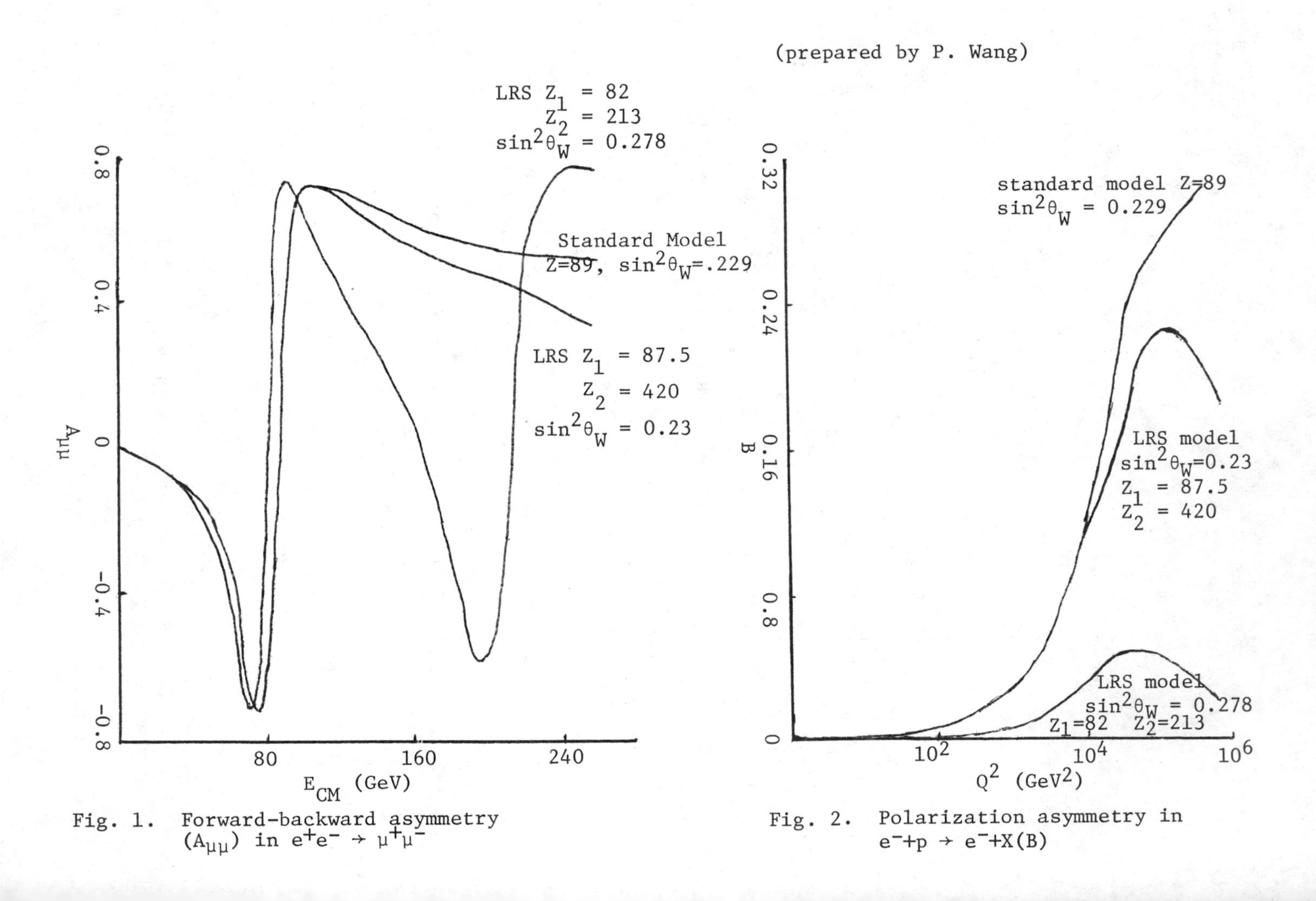

Fig. 1. Forward-backward asymmetry ($A_{\mu\mu}$) in $e^+e^- \to \mu^+\mu^-$

Fig. 2. Polarization asymmetry in $e^-+p \to e^-+X(B)$

this may well be true unless the intensities become larger at the lower energies.

The prediction by the LRS model of light (ν_L) and heavy (N_R) Majorana neutrinos may be more amenable to experimental tests in the near future. The LRS model predicts for the neutrino masses [12]:

$$m_{\nu_L} \sim \frac{m_\ell^2}{m_{N_R}} \text{ with } m_{N_R} \sim m_{W_R} \quad (m_\ell \text{ is charged lepton mass}) \qquad (8)$$

Apart from trying to measure a mass of the ν_e in the neighborhood of 0.1 - 1 eV, a heavy Majorana neutrino with a mass in the neighborhood of 100 GeV (see Eq. (7)) would have interesting consequences for several rare decay processes under investigation for some time [17], such as neutrinoless double β decay, $\mu \to e\gamma$, $\mu^- + A(Z) \to e^- + A'(Z)$, etc. For $(\beta\beta)_0$ of a typical double β emitter (e.g. Ge^{76}), a lifetime in the range $10^{23\pm2}$ yr. is predicted. For the muon decay and muon conversion processes, branching ratios of 10^{-12} are possible. Because of the Majorana character of the heavy neutrino, the muon conversion process $\mu^- + A(Z) \to e^+ + A(Z-2)$ could occur with a comparable branching ratio to that of the regular muon conversion process. It should be emphasized that diagrams resulting from the presence of a light Majorana neutrino in the mass range indicated above would yield much smaller transition probabilities for these rare decay processes.

We proceed with our discussion of B-L gauge models by adjoining $SU(3)_C$ to the LRS group; the rank 5 B-L group $SU(3)_C \times SU(2)_L \times SU(2)_R \times U(1)_{B-L}$ is now an enlargement of the rank 4 standard group $SU(3)_C \times SU(2)_L \times U(1)_Y$. The standard electroweak and strong interaction group $SU(3)_C \times SU(2)_L \times U(1)_Y$ is the maximal subgroup of the rank 4 simple group SU(5) and for this reason has been studied in considerable detail. The failure of SU(5) GUT to correctly predict the proton lifetime via the gauge boson mechanism is deeply troubling and may require a reassessment of the grand unification program. However, one may be more conservative and try to cope with a different unification group.

Let me now explain how a straightforward enlargement of the B-L LRS and strong interaction group $SU(3)_C \times SU(2)_L \times SU(2)_R \times U(1)_{B-L}$ through successive stages to the rank 5 B-L Pati-Salam group and the rank 5 B-L SO(10) group has some interesting consequences for proton decay and neutron oscillations. Eq. (9) shows how B-L SO(10) GUT breaks down into the B-L Pati-Salam group which in turn breaks down into the B-L LRS electroweak and strong interaction group, then into the non-(B-L) standard electroweak and strong interaction group and finally into the unbroken non-(B-L) $SU(3)_C \times U(1)_{EM}$ group. Eq. (10) repeats the process starting with the Pati-Salam group so that the Pati-Salam Higgs representations can be

$$SO(10) \xrightarrow[\{54\}]{M_X} SU(4)_C \times SU(2)_L \times SU(2)_R \xrightarrow[\{45\}]{M_C} SU(3)_C \times SU(2)_L \times$$

$$SU(2)_R \times U(1)_{B-L} \xrightarrow[\{126\}]{M_{W_R}} SU(3)_C \times SU(2)_L \times U(1)_Y$$

$$\xrightarrow[\{10\}]{M_{W_L}} SU(3)_C \times U(1)_{EM} \tag{9}$$

$$SU(4)_C \times SU(2)_L \times SU(2)_R \xrightarrow[\Sigma(15,\ 1,\ 1)]{M_C} SU(3)_C \times SU(2)_L \times$$

$$SU(2)_R \times U(1)_{B-L} \xrightarrow[\Delta_R(\overline{10},1,3)]{M_{W_R}} SU(3)_C \times SU(2)_L \times U(1)_Y$$

$$\xrightarrow[\phi(1,2,2)]{M_{W_L}} SU(3)_C \times U(1)_{EM} \tag{10}$$

identified. In Eq. (9), I have denoted above each arrow the mass of the appropriate gauge boson and below each arrow the SO(10) representation of the Higgs that is responsible for the breaking. Eq. (10) shows the sequence of symmetry breakings that starts with the Pati-Salam group but continues in precisely the same fashion as in Eq. (9). Again, the masses above the arrows represent the masses of the gauge bosons and the Higgs representations are given below the arrows with the quantum numbers now of the Pati-Salam group.

Let us first consider the implications of the Pati-Salam group. From Eq. (10), we can say that the ordering of the masses is $M_C > M_{W_R} > M_{W_L}$ where $M_C > 10^4$GeV from the absence of $K_L \to \mu^+ + e^-$ decay [5] and $M_{W_L} = 80$ GeV. It can be shown that the Higgs Δ_R $(\overline{10}, 1, 3)$ - which breaks the LRS electroweak and strong interaction group down to the standard electroweak and strong interaction group (see Eq. (10)) - can give rise to a six-quark diagram that allows neutron oscillations (i.e. the $\Delta B = 2$ transition $n \to \bar{n}$). The six-quark diagram is due to a combination of the Yukawa interactions between the Δ_R Higgs and the quark pairs and the quartic interaction among the Higgs themselves; for details see the paper by Mohapatra and myself [11]. The resulting strength of the n-$\bar{n}$ transition is of the order of $\lambda h^3 M_{W_R}/M^6_{\Delta_R}$ where λ is the quartic coupling constant of the Higgs and h is the Yukawa coupling constant. An interesting value of the mixing time $t_{n-\bar{n}}$ ($\sim 10^7$ sec) is obtained if $M_{\Delta R} \sim 10^5$GeV, a value which is consistent with our present knowledge. It is noteworthy that the Pati-Salam group with the Higgs scalars chosen as in Eq. (10) leads to an accidental global symmetry that forbids proton decay [18]. This feature of the Pati-Salam group depends on the fact that the breaking of B-L local symmetry is related to the breaking of parity (see Eq. (5)).

It must be said that this rosy picture of the detectability of neutron oscillations on the basis of the Pati-Salam group has been questioned by Mohapatra and Senjanovic [19]. They point out that the decomposition of $\Delta_R(\overline{10}, 1, 3)$ is $\Delta_R(\overline{6}, 1, 3) + \Delta_R(\overline{3}, 1, 3) + \Delta_R(1, 1, 3)$ and that whereas $\Delta_R(1, 1, 3)$ gives rise to M_{W_R}, it is $\Delta_R(6, 1, 3)$ which is responsible for the neutron oscillation effect. They then argue that a large value of M_{Δ_R} (at the M_C level) would be required by the principle of minimal fine tuning that they introduce. Such a large value of M_{Δ_R} would suppress neutron oscillations within the framework of the Pati-Salam group and would compel us to move down to the group $SU(3)_C \times SU(2)_L \times SU(2)_R \times U(1)_{B-L}$ as the source of neutron oscillations if they occur. However, in my view, it would be premature to accept the principle of minimal fine tuning since it assumes the heavier Higgs know what the lighter are doing and fine tuning itself is so little understood. It is precisely because of this lack of clarity on the issue and to gain insight into the Higgs mechanism (proton decay occurs through the gauge bosons) that I would urge doing a good neutron oscillation experiment.

Moving on to the grand unification group SO(10), we first note that the basic representation (16) of this B-L group can accommodate the quarks and leptons of a single generation (including a right-handed neutrino); let us remind ourselves that this is in contrast to SU(5) GUT where two representations $\overline{5}$ and 10 are required to accommodate all quarks and leptons of a single generation (without a right-handed neutrino). Insofar as neutron oscillations are concerned, the principle of minimal fine tuning would further suppress them because of the elevation of the mass scale ($M_X > M_C$ - see Eq. (9)). Even accepting this principle (which I do not - see above), it has recently been shown [20] that the introduction of supersymmetry into SO(10) GUT creates new opportunities for neutron oscillations. The point is that with supersymmetry the dimension 9 of the six-quark diagram necessary for $n \rightarrow \bar{n}$ is reduced to dimension 7 so that the mass scale need not be so low to achieve observability of the process. Indeed, reasonable estimates of the parameters [20] lead to a mixing time $t_{n \rightarrow \bar{n}} \sim 10^7$ sec.

Apart from neutron oscillations, SO(10) GUT (without supersymmetry) is attractive for another reason; it turns out that the breakdown of SO(10), proceeding through the succession of B-L subgrounds shown in Eq. (9), gives a unique prediction of a longer proton lifetime (as required by experiment [2]) than that predicted by SU(5). This lifetime depends on M_{W_R} but not on the other intermediate mass scales nor on $\sin^2\theta_W$ in the limit where Higgs contributions are neglected (which is a very good limit). More explicitly, Tosa, Branco and I have found [21] that:

$$M_X = M_5 \left(\frac{M_5}{M_{W_R}}\right)^{1/2} \tag{11}$$

where M_5 is the grand unification mass for SU(5) GUT. Eq. (11) tells us that there is a simple connection between the mass scale of right-handed weak currents and the SO(10) unification mass which requires M_X to be larger than M_5 so that a longer proton lifetime follows. It is well known that when SO(10) breaks down to SU(5) x U(1) (a non-(B-L) group), the predicted proton lifetime is shorter than that given by SU(5) GUT. And so we have another test of the role of B-L local symmetry.

In concluding my talk, I should like to make two more remarks about the special character of B-L gauge groups. The first remark has to do with the answer to the question of what grand unified groups have B-L as a generator under very general assumptions about the color and electroweak classification schemes for the quarks and leptons. Assuming that the grand unified group contains $SU(2)_L$ and B-L as local symmetries and that there are no mirror fermions, one finds [22] that only our familiar groups SO(10) and $SU(4)_C \times SU(2)_L \times SU(2)_R$ are the only ones to qualify. It is surprising how restrictive the condition of having B-L as a generator is on the possible choice of unification groups.

Further support for the idea that B-L local symmetry has a very special role to play (comparable to $SU(2)_L$) in constraining unification models comes from a completely different direction, namely in searching for a preon model that predicts at least three generations of quarks and leptons. Tosa and I [23] started our search by considering a variety of preon models (with the composite quarks and leptons postulated to consist of either fermions alone or a combination of fermions and bosons) where the preons were assigned strong $SU(3)_C$ quantum numbers and either the standard electroweak or the LRS electroweak quantum numbers. The condition imposed on these models was the absence of exotics and anomalies in the strong and electroweak sectors. We were able to establish the impossibility of having two or more generations of quarks and leptons under this condition. It was clearly necessary to enlarge the preon group.

We continued this program [24] by allowing the color-flavor preon group, G(CF), to be any arbitrary partial unification or GUT group, and we also introduced a metacolor group, G(MC), that would confine the preons inside the quarks or leptons, in the same way the SU(3) color group confines quarks within hadrons. We limited ourselves to three-fermion preon models (since these seem to be the most interesting) and imposed the following set of plausible conditions: (1) the preons are (massless) Weyl spinors and belong to low-dimensional chiral representations of the gauged symmetry group $G(MC) \otimes G(CF)$; (2) G(MC) is an asymptotically free simple group so that one can think of the preons as having an existence in the asymptotic limit; (3) the Pauli principle can be generalized to the metacolor degree of freedom just as the Pauli principle, in the case of quarks, was generalized to the color degree of freedom; (4) no anomalies exist in the metacolor and color-flavor sectors; (5) the composite quarks and leptons are metacolor singlets (with the correct color and flavor quantum numbers) and are massless on the metacolor scale; (6) there are no low representation exotics nor mirror fermions.

Remarkably enough, it turns out that the only color-flavor preon groups that satisfy these conditions belong to the small class of B-L gauge groups that we have already discussed, namely SO(10) and $SU(4)_C \times SU(2)_L \times SU(2)_R$. (For example, the SU(5) color-flavor group is eliminated by its failure to satisfy condition (4) requiring anomaly cancellation.) As far as the metacolor group is concerned, we show that the only possible metacolor group with the property that the triple product of its basic representation yields a metacolor singlet quark or lepton, are SU(3), E_6, E_8, G_2 and F_4. The final result is that the only GUT preon models satisfying the above six conditions are SU(3)(MC) $\otimes$ SO(10)(CF) with 4 generations of quarks and leptons, F_4(MC) $\otimes$ SO(10)(CF) with 3 generations, and E_6(MC) $\otimes$ $SU(4)_C \times SU(2)_L \times SU(2)_R$ with 3 generations. The first two preon groups have only marginal asymptotic freedom whereas the third has very satisfactory asymptotic behavior. It is most interesting that all three permissible preon models contain B-L color-flavor gauge groups and that the most promising of the three contains the Pati-Salam color-flavor group and predicts exactly three generations. This last preon model even gives a proper ordering of the masses of Higgs doublets and Higgs triplets treated as preon condensates. The next difficult step is to understand the large scaling factors in moving from the leptons to the quarks and from one generation to the next.

I hope that I have communicated to you some sense of why I believe that B-L gauge models will play a larger role in the future development of particle theory.

REFERENCES

1. C. Rubbia (report at Singapore Conference, June 1983) has given the following numbers from the UA(1) experiment: $M(W_L) = 81 \pm 2$ GeV; $M(Z^0) = 95.2 \pm 2.5$ GeV, $\Gamma(Z^0) \lesssim 4.1$ GeV.
2. The (evaluated) running time of the IMB experiment as of June 1983 is 130 days without a single p decay event, yielding a minimum lifetime of 10^{32} yrs. (M. Goldhaber - private communication).
3. G. 't Hooft, Phys. Rev. Lett. 37, 8 (1976).
4. E. C. G. Sudarshan and I used $m_\nu = 0$ and the implied γ_5 invariance to motivate our original derivation of the V-A interaction [see E. C. G. Sudarshan and R. E. Marshak, Proc. Padua-Venice Conf. on Mesons and Newly Discovered

Particles (1957), reprinted in P. K. Kabir, Development of Weak Interaction Theory (Gordon and Breach, 1963), pp. 118-128.] However, for the reasons given in the text, this assumption may have to be modified slightly and a small mass assigned to the neutrino.
5. J. C. Pati and A. Salam, Phys. Rev. D10, 275 (1974); R. N. Mohapatra and J. C. Pati, Phys. Rev. D11, 566 (1975); G. Senjanovic and R. N. Mohapatra, ibid 12, 1502 (1975).
6. A. Gamba, R. E. Marshak and S. Okubo, Proc. Nat. Acad. of Sci. 45, 881 (1959); at the 1959 Kiev Conference, I suggested that a second (muon) neutrino would require another basic constituent p' (having the same charge as the proton) (See Proc. of Ninth Conf. on High Energy Phys., Acad. of Sci., Moscow 1960, Vol. II, p. 269). Cf. J. D. Bjorken and S. L. Glashow, Phys. Lett. 11, 255 (1964).
7. S. Sakata, Prog. Theor. Phys. 16, 686 (1956).
8. R. E. Marshak and R. N. Mohapatra, Phys. Lett. 91B, 222 (1980).
9. T. D. Lee and C. N. Yang [Phys. Rev. 104, 254 (1956)] have shown the Eötvos experiment requires B-L local symmetry to be broken.
10. A. Davidson, Phys. Rev. D20, 776 (1979); this paper contains an independent derivation of B-L as the fourth color; cf. V. Barger, W. Keung and E. Ma, Phys. Rev. Lett. 44, 1169 (1980).
11. R. N. Mohapatra and R. E. Marshak, Phys. Rev. Lett. 44, 1316 (1980); see also V. A. Kuz'min, JETP Lett. 12, 228 (1970) and S. L. Glashow, Recent Developments in Gauge Theory, Plenum 1980).
12. R. N. Mohapatra and G. Senjanovic, Phys. Rev. Lett. 44, 912 (1980); see also M. Gell-Mann, P. Ramond and R. Slansky (unpublished) and E. Witten, Phys. Lett. B91, 81 (1980).
13. R. N. Mohapatra and G. Senjanovic, Fermilab 80/61 THY (1980).
14. Cf. T. Rizzo and G. Senjanovic, Phys. Rev. D24, 704 (1981); ibid D25, 235 (1982).
15. These calculations have been done by P. Wang (VPI); cf. also W. Hollik, Z. Phys. C8, 149 (1981).
16. Cf. H . Fritsch and P. Minkowski, Phys. Reports 73, 126 (1981).
17. Riazuddin, R. E. Marshak and R. N. Mohapatra, Phys. Rev. D24, 1310 (1981).
18. See fn. 15 of ref. 11.
19. Cf. R. N. Mohapatra and G. Senjanovic, Phys. Rev. D27, 1601 (1983) and earlier work by the same authors.
20. D. Lüst, Univ. of Munich preprint (1983); see also Univ. of Munich Thesis (1982).
21. Y. Tosa, G. Branco and R. E. Marshak (to be published in Phys. Rev. D).
22. Y. Tosa, S. Okubo and R. E. Marshak, Phys. Rev. D27, 444 (1983).
23. Y. Tosa and R. E. Marshak, Phys. Rev. D26, 303 (1982).
24. Y. Tosa and R. E. Marshak, Phys. Rev. D27, 616 (1983).

LOW-ENERGY EFFECTS OF MIRROR LEPTONS

Presented by M. Roos

K. Enqvist[+], J. Maalampi[*,++], K. Mursula[++] and M. Roos[++,**]

[+] Research Institute for Theoretical Physics
University of Helsinki, Helsinki, Finland

[++] Department of High Energy Physics
University of Helsinki, Helsinki, Finland

[*] CERN, Genève, Switzerland

[**] Max-Planck-Institut für Physik und Astrophysik
Munich, Fed. Rep. Germany

1. INTRODUCTION

Mirror leptons are leptons with V+A interactions, all other quantum numbers being the same as for ordinary leptons. The charged mirror leptons must be heavier than about 20 GeV since they have not been pair-produced at PETRA[1], and lighter than a few hundred GeV, since their mass originates in the breaking of $SU(2)_W \times U(1)_Y$ and also because they can affect radiatively the value of the parameter $\rho = (M_W/M_Z\cos\theta_W)^2$, known to be close to 1.0[2].

Mirror leptons appear in GUTs such as $SO(n > 10)$[3,4], $SU(n < 5)$ and E_8[6]. They can also appear in composite models[7], and may also be needed in globally supersymmetric theories with an extra U(1)-group à la Fayet[8] to cancel axial anomalies connected with that group .

When mirror leptons mix with ordinary leptons, either via explicit Yukawa terms or radiatively[3a], they change the chiral structure of weak currents. Charged currents (CC) acquire a small V+A component, while for neutral currents (NC) only the axial vector part will be modified. These changes might be detectable already at present energies. In such "mirror mixing models" there naturally appear also mirror quarks. However, the quark-mirror quark mixing angles must be so small that they have no observable effect on the weak processes because the mirror quark mixing would otherwise spoil the subtle cancellation of flavor changing neutral currents (FCNC). In the most general case FCNC arise already at the tree level. According to a recent general analysis[10] of the first two generations, the observed absence of FCNC sets an upper bound of the order of 10^{-3} to the mirror quark mixing angles. A more stringent limit is obtained from the fact that FCNC are also suppressed at the one-loop level[11]. Therefore we will neglect mirror mixings in the quark sector.

In contrast, the effects of mixing of ordinary leptons and mirror leptons can be much larger. For this reason we make a statistical analysis of all the relevant CC and NC reactions to determine the range of mixing parameters allowed by recent data.

2. THE MIRROR MIXING MODEL[12-16]

An example of a GUT based on a symmetry group containing mirror fermions is SO(11)[3b]. The fundamental spinor representation is real, 32-dimensional, and decomposes under SO(10) as $\underset{\sim}{32} = \underset{\sim}{16} + \underset{\sim}{16}^*$, where the conventional lefthanded fermions of one family populate the $\underset{\sim}{16}$ spinor of SO(10). The $\underset{\sim}{16}^*$ contains a corresponding set of mirror fermions. Analogously to the ordinary fermions, the mirror electron L_e and its mirror neutrino N_e go into an $SU(2)_L$ doublet $\begin{pmatrix} N'^c_e \\ N'^c_e \end{pmatrix}_L$, the mirror quarks U, D form doublets $\begin{pmatrix} U'^c \\ D'^c \end{pmatrix}_L$ of three colors, and the remaining lefthanded mirror fermions of $\underset{\sim}{16}^*$, D'_L, U'_L, L'_e, N'_{e_L} form 8 singlets (primes denote weak interaction eigenstates). Consequently the coupling of the ordinary fermions to the W-boson is lefthanded, e.g. $\bar{\nu}'\gamma^\alpha(1-\gamma_5)e'W_\alpha$, whereas the mirror fermions couple righthandedly to the same W-boson, e.g. $\bar{N}_e'\gamma^\alpha(1+\gamma_5)L_e'W_\alpha$.

Let us denote the mass eigenstates[15]

$$\psi_\ell = \begin{pmatrix} \ell^- \\ L_\ell^- \end{pmatrix}, \quad \psi_{\nu_\ell} = \begin{pmatrix} \nu_\ell \\ N_\ell \end{pmatrix}, \tag{1}$$

where $\ell = e, \mu, \tau$. Allowing for the mixing of leptons with the corresponding mirror leptons through the mass matrix, the weak Lagrangian in the mass eigenstates basis takes the following form

$$\begin{aligned} \mathcal{L}^{CC} &= -\frac{g}{2\sqrt{2}} W_\alpha^+ \sum_\ell \{\bar{\psi}_{\nu_\ell}\gamma^\alpha[R(\varphi_\ell-\theta_\ell) - R(\varphi_\ell+\theta_\ell)\gamma_5\tau_3]\psi_\ell\} + h.c., \\ \mathcal{L}^{NC} &= -\frac{g}{4\cos\theta_W} Z_\alpha \sum_\ell \{\bar{\psi}_{\nu_\ell}\gamma^\alpha[1-R(2\varphi_\ell)\gamma_5\tau_3]\psi_{\nu_\ell} + \bar{\psi}_\ell\gamma^\alpha[4\sin^2\theta_W - \\ &\qquad -1 + R(2\theta_\ell)\gamma_5\tau_3]\psi_\ell\} \end{aligned} \tag{2}$$

where the $R(\theta)$ are the 2x2 orthogonal mixing matrices diagonalizing the respective mass matrix. The particular V,A-structure of the Lagrangian (2) is a direct consequence of the rotation of chirality eigenstates to mass eigenstates. If we restrict ourselves to the e and μ families, the Lagrangian (2) depends on the ℓ-L mixing angles θ_e, θ_μ and on the ν_ℓ-N_ℓ mixing angles φ_e, φ_μ.

In contrast to charged mirror fermions, mirror neutrinos may be light ($< m_e$) or heavy ($> m_K$). Their couplings to the muon are known to be severely restricted for masses between 10 MeV and 200 MeV .
Therefore we will consider the following cases:

model a) m_{N_e} , $m_{N_\mu} < m_e$ model b) $m_{N_e} < m_e$, $m_{N_\mu} > m_K$

model c) m_{N_e} , $m_{N_\mu} > m_K$ model d) $m_{N_\mu} < m_e$, $m_{N_e} > m_K$.

Model c) with heavy mirror neutrinos decoupled from low energy phenomena also corresponds to the more general case that only ordinary fermions are present but with general V,A CC couplings and modified NC couplings (cf. Eq. 17).

The limit of the standard GWS model is obtained when all the mixing angles vanish. The structure of the CC constraints is, however, such that the V-A limit may also be obtained for some choices of non-vanishing mixing angles[12,13]. For instance in model c, this is true if two of the angles vanish and the two other angles are equal but non-vanishing. This degeneracy is removed when both CC and NC constraints are used[15] (cf. Sec. 6).

In models with light mirror neutrinos the neutrino beam produced in pseudoscalar or nuclear beta decay is a coherent superposition of the two mass eigenstates[12,14]. This leads to an oscillating probability for finding an interacting neutrino or mirror neutrino. The phenomenon is analogous to the well known case of flavor oscillations, but the additional novel feature is that now both chiralities interact. The rate of oscillations depends on $\Delta m^2_\nu = |m^2_N - m^2_\nu|$.

The main decay mode of light mirror neutrinos as well as neutrinos is radiative decay $N_\ell \to \nu_\ell \gamma$ (or $\nu_\ell \to N_\ell \gamma$, whichever is kinematically allowed). For neutrinos $\lesssim 0$ (100 eV), however, there exist severe astrophysical constraints on the lifetime of unstable neutrinos[18]: the observed cosmic photon spectrum and the 3°K background radiation do not allow for the existence of light neutrinos with lifetimes $\lesssim 0\,(10^{20}$ s). For longlived neutrinos in the mass range 10-100 eV there are also limits coming from the observed galactic UV radiation[19].

One possibility to confirm with these limits is to set $\theta_\ell = 0$. Since φ_ℓ can still be large and since there are no very strict bounds on the masses of neutrinos and mirror neutrinos, neutrino-mirror neutrino oscillation is possible. The scattering cross sections are then dependent on the oscillating term which includes the unknown mass factor Δm^2_ν. Assuming that this factor is such that, in all scattering processes to be considered, the oscillations are already well developed before the beam hits the detector, it follows that the oscillating term is averaged and the mass dependence drops out. The beam is then effectively an incoherent mixture of neutrino and mirror neutrino mass eigenstates. Accordingly, we call this incoherent scattering. This is justified e.g. if m_N and m_ν are not accidentally degenerate and if the larger of them is at least of the order of 10 eV.

The second possibility which we consider here more thoroughly, is to set a strict upper bound on all light neutrino masses ($N=N_\ell, \nu_\ell$)

$$m_N < 2\cdot 10^{-2} \text{ eV},$$

in which case oscillations turn on very slowly. Therefore the scattering cross sections are now independent of the arbitrary mass difference and also of the respective neutrino mixing angle ϕ_ℓ. We call this coherent scattering.

A third possibility, to which we shall return briefly in Sec. 7, is to have mirror neutrinos heavier than about 1 keV.

Because of the scarcity of data available, we cannot afford to keep the ν_ℓ and N_ℓ masses as free parameters in the fits. The models a-d are therefore extreme cases which permit us to neglect these masses.

3. CHARGED CURRENT REACTIONS

3.1 Muon decay

In contrast to the V-A theory, the muons produced in $\pi_{\mu 2}$ and $K_{\mu 2}$ decay no longer have the trivial longitudinal polarization $P_{\mu\pm} = \mp 1$. Moreover, the muons have four possible decay channels.

$$\text{(i)} \quad \mu \to e\nu_\mu\bar{\nu}_e \qquad\qquad \text{(ii)} \quad \mu \to e\nu_\mu\bar{N}_e$$
$$\text{(iii)} \quad \mu \to eN_\mu\bar{\nu}_e \qquad\qquad \text{(iv)} \quad \mu \to eN_\mu\bar{N}_e \tag{3}$$

Kinematically allowed processes in model a are all the channels (i) to (iv), in model b: (i) and (ii), in model c: (i), and in model d: (i) and (iii). The energy spectrum of the decay electron can be parametrized in the usual form[20]. The spectrum determines the parameters ρ, δ, ξ and the Fermi coupling constant G_F. The different subprocesses contribute additively to these parameters.

We use the experimental values $\rho = 0.7517 \pm 0.0026$[21], V-A value $\frac{3}{4}$; $\delta = 0.7551 \pm 0.085$[21], V-A value $\frac{3}{4}$; $-\xi P_{\mu^+} = 0.975 \pm 0.015$[21], V-A value 1.

The theoretical expressions for the above quantities in the different models are detailed elsewhere[12].

3.2 Nuclear beta decay

Measurement of the longitudinal polarization of the electron in the nuclear ß-decay provides a well-known test for the V-A hypothesis. Pure V-A predicts the value $P_{\beta^-} = -v/c$ (for the most general case see Ref. 23). In the models we are considering this reduces to the same quantity P_{e^-} as in muon decay[12]. Thus we can average the measurement $-P_{e^+} = 1.008 \pm 0.054$[24] from muon decay, and $-(c/v)P_\beta$(Gamov-Teller) $= 1.001 \pm 0.008$[25] from nuclear beta decay. The V-A value is 1.

3.3 Inverse muon decay

The process

$$\text{(i)} \quad \nu_\mu + e^- \to \mu^- + \nu_e \tag{4}$$

provides information about the V,A structure of the leptonic CC, beyond what μ-decay without observation of the decay neutrinos gives.

In the fermion-mirror fermion mixing model there are three further processes that may contribute to the inverse muon decay:

$$\text{(ii)} \quad \nu_\mu + e^- \to \mu^- + N_e \qquad \text{(iv)} \quad N_\mu + e^- \to \mu^- + N_e$$

$$\text{(iii)} \quad N_\mu + e^- \to \mu^- + \nu_e$$

If N_μ is light enough to be produced in π and K decays, the neutrino beam scattering on electrons consists of two components ν_μ and N_μ, with well known relative probabilities. The allowed processes in the different models are: model a (i)-(iv), model b (i) and (ii), model c (i), and model d (i) and (iii).

In the experimental studies the differential cross section is presented in the units of $G_F^2 s/2$, the value of the cross section in the V-A limit. The expressions for this quantity which we denote S, are in the different models given in Ref. 12. Experimentally we have $S = 0.98 \pm 0.12$.

3.4 Pseudoscalar meson leptonic decay

For the muon polarization in pion decay we have $-P_{\mu^+} = 0.99 \pm 0.16$[27], V-A value 1.

We use in our analysis the ratios of the total electronic and muonic 2-widths for the pseudoscalars P = π, K:

$$R_P = \frac{\Gamma\ \text{total}\ (Pe_2)}{\Gamma\ \text{total}\ (P_{\mu 2})} = \frac{m_e^2(m_P^2 - m_e^2)^2}{m_\mu^2(m_P^2 - m_\mu^2)^2}(1 + \Delta_P)r \tag{5}$$

where Δ_P is the $O(\alpha)$ radiative correction[28]. The factor r is a ratio of terms depending on the parameters of the lepton currents. In models a, b and d also the nondiagonal decay $P \to \ell \bar{N}_\ell$ is allowed. The expressions for R_P in the different models are given in Ref. 12. We use an average of R_π and R_K: $R_{\pi,K} = 1.016 \pm 0.017$[21].

4. LEPTONIC NEUTRAL CURRENT REACTIONS

4.1 Elastic $\bar{\nu}_e e^-$-scattering

This reaction proceeds through both CC W^- exchange and NC Z^0 exchange. In models c and d N_e is heavy and thus decoupled from low energy phenomena. In models a and b the N_e is light, so we have both coherent and incoherent scattering.

Let us first consider incoherent scattering. There the beam is an incoherent mixture of the two mass eigenstates $\bar{\nu}_e$ and $\bar{N}_e$. We then have the following four subprocesses which contribute to "elastic" $\bar{\nu}_e e^-$-scattering:

$$\bar{\nu}_e e^- \to \bar{\nu}_e e^- , \qquad \bar{N}_e e^- \to \bar{\nu}_e e^- ,$$
$$\bar{\nu}_e e^- \to \bar{N}_e e^- , \qquad \bar{N}_e e^- \to \bar{N}_e e^- . \tag{6}$$

The coupling constants for each of these subprocesses can be read off from the Lagrangian (2) with $\theta_e = 0$.

In the case of coherent scattering the beam can be regarded effectively as an incoherent mixture of weak eigenstates $\bar{\nu}'_e$ and $\bar{N}'_e$, whose weak currents written in a mixed basis of neutrino weak eigenstates and charged lepton mass eigenstates are:

$$J^\mu_{CC} = -\frac{g}{2\sqrt{2}}\{\cos\theta_\ell\, \bar{\nu}'_\ell \gamma^\mu(1-\gamma_5)\ell - \sin\theta_\ell\, \bar{N}'_\ell \gamma^\mu(1+\gamma_5)\ell + \sin\theta_\ell\, \bar{\nu}'_\ell \gamma^\mu(1-\gamma_5)L_\ell - \cos\theta_\ell\, \bar{N}'_\ell \gamma^\mu(1+\gamma_5)L_\ell\}, \tag{7}$$

$$J^\mu_{NC} = -\frac{g}{4\cos\theta_W}\{\cos\theta_\ell\, \bar{\nu}'_\ell \gamma^\mu(1-\gamma_5)\nu_\ell - \sin\theta_\ell\, \bar{N}'_\ell \gamma^\mu(1+\gamma_5)\nu_\ell + \sin\varphi_\ell\, \bar{\nu}'_\ell \gamma^\mu(1-\gamma_5)N_\ell + \cos\varphi_\ell\, \bar{N}'_\ell \gamma^\mu(1+\gamma_5)N_\ell\}. \tag{8}$$

All the subprocesses (6) with the incoming neutrinos being now the weak interaction eigenstates $\bar{\nu}'_e$ and $\bar{N}'_e$, add up incoherently to the total cross section. The cross sections for both coherent and incoherent scattering have been given in Ref. 11. We note here that the cross section is proportional to the oscillation factor $P^e_{\nu\nu}$ given by

$$P^e_{\nu\nu} = 1 \quad \text{(coherent case)},$$
$$P^e_{\nu\nu} = 1 - \tfrac{1}{2}\sin^2 2\varphi_e \quad \text{(incoherent case)}. \tag{9}$$

Thus for models a and b in the incoherent case this reaction furnishes information on φ_e, whereas no other reaction does.

We note also that the cross section for this reaction in model b does depend also on φ_μ and θ_μ, which might seem surprising, since no muons or muon neutrinos participate in the reaction. The reason is that the scale of the cross section is given by a modified Fermi constant $\ddot{G}$, determined in muon decay and thus dependent (differently in the four models) on some of the four mixing angles[12]. This dependence is, however, rather weak.

We use the experimental cross section for low energy neutrinos ,

$$\langle \sigma(\bar{\nu}_e e^-)\rangle_{low} = (7.6 \pm 2.2)\ 10^{-46}\ \mathrm{cm}^2 \quad \text{(standard model: 5.58)}$$

together with the Davis antineutrino spectrum[30].

4.2 Elastic $\overset{(-)}{\nu}_\mu e^-$-scattering

These cross sections follow easily from the previous case by changing φ_e to φ_μ and by neglecting the electron mass terms. Now it is models a and d, containing a light N_μ, which allow to distinguish between coherent and incoherent beams.

We take the experimental values from Mo[31], however, subtracting out the (then preliminary) cross sections of the CHARM experiment:

$$\sigma(\bar{\nu}_\mu e^-) = (1.54 \pm 0.67)\cdot 10^{-42} E_{\bar{\nu}}\ [\mathrm{cm}^2/\mathrm{GeV}] \quad \text{(standard model: 1.38)},$$

$$\sigma(\nu_\mu e^-) = (1.46 \pm 0.24)\cdot 10^{-42} E_{\nu}\ [\mathrm{cm}^2/\mathrm{GeV}] \quad \text{(standard model: 1.50)}.$$

From the CHARM experiment[32] we take the cross section ratio

$$\frac{\sigma(\nu_\mu e^-)}{\sigma(\bar{\nu}_\mu e^-)} = 1.37 \begin{array}{l} +0.65 \\ -0.44 \end{array} \quad \text{(standard model: 1.09)}.$$

4.3 NC effects in $e^+e^- \to \mu^+\mu^-$

The formal expression for the differential cross section $\frac{d\sigma}{d\Omega}(e^+e^- \to \mu^+\mu^-)$ is the same for all the models considered in this paper. It is given by

$$\frac{d\sigma}{d\Omega}(e^+e^- \to \mu^+\mu^-) = \frac{\alpha^2}{4s}\left[F_1(s)(1+\cos^2\theta) + F_2(s)\cos\theta\right], \tag{10}$$

where $F_1(s)$ and $F_2(s)$ are functions of the vector and axial vector NC coupling constants of the leptons, depending on the charged lepton mixing angles θ_e and θ_μ as well as on $\sin^2\theta_W$[11].

As experimental input for the forward-backward asymmetry $A^{FB}=3F_2/8F_1$ we use the averaged values[33] given for three energy regions:

A^{FB} (14 GeV) = (2.6 ± 3.0) % (standard model: -1.4),
A^{FB} (22 GeV) = (-6.9 ± 3.4) % (standard model: -3.7),
A^{FB} (34 GeV) = (-11.9 ± 1.5) % (standard model: -10.5).

There also exists separate information on the coefficient of the $(1 + \cos^2\theta)$-term in Eq. (10), determining

$$h_{VV} = \frac{\tilde{G}}{G_F} \cdot \frac{1}{4} \left(1 - 4\sin^2\theta_W\right)^2 . \tag{11}$$

We use the experimental average of PETRA measurements[34] $h_{VV}^{exp} = 0.009 \pm 0.040$ (standard model: 0.0005) as an additional constraint furnishing information on θ_W.

5. SEMILEPTONIC NEUTRAL CURRENT REACTIONS

5.1 Neutrino hadron scattering

As we have argued in Sec. 1 (and more fully in Ref. 11), the mixing between quarks and mirror quarks can be neglected. This is fortunate because otherwise the proliferation of parameters would make all semileptonic data useless. Instead we find that some of the cross sections of ν and $\bar{\nu}$ scattering on hadrons can be modified for the mirror mixing model when the leptonic mixing angles are small. To second order in the mixing angles the cross sections for nuclear target N become:

$$\begin{aligned}
\sigma_{CC}(\overset{(-)}{\nu}_\mu N) &= \frac{\tilde{G}^2}{G_F^2} k_{CC}\, \sigma_{CC}^{V-A}(\overset{(-)}{\nu}_\mu N) , \\
\sigma_{NC}(\nu_\mu N) &= \frac{\tilde{G}^2}{G_F^2} \left\{ k_{NC}^1\, \sigma_{NC}^{V-A}(\nu_\mu N) + k_{NC}^2\, \sigma_{NC}^{V-A}(\bar{\nu}_\mu N) \right\} , \\
\sigma_{NC}(\bar{\nu}_\mu N) &= \frac{\tilde{G}^2}{G_F^2} \left\{ k_{NC}^1\, \sigma_{NC}^{V-A}(\bar{\nu}_\mu N) + k_{NC}^2\, \sigma_{NC}^{V-A}(\nu_\mu N) \right\} ,
\end{aligned} \tag{12}$$

where the correction factors e.g. in models b and c are

$$k_{CC} \simeq 1 - \theta_\mu^2 - \varphi_\mu^2 , \quad k_{NC}^1 \simeq 1 - 2\varphi_\mu^2 , \quad k_{NC}^2 \simeq 0 . \tag{13}$$

We use the data on isoscalar targets $\bar{N}$ in the form[35,36]

$$R^{\pm} = \frac{\sigma_{NC}(\nu_\mu \bar{N}) \pm \sigma_{NC}(\bar{\nu}_\mu \bar{N})}{\sigma_{CC}(\nu_\mu \bar{N}) \pm \sigma_{CC}(\bar{\nu}_\mu \bar{N})} . \tag{14}$$

For the Paschos-Wolfenstein ratio R^- the results of four experiments[38,39] can be combined to yield $R^- = 0.264 \pm 0.007$ (standard model: 0.261). The CFRR group[39] has also reported a value of R^+ which we can use: $R^+ = 0.315 \pm 0.009$ (standard model: 0.325).

We also use some data on inclusive $\nu(\bar{\nu})$ scattering on protons and neutrons[37]: $R_p = 0.47 \pm 0.064$ (standard model: 0.40), $R_n = 0.22 \pm 0.031$ (standard model: 0.24).

5.2 Charge asymmetry: (i) polarized electrons on deuterium

The asymmetry can be written in the form $A_D(x,y)/Q^2 = a_1(x) + a_2(x)F(y)$ where a_1 and a_2 depend on $\sin^2\theta_W$ and $\cos 2\theta_e$, and $F(y)$ is a known function. We can then compare the theoretical formulas with the results of the SLAC experiment[40]: $a_1 = (-9.7 \pm 2.6) \times 10^{-5}$ GeV^{-2} (standard model: -7.6), $a_2 = (4.9 \pm 8.1) \times 10^{-5}$ GeV^{-2} (standard model: -0.7).

(ii) muons on carbon

There also exists a result for the asymmetry B from μC-scattering[41]. The expression valid for B in all mixing models is

$$B(-P,P) = -\frac{3}{4}\cdot\frac{3\tilde{G}}{5\sqrt{2}\pi\alpha}\left[\cos 2\theta_\mu + P(4\sin^2\theta_W - 1)\right]F(y)Q^2. \quad (15)$$

The experimental value is (with P = 0.81):

$$\frac{B(-P,P)}{F(y)Q^2} = -(1.40 \pm 0.35) \times 10^4 \text{ GeV}^{-4} \text{ (standard model: } -1.56).$$

6. RESULTS OF THE FITS

The values of the mixing angles have been determined previously[12,13] using leptonic CC data alone. Extending the analysis to neutral currents, they help to determine the mixing parameters at the cost of introducing only one additional parameter, $\sin^2\theta_W$. For example e^+e^--asymmetry measures essentially the product $\cos 2\theta_\mu \cos 2\theta_e$, and the $\overset{(-)}{\nu}_\mu$e-scattering depends on $\cos 2\varphi_\mu$.

Furthermore, as noted in Ref. 12, since the CC data do not involve any scattering experiment with $\overset{(-)}{\nu}_e$ beams, the corresponding neutrino mixing angle φ_e is left arbitrary in models a and b. Including now the elastic $\bar{\nu}_e$e-scattering in the analysis we get also φ_e constrained in models a and b with incoherent scattering.

In all models we find that the mirror mixing angles are consistent with zero, although their best values may differ from zero.

For example, in the general 5-parametric model c we find with 68 % confidence: $\theta_e < 16.4°$, $\varphi_e < 20.5°$, $\theta_\mu < 13.2°$, $\varphi_\mu < 21.9°$ at $\sin^2\theta_W = 0.218$.

We note that the effect of mixing on $\sin^2\theta_W$ is to decrease it with respect to its standard model value $\sin^2\theta_W = 0.239 \pm 0.007$, the reason for this value being about 1σ higher than in other global fits of NC data[2,42] is that we have used the data on inclusive $\nu(\bar{\nu})$ scattering on isoscalar target in the Paschos-Wolfenstein form.

The θ_ℓ are much more constrained in coherent models ($\theta_e < 4°$, $\theta_\mu < 9°$) than the φ_ℓ in coherent models ($\varphi_e < 36°$, $\varphi_\mu < 18°$, θ_e and/or θ_μ zero). This leaves the possibility of large neutrino-mirror neutrino oscillations valid. Note that in coherent models one or two φ_ℓ's remain totally unrestricted.

The inclusion of NC data in addition to CC only slightly modifies the values and limits of the mixing angles[12] of models a, b and d. However, model c is now better determined, due to the resolution of the above mentioned multiple V-A limit problem of CC data.

7. BEAM DUMP EXPERIMENTS

The observation in the beam dump experiments[43,44] is that the prompt neutrino flux from the beam dump produces electrons and muons in the ratio

$$R_1 = \frac{e^+ + e^-}{\mu^+ + \mu^-} \simeq 0.52 \pm 0.15 \tag{16}$$

This is in contrast to the expectation, $R_1 = 1$, if universality holds and if the main source of prompt neutrinos is the decay of charmed hadrons.

Various suggestions have been made to explain this e-μ asymmetry. If the electron neutrino oscillates to a neutrino of another flavour, say[45], the amount of produced electrons would decrease correspondingly. Also speculations on charged Higgses with non-universal couplings or an enhanced purely leptonic branching ratio of the charmed particles[47] have been proposed.

None of these alternatives, however, has reached the observed level of asymmetry without introducing other undesirable features. The flavor oscillation required surpasses by far the oscillations allowed by present dedicated experiments. Enhancing the leptonic branching ratio of D and F leads at most to $R_1 = 0.9$[47]. On the other hand, the possible mixing of mirror and ordinary fermions naturally leads to a modified non-universal V,A structure of weak currents, which might be the cause of the leptonic asymmetry[16].

Let us first consider the status of e-μ universality in charged weak currents, independently of the mirror mixing model. Modifying the conventional V-A structure to a general mixture of (real) V and A couplings[11,12,14] of fermions f, f', we write

$$J^{\alpha} = \bar{f}\gamma^{\alpha}(V_f - A_f\gamma_5)f' . \tag{17}$$

We can obtain limits on the ratios of lepton couplings,

$$\lambda_i \equiv \frac{A_i}{V_i} \ , \ i = e, \mu \quad ; \quad \kappa \equiv \frac{V_\mu}{V_e} . \tag{18}$$

We find[11,15] the following best fit values (and 1σ-limits) for them:

$$\begin{aligned} \lambda_e &= 1.085 \ (< 1.15) , \\ \lambda_\mu &= 1.00 \ (< 1.115) , \\ (0.905 <) \ \kappa &= 1.047 \ (< 1.115) .^{*)} \end{aligned} \tag{19}$$

The constraints used (except R_p) depend only on λ_e and λ_μ and are symmetrical with respect to the replacements $\lambda_i \to 1/\lambda_i$. The parameter κ which directly measures e-μ universality is constrained only by the remarkably accurate pseudoscalar rates R_p, cf. Section 3.4.

The beam dump ratio can now be expressed[16] by

$$R_1 \simeq \frac{1}{\kappa^4} \cdot \frac{1 + 6\lambda_e^2 + \lambda_e^4}{1 + 6\lambda_\mu^2 + \lambda_\mu^4} . \tag{20}$$

Inserting the limits (19) one finds that the maximum non-universality allowed yields $R_1 \gtrsim 0.96$. Thus non-universality of the charged currents alone cannot explain the beam dump result in this general formalism.

Consider next the mirror lepton models c and d where N_e is heavy enough not to be produced in weak decays of either light or heavy particles ($m_{N_e} > m_D$). Then the situation is similar to the previous general case and $R_1 \gtrsim 0.96$.

Models a and b, however, with a light mirror neutrino N_e, offer a new situation: there appears an oscillating pattern in the cross sections due to neutrino-mirror neutrino oscillations. In particular, the cross sections of the $\overset{(-)}{\nu}_e$ beam are modified (see Refs. 11,12,14,16 for details) with an oscillation factor at the distance x

$$P^e_{\nu\nu}(x) = 1 - \frac{1}{2}\sin^2 2\varphi_e \left(1 - \cos\frac{2\pi x}{L}\right) . \tag{21}$$

*)This limit is further tightened by the new measurement of Bryman et al.[48] of R .

The oscillation length is, as usual, $L=4\pi E_\nu/\Delta m_e^2$, where $\Delta m_e^2 = |m^2_{N_e} - m^2_{\nu_e}|$. For Δm_e^2 sufficiently large $P^e_{\nu\nu}(x)$ is averaged to $\bar{P}^e_{\nu\nu} = 1 - \frac{1}{2}\sin^2 2\varphi_e$ and an effectively incoherent scattering follows.

In an experiment with reactor antineutrinos it is enough to take $\Delta m_e^2 \gtrsim 0.1 - 0.2\ eV^2$ in order to have the beam flux incoherent and thus depleted by the factor $\bar{P}^e_{\nu\nu}$, while in all beam dump experiments a somewhat heavier mass scale is needed (in most of them[43,44a,44b], $\Delta m^2_e \gtrsim 20\ eV^2$). This means that if the observed electron deficiency in beam dump experiments is due to neutrino-mirror neutrino oscillations it would also lead to a diminished flux in reactor experiments. However, the recent reactor experiments[49] tell us that the reactor antineutrino flux is known to an inaccuracy at 5-10 %. Accordingly, the ratio R_1 can only be allowed to decrease to the value 0.9 in models a and b with light ($< m_e$) mirror neutrinos.

One can thus conclude that neither very light ($< m_e$) nor very heavy ($> m_p$) mirror neutrinos can provide an explanation for a small ratio R_1. For the mass range $m_e < m_{N_e} < (m_K - m_\pi)$ there is an experimental limit from K^+_{e3}-decay which tells that no sizeable depletion of the ν_e-flux is allowed. Since the experimental arrangement is similar to the beam dump experiments, mirror neutrinos with a mass <200 MeV, small enough to be produced in K^+_{e3}-decay cannot have mixing with ν_e's large enough to make the R_1-ratio small. However, there still remains the possibility that N_e is heavier than 200 MeV and smaller than about 1 GeV. This is discussed in more detail in Ref.16.

ACKNOWLEDGEMENTS

It is a pleasure to acknowledge the hospitality and generosity of the Max-Planck-Institut, where one of us (M.R.) was staying during the preparation of this work. We also want to acknowledge the financial support of the Emil Aaltonen Foundation (to K.E.) and of the Academy of Finland (to J.M. and M.R.).

REFERENCES

1. For a recent review see K.H. Mess and B.H. Wiik, preprint DESY 82011 (1982), unpublished.
2. I. Liede and M. Roos, Nucl. Phys. B167, 397 (1980).
3a. K. Enqvist and J. Maalampi, Nucl. Phys. B191, 189 (1981).
b. J. Maalampi and K. Enqvist, Phys. Lett. 97B, 217 (1980).
4. M. Ida, Y. Kayama and T. Kitazoe, Prog. Theor. Phys. 64, 1745, (1980); R.N. Mohapatra and B. Sakita, Phys. Rev. D21, 1062,

(1980); H. Sato, Phys. Lett. 101B, 233 (1981); F. Wilczek and A. Zee, Phys. Rev. D25, 553 (1982).
4. N.S. Baaklini, Phys. Rev. D21, 343 (1980); J. Chakrabarti, M. Popović and R.N. Mohapatra, Phys. Rev. D21, 2312 (1980); C. W. Kim and C. Roiesnel, Phys. Lett. 93B, 343 (1980); Z.-q. Ma, T.-s. Tu, P.-y. Xue and X.-j. Zhou, Phys. Lett. 100B, 399 (1981); I. Umemura and K. Yamamoto, Phys. Lett. 100B, 34 (1981); M. Chaichian, Yu. N. Kolmakov and N.F. Nelipa, Phys. Rev. D25, 1377 (1982); J.C. Pati, A. Salam and Strathdee, Phys. Lett. 108B, 121 (1982).
6. I. Bars and M. Günaydin, Phys. Rev. Lett. 45, 859 (1980).
7. M. Chaichian, Yu. N. Kolmakov and N.F. Nelipa, Helsinki Univ. preprint HU-TFT 82-15 (1982) (unpublished).
8. P. Fayet, Phys. Lett. 69B, 489 (1977); 84B, 416 (1979).
9. P. Fayet, Proc. XVII Rencontre de Moriond on Elem. part., Ed. Tran Thanh Van, Editions Frontieres, Gif-sur-Yvette (1982) p. 483),
10. I. Umemura and K. Yamamoto, Phys. Lett. 108B, 37 (1982).
11. K. Enqvist, K. Mursula, and M. Roos, Helsinki University preprint HU-TFT-82-51 (1982) (to be publ. in Nucl. Phys. B).
12. J. Maalampi, K. Mursula and M. Roos, Nucl. Phys. B207, 233 (1982).
13. S. Nandi, A. Stern and E.C.G. Sudarshan, Phys.Rev. D26,2522 (1982).
14. J. Maalampi and K. Mursula, Z. Phys. C16, 83 (1982).
15. K. Enqvist, K. Mursula, J. Maalampi and M. Roos, Helsinki Univ. preprint HU-TFT-81-18 (1981) (unpublished).
16. K. Enqvist, K. Mursula, and M. Roos, Helsinki Univ. preprint HU-TFT-83-10 (1983), revised version.
17. Y. Asano et al., Phys. Lett. 104B, 84 (1981); R. Abela et al., SIN preprint PR 81-06 (1981).
18. See eg. F.W. Stecker and R.W. Brown, NASA preprint 83873 (1981) (unpublished) and references therein.
19. A. de Rújula and S.L. Glashow, Phys. Rev. Lett. 45, 942 (1980).
20. F. Scheck, Phys. Reports 44, 187 (1978).
21. M. Roos et al., Review of Particle Properties, Phys. Lett. 111B, 1, (1982).
22. V.V. Akhmanov et al., Soviet J. of Nucl. Phys. 6, 230 (1968).
23. R.E. Marshak, Riazuddin, and C.P. Ryan, Theory of weak interactions in particle physics (Wiley-Interscience, 1969).
24. F. Corriveau et al., Phys. Rev. D24, 2004 (1981).
25. F.W. Koks and J. von Klinken, Nucl. Phys. A272, 61 (1976).
26. M. Jonker et al., Phys. Lett. 93B, 203 (1980), and F. Bergsma et al. Phys. Lett. 122B, 465 (1983).
27. R. Abela et al., Nucl. Phys. A395, 413 (1983).
28. T. Goldman and W.J. Wilson, Phys. Rev. D15, 709 (1977).
29. F. Reines, H.S. Gurr and H.W. Sobel, Proc. Int. Neutrino Conf., Aachen, 1976 (Vieweg, Braunschweig, 1977) p. 217; Phys. Rev. Lett. 37, 315 (1976).
30. B.R. Davis et al., Phys. Rev. C19, 2259 (1979).
31. L.W. Mo, in "Neutrino Physics and Astrophysics", ed. Ettore Fiorini, Plenum Press, New York, 1982, p. 191.

32. M. Jonker et al., Phys. Lett. 105B, 242 (1981), and CERN preprint CERN-EP/82-109 (1982) (unpublished).
33. P. Steffen, DESY preprint DESY 82-039 (1982)(unpublished).
34. F. Niebergall, Proc. Int. Conf. Neutrino '82, Balatonfüred, 1982, Vol. II, p. 62 (Budapest, 1982).
35. E.A. Paschos and L. Wolfenstein, Phys. Rev. D7, 91 (1972).
36. P.Q. Hung and J.J. Sakurai, Phys. Lett. 63B, 295 (1976).
37. T. Kafka et al., Phys. Rev. Lett. 48, 910 (1982).
38. M. Holder et al., Phys. Lett. 72B, 254 (1977); C. Geweniger, Proc. Int. Neutrino Conf., Bergen, 1979 (Åvstedt Industrier AIs, 1979) Vol. II, p. 392; M. Jonker et al., Phys. Lett. 99B, 265 (1981).
39. R. Blair et al., Proc. Int. Neutrino Conf., Hawaii, 1981, Vol. I, p. 311.
40. C.Y. Prescott et al., Phys. Lett. 84B, 524 (1979).
41. A. Argento et al., Phys. Lett. 120B, 245 (1983).
42. J.E. Kim, P. Langacker, M. Levine and H.H. Williams, Rev. Mod. Phys. 53, 211 (1980).
43. P. Fritze et al., Phys. Lett. 96B, 427 (1980).
44a. M. Jonker et al., Phys. Lett. 96B, 435 (1980);
b. H. Abramowicz et al., Z. Phys. C13, 179 (1982);
c. R.C. Ball et al., Proc. Int. Conf. Neutrino Conf. Neutrino '82, A. Frenkel and L. Jenik (eds.), Budapest (1982), vol. I, p. 89.
45. A. De Rujula et al., Nucl. Phys. B168, 54 (1980).
46. V. Barger, F. Halzen, S. Pakvasa and R.J.N. Phillips, Phys. Lett. 116B, 357 (1982).
47. E.L. Berger, L. Clavelli and N.R. Wright, Argonne preprint ANL-HEP-PR-82-32 (1982).
48. D. Bryman et al., Phys. Rev. Lett. 50, 7 (1983).
49. H. Kwon et al., Phys. Rev. D24, 1097 (1981);
J.L. Vuilleumier et al., Phys. Lett. 114B, 298 (1982).
50. H. Deden et al., Phys. Lett. 98B, 310 (1981).

ELECTROWEAK PHENOMENOLOGY IN COMPOSITE MODELS

Dieter Schildknecht

Max-Planck-Institut für Physik und Astrophysik
Munich, Fed. Rep. Germany, and (permanent address)
Department of Theoretical Physics, University of Bielefeld
Bielefeld, Fed. Rep. Germany

ABSTRACT

We review the phenomenology of electroweak theories based on composite leptons and quarks. In particular, we concentrate on a concept of compositeness, in which the short range weak interaction between leptons and quarks is treated in close analogy to the short range strong interaction between hadrons. The standard neutral current structure as well as the standard boson mass relations appear as dynamical low energy approximations based on W-dominance without invoking $SU(2)_I \times U(1)_Y$ symmetry at short distances.

1. INTRODUCTION

1.1. The Standard Model Confronted with Experiment

The standard $SU(2)_I \times U(1)_Y$ spontaneously broken electroweak gauge theory[1] is empirically strongly supported[2] by the neutral current structure observed at "low" energies. More direct empirical support is due to the first indication for the existence of the postulated charged W boson reported[3,4] by the UA1 and UA2 collaborations at this meeting. From the observed five candidates for $W^\pm$ production, $p\bar{p} \rightarrow W^\pm + X$, the UA1 collaboration extracts a lower limit of $M_W > 73$ GeV and masses of $M_W = 74 \pm 4$ GeV and $M_W = 81 \pm 5$ GeV, depending on the analysis applied to their data. The observed masses are consistent with the expectation of the $SU(2)_I \times U(1)_Y$ gauge theory, which (upon applying radiative corrections[5]) leads to $M_W = 82 \pm 2.4$ GeV.

All our present empirical information is thus consistent with the spontaneously broken $SU(2)_I \times U(1)_Y$ gauge theory. Nevertheless, there is a large part of this theory which has not been empirically verified so far. In particular:

i) We do not yet know whether it is really spontaneous symmetry breaking which is responsible for the charged W mass as well as the $W^{(o)}$ - B mass mixing matrix, which upon diagonalization yields the Z mass and the massless photon. Breaking $SU(2)_I \times U(1)_Y$ local gauge invariance explicitly by introducing boson mass terms as well as mass mixing terms yields all consequences of the theory which have been empirically verified so far, including the boson mass formulae (without radiative corrections). This is related to the fact that $SU(2)_I \times U(1)_Y$ symmetry at short distances is implicitly required by restricting symmetry breaking to mass mixing terms, while dropping additional current mixing terms, which a priori would be allowed. Observing the Higgs scalar particle and empirically establishing its properties are thus absolutely essential for experimental confirmation of spontaneous symmetry breaking.

ii) We do not know the empirical high energy behaviour (in the TeV region) of lepton, quark and weak boson interactions, and we do not yet have a clear empirical test of the renormalizability of the theory, which is closely connected with the high energy behaviour.

From i) and ii), I conclude that a world is not empirically excluded in which lepton and quark masses are generated without invoking spontaneous symmetry breaking (e.g., by a confinement mechanism analogous to QCD), and lepton (quark) interactions at high energies (TeV) show features which are similar to hadron interactions (QCD). This is the world of composite intermediate bosons, leptons and quarks about which I will be talking. We will see that all electroweak phenomena so far observed are accounted for within such a framework, whereas the high energy (TeV) phenomena are expected to differ from the standard scenario.

1.2. Motivations for Compositeness

Apart from not being ruled out by experiment, there are several motivations to contemplate on further compositeness[6,7] on the quark, lepton and weak boson level at distances smaller than 10^{-16} cm:

i) The existence of apparently identical families of leptons and quarks with a mass pattern, which has withstood a convincing explanation so far.

ii) Connected with i), the existence of quark (lepton?) mixing angles. The family pattern and the mixing angles may be the outgrowth of a rich dynamical structure at short distances ($\lesssim 10^{-16}$ cm).

iii) The "desert syndrome" (an expression borrowed from A. Salam[8]), i.e., the assumption of no further structures at short distances. According to grand unified theories this assumption is upheld in a self-consistent manner from 100 GeV to 10^{15} GeV. I believe that nature may in fact be richer than the GUTs' imagination.

iv) The possibility that all forces in nature are due to unbroken gauge interactions. Such a conjecture must not be necessarily realized in nature, but it may be worthwhile to be explored.

There may be additional motivations. While it is fair to say that no composite model has so far succeeded in providing us with much insight into such basic questions as to the origin of the family structure, I still believe that the idea of further structure at very short distances ($\lesssim 10^{-16}$ cm) may open up a road for finding answers to such questions.

2. COMPOSITENESS AND W-DOMINANCE

2.1 The QCD-QHD Parallelism

I am advocating a point of view[9] according to which compositeness of weak bosons, leptons and quarks is realized in a manner which shows certain analogies to QCD and hadron physics.[10] Leptons, quarks and weak bosons are assumed to be composed of subunits, which are bound by some kind of confining hypercolor force. The short range ($\lesssim 10^{-16}$ cm) weak interaction appears as a residue of the hypercolor force,[11] just as strong interactions appear as a residue of the color force. With elementary photons and gluons the theory should thus be based on the gauge group $U(1)_{EM} \times SU(3)_C \times G_{HC}$, where the subscripts refer to electromagnetic, color and hypercolor, respectively. Of course, the hypercolor dynamics cannot be written down and solved from first principles. This has not even been carried through for QCD in a satisfactory manner. In developing an effective low energy description of weak interactions, I will thus mainly be guided by analogies to low energy hadron interactions, which explicitly or implicitly are consequences of the underlying color interactions of QCD. I will start from a parallelism between color (hadron) and hypercolor (weak) interactions.

a) QCD

i) Strong interactions are a residue of the color force. In the low energy approximation a process such as, e.g., pion-pion scattering is described by the quark line diagram of Fig. 1, which is dominated by ρ^0 exchange (with an additional scalar (ε) contribution in the neutral channel). The effective interaction of the bound state ρ mesons is invariant under strong $SU(2)_I$ transformations.

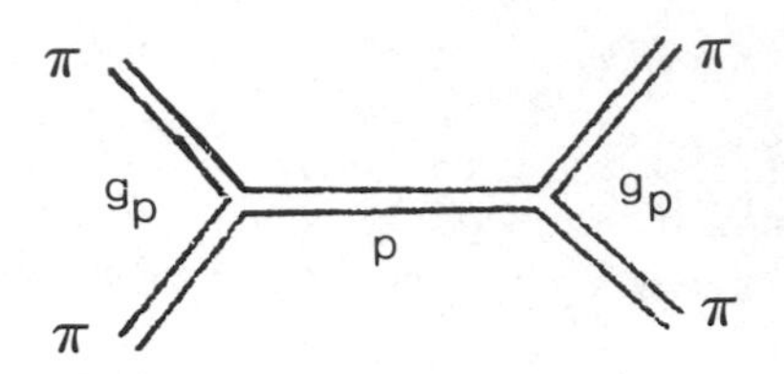

Fig. 1. Quark line diagram for pion-pion scattering.

ii) The photon quark interaction leads to transition of the photon to ρ^o, ω, ... mesons, which break strong isospin invariance (see Fig. 2).

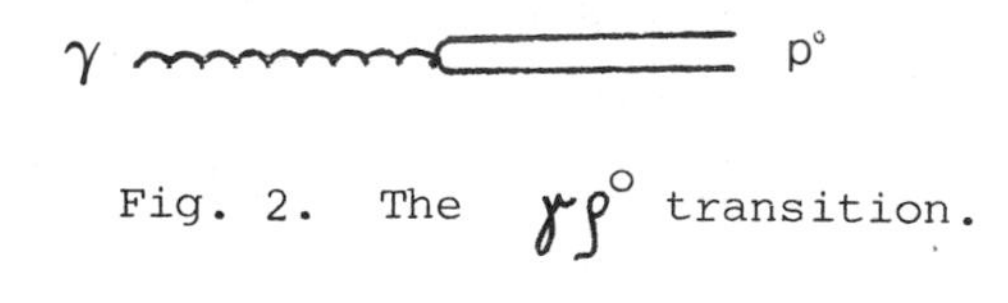

Fig. 2. The $\gamma\rho^o$ transition.

iii) The photon interacts with quarks in hadrons. As a consequence, it interacts with hadrons via electromagnetic form factors. These are dominated by the least massive mesons, i.e., the pion electromagnetic form factor is dominated by the ρ^o ("ρ^o-dominance").

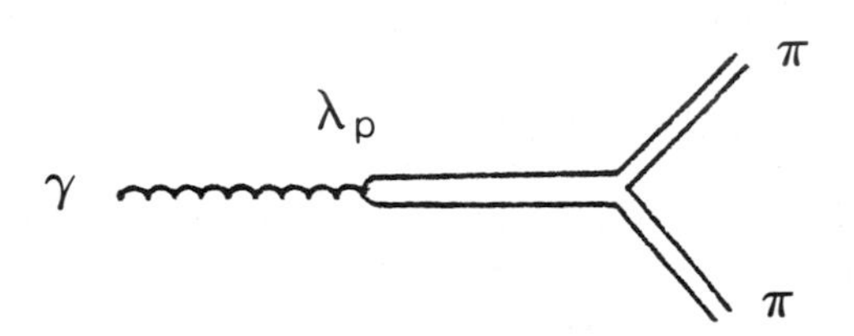

Fig. 3. ρ -dominance of the pion form factor.

b) QHD: We are now going to translate the above features of low energy hadron-hadron and photon-hadron interactions into QHD ("Quantumhypercolordynamics" or "Quantumhaplodynamics") as follows:

i) The short range weak interaction is the residue of a hypercolor force. In the low energy approximation a process such as lepton-lepton scattering is described by the diagrams of Fig. 4. The substructure of leptons, quarks and weak bosons is assumed to generate an effective interaction, which is invariant under weak isospin transformations $SU(2)_W$.

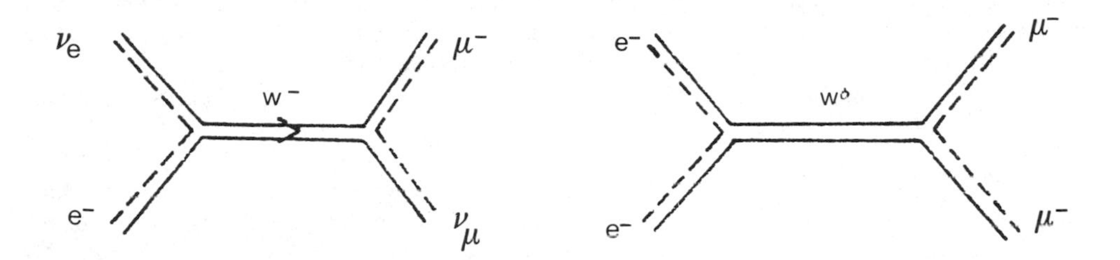

Fig. 4. Charged current and neutral current weak interactions in terms of subconstituents.

ii) Photons interact with subconstituents. As a consequence, (virtual) photons make transitions to neutral bosons, i.e., to the neutral member W^o of the isotriplet W^+, W^-, W^o (Fig. 5).

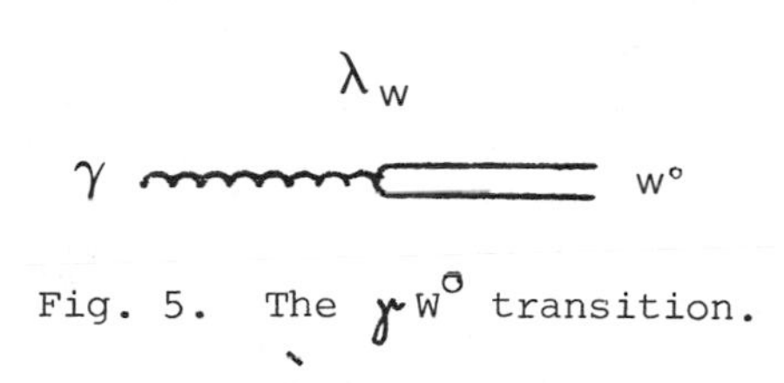

Fig. 5. The γW^o transition.

iii) Photons interact directly with subconstituents and thus they interact with leptons and quarks via electromagnetic form factors (Fig. 6). In the simplest dynamical approximation, the form factors are to be saturated by a single W^o boson. (W-dominance, as introduced in Ref. 9.) Subsequently, we will discuss the implication of more massive additional contributions to the form factors.

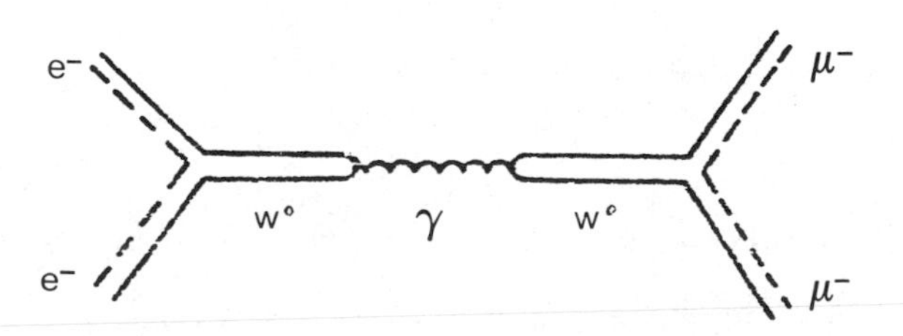

Fig. 6. The electromagnetic interaction between an electron and a muon.

The above diagrams are somewhat schematic and several remarks on them are in order. The full lines denote the constituents of W bosons, which are also assumed to be constituents of leptons and quarks. In the "haplon" model of Ref. 12, these constituents carry spin 1/2 in full analogy to the quark structure of hadronic vector mesons. The dotted lines in the above diagrams denote constituents of quarks and leptons which are not constituents of the bosons and allow one to endow quarks with color in contrast to leptons and to also take care of the difference between different generations. In the model of Ref. 12, the additional constituents of the fermions are spin zero particles.

The parallelism between hadrons as composite states of quarks, and leptons (quarks) as composite states of subconstituents, finds its limitation in parity violation by the weak interactions. While the ρ meson in nucleon-nucleon interactions couples to a parity conserving vector current, the least massive W boson has to couple to the usual left-handed weak isospin current. In composite models, it seems natural to start from a left-right symmetric theory and assume that the parity breaking mechanism pushes the masses of the bosons coupled to right-handed currents to sufficiently high values. The lack (or strong suppression) of scalar and pseudoscalar exchanges may also find its explanation in connection with parity violation.

In order to correctly reproduce the electric charge assignment of leptons and quarks and the parity conserving character of photon interactions, there must be bosons and corresponding currents present, other than just the neutral member, W^o, of a triplet of bosons coupled to the left-handed weak isospin current. The existence of additional isoscalar bosons seems to be natural in composite models. We describe the additional interactions in an effective manner by introducing a Y boson (of sufficiently high mass) coupled to a current of the form of the usual weak hypercharge current, $j^{(Y)} \equiv j_{EM} - j^{(3)}$. For details we refer to Ref. 9. Also, we will replace the photon propagator in Fig. 6 by the full propagator of Fig. 7, since the $W^{(o)}$ transition will turn out to have considerable strength ($\lambda_W^2 = \sin^2\Theta_W$).

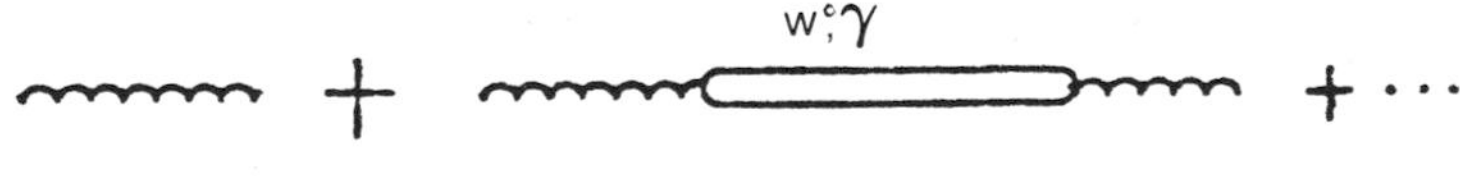

Fig. 7. The full photon propagator.

2.2 Single W,Y-Dominance

To start with, we now assume that the interactions of Fig. 4 and Fig. 6 at low energies are dominated by a single triplet of bosons $W^{\pm}$ and W^{o}, which is coupled in a globally SU(2) invariant manner to the usual* left-handed weak isospin current $j^{(\pm)}, j^{(o)}$, and let us introduce an additional (more massive) boson coupled to the weak hypercharge current, $j^{(Y)}$. The charged current interaction (Fig. 4) yields as usual

$$\frac{g_W^2}{m_W^2} = \frac{8 G_F}{\sqrt{2}}, \tag{2.1}$$

where $G_F \cong 1/(296 \text{ GeV})^2$ is the Fermi coupling and g_W the coupling of the boson triplet of mass m_W. In the neutral current sector (Figs. 4 and 6) weak interactions and the full electromagnetic one are described by

$$\begin{aligned} -2\mathcal{L}_{eff}^{EM+NC} &= \frac{-g_W^2}{q^2 - m_W^2} j^{(3)2} + \frac{-g_Y^2}{q^2 - m_Y^2} j^{(Y)2} \\ &+ D(q^2)\left(g_W \lambda_W \frac{m_W^2}{q^2 - m_W^2} j^{(3)} + g_Y \lambda_Y \frac{m_Y^2}{q^2 - m_Y^2} j^{(Y)} \right)^2, \end{aligned} \tag{2.2}$$

where the full photon propagator[9]

$$D(q^2) = -\frac{1}{q^2}\left(1 - q^2 \sum_i \frac{\lambda_i^2}{q^2 - m_i^2}\right)^{-1} \tag{2.3}$$

has to be inserted. In the limit of $q^2 \rightarrow 0$, to be taken below, it takes the form

$$D(q^2) \cong -\frac{1}{q^2}\left(1 - q^2 \sum_i \frac{\lambda_i^2}{m_i^2} + \cdots\right). \tag{2.4}$$

Boson masses and couplings have been denoted by $m_{W,Y}$, and the photon boson transition strengths by $\lambda_{W,Y}$. As explicitly seen in (2.2), the photon interacts via W-dominated form factors with leptons and quarks.

---

*In the low energy approximation, composite leptons and quarks are effectively described by local fields.

In the limit $q^2 \to 0$, Ansatz (2.2) first of all contains the photon pole term with the correct coupling strength, provided we identify the electromagnetic current with

$$j_{EM} \equiv j^{(3)} + j^{(Y)} \tag{2.5}$$

and require

$$\lambda_W g_W = \lambda_Y g_Y = e, \tag{2.6}$$

where e denotes the unit of electric charge. The constraint (2.6) is a consequence of W-dominance for the isovector part (and of Y dominance for the additional piece) of the electromagnetic lepton (quark) form factor. It may be called "compositeness" or "saturation" condition* within the present context. Inserting the identity

$$\frac{m_W^2}{q^2 - m_W^2} \equiv \frac{q^2}{q^2 - m_W^2} - 1, \tag{2.7}$$

the full expression for the neutral current interaction Lagrangian in the $q^2 \to 0$ limit is easily obtained from (2.2),

$$\begin{aligned} -\mathcal{L}_{eff}^{EM+NC}(q^2 \to 0) &= -\frac{e^2}{2q^2}\, j_{EM}^2 \\ &+ \rho\, \frac{4G_F}{\sqrt{2}} \left(\left(j^{(3)} - \sin^2\Theta_W\, j_{EM} \right)^2 + C\, j_{EM}^2 \right), \end{aligned} \tag{2.8}$$

where

$$\rho = 1 + \frac{g_Y^2}{m_Y^2} \Big/ \frac{g_W^2}{m_W^2} \gtrsim 1, \tag{2.9}$$

*Relation (2.6) coincides with what has been called[13] the "unification condition". Here, it arises as a consequence of the dynamical W dominance approximation and not as a consequence of a unification requirement.

and

$$C = \frac{(\rho-1)}{\rho^2}\left(1 - \lambda_W^2 - \lambda_Y^2\right)^2, \tag{2.10}$$

and $\sin^2\Theta_W \cong 0.23$ has been identified with

$$\sin^2\Theta_W = \rho^{-1}\left(\lambda_W^2 + (\rho-1)(1-\lambda_Y^2)\right). \tag{2.11}$$

In the limit of $m_Y \to \infty$, one obtains $\rho = 1$ and $C = 0$ and (2.8) reduces exactly to the effective Lagrangian of the standard model. Relation (2.11) reduces to

$$\lambda_W^2 = \sin^2\Theta_W. \tag{2.12}$$

In order to compare with experimental limits on ρ and C, let us assume $\lambda_Y^2 \cong \lambda_W^2$. This assumption is reasonable, as the transition strengths of the photon to the W and Y bosons should be of comparable magnitude as a consequence of similar bound state dynamics. From (2.6) we then obtain $g_Y \cong g_W$ and thus

$$\rho \cong 1 + \frac{m_W^2}{m_Y^2}. \tag{2.13}$$

Empirically, on the level of one standard deviation, an experimental value as large as $\rho = 1.05$ is not definitely ruled out.[14] Inserting this value in (2.13) yields

$$m_Y \gtrsim 5\, m_W, \tag{2.14}$$

i.e., the small deviation from $\rho = 1$ naturally implies a large spacing of the weak isoscalar and isovector bosons.

For $\lambda_W = \lambda_Y$, the parameter C becomes a function of ρ and $\sin^2\Theta_W$. One easily derives from (2.10) and (2.11)

$$C = \frac{(\rho-1)}{(2-\rho)^2}\left(1 - 2\sin^2\Theta_W\right)^2. \tag{2.15}$$

Experimentally, from PETRA,[16] the parameter C is bounded by $C \lesssim 0.02$, which implies $\rho \lesssim 1.07$ according to (2.15), while $C \lesssim 0.01$ yields $\rho \lesssim 1.04$. This bound on ρ obtained from e^+e^- annihilation under the assumptions mentioned is thus roughly as stringent as the one from deep inelastic scattering.

Let us finally combine the charged current normalization (2.1) with the saturation condition (2.6) and the identification of λ_W^2 with $\sin^2\Theta_W$ in (2.12) to obtain for the charged W mass, m_W,

$$m_W = M_W \equiv \frac{e^2\sqrt{2}}{8\, G_F \sin^2\Theta_W} \cong (79\, GeV)^2. \tag{2.16}$$

Next, from the zeros of the photon propagator (2.3), for $m_Y^2 \gg m_W^2$, we derive for the lowest neutral boson the relation

$$m_Z^2 = \frac{m_W^2}{1-\lambda_W^2} = \frac{M_W^2}{\cos^2\Theta_W} \equiv M_Z^2 \cong (90\, GeV)^2. \tag{2.17}$$

Both mass formulae, (2.16) and (2.17), coincide with the canonical ones of the $SU(2)_I \times U(1)_Y$ spontaneously broken gauge theory,[1] which contains $SU(2)_I \times U(1)_Y$ symmetry in the large energy (short distance) limit. Within the present context of compositeness, the mass formulae appear as a consequence of the dynamical approximation of W-dominance for the electromagnetic interaction of leptons and quarks in the low q^2 limit. No symmetry assumption, such as $SU(2)_I \times U(1)_Y$ at short distances or asymptotic energies, as contained in the standard framework, is needed. Possible changes of the boson mass formulae (2.16) and (2.17) as a consequence of additional more massive contributions[15] to the lepton (quark) form factors will be discussed below and will be found to be fairly small for large level spacing, which will be shown to be expected due to the large value of $\sin^2\Theta_W \cong 0.23$.

When writing down the mass formulae (2.17) for m_Z, as well as in deriving the effective Lagrangian (2.8) and the expression for ρ_0 in (2.9), we assumed that the charged and neutral bosons $W^\pm$ and W^0 are degenerate in mass (global SU(2)). This assumption may be violated due to different Coulomb interactions of the subconstituents within neutral and charged bosons. If the mass of the unmixed neutral boson, m_{W^0}, is slightly smaller than the mass of the charged one, $m_{W^\pm}$, the neutral boson mass will be less strongly shifted, with respect to the charged one, than indicated by (2.17), i.e., (2.17) must be replaced by

$$m_Z = \frac{m_{W^0}}{\cos\Theta_W} < \frac{M_W}{\cos\Theta_W}, \tag{2.18}$$

and ρ in (2.9) becomes ρ',

$$\rho' = \frac{m_{W^+}^2}{m_Z^2 \cos^2\Theta_W} \left(1 + \frac{g_Y^2}{m_Y^2} \Big/ \frac{g_W^2}{m_W^2}\right) \gtrsim 1. \tag{2.19}$$

ρ' may thus be larger than unity, even in the limit $m_Y \rightarrow \infty$.

In our analysis of the Ansatz (2.2) for the effective Lagrangian of the electroweak interaction we have emphasized the $q^2 \rightarrow 0$ behaviour. A form of the effective Lagrangian, which explicitly displays the Z-boson and Y-boson poles besides the photon pole has been given in Ref. 9.

Let us summarize: We assumed that quarks, leptons and weak bosons form composite objects. The coupling of the elementary photon to the charged constituents of the W^0 boson leads to γW^0 transitions (mixing) analogous to $\gamma\rho^0$ transitions in QCD. The coupling of the photon to leptons and quarks originates from the coupling to the same charged constituents, which are constituents of the vector bosons. Leptons and quarks thus develop electromagnetic form factors, which are analogous to the hadron form factors in QCD. Dominating these form factors by the lowest lying bosons, the low energy effective interaction in the limit $m_Y \gg m_W$, and $m_{W^0} = m_{W^\pm}$, turns out to be identical to the one of the standard $SU(2) \times U(1)$ theory. The magnitude of the resulting charged boson mass in the W-dominance (low energy) approximation agrees with the canonical value. The neutral boson is shifted upward relative to the charged one due to γW^0 mixing. For $m_{W^\pm} = m_{W^0}$, the neutral boson mass, m_Z, agrees with the canonical value.

2.3. Universality of Weak Interactions and Current Algebra for Subconstituent Currents

Writing (2.6) explicitly for different fermion doublets, ν_e, e^-; ν_μ, μ^-; ... etc., yields

$$\lambda_W g_{Wee} = \lambda_W g_{W\mu\mu} = \cdots = e, \tag{2.20}$$

or

$$g_{Wee} = g_{W\mu\mu} = \cdots \equiv g_W , \tag{2.21}$$

i.e., the W boson has to couple universally to all leptons and quarks. Universality of the weak interactions (which in the SU(2) x U(1) gauge theory is based on gauge invariance) appears as a consequence[9] of W-dominance of (isovector) photon lepton (quark) interactions.

In arriving at (2.6) and (2.20) from (2.2), we made use of the fact that the different generations of leptons (quarks) have the same electric charge, e, to which the form factors at $q^2 = 0$ have to be normalized. Without referring to electromagnetism, the normalization may be deduced as a property of the matrix elements of the weak isovector current of the subconstituents, provided these currents fulfill[17] a local current algebra. This is the case in the composite model of Ref. 12.

Let us consider the matrix elements of leptons (quarks) of the weak current $F^W_{\mu,i}(x)$. The current algebra requirement then yields universal normalization for the corresponding form factors for $q^2 = 0$, i.e., $F_e(q^2 = 2) = F_\mu(q^2 = 0) = \ldots = 1$. Requiring W-dominance of the form factor of any lepton or quark f,

$$F_f(q^2) = \frac{m_W^2}{f_W} \frac{g_{Wff}}{m_W^2 - q^2} , \tag{2.22}$$

gives

$$f_W = g_{Wee} = g_{W\mu\mu} = \cdots \equiv g_W \tag{2.23}$$

as in (2.13). Universality of weak interactions thus appears as a consequence of current algebra (which provides the normalization) and W-dominance.

Let us add the remark that more massive vector boson contributions to the form factors (2.22) may in fact spoil the universality of the couplings to the least massive boson, while approximate universality of the Fermi coupling G_F may still be valid.

3. HIGHER MASS CONTRIBUTIONS

3.1. Duality, QHD Sumrules

In a bound state model of weak bosons, one expects a spectrum of such states eventually followed by a continuum of subconstituents similar to, e.g., the situation with the $c\bar{c}$ system of QCD. In the case of e^+e^- annihilation into hadrons the concept of q^2 duality[18] successfully allows one to relate vector meson parameters, such as photon couplings, masses and level spacings to the asymptotic quark antiquark production cross section. More recently the duality concept has been reformulated[19] within the framework of QCD. For our purposes the simple original formulation is sufficient. Translated into QHD, it says[20,9]

$$\lambda_W^2 = \sin^2\Theta_W = \frac{\alpha}{3\pi} N Q^2 \frac{\Delta m_W^2}{m_W^2} . \tag{3.1}$$

The magnitude of the $\gamma W_2^{(o)}$ transition strength is thus related to the level spacing m_W^2 of the neutral boson spectrum, the product of the number of colors and hypercolors of the subconstituents N, and the average charge Q (in units of e) of the subconstituents within a boson bound state. If one assumes three colors and three hypercolors of the haplons, and $Q^2 = 1/2$ for the isovector charge, one obtains $m^2 \cong 0.4\ \text{TeV}^2$ and $M_Z' \cong 0.6$ TeV for the first excited neutral boson state. The large empirical value of $\sin^2\Theta_W \cong 0.23$ thus implies a level spacing, which is large compared with the least massive boson. This may be connected with the strength of the con-

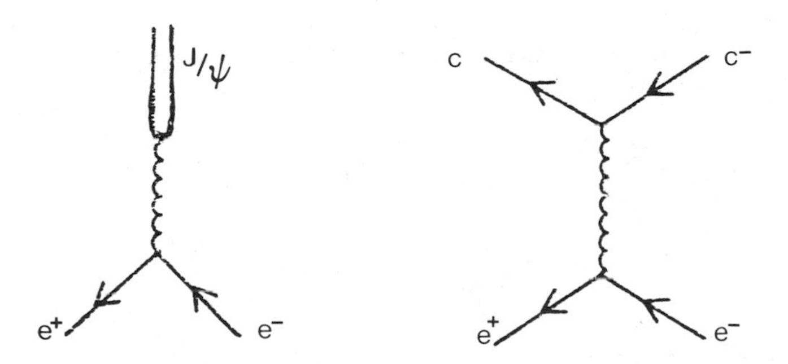

Fig. 8. Duality of J/ψ production to $c\bar{c}$ production.

fining force. Phenomenologically, the least massive boson state in QHD in this respect shows similarity to the case of the pion in QCD. It is useful to think of the boson spectrum in terms of the substitution

$$GeV|_{QCD} \rightarrow TeV|_{QHD} \tag{3.2}$$

We note that the pointlike coupling of photons to subconstituents combined with duality, i.e., the smooth averaging of the low lying resonances by the asymptotic cross section embodied in (3.1), implies a relation between couplings and masses of successive resonances, which is given by

$$\lambda'_W = \lambda_W \frac{m_W}{m'_W} . \tag{3.3}$$

3.2. Influence on the Masses of the Lightest Bosons

In collaboration with Masaaki Kuroda, I have investigated how strongly W-dominance predictions for the charged and neutral boson masses are affected[15] by the presence of additional more massive contributions to lepton (quark) form factors. We simply considered the effect of a single additional isovector triplet, W', which yields a sufficiently accurate estimate of the influence of more massive states on the masses of the lowest one. The restrictions due to the empirical values of G_F, (2.1), $\sin^2{}_W$, (2.12), and e, (2.6), have been appropriately generalized and the duality relation (3.3) has been incorporated. For details we refer to Ref. 15 and the reviews of Ref. 7. The most stringent limits are obtained if the W' coupling to leptons (quarks) is subjected to the reasonable constraint $|g_{W'}| \lesssim |g_W|$. These limits will be referred to in what follows.

From the arguments of subsection 3.1 we know that the level spacing of W bosons should be large, i.e., $m_{W'} \gtrsim 5\, m_W$. Under this assumption we found[15] that m_W should differ from the canonical value of 79 GeV by $\pm$ 10% at most, i.e.,

$$71\,GeV \lesssim m_W \lesssim 87\,GeV, \tag{3.4}$$

where the bounds are actually very conservative ones. The mass of the least massive neutral boson is then given by

$$m_Z^2 = \frac{m_W^2}{1 - \lambda_W^2} \tag{3.5}$$

with

$$\lambda_W^2 = \sin^2\Theta_W \left(\frac{m_W^2}{M_W^2} + \frac{(m_W^2 - M_W^2)^2}{m_W^2 M_W^2} \right). \tag{3.6}$$

In the relevant interval (3.4), relation (3.6) is well approximated by

$$\lambda_W^2 = \sin^2\Theta_W \frac{m_W^2}{M_W^2} \tag{3.7}$$

Numerically, with $\sin^2\Theta_W = 0.23$ and $M_W = 79$ GeV, one obtains the following results for the neutral boson mass, m_Z, as a function of the charged one, m_W:

Table 1. The neutral boson mass m_Z as a function of the charged one (or the unmixed neutral one), m_W.

m_W (GeV)	m_Z (GeV)	$m_Z - m_W$ (GeV)
71	79	8
75	84	9
79	90	11
83	96	13
87	102	15

The mass shift rises from 8 to 15 GeV when the charged boson changes within the limits of 71 GeV and 87 GeV. For $m_W = 79$ GeV, m_Z coincides with the canonical value of 90 GeV. Let us note that in (3.6) and (3.7) the mass m_W is the unmixed neutral boson mass, $m_W \equiv m_{W^0}$. As mentioned in Section 2.2, if, because of Coulomb corrections in the bound state dynamics, m_{W^0} is actually smaller than $m_{W^\pm}$, the Z-mass will be less strongly shifted relative to the charged boson mass than inferred from the above table with $m_W = m_{W^\pm}$.

In summary, from the expected large level spacing in the spectrum of more massive W bosons, one infers that the canonical mass relation for the lightest charged boson, $m_{W^\pm} = 79$ GeV, is only little affected. Again under the assumption of large level spacing

($m_{W'} \gtrsim 5\, m_W$), due to the γW^o mixing, the neutral boson mass is exactly predicted as a function of the charged one. This prediction is made under the caveat, however, that the unmixed neutral boson is exactly degenerate in its mass with the charged one.

4. EXPERIMENTAL TESTS OF COMPOSITENESS

In discussing how to test the ideas on electroweak interactions, elaborated upon in the preceding sections, let me discriminate between the "low energy" domain, which is being explored with lepton beams and in e^+e^- annihilation at present, the high energy region with the possibility of producing weak bosons (explored at the $p\bar{p}$ collider and future e^+e^- devices), and, finally, the region of very high energies of the order of 1 TeV.

If the hypothesis of further compositeness and complexity makes sense, one would expect unconventional weak interaction effects to show up at some level of accuracy in the low energy region. Search for, and constraints on effects due to isoscalar exchange as well as right-handed or scalar and pseudoscalar bosons will be valuable. Effects in e^+e^- annihilation due to the C term in (2.8) should be further looked for.

In the high energy region, establishing masses and coupling constants of the least massive $W^\pm$ and Z bosons is of obvious importance. Deviations from the SU(2) x U(1) gauge theory predictions may be small, however. For the boson masses, this has been discussed in the previous section. If a second neutral boson exists, the forward-backward asymmetry in $e^+e^- \rightarrow \mu^+\mu^-$ beyond the first Z peak will be affected.[21] As pointed out[22] by Renard, compositeness implies enhancement factors of the order of 10^3 to 10^4 for rare Z decays, such as $Z \rightarrow 3\gamma$, $Z \rightarrow 3$ gluons. The enhancement is a consequence of replacing annihilation via a fermion loop in the SU(2) x U(1) gauge theory (Fig. 9) by pointlike annihilation, determined by the wave function of the bound state at zero distance, which is proportional to $\sin^2\Theta_W$.

Finally, in the very high energy domain, the substructure should lead to dramatic effects, such as multilepton production.

5. CONCLUSIONS

Let us end with the following summarizing and concluding remarks:

i) The notion of compositeness adopted implies an interpretation of $\sin^2\Theta_W$ as γW^o transition strength analogous to the $\gamma\rho^o$ transitions in QCD.

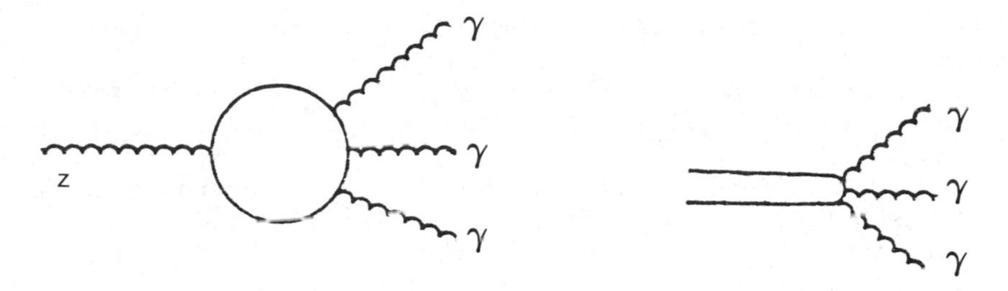

Fig. 9. $Z \rightarrow 3\gamma$ in the standard model and in composite models.

ii) Photons interact with subconstituents of leptons and quarks. As a consequence, leptons and quarks develop electromagnetic form factors. Single W-dominance for the form factors implies[9] the canonical mass relations[23] as a low energy dynamical approximation. No assumption of short distance (high energy) SU(2) x U(1) symmetry is required, as it is embodied in the standard model. The true high energy (TeV) scenario may be completely different from the standard one.

iii) The $W^{\pm}$ boson, for which the first experimental evidence[24] is available now, may be a nonstandard composite boson. Due to the large level spacing of possible excited bosons, which has been inferred from the large value of $\sin^2\theta_W$, deviations from the canonical values of masses and couplings are expected to be small, however. Precise measurements are needed. Nevertheless, it may require energies in the TeV scale to prove or disprove the conjecture of compositeness described here.

iv) If one finds the notion of compositeness we have adopted to be an unlikely one, he may be reminded of the history of strong interactions; the ρ^0, ω and ϕ mesons were originally predicted[25] as gauge bosons.

REFERENCES AND FOOTNOTES

1. S.L. Glashow, Nucl. Phys. 22 (1961) 579;
   S. Weinberg, Phys. Rev. Lett. 19 (1967) 1264;
   A. Salam, Elementary Particle Theory, ed. N. Svarthold (Almquist and Wiksell, Stockholm, 1968) p. 367.
2. P.Q. Hung and J.J. Sakurai, Ann. Rev. Nucl. Part. Sci. 31 (1981) 375.
3. UA1 Collaboration, G. Arnison et al., Phys. Lett. 112B (1983) 103, and report to this conference.

4. UA2 Collaboration, CERN-EP/83-25 (1983), and report to this conference.
5. W.J. Marciano and A. Sirlin, Nucl. Phys. B189 (1981) 442; C.H. Llewellyn Smith and J.F. Wheater, Phys. Lett. 105B (1981) 486.
6. For a fairly complete list of references on various composite models we refer to the review by L. Lyons, Oxford University preprint, Ref. 52/82 (1982). Further recent reviews include: H. Harari, Weizmann Institute preprint, WIS-82/60; R.D. Peccei, MPI-PAE/PTh 69/82 (1982); M.E. Peskin, Proceedings of the 1981 International Symposium on Lepton and Photon Interactions (Bonn), p. 880.
7. For recent reviews on phenomenological implications of compositeness based on the standpoint taken in the present paper see: R. Kögerler, BI-TP 83/01, to appear in "Electroweak Interactions at High Energies", Proceedings of the 1982 DESY Workshop, Hamburg, ed. R. Kögerler and D. Schildknecht, (World Scientific Publishing Co., Singapore); D. Schildknecht, MPI-PAE/PTh 81/82 (1982), to appear in the Proceedings of the Eighth Interantional Workshop on Weak Interactions and Neutrinos, Javea (Alicante), Spain, September 1982, ed. A. Morales; compare also the contribution of H. Fritzsch to this conference.
8. A. Salam, Trieste preprint, IC/82/215.
9. R. Kögerler and D. Schildknecht, CERN-TH-3231 (January 1882).
10. The conjecture that the short range weak interaction shows features, which are analogous to hadron interactions, e.g., a full spectrum of weak bosons as mediators of the weak interactions, has been advocated by: J.D. Bjorken, Phys. Rev. D19 (1979) 335; E.H. de Groot and D. Schildknecht, Z. f. Physik C10 (1981) 55; see also D. Schildknecht, BI-TP 81/12 (1981), and Proceedings of the International School of Subnuclear Physics, Erice, Sicily, 1980, ed. A. Zichicchi.
11. Specific variants of the conjecture that the weak interactions are residual effects of a confining force are given in: H. Harari and N. Seiberg, Phys. Lett. 98B (1981) 269; O.W. Greenberg and J. Sucher, Phys. Lett. 99B (1981) 339; H. Fritzsch and G. Mandelbaum, Phys. Lett. 102B (1981) 319; L. Abbot and E. Farhi, Phys. Lett. 101B (1981) 69; R. Barbieri, R.N. Mohapatra and A. Masiero, Phys. Lett. 105B (1981) 369.
12. H. Fritzsch and G. Mandelbaum, Ref. 11.
13. P.Q. Hung and J.J. Sakurai, Nucl. Phys. B143 (1978) 81.
14. See, e.g., the analysis of M. Roos presented at this conference.
15. M. Kuroda and D. Schildknecht, Phys. Lett. 121B (1983) 173.
16. J.G. Branson, DESY 82-066 (1982); M. Davier, Rapporteur's talk at the 21st International Conference on High Energy Physics, Paris, 1982.

17. H. Fritzsch, D. Schildknecht and R. Kögerler, Phys. Lett. 114B (1982) 157.
18. A. Bramon, E. Etim and M. Greco, Phys. Lett. 41B (1972) 609; J.J. Sakurai, Phys. Lett. 46B (1973) 207; D. Schildknecht and F. Steiner, Phys. Lett. 56B (1975) 36.
19. V.A. Novikov, L.B. Okun, M.A. Shifman, A.I. Vainshtein, M.B. Voloshin and V.I. Zakharov, Phys. Rep. 41 (1970) 1; M.A. Shifman, A.I. Vainshtein, V.I. Zakharov, Nucl. Phys. B147 (1979) 385, 448; for applications to QHD compare S. Narison, Phys. Lett. 122B (1983) and LAPP-TH-68 (1982).
20. P. Chen and J.J. Sakurai, Phys. Lett. 110B (1982) 481.
21. M. Kuroda and D. Schildknecht, Phys. Rev. D26 (1982) 3167.
22. F.M. Renard, Phys. Lett. 116B (1982) 269.
23. The prediction[9] of the canonical mass relations from compositeness and W-dominance has been at variance with the conjecture of some of the authors of Ref. 11 that compositeness implies W-boson masses of at least 120 GeV.
24. The forward-backward asymmetry expected for W-production still needs empirical confirmation.
25. J.J. Sakurai, Ann. Phys. 11 (1960) 1.

THE INTERNAL STRUCTURE OF W-BOSONS

Harald Fritzsch

Sektion Physik, Universität München and

Max-Planck-Institut für Physik und Astrophysik

München, Germany

Recently the weak intermediate bosons have been found in p-$\bar{p}$-collisions at high energies[1]. The masses observed are in good agreement with the values predicted within the standard SU(2) x U(1)-theory. Nevertheless a number of questions remain to be answered, e.g.:

Is the SU(2) x U(1) gauge theory a microscopic theory of the electroweak interactions, or is it merely an effective theory, describing the low energy properties of a more fundamental theory?

Are the weak bosons basic gauge particles like the photon, or are they bound states of certain constituents?

Are leptons and quarks composite and are they made of the same constituents as the weak bosons?

In this talk I shall address these and other related questions. Before doing so, let me emphasize that in particle physics there exist two different ways to generate masses:

a) DYNAMICAL MASS GENERATION BY CONFINEMENT

Here the mass of a particle is generated by a force between constituents, which becomes strong at sufficiently large distances. Examples: the ρ-meson or nucleon masses in QCD. The masses of the particles are measures of the energy of the fields confined inside a finite volume.

b) MASS GENERATION BY SPONTANEOUS SYMMETRY BREAKING

Here a mass scale is introduced by a scalar field which acquires via its self interaction a non-zero vacuum expectation value. For example, in the standard SU(2) x U(1) theory the W-boson mass is introduced via a scalar field whose vacuum expectation value v is of the order of 300 GeV:

$$M_W = \frac{1}{2} g \cdot v \tag{1}$$

$$\frac{G}{\sqrt{2}} = \frac{g^2}{8 M_W^2} = \frac{1}{2v^2} \qquad (v = 246 \text{ GeV}).$$

(g: $SU(2)^W$ - coupling constant; G: Fermi constant).

The question arises whether the W-boson mass is due to a spontaneous breakdown of the SU(2) x U(1) gauge symmetry, or whether it has a dynamical origin. If the first possibility is realized, the SU(2) x U(1) gauge theory may well be the correct theory not only of the electroweak interactions at relatively small energies, but also at energies much above the energy scale exhibited by the Fermi constant (of the order of 300 GeV).

If the W mass is generated dynamically, one expects the W-bosons to be bound states of at least two constituents. No unification of the electromagnetic and weak interactions is achieved. The W-boson will be the ground state of a complicated spectrum of states, just like the ρ-meson is the ground state of the spectrum of infinitely many states in the $\bar{q}q(J^P = 1^-)$ channel. It is this possibility which I would like to address here.

During the progress of physics within the past hundred years it has happened twice that observed short range forces were recognized as indirect consequences of an underlying substructure of the objects considered. Thus the short-range molecular and van-der-Waals forces turned out to be indirect consequences of the substructure of atoms; they are remnants of the long range electromagnetic forces. Since 1970 something similar has happened to the short-range nuclear force, which has turned out to be a relict of the quark substructure of hadrons and the strong long range color forces between the quarks.

The only short range interaction left in physics which has not been traced back to a substructure and to a fundamental long range force between constituents is the weak interaction. Recently a number of authors has become interested in interpreting the weak force as some kind of "Van der Waals" remnant of an underlying lepton-quark substructure[3]. The lepton- quark constituents for which I will use the name "haplons"[x] are supposed to be bound together by very strong so-called hypercolor forces which are supposed to be confining forces, presumably described by a non- Abelean gauge theory (although other types of forces are not excluded). The short range character of the weak interaction arises since the leptons and quarks are hypercolor singlets, but have a finite size. The energy scale provided by the Fermi constant is of the order of 300 GeV; the inverse size of the leptons, quarks and weak bosons is expected to be of the same order, i.e. their radii are of the order of 10^{-17} cm.

If the weak interaction turns out to be a remnant of the hypercolor force, a new interpretation of the relationship between the electromagnetic and weak interaction is required. The W- and Z- bosons cease to be fundamental gauge bosons, but acquire the less prestigious status of bound states of haplons. However the photon remains an elementary object (at least at the scale of the order of 10^{-17} cm, discussed here). As a whole, the SU(2) x U(1)-theory cannot be regarded anymore as a fundamental microscopic theory of the electroweak interactions, but at best can be interpreted as an effective theory, which is useful only at distances larger than the hypercolor confinement scale. It acquires a status comparable to the one of the σ-model in QCD, which correctly describes the chiral dynamics of π-mesons and nucleons at relatively low energies, but fails to be a reasonable description of the strong interaction at high energies.

However I would like to emphasize that at the present time no indication whatsoever comes from the experimental side that leptons, quarks and weak bosons may be bound states of yet smaller constituents. It may well be that the weak force will turn out in the future as a fundamental gauge force, as fundamental as the electromagnetic one and the color force. In fact, interpreting the weak forces as effective forces poses a number of problems which have not been solved in a satisfactory manner. First

[x] derived from Greek "haplos" (simple)

of all, the weak interactions violate parity, and they do that not in an uncontrolled way, but in a very simple one: only the lefthanded leptons and quarks take part in the charged current interactions. If we interpret the weak interactions as Van der Waals type interactions, the parity violation is a point of worry. How should one interpret the observed parity violation? Does it mean that the lefthanded fermions have a different internal structure than the righthanded ones? Or are we dealing with two or several different hypercolor confinement scales, for example one for the lefthanded fermions, and one for the righthanded fermions, such that the resulting effective theory is similar to the left - right symmetric gauge theory, based on the group $SU(2)_L \times SU(2)_R$?

Another point of concern is the fact that the weak interactions show a number of regularities, e.g. the universality of the weak couplings, which one would not a priori expect if the weak interaction is merely a hypercolor remnant. On the other hand it is well - known that the interaction of pions or ρ-mesons with hadrons shows a number of regularities which can be traced back to current algebra, combined with chiral symmetry or vector meson dominance. Despite the fact that both the ρ-mesons and the pions are quark - antiquark bound states for which one would not a priori expect that their interaction with other hadrons exhibits remarkable simple properties (e.g. the universality of the vector meson couplings), the latter arise as a consequence of the underlying current algebra, which is saturated rather well at low energies by the lowest lying pole (either the pion pole in the case of the divergence of the axial vector current, or the ρ- or A_1-pole in the case of the vector or axial vector current).

In the case of chiral SU(2) x SU(2) the pole dominance works very well - the predictions of current algebra and PCAC seem to be fulfilled within about 5 %. The universality of the weak interactions is observed to be valid within 1 % in the case of the couplings of the weak currents to electrons, myons, as well as u,d and s- quarks. Much weaker constraints exist for the heavy quarks. It remains to be seen whether the observed universality of the weak interactions will find an explanation along lines similar to the ones used in hadronic physics.

Here I shall concentrate on models in which the W-bosons consist of a haplon and an antihaplon. Since the observed weak interactions exhibit a symmetry $SU(2)_L$, we assume the existence of two haplons, denoted by the doublet $\binom{\alpha}{\beta}$, which carry hypercolor and electric charge, but no ordinary color. The electric charges are assumed to be: $Q(\alpha) = +1/2$, $Q(\beta) = -1/2$ (see ref. (4)).

The spectral functions at energies much above the hypercolor confinement scale are supposed to be described by a continuum of haplon- antihaplon pairs. At low energies the weak amplitudes will be dominated by the lowest lying poles, which are identified with the W-particles. The latter form the triplet:

$$\begin{pmatrix} W^+ \\ W^3 \\ W^- \end{pmatrix} = \begin{pmatrix} \bar{\beta}\alpha \\ \frac{1}{\sqrt{2}}(\bar{\alpha}\alpha - \bar{\beta}\beta) \\ \bar{\alpha}\beta \end{pmatrix}$$

The experimental data on the neutral current interaction require a mixing between the photon and the W_3 boson (the neutral, isovector partner of W^+ and W^-), which in the standard SU(2) x U(1) scheme is caused by the spontaneous symmetry breaking. Within our approach this mixing is due to the $W_3 - \gamma$ transitions, generated dynamically like the ρ-γ transitions in QCD (for an early discussion, based on vector meson dominance, see ref. (5,6)). The magnitude of $\sin^2\theta_W$ is directly related to the strength of the γ-W_3 transition. The latter is determined by the electric charges of the W-constituents and by the W wave function near the origin. We suppose that in the absence of electromagnetism the weak interactions are mediated by the triplet (W^+, W^-, W^3), where $M(W^+) = M(W^-) = M(W^3) = O(\Lambda_H)$.

After the introduction of the electromagnetic interaction the photon and the W^3- boson mix. We denote the strength of this mixing by a parameter λ, following ref. (6), which is related to g (W-fermion coupling constant) and the effective value of $\sin^2\theta_W$

$$\sin^2\theta_W = \frac{e}{g} \cdot \lambda$$

Furthermore one has:

$$M_W = g \cdot 123 \text{ GeV}$$

$$M_Z^2 = \frac{M_W^2}{1-\lambda^2}$$

The mixing parameter λ is determined by the decay constant F_W of the W-boson, which we define in analogy to the decay constants of the ρ_o-meson (F_ρ):

$$|W^3\rangle = \frac{1}{\sqrt{2}} \frac{1}{\sqrt{n}} \sum_{j=1}^{n} (\bar{\alpha}_j\alpha_j - \bar{\beta}_j\beta_j)\ \phi(x)$$

($\phi(x)$: wave function in coordinate space, j: hypercolor index). The current matrix element can be written as

$$\langle o|j_\mu{}^3|W^3\rangle = \varepsilon_\mu\sqrt{n} \cdot \sqrt{2M_W} \cdot \phi\ (o) = \varepsilon_\mu \cdot M_W\ F_W$$

$$F_W = \sqrt{n} \cdot /2\sqrt{M_W}\ \phi(o)$$

$$\sin^2\theta_W = \frac{e^2}{g} \sqrt{n} \cdot \sqrt{2/M_W{}^3}\ \phi(o)$$

$$= e^2\ /\ g \cdot F_W\ /\ M_W.$$

e.g. $\sin^2\theta_W$ is proportional to the coordinate space wave function of the W-boson at the origin. Taking for example $g = 0.65$ and $M_W = 79$ GeV, one obtains $F_W = 123$ GeV, a value which seems not unreasonable for a bound state of the size 10^{-16} cm.

In the SU(2) x U(1) gauge theory the SU(2) coupling constant g is related to e by the relation $g = e/\sin\theta_W$. In bound state models of the weak interactions discussed here this relation need not be true in general. However it has been emphasized recently[7)] that this relation is approximately fullfilled if the lowest lying W-pole dominates the weak spectral function at low energies. This leads to the relation

$$g = F_W\ /\ M_W = e/\sin\theta_W \approx 0.65$$

(we have used $\sin^2\theta_W = 0.22$).

It is interesting to note that many aspects of the bound state models can be derived from a local current algebra of the weak currents. We observe that the left-handed leptons and quarks form doublets of the weak isospin. The weak isospin charges $F_i^W (i = 1, 2, 3)$ obey the isospin charge algebra

$$[F_i^W, F_j^W] = i\ \varepsilon_{ijk}\ F_k^W$$

Let us assume that these charges can be constructed as integrals over local charge densities $F_{oi}^W(x)$, i.e.,

$$F_i^W(x^o) = \int F_{oi}^W(x)d^3x.$$

Furthermore we suppose that the charge densities obey at equal times the local current algebra

$$[F_{oi}^W(x), F_{oj}^W(y)]_{x^o=y^o} = i\varepsilon_{ijk}F_{o\kappa}^W \delta^3(\vec{x}-\vec{y}).$$

The local algebra is trivially fulfilled in a model in which leptons and quarks are pointlike objects and the weak currents are simply bilinear in the lepton and quark fields. However, if leptons and quarks are extended objects, the situation changes drastically. Currents, which are bilinear in the (composite) lepton and quark fields would not obey the local algebra, just like the currents, which are bilinear in nucleon fields, do not obey the local current algebra of QCD. Thus the local algebra becomes a highly non-trivial constraint. It is fulfilled in the haplon models discussed above, in which the currents are bilinear in α and β.

We consider matrix elements of the weak currents between the various fermion fields. In order to do so, we shall assume that the higher families composed of μ, τ, ... etc. are dynamical excitations of the first family (ν_e, e^-, u, d), without specifying in detail the dynamical structure of these states.

Let us look at the form factors of the left-handed weak neutral current $F_{\mu 3}^L(x)$, i.e., the matrix elements of this current between different lefthanded lepton or quark states, e.g., $\langle e_L^- | F_{\mu 3}(0) | e_L^- \rangle$. Denoting these form factors by $F_e(t)$, $F_\mu(t)$, $F_\tau(t)$, etc., the weak isospin algebra requires a universal normalization at t = 0, i.e., $F_e = F_{\nu_e} = F_\mu = F_\nu = \ldots = 1$. Assuming W dominance to be a reasonably good approximation, we may write for the dependence on the four-momentum transfer t,

$$F_f(t) = \frac{m_W^2}{f_W} \frac{f_W^{ff}}{m_W^2 - t},$$

where f denotes any one of the fermions e^-, ν_e, μ^-, ν_μ, etc. From $F_f(0) = 1$, we obtain the universality relation

$$f_W = f_W^{ff} \equiv g .$$

The neutral W and, because of the weak isospin algebra, also the charged W bosons couple universally to leptons and quarks, $f_W^{ff} \equiv g$. Thus the universality of the weak interactions follows from the W-dominance.

Let us stress that W dominance is a dynamical approximation of the underlying hypercolour dynamics, which is expected to be valid in the low energy region. At high energies (~1 TeV) lepton-lepton (quark) scattering is expected to show completely new phenomena, like multiple lepton (quark) production, similar to hadron-hadron interactions at high energies. Since universality is valid, we expect W dominance to be a very good approximation in the low energy region. As a consequence, deviations from the predicted charged and neutral boson masses should be small.

If the weak interactions are a manifestation of hypercolor dynamics, the question of SU(2) breaking is again open. One may wonder why the large violation of the weak isospin in the lepton - quark spectrum does not imply a large breaking of the symmetry in dynamical parameters like the W- fermion- coupling constants.

We should like to study the violation of the isospin in the pion dynamics. It is well - known that there exist two different sources of the violation of the isospin in strong interaction: the difference of the quark masses m_u and m_d, and the electromagnetic interaction. Only the second source contributes to the $\pi^+ - \pi^o$ mass difference. If we set $m_u = m_d = m$ and let m go to zero, the pion mass approaches zero as well, provided we neglect the electromagnetic interaction.

Following the laws of chiral symmetry breaking, one finds:

$$M_\pi^2 = (m_u + m_d) \cdot B + O(m_q^2 \ln m_q)$$

$$B = - \frac{2}{F_\pi^2} \langle o|\bar{u}u|o\rangle$$

where u, d denote the light quark flavors, m_u, m_d the quark masses, $|o\rangle$ the QCD vacuum, F_π the pion decay constant. Typical values are $(m_u + m_d) \cong 14$ MeV, B = 1300 MeV[8].

The electromagnetic self energy of the π^o vanishes in the chiral limit. The electromagnetic self energy of the charged pion can be calculated to order α in terms of the vector and axial vector spectral functions[8]:

$$(\Delta M_{\pi^+}^2)_{el.} = \frac{3\alpha}{4\pi \cdot F_\pi^2} \int_0^\infty ds \cdot s \ln\left(\frac{\mu^2}{s}\right)[\rho^V(s) - \rho^A(s)]$$

(ρ_V, ρ_A: vector - and axial vector spectral functions, μ: arbitrary renormalization point).

Saturating the integral above with the ρ- and A_1-poles and using the spectral function sum rules one obtains:

$$(\Delta M_{\pi^+}^2)_{el.} = \frac{3\alpha}{4_\pi} \cdot M_\rho^2 \cdot \left(\frac{F_\rho}{F_\pi}\right)^2 \cdot \ln\left(\frac{F_\rho^2}{F_\rho^2 - F_\pi^2}\right).$$

Using the measured values $F_\pi = 132$ MeV and $F_\rho = 204$ MeV, one obtains $(\Delta M_{\pi^+}^2)_{el.} \cong (36.4 \text{ MeV})^2$, which is close to the observed mass difference $\Delta M_\pi^2 = (35.6 \text{ MeV})^2$.

Combining the two relations, denoted above, one finds:

$$M_{\pi^0}^2 = (m_u + m_d) \cdot B + \ldots$$

$$M_{\pi^+}^2 = (m_u + m_d) \cdot B + \alpha\ M_\rho^2 \cdot 0.31 + \ldots$$

In the chiral limit $m_u = m_d = 0$ we obtain $M_{\pi^0} = 0$, $M_{\pi^+} \approx 36$ MeV. As an illustration we consider the case $m_u = m_d = 1$ KeV. One finds $M_{\pi^0} = 1.6$ MeV, $M_{\pi^+} = 36.4$ MeV, i.e. the neutral and charged pion mass differ by a factor of about 23.

We have just found a situation in QCD, which resembles the one in the lepton- quark- spectrum, namely a large isospin breaking despite the fact that for $m_u = m_d$ the isospin is an exact symmetry of QCD. In the chiral limit the π-mesons are particles, which in the absence of electromagnetism have zero mass, but have a finite size. Their inverse size is of order Λ.

Including the electromagnetic interaction has the effect of lifting the charged pion mass from zero to the finite value $M_{\pi^+} \approx 0.16 \cdot e \cdot M\rho \approx 36$ MeV. The neutral pion stays massless. The charged pion mass is of order $e \cdot \Lambda[QCD]$, i.e. $e \cdot$ (inverse size of pion).

We note that the π^+ - mass is of electromagnetic origin. The self energy diagram consists of a charged pion emitting a virtual photon and turning itself into a massive state (ρ, A_1, ...). Due to the chiral symmetry the sum of all these contributions is finite and of order $e \cdot \Lambda(QCD)$.

With these preparations in mind, we are ready to consider the lepton quark spectrum. The leptons and quark flavors observed thus far seem to come in three families:

$$\text{I.} \qquad \left(\begin{array}{l|l} \nu_e & u\ (5) \\ e^-\ (0.5) & d\ (9) \end{array}\right)$$

$$\text{II.} \qquad \left(\begin{array}{l|l} \nu_\mu & c\ (1{,}200) \\ \mu^-(106) & s\ (180) \end{array}\right)$$

$$\text{III.} \qquad \left(\begin{array}{l|l} \nu_\tau & t\ (>18{,}000) \\ \tau^-(1784) & b\ (4{,}800) \end{array}\right)$$

The numbers in brackets denote the masses of the particles in MeV. For the light quarks we used typical "current algebra" masses. Experimental astrophysical constraints require the neutrino masses to be < 100 eV. Looking closer at the fermion mass spectrum one observes the following facts:

a) There exists a well-obeyed hierarchical structure. The mass ratios of fermions belonging to different generations are given by:

$$\frac{m_s}{m_d} \approx 20, \qquad \frac{m_b}{m_s} \approx 27, \qquad \frac{m_\tau}{m_\mu} \approx 17$$

$$\frac{m_c}{m_u} \approx 240, \qquad \frac{m_t}{m_c} \approx 15, \qquad \frac{m_\mu}{m_e} \approx 207$$

All charged fermions of the first family (e, u, d) are lighter than the ones of the second family (μ, c, s), and those are lighter than the members of the third family (τ, t, b). Note that, with exception of $\frac{m_c}{m_u}$ and $\frac{m_\mu}{m_e}$ all mass ratios are of order 20.

b) inside families the mass ratios of charge $\frac{2}{3}$ to charge $(\frac{-1}{3})$ quarks are given by:

$$\frac{m_u}{m_d} \approx 0.6, \qquad \frac{m_c}{m_s} \approx 6.5, \qquad \frac{m_t}{m_b} > 4$$

i.e. the charge $\frac{2}{3}$ quarks of the second and third family are heavier than the corresponding charge $(-\frac{1}{3})$ quarks. This pattern is broken by the quarks of the first generation.

c) The mass eigenstates are not eigenstates of the weak interactions. The corresponding mixing angles (three angles in case of three families) all seem to be relatively small.

d) The lepton-quark mass spectrum looks rather arbitrary. It lacks any approximate symmetry. Wide fluctuations of the mass parameters are observed. Nevertheless it seems that the masses are somewhat correlated with the electric charge. Neutrinos, being electrically neutral, have no (or an exceedingly small) mass; the masses of the quarks of charge 2/3 are much larger that the masses of the quarks of charge -1/3 (except the first family, see b)).

A solution of the fermion mass problem is required to give answers to the following two questions:

a) Why are the fermion masses much smaller than 1 TeV?

b) What is the mechanism responsible for the generation of mass?

We suppose that the answer to the first question is given by a symmetry. On the scale of the hypercolor interaction, both color and electromagnetism can be viewed as small perturbations. We suppose that in the limit where those interactions are switched off the leptons and quarks are massless; one is dealing with 24 massless states. A chiral symmetry (either a continous or a discrete chiral symmetry) is supposed to provide the reason for the absence of mass.

Besides the 24 massless fermions an infinite number of other fermions with mass of order 1 TeV or larger exist. For those states we see no reason why a chiral symmetry should be valid. Such a symmetry if it remained unbroken would require a parity doubling of all massive states. We suppose that the chiral symmetry is strongly broken by the hypercolor dynamics in the heavy fermion sector. The observed SU(2) symmetry of the weak interactions is supposed to be a flavor symmetry of the hypercolor interaction.

If the electromagnetic interaction is introduced, the leptons and quarks aquire a finite electromagnetic self energy, where the heavy fermions (mass $\sim$ 1 TeV) serve as intermediate states. The self energy is finite due to the cut off of order Λ_h provided by the finite sizes (of order Λ_h^{-1}) of the leptons and quarks. Note that the leptons and quarks have in particular a finite charge radius of order of Λ_h^{-1}.

Thus far we have mentioned only the electromagnetic perturbation of hypercolor dynamics. In a similar way one may consider the QCD interaction. However the gluonic self energy of a quark is strongly model dependent. In all bound state models of the type considered here the fermions do have a finite charge radius. No finite color radius is implied. In these schemes the color resides on one of the constituents (often a scalar object) of the leptons and quarks. In this case the color self energy can be represented by the renormalized mass of the corresponding constituent, which in general is arbitrary; it may vanish as a result of a symmetry (e.g. as a consequence of an underlying supersymmetry). If the QCD interaction would be an important interaction for the generation of mass for leptons and quarks, the mass difference inside weak doublets would be small compared to the average mass (e.g. $m_t - m_b << \frac{1}{2} (m_t + m_b)$). This is not the case. Instead a very strong dependence of the masses on the electric charges is observed. We conclude: the color force is either excluded from contributing to the lepton or quark masses, or contributes only very little. This has implications for the experimental search for lepton-quark substructure: leptons and quarks are expected to have a charge radius of the order of 10^{-17} cm, but no color radius of this order. Subsequently we shall assume that only the QED interaction is responsible for the fermion mass generation[11].

We analyze the fermion electromagnetic self energy diagram somewhat more in detail[12]. In principle, infinitely many heavy (mass > 1 TeV) states will contribute as intermediate states. Since we have no detailed information about the heavy states there is no way to compute the electromagnetic self energy exactly. However, we expect that the lowest contributing intermediate state will dominate. (Something similar is true in hadron physics: the electromagnetic selfenergy of the pion or the proton is dominated by the lowest intermediate state). Under this assumption the resulting fermion mass matrix reads:

$$M = \frac{\alpha}{\pi} Q^2 K \Lambda_h \begin{pmatrix} |f_1^{(n)}|^2 & f_1^{(n)+} f_2^{(n)} & f_1^{(n)+} f_3^{(n)} \\ f_1^{(n)} f_2^{(n)+} & |f_2^{(n)}|^2 & f_2^{(n)+} f_3^{(n)} \\ f_1^{(n)} f_3^{(n)+} & f_2^{(n)} f_3^{(n)+} & |f_3^{(n)}|^2 \end{pmatrix}$$

$$+ O(\alpha^2)$$

Q denotes the electric charge of the fermion and K is a parameter of order one, depending on the transition form factors. The elements of the vectors

$$\vec{f}^{(n)} = \begin{pmatrix} f_1^{(n)} \\ f_2^{(n)} \\ f_3^{(n)} \end{pmatrix}$$

are measures of the transitions $\langle i|j_\mu|n\rangle$, where $|i\rangle$ denotes a quark or lepton state, $|n\rangle$ the intermediate heavy state, and j_μ the electromagnetic current. It is useful to rewrite M in terms of the matrix

$$A = \begin{pmatrix} f_1^{(n)} & f_2^{(n)} & f_3^{(n)} \\ 0 & 0 & 0 \\ 0 & 0 & 0 \\ \vdots & \vdots & \vdots \end{pmatrix}$$

The result is

$$M = \frac{\alpha}{\pi} Q^2 K \Lambda_h (A^+ A) + O(\alpha^2).$$

This shows that M is of rank 1, and the diagonalization gives:

$$M = \frac{\alpha}{\pi} Q^2 K \Lambda_h |\vec{f}^{(n)}|^2 \begin{pmatrix} 0 & 0 & 0 \\ 0 & 0 & 0 \\ 0 & 0 & 1 \end{pmatrix} + O(\alpha^2)$$

Thus in the approximation made above (one intermediate state dominates, $O(\alpha^2)$ terms are neglected) only one family of quarks and leptons aquires a mass. The latter is identified with the third family (t, b, τ). The hierarchical pattern of fermion masses starts to emerge.

As soon as we give up the assumption of the dominance of the fermion self energy by one intermediate state, M ceases to be of rank 1. In the case of 2 intermediate states ($|n>$ and $m>$) the second row of the matrix A ceases to be zero and one finds:

$$A = \begin{pmatrix} f_1^{(n)} & f_2^{(n)} & f_3^{(n)} \\ f_1^{(m)} & f_2^{(m)} & f_3^{(m)} \\ 0 & 0 & 0 \\ \vdots & \vdots & \vdots \end{pmatrix}$$

Now M is of rank 2. Still one eigenvalue of M is exactly zero which means that the third and second family are massive. The magnitude of the fermion mass hierarchy is related to the quality of the dominance of the fermion self energy by one intermediate state, i.e. to the magnitude of $f_i^{(m)}/f_i^{(n)}$. If this quantities are << 1 the diagonalization of the mass matrix gives:

$$M \simeq \frac{\alpha}{\pi} Q^2 K \Lambda_h |\vec{f}^{(n)}|^2$$

$$\mathrm{diag}\left(0, \frac{|\vec{f}^{(n)} x \vec{f}^{(m)}|^2}{|\vec{f}^{(n)}|^4}, 1 + \frac{|\vec{f}^{(m)}|^2}{|\vec{f}^{(n)}|^2} - \frac{|\vec{f}^{(n)} x \vec{f}^{(m)}|^2}{|\vec{f}^{(n)}|^4}\right) + O(\alpha^2)$$

Finally, if more than two intermediate states contribute, also the third row etc. of A is nonzero, M is of rank 3, and the first generation aquires a mass.

The situation is as follows: We found that the fermion mass hierarchy is related to the quality of the dominance of the fermion self energy by one or several intermediate states. The neutrinos being neutral remain massless to $O(\alpha)$. The quarks obey the following mass relations:

$$\frac{m_u}{m_d} = \frac{m_c}{m_s} = \frac{m_t}{m_b} = \frac{(\frac{2}{3})^2}{(-\frac{1}{3})^2} = 4$$

Thus the mass matrices for up and down-type quarks are, up to a factor 4, the same. Therefore, no weak interaction mixing exists in this approximation. The factor 4 between u- type and d- type quark masses as well as the absence of weak mixing is a consequence of the SU(2)-invariance of the transition matrix elements $\langle i|j_\mu|n\rangle$. In order to introduce mixing and departures from relation one has to allow for a violation of the SU(2)-invariance which is caused by electromagnetic effects of order α^2.

An attempt to estimate these effects was made in ref. (12). As a general result we mention that the departures from the relations m_t: m_b = 4: 1 etc. are related to the weak mixing angles. The latter are of order α, e.g. $\sin\theta_c = \frac{3\alpha}{\pi} \cdot (m_s / m_d) \cdot c$ (c: parameter of order one, depending on the quark formfactors). It is interesting to observe that the weak interaction mixing of heavy quarks is expected to be very small. The heavy quarks t and b are essentially inert with respect to the mixing effects. On the other hand the masses of the light quarks u and d are strongly perturbed by mixing effects, and correspondingly the Cabibbo angle is exceptionally large. The b - c transition element in the weak mixing matrix is expected to be of the order of 0.05, implying a life time for the B-meson of the order of 10^{-12} s.

In lowest order of α one finds m_t: m_b = 4: 1, i.e. $m_t \approx 19$ GeV (we use $m_b \approx 4.75$ GeV). The higher order corrections cause m_t/m_b to be slightly larger than 4, but significently less than m_c: $m_s \approx 6...7$. We expect m_t: $m_b \lesssim 5$, i.e. $m_t \lesssim 25$ GeV. If it should turn out that m_t is much larger than 25 GeV, the obvious conclusion would be that also the (t, b)-system is influenced strongly by weak interaction mixing effects, involving one further family of heavy quarks(t', b') with masses significantly larger than the (t,b)-masses. In that case the relation m(t'): m(b') $\approx$ 4...5 should hold, and the B-meson lifetime should be significantly less than the lifetime of the order of 10^{-12} s quoted above.

Finally let me emphasize that the strongly interacting system of W-bosons described here is in its consequences for weak interaction physics in the region 0 ... 100 GeV essentially identical to the SU(2) x U(1) gauge theory. Nevertheless small deviations from the latter are expected, for example in the W- Z- mass spectrum and in the universality of the weak currents. I hope that such small deviations are discovered soon. I do not expect that there will be any important progress in high energy physics, unless this has happened.

REFERENCES

1. G. Arnison et al., Phys. Lett. 112 B (1983) 103.

2. See also: D. Schildknecht, these proceedings.

3. H. Harari and N. Seiberg, Phys. Lett. 98 B (1981) 269
   L. Abbott and E. Farhi, Phys. Lett. 101 B (1981) 69
   H. Fritzsch and G. Mandelbaum, Phys. Lett. 102 B (1981) 319,
   R. Barbieri, A. Masiero, and R. N. Mohapatra, Phys. Lett. 105 B (1981) 369.

4. H. Fritzsch and G. Mandelbaum, see ref. (3), and Phys. Lett. 109 B (1982) 224.

5. J. D. Bjorken, Phys. Rev. D 19 (1979) 335.

6. P. Hung and J. Sakurai, Nucl. Phys. B 143 (1978) 81.

7. H. Fritzsch, D. Schildknecht and R. Kögerler, Phys. Lett. 114 B (1982) 157. See also: R. Kögerler and D. Schildknecht, CERN preprint TH 3231 (1982).

8. See e.g.: J. Gasser and H. Leutwyler, Physics Reports 87 (1982) 77.

9. See e.g.: G.'t Hooft, in: "Recent developments in Gauge Theories", Plenum Press, N.Y. (1980), p. 135.

10. See e.g. the class of models, discussed in ref. (4).

11. See also: H. Fritzsch, in: Proceedings of the Int. Conf. Neutrino 82 (June 1982), Balatonfüred, Hungary, A. Frenkel, L. Jenik ed. (Budapest, 1982).

12. U. Baur and H. Fritzsch, The masses of Composite Quarks and Leptons as Electromagnetic Self Energies, Munich preprint MPI - PAE/PTh 37/83 (June 1983).

CLASSICAL RADIATION SYMMETRY IN GAUGE THEORIES

Robert W. Brown

Physics Department
Case Western Reserve University
Cleveland, Ohio 44106

INTRODUCTION

There are spin-independent zeros in every tree-approximated radiation amplitude in gauge theories. This theorem[1,2] applies to the emission/absorption of any number of massless gauge bosons whose gauge-group symmetry is unbroken. Under certain circumstances the "radiation zeros" lie in physical regions, and may lead to measurable experimental tests of gauge theories. In any case, they imply constraints on the form of the tree amplitudes, and do lead to a new radiation representation for theoretical analysis.

We will discuss the experimental situation, particularly with reference to electroweak physics, in due course. The underlying symmetry and its relationship to gauge theory is considered first and at some length.*

For definiteness, consider single-photon emission in the general scattering process, k particles $\rightarrow$ n - k particles + photon, where the photon has momentum $q = \omega n$, $n \equiv (1,\hat{n})$, and the other particles have charges Q_i and momenta p_i. Then the spin-independent condition for such null zones is simply that all particles have the same $Q/p \cdot q$ ratio. This corresponds to the $n - 2$ equations (one is eliminated by charge and momentum conservation),

$$\frac{Q_i}{p_i \cdot q} = \frac{Q_1}{p_1 \cdot q} , \tag{1}$$

*Background material for this paper may be found in Refs. 1, 2.

standardizing to $i=1$ as the common (but not necessarily constant) value.

For the null zone to overlap with the physical region, (1) requires that all nonzero charges have the same sign and that neutral particles must be massless and and travel in the photon's direction. In Ref. 2, examples of physical null zones are discussed and a physical-null-zone theorem has been given which states that we may always find physical values in the solutions of (1) if all particles are either massless or have the same Q_i/m_i ratio (the ratio for initial and final states may differ). The null zone constraint has also been studied by Samuel[3] and Passarino.[4] Naculich[5] has analyzed the physical-null-zone theorem in more detail and has developed geometrical pictures for the massless case.

We may write the complete single-photon tree amplitude as

$$M = \sum \frac{Q_i I_i}{p_i \cdot q} , \tag{2}$$

by linearity in the charges (the internal charges are fixed by the external charges). Hence the radiation theorem amounts to the current sum rule,

$$\sum I_i = 0 \quad , \tag{3}$$

whose proof will be discussed later.

A complementary theorem is evident from (2). If all particles have the same $I/p \cdot q$ ratio, then $M=0$ by conservation of charge,

$$\sum Q_i = 0 \quad , \tag{4}$$

defining all particles as outgoing for convenience. The $n-2$ equations analogous to (1),

$$\frac{I_i}{p_i \cdot q} = \frac{I_1}{p_1 \cdot q} , \tag{5}$$

are spin dependent and complicated, and do not have interesting physical solutions in general.

The (double) zeros in M, implied by (3) and (4) for the conditions (1) and (5), respectively, can be explicitly exhibited by rewriting (2),

$$M = \sum p_i \cdot q \left(\frac{Q_i}{p_i \cdot q} - \frac{Q_j}{p_j \cdot q}\right) \left(\frac{I_i}{p_i \cdot q} - \frac{I_k}{p_k \cdot q}\right) \quad . \tag{6}$$

Any choice $j \neq k$ reduces (6) to the expected sum of $n-2$ terms. It is easy to verify (6) using (3), (4), and the conservation of momentum,

$$\sum p_i \cdot q = 0 \quad . \tag{7}$$

We may describe the "radiation representation" (6) and its zeros as due to "radiation symmetries." The sum rule, which will be seen to be related to Poincaré invariance, can be restated as a symmetry (in M) under

$$\frac{Q_i}{p_i \cdot q} \rightarrow \frac{Q_i}{p_i \cdot q} + C \quad . \tag{8}$$

Similarly, (4) is equivalent to a symmetry under

$$\frac{I_i}{p_i \cdot q} \rightarrow \frac{I_i}{p_i \cdot q} + C' \quad . \tag{9}$$

By (7), we may combine (8) and (9); the arbitrariness in C and C' lead to (6).

We thus have an alternative description of the radiation theorem: Radiation symmetries are present in tree amplitudes for all theories (spin ≤ 1) whose derivative couplings, if any, follow the gauge prescription. Single derivatives of vector fields can occur in Yang-Mills trilinear form; Dirac derivatives are avoided; single derivatives of scalar fields are allowed generally. Double derivatives of scalar fields can occur as they are obtained by replacing a vector field by a derivative of a scalar field, á la Higgs, in the Yang-Mills coupling.

Since interactions with arbitrary products of fields (and their gauge derivative forms) preserve the tree symmetries as well, we do include non-renormalizable theories. We could think* of this class as quasi-renormalizable in that the products may be considered as limits of heavy particle exchange, leading to tree segments of zero length. But aside from such products, the derivatives - and their gauge covariantization - must occur only as seen in renormalizable gauge theories. In particular, the electromagnetic couplings must correspond to the gyromagnetic value $g = 2$ for all particles.

CLASSICAL BASIS

We can see that the radiation zeros are the relativistic generalization of the well-known absence of electric and magnetic dipole

*I thank C. Goebel for this suggestion.

radiation for nonrelativistic collisions involving particles with the same charge/mass ratio and g-factor. First note that the nonrelativistic limit of (1) is

$$\frac{Q_i}{m_i} = \frac{Q_1}{m_1} \quad . \tag{10}$$

If (10) holds and all spinning particles have identical g_i, then the absence of external forces implies $\ddot{\vec{d}} = \ddot{\vec{\mu}} = 0$ where $\vec{d}$ ($\vec{\mu}$) is the electric (magnetic) dipole moment.

A direct connection to the infrared piece of quantum tree graphs can be made by looking for a relativistic generalization of the dipole result. As we will see, this leads to (1).

The sudden disappearance/appearance of uniformly moving charges is an adequate picture for radiation in the infrared limit ($\omega \to 0$). The corresponding classical amplitude is[1,2]

$$A_{IR} = -\sum \frac{Q_i}{\omega(1 - \hat{n}\cdot\vec{v}_i)} \vec{v}_i \cdot \vec{\varepsilon} \quad . \tag{11}$$

Indeed, A_{IR} vanishes under (10) in the nonrelativistic (electric dipole) limit, by conservation of momentum.

In covariant notation, (11) reads

$$A_{IR} = \sum \frac{Q_i}{p_i\cdot q} p_i \cdot \varepsilon \quad . \tag{12}$$

All the zeros and symmetries described previously are evident in (12) where $I_i = p_i\cdot\varepsilon$. Under (1) there is complete destructive interference of the long wavelength (external classical lines) radiation in relativistic collisions. The incorporation of spin can best wait until the covering field theory calculations.

PROOF

The proof of the radiation theorem shows how the classical result for particle orbits is extended to quantum tree amplitudes (classical field solutions), and illustrates the underlying symmetry seen for gauge couplings. In addition, a useful reorganization for radiation problems emerges.

Consider first a vertex (no internal lines) graph as a source to which we attach a photon in all possible ways. The same expression (12) is found for the scattering of scalar particles at a point with no derivative couplings.* Notice that the photon can now be hard.

The inclusion of spin is remarkable in that, for $g = 2$, the spin currents, linear in q, generate a first-order Lorentz transformation of the associated leg (wave function). In the null zone (1) this transformation is the same (universal) for all vertex legs, so that the spin currents must cancel out by Lorentz invariance just as the convection currents (present for all spins) cancel by momentum conservation.

Equally remarkably (when derivative couplings are included), the lowest-order (in q) contact currents, each comprised of a sea-gull term plus a momentum-shift term, provide the analogous Lorentz transformations of any derivative associated with a given leg. Thus the photon attachments lead to the transformations of the new tensors (the derivatives) in the problem necessary to maintain the first-order null zone cancellation.

The radiation amplitude M derived from a vertex source graph V_G will therefore vanish in the null zone if there are no higher-order currents. This is why we must stay with the gauge prescription given earlier. Although the Yang-Mills vertex yields a quadratic current, it cancels in the null zone due to its cyclic nature.

From this diagrammatic analysis, our result is

$$M(V_G) = \sum_{\text{legs}} \frac{Q_i J_i}{p_i \cdot q} \tag{13}$$

where, for gauge derivatives,

$$\sum_{\text{legs}} J_i = 0 \quad , \tag{14}$$

by Poincaré and Yang-Mills symmetry. We obtain J_i by inserting j_i into leg i of V_G where

$$j = j_{conv} + j_{spin} + j_{cont} + j_{YM} \quad , \tag{15}$$

with

$$j_{conv} = \pm\, p \cdot \varepsilon \text{ for outgoing } (+) \text{ or incoming } (-) \; , \tag{16}$$

$$j_{spin} = \{0; +\tfrac{i}{4}\,\sigma^{\alpha\beta}\omega_{\alpha\beta}; -\tfrac{i}{4}\,\sigma^{\alpha\beta}\omega_{\alpha\beta}; g_{\alpha\beta} \to \omega_{\alpha\beta}\} \text{ for } \{\text{scalar; spinor } \bar{u},\bar{v}\text{; spinor } u,v\text{; vector } \eta_\alpha = g_{\alpha\beta}\eta^\beta, \eta_\alpha^+ = g_{\alpha\beta}\eta^{+\beta}\}, \tag{17}$$

*This scalar calculation has also been performed independently by M. Samuel (private communication).

$$j_{cont} = \text{replacement } g_{\alpha\beta} \to \omega_{\alpha\beta} \text{ for } p_\alpha = g_{\alpha\beta}p^\beta \quad , \tag{18}$$

$$j_{YM} = \text{product of the spin current and momentum shift for Yang-Mills trilinear coupling,} \tag{19}$$

in terms of Lorentz generator

$$\omega_{\alpha\beta} = q_\alpha \varepsilon_\beta - \varepsilon_\alpha q_\beta \quad . \tag{20}$$

To finish the proof we consider general tree source graphs T_G which involve photon couplings to (fixed) internal lines. If Γ is the (lowest-order) photon vertex corresponding to $g = 2$, we have available the decomposition identity,

$$D(p-q)\Gamma D(p) + \text{seagulls (if any)} = D(p-q)j\frac{Q}{p\cdot q} + \frac{Q}{p\cdot q}j\,D(p), \tag{21}$$

for an internal particle (spin ≤ 1) with propagator D. We get exactly the $Qj/p\cdot q$ factor for each internal leg of every vertex that we found previously for the external legs. With the aid of (21), we obtain a reorganized "vertex expansion,"

$$M(T_G) = \sum M(V_G)\ R(V_G) \quad , \tag{22}$$

with all propagators included in the R factors.* Eqs. (13) and (14) still apply. In the null zone (1), all internal lines of a tree graph also have the same $Q/p\cdot q$ ratios, giving $M(T_G) = 0$.

NEUTRAL PARTICLES

We have already noted that neutral particles must be massless for a physical null zone, although the radiation representation is independent of this question. When $p_r \propto q$ for $Q_r = 0$, then it can be shown that $J_r = 0$ (using the fact that massless vector bosons must be coupled to conserved currents) so that (14) is still relevant even though the rth term is absent in (13). An exception is Compton scattering whose null zone is in the forward direction. The spin terms transfer no momentum through the photon-charge coupling and are not eliminated by current conservation.

VIOLATIONS

Anomalous intrinsic magnetic moments ($g \neq 2$) modify the identity (21), contributing additional spin currents that cannot be described by a universal Lorentz transformation. Other higher derivative

*The expansion (22) is a gauge-invariant decomposition and is less cumbersome than (2). The radiation representation can be used for (13).

couplings (beyond the gauge couplings detailed in the Introduction) have no general mechanism for their cancellation, leading to non-vanishing $O(q^2)$ terms on the right-hand-side of (14). These non-gauge couplings violate the radiation theorem.

It may therefore be expected that closed loops also violate the theorem since they give corrections to the electromagnetic moments and they imply an infinite derivative-coupling series. These violations may be anticipated from the uncertainty principle: An exact interference cancellation should be spoiled by short-range quantum correlations among the particles. Except for scalar self-energies or overriding mechanisms such as angular momentum conservation or the exclusion principle, radiation zeros are not sustained by loops; see Refs. 1,2. Indeed, explicit calculations[6,7] by Laursen, Samuel, Sen, and Tupper show that, while first-order scalar bubbles preserve zeros, triangle and seagull closed loops do not.

CANCELLATIONS IN GENERAL AMPLITUDES

From the previous remarks it is clear that, even though a general radiation amplitude which includes higher-order corrections may not vanish in the null zone, its infrared-divergent term will. The next-order $O(q^o)$ term has the same zeros if $g = 2$ for all external particles. Consequently, low-energy theorems naturally separate out the contributions that satisfy the radiation theorem.

It follows from the standard infrared analysis that the infrared-divergent virtual corrections in the elastic cross section that usually cancel against the bremsstrahlung divergence must vanish in the region obtained by a $q \to 0$ limit of (1).* We see that radiation symmetry has implications for virtual photons as well. For every photon reaction, the infrared limit of its null zone yields a region of suppression for the corrections to the corresponding reaction without the photon.

Any higher-order correction that is proportional to the lowest-order tree graph obviously vanishes along with the tree graph. Hence all the soft-photon corrections to a tree radiation amplitude share its radiation zeros. Consistently, the infrared terms in the cross section corrections, which are normally cancelled by the soft-photon divergences, are themselves proportional to the lowest-order tree graph.

In recent years we have learned how to factorize mass singularities in the all-orders, all-logarithms proof of the parton model.[8] Such singularities should therefore cancel out in the null zone of

*The remaining discussion in this section has been inspired by conversations with J. Smith and G. Sterman.

the lowest-order (hard scattering) amplitude. (This remark is relevant to the cancellations seen recently by Laursen et al.[7]) In a by-now familiar story, the mass singularities against which these cancel, under the conditions of the KLN theorem,[9] must themselves vanish in the appropriate null zone limit.

NON-ABELIAN RADIATION

The extension to the emission/absorption of any gauge boson g of a general gauge group G is straightforward. We replace Q by Q_g in the null zone condition where Q_g is the contraction of Clebsch-Gordon coefficients appropriate to the leg and the source vertex. Conservation of charge now corresponds to the fact that g is in the adjoint representation so that the sum of its couplings around a G-invariant vertex is zero. Using Feynman rules in factored form we obtain the same currents J_i as before.

We observe that the most obvious candidate, QCD, hides its radiation zeros, since the color-singlet physical states are connected to quark and gluon particles only through color averaging and summing. The radiation representation still applies, of course.

DUAL SYMMETRY GENERATOR

The radiation symmetries, (8) and (9), show a dual role for the electromagnetic or other gauge theory current: The current generates transformations both in G space and, in effect, in space-time. For identical $Q/p{\cdot}q$, the convective current effectively generates a universal displacement of the associated wave function; the spin and contact currents effectively generate a universal Lorentz transformation of the wave function and its derivative, respectively. We view the massless gauge boson as having joint membership in the adjoint representations of both the internal group and a Poincaré little group.

EXPERIMENTAL IMPLICATIONS

The relativistic null zone condition is sufficiently severe that it was only a few years ago that a theoretical example was found by accident. (The Mitsubishi Zeros* of the 1940's were eventually shot down.) In the context of the program[10,11] for the study of weak-boson self-couplings in pp and $p\bar{p}$ collisions, Mikaelian, Samuel, and Sahdev[12] first pointed out that a zero exists in the $q\bar{q} \to W\gamma$ angular distribution for gauge couplings. Another example is $\nu e \to W\gamma$; related zeros appear in W radiative decay.[13]

*N.Kaplan may be forgiven for reminding me of these zeros.

The MSS zero implies that each helicity amplitude for these $W\gamma$ reactions must factorize, the algebra for which has been developed very nicely by Goebel, Halzen, and Leveille[14] and used by Dongpei.[15] The phenomenon to this point historically seemed to be restricted to 3-vertex sources, and the absence of an explanation provided the motivation for the work described in Refs. 1 and 2. We now recognize Mikaelian factorization, the MSS zero and the CHL algebra as $n = 3$ examples of the radiation zeros (and representation) which are the relativistic generalization of the textbook dipole result.

Thus we have more possibilities for the probe of gauge theory couplings.* The investigation of bremsstrahlung null zones in hard quark scattering, $q\,q \to q\,q\,\gamma$ and $e\,q \to e\,q\,\gamma$, as well as of radiative decay zeros may give a measure of heavy quark and heavy lepton parameters. In general, the deviations from zero provide bounds on magnetic moments, on higher-order corrections, on masses (the larger the mass the smaller the null zone) and so forth. We may study standard reactions (the zero in $e^-e^- \to e^-e^-\gamma$ was not known before this) or exotic-particle processes.

The most promising experiment remains $p\,\bar{p}\,(p\,p) \to W\,\gamma\,X$, possibly measurable in the near future[16] at the SPS. The hard transverse photon recoils against the W, coupling in leading twist only to the hard-scattering subprocess. There are quark transverse momentum smearing and gluon radiative corrections [of order $\alpha_s(M_W^2)/\pi$]. Monyonko and Reid[17] are in the midst of a thorough calculation of the α_s corrections where (apropos the remarks in the previous section) it is noted that, since the leading contribution of the virtual corrections comes from interference with the lowest order amplitude, they also vanish at the $W\gamma$ zero. In the related W radiative decay, Passarino[4] has proposed that the null zone serve as a filter which lets only heavier quarks through.

STREAMLINING

It is of interest to know the extent to which radiation symmetries can be used to simplify computations. Heroic CALKULators[18] in QED and QCD have recently found that lowest-order radiative amplitudes reduce to simple forms, where in particular they show how 5-body single-photon amplitudes factorize in the massless limit. We have verified that radiation zeros are present in these forms. Also, Passarino[5] has combined the helicity techniques of CALKUL and the radiation representation in examples of amplitude analysis.

---

*I thank R. Brock, R. Decker, E. Paschos, and D. Stump for discussions concerning these possibilities.

GAUGE-FIELD DECOUPLING

Our most recent work[19] in radiation symmetry is motivated by the multi-photon null zones and the finite Lorentz transformation described in Ref. 2. We have been able to show that the radiation zero is itself a special case of a general decoupling theorem: To all orders in classical gauge theory, an external plane wave, $e^{iq\cdot x}$, $q^2 = 0$, decouples from scattering amplitudes in the null zone. The decoupling mechanism is again a combination of Poincaré and gauge invariance; the external field can be eliminated in the momentum-space region, given by (1) or its non-Abelian equivalent, using finite Poincaré and gauge transformations. These same transformations can be used to find solutions of the classical particle and matter field equations for external plane wave potentials.

OUTLOOK

Radiation symmetry is another distinctive characteristic of gauge theories. We look for it to provide experimental tests of particle parameters, directions in the organization of computations, and new understanding of theoretical issues such as renormalizability.

ACKNOWLEDGMENTS

I am indebted to my collaborators, Professors Stanley J. Brodsky and Kenneth L. Kowalski, for many helpful discussions. This work has been supported by the National Science Foundation.

REFERENCES

1. S.J. Brodsky and R.W. Brown, Phys. Rev. Lett. 49, 966 (1982).
2. R.W. Brown, K.L. Kowalski and S.J. Brodsky, Phys. Rev. D1 (to be published).
3. M.A. Samuel, Oklahoma State Preprint 133 (1982).
4. G. Passarino, SLAC-PUB-3024 (1982).
5. S. Naculich, Case Western Reserve Senior Thesis, CWRUTH-83-4.
6. M.L. Laursen, M.A. Samuel, A. Sen and G. Tupper, Oklahoma State Preprint 137 (1982); M.A. Samuel, ibid. 139 (1982).
7. M.L. Laursen, M.A. Samuel and A. Sen, ibid. 142 (1983).
8. See, for example, the discussion and references in A.H. Mueller, Phys. Rep. 73C, 237 (1981).
9. T. Kinoshita, J. Math. Phys. 3, 650 (1962); T.D. Lee and M. Nauenberg, Phys. Rev. 133, 1549 (1964).
10. R.W. Brown and K. O. Mikaelian, Phys. Rev. D19, 922 (1979).
11. R.W. Brown, D. Sahdev, and K.O. Mikaelian, Phys. Rev. D20, 1164 (1979). For a review, see R.W. Brown, in Proton-Antiproton

Collider Physics - 1981, edited by V. Barger, D. Cline, and F. Halzen (AIP, New York, 1982), p. 251.

12. K.O. Mikaelian, M.A. Samuel, and D. Sahdev, Phys. Rev. Lett. 43, 746 (1979). See also Ref. 11.
13. T.R. Grose and K.O. Mikaelian, Phys. Rev. D23, 123 (1981).
14. C.J. Goebel, F. Halzen, and J. P. Leveille, Phys. Rev. D23, 2682 (1981). Cross section factorization was first shown for $\gamma q \to W q$ by K.O. Mikaelian, Phys. Rev. D17, 750 (1978).
15. Z. Dongpei, Phys. Rev. D22, 2266 (1980). See also K.O. Mikaelian, Phys. Rev. D25, 66 (1982).
16. D. Cline and C. Rubbia (private communication).
17. N.M. Monyonko and J.H. Reid, Simon Fraser Preprint (1983).
18. F.A. Berends, R.K. Kleiss, P. De Causmaecker, R. Gastmans, W. Troost and T.T. Wu, Nucl. Phys. B206, 61 (1982) and references therein.
19. R.W. Brown and K. L. Kowalski, Case Western Reserve Preprint, CWRUTH-83-5.

NEW TECHNIQUES AND RESULTS IN GAUGE THEORY CALCULATIONS*

Presented by R. Gastmans

Calkul Collaboration

F.A. Berends[a], D. Danckaert[b], P. De Causmaecker[b,1],

R. Gastmans[b,2], R. Kleiss[a,3], W. Troost[b,4] and T.T. Wu[c,5]

[a] Instituut-Lorentz, Leiden, The Netherlands
[b] Instituut voor Theoretische Fysica, University of Leuven B-3030 Leuven, Belgium
[c] Deutsches Elektronen-Synchrotron DESY, Hamburg, Germany and Gordon McKay Laboratory, Harvard University Cambridge, MA 02138, USA

ABSTRACT

It is shown how the introduction of explicit polarization vectors of the radiated gauge particles leads to great simplifications in the calculation of bremsstrahlung processes at high energies. Applications of this technique are given for single and multiple bremsstrahlung in QED and QCD, in particular for $e^+e^- \to 4$ jets.

---

*Work supported in part by the NATO Research Grant No. RG 079.80.
[1] Navorser, I.I.K.W., Belgium.
[2] Onderzoeksleider, N.F.W.O., Belgium.
[3] Work supported in part by the Stichting F.O.M., The Netherlands.
[4] Bevoegdverklaard Navorser, N.F.W.O., Belgium.
[5] Work supported in part by the United States Department of Energy under Contract No. DE-AS02-76-ER03227.

I. INTRODUCTION

As a rule, the final results in gauge theory calculations are always much simpler than the intermediate formulae. The intermediate expressions are usually lengthy, and they implicitly contain cancellations which ultimately yield the simple final answers.

It has been said many times that simple formulae should be obtained in a simple way. In this paper, we present a method for obtaining bremsstrahlung cross sections in gauge theories which naturally leads to simple final answers without generating long and untransparent intermediate formulae. We show that the calculation of the various helicity amplitudes for a bremsstrahlung process in the high energy limit, where fermion masses can be neglected, is greatly simplified by the introduction, in a covariant way, of explicit polarization vectors for the radiated gauge particles.

In this way, multiple bremsstrahlung of hard photons and/or gluons becomes calculable, although these processes are described by many Feynman diagrams. Because of the relative simplicity of the formulae, certain properties of cross sections, like e.g. for $e^+e^- \to$ 4 jets, could easily be established.

We explain the method in Sec. II and give applications to single bremsstrahlung processes in Sec. III. Multiple bremsstrahlung is treated in Sec. IV, and Sec. V summarizes our results.

We hope that our technique can lead to similar simplifications in the calculation of virtual radiative corrections. Recently, however, it has been proposed that the use of the Gegenbauer polynomial technique in x-space[1] could be very useful for massless theories. In the Appendix, we present some results in the calculation of two-loop vertex corrections.

II. HELICITY FORMALISM [2,3]

Consider, for example, a photon with four-momentum k which is radiated from a charged fermion line for which q_- and q_+ are the momenta of the outgoing fermion and anti-fermion. We can then construct two photon polarizations orthogonal to k and to each other:

$$\varepsilon_\mu^{\parallel} = N_q \, [(q_+k)q_{-\mu} - (q_-k)q_{+\mu}] \,, \qquad (1)$$

$$\varepsilon_\mu^{\perp} = N_q \, \varepsilon_{\mu\alpha\beta\gamma} \, q_+^\alpha \, q_-^\beta \, k^\gamma,$$

where the normalization factor is

$$N_q = [2(q_+q_-)(q_+k)(q_-k)]^{-1/2} \,. \qquad (2)$$

Equivalently, one can construct two circular polarization

vectors

$$\varepsilon_\mu^\pm = (\varepsilon_\mu^\parallel \pm i\,\varepsilon_\mu^\perp)/\sqrt{2}\ . \tag{3}$$

But, in QED, only the combination $\not\varepsilon^\pm$ appears, which can effectively be written as

$$\not\varepsilon^\pm = -\frac{1}{2\sqrt{2}}\, N_q\, [\not k\, \not q_- \not q_+ (1 \pm \gamma_5) - \not q_- \not q_+ \not k (1 \mp \gamma_5)]\ . \tag{4}$$

This equation is obtained after dropping a term proportional to $\not k\,\gamma_5$, which in massless QED gives a vanishing contribution because of axial current conservation.

From this expression it is immediately clear that if the helicities of the fermions are also fixed, only one of the terms can contribute. Furthermore, if $\not\varepsilon^\pm$ stands next to a spinor $\bar{u}(q_-)$ or $v(q_+)$, this one term will either give zero because of the Dirac equations $\bar{u}(q_-)\not q_- = 0$ or $\not q_+ v(q_+) = 0$, or give rise to a factor $2(q_\pm k)$ which cancels the denominator of the adjacent fermion propagator in the Feynman diagram. These properties are at the origin of the simplifications which were announced in the Introduction.

Often, there is more than one charged fermion line in the Feynman diagrams of the process. In order to generate the simplifications, it is however essential to use for the photon polarizations, Eq. (4), the external momenta of the fermion line to which the photon is attached. Since a photon can only have two polarizations, the polarization vectors expressed in terms of the momenta $q_\pm$ are related to the polarization vectors in terms of momenta $p_\pm$ by the equation

$$\varepsilon_\mu^\pm(q_+,q_-) = e^{\pm i\phi}\, \varepsilon_\mu^\pm(p_+,p_-) + \beta_\pm k_\mu\ . \tag{5}$$

The constants $\beta_\pm$ are of no relevance, but the phase ϕ must be properly taken into account. It is given by the dot product of the ε's.

With some minor modifications, this technique can also be applied advantageously to the case of gluon bremsstrahlung.

III. SINGLE BREMSSTRAHLUNG [4]

Consider first the process

$$e^+(p_+) + e^-(p_-) \to \mu^+(q_+) + \mu^-(q_-) + \gamma(k)\ , \tag{6}$$

where the momenta of the particles are given between parentheses, and let us introduce the following notation:

$$s = (p_+ + p_-)^2 \ , \ t = (p_+ - q_+)^2 \ , \ u = (p_+ - q_-)^2 ,$$
$$s' = (q_+ + q_-)^2 \ , \ t' = (p_- - q_-)^2 \ , \ u' = (p_- - q_+)^2 \ . \tag{7}$$

All particles being massless, we have furthermore

$$s + s' + t + t' + u + u' = 0. \tag{8}$$

To simplify the discussion, we shall assume first that the photon is emitted by the muons only. We then have to consider the two top Feynman diagrams of Fig.1. Their corresponding Feynman amplitudes are given by

$$M_1 = \frac{ie^3}{2s(q_-k)} \bar{v}(p_+)\gamma^\mu u(p_-) \, \bar{u}(q_-) \not{\epsilon} \, (\not{q}_- + \not{k})\gamma_\mu v(q_+),$$
$$M_2 = \frac{ie^3}{2s(q_+k)} \bar{v}(p_+)\gamma^\mu u(p_-) \, \bar{u}(q_-)\gamma_\mu \, (-\not{q}_+ - \not{k}) \not{\epsilon} \, v(q_+). \tag{9}$$

Suppose we want to calculate the helicity amplitude $M(+,-,+,-,+)$, where the arguments indicate the helicities of the e^+, e^-, μ^+, μ^-, and γ . Because of the $(1-\gamma_5)/2$ helicity projection operator for the spinor $v(q_+)$, only the second term of $\not{\epsilon}^+$ [Eq.(4)] can contribute. Inserting the second term of $\not{\epsilon}^+$ in M_1 gives zero because of the Dirac equation, hence only M_2 contributes yielding

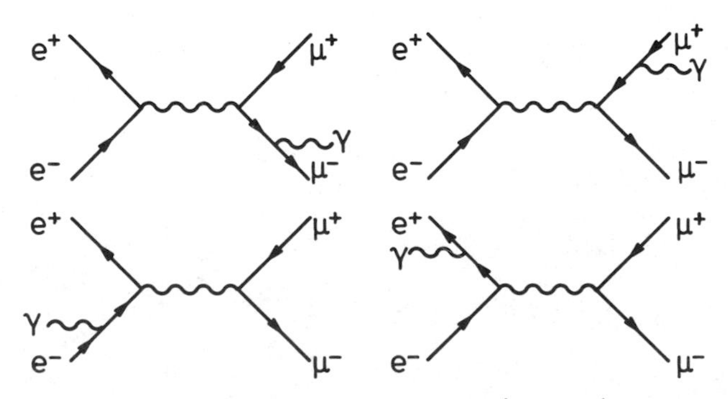

Fig.1. Feynman diagrams for $e^+e^- \to \mu^+\mu^-\gamma$.

$M(+,-,+,-,+)$

$$= -\frac{ie^3}{4s\sqrt{2}} N_q \bar{v}(p_+)\gamma^\mu(1-\gamma_5)u(p_-)\, \bar{u}(q_-)\gamma_\mu(\not{p}_+ + \not{p}_-)\not{q}_-(1-\gamma_5)v(q_+). \tag{10}$$

This formula can further be simplified by eliminating the repeated index μ. To this end, we observe that, for any lightlike vector q,

$$(1 \mp \gamma_5)v(q)\bar{v}(q)(1 \pm \gamma_5) = (1 \mp \gamma_5) \sum_{hel} v(q)\bar{v}(q)(1 \pm \gamma_5) = 2(1 \mp \gamma_5)\not{q} . \tag{11}$$

Hence,

$$M(+,-,+,-,+) = -\frac{ie^3}{4s\sqrt{2}} N_q$$

$$\times \frac{\bar{v}(p_+)\gamma^\mu(1-\gamma_5)u(p_-)\bar{u}(p_-)\not{a}(1-\gamma_5)u(q_-)\bar{u}(q_-)\gamma_\mu(\not{p}_+ + \not{p}_-)\not{q}_-(1-\gamma_5)v(q_+)}{\bar{u}(p_-)\not{a}(1-\gamma_5)u(q_-)}$$

$$= -\frac{ie^3}{s\sqrt{2}} N_q \frac{\bar{v}(p_+)\gamma^\mu \not{p}_- \not{a}\, \not{q}_-\gamma_\mu(\not{p}_+ + \not{p}_-)\not{q}_-(1-\gamma_5)v(q_+)}{\bar{u}(p_-)\not{a}(1-\gamma_5)u(q_-)} \tag{12}$$

$$= \frac{ie^3}{s}\sqrt{2}\, N_q \frac{\bar{v}(p_+)\not{q}_-\not{a}\, \not{p}_-\not{p}_+\not{q}_-(1-\gamma_5)v(q_+)}{\bar{u}(p_-)\not{a}(1-\gamma_5)u(q_-)} .$$

Choosing $a = p_+$ yields

$$M(+,-,+,-,+) = -\sqrt{2}ie^3 N_q\, u \frac{\bar{v}(p_+)\not{q}_-(1-\gamma_5)v(q_+)}{\bar{u}(p_-)\not{p}_+(1-\gamma_5)u(q_-)}$$

$$\hat{=} \sqrt{2}\, e^3\, s' N_q \frac{u}{(ss')^{1/2}} , \tag{13}$$

where in the last line we neglected an irrelevant overall phase factor.

If, in addition to radiation from the muon line, one includes radiation from the electron line, one has to use an expression $\not{\epsilon}^+$ in terms of the momenta p_+ and p_-. Including the Feynman diagrams M_3 and M_4 then yields

$$M(+,-,+,-,+) \hat{=} \sqrt{2}\, e^3 [s' N_q + s\, N_p\, e^{i\phi}] \frac{u}{(ss')^{1/2}} , \tag{14}$$

with

$$N_p = [2(p_+p_-)(p_+k)(p_-k)]^{-1/2} , \quad (15)$$

$$e^{i\phi} = -(\varepsilon_p^- \varepsilon_q^+) = \frac{1}{8} N_p N_q \ \mathrm{Tr}\ [\not{p}_+ \not{p}_- \not{k}\ \not{q}_- \not{q}_+ \not{k}(1-\gamma_5)] .$$

One thus finds that the helicity amplitude (14) factorizes:

i) the first factor, in the brackets, has a squared absolute value which is easily seen to be proportional to the well-known "infrared factor" [5]:

$$S(k) = -e^2 \left[\frac{p_+}{(p_+k)} - \frac{p_-}{(p_-k)} - \frac{q_+}{(q_+k)} + \frac{q_-}{(q_-k)}\right]^2 . \quad (16)$$

ii) the second factor is given by the particular helicity configuration one considers. Indeed, for different helicities, one finds that the quantity u in Eq. (14) has to be replaced by u', t, or t'. Thus,

$$\sum_{pol} |M_1 + M_2 + M_3 + M_4|^2 = S(k)\ 4\ e^4\ \frac{t^2+t'^2+u^2+u'^2}{ss'} . \quad (17)$$

The last factor in this formula can be seen as a suitable extension of the non-radiative cross section, which in this case is simply $8e^4(t^2+u^2)/s^2$.

The factorization property is by no means particular for $e^+e^-\to\mu^+\mu^-\gamma$. It is found that also for $e^+e^-\to\gamma\gamma\gamma$ and $e^+e^-\to e^+e^-\gamma$, the helicity amplitudes exhibit a structure similar to Eq. (14).

In the case of QCD, the "infrared factors" have a more complicated structure, but we find that, mutatis mutandis, the single bremsstrahlung processes $qq\to q'q'g$, $qq\to qqg$, $q\bar{q}\to ggg$, and $gg\to ggg$ have cross sections very similar to Eq. (17). In particular, it is seen that these cross sections never have double poles in the invariants.

IV. MULTIPLE BREMSSTRAHLUNG [2,6]

Certain helicity amplitudes for multiple bremsstrahlung processes are straightforward generalizations of single bremsstrahlung helicity amplitudes. E.g., for

$$e^+(p_+) + e^-(p_-) \to \mu^+(q_+) + \mu^-(q_-) + \gamma(k_1) + \ldots + \gamma(k_n), \quad (18)$$

one finds that

$$|M(+,-,+,-,+,\ldots,+)|^2 = S(k_1)\ldots S(k_n)\ 2e^4\ \frac{u^2}{ss'} . \quad (19)$$

In this case, the simplicity is due to the fact that all the photons have the same helicity.

When the photon helicities are different, the helicity amplitudes have a more complicated structure. The best one can do is to choose a specific representation for the spinors and to evaluate the helicity amplitudes as complex numbers for a given point in phase space. This is done as follows:

We go to the e^+e^- c.m. frame with the z-direction along $\vec{p}_+$, and introduce the notation

$$k_\pm = k_o \pm k_z, \; k_\perp = k_x + i\, k_y = |k_\perp| e^{i\phi_k} , \tag{20}$$

for any light-like vector k. Choosing a representation for the γ-matrices for which

$$\gamma_5 = \begin{pmatrix} 1 & 0 \\ 0 & -1 \end{pmatrix} , \tag{21}$$

we can take

$$u_+(k) = v_-(k) = \begin{pmatrix} \sqrt{k_+} \\ \sqrt{k_-}\, e^{i\phi_k} \\ 0 \\ 0 \end{pmatrix} , \; u_-(k) = v_+(k) = \begin{pmatrix} 0 \\ 0 \\ -\sqrt{k_-}\, e^{-i\phi_k} \\ \sqrt{k_+} \end{pmatrix} , \tag{22}$$

where the first spinor is an eigenstate of $1 + \gamma_5$, and the second one of $1 - \gamma_5$.

With these formulae, it is easily seen that

$$\bar{u}(k_i)(1+\gamma_5)u(k_j) = [\bar{u}(k_j)(1-\gamma_5)u(k_i)]^\star = 2 \, \frac{k_{j\perp} \, Z^\star_{ij}}{k_{j-}\sqrt{k_{i+}k_{j+}}} , \tag{23}$$

with

$$Z_{ij} = k_{i+}k_{j-} - k^\star_{i\perp} k_{j\perp} . \tag{24}$$

Once the repeated indices are eliminated from the helicity amplitudes, as explained in Sec. III, one can reduce any spinorial expression to a product of elementary expressions, like in Eq.(23), by using Eq. (11) as often as necessary. In this way, helicity amplitudes are expressed in terms of the quantities Z_{ij}, which are complex functions of the various momenta in the process.

We have applied these procedures to the calculation of the helicity amplitudes for $e^+e^- \to q\bar{q}gg$ and $e^+e^- \to q\bar{q}\, q\bar{q}$, which are the subprocesses for $e^+e^- \to 4$ jets.[6] Because of the relative simplicity of our formulae, we were able to establish that the 4-jet cross section is proportional to $R = \Sigma\, Q_f^2$, where Q_f are the fractional charges of the quarks and where the summation runs over the different quark flavors. The terms proportional to $(\Sigma\, Q_f)^2$ dropped out after momentum symmetrization.

V. CONCLUSIONS

In principle, our helicity formalism can be applied systematically to all multiple bremsstrahlung processes in gauge theories at high energies. It is especially useful for those processes for which the standard manipulations of covariant summation over polarization degrees of freedom would become prohibitively lengthy (Just imagine that $e^+e^- \to 6$ jets has to be calculated!).

The main advantages of our procedure are that

1) only a limited number of Feynman diagrams contribute to a given helicity amplitude;
2) the introduction of explicit polarization vectors for the gauge particles can be done in a covariant way;
3) unphysical ghost contributions do not have to be considered;
4) repeated indices can be eliminated;
5) factorization properties can be found at the level of the helicity amplitudes;
6) multiple bremsstrahlung becomes calculable by the introduction of explicit spinors.

ACKNOWLEDGEMENT

One of us (RG) would like to thank Professor Harvey Newman for his kind invitation and for providing the stimulating environment of this Europhysics Study Conference.

APPENDIX

Since the massless limit leads to such simplifications in the calculation or real bremsstrahlung processes, one might wonder whether massless theories also have virtual corrections which can be calculated in a simple way. Chetyrkin et al.[1] advocate that the use of the Gegenbauer polynomial method in x-space is ideally suited for this purpose. They illustrate the technique for the 3-loop β-function in QED.

Essentially, the method consists in Fourier transforming the momentum space propagators

$$\frac{1}{k^2} = \frac{\Gamma(\lambda)}{\pi^{\lambda+1}} \int \frac{\exp\,[2i(kx)]}{(x^2)^\lambda} d^D x, \qquad (A.1)$$

where $\lambda = (D-2)/2 = 1 + \varepsilon$.

The integration over the virtual momenta in the Feynman diagram then yields δ-functions, which can be used to integrate over some of the x-variables.

The next step consists in expanding the x-space propagators in Gegenbauer polynomials:

$$\frac{1}{(x_1-x_2)^{2\lambda}} = \frac{1}{(\max\, r_1, r_2)^\lambda} \sum_{n=o}^{\infty} C_n^\lambda(\hat{x}_1 . \hat{x}_2) \left\langle \frac{r_1}{r_2} \right\rangle^{n/2}, \qquad (A.2)$$

where $r_1 = x_1^2$, $r_2 = x_2^2$, $\hat{x} = \vec{x}/\sqrt{r}$, and

$$\left\langle \frac{r_1}{r_2} \right\rangle = \begin{cases} r_1/r_2 & \text{if } r_1 \leqslant r_2 \\ r_2/r_1 & \text{if } r_2 \leqslant r_1 \end{cases} . \qquad (A.3)$$

Analogously, the remaining exponentials can be expanded:

$$\exp[2i(kx)] = \Gamma(\lambda) \sum_{n=o}^{\infty} i^n (n+\lambda) C_n^\lambda(\hat{x}.\hat{k})(k^2 r)^{n/2} j_{\lambda+n}(k^2 r),$$

$$j_\alpha(z) = \sum_{n=o}^{\infty} \frac{(-z)^n}{n!\,\Gamma(n+\alpha+1)} . \qquad (A.4)$$

Most of the time, the integration over the angular variables $\hat{x}_i$ can be done using the orthogonality relation of the Gegenbauer polynomials, and the remaining radial integrations can then be performed using

$$\int_0^\infty z^b j_a(z) dz = \Gamma(b+1)/\Gamma(a-b), \qquad \mathcal{R}e\, b > -1, \quad \mathcal{R}e\, a > 2\, \mathcal{R}e\, b + 1/2. \qquad (A.5)$$

The final expression for the Feynman diagram is then obtained by evaluating the remaining (multiple) sums in the limit $\varepsilon \to 0$.

We verified that the method could also be used for evaluating 2-loop vertex corrections in massless theories. For the Feynman diagrams of Fig.2, one has to know the integrals

$$I_1 = \int d^D k d^D \ell \, [(p-k)^2 (k-Q)^2 (k+Q)^2 (k-\ell)^2 (\ell-Q)^2 (\ell+Q)^2]^{-1} ,$$

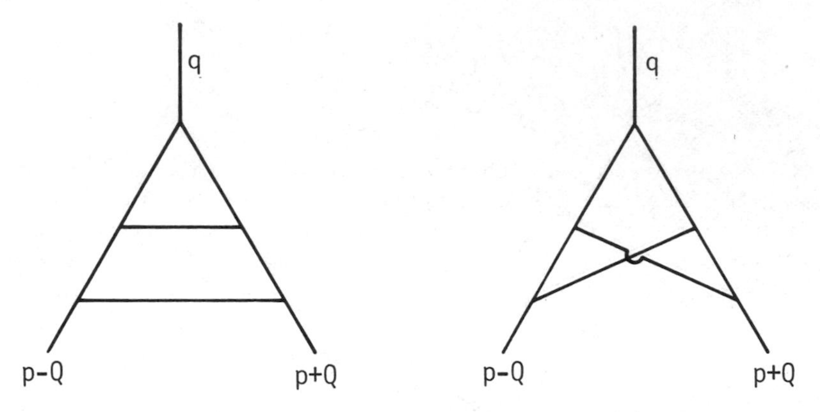

Fig. 2. Two-loop vertex corrections.

$$I_2 = \int d^Dk d^D\ell \ [(p-k)^2(k-Q)^2(k-\ell)^2(\ell-k+p+Q)^2(\ell-Q)^2(\ell+Q)^2]^{-1}, \tag{A.6}$$

in the limit

$$(p-Q)^2 = 0 , \quad (p+Q)^2 = 0 , \\ p^2 = - Q^2 , \quad (pQ) = 0 , \quad q^2 = 4Q^2 . \tag{A.7}$$

Again, the answers are surprisingly simple:

$$I_1 = \frac{1}{4} A \ [1 + 5\zeta(2)\varepsilon^2 - \frac{58}{3}\zeta(3)\varepsilon^3 + \frac{135}{4}\zeta(4)\varepsilon^4 + \ldots] , \\ I_2 = A \ [1 - 8\zeta(2)\varepsilon^2 + \frac{89}{3}\zeta(3)\varepsilon^3 - \frac{165}{4}\zeta(4)\varepsilon^4 + \ldots] , \tag{A.8}$$

where

$$A = \pi^{4+2\varepsilon}(q^2)^{-2+2\varepsilon} \ e^{2\gamma\varepsilon}\varepsilon^{-4}. \tag{A.9}$$

In these formulae, γ is Euler's constant and $\zeta(n)$ are the Riemann zeta-functions.

It should be said, however, that the evaluation of I_1 and I_2 can also be done directly in momentum space and that it is purely a matter of taste as to which method is considered the simplest. A very preliminary investigation of 3-loop vertex corrections did not lead us to a case where one of the methods would be clearly more indicated than the other.

REFERENCES

1. K. G. Chetyrkin, A. L. Kataev, and F. V. Tkachov, Nucl.Phys.B174: 345 (1980).
2. P. De Causmaecker, R. Gastmans, W. Troost, and T. T. Wu, Phys. Lett. 105B:215 (1981).
3. P. De Causmaecker, R. Gastmans, W. Troost, and T. T. Wu, Nucl. Phys. B206:53 (1982).
4. F. A. Berends, R. Kleiss, P. De Causmaecker, R. Gastmans, and T. T. Wu, Phys. Lett. 103B:124 (1981);
   F. A. Berends, R. Kleiss, P. De Causmaecker, R. Gastmans, W. Troost, and T. T. Wu, Nucl.Phys. B206:61 (1982).
5. D. R. Yennie, S. C. Frautschi, and H. Suura, Ann. Phys.(N.Y.) 13: 379 (1961).
6. D. Danckaert, P. De Causmaecker, R. Gastmans, W. Troost, and T. T. Wu, Phys. Lett. 114B:203 (1982).

SEARCH FOR COSMIC RAY MONOPOLE FLUX

USING SUPERCONDUCTIVE DETECTORS

Blas Cabrera

Physics Department
Stanford University
Stanford, California 94305

INTRODUCTION

On February 14, 1982 in our laboratory a prototype superconductive detector designed to search for magnetic monopoles observed a single candidate event. Even though it has not been possible to rule out a spurious cause, the possible existence of superheavy magnetically charged particles passing through the earth's surface has produced enormous interest within the scientific community. During this last year, we have designed, built and are now operating a new superconductive detector with three larger independent sensing loops used in coincidence. In addition we are planning an order of magnitude yet larger detector utilizing a superconducting sheet which will be periodically scanned with a magnetometer. This paper summarizes our work on these detectors.

Serious interest in magnetic monopoles began in 1931, when P.A.M. Dirac[1] proposed the existence of magnetically charged particles to explain the observed quantization of electric charge. He showed that only integer multiples of a fundamental magnetic charge g (Dirac charge) are consistent with quantum mechanics. Many years of experimental searches produced no convincing candidates.

Over the last decade, work on unification theories has unexpectedly yielded strong renewed interest in monopoles. In 1974, 't Hooft and independently Polyakov[2] showed that in true unification theories (those based on simple or semi-simple compact groups) magnetically charged particles are necessarily present. These include the standard SU(5) grand unification model. The modern theory predicts the same long-range field and thus the same charge g as the Dirac solution; now, however, the near field is also specified leading to a calculable mass. The standard SU(5) model predicts a monopole mass of 10^{16} GeV/c^2, horrendously heavier than had been considered in previous searches.

Such supermassive magnetically charged particles would possess qualitatively different properties from those assumed in earlier searches. These include necessarily nonrelativistic velocities from which follow weak ionization and extreme penetration through matter. Thus such particles may very well have escaped detection in earlier searches.

Superconductive technologies, many developed at Stanford University over the last decade, have led naturally to very sensitive detectors for magnetically charged particles. These superconductive detectors directly measure the magnetic charge independent of particle velocity, mass, electric charge and magnetic dipole moment. In addition, the detector response is based on simple and fundamental theoretical arguments which are extremely convincing. Because of their velocity independent response, these detectors are a natural choice in searches for a particle flux of supermassive (thus slow) magnetically charged particles.

Before describing our experimental work on superconductive detectors, we briefly summarize a now vast body of literature on monopole particle flux limits based on astrophysical arguments and on the response of conventional detectors (e.g., scintillators) to the passage of a magnetic charge.[3]

ASTROPHYSICAL LIMITS AND CONVENTIONAL DETECTORS

Cosmological theories based on GUTs lead to predictions for monopole particle flux limits which are impossibly high or unobservably low. However, the latter results are exponentially model dependent and thus are not inconsistent with observable levels. Thus we turn to astrophysical arguments for more concrete observational limits. In the discussion that follows we assume a particle mass of 10^{16} GeV/c^2 to obtain representative numbers. Then an absolute upper bound for the galactic monopole particle flux of 4×10^{-10} cm^{-2} s^{-1} sr^{-1} is obtained from the limits on the local galactic dark mass.

A much smaller upper bound of 10^{-15} cm^{-2} s^{-1} sr^{-1} is obtained assuming an isotropic flux from arguments based on the existence of the 3 microgauss galactic magnetic field. However, several authors have demonstrated that models incorporating monopole plasma oscillations may allow a much larger particle flux, approaching in some cases the local galactic dark mass limit. All of these galactic bounds suggest particle velocities in gravitational virial equilibrium, thus very near 10^{-3} c.

An enhanced monopole density, gravitationally bound to our own solar system, has been suggested and would allow much smaller average galactic flux levels and leads to lower particle velocities near 10^{-4} c. Although a mechanism for the formation of such a cloud remains obscure, its possible existence is not ruled out.

Perhaps the most important question regarding direct experimental detection is: can conventional ionization or scintillation devices with their much larger sensing areas than superconductive devices detect the passage of single Dirac charges with velocities of order 10^{-4} to 10^{-3} c? Several phenomenological theories suggest the answer is yes, however, experimental tests of these theories are not possible. Recently, new calculations based on fundamental quantum mechanical arguments suggest that helium gas devices would provide such a sensitivity.

We feel that experimental efforts with both conventional and superconductive detectors should be encouraged. These efforts are complementary, with the conventional detectors providing up to 1000 times larger sensing areas and convincing detection for any slow moving electric or magnetic particles with velocities above about 10^{-3} c, and the superconductive detectors providing definitive identification of magnetic versus electric charge for any particle velocity.

Finally, it has been shown theoretically that the supermassive monopoles arising from grand unification theories will catalyze nucleon decay processes. If the cross section for such events is of order the hadron cross section, as has been suggested, then all attempts at direct detection of these monopoles may be doomed to failure. Arguments based on x-ray flux limits form galactic neutron stars and which assume a strong cross section lead to an upper bound for a magnetic particle flux of about 10^{-22} cm^{-2} s^{-1} sr^{-1}. However, there remain unanswered questions both on the catalysis cross section and on the astrophysical arguments based on our incomplete understanding of neutron stars.

Even though some of these theoretical and astrophysical arguments suggest particle flux levels well below one per year in our detectors, we see no compelling reasons for abandoning searches with superconductive detectors and, to the contrary, we continue to be attracted by the elegance and simplicity of the technique. By far, the most definitive positive identification of a magnetic charge would be provided by a superconductive detector.

MONOPOLE COUPLING TO SUPERCONDUCTING RING

The theoretical similarities between flux quantization in superconductors and Dirac magnetic monopoles make superconductive systems natural detectors for these elusive particles. Alvarez and coworkers at Berkeley were the first to use superconductive monopole detectors in static matter searches.[4]

The magnetic flux emanating from a Dirac charge, $4\pi g = hc/e$, is exactly twice the flux quantum of superconductivity, $\phi_o = hc/2e$. Consider a magnetic charge g passing through a superconducting ring, as shown schematically in Fig. 1. Since no magnetic field line may pass through the superconductor, every field line emanating from the pole must leave a closed loop around the ring wire as the particle moves through. Thus the net change of flux through the ring must exactly equal the total flux from the pole - two ϕ_o for a monopole of unit Dirac charge.

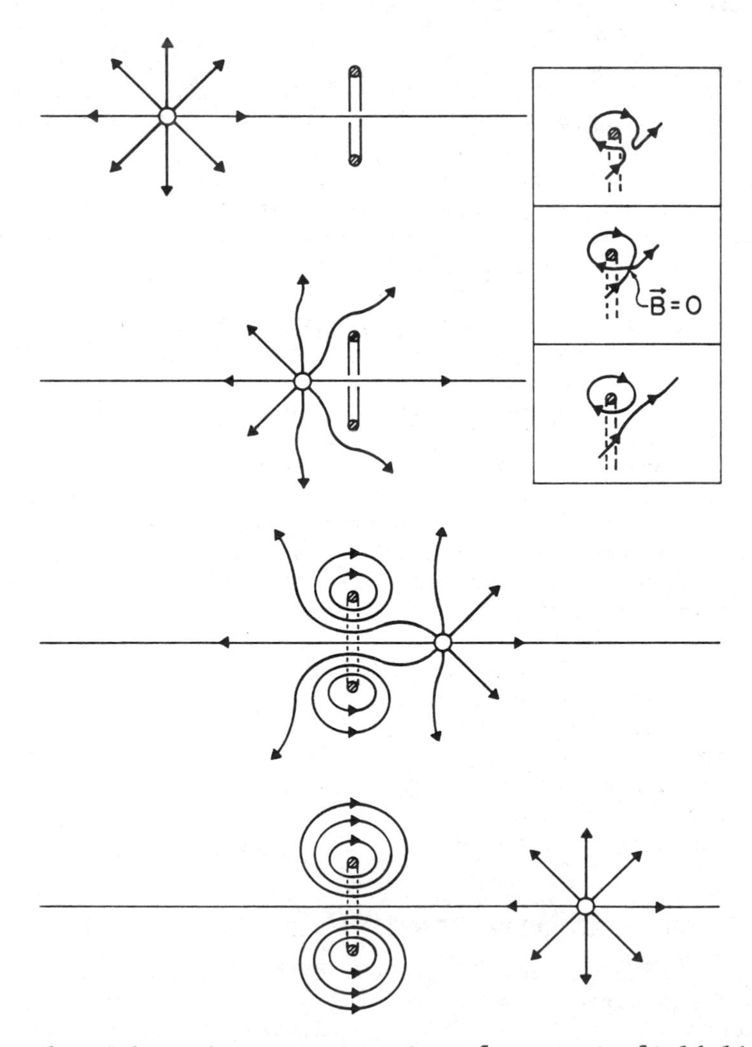

Fig. 1. Schematic representation of magnetic field lines as monopole passes through superconducting ring.

The same result is obtained for all particle trajectories which pass through the ring, whereas no net flux change results from trajectories which miss the ring. To illustrate, in Fig. 2 the induced supercurrent resulting from each of four different trajectories has been computed. Note that the rise times are similar and are of order the ring radius divided by the particle velocity. For trajectories which miss the ring only transient currents are obtained with a maximum peak to peak excursion of one ϕ_o. Unfortunately, present instrument bandwidths will not resolve the rise times expected from a magnetic charge moving at 10^{-3} or 10^{-4} c, thus only the resulting dc current change is sensed.

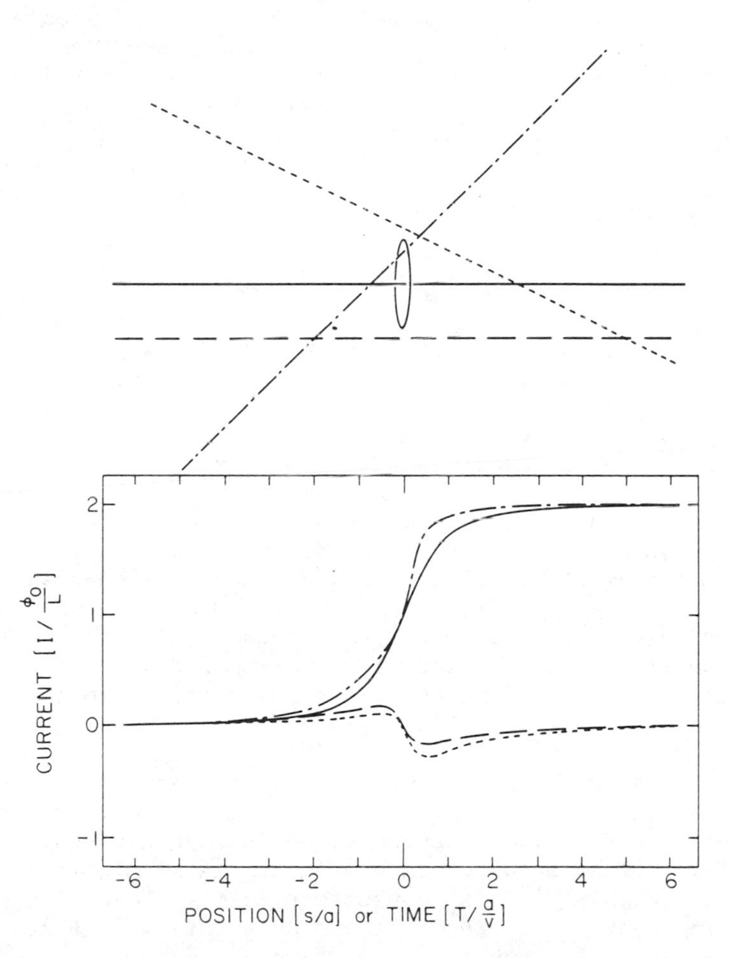

Fig. 2. The induced current in a superconducting ring of radius "a" is plotted against time (or position) for each of four different monopole trajectories. The particle travels at constant velocity v.

Thus if a magnetic charge g passes through the ring, the number of flux quanta threading the ring will change by two; whereas, if the particle does not pass through, the flux will remain unchanged. Finally if the magnetic charge passes through the wire of the ring it would leave a trapped doubly quantized vortex, and some intermediate total current would persist. Any solid or sheet superconductor can be thought of as an infinite array of filament loops. Two ϕ_o would be present through each filament which surrounded the monopole trajectory. The resulting magnetic flux penetrates the superconductor through a non-superconducting core typically 1000 A in diameter, which is surrounded by a supercurrent vortex.

Any superconductive system based on these properties is sensitive only to magnetic charges and thus makes a natural detector. The passage of any presently known particle possessing only electric charge or a magnetic dipole moment would cause very small transient signals but no dc shifts. Thus, a cosmic flux of magnetically charged particles such as the predicted supermassive GUT monopoles can be detected by monitoring the current in a superconducting loop or scanning the surface of a superconducting sheet for spontaneously appearing doubly quantized vortices. Our work on several such devices is described below.

PROTOTYPE SINGLE LOOP DETECTOR

To test the feasibility of such devices an existing instrument, originally built as an ultra sensitive magnetometer, was used in a prototype search for a particle flux of supermassive monopoles.[5] This instrument has been operated as a monopole detector for a total of 382 days. It consists of a four turn 5 cm diameter loop made of 0.005 cm diameter niobium wire. As shown in Fig. 3, the coil is positioned with its axis vertical and is connected to the superconducting input coil of a SQUID. The passage of a single Dirac charge through the loop would result in an 8 ϕ_o change in the flux through the superconducting circuit, comprised of the detection loop and the SQUID input coil (2 ϕ_o couple to each of the four turns). The SQUID and the loop are mounted inside an ultra low field shield in an ambient field of 5 x 10^{-8} gauss. Such low field regions are produced using expandable superconducting shields.

Several intervals throughout a continuous one month time period are shown in Fig. 4a, where no adjustment of the dc level has been made. Typical disturbances caused by daily liquid nitrogen and weekly liquid helium transfers are evident. A single large event was recorded (Fig 4b). It is consistent in magnitude with the passage of a single Dirac charge within an uncertainty of $\pm$ 5%. It is the largest event of any kind in the record.

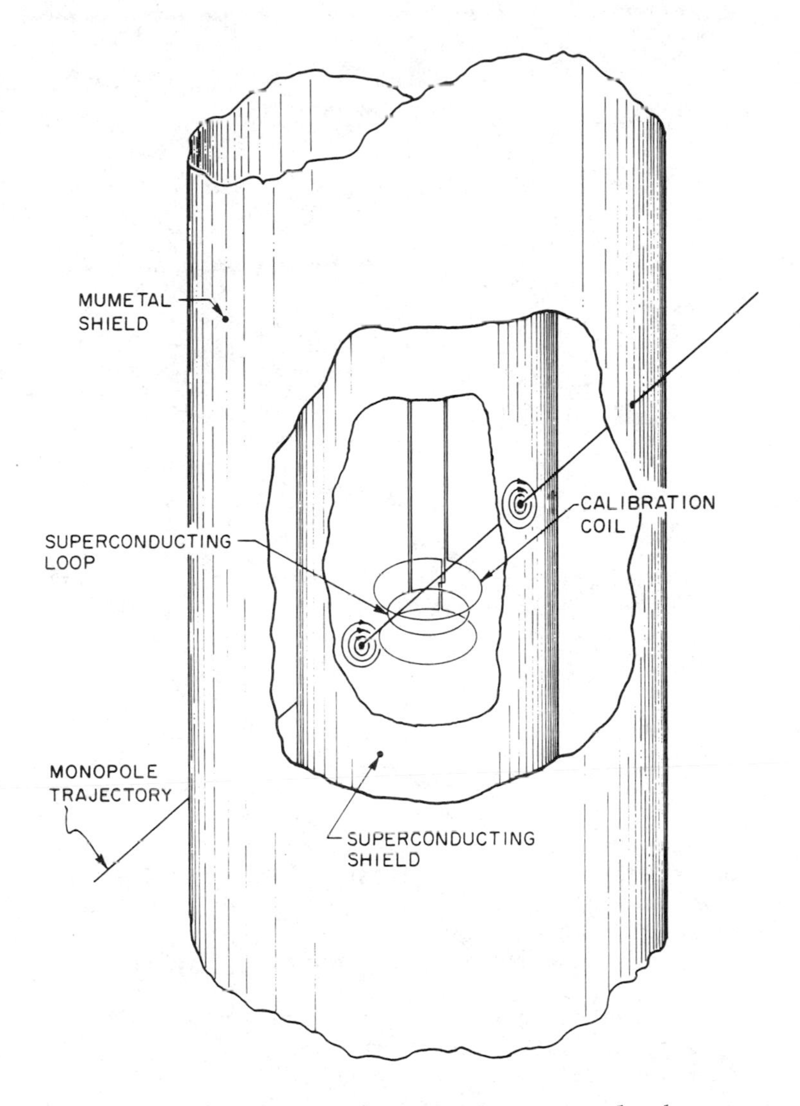

Fig. 3. Schematic of prototype monopole detector.

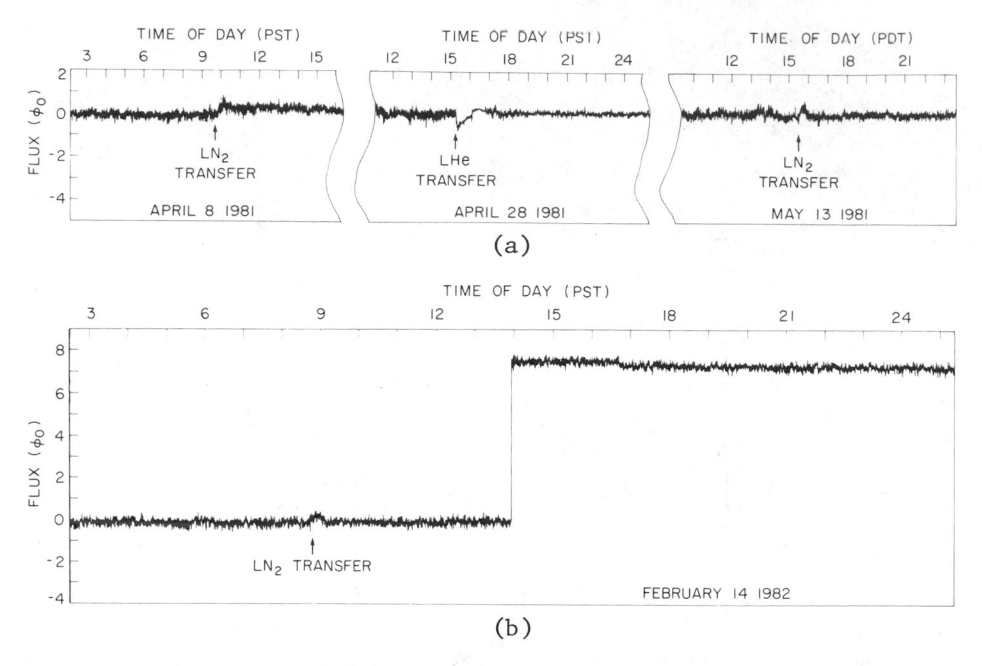

Fig. 4. Data of (a) typical stability during one month, and (b) the candidate event.

The most likely sources for spurious signals are associated with the mechanical sensitivity of the apparatus. Mechanically induced offsets have been intentionally generated. These could be caused by shifts of the four turn loop wire geometry, which would produce inductance changes and inversely proportional current changes (since the flux remains constant). Alternatively, trapped supercurrent vortices in or near the SQUID could move from mechanically induced local heating. In any case, sharp raps with a screwdriver handle against the detector assembly cause offsets.

Although we have not been able to produce such events which appear as clean as the candidate event, magnitudes close to the candidate event magnitude have been generated. If we further assume that mechanical stresses frozen into the detector upon initial cooling from room temperature could spontaneously release or be triggered by a small initial external disturbance, it would be possible to imagine a large spurious signal. The difficulty with easily accepting this scenario as a plausible explanation for the candidate event is that the data from this detector has been accumulated over five separate runs, each commencing with a cooldown of the apparatus from room temperature, and no other such signals were ever seen. Nevertheless, we feel IT IS NOT POSSIBLE TO RULE OUT A SPURIOUS CAUSE FOR THE FEBRUARY CANDIDATE EVENT.

We are planning a series of tests on this detector aimed at reducing the uncertainty in our calibration and further clarify the origin of the mechanically induced offsets. These tests will include repeated thermal cycling of the apparatus.

Since July, 1982, our primary effort has gone into building a new detector with a larger sensing area and with sufficient cross checks to rule out all spurious causes for any new events. It is described in the next section.

NEW THREE LOOP DETECTOR

In July of 1982, a group composed of M. Taber, J. Bourg, S. Felch, R. Gardner, R. King and myself began work on a larger detector. We have completed the construction and begun operating a new detector based on the same principles as the prototype system but with improved mechanical stability, a greater sensing area and much better spurious signal discrimination. As shown schematically in Fig. 5, it consists of three mutually orthogonal two turn superconducting loops, each 10.2 cm in diameter (twice that of the original device). In addition to a sevenfold increase (70.9 cm^2) in direct average isotropic sensing area for trajectories passing through at least one loop (4 ϕ_o signal), this new instrument is sensitive to near miss trajectories which traverse the shield but do not intersect the three loops. The near miss category provides and additional effective sensing area for single Dirac charges 53 times greater (540 cm^2) than that of the prototype detector for signals of magnitude 0.1 ϕ_o or larger. Because of the smaller signal levels, this near miss sensitivity can provide lower flux bounds if no candidate events are seen, however such a near miss candidate would not be as definitive as one corresponding to a trajectory passing through at least one loop.

To monitor the detector, we have designed a computer based data acquisition system. The output signals from the fluxgate and the accelerometer together with the outputs from the three independent SQUID loop current sensors are continuously sampled at 200 readings per second per channel. These data are temporarily stored onto a computer ring buffer (about 30 seconds long) and digitally filtered to 0.1 hz bandwidth (one point every 5 seconds) in real time. The filtered readings are then stored on hard disc to form a permanent continuous record. In addition, these filtered data are constantly compared to detect changes in the SQUID sensor levels above a 0.1 ϕ_o threshold. Such an event then triggers in software the permanent storage of the ring buffer contents around the time of the event. The high bandwidth information allows convincing discrimination against mechanically induced spurious events of any kind. However, the 10 millisecond resolution us NOT capable of observing supercurrent rise times of order 0.1 to 1 microsecond expected from the passage of a Dirac charge with velocity between 10^{-3} and 10^{-4} the speed of light.

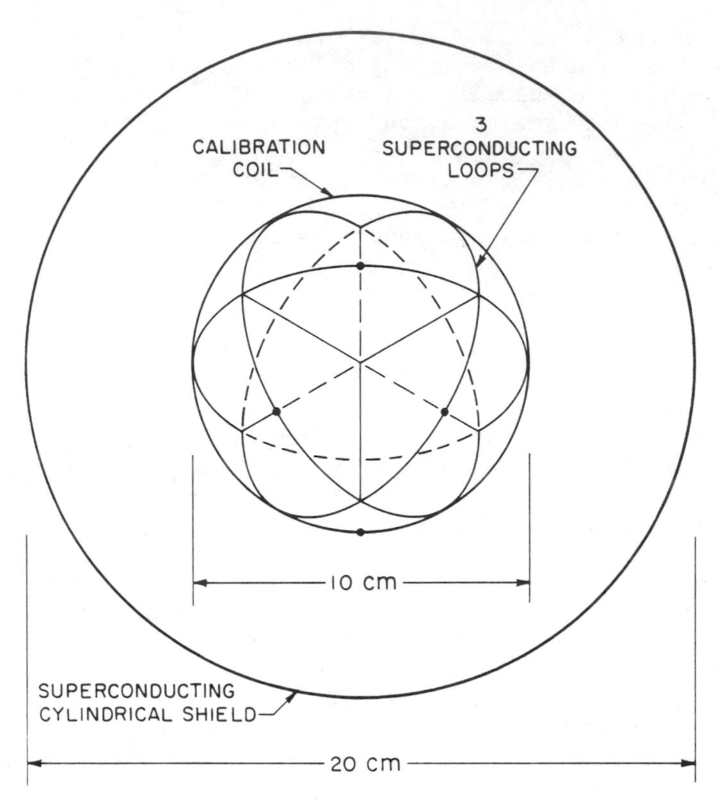

Fig. 5. Schematic top view of three loop monopole detector.

This data acquisition system has been assembled and soon is expected to become fully operational.

As a backup to the computer the detector is also continuously monitored on a six channel strip chart recorder at 0.1 hz bandwidth. As of January, 1983 we have operated the system for nearly six months. Fig. 6 shows ten hours of typical recent strip chart data. We have accumulated 125 days of data primarily on the strip chart recorder with SQUID noise levels of 1 ϕ_o or better. The signal from any Dirac magnetic charge passing through at least one loop would have been detectable. NO SUCH SIGNAL HAS YET BEEN SEEN. That noise level did not allow clear detection of near miss signals. However, using only the direct loop sensing area the dominant contribution to our experimental particle flux bound now comes from our new three axis detector.

As of January 3, 1983, the limit on the particle flux of magnetic charges passing through earth's surface at any velocity and of any mass is less than 7.3×10^{-11} cm^{-2} s^{-1} sr^{-1}. This limit is derived combining 382 days of data from the prototype detector with its

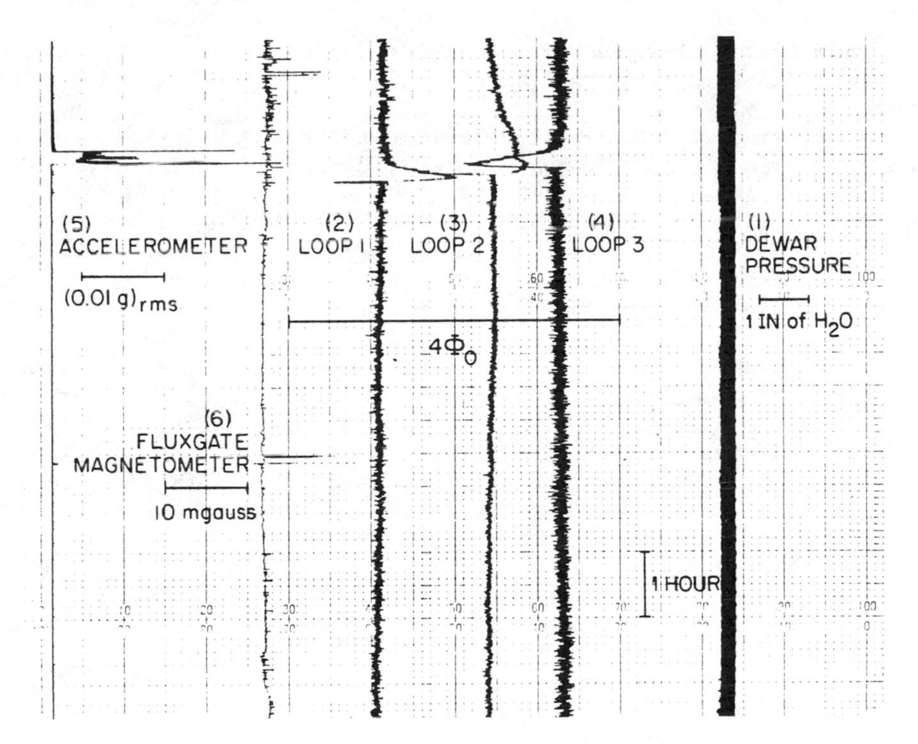

Fig. 6. Recent typical data taken at 0.1 hz bandwidth on backup six pen strip chart recorder. It includes a liquid nitrogen transfer cycle. The passage of a Dirac magnetic charge through the loop would cause a $4\ \phi_o$ change in one or more loops.

10.1 cm^2 of average loop sensing area and 125 days from the new three axis detector with its 70.9 cm^2 loop area. The limit is more than a factor of eight lower than that published in the May, 1982 Physical Review Letter.[5]

FUTURE SCANNING DETECTORS

At present, superconductive detectors are the only known monopole detectors which are sensitive to arbitrarily low particle velocities. Their response is exactly calculable using simple fundamental concepts. Thus an important quesiton is how large can the sensing areas of such devices be made using existing technologies?

Larger diameter loops with their larger self inductances suffer from ever-decreasing signal-to-noise ratios and for loop diameters larger than about 10 cm cannot be effectively used as monopole detectors.

To overcome this limitation we are designing a new generation of supercodnuctive detector which will use up to several square meters of a thin superconducting sheet in the form of a cylinder for recording magnetically charged particle tracks. It is analogous to a photographic emulsion which records electrically charged particle tracks. As discussed in the third section, a magnetic charge traversing the cylinder would record a signature consisting of doubly quantized trapped flux vortices in the walls (Fig. 7). As long as the sheet remained superconducting, the strong flux pinning due to lattice and surface defects would prevent any motion.

The ambient trapped flux pattern, about one quantized vortex per cm^2 in a 10^{-7} gauss field, would be periodically recorded using a small scanning coil coupled to a SQUID. We have calculated the coupling of a trapped quantized vortex to a 38 turn and 3 mm diameter coil. We assume that the vortex is located in an infinite flat sheet and that the coil is held a constant 1.5 mm above the surface with its axis perpendicular to the surface, but is free to move parallel to the surface. The inductive coupling is plotted in Fig. 8 as a function of the distance between the coil axis and a parallel line perpendicular to the surface and through the vortex.

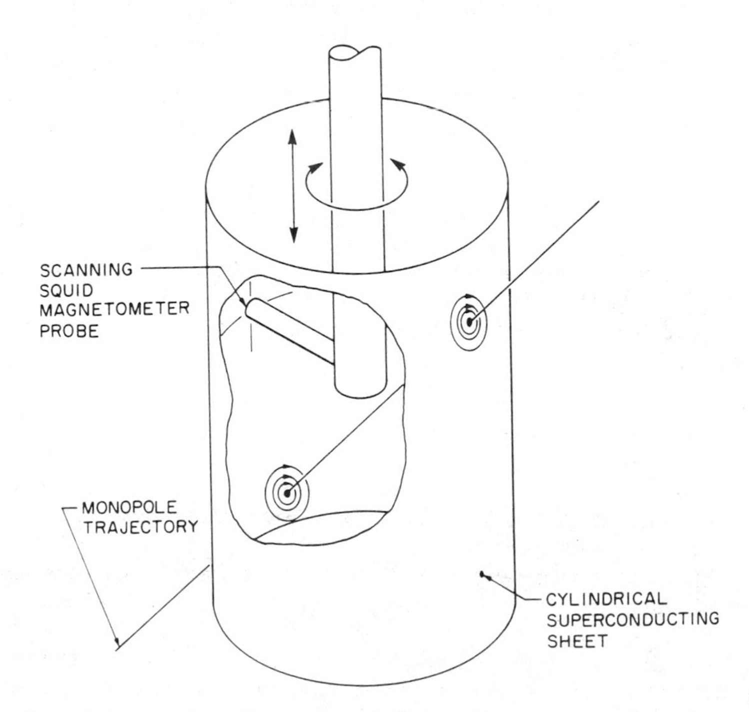

Fig. 7. Schematic of proposed large area scanning detectors.

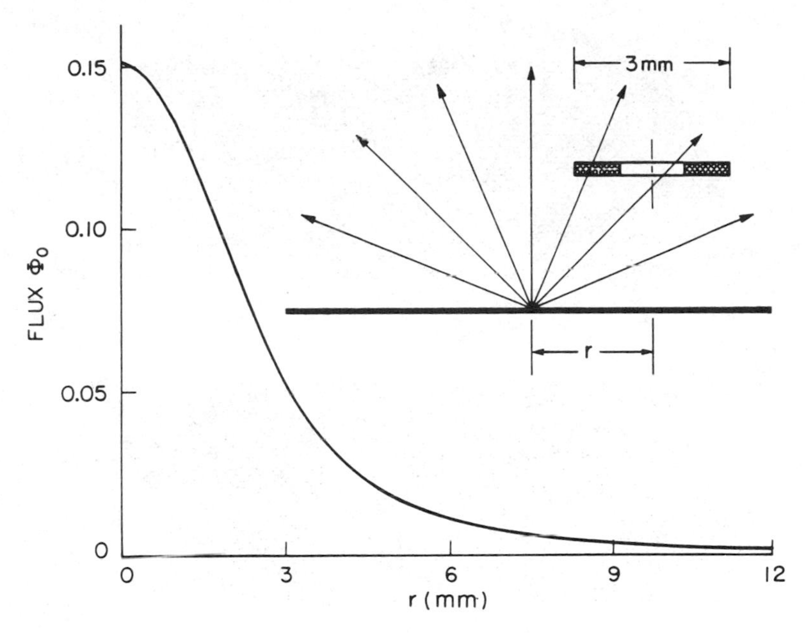

Fig. 8. Coupling of single trapped vortex to the scanning coil.

To better visualize the data obtained from such a scanning procedure, we have used the calculated coupling shown in Fig. 8 to generate simulated data. Two simulated scans, over a 10 cm by 10 cm area, are shown in Fig. 9(a) and (b). We assume that the scanning coil would be connected to one of our existing SQUID systems, thus a typical experimental SQUID noise level has been added. In (b) a new doubly quantized vortex has appeared and is most clearly seen when (a) is subtracted from (b) as shown in (c). With our prototype scanning device, we expect a factor of three greater signal to noise obtained primarily from a closer coil to surface spacing.

The advantages of such scanning detectors over loop detectors include:

(1) a signal to noise for the observation of a Dirac magnetic charge which is independent of the detector size. It depends only on the coil design and coil to surface spacing.

(2) direct quantum mechanical calibration of the detector provided by the singly quantized trapped flux vortices present in the niobium film.

(3) insensitivity to external magnetic field changes. A new quantized vortex appearing in the film can only have been produced by the passage of a magnetic charge as long as the superconducting sheet is kept below its transition temperature.

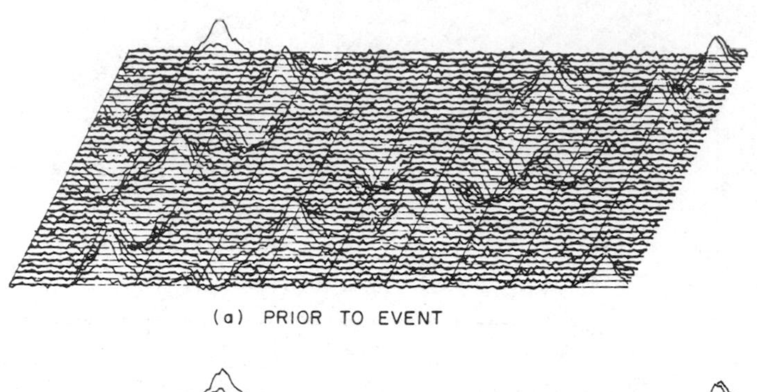

(a) PRIOR TO EVENT

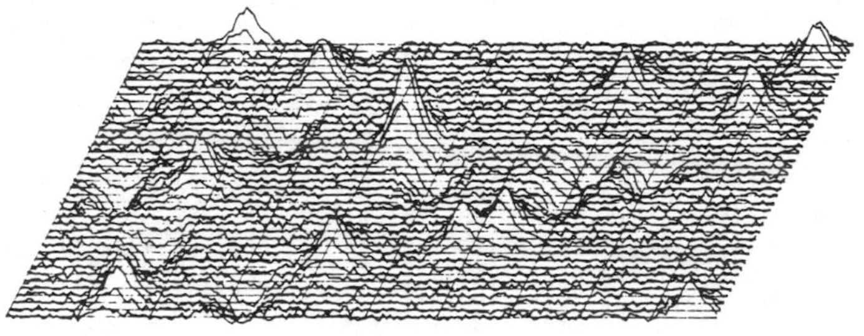

(b) AFTER EVENT

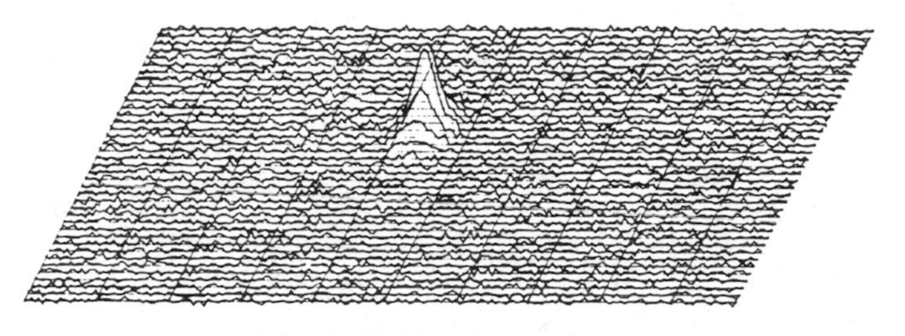

(c) BACKGROUND SUBTRACTION

Fig. 9. Simulated data from scanning detectors.

(4) a spatial rather than temporal record. The appearance of a new vortex can be checked and rechecked by repeatedly scanning the region of interest on the cylinder. If existing vortices have moved this will be clearly indicated by the subtraction technique described above.

(5) double coincidence signals when particle trajectory intersects the cylinder wall twice. Two doubly quantized vortices of opposite polarity would simultaneously appear between two successive scans.

and (6) high precision directional information, should convincing candidates be found which intersect the cylinder walls twice.

To aid us in optimizing the design for a given scanning detector, we have computed the sensing cross section averaged over all solid angle for a cylinder of diameter D and length L. The results are summarized in Fig. 10 where σ_1 and σ_2 are the average sensing areas for single and double intersection by the particle trajectory with the cylinder walls. The total σ_{tot} is the sum of σ_1 and σ_2. These average sensing areas are equal to DL (cylinder diameter times length) times a unitless factor which depends only on the ratio D/L. and which is plotted in the figure. Diameter to length ratios (D/L) between 0.2 and 0.4 are planned both for easy interfacing with our ultra low field technology and for a high double coincidence sensing area (σ_2).

We will first build and operate a prototype scanning detector which will fit within our existing ultra low field shields and have an average sensing area approaching 0.1 m^2. Once having demonstrated feasibility, we intend to build and operate a scanning detector with more than 1 m^2 of average sensing area. Large dewar technology has been developed at Stanford for more than a decade on the superconducting accelerator work and the gravitational radiation detector.[6] The latter program has built and operated a large liquid helium dewar with a cryogenic region 1 m in diameter by 3 m long. Such a dewar could house a scanning monopole detector with nearly 2 m^2 sensing area averaged over a solid angle.

If one or more convincing candidates are seen, we will design probes to measure or set limits on the magnetic particle velocity and mass. With our new three loop detector, or later with the larger sensing area scanning detectors, either we will begin to see more events which would be very convincing, or we will set an upper limit below 10^{-12} cm^{-2} s^{-1} sr^{-1} on a cosmic ray magnetic particle flux - more than 100 times smaller than our present value. Either way, this is a very exciting time for our group.

ACKNOWLEDGEMENTS

I would like to thank Harvey Newman and his organizing team along with the other conference participants for an extremely stimulating and productive conference. I also thank the "Ettore Majorana" Centre for Scientific Culture and its fine staff for the remarkable hospitality we received during our stay in Erice.

The research described in this paper has been funded in part by DOE contract DE-AM03-76SF00-326 and NSF grant DMR 80-26007.

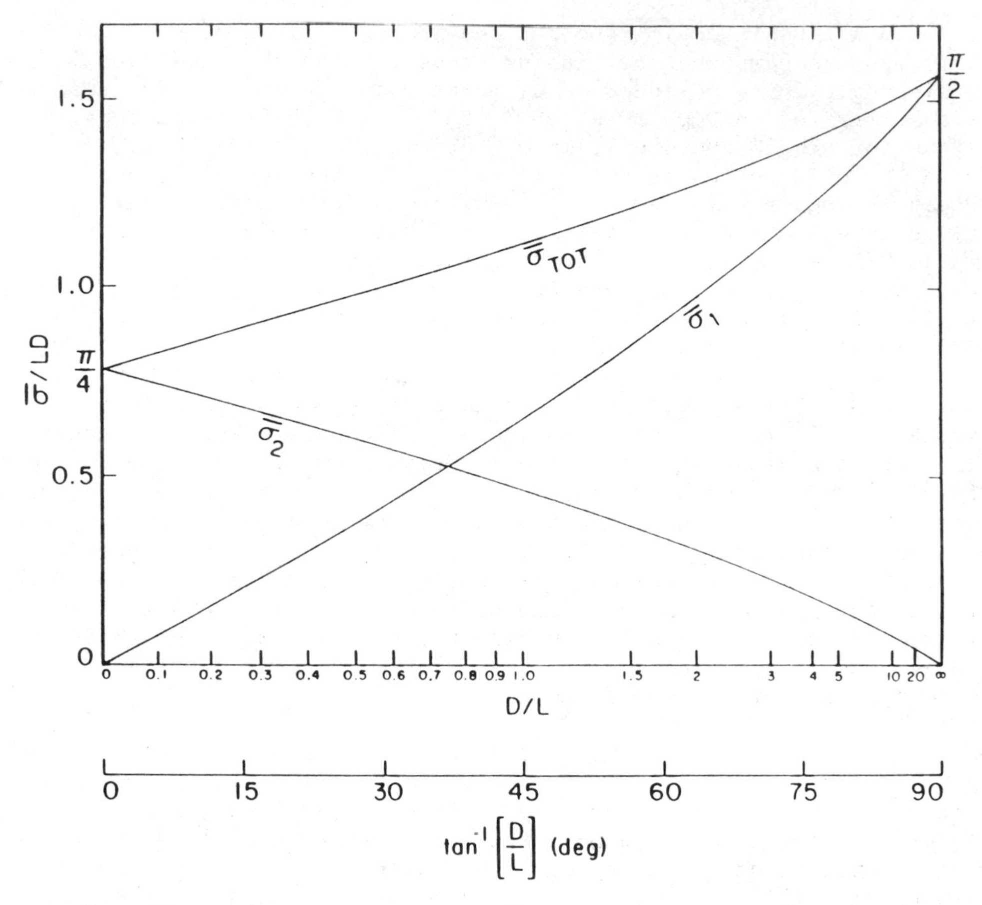

Fig. 10. Sensing area averaged over solid angle for single intersection by trajectory (σ_1), double intersection by trajectory (σ_2) and for either (σ_{tot}).

REFERENCES

1. P. A. M. Dirac, Proc. Roy. Soc. A 133, 60 (1931); Phys. Rev. 74, 817 (1948).
2. G. 't Hooft, Nucl. Phys. 79B, 276 (1974) and Nucl. Phys. 105B, 538 (1976); A. M. Polyakov, JETP Letts. 20, 194 (1974).
3. For a detailed review of recent theory and experiment, see Proceedings of Magnetic Monopole Workshop, held at Wingspread in Racine, Wisconsin, October 14-17, 1982; eds. W. P. Trower and R. A. Carrigan (in press by Plenum).

4. L. W. Alvarez, Lawrence Radiation Laboratory Physics Note 470, 1963 (unpublished); P. Eberhard, Lawrence Radiation Laboratory Physics Note 506, 1964 (unpublished); L. J. Tassie, Nuovo Cimento 38, 1935 (1965); L. Vant-Hull, Phys. Rev. 173, 1412 (1968); P. Eberhard, D. Ross, L. Alvarez and R. Watt, Phys. Rev. D 4, 3260 (1971); for more recent work, see papers by C. C. Tsuei and D. Cline in ref. 3.
5. B. Cabrera, Phys. Rev. Lett. 48, 1378 (1982); see also paper by B. Cabrera in ref. 3.
6. W. O. Hamilton, P. B. Pipes, S. Kleve, T. P. Bernat, D. G. Blair, D. H. Darling, D. Dewitt, M. S. McAshan, R. Taber, S. P. Boughn, W. M. Fairbank, W. P. Montgomery and W. C. Oelfke, Cryogenics, March 1982, p. 107.

EXPERIMENTAL SEARCHES FOR SLOWLY-MOVING MAGNETIC MONOPOLES

Donald E. Groom

Physics Department, University of Utah

Salt Lake City UT 84112 USA

INTRODUCTION

The mythological Dirac monopole has been the object of experimental searches for nearly 50 years. In its modern context (since grand unified theories) it is massive and primordial. Experimentally, such an object should be slow (with solar or galactic escape velocity) and possibly very lightly ionizing. In the flurry of excitement surrounding B. Cabrera's observations of last spring,[1] many groups attempted to observe or set limits on the flux of slow monopoles with "traditional" electronic particle detectors--plastic scintillators or proportional wire chambers. These experiments are the main subject of this talk, since older searches for relativistic monopoles, for monopoles at rest, etc., have recently been reviewed elsewhere.[2,3] There have also been improvements in the theory of ionization by slowly-moving monopoles, and new looks at astrophysical limits. As a result, a new generation of electronic detectors is being developed which take advantage of the theoretical improvements and with sufficient aperture to improve upon astrophysical limits.

MONOPOLES

Grand unified theories (GUT's) of various kinds all require the existence of a magnetic monopole with mass $M \approx M_X/\alpha_X$, where M_X is the mass of the lepto-quark boson (about 10^{14} GeV), and α_X is the GUT fine structure constant (about 1/50).[4,5] Presumably at least one monopole/event horizon was produced in connection with the transition from the GUT symmetry to SU(3)×SU(2)×U(1) when the age of the universe was about 10^{-35}s.[6,7] At a very much later epoch (10^{-6}s) nucleons condensed from the quark plasma. If the expansion proceeded in a simple way between these epochs, we should expect about 10^{-5} monopoles/nucleon at the present time. Most of us have already noticed that it is not true that all but 10^{-10} of the mass of the universe is in the form of monopoles. Salvation from this dilemma has apparently come with the discovery of inflation by A.H. Guth:[8] in connection with the phase change from SU(5) (or your favorite GUT symmetry) to the SU(3)×SU(2)×U(1) of the next epoch, the radius of the universe (in the Robertson-Walker sense) increased dramatically before resuming a "normal" expansion. As a result, our present information horizon may not have yet reached the size of the horizon containing "our" monopole. To slightly overstate the theoretical situation, GUT-based estimates of the present flux of primordial monopoles vary by a factor of 10^{75}!

Because of the size of the proton's magnetic moment, a bound monopole-proton state should exist.[9] Indeed, nearly every monopole may have accumulated such a proton during its long journey. Even more interestingly, Rubakov[10] and Callan[11] have pointed out that monopoles might catalyze nucleon decay: A nucleon's close encounter with a monopole and its surrounding quark condensate results in the decay of the nucleon. The cross section is expected to lie between geometrical and 10^{-5} or so of geometrical.[12] The very interesting implications of this process for monopole flux limits, or more properly the product of catalysis cross section and flux,[13-15] will be explored by Prof. Peccei in the next talk.

As experimentalists making direct searches for the monopole flux, however, it is advisable to ignore these possibilities. Catalysis may or may not occur, and if we are sensitive to a "naked" monopole we are likely to be even more sensitive to a monopole with an attached electric charge. We must also include the possibility that intermediate mass monopoles predicted by other theories exist,[16] and not limit our searches to "slow" monopoles.

ASTROPHYSICS

While astrophysical arguments concerning monopole flux are not the subject of this talk, they form an essential context for the evaluation and design of experiments. In the first place, we may tabulate expected velocities for monopoles in various "habitats:"

1. Not gravitationally bound to our galaxy or supercluster: The primordial monopole gas is now quite cold, so that individual monopoles would arrive at the earth with at least the escape velocity of the galaxy, or $v \geqslant 10^{-3}c$.

2. Bound to the galaxy or supercluster: $v \sim 10^{-3}c$. Circular velocity at the sun's distance from the galactic center is over 80% of escape velocity from the galaxy, which is in turn is not much below supercluster escape velocity. Both the sun and a bound monopole

thus move with $v \sim 10^{-3}c$ with respect to a galactic rest frame, so our encounter velocity with the monopole must be quite close to $10^{-3}c$. As will be seen, this possiblity is the most attractive from experimental and theoretical points of view.

3. Bound to stellar systems: $v \sim 10^{-4}c$ for terrestrial observations, since $v = 10^{-4}c$ in the solar rest system. In spite of initial arguments to the contrary,[17] it now appears that no large flux enhancement over the galactic halo case (above) would occur.[18]

4. Bound to planets, e.g., the earth: $v < 4\times10^{-5}c$ at the earth's surface. We view this possibility as somewhat farfetched because of the lack of a credible mechanism for capturing monopoles in this way. In addition, M.S. Turner[15] has shown that the catalysis mechanism would lead to unacceptably large terrestrial heating unless the catalysis cross section-flux product were exceedingly small.

Known limits on the average density of matter in the universe lead to severe flux limits on GUT monopoles under possibility (1), as shown by the curves labled "Uniform" in Fig. 1. Limits on the galactic halo mass permit a more generous limit by a factor of 10^5 or so under possibility (2), as is shown by the curve labled "Clustered."

Parker has long since noted that galactic magnetic field survival considerations then limit the monopole flux.[19] This problem has recently been reconsidered by Turner, Parker, and Bogden (TPB),[20] and by others.[21] Because of the relevance of their limits to existing and proposed monopole searches, we summarize them here.

Interstellar fields of 2-5 μG have been measured.[22] Although the distribution is fairly chaotic, there is a rough tendency of lines to follow the spiral structure; indeed, differential rotation of the galaxy "stretches" them in this direction and creates new field. The regeneration time is thus about the rotation period, or 2×10^8y. Fields which are coherent over large distances have a maximal effect on monopoles (and conversely!), so TPB consider the most favorable case for large monopole flux which is consistent with the data,[23,24] or fields which are coherent over small "domain" sizes of ~300 pc but are otherwise random. A monopole at rest will acquire v_{mag} in crossing one domain. A monopole entering a domain with velocity $|\vec{v}| >> v_{mag}$ will emerge with velocity $\vec{v} + \Delta\vec{v}$, or energy change

$$\Delta E = \frac{1}{2} M(\vec{v} + \Delta\vec{v})^2 - \frac{1}{2} Mv^2 \quad \Rightarrow \quad \langle\Delta E\rangle = \frac{1}{2} M(\Delta v)^2$$

As one might expect, velocity changes in the direction of the original velocity average to zero, while transverse "kicks" are cumulative, slowly accelerating the monopoles and draining field energy. From these considerations a number of interesting monopole mass limits and flux limits can be inferred. For example, monopoles would eventually escape if their masses were low enough that they slowly acquired

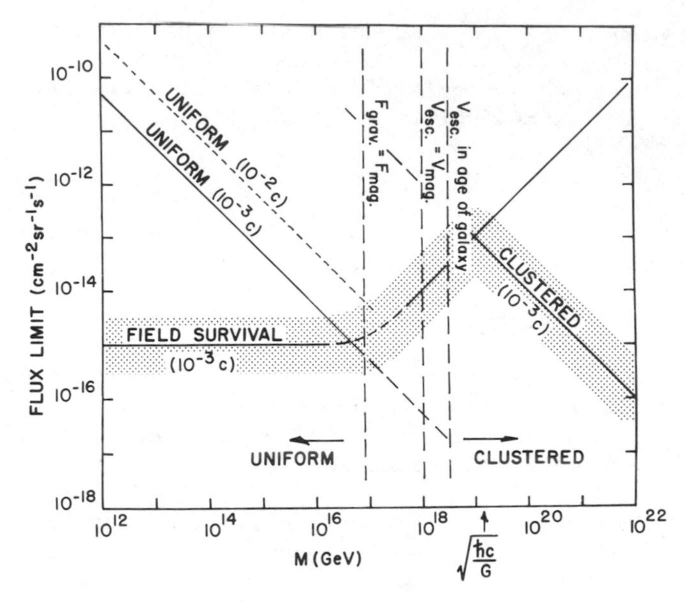

Fig. 1. Astrophysical limits on the monopole flux, calculated in the spirit of Ref. 20. The net optimistic upper limit as a function of monopole mass, set by density and field survival considerations, is indicated by the shaded band. The width indicates a general uncertainty but has no quantitative significance. Vertical lines indicate lower mass limits under various scenarios for monopoles gravitationally bound to the galaxy.

escape velocity during the lifetime of the galaxy, if they acquired escape velocity in crossing one domain, or if the magnetic force exceeded the gravitational force. All of these mass limits lie between 10^{17} GeV and 3×10^{18} GeV, as shown in Fig. 1. Similarly, the energy transfer rate should not exceed the field energy divided by the regeneration time, as shown for the "field survival" line in the Figure, drawn for $v = 10^{-3}c$. Finally, the halo mass limit sets the "clustered" flux limit shown in the Figure. The net result is a mass-dependent flux limit indicated by the shaded band. For a monopole mass of 10^{19} GeV (which by accident is the Planck mass!) the flux could be as high as $10^{-13}cm^{-2}sr^{-1}s^{-1}$, but it is otherwise difficult to exceed $10^{-15}cm^{-2}sr^{-1}s^{-1}$--and TPB have strained the astrophysical quantities as far as they thought possible in the direction of high monopole flux.

Last year it was reasonable to look for fluxes at the $10^{-10}cm^{-2}sr^{-1}s^{-1}$ to $10^{-11}cm^{-2}sr^{-1}s^{-1}$ level, under the stimulus of Cabrera's tantalizing results. Now that lower flux limits exist (if our detectors really work!), the neo-Parker limit or better would be a reasonable goal for future experiments. The various velocity regimes should serve as design guides. For example, a search for $v > 10^{-3}c$ massive monopoles with a flux limit above $10^{-15}cm^{-2}sr^{-1}s^{-1}$ is unlikely to yield positive results on the basis of field survival arguments.

IONIZATION AND EXCITATION BY SLOW MONOPOLES

The best monopole detector is undoubtedly a superconducting loop or similar device, since the detection scheme then employs the one attribute of a monopole of which we are certain. On the other hand, areas of 1000 m^2 or greater are necessary to probe the limits discussed above. We are thus led to consider the cheap, large-area detectors normally used for particle detection, plastic scintillators and proportional wire chambers (PWC's). But can they detect slow monopoles? Our cautiously affirmative answer follows from a four-step argument:

1. Energy loss mechanisms for charged particles (protons and heavier ions) are well understood both theoretically and experimentally down to $v < 10^{-3}c$. Following work by Fermi and Teller,[25] Lindhard and various collaborators[26,27] approximated the electronic structure of solids by a Fermi distribution in constructing a semi-phenomenological theory. A recent theoretical treatment is given by Ahlen and Kinoshita (AK),[28] and the lowest velocity experiments known to us extend to $6\times10^{-4}c$.[29] The theoretical prediction that the ionization part of dE/dx is linear in β probably becomes invalid below $v = 10^{-3}c$, because the electronic structure of real detectors, which are made of insulators, is inadequately described by a Fermi distribution.

2. The step to energy loss by monopoles may be made in a minimal way by considering only the monopole's induced electric field. An electron experiences an electric field proportional to gv times geometrical factors. For a slow monopole the relative velocity v is very close to the electron's velocity, which for an atomic electron is about $\alpha c = c/137$. Since $g = e/2\alpha$ for a Dirac monopole, $gv \approx e/2$. In other words, a slow monopole's interaction with an electron is the same as that of an electric charge e/2, and we should thus expect a slow monopole to be 1/4 as ionizing as a proton with the same velocity. AK use a slightly more sophisticated version of this argument, with a monopole version of the Rutherford cross section,[30] in calculating the energy loss curve shown in Fig. 2.

3. Extension to $v = 10^{-4}c$ is complicated by the energy gap of a scintillator or the ionization potential of a PWC gas. Ritson[31] has taken such effects into account in a rough way by (a) approximating the electronic structure by a Thomas-Fermi distribution and (b) requiring a collision energy transfer in excess of some gap energy G_o. In Fig. 2 we show the effect of extending AK's curve for a decade by this method. Curves for G_o = 16 eV (ionization) in argon and

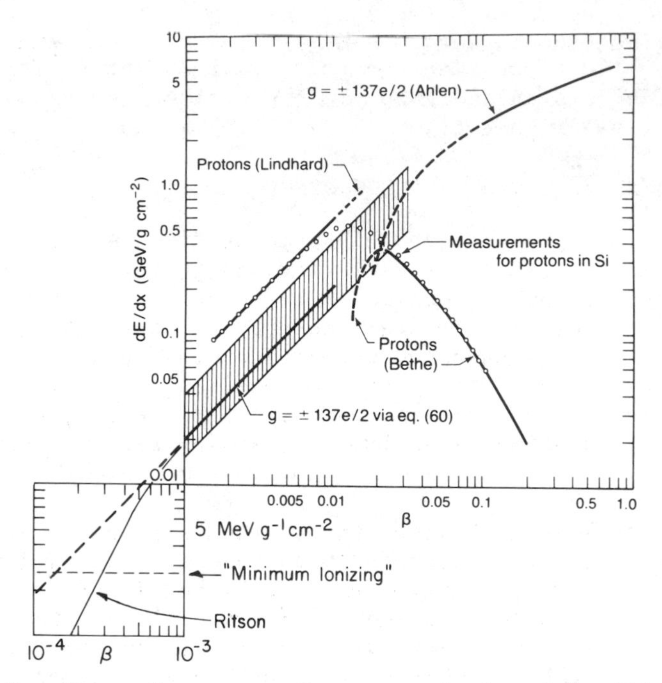

Fig. 2. Monopole energy loss as calculated by Ahlen and Kinoshita,[28] and as extended by Ritson.[31] Except for the the lower decade, the figure is as taken from Ref. 28.

G_o = 4 eV (excitation) in carbon (scintillator) are given in Ritson's paper; a similar curve for G_o = 3 eV in carbon has been published elsewhere.[32] We believe that these results represent a pessimistic lower bound on ionization/excitation energy loss by monopoles. (Ahlen and Tarlé[33] have recently claimed that a kinetic cutoff to excitation in scintillator exists at $v \sim 6\times10^{-4}c$, but this conclusion results from the sharp momentum cutoff of their assumed Fermi distribution.)

4. Drell et al.[34] have recently published an elegant and thorough treatment of slow monopole energy loss in atomic hydrogen and helium. They find energy loss rates an order of magnitude or so higher than is found in the "induced E field" method of AK and its extension by Ritson. Level mixing results in a high cross section for leaving helium in its metastable state; one can surmise that the same effect will be important in argon and other noble gases. Since energy transfer to quenching gases is known to be efficient,[35] we might expect large ionization effects if the ionization potential of the quenching gas is less than the deexcitation energy of the noble gas. This condition is unfortunately not met in any PWC array monopole search yet reported, but PWC's for new experiments are being designed to utilize the Drell et al. mechanism.

EXPERIMENTS

With these considerations in mind we are at last in a position to examine the dozen-odd experiments that have been reported or proposed within the past year or so. To deal with the Babel of units and confidence levels reported by various workers, we have simply reduced all reported limits to be reciprocal of the aperture-solid angle time product, in $cm^{-2}sr^{-1}s^{-1}$. The reader can multiply by 2.3 if he prefers 90% confidence levels.

Results are summarized in Fig. 3. The horizontal axis covers the velocity range of interest, if we discount the possibility of a swarm of monopoles traveling with the earth. (As discussed in the last section, the detectors under discussion are very likely to be insensitive for velocities under a few times $10^{-4}c$.) The vertical scale ranges from the loose astrophysical upper limit to the flux limit set by Cabrera, which as of this meeting is $7.1\times10^{-11}cm^{-2}sr^{-1}s^{-1}$. We have also shown the older Berkeley limit[36] obtained with Lexan detectors, at $6.7\times10^{-14}cm^{-2}sr^{-1}s^{-1}$ for $v > 0.02c$.

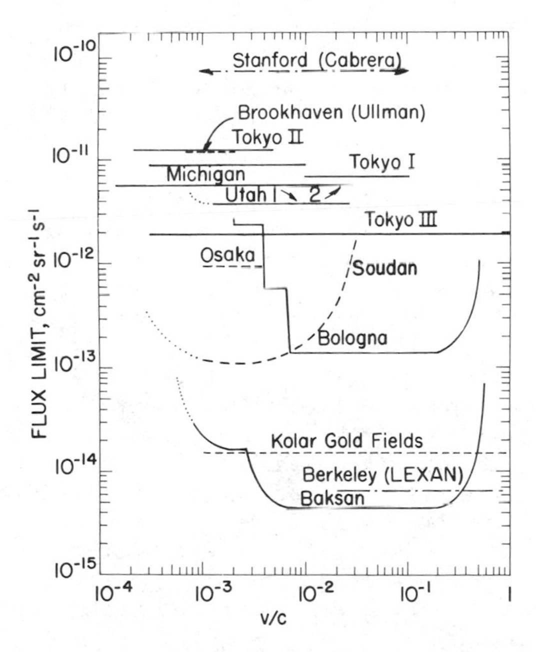

Fig. 3. Experimental monopole flux bounds obtained with scintillator arrays (solid lines) and proportional wire chamber arrays (dotted lines). Also shown are Cabrera's limit as of 02/82 and the Berkeley limit for relativistic monopoles (Ref. 36).

As a consequence of the above dE/dx discussion it is appropriate to group experiments by detector type. Within types, experiments ments are arranged chronologically, with published results first:

1. Scintillator experiments. After some electronic modification by two of the groups, the low-velocity cutoff in the five reported experiments is now determined by the problematical matter of scintillator response at low velocities. As discussed above, we use Ritson's curves as a conservative lower limit, even though agreement on this point is far from unanimous. High-velocity limits are usually imposed by the necessity of vetoing cosmic rays, although one array operated by the Tokyo group, (as well as the Kolar PWC array) is far enough underground that this constraint can be relaxed. Individual comments follow:

(a) Bologna.[37] For reasons of background rejection, the discrimination threshold was at about 25 times the ionization produced by relativistic cosmic ray muons, or 25 I_{min}. The detector was enlarged after the first publication to 38 m^2sr. Lower velocity limits varied in different runs from 0.002c to 0.007c. Both contribute to the v-dependent flux limit shown in Fig. 3, which summarizes the results as of 10 May 1982.

(b) Tokyo.[38] The original publication has been supplemented by an undated preprint. A small (1.1 m^2sr) 6-layer surface detector with a 1.2 I_{min} threshold was used to obtain the data shown as "Tokyo I" and "Tokyo II"; the difference was that in the second case the velocity window was $2\times10^{-4}c < v < 5\times10^{-3}c$ rather than $0.01c < v < 0.25c$. We would expect a dE/dx-imposed threshold at about $3\times10^{-4}c$ on the basis of Ritson's curves. A much larger array ($A\Omega$ = 22 m^2sr) located 250 m underground was used to obtain the "Tokyo III" limit. It had a threshold of 0.06 I_{min} for $v > 0.001c$, and 0.25 I_{min} for $3\times10^{-4}c < v < 0.001c$. All things considered, we regard this limit as the best low-velocity result available.

(c) Baksan.[39] This enormous ($A\Omega \sim 1850$ m^2sr) detector located 850 hg/cm^2 underground in the Baksan Mountains has been used to set the best available flux limit on galactic monopoles. While the original velocity window was too high for relevance to such objects, a later electronic modification reduced the threshold to $\sim7\times10^{-3}c$, with the new limit due to increased pulse width at lower velocities. (Discriminators are voltage sensitive and no pulse integrators are used.) Thresholds were set at 0.25 I_{min}. The curve shown in Fig. 3 is from a communication from A. Chudakov received in February 1983.

(d) Utah-Stanford.[32] A small 3-layer detector 507 hg/cm^2 underground operated with ~0.25 I_{min} thresholds to obtain the curve labeled "Utah 1". Pulse integrators were then introduced to permit velocity-independent triggering, and thresholds were reduced to

0.12 I_{min} ("Utah 2"). The estimated velocity lower limit (1.4×10^{-4}c) is based upon Ritson's excitation curves.

(e) Michigan.[40] A 5-layer surface array was set to trigger at 0.01 I_{min}. Since one usually collects less than 100 photoelectrons per muon traversal, we do not completely understand this statement. In any case, the experiment certainly has the lowest threshold of any yet reported.

2. PWC experiments: In principle, the comparatively low cost of PWC detectors make them attractive choices for a large-aperture array. There is one serious problem: The questions we raised concerning the sensitivity of scintillator to slow monopoles are replaced by pessimism in the case of commonly used PWC gases. For example, Ritson's curves for ionization in argon suggest that Ullman's pioneering experiment[41] was insensitive to monopoles over the entire velocity window as set by the electronics. If the Drell et al. mechanism were operative the experiment would still be insensitive because the ionization potential of methane (13.0 eV) is greater than the argon metastable excitation energy (11.6 eV). We do not wish to imply that PWC search results are invalid for $v < 10^{-3}c$, but merely that there are problems in the context of present first steps toward understanding energy loss by slow monopoles. Results from four such experiments are available:

(a) Brookhaven (Ullman).[41] The modest limit set by this experiment is offset by the fact that it was by far the earliest, even antedating Cabrera's. A 3-layer surface array of argon-methane chambers subtended 1.8 m^2sr for 2-way incidence. The threshold was at 2 I_{min}, and the velocity window was $3.3\times10^{-4}c < v < 1.2\times10^{-3}c$.

(b) Soudan.[42] This underground (1800 hg/cm^2) experiment makes use of a massive proton decay array ($A\Omega$ = 72 m^2sr) with 48 layers of argon-CO_2 filled PWC's. The electronic velocity window is $3\times10^{-4}c < v < 1.2\times10^{-3}c$, but with their thresholds at 0.5 I_{min} the authors do not claim acceptance below 10^{-3}c. Details concerning time resolution and trigger requirements result in the velocity-dependent limits shown in Fig. 3.

(c) Kolar Gold Fields.[43] We have only the evidence of a private communication of last spring (1982) concerning this large (208 m^2sr) PWC detector operated very deep underground in the Kolar Gold Fields (India) by a Japanese-Indian collaboration. The flux limit was $1.5\times10^{-14}cm^{-2}sr^{-1}$ between 10^{-3}c and c, with a threshold at 2.5 I_{min}. This limit is comparable to the Baksan result for $v \sim 10^{-3}c$.

(d) Osaka.[44] A surface detector with five layers of A-CH_4 filled PWC's is operated by physicists from two universities in Osaka,

Japan. The threshold of the 4.6 m^2sr detector is at 0.05 I_{min}, and the velocity window is between $10^{-3}c$ and $4\times10^{-3}c$. The limit shown in Fig. 3 is based upon 367 days of running.

3. Proposed experiments. There are probably important omissions in our list of planned experiments, which includes two important "classes": Large neutrino detectors at accelerators with re-worked trigger logic, and new PWC arrays using gas mixtures for which the Drell et al. mechanism can be utilized. Our order is fairly arbitrary:

(a) The CHARM Detector (CERN).[45] The aperture of this CERN neutrino detector is about 400 m^2sr. A monopole trigger consists of any six layers of scintillators triggering above 0.3 I_{min}. Data were taken for 10 days prior to February 1983 (but not analyzed:); more may be expected in the fall.

(b) The EAGLE 3 Detector (Fermilab).[46] This ~330 m^2sr detector will trigger if more than 20 I_{min} is deposited in each of four PWC planes.

(c) Homestake (U. of Pennsylvania and BNL).[47] A liquid scintillator "box" surrounding the ^{37}Cl (Davis) solar neutrino detector 4200 hg/cm^2 underground should be in operation by the fall of 1983. The ~1500 m^2sr detector threshold will be 0.1 I_{min} or lower, and should produce flux limits of $10^{-15}cm^{-2}s^{-1}s^{-1}$ in two years. The velocity window will extend from $10^{-4}c$ to nearly c.

(d) Berkeley (Ahlen, Liss, Price, and Tarlé).[48] A single thick slab of napthalene-based scintillator subtending 17.5 m^2sr is viewed by 52 photomultipliers. Pulse shape discrimination is used to eliminate cosmic rays and measure monopole velocity in the range $5 \times 10^{-4}c$ to $4 \times 10^{-3}c$. Preliminary data is available; a planned Cherenkov cosmic-ray veto counter was unnecessary.

(e) Texas (McIntyre and Webb).[49] A box of plastic scintillation counters with $A\Omega$ = 900 m^2sr will be located in a salt mine 500 hg/cm^2 underground. The threshold will be 4 I_{min} and the velocity window 0.004c to c. Its flux limit will be above that shown for unbound monopoles in Fig. 1.

(f) Utah-Stanford (Ritson and Loh).[50] PWC's containing helium plus a quencher are being developed for a large-area 8-layer detector. Making use of the Drell et al. mechanism in a known way, it should set limits below $10^{-15}cm^{-2}sr^{-1}s^{-1}$ and provide a useful compliment to the Homestake limit.

(g) La Jolla (Macek et al.).[51] Similar $He\text{-}CO_2$ PWC's are being developed for a proposed 4-layer detector which may eventually reach 6300 m^2sr.

CONCLUSIONS

The flux of monopoles gravitationally bound to the sun is less than $5\times10^{-12}cm^{-2}sr^{-1}s^{-1}$, on the basis of experiments performed in Tokyo and Utah. Kolar Gold Field and Baksan Mountains experiments limit the flux of galactic halo monopoles to $10^{-14}cm^{-2}sr^{-1}s^{-1}$, while older Berkeley experiments provide limits on relativistic monopoles at just over half this value. Proposed experiments will set limits below the astrophysical bounds near $10^{-15}cm^{-2}sr^{-1}s^{-1}$, and at the same time avoid interpretative questions concerning the ability of existing experiments to detect monopoles.

ACKNOWLEDGEMENTS

I have received private communications or conversed with the authors of more than half the experiments described above. Conversations with B. Cabrera and R. Peccei have been exceedingly helpful, as has the recent review by E.C. Loh.[52]

REFERENCES

1. B. Cabrera, Phys. Rev. Lett. 48, 1378 (1982); flux limits updated by private communication, February 1983.
2. R.A. Carrigan, Jr., Nature 288, 348-350 (1980).
3. M.J. Longo, Phys. Rev., D25, 2399-2405 (1982).
4. G. t'Hooft, Nucl. Phys. B79, 276 (1974).
5. A. Polyakov, Pis'ma Zh. Eksp. Theor. Fiz. 20, 430 (1974) [JETP Lett. 20, 194 (1974)].
6. Ya.B. Zeldovich and M. Yu. Khlopov, Phys. Lett. 79B, 239 (1978).
7. J.P. Preskill, Phys. Rev. Lett. 43, 1365-1368 (1979).
8. A.H. Guth, Phys. Rev. D23, 347-356 (1981).
9. D. Sivers, Phys. Rev. D2, 2048-2054 (1970).
10. V.A. Rubakov, Zh. Eksp. Teor. Fiz. 33, 658 (1981) [JETP Lett. 33, 644 (1981)].
11. C.G. Callan, Princeton Univ. Preprints "Disappearing Dyons" and "Dyon-Fermion Dynamics" (1982); see also F.A. Wilczek, Phys. Rev. Lett. 48, 1146 (1982).
12. F.A. Bais, J. Ellis, D.V. Nanopoulos, and K.A. Olive (CERN), CERN-TH-3383, Aug. 1982.
13. E.W. Kolb, S.A. Colgate, and J.A. Harvey, Phys. Rev. Lett. 49, 1373-1375 (1982).
14. S. Dimopoulos, J.R. Preskill, and F. Wilczek, Phys. Let. 119B, 320-322 (1983).
15. M.S. Turner, Nature 302, 804-806 (1983).
16. Q. Shafi, private communication 02/83.
17. S. Dimopoulos, S.L. Glashow, E.M. Purcell, and F. Wilczek, Nature 298, 824 (1983).
18. K. Freese and M.S. Turner, Phys. Lett. B (in press).
19. E.N. Parker, "Cosmical Magnetic Fields,"(Clarendon, Oxford, 1979).
20. M.S. Turner, E.N. Parker, and T.J. Bogdan, Phys. Rev. D26, 1296 (1982).

21. E.E. Salpeter, S. L. Shapiro and I. Wasserman, Phys. Rev. Lett. 49, 114-117 (1982).
22. C. Heiles, Ann. Rev. Astron. Astrophys. 14, 1-22 (1976).
23. J.R. Jokipii and E.N. Parker, Ap. J. 155, 799 (1969).
24. J.R. Jokipi, I. Lerche, and R. A. Schommer, Ap. J. Lett. 157, L119 (1969).
25. E. Fermi and E. Teller, Phys. Rev. 72, 399 (1947).
26. J. Lindhard, Mat. Fys. Medd. Dan. Vit. Selsk. 28, No. 8 (1954).
27. J. Lindhard and M. Scharff, Phys. Rev. 124, 128 (1961).
28. S.P. Ahlen and K. Kinoshita, Phys. Rev. D26, 2347 (1982).
29. S.H. Overbury, P.F. Dittner, S. Datz, and R.S. Thoe, Rad. Eff. 41, 219(1979). Also, see J.D. Garcia, Phys. Rev. A1, 280(1970) for a summary of experimental results of K-shell ionization by slow protons.
30. Y. Kazama, C.N. Yang and A.S. Goldhaber, Phys. Rev. D 15, 2287 (1977).
31. D. M. Ritson, SLAC-Pub-2950 (1982).
32. D.E. Groom, E.C. Loh, H.N. Nelson, and D.M. Ritson, Phys. Rev. Lett. 50, 573 (1983).
33. S.P. Ahlen and G. Tarlé, Phys. Rev. D (Rapid Communications) 27, 688(1983).
34. S. D. Drell, N. M. Kroll, M. T. Mueller, S. J. Parke, and M.A. Ruderman, Phys. Rev. Lett. 50, 644-648 (1983).
35. C.f. for example W.P. Jesse, J. Chem. Phys. 41, 2060 (1964).
36. K. Kinoshita and P.B. Price, Phys. Rev. D24, 1707 (1981).
37. R. Bonarelli et al., Phys. Lett. 112B, 100-102 (1982), and preprint dated 10 May 1982.
38. T. Mashimo, K. Kawoge, and M. Koshiba, J. Phys. Soc. Japan 51, 3065-3066 (1982), and undated preprint obtained May 1982.
39. E.N. Alexeyev, M.M. Boliev, A.E. Chudakov, B.A. Makoev, S.P. Mikheyev, and Yu.V. Sten'kin, Lett. Nuovo Cimento 35, 413 (1982), and private communication from A.E. Chudakov, February 1983.
40. J.K. Sokolowski and L.R. Sulak, Proc. 21st Conf. on High Energy Phys., Paris, 1982 (to be published), and L.R. Sulak, private communication 25 January 1983.
41. J.D. Ullman, Phys. Rev. Lett. 47, 289-292 (1981).
42. J. Bartelt et al., Phys. Rev. Lett. 50, 655-658 (1973).
43. N.K. Mondal, private communication to E.C. Loh May 1982.
44. S. Higashi, S. Ozaki, T. Takahashi, and K. Tsuji, undated preprint obtained from S. Ozaki in January 1983.
45. K. Winter, private communication February 1983.
46. E.C. Loh, private communication December 1982.
47. M.L. Cherry et al., proposal dated 08/82, and M.L. Cherry, private communication April 1983.
48. G. Tarlé, private communication 29 April 1983.
49. P.M. McIntyre and R.C. Webb, proposal dated 01 February 1982.
50. E.C. Loh and D.M. Ritson, private communications Feb.-May 1983.
51. G. Masek, private communication and note dated 23 March 1983.
52. E.C. Loh, in Proc. of the Wingspread Workshop, Racine WI USA, (1982), to be published.

THEORETICAL REVIEW OF MONOPOLE BOUNDS

Roberto D. Peccei

Max-Planck-Institut für Physik und Astrophysik
- Werner Heisenberg Institut für Physik -
Munich, Fed. Rep. Germany

ABSTRACT

A brief description of the principal features of Grand Unified Monopoles is given. This is followed by an extended overview of the current terrestrial, astrophysical and cosmological bounds on their existence.

In one of the classic papers of modern physics, Dirac[1] in 1931 showed that quantum mechanics plus the existence of magnetic charge implies charge quantization. Let m be the magnetic charge, then Dirac's result is that

$$em = 2\pi N \qquad (N \text{ integer}) \tag{1}$$

(I am using natural units $\hbar = c = 1$ and $\alpha = \frac{e^2}{4\pi} \simeq \frac{1}{137}$)

Grand Unified Theories (GUTs)[2] which unify the forces which describe strong and electroweak interactions into a simple group G automatically have charge quantization, since the charge is one of the generators of G. The Grand unified group G is not, however, respected by the vacuum and the symmetry is spontaneously broken down to SU(3) x SU(2) x U(1) (or an equivalent "low-energy" theory) at a scale M_x. Remarkably, as 't Hooft[3] and Polyakov[4] showed in 1974, in theories where a simple group G breaks down to $U(1)_{em}$ one necessarily has magnetic monopoles. Thus the circle closes ! This is illustrated pictorially in Fig. 1.

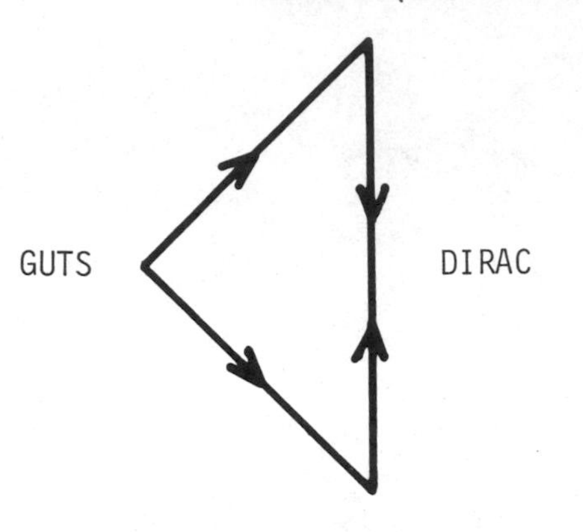

Fig. 1. The circle closes

CHARACTERISTICS OF THE GUT MONOPOLE

I will be here concerned only with GUT monopoles and to be specific, I shall consider that the relevant GUT is SU(5)[5]. Before I embark into a discussion of possible bounds on GUT monopoles I will need to recall certain of their features[6]. GUT monopoles arise as G suffers a breakdown to $U(1)_{em}$. For the particular case of SU(5) this occurs through the sequence

$$SU(5) \underset{M_x}{\rightarrow} SU(3) \otimes SU(2) \otimes U(1) \underset{M_w}{\rightarrow} SU(3) \otimes U(1)_{em} \qquad (2)$$

with $M_x \simeq 10^{14}$ GeV, $M_W \simeq 10^2$ GeV. Monopoles ensue when the Higgs field configurations, responsible for the breakdown, approach as $\mathbf{r} \rightarrow \infty$, different orientations in group space correlated with physical space directions. This physical space-group space correlation is also exhibited by the gauge fields. The monopole corresponds to a complicated, static, but non-singular, gauge field configuration within a core of size $\sim \frac{1}{M_x}$, where the Higgs fields tend to zero. Outside the core, the Higgs fields magnitude grows to that given by their vacuum expectation values ($H \rightarrow \langle H \rangle$) and the non-Abelian gauge field configurations, which correspond to the monopole, approach that of an Abelian (Dirac) monopole. More precisely, for $r \gg \frac{1}{M_x}$, one has[7]

$$g\lambda_a W_a^i(r) \simeq T W_{\rm Dirac}^i(r) \quad (3)$$

Here g is the gauge coupling constant of the GUT group, and $\lambda_a W_a^i(r)$ is a matrix of the gauge fields. The matrix T is given by the sum of the diagonal generators of the unbroken group (e.g. for SU(5), a sum of the diagonal generators of SU(3) x $U(1)_{em}$). $W^i_{\rm Dirac}(r)$ is the vector potential for an Abelian monopole

$$\vec{W}_{\rm Dirac}(r) = \frac{\hat{r}\times\hat{n}}{r(1-\hat{n}\cdot\hat{r})} \quad (4)$$

where $\hat{n}$ is a unit vector.

For the SU(5) case one can show[6,7] that the 5x5 matrix T takes the form

$$T = \begin{bmatrix} a & & & & 0 \\ & a & & & \\ & & b & & \\ & & & -b-2a & \\ 0 & & & & 0 \end{bmatrix} \quad (5)$$

The quantization condition which replaces (1) for the non-Abelian case is

$$e^{2\pi i T} = 1 \quad (6)$$

which implies that a and b must be integers. Stability of the monopole configuration[6], furthermore, implies that only the values

$$a - b = 0, \pm 1 \quad (7)$$

are allowed. Finally, the topological charge associated with the monopole can be seen to be given by

$$Q_t = 2a + b \quad (8)$$

Thus the lowest topological charge monopole (Q_t = 1) has a = 0, b = 1, It will be seen below that this monopole carries precisely one Dirac unit of magnetic charge, $m = \frac{2\pi}{e}$. For this case, the matrix T takes the simple form

$$T = \begin{bmatrix} 0 & & & & 0 \\ & 0 & & & \\ & & 1 & & \\ & & & -1 & \\ 0 & & & & 0 \end{bmatrix} = \begin{bmatrix} 1/3 & & & & 0 \\ & 1/3 & & & \\ & & 1/3 & & \\ & & & -1 & \\ 0 & & & & 0 \end{bmatrix} + \begin{bmatrix} -1/3 & & & & 0 \\ & -1/3 & & & \\ & & 2/3 & & \\ & & & 0 & \\ 0 & & & & 0 \end{bmatrix} \tag{9}$$

The second equality above shows that T is a sum of the electric charge generator and of a color hypercharge generator. Thus the simplest monopole has, away from the core, both magnetic and chromomagnetic fields.

Using Eq. (3), it is straightforward to calculate the electromagnetic potential for the $Q_t = 1$ monopole. Outside the core

$$\vec{A}(r) = \frac{1}{2g} Tr(\lambda_Q T)\vec{W}_{\text{Dirac}}(r) \tag{10}$$

Here

$$\lambda_Q = \sqrt{3/2} \begin{bmatrix} 1/3 & & & & 0 \\ & 1/3 & & & \\ & & 1/3 & & \\ & & & -1 & \\ 0 & & & & 0 \end{bmatrix} \tag{11}$$

is the appropriately normalized charge generator ($Tr\,\lambda_Q \lambda_Q = 2$). Because for SU(5) $\sin^2\theta_W = 3/8$ one has that $g = \sqrt{\frac{8}{3}}\, e$. Whence, it follows from Eq. (10) that

$$\vec{A}(r) = \frac{2\pi}{e}\left\{\frac{1}{4\pi r}\,\frac{\hat{r}\times\hat{n}}{(1-\hat{r}\cdot\hat{n})}\right\} \tag{12}$$

which demonstrates that the magnetic charge is indeed

$$m = \frac{2\pi}{e} \tag{13}$$

The monopole mass can be estimated by computing the volume integral of the field energy density outside the core. In this region

$$\theta^{00} = \frac{1}{4} Tr\lambda_a \vec{B}_a \cdot \lambda_b \vec{B}_b = \frac{1}{2g^2}(b^2 + 2ab + 3a^2)\vec{B}^2_{\text{Dirac}} \tag{14}$$

where we have used that $\lambda_a \vec{B}_a = \frac{1}{g} T \vec{B}_{\text{Dirac}}$, with $\vec{B}_{\text{Dirac}} = \frac{\hat{r}}{r^2}$. Thus

$$M_{\text{Mon.}} \geq \int_{1/M_x}^{\infty} 4\pi r^2 \theta^{00} dr = \frac{1}{2\alpha_{\text{GUT}}}(b^2 + 2ab + 3a^2)M_x \tag{15}$$

For the simplest monopole ($b = 1$, $a = 0$) and using[8] $M_x \simeq 2\times10^{14}$ GeV, $\alpha_{\text{GUT}} = \frac{g^2}{4\pi} \simeq \frac{1}{45}$ one has

$$M_{\text{Mon.}} \geqslant 5\times10^{15}\text{GeV} \tag{16}$$

For the purposes of this paper I shall adopt for M_{Mon} the value $M_{Mon} = 10^{16}$ GeV.

The last important property of the GUT monopole I shall need is connected with the possibility that, in the presence of the monopole field, baryon violating decays may occur very rapidly. This remarkable property of monopoles has been discussed very recently by Rubakov[9] and Callan[10], but it is not without controversy. What Rubakov and Callan show (argue ?) is that surrounding the monopole core there exist fermion-antifermion condensates whose density drops off only geometrically. Some of these condensates, furthermore, have $\Delta B \neq 0$ and hence could induce $\Delta B \neq 0$ processes at hadronic rates.

Consider one fermion generation ($\bar{5}$ + 10) of SU(5) in the presence of the $Q_t = 1$ monopole. With respect to the SU(2) subgroup of SU(5) defined by T in Eq. (9), the quarks and leptons appear as the following doublets:

$$\begin{pmatrix} d_3 \\ e^- \end{pmatrix}_L \begin{pmatrix} e^+ \\ d_3 \end{pmatrix}_L \begin{pmatrix} u_1 \\ \bar{u}_2 \end{pmatrix}_L \begin{pmatrix} u_2 \\ \bar{u}_1 \end{pmatrix}_L \tag{17}$$

where the indices 1,2,3 are color indices. The fermion condensates that form in the presence of the monopole contain two $T_3 = +1$ and two $T_3 = -1$ states. It is clear from (17) that some of these condensates have $\Delta B \neq 0$. In a model calculation[9,10] one finds that, for example,

$$\langle M|\bar{d}_3 e^+ \bar{u}_2 \bar{u}_1|M\rangle \simeq \frac{1}{(2\pi)^4}\frac{1}{r^6} \tag{18}$$

The purely geometric cutoff in Eq. (18) should hold up to distances where color gets confined $r \sim \frac{1}{\Lambda_{QCD}}$.

The presence of a condensate like (18) surrounding a monopole can induce in matter proton decay. The catalysis reaction $M + p \rightarrow e^+ + M + X$ can be imagined pictorially as in Fig. 2. However, the magnitude of the cross section is difficult to calculate. A guess is that it should be of typical strong interaction size

$$\sigma_{\Delta B\neq 0} = \frac{1}{\beta}\frac{1}{(1\,\text{GeV})^2}\sigma_0 \tag{19}$$

Here ß is the monopole velocity and is a purely kinematical flux factor. σ_0 is a fudge factor which one would expect to be of O(1), if $\sigma_{\Delta B \neq 0}$ is indeed of strong interaction size. Bais, Ellis, Nanopoulos and Olive[11], have tried to be a little careful with (2π)-factors and argue that σ_o could perhaps be of order $10^{-4} - 10^{-5}$. Because, the existence of the fermion condensates and of the catalysis has not been totally satisfactorily demonstrated - indeed there exists controversy whether it is suppressed by powers of the weak interaction scale[12] or whether it only holds in the truly zero fermion mass limit[13] - one should keep in mind that σ_o could possibly be vanishingly small ! Hence, in what follows I shall quote limits for monopoles, leaving σ_o as a free parameter.

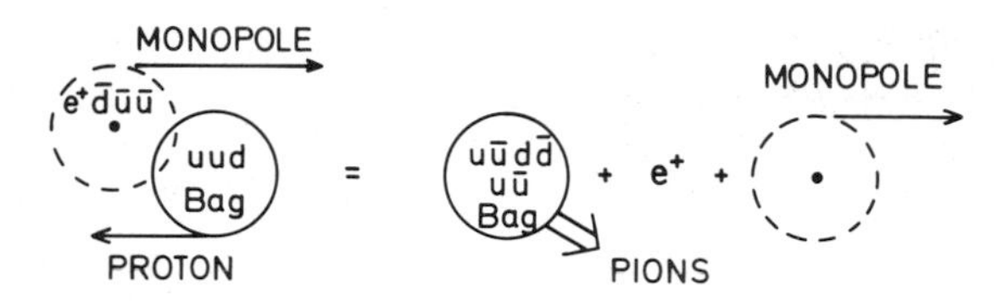

Fig. 2. Schematic monopole catalysis of proton decay

MONOPOLES ON EARTH

I shall begin my overview of bounds on monopoles at the center of the earth; then I shall move outwards to the earth surface and subsequently to the cosmos. Finally, I shall go backwards in time and discuss what the early universe can tell us about monopoles. Throughout, I shall assume that $M_{Mon} = 10^{16}$ GeV and $m = \frac{2\pi}{e}$.

The gravitational force on a GUT monopole on the surface of the earth is of the order of 0.1 eV/A°. Thus, it is clear that monopoles in general will fall to the earth's core. However, if they are fast enough they will have enough kinetic energy to sail through. Roughly speaking, monopoles with $\beta \gtrsim 3 \times 10^{-5} - 10^{-4}$ and therefore $T \gtrsim 10^7 - 10^8$ GeV will not stop in the earth. Carrigan[14] has put a bound on the number of monopoles trapped in the earth, by estimating the amount of heat flowing out of the earth arising from monopole-antimonopole annihilation in the earth's core during periods of magnetic field reversal. He obtains in this way the following bound on the number of monopoles to nucleons:

$$\frac{n_M}{n_N} \lesssim 2\times10^{-28} \tag{20}$$

It is amusing to note that this is the same kind of bound one obtains also from searches for monopoles in moon rocks, ocean sediments etc.[15] . However, these latter (surface) bounds are not expected to be valid for the superheavy GUT monopoles under discussion.

One can turn the bound (20) into a bound on the monopole flux of slow monopoles ($\beta \lesssim 3 \times 10^{-5} - 10^{-4}$), assumed to have accumulated in the earth's core throughout its lifetime ($T_e \simeq 10^{10}$ years):

$$\Phi^{\mathrm{Carrigan}} = \frac{n_M}{4\pi^2 R^2 T_e} \lesssim 3\times10^{-13}\mathrm{cm}^{-2}\mathrm{sec}^{-1}\mathrm{st}^{-1} \qquad (\beta \lesssim 3\times10^{-5}-10^{-4}) \tag{21}$$

Actually, there exists a potentially much more stringent bound than the above, arising from the effects that monopoles in the core would have on the earth magnetic field. This bound, which is discussed by Dimopoulos, Glashow, Purcell and Wilczek[16], follows from considerations first put forward by Parker[17]. The magnetic energy dissipated by the monopoles trapped in the earth's core, by their doing work agains the earth's magnetic field, must be less than the total available magnetic energy stored during the characteristic growth time of the geomagnetic field, τ_g. If d_M is the monopole density and v_c the typical velocity for core monopoles, then one must have that

$$m d_M v_c B \leqslant \tfrac{1}{2} B^2/\tau_g \tag{22}$$

or

$$d_M \lesssim \frac{1.2\times10^6 B(\mathrm{Gauss})}{[v_c\tau_g](\mathrm{cm})}\ \mathrm{cm}^{-3} \tag{23}$$

Using as typical numbers $\tau_g \simeq 10^4$ years. $B \simeq 10^2$ Gauss and $v_c \simeq 7 \times 10^5$ cm/sec (This value follows from energetic considerations: $\frac{1}{2} Mv_c^2 \simeq mBR_{core}$, $R_{core} \simeq \frac{1}{2} R_e$) one has

$$d_M \lesssim 5\times10^{-10}\text{cm}^{-3} \tag{24}$$

which corresponds to

$$\frac{n_M}{n_N} \lesssim 8\times10^{-35} \tag{25}$$

and a flux limit, if one assumes that the earth's magnetic field properties have not varied substantially in time:

$$\Phi^{DGPW} \lesssim 2\times10^{-19}\text{cm}^{-2}\text{sec}^{-1}\text{st}^{-1} \quad (\beta \lesssim 3\times10^{-5}\text{--}10^{-4}) \tag{26}$$

Cabrera[18], in elegant experiment designed to detect magnetic monopoles by observing quantized current rises in a 20 cm² superconducting loop, reported last year a candidate event. The current rise was precisely of the magnitude expected for a unit magnetic charge monopole ($m = \frac{2\pi}{e}$). This event obtained during a run of 151 days, if taken at face value, would correspond to a flux:

$$\Phi^{Cabrera} = \frac{1}{2\pi A t_{exp}} = 6.1\times10^{-10}\text{cm}^{-1}\text{sec}^{-1}\text{st}^{-1} \tag{27}$$

Since that time, as Cabrera told us in this meeting[19], a new three ring apparatus has been built, which has a much larger effective area. Since no new event has been detected, during the limited running time of this new device, this lowers the limit of Eq. (27) to[19]:

$$\Phi'^{Cabrera} = 7.1\times10^{-11}\text{cm}^{-2}\text{sec}^{-1}\text{st}^{-1} \tag{28}$$

These values are so much greater than the limits obtained for the very slow monopoles which can be captured by the earth, that they suggest that, if Cabrera's event is real, then its flux is that characteristic of monopoles with $\beta \gtrsim 3 \times 10^{-5} - 10^{-4}$. However, as I shall discuss below, monopole fluxes for $\beta \gtrsim 10^{-3} - 10^{-4}$ can be severely bounded by ionization experiments. The combination of both geological and ionization bounds leaves a very narrow velocity range for Cabrera's event.

Moving monopoles traversing matter transfer energy, to the (nearly) free electrons because of the induced electric fields that they set up. Since $\vec{E} = \vec{\beta} \times \vec{B}$, it is clear that the energy loss of a monopole will be the same as that of a charged particle, save for an overall factor of $(\frac{m\beta}{e})^2 = \frac{\beta^2}{4\alpha^2}$. Thus

$$\left.\frac{dE}{dx}\right|_{Mon.} = \beta^2/u\alpha_2 \left.\frac{DE}{dx}\right|_{Ch.\ Part.} \tag{29}$$

The above formula is reliable for sufficiently fast monopoles, where electron binding can be safely neglected. The $1/\alpha^2$ factor, of course, indicates that fast monopoles are expected to ionize very strongly. Eq. (29) ceases to be valid when the time the monopole spends in the vicinity of a given atom - the collision time - is comparable to the typical orbit time of the bound electrons. To be able to neglect binding one must have that

$$(\Delta t)_{\text{Coll.}} \simeq \frac{\langle r \rangle_{\text{atom}}}{\beta} \lesssim (\Delta t)_{\text{orbit}} \simeq \frac{1}{(\Delta E)\text{spacing}} \tag{30}$$

The above, for hydrogenic atoms, implies that

$$\beta \gtrsim \beta_0 = \frac{3}{8} z\alpha = 2\times10^{-3} z \tag{31}$$

For $\beta \lesssim \beta_0$ the energy transfer from the monopole to matter should begin to be ineffective and ultimately the energy loss will vanish exponentially.

To actually obtain the energy loss of slow moving monopoles requires detailed calculation. Such detailed calculations have been performed recently by a number of authors, notably Ahlen and Kinoshita[20] and Ritson[21], who make use of a modified Thomas-Fermi model. This complicated subject was reviewed by D.H. Groom[22] at this meeting and I inferred from his talk that one may begin to have confidence in monopole ionization formulas, down to perhaps as low a β as 10^{-4}.

Matters, however, are somewhat complicated by another effect caused by the monopole's passage through matter. The strong B field of the monopole causes tremendous level mixing and crossings and this provides an additional source of ionization. The typical Zeeman shift due to the presence of the monopole is

$$(\Delta E)_{\text{Zeeman}} \simeq \frac{e}{2m_e} \frac{m}{4\pi \langle r \rangle^2_{\text{atom}}} \simeq \frac{m_e \alpha^2 z^2}{4} \tag{32}$$

which is of the order of the binding energy and thus it clearly cannot be neglected[23]. The importance of this source of monopole energy loss is discussed in detail in a recent paper by Drell et al.[24]. These authors find that, for low β, the main energy loss is due to these magnetic effects, rather than through the more conventional electric effects. Thus, it may well be that, for some materials, monopoles are above minimum ionizing down to $\beta \simeq 10^{-4}$. This issue is far from settled and is discussed further by Groom[22].

In the last year, spurred in part by Cabrera's paper, many ionization experiments were set up to obtain limits on the ambient monopole flux. A compilation of these experimental results was given by Giacomelli[25] at the Wingspread Meeting. These bounds have been updated and revised by Groom[22] and I display them in Fig. 3, which is essentially adapted from Groom's compilation. As can be seen from the figure, the Cabrera flux is in contradiction with these limits and the geological bounds except for a narrow region about $\beta \simeq 10^{-4}$. However, one should remember that there are still some uncertainties in the calculations of the amount of ionization of slow monopoles ($\beta \lesssim \beta_o$), so that the region much below 10^{-3} is still a bit questionable for ionization experiments.

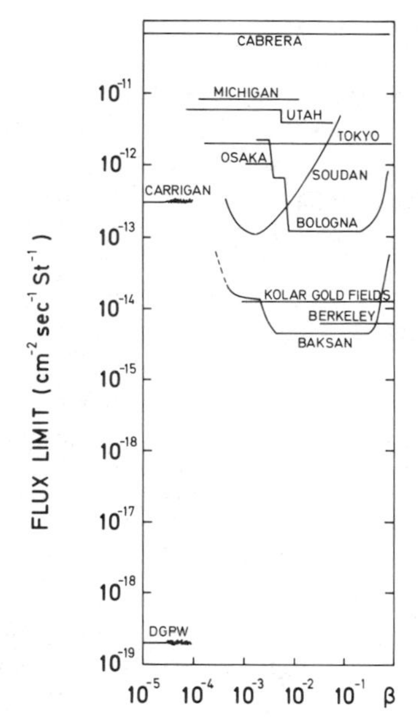

Fig. 3. Terrestrial Monopole Bounds (excluding catalysis)

If monopole catalysis of proton decay occurs at a strong interaction rate, then the present generation of proton decay experiments can also put some bounds on the monopole flux on earth[11,26]. Using Eq. (19) for the catalysis cross section, it follows that the mean free path between decays is

$$\lambda \simeq \frac{43\beta}{\sigma_0\rho(\mathrm{g/cm^3})} \text{ meters} \tag{33}$$

For reasonably slow monopoles ($\beta \simeq 10^{-3} - 10^{-4}$) and if σ_o is of 0(1), one would therefore expect multiple decays in a typical proton decay detector. Accepting that $\tau_P \gtrsim 10^{31}$ years and using simply

$$\sigma_{\Delta B \neq 0} \Phi \lesssim \frac{1}{\tau_p} \tag{34}$$

gives the bound

$$\sigma \Phi_{Cat.} \lesssim 8 \times 10^{-12} \beta \, cm^{-2} sec^{-1} st^{-1} \tag{35}$$

This is a better bound, for $\beta \lesssim 10^{-3}$, on the flux of monopoles than those obtained in ionization experiments if σ_o is indeed of 0(1). On the other hand, if one requires the above to be consistent with Cabrera's flux at $\beta \simeq 10^{-4}$ one would need $\sigma_o \simeq 10^{-5}$.

Turner[27] recently has used monopole nucleon decay catalysis to provide a further strong bound on the flux of monopoles trapped in the earth. Simply put, Turner requires that the total energy release by catalysis should not exceed the surface heat flow. This bounds the total number of monopoles in the earth and hence provides directly a limit on the flux of very slow monopoles. The rate of nucleon decays induced by monopoles is

$$\Gamma_{Cat.} = d_n \beta \sigma_{\Delta B \neq 0} \simeq 4 \times 10^9 \sigma_0 sec^{-1} \tag{36}$$

where in the above d_n is the earth's nucleon density ($d_n \simeq 3.3 \times 10^{24} cm^{-3}$). Assuming that roughly 1 GeV of energy is released per decay this gives a rate of energy produced by catalysis:

$$\left. \frac{dE}{dt} \right|_{Cat.} = n_M (1\,GeV) \Gamma_{Cat.} \tag{37}$$

where n_M is the number of monopoles on earth. From the relation between the monopole flux and n_M (Eq. 21) and demanding that $\left. \frac{dE}{dt} \right|_{cat.}$ be less than the surface heat flow ($(\frac{dE}{dt})_{Heat} \lesssim 3 \times 10^{20}$ ergs/sec.) one obtains the bound

$$\sigma_0 \Phi_{Cat.}^{Turner} \lesssim 2 \times 10^{-21} cm^{-2} sec^{-1} st^{-1} \quad (\beta \lesssim 3 \times 10^{-5} - 10^{-4}) \tag{38}$$

For $\sigma_o \simeq 0(1)$ this is a very strong bound. Indeed, even if $\sigma_o \simeq 10^{-2}$ this bound is of the same order as that given in Eq. (26). Fig. 4 displays the catalysis bounds for some values of σ_o, along with some of the ionization bounds of Fig. 3.

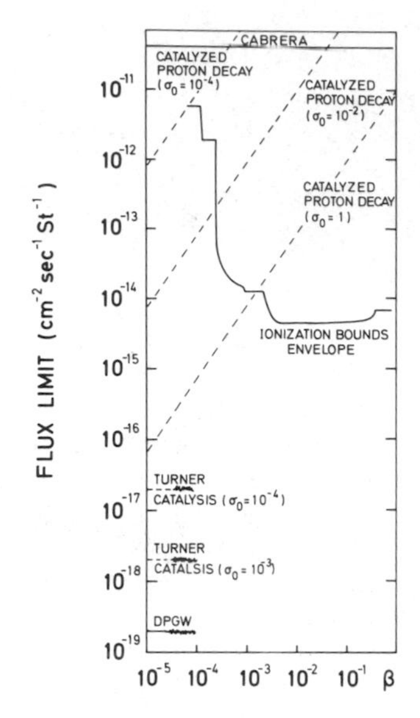

Fig. 4. Catalysis Bounds on Earth

ASTROPHYSICAL BOUNDS ON MONOPOLES

Perhaps the best known astrophysical bound on the monopole flux is the, so called, Parker limit[28]. This bound, which applies to the average flux of monopoles in a typical galaxy, limits the monopole flux by demanding that monopoles should not erase the Galactic B-field. The argument is similar to the one I discussed earlier for the persistence of the earth's magnetic field. One requires that the rate of energy loss per unit volume, due to the presence of monopoles in the galaxy:

$$\left.\frac{dE}{dt}\right|_{\text{Mon.}} = \vec{J}_m \cdot \vec{B} \tag{39}$$

where

$$\vec{J}_m = m d_M \vec{v} \tag{40}$$

with d_M being the monopole density, be less than the magnetic power, per unit volume, stored in the Galaxy. If τ_{gal} is the typical time to regenerate the Galactic magnetic field - believed to be of the order of 10^8 years - then one asks that

$$\vec{J}_m \cdot \vec{B} \leqslant \tfrac{1}{2}\vec{B}^2/\tau_{gal.} \tag{41}$$

The monopole flux is

$$\Phi - d_M |v|/4\pi \tag{42}$$

Thus if one approximates $\vec{v}.\vec{B}$ by $|v||B|$, Eq.(41) implies the bound

$$\Phi_{gal.} \leqslant \frac{e|B|_{gal.}}{(4\pi)^2\tau_{gal.}} \leqslant 10^{-16}\text{cm}^{-2}\text{sec}^{-1}\text{st}^{-1} \quad \text{(Parker limit)} \tag{43}$$

where the numerical result follows by taking for the typical galactic field $B_{gal} \simeq 3 \times 10^{-6}$ Gauss .

The Parker limit (43) is velocity independent (see, however, below) and almost 6 orders of magnitude smaller than the Cabrera flux. Thus, if Cabrera's event was real, it follows that the flux to which it corresponds is atypical for the Galaxy as a whole, but must correspond to some local enhancement. Dimopoulos, Glashow, Purcell and Wilczek[16] suggested that such a large local enhancement in the flux at the earth would occur if there were a cloud of gravitationally captured monopoles in orbit around the sun. Such a monopole cloud, furthermore, would have typical orbital velocities of around $\beta \simeq 10^{-4}$, falling in the narrow band where the existence of monopoles is not otherwise ruled out. This important issue has been reexamined recently by Freeze and Turner[29], with different conclusions than those of Dimopoulos et al.[16].

Freeze and Turner calculate the flux on earth of solar captured monopoles as follows (I am slightly oversimplifying their arguments; for more details see Ref. 29):

$$\Phi_{earth} \simeq \frac{1}{4\pi V_{cloud}} \beta \, [4\pi^2 R_s^2 \tau\varepsilon\Phi_{gal.}] \tag{44}$$

Here V_{cloud} is the volume swept by the monopole cloud and the quantity in the square brackets represents the number of monopoles captured by the sun (radius R_s) in a time τ with a presumed capture efficiency ϵ. If for V_{cloud} one uses the volume of the earth-sun system, V_{e-s} and takes $\beta \simeq 10^{-4}$ and a "typical" capture time $\tau \simeq \tau_{capt} \simeq 10^9$ years, then one finds

$$\Phi_{earth} \simeq 10^5\varepsilon\Phi_{gal.} \tag{45}$$

which, indeed, corresponds to a large local enhancement of the galactic flux for efficiencies, ϵ , of monopole capture of order unity. However, Freeze and Turner[29], argue that this is a bad overestimate. First of all τ should be more of order $\tau = \tau_{pert} \simeq 10^8$ years, which is the time by which the sun's B field would sufficiently perturb the monopole gravitational orbits, rendering them unstable. Secondly, by direct computation, they show that the extent of the monopole cloud which impinges on earth has a much bigger size than the naive geometrical one. They find $V_{could} \simeq 70\, V_{e-s}$. Finally, they estimate that the efficiency of galactic monopoles being captured by the sun is at most of the order of 10 %. Whence, the enhancement expected by Freeze and Turner is small:

$$\Phi^{FT}_{earth} \simeq 50\Phi_{gal.} \tag{46}$$

In fact, even to sustain such an enhancement requires mechanisms which continually pump angular momentum into the monopole cloud, if not it will inexhorably spiral into the sun[29].

The conclusions I draw from the above discussion is that large local enhancements of monopole fluxes, beyond the Parker limit are probably unlikely. However, before ruling out the possibility of having monopole fluxes at the Cabrera level from astrophysical considerations, one must examine carefully some of the assumptions behind the Parker bound. There is, of course, the whole issue of the reliability of the estimation of the origin, size and time scale needed to set up the galactic field[17]. Indeed, it has been suggested recently[30] that one could bypass the Parker bound altogether if the origin of the galactic magnetic fields is due to magnetic plasma oscillations. The monopoles and antimonopoles oscillate back and forth alternatively draining and restoring the galactic B field. However, whether this phenomena has all the necessary astrophysical characteristics is rather unclear still[30]. Even granting the usual scenario, there is a simpler worry one might have with the derivation of the Parker limit. The approximation $\vec{v}\cdot\vec{B} \simeq |v||B|$ used to derive the bound corresponds, roughly speaking, to assuming that the monopoles are always removing energy from the magnetic field. Does this not lead to an overestimate of the energy loss ?

Turner, Parker and Bogdan[31] have examined this last issue recently They find that, as long as the monopole velocities are below a critical velocity, β_c, the approximation of neglecting the directionally of the monopoles is reasonable. Above β_c, however, it is no longer true that the monopoles always effectively allign themselves so as to take energy out of the magnetic field. For $\beta > \beta_c$ the Parker limit is correspondingly reduced. The critical velocity β_c can be estimated as the velocity that a monopole acquires in a typical galactic coherence length $L \simeq 10^{21}$ cm by magnetic acceleration. For a GUT monopole ($M = M^{16}$ GeV), using $B_{gal} \simeq 3 \times 10^{-6}$ Gauss, one has $\beta_c \simeq 3 \times 10^{-3}$. The result of the more refined analysis of Ref. 31 is

$$\Phi^{TPB}_{gal.} \leqslant 10^{-15}\text{cm}^{-2}\text{sec}^{-1}\text{st}^{-1} \cdot \begin{cases} (\beta/\beta_c)^L & 1 \\ \beta > \beta_c & \beta < \beta_c \end{cases} \tag{47}$$

This result, at least for low ß, is essentially the same as that of Eq. (43). The one order of magnitude change between Eqs. (47) and (43), for low ß, is not to be worried about. The simple analysis I presented, which lead to Eq. (43), is certainly not expected to be better than that.

It is possible to obtain even more restrictive bounds on the monopole flux than the Parker limit, but these bounds involve in general more speculative assumptions. For instance, Ritson[32], has bounded the monopole flux from the fact that in certain stars one still observes what are believed to be primordial magnetic fields. The, so called, peculiar rotating A_4 stars have coherent magnetic fields ranging from 2×10^2 to 3×10^4 Gauss. These fields are in the direction opposite to that expected from rotation and they are therefore presumed to have been frozen in at the stars formation time ($T_{star} \simeq 5 \times 10^8$ years). The energy loss of a typical galactic GUT monopole ($\beta \simeq 10^{-3}$, $T = 10^{10}$ GeV) as it traverses the star is such that it is likely to be trapped. The monopole density in the star - and therefore the monopole flux in the galaxy - is bounded by the by now familiar Parker argument, which leads to the formula

$$d_M \leqslant \frac{B_{star}e}{4\pi\beta_{drift}T_{star}} \tag{48}$$

The drift velocity can be estimated by equating the rate of change of the monopole energy with distance ,

$$\frac{dE}{dx} \simeq 3\times10^2\beta_{drift}\ \text{GeV/cm} \tag{49}$$

obtained by treating the star as an electron gas, with the magnetic force mB_{star} experienced by the monopole. This estimate in (48) gives

$$d_M \leqslant \frac{2\times10^{-5}}{T_{star}(\text{years})}\ \text{cm}^{-3} \tag{50}$$

Translating the above to a flux bound, and using $T_{star} = 5 \times 10^8$ years, $R_{star} = 10^{11}$ cm, gives

$$\Phi^{\text{Ritson}} = \frac{R_{star}d_M}{3\pi T_{star}} \leqslant 2.5\times10^{-20}\text{cm}^{-2}\text{sec}^{-1}\text{st}^{-1} \quad (\beta \lesssim 10^{-2}) \tag{51}$$

The ß restriction above follows since very fast monopoles would sail through the star. One should note that this is a such a strong bound primarily because $\Phi \sim \frac{1}{T^2_{star}}$ and one uses the fact that the B-field has been frozen in for a long time. Whether this assumption is really warranted, however, is the principal uncertainty concerning this limit.

A stronger bound that the Parker limit (47) for the flux of intergalactic monopoles has been derived recently by Raphaeli and Turner[33]. This is another example of a stronger limit which follows from perhaps more questionable assumptions. Raphaeli and Turner find the following bound for the intercluster (IC) flux of monopoles:

$$\Phi_{IC} \leqslant 2\times10^{-19}\text{cm}^{-2}\text{sec}^{-1}\text{st}^{-1}\begin{cases}(\beta/10^{-3})^2 & \beta > 10^{-3} \\ 1 & \beta < 10^{-3}\end{cases} \tag{52}$$

This bound, which is again obtained by Parker-like considerations, is stronger than that of Eq. (47) principally because the intercluster magnetic field is thought to be smaller (by two orders of magnitude) than the galactic field and have a longer lifetime, $\tau_{IC} \simeq 10^9$ years. However, it has to be said that the physics which is responsible for the origin of the intercluster fields is still more uncertain than that which gives rise to the galactic magnetic fields.

It was immediately realized by a number of authors[34] that if monopoles catalyze baryon number violating processes, then one could get a good bound on the galactic monopole flux from neutron stars. The argument is similar to the one exposed earlier, relating to the expected heat flow from the earth due to catalysis. Only the relevant parameters change. A typical neutron star has a radius $R_N \simeq 10^6$ cm; a nucleon density $d_n \simeq 2 \times 10^{38}$ cm^{-3}; and an escape velocity $\beta_e \simeq 1/2$. The expected monopole energy loss in the neutron star is[34]

$$\frac{dE}{dx} \simeq 10^{11}\beta\,\text{GeV/cm} \tag{53}$$

which means that a typical galactic GUT monopole ($\beta \simeq 10^{-3}$, $T \simeq 10^{10}$ GeV) will always be stopped by the star and captured. Because typically galactic monopoles are much slower than the neutron star escape velocity, the number of monopoles captured for a given flux will be larger than that estimated by naive geometrical considerations (cf. Eq. (21)). The capture cross section for monopoles of velocity β is given by

$$\sigma_{\text{Capt.}} = \pi R_N^2\left[1 + \frac{2M_N G}{R_N\beta^2}\right] \simeq \pi R_N^2\left(\frac{\beta_e}{\beta}\right)^2 \simeq \frac{8\times10^{11}}{\beta^2}\text{cm}^2 \tag{54}$$

Using the above, it follows that the number of monopoles accumulated in an old neutron star - $T \simeq 10^{10}$ years - is[34]

$$n_M = 4\pi\sigma_{\text{Capt.}}T\Phi = \frac{3.2\times10^{30}}{\beta^2}\,\Phi(\text{cm}^{-2}\text{sec}^{-1}\text{st}^{-1}) \tag{55}$$

From Eq. (55) one can immediately calculate the luminosity (Energy released/sec.) due to monopole induced baryon decay. The expected luminosity arising from the release of approximately one GeV of energy per decay is just

$$L = n_M(1\,\text{GeV})\Gamma_{\text{cat}} \tag{56}$$

Using the, typical neutron star, catalysis rate

$$\Gamma_{cat} = d_n \beta \sigma_{\Delta B \neq 0} = 2.4\times10^{21} \sigma_0 \sec^{-1} \tag{57}$$

and Eq. (55) yields a luminosity

$$L = 1.2\times10^{49} \frac{1}{\beta^2} \sigma_0 \Phi(\mathrm{cm}^{-2}\mathrm{sec}^{-1}\mathrm{st}^{-1}) \frac{\mathrm{ergs}}{\mathrm{sec}} \tag{58}$$

The total luminosity arising from neutron stars can be bounded, principally by looking at the photon spectra. If one adopts the most conservative estimate in Ref. 34, which is that of Kolb, Colgate and Harvey, one gets the flux bound:

$$\sigma_0 \Phi \leqslant 10^{-16} \beta^2 \mathrm{cm}^{-2} \mathrm{sec}^{-1} \mathrm{st}^{-1} \tag{59}$$

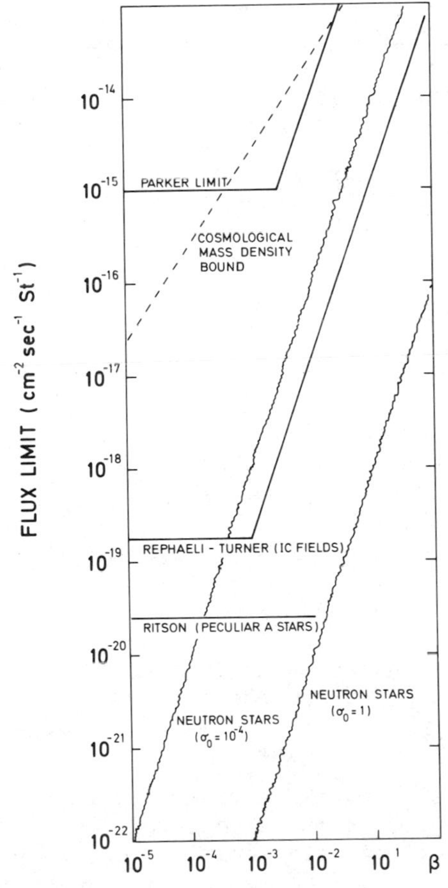

Fig. 5. Astrophysical Bounds on the Monopole Flux

If σ_o is indeed of O(1) this is, for typical galactic monopoles, a bound many orders of magnitude stronger than the Parker limit (47). I should note that Harvey[35] has worried about whether monopole-antimonopole annihilations may reduce n_M in neutron stars and hence weaken the above flux bound. His conclusion is that this effect is unlikely to be significant. Thus the bound (59) should remain reliable. It is a significant improvement over (47) as long as $\sigma_o \gtrsim 10^{-5}$.

The astrophysical bounds on monopole fluxes are summarized in Fig. 5 above. Included in this figure is also the bound on the monopole flux obtained by demanding that the monopole mass density should not exceed the critical density ρ_c for which the universe closes. This bound, as well as other cosmological considerations on monopoles are the subject of the following section.

COSMOLOGICAL BOUNDS AND PROBLEMS

Because GUT monopoles are so heavy, their total density in the universe cannot be too great. A simple bound is obtained by requiring that the monopole mass density should not exceed the total mass density of the universe, which is known to be near the critical density[36]:

$$\rho_c = \frac{3H_0}{8\pi G} = 2\times10^{-29}\text{g/cm}^3 \tag{60}$$

From

$$\rho_M = M_{\text{Mon.}} d_M \leqslant \rho_c \tag{61}$$

one gets the bound

$$d_M \leqslant 10^{-21}\text{cm}^{-3} \tag{62}$$

This in turn means that the average flux of monopoles in the universe as a whole is bounded by

$$\Phi_{\text{univ.}} = \frac{1}{4\pi} d_M |v| \leqslant 2.4\times10^{-12}\beta\,\text{cm}^{-2}\text{sec}^{-1}\text{st}^{-1} \tag{63}$$

This is the limit plotted in Fig. 5.

For cosmological purposes it is better instead of quoting flux limits to quote values for the ratio

$$r = d_M/s \tag{64}$$

of the monopole density to the entropy density. If the universe's expansion is adiabatic, as it is believed, and if one can neglect possible monopole-antimonopole annihilation, then r is a constant in time. At present[36] $s \simeq 7\, n_\gamma \simeq 10^3\ \text{cm}^{-3}$ so that the limit (62) implies

$$r(\text{now}) \leqslant 10^{-24} \tag{65}$$

It is possible to derive a weaker bound on r, but at a much earlier stage in the universe's history, by requiring that at the time of nucleosynthesis (Temperature $T \simeq 1$ MeV) ρ_M did not exceed the energy density of the universe ρ. In this epoch the universe was radiation dominated and thus $\rho/s = 3/4$ T. Hence:

$$r(1\,\text{MeV}) = \frac{d_M}{s} = \frac{\rho_M}{M_{\text{Mon.}} s} \leqslant \frac{1}{M_{\text{Mon.}}} \frac{\rho}{s} = \frac{3}{4} \frac{T}{M_{\text{Mon.}}} \simeq 10^{-19} \tag{66}$$

The cosmological monopole problem[37,38] is that both r (now) and r(1 MeV) are very much smaller than what might be naively expected from a calculation of this number at the time of the GUT phase transition, when the monopoles are supposed to be created. The problem is exacerbated because, as Preskill has shown[37], the probability of monopole-antimonopole annihilation after production is very small.

In the standard cosmological scenario the universe cools in a symmetric phase up to temperatures of order $T_c \simeq 10^{15}$ GeV. At this temperature the GUT phase transition takes place and certain Higgs fields (H) acquire non zero vacuum expectation values, thereby breaking the original GUT symmetry down. Near $T \simeq T_c$ not all regions of space have $\langle H \rangle \neq 0$. Typically there are domains of size of order T_c^{-1} where $\langle H \rangle \neq 0$ among regions where $\langle H \rangle = 0$. Furthermore, the direction of $\langle H \rangle$ in the domains are uncorrelated so that monopoles, as topological knots formed between the domains, can appear. According to this very simple picture, the density of monopoles at formation is inversely proportional to the volume occupied by the domains. Since the typical correlation length ξ, for T near T_c, is $\xi \sim T_c^{-1}$, one expects

$$d_M(T_c) \simeq \frac{1}{\xi^3} \simeq T_c^3 \tag{67}$$

and, since $s \simeq T_c^3$ also, it follows that $r(T_c)$ should be of O(1).

Actually this discussion is a bit too rough[39]. The relevant temperature that sets the scale for the correlation length ξ is not the transition temperature, but the Ginzburg temperature ($T_G < T_c$) - the temperature at which the hopping from symmetric to asymmetric phase becomes prevalent. This has the effect of reducing r somewhat, and it has been estimated that[40]

$$r(T_G) \simeq 10^{-5} \tag{68}$$

This number is still very large compared to the bounds of Eqs. (65) and (66) and one might worry that the estimate above might unaccountably be a gross overestimate. Unfortunately, Einhorn[41] showed indirectly, that (68) is probably a reasonable estimate by obtaining a lower bound for $r(T_G)$ which is far in excess of the upper bounds (65) and (66) ! Einhorn argues, simply and convincingly, that the biggest that the domain correlation length can be is the size of a causally connected region at $T = T_G$

$$\xi_{Max} \simeq \frac{M_{Planck}}{T_{G^2}} \simeq \frac{10^4}{T_G} \tag{69}$$

This then gives the lower bound

$$r(T_G) \gtrsim 10^{-12} \tag{70}$$

In essence the lower bound (70) (or even worse the estimate (68)) coupled with the calculated paucity of $M\text{-}\bar{M}$ annihilations[37] during the universe's history, totally contradicts the limits of Eqs. (65) and (66). This is the cosmological monopole problem. Somewhere some assumptions made must have been wrong ! Theorists have, of course, suggested many (mostly unsatisfactory) solutions to the cosmological monopole problem[42]. Of these, I would like to briefly discuss what is perhaps the nicest among them, which is the inflationary universe scenario.

Guth[43] suggested that if the GUT phase transition instead of being of 2nd order was of 1st order, then many outstanding cosmological problems, like the horizon and flatness problems, could be solved. In a 1st order transition, after T_c, the universe remains in the false vacuum for a while. During this time the energy density of the vacuum dominates over the radiation energy density and the universe expands exponentially. The phase transition to the true vacuum proceeds eventually because, through quantum mechanical tunnelling, bubbles of the true vacuum form. In the "old inflationary" scenario of Guth the tunnelling rate was low and as a result the number of monopoles formed was reduced[44] (r is no longer just a function of the correlation length but also of the formation rate of bubbles). Unfortunately, since the tunnelling rate is low it also turns out that the bubbles of true vacuum never coalesce and the resulting universe is too inhomogenous to be acceptable cosmologically.

A "new inflationary" scenario has been proposed by Linde[45] and Albrecht and Steinhardt[46] which remedies this defect. Basically these authors achieve a high tunnelling rate for the bubbles because the Higgs potential chosen is very flat between the false and true vacuum minima. However, because of the flatness of the potential it turns out that, after the tunnelling, the bubble still has $\langle\phi\rangle \simeq 0$ and that the "roll over" to $\langle\phi\rangle = \langle\phi\rangle_{min}$ is slow. This means that, in effect, the bubble itself will expand exponentially and that all the universe is within one bubble ! Clearly in this scenario there are no monopoles produced as defects between domains. However, there may be some monopole pair production by thermal fluctuations, when the universe reheats after the phase transition is completed. Unfortunately, any estimate of r here depends exponentially on the reheating temperature, through the usual Boltzman factor e^{-M/T_r}, and is extremely sensitive. My conclusion is that the (new) inflationary scenario may indeed solve the cosmological monopole problem, in that no contradiction need exist with the limits (65) and (66). But this scenario is totally incapable of providing a reliable number for the flux of monopoles to be expected today.

CONCLUSIONS AND SUMMARY

I draw three main conclusions from this overview of the "state of the art" in monopole bounds:

1) Earth based ionization experiments plus geological considerations exclude sizable fluxes of monopoles for all ß, except for the region around $ß \simeq 10^{-4}$. The Cabrera event, if real, is limited to this narrow velocity range.

2) Astrophysical considerations give several order of magnitude lower monopole flux limits than that attributable to the Cabrera event (If catalysis of proton decay by monopoles proceeds at strong rates the discrepancy between these limits and the Cabrera's flux can be as much as ten orders of magnitude!). This argues for a local enhancement of the monopole flux. However, solar capture of galactic monopoles does not seem to produce a sufficiently large local flux to explain Cabrera's event.

3) Cosmological limits for the expected monopole flux are useless. Either one obtains embarassing large fluxes (the monopole problem) or one has scenarios where the number of monopole produced is essentially negligible.

These observations suggest that GUT monopoles, if they exist at all in our universe, may be beyond detection on earth. However, the only ways we shall ever be able to settle this issue is by:

(a) Making a concerted push with ionization experiments to get down to $ß \lesssim 10^{-4}$ and to obtain flux limits $\Phi \lesssim \Phi_{Parker}$

and/or

(b) Having Cabrera (or someone else) find another event !

ACKNOWLEDGEMENT

I would like to thank B. Cabrera and D. Groom for some helpful discussions on monopoles. I am also indepted to D. Groom for making available to me his compilation of ionization experiment bounds.

REFERENCES

1. P.A.M. Dirac, Proc. Roy. Soc. London A133, 60 (1931).
2. For a review, see for example P. Langacker, Phys. Rept. 72, 185 (1981).
3. G. 't Hooft, Nucl. Phys. B79, 276 (1974).
4. A.M. Polyakov, JETP Lett. 20, 194 (1974).
5. H. Georgi and S.L. Glashow, Phys. Rev. Lett. 32, 438 (1974).
6. For a review, see for example S. Coleman, Lectures at the 1981 Ettore Majorana School of Subnuclear Physics (HUTP-82/A032).
7. C. Dokos and T. Tomaras, Phys. Rev. D21, 2940 (1980).
8. W. Marciano, Proceedings of the 1982 Orbis Scientiae, Coral Gables, Florida.
9. V.A. Rubakov, JETP Lett. 33, 644 (1981); Nucl. Phys. B203 311 (1982).
10. C.C. Callan, Phys. Rev. D26, 2058 (1982) and Princeton preprints 1982.
11. F. Bais, J. Ellis, D. Nanopoulos and K. Olive, CERN preprint TH-3383.
12. F. Wilczek, Phys. Rev. Lett. 48, 1142 (1982).
13. T. Walsh, P. Weisz and T.T. Wu, DESY preprint (1983).
14. R.A. Carrigan, Nature 288, 348 (1980).
15. P.H. Eberhard, R.R. Ross, L.W. Alvarez and R.D. Watt, Phys. Rev. D4, 3260 (1971); R.R. Ross, P.H. Eberhard and L.W. Alvarez, Phys. Rev. D8, 698 (1973).
16. S. Dimopoulos, S.L. Glashow, E.M. Purcell and F. Wilczek, Nature 298, 824 (1982).
17. E.N. Parker, Ap. J. 160, 383 (1970); Cosmical Magnetic Fields (Clarendon Press, Oxford 1979).
18. B. Cabrera, Phys. Rev. Lett. 48, 1378 (1982).
19. B. Cabrera, these Proceedings.
20. S.P. Ahlen and K. Kinoshita, Phys. Rev. D26, 2347 (1982).
21. D. Ritson, SLAC-PUB 2950 (1982).
22. D.H. Groom, these Proceedings.
23. G. Lazarides, R. Shafi and T. Walsh, Phys. Lett. 100B, 21 (1981); P.M. McIntyre and R.C. Webb, Texas ASM preprint (1982).
24. S. Drell, M. Kroll, M. Mueller, S. Parke and M. Rudermann, Phys. Rev. Lett. 50, 644 (1983).
25. G. Giacomelli, Proceedings of the Magnetic Monopole Workshop, Wingspread, Wisconsin 1982.
26. T. Walsh, Proceedings of the XXI International Conference on High Energy Physics, Paris 1982.
27. M. Turner, Chicago preprint 1982 (EFI 82-55).
28. E.N. Parker ref. 17; S. Bludman and M. Ruderman, Phys. Rev. Lett. 36, 840 (1976); G. Lazarides, R. Shafi and T. Walsh, ref. 23.
29. K. Freeze and M. Turner, Phys Lett. B123, 293 (1983).
30. P. Eberhard, LBL preprint 1982; J. Arons and R.D. Blauford, ITP-UCSB preprint 1982; M.S. Turner, E.N. Parker and T.J. Bogdan, Phys. Rev. D26, 1296 (1982); E. Salpeter, S. Shapiro and I. Wasserman, Phys. Rev. Lett. 49, 1114 (1982).

31. M. Turner, E.N. Parker and T.J. Bogdan, ref. 30.
32. D. Ritson, SLAC-PUB 2977 (1982).
33. Y. Raphaeli and M. Turner, Phys. Lett.
34. F. Bais, J. Ellis, D. Nanopoulos and K. Olive, ref. 11; T. Walsh, ref. 26. E.W. Kolb, S.A. Colgate and J. Harvey, Phys. Rev. Lett. 49, 1373 (1982); S. Dimopoulos, J. Preskill and F. Wilczek, Phys. Lett. B119, 320 (1982).
35. J. Harvey, Princeton preprint 1982.
36. See for example, S. Weinberg, Gravitation and Cosmology (John Wiley and Sons, Inc. New York 1972).
37. J. Preskill, Phys. Rev. Lett. 43, 1365 (1979).
38. Y. Zeldovich and M. Kholopov, Phys. Lett. 79B, 239 (1979).
39. For more detailed discussion see T. Kibble, Proceedings of the Monopoles in Quantum Field Theory Conference, Trieste 1981; N. Straumann, Lectures at the SIN Spring School ZVOZ, Switzerland 1982.
40. T. Kibble, J. Phys. A9, 1387 (1976); M. Einhorn, D.L. Stein and D. Toussaint, Phys. Rev. D21, 3295 (1980).
41. M. Einhorn, Proceedings of the Europhysics Conference on Unification of the Fundamental Interactions, Erice 1980.
42. For a review see G. Lazarides, in Proceedings of the Magnetic Monopole Workshop, Wingspread, Wisconsin 1982.
43. A. Guth, Phys. Rev. D23, 347 (1981).
44. A. Guth and S.H. Tye, Phys. Rev. Lett. 44, 631 (1980); A. Guth and E. Weinberg, Phys. Rev. D23, 876 (1981); G.B. Cook and K.T. Mohanthappa, Phys. Rev. D23, 1321 (1981); M. Einhorn and K. Sato, Nucl. Phys. B180, 395 (1981).
45. A.D. Linde, Phys. Lett. 108B, 389 (1982).
46. A. Albrecht and P. Steinhardt, Phys. Rev. Lett. 48, 1220 (1982).

REVIEW OF PROTON DECAY EXPERIMENTS

M. Rollier

Dipartimento di Fisica and Sezione INFN

Milano, Italy

ABSTRACT

The latest results on the nucleon decay experiments are discussed and compared with the theoretical predictions. Up to now no clear indication for the existence of the proton decay has been found. The predicted lifetime in the standard SU_5 model is now close to the experimental limit of about 2 10^{31} years.

INTRODUCTION

In the last two years a big experimental effort has been dedicated to experiments on the nucleon stability. This effort has been stimulated by the fact that the "Standard Model" predicts that the nucleons should decay with a lifetime accessible to experiments.

All the recent experimental results mainly from the e^+e^- interactions at PETRA and SLAC and also the $W^{\pm}$ discovery at CERN are very strong supports for the so called "Standard Model", but, just for this reason, the proton decay remains one of the more important tests for the model.

Three new experiments (KGF, NUSEX, IMB) have already presented limits on the proton lifetime. In this review I will report the new data of these experiments as they have been presented at the recent ICOMAN 83 Conference in Frascati[1].

THE THEORETICAL PREDICTIONS

In the general framework of the so called "grand unified models" many different predictions can be obtained depending on the different assumptions. In Table I[1] some predictions are presented for

TABLE I [1] Class of theoretical models predicting baryon number violation. F refers to the number of light fermions families, η_H to the number of light Higgs doublets and M_H to the mass of superheavy Higgs particles.

Class	τ_p(yr)	Modes
Minimal SU_5 ($F=3$, $n_H=1$, $M_H=M_X$)	$3.2\ 10^{29\pm2}$ for $\Lambda = 160^{+100}_{-80}$	$\Delta(B-L)=0$ $\Delta S/\Delta B=1,0$ (mainly $p\to e^+\pi^0$)
($F\neq3$, $n_H\neq1$, $M_H\neq M_X$	Same $10^{\pm2}$	Same
Superheavy Higgs predicted decays	No prediction	μ,K enhanced
3 mass scales ($M_{X1}, M_{X2} >>> M_W$)	No prediction	Same
Susy GUTs	$10^{30\pm2}$	$\mu K\bar{\nu}$ ($p\to K^+\bar{\nu}$, $K^0\mu^+$)
Low mass scales e.g. (moderate mass) Higgs-mediated composite fermions Pati-Salam	No prediction	Selection rules depend on mass scale

different classes of models.

I just want to point out that for minimal SU_5 it is difficult to bring up the proton lifetime to a limit higher than $3\ 10^{31}$ years which roughly is the actual experimental limit.

Also for the possible decay modes and branching ratios the predictions are strongly dependent from the assumptions in the different models.

In Table I we can see that the ($p\to e^+\pi^0$) and ($n\to e^+\pi^-$) modes are dominant in minimal SU_5 but in the SUSY models and also in other GUT's predictions the ($p\to K^+\bar{\nu}$) and ($p\to K^0\mu^+$) decays can be important.

THE KOLAR GOLD FIELD EXPERIMENT

Results from the experiment carried out by the Indian and Japanese Collaboration of Tata Institute and Osaka University in the Kolar Gold Field mine in India have been already published[2]. I will only summarize the main features of the experiment: the detector is a calorimeter mode with iron plates 1.2 cm thick interleaved with 1500 proportional tubes of (10 x 10 cm^2) section and 6 meters long. The total mass is 140 tons, but the fiducial mass for confined events is 60 tons. The background due to cosmic rays is low and the number of atmospheric muons crossing the detector is only ∿2 per day.

The experiment has been running now for 1.72 years and only 3 completely confined events have been found which can be interpreted as proton decays (see Fig. 1):

Event 587 can be $p \to e^+ \pi^0$ with visible energy $E \simeq 1 \pm 0.4$ GeV
Event 869 can be $p \to \bar{\nu} \pi^+$ with visible energy $E \simeq 0.435 \pm 0.2$
Event 877 can be $p \to K^0 \mu^+$ with visible energy $E \simeq 0.950$

The energy can be calculated from the signals in the proportional tubes.

Clearly it is difficult for these three events to completely rule out the hypothesis that they are due to atmospheric ν background. In any case the KGF Collaboration, assuming these events as proton decays, and assuming a detection probability in their detector of 50 %, give a value for the nucleon lifetime divided by the branching ratio:

$$\tau/\text{B.R.} \simeq 1 \times 10^{31} \text{ years}$$

THE NUSEX EXPERIMENT

Also the first results from the experiment carried out by a Frascati, Milano, Torino, CERN Collaboration in the Mont Blanc Tunnel Laboratory have been already published[3].

The detector is a calorimeter and in Table II the main parameters of the experiment are summarized.

In Fig.2 a general sketch of the apparatus with the details of the detectors and the read-out system is shown.

The experiment has been running from June 1982 with high efficiency (∿3700 hours of effective operation) with a fiducial mass of 120 tons that correspond to 64 tons x year and $3.8\ 10^{31}$ nucleons x year. The measured muon background is of the order to 1 muon x hour as expected. In total 4350 muons crossing the detector have been seen of which 38 are stopping (see Fig. 3).

Also 52 "muons bundles" have been seen with two or more parallel muons at the same time (see Fig. 3). Five totally confined events (Fig. 4) have been found (TABLE III) of which 4 are easily interpreted as neutrino background events. From the calculated fluxes of atmospheric neutrinos about 10 ν events per year are expected in the detector, corresponding to ∿ 4 events in the running time of the experiment.

However event 19-503 shown in Fig. 4 looks like a 3 prong event apparently balanced in momentum. It is difficult to consider it as a ν events but, if we take the vertex in A, it is possible to interpret it as a proton decay giving a stopping positive muon (track AB) and a decaying $K^0 \to \pi^+ \pi^-$ at the same point A (tracks AC and AD). All the kinematical quantities are compatible with the expectations for a proton decaying: $p \to K^0 \mu^+$ (Table IV).

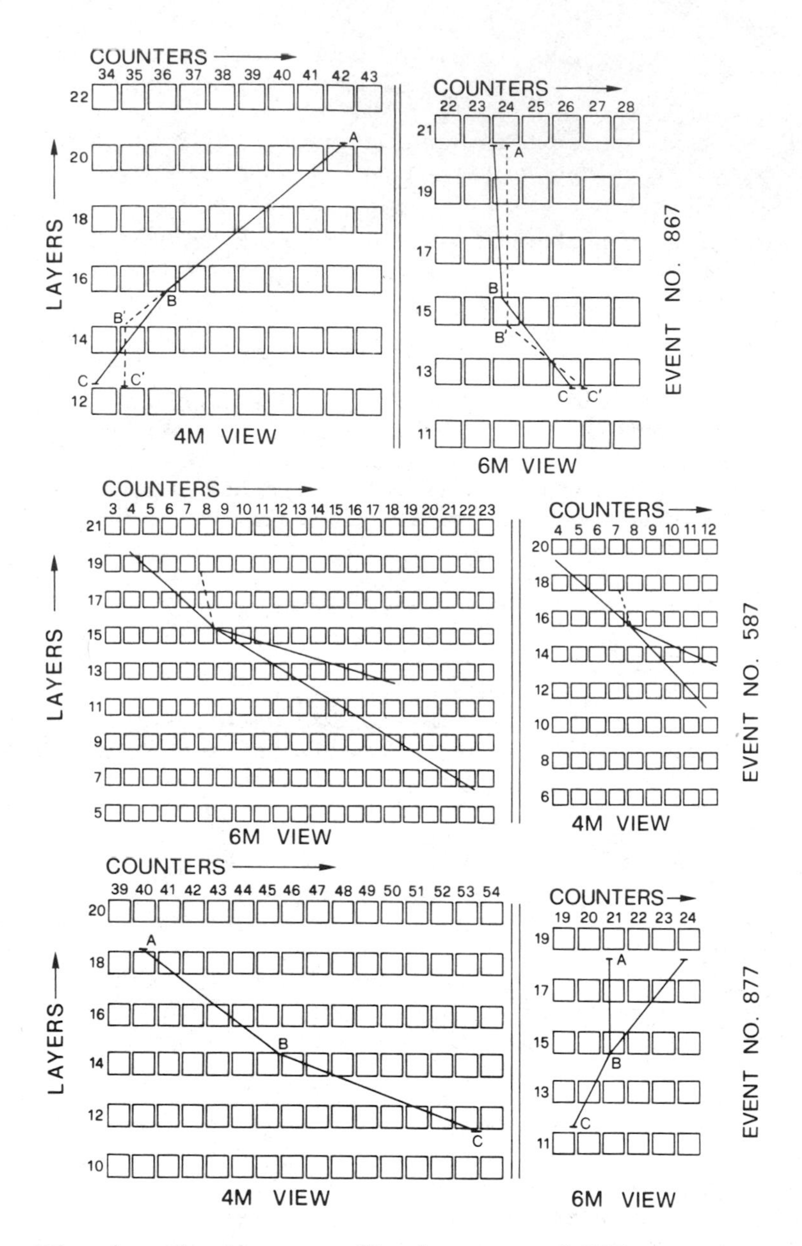

Fig. 1. The three confined events of KGF experiment.

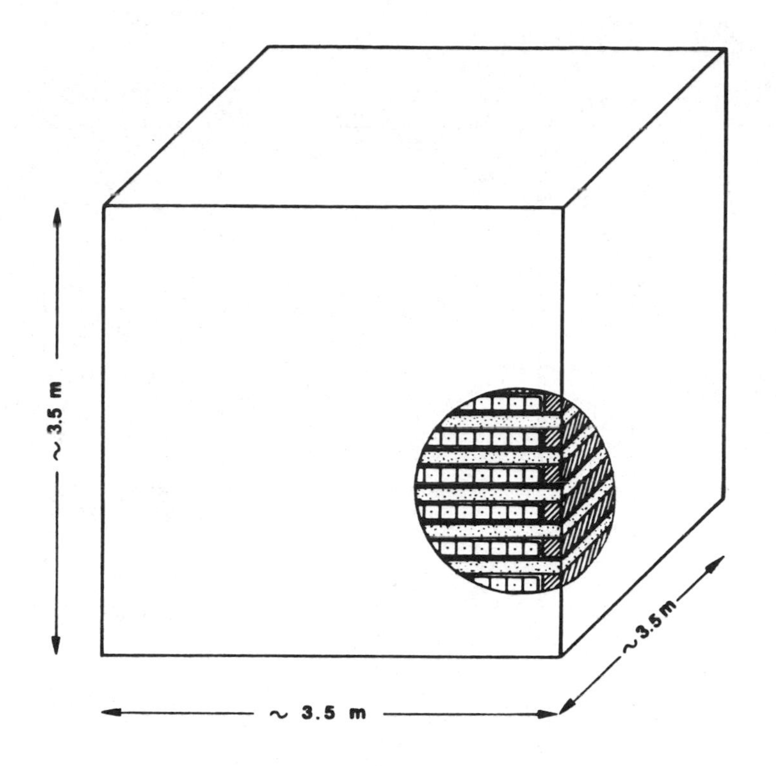

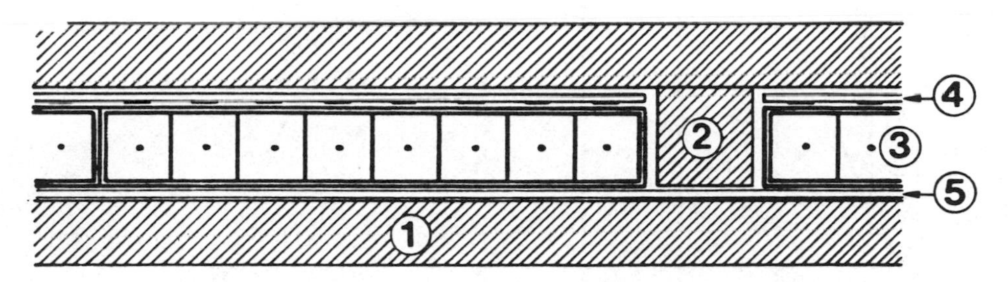

Fig. 2. The Nusex detector. 1) iron plates of 1.5 cm.
2) iron spacer
3) streamer tubes
4) - 5) strips for the read-out

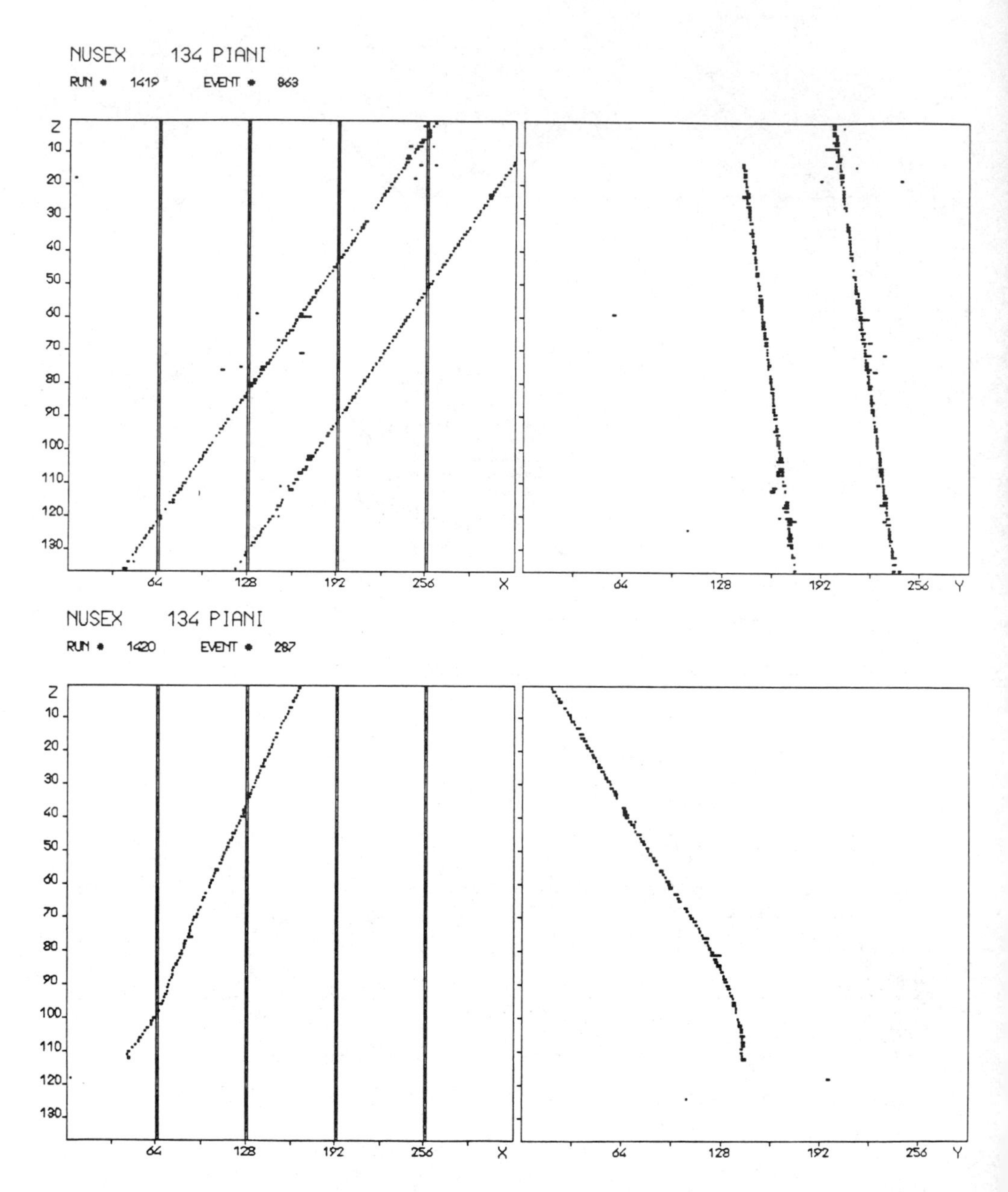

Fig. 3. A muon "bundle" and a stopping muon in the Nusex experiment.

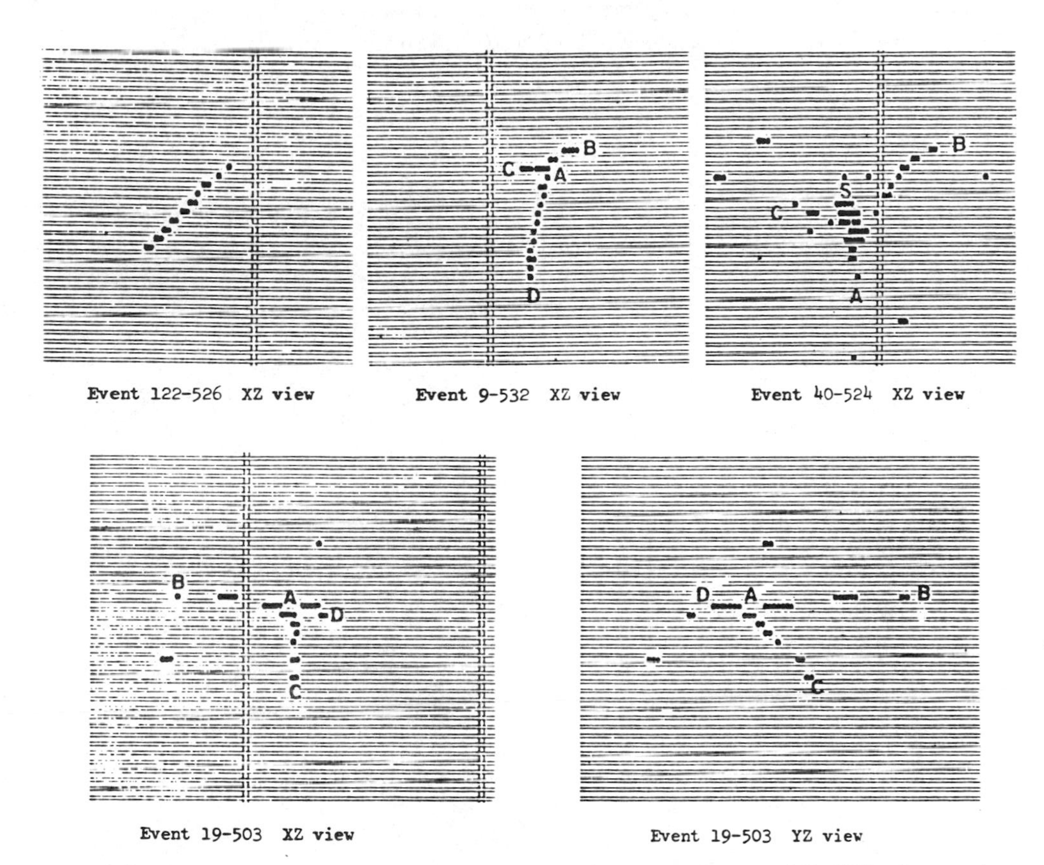

Fig. 4. Four events from Nusex experiment. Event 19-503 shown in both projections can be a proton decaying in a μ^+ and a K^0_s

TABLE II

NUSEX DETECTOR

Size:	3.5 x 3.5 x 3.5 m^3
Site:	Mont Blanc Tunnel (Garage no.17)
Background expected:	1μ crossing the detector per hour
	∿ 12 ν events E_ν>300 MeV per year in the detector
Cover:	∿ 5000 m w.e.
Total mass:	150 tons
Average density:	3.6 gr/cm^3
Mean radiation length:	4.5 cm
Target:	136 iron planes 1 cm thick interleaved with detectors
Detectors:	Plastic streamer tubes with 1. x 1. cm^2 section and 3.5 meters long.
Read out:	on two sets of orthogonal strips
Trigger:	obtained by the "OR" of each plane requiring as a minimum 4 consecutive planes hitted

TABLE III

no. event	Topology	Visible energy	Interpretation
19 - 503	3 prongs ?	1.0 ± 0.2 GeV	$\nu_\mu \to \mu\pi$?proton decay?
40 - 524	3 prongs	1.5 ± 0.4 GeV	$\nu_e \to e\pi\pi$
122- 526	1 prong	0.4 ± 0.1 GeV	$\nu_\mu \to \mu$
9 - 532	1 prong	0.33± 0.15	$\nu_\mu \to \mu$
133-1062	1 prong	1.5 ± 0.3	$\nu_\mu \to \mu$

TABLE IV

	Expected	Measured
Visible energy	0.94 GeV	1. ± 0.2 GeV
μ^+ momentum	0.33 GeV	0.38 ± 0.15 GeV
K^0 mass	0.498 GeV	0.55 ± 0.08 GeV
K^0 momentum	0.33 GeV	0.34 ± 0.10 GeV
Unbalanced momentum (Fermi motion)	0 ↔ 0.3 GeV	0.4 ± 0.2 GeV

The same event can also be interpreted as a neutrino event in two ways:

i) as a 3 prongs ν event with vertex at point A. In this case the kinematical fit is bad, the total visible energy $E_\nu = 1.0 \pm 0.2$ is not compatible with the total momentum $P_\nu = 0.4 \pm 0.2$. From the analysis of 400 ν events in a test run done at CERN with a reduced module of the same detector only 1 event has been found with 1 track with an angle with the muon direction larger than $140°$, so that the probability that the event be neutrino induced is < 1 % at 90 % C.L.

ii) as a neutrino interacting at point D and giving two prongs ($\nu_\mu \to \mu\pi$). In this case the pion track DA makes a large angle scattering at point A. Again from the 400 ν events of the test run, as <u>no</u> two prong event has been found with such a small opening angle (less than $15°$), also for this topology at 90 % C.L. less than 0.04 ν events are expected.

The conclusion is that clearly it is difficult to explain this event with the ν background.

Summarizing the NUSEX results:

- 4 neutrino interactions have been found in agreement with expectation
- 1 event can be a proton decay in the $p \to K^0\mu^+$ channel.

With 80 % detection probability one event correspond to a lifetime for the proton decay in the $K^0\mu^+$ mode of τ/B.R. $\simeq 1.0 - 2.8\ 10^{31}$ years.

THE IRVINE-MICHIGAN-BROOKHAVEN EXPERIMENT

First results on this experiment have been presented in January 1983 at the ICOMAN Conference in Frascati . The experiment uses a completely different technique: the detector is a very large cubic tank (21 x 21 x 21 m^3) filled with purified water in which the Cerenkov light emitted by low mass particles is seen by a grating of phototubes put on the 6 sides of the cube (Fig. 5). The PM tubes (5 diameter) are displayed in a 16 x 16 array on each face at a distance of 1.2m one from the other. The total volume corresponds to $\sim$ 8 Ktons of water but the fiducial volume (16 x 16 x 16 m^3) has been limited to 3.3K tons corresponding to $2\ 10^{33}$ nucleons.

The experiment has been carried out in a salt mine in Ohio at a depth of 1600 m w.e. where the background of muons is still 500 muons/m^2 per day. The apparatus is operational since December 1981 but the first results presented refer only to 80 days of effective run. The event analysis is rather difficult since the number of triggers is very high (240.000 per day). The selection has been carried out separately by different methods but ending with about the same sample.

Out of $18.6\ 10^6$ triggers only 72 events have been selected satisfying the following criteria:

- ≥ 1 track with an origin in the fiducial volume
- Cerenkov energy in these limits: $300\ MeV < E_C < 1600$ MeV.

These events can be reasonably explained as neutrino background: the vertex position and the incoming neutrino directions are isotro-

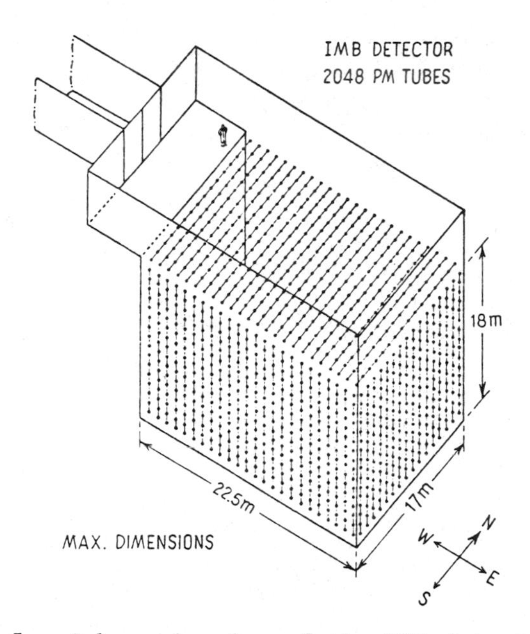

Fig. 5. Schematic view of the IMB detector.

TABLE V

Event number	Number of hits	Cerenkov energy	Track energy E_1	E_2	Opening angle (degree)	Muon decay signature
151-35037	188	1150	675	475	135 ± 7	1
125 7797	166	1700	900	300	140 ± 10	1
247-12891	225	1170	750	420	120	0
388-19376	340	1770	1080	690	115 ± 15	0

pically distributed (Fig.6a) and thc energy distribution (Fig.6b) is in agreement with predictions from atmospheric neutrinos.

To select the possible candidate for a nucleon decay in the channel $p \to e^+\pi^0$ only events with two tracks having an opening angle larger than 110° have been considered: only 4 events are left (see TABLE V).

In the last column 1 means that one track has a signature for a muon decay recognized by a Cerenkov ring delayed by few microseconds due to the electron.

The two events having the muon decay signature are excluded and also the last two events have been ruled out for the opening angle, the total energy and the energy distribution between the two tracks.

In conclusion in the first run of 80 days no candidate for a proton decay in the $e^+\pi^0$ channel has been found. Taking into account the absoption probability for pions in a nucleus (32 %) and the detection efficiency of 90 %, the limit on the proton lifetime for this channel is:

$$\tau/\text{B.R.} > 6.5\ 10^{31}\ \text{years}$$

If only the protons of hydrogen in water are considered a lower limit only for "free" protons lifetime can be given:

$$\tau/\text{B.R. on free protons} > 1.9\ 10^{31}\ \text{years}$$

CONCLUDING REMARKS

If we compare the results from these three experiments (KGF, NUSEX, IMB) we can see that in thc $e^+\pi^0$ channel the IMB experiment with only 80 running days gives the higher limit. If we consider the $e^+\pi^0$ candidate from the KGF Collaboration as a good event clearly there is disagreement with the IMB result.

On the other side the μ^+K^0 event from the NUSEX Collaboration looks as a "possible candidate for a proton decay". However a single event can only represent an indication for proton decay.

At present we can only say that no clear definite evidence for the proton decay have been seen. Probably it will be necessary to improve the detectors in order to have a better understanding of

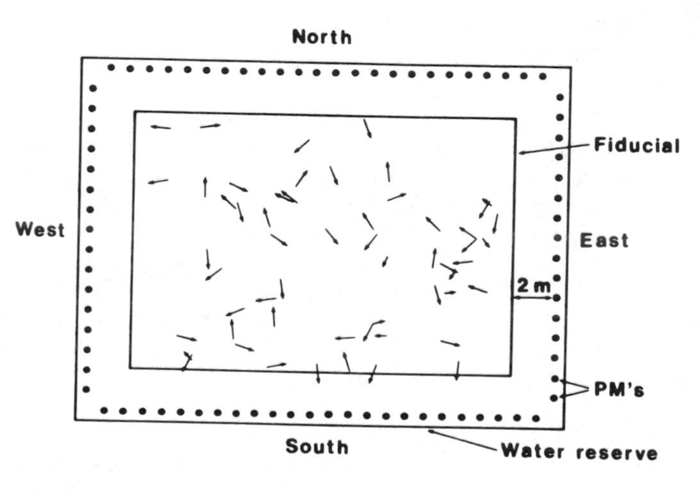

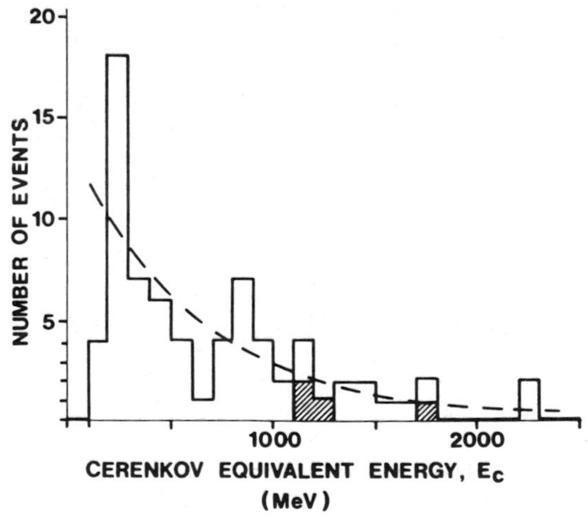

Fig. 6. IMB experiment. a) spatial distribution of the 72 events selected.
b) energy distribution.

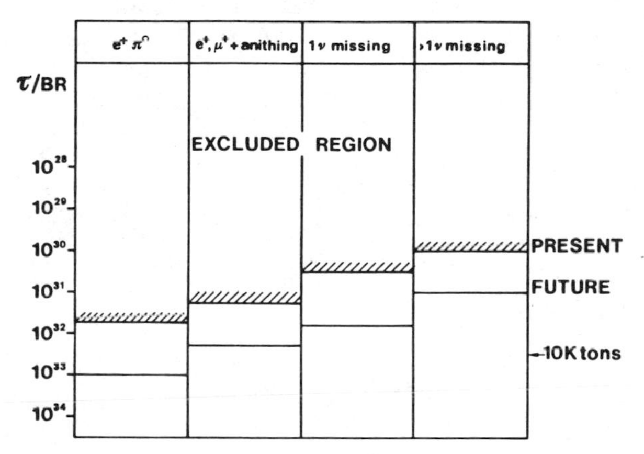

Fig. 7. Experimental limit on the nucleon lifetime in present and future experiments for different decay modes.

all background events. It is also important to have different techniques (calorimeters and water Cerenkov) in order to compare the results.

FUTURE EXPERIMENTS

Very soon new results will improve the present limit on the proton lifetime. The IMB experiment can easily increase the statistic and set a new limit in the $e^+\pi^0$ channel and probably also will try to look for other decay modes.

Other experiments will also start taking data in a short time:

- a calorimeter detector, with a better granularity than Nusex of $\sim$ 1 Kton will be ready next year in the Frejus Tunnel.
- a water Cerenkov of 3.5 Ktons of water, with improved PM tubes will start taking data next month in Kamioka mine in Japan.

Also other detectors of very large masses have been proposed and will be built in next years (SOUDAN II, GRAN SASSO, ...).

One question can be raised: what is the maximum limit on the proton lifetime that any experiment can achieve? The limit is due to the ν-background which cannot be avoided. In Fig. 7 the result of a simple calculation[1] is shown on the maximum limit for different possible decay modes.

REFERENCES

1. D. Ayres, Proceedings of the 1982 Summer Workshop on proton decay experiments, June 7-1 1982 A.N.L.
2. M.R. Krishnaswamy, M.G.K. Menon, N.K. Mondal, V.S. Narasimham, B.V. Sreekantan, Phys. Lett. 115B, 349 (1982);
   S. Miyake, Communication at the ICOMAN Conference Frascati 1983.
3. G. Battistoni, E. Bellotti, G. Bologna, P. Campana, C. Castagnoli, V. Chiarella, D.C. Cundy, B. D'Ettorre Piazzoli, E. Fiorini, E. Iarocci, G. Mannocchi, G.P. Murtas, P. Negri, G. Nicoletti, P. Picchi, M. Price, A. Pullia, S. Ragazzi, M. Rollier, O. Saavedra, L. Trasatti, L. Zanotti, Phys. Lett. 118B, 461 (1982)
   G. Battistoni, Communication at the ICOMAN Conference, Frascati 1983.
4. L.R. Sulak, J.C. Van der Velde, Communication at the ICOMAN Conference, Frascati 1983.

THEORETICAL ASPECTS OF PROTON DECAY

F. J. Ynduráin

Departamento de Física Teórica
Universidad Autónoma de Madrid
Cantoblanco, Madrid-34, SPAIN

ABSTRACT

An informal discussion is presented on what the new existing bounds for the proton lifetime implies on minimal grand unified theories. It is argued that they probably finish minimal SU(5). Ways out (via extra Higgses on extensions to supersymmetric SU(5)) are then briefly discussed. A few words are also said about other p-decay producing mechanisms: black holes, instantons and monopoles.

As you have heard in the lecture of Rollier in this symposium, the IMB experiment implies a negative result on the mode $p \to e^+\pi^\circ$ for proton decay:

$$\tau_p \to e^+\pi^\circ > 6 \times 10^{31} \text{ yr.} \qquad (1)$$

From the Mont Blanc experiment we have a bound on all decays producing e^+ or μ^+:

$$\tau_p \to e^+/\mu^+ + \text{mesons} \gtrsim 10^{31} \text{ yr.} \qquad (2)$$

We have the possibility for equality in (2) because there is a candidate event for $p \to \mu^+K^\circ$ at a level of $\sim 10^{31}$ yr. In this seminar I will discuss some of the theoretical implications of proton decay, especially in connection with the bounds (1), (2). The theoretical framework will be that of minimal GUTs (grand unified theories); and I will thus start by explaining what I mean by this description--minimal GUTs.

I think that the evidence that the standard SU(3) x SU(2) x U(1) gauge theory describes low energy (< 1 TeV) phenomenology is overwhelming. It is perhaps hard to appreciate nowadays the fact that the Weinberg-Salam SU(2) x U(1) piece is only one among many possible models: the only property that singles it out is that of its minimality in the sense that it has only the observed fermions, and the symmetry is broken by a minimal Higgs system. Since this has worked once, we may try it again and see how far we can go with the minimal structure that contains the known families of fermions and which unifies SU(3) x SU(2) x U(1) inside a simple group. Of course the answer is the SU(5) model of Georgi and Glashow. One may note that nobody has seen yet the t quark, and choose instead E_6; or we may want to incorporate right handed neutrinos, and select SO(10). I will concentrate on SU(5); but, to a very large extent, the conclusions will be valid for the other groups.

If we want to fit the known fermions into representations of SU(5) we are forced to put quarks, and antiquarks, in the same multiplet. For example, we have the three decuplets

$$\begin{pmatrix} \bar{e} \\ u_i \\ d_i \\ \bar{u}_i \end{pmatrix}, \quad \begin{pmatrix} \bar{\mu} \\ c_i \\ s_i \\ \bar{c}_i \end{pmatrix}, \quad \begin{pmatrix} \bar{\tau} \\ t_i \\ b_i \\ \bar{t}_i \end{pmatrix}, \qquad i = 1, 2, 3. \tag{3}$$

Therefore bayon (N_B) and lepton (N_L) numbers are not separately conserved and we may have proton decay. Before embarking upon it, however, let me list the successes and failures of SU(5) with the minimal system of Higgses necessary to break the symmetry in two stages: at the grand unification energy $M_X \simeq 2 \times 10^{14}$ GeV, and at the Weinberg-Salam scale, $G_F^{-1/2} \simeq 3 \times 10^2$ GeV. The most startling success is doubtlessly the prediction $\sin^2\theta_W \simeq 0.22$. This is, to a large extent, independent of the number of families or the Higgs structure. With a minimal Higgs structure and just three or four generations we also have $m_b \simeq 3m_\tau$; but the same assumptions imply the disastrous prediction

$$\frac{m_\mu}{m_e} = \frac{m_s}{m_d} \tag{4}$$

SU(5) with three or more families features CP violation and N_B non-conservation: hence the necessary ingredients to explain the generation of the observed baryon assymmetry of our universe. However, the actual calculations fail to reproduce the observed baryon to photon ratio N_B/N_γ, and, to get it, we have to introduce extra Higgses. Finally, we have the serious problem of the hierarchy: the GUT breaking would, via radiative corrections, seep into the Weinberg-

Salam breaking giving $G_F^{1/2} \sim 10^{11}$ GeV unless some fantastic cancellations take place.

How about the predictions for proton decay? There are two types of calculations for this process: exclusive or inclusive ones. In the exclusive calculations one considers a pair $q_1 q_2$ of quarks in the proton which fuse to give $q_3 e^{\mp}$ (for example). The effective Lagrangian is

$$\mathcal{L}_{p \to e^+} + \ldots = \frac{g_{GUT}^2}{2M_x^2} \sum \varepsilon_{ijk} \bar{u}_i^c \gamma_\mu \frac{1-\gamma_5}{2} u_j \bar{e}^+ \gamma^\mu \frac{1-\gamma_5}{2} d_k$$

Concentrating upon $p \to e^+ \pi^\circ$ we then find

$$\tau_{p \to e^+ \pi^\circ} = \frac{\pi M_x^4}{6 g_{GUT}^4 E a(f) |\phi(o)|^2} (BR)^{-1} \tag{5}$$

Here M_x is the grand unified mass, g_{GUT} the coupling at GUT energies, $g_{GUT}^2/4\pi = 0.022$, E the energy release, $E \sim 0.8$ GeV, a(f) an enhancement factor of the order of 16 (if we normalize ϕ at 1 GeV), $\phi(o)$ the wave function for two quarks at the same point, and (BR) the branching ratio. $|\phi(0)|^2$ may vary, in different models, by a factor 2 or 1/2, and (BR) is of some 35 to 70%. To obtain M_X we require only α -- a very well known quantity -- and Λ_{QCD}. For the last I will take the world average, with reasonable error estimates: $\Lambda_{QCD} \simeq 140^{+80}_{-50}$ MeV. g_{GUT} is fairly stable. What does this give for the rate? One has to be careful with errors, as some of the quantities in (5) are strongly correlated. Being conservative, we would write

$$\tau_{p \to e^+ \pi^\circ} \simeq 1.5 \times 10^{30 \pm 1.5} \text{ yr.} \tag{6}$$

This is not yet fully incompatible with (1); but there is more. One can try and calculate not a particular channel, but $p \to e^+ + \text{all}$, and take into account that most protons are bound inside nuclei. This can be done in a variety of ways; all the calculations agree within a factor of 2 and we find

$$\tau_{p \to e^+ + \text{all}} \simeq 2 \times 10^{29 \pm 2} \text{ yr.} \tag{7}$$

Note the extra error. This again is barely compatible with (2); what is more, all calculations agree that $\tau_{p \to e^+} / \tau_{p \to \mu^+} < 2$ to 10, so we cannot believe the event $p \to \mu^+ + K$ together with the absence of $p \to e^+ + \pi$, within the model. It may be that p decay is just lurking around the corner, but I will now take the attitude that (6), (7) are really incompatible with (1), (2) and see what would have to be done to the model to save the day.

We have seen before that a minimal Higgs structure has some unwanted consequences: relation (4) and too low a value for N_B/N_γ, not to mention the problem of the CP violating $\Theta\tilde{G}G$ term and eventual solutions involving axions. However, once one gives up minimality of Higgs systems there are few limits to what we can do: I will exemplify this with an amusing (if artificial) possibility.

We know that the (u,d); (c,s); (t,b) go together, up to small Cabibbo-like mixings, for SU(2) x U(1) interactions: but there is no reason why the same will hold true for interactions with proton-decay producing vector mesons. For example we could modify (3) to

$$\begin{pmatrix} \bar{e} \\ u_i \\ d_i \\ \bar{c}_i \end{pmatrix}, \quad \begin{pmatrix} \bar{\mu} \\ c_i \\ s_i \\ u_i \end{pmatrix},$$

and leave the third family as it stands. This would suppress decays $p \to e^+\pi$, but would allow $p \to \mu^+ K$ at a rate compatible with the Mont Blanc event. By replacing $\bar{u}_i$ in (3) with $\bar{t}_i$ one may rotate p decay entirely away. This possibility, discussed first by C. Jarlskog, requires weird Higgs systems: but, since we do not understand why $\theta_{cabibbo} \sim 0$, there is no reason why the new angles could not be $\theta_x \sim \pi/2$.

Another possibility is that the family structure breaks down for superheavy fermions and we get representations other than 5 and 10: or we could have partial unifications well below 10^{14} GeV. The range of possibilities is infinite; but there is one particularly interesting, which allows us to retain the SU(5) and family structure. This is to extend the symmetry to a supersymmetry (SUSY). This presents several advantages. First of all it may eventually allow us to include gravity in the unification scheme. A less speculative point in favour of SUSY ideas is that it leads naturally to a possible way out of the hierarchy problem and, in some versions, to dispense with one mass scale which would appear via radiative corrections.

One can repeat the calculation of, say, $\tau_{p \to e^+\pi^\circ}$ in SUSY - SU(5). The changes are as follows: M_x increases to 10^{15} GeV because, as there are more fermions and scalars, unification is slower. $g^2_{GUT} / 4\pi$ increases to 0.04; and a(f) decreases to ~ 6. The rest is as before: the net result is that the proton lifetime increases to $\tau_p^{SUSY} \sim 10^{35}$ yr., well beyond the experimental reach. However, in SUSY-SU(5) the proton may decay via other mechanisms. For example, you may have u + d going to gluino + shiggs, with exchange of squark and, with another exchange of squark, materialization into $\bar{\nu}_\mu \bar{s}$ or $\bar{\nu}_e \bar{d}$. Because some couplings are Higgs couplings proportional to the mass, the first mode is much favoured. The effective Lagrangian is

$$\mathcal{L}^{SUSY-5}_{eff} \sim \frac{f_u f_s m_\lambda \alpha_s}{2\pi\, m_\psi m_{\Psi_H}} \, uds\, \nu_\mu \,,$$

where $f_{u,s}$ are the Higgs couplings, λ the gluino, ψ the squark and Ψ_H the shiggs. α_s is the QCD coupling constant. As will be seen, the predictions will be affected by errors much larger than in the previous cases: all one can say is that τ_p may vary between 10^{26} and 10^{35} yr. A peculiarity of this mechanism (technically known as "decay via dimension five operators" because the decay-inducing interaction is $u\,\nu_\mu \phi_d \phi_s$) is that it favours decays into strange particles. This comes about because the Higgs coupling f_a for flavour a is proportional to the a-quark mass. Hence,

$$\frac{f_s}{f_d} = \frac{m_s}{m_d} \sim 20:$$

decay into strange particles is favoured. In the simplest SUSY models the preferred decay is $p \to K^+ \nu_\mu$ but you may cook up things to get $p \to K^\circ \mu^+$. The contribution of Nanopoulos to these proceedings contains a discussion of such possibilities, and I will send you there for details.

This is all a bit depressing. The (reasonably) safe prediction of the minimal model is getting in trouble with the experiment: if we enlarge the model we get almost no prediction at all.

I will now finish this brief survey with a quick review for other possible mechanisms for p decay. First we have decay via virtual black holes first discussed by Zeldovish. The rate may be easily obtained from e.g. eq. (5) replacing M_x by M_{Planck} and dropping g^4_{GUT}. One gets $\tau_p \sim 10^{45}$ yr. I will discuss in somewhat more detail decays via instantons since they are very similar to monopole-catalyzed decays which may have phenomenological relevance.

Let us consider a simple model with a doublet of left-handed quarks $(q_1\,, q_2)_L$ and a triplet of massless $\vec{W}_\mu$. The quark number operator is

$$N_q = \sum_i \int d\vec{x}\; q^+_{iL}(x)\, q_{iL}(x) -$$

$$= \sum_i \int d\vec{x}\; \bar{q}_i(x) \gamma^\circ \frac{1-\gamma_5}{2} q_i(x) \quad .$$

One can write its time derivative as

$$\partial_o N_q = \sum_i \int d\vec{x}\ \partial_\mu \sum_i \bar{q}_i(x)\gamma^\mu \frac{1-\gamma_5}{2} q_i(x)$$

$$+ \sum_i \int d\vec{x}\ \vec{\partial}\ \bar{q}_i(x)\vec{\gamma} \frac{1-\gamma_5}{2} q_i(x) \ .$$

the term $\partial_\mu\ \bar{q}\gamma^\mu q$ drops because the vector current is conserved; the $\vec{\partial}\ \bar{q}\ \vec{\gamma}(1-\gamma_5)/2\ q$ because it is a divergence. The net change in quark number is then

$$\Delta N_q \equiv N_q\ (+\infty) - N_q\ (-\infty) = \int_{-\infty}^{+\infty} dx^\circ \partial_o\ N_q$$

$$= -\ \frac{1}{2} \int d^4x\ \partial_\mu \sum_i \bar{q}_i\ \gamma^\mu \gamma_5\ q_i.$$

the divergence of the axial current is given by the Adler-Bell-Jackiw anomaly so

$$\Delta N_q = -\ \frac{g^2}{32\pi^2} \int d^4x\ \tilde{F}_W^{\ \mu\nu}\ F_{W\mu\nu}$$

Now, as noted by 't Hooft, the point is that there exist <u>classical</u> field configurations (instantons) for which the r.h.s. is nonzero, thus leading to violation of quark number. In realistic models this is compensated by a change in the number of leptons; in SU(2) x U(1) one still gets $N_B - N_L$ = conserved.

To violate baryon number you have to create the instanton. The action for an instanton is

$$A = -\ \frac{1}{4} \int d^4x\ \tilde{F}_W^{\mu\nu}\ F_{W\mu\nu};$$

so you pay a penalty of exp(-A) (tunnelling effect: the instanton connects different vacua) which is of the order of $\exp(-\ 8\pi/g_W^2)$. This is very small and, in the standard model, leads to $\tau_p \sim 10^{80}$yr.

A 't Hooft-Polyakov monopole is essentially an SU(5) instanton: thus it will produce proton decay. However, if a monopole comes near a proton the monopole is already there: as noticed by Callan and Rubakov you do not have to pay the exponential penalty for creating it. The monopole does not get destroyed in the process of disintegrating baryons so it should produce a shower of events if it came through a proton decay detector. Indeed, if we were lucky we could find proton decay <u>and</u> monopoles at the same time. Unfortunately, as you may check in the contributions of Cabrera, Peccei and

Groom in these proceedings, the existing bounds on monopole fluxes are such that the chances of success are rather slim.

In conclusion, and to lift a bit the gloomy impression that this seminar may have caused, I would like to say that, although present experiments may fail to detect p decay, I believe they are certainly worth the while. τ_p is a basic quantity: any effort to know it better is justified. Even if it only serves, from the theoretical point of view, to finish off already falling models.

Bibliographical note

I have not tried here to give any references. For "classical" proton decay calculations, cf. the review of P. Langacker, Phys. Rep. 72C, 815 (1981). SUSY decays are reviewed by D. Nanopoulos and monopole ones by R. Peccei, both in these proceedings.

NEUTRINO OSCILLATIONS AND $n\bar{n}$ OSCILLATIONS

Felix Boehm

California Institute of Technology
Pasadena, CA 91125

INTRODUCTION

This review will cover two topics which are only marginally related. Neutrino oscillations and their implications on neutrino mass will be discussed first. It will be followed by an overview of the status of neutron-antineutron oscillations. While the first process is a sensitive source of information on lepton number conservation and neutrino mass, the second topic has to do with a possible $\Delta B = 2$ baryon number violation process.

NEUTRINO OSCILLATIONS

I shall start this part of my talk with an overview and a conclusion. The more general questions of lepton number conservation and neutrino mass have been marred by controversy during the past two years. Several experiments performed around 1980 have staked claims for evidence for neutrino mixing and finite neutrino mass. Much of the evidence, however, did not stand up to recent rigorous experimental tests. Some of the principal sources of information on neutrino mass and their current status are outlined below.

We begin with neutrino oscillations. This process, to be discussed here in detail, constitutes a test for neutrino mixing and neutrino mass[1]. Early results by Reines[2], based on measurements of the relative yields of the charged current reaction $\bar{\nu}_e d \rightarrow e^+ nn$ and the neutral current reaction $\bar{\nu}_e d \rightarrow \bar{\nu}_e pn$ were presented as evidence for neutrino oscillations with large mixing. Recent work by the Caltech-Munich-SIN group[3] however disagrees with the mentioned findings. No evidence for oscillations could be found and stringent limits are set for the oscillation parameters.

Supporting evidence for oscillations was also inferred from CERN beam dump data of 1980[4]. This was based on the ν_e/ν_μ ratio which was found to be 0.5 to 0.6 with an error of 0.2, instead of 0.9 to 1.0 for no oscillations. Recent results from Fermilab[5] however find a ν_e/ν_μ ratio consistent with no oscillations.

One remaining suggestive piece of evidence for oscillations is the solar neutrino puzzle[6]. The discrepancy between the experimentally observed solar ν_e flux and that calculated by the accepted solar model lends support to neutrino oscillations with maximum mixing. Possible mass parameters Δm^2 between 10^{-2} and 10^{-11} eV^2 required here are not in conflict with present laboratory data. Some of the reaction cross sections responsible for the ^8B neutrinos, however, have come under fire recently[6] and it remains to be seen how much of the puzzle persists after all the low energy cross sections are verified.

The other important source for neutrino mass, equally cast in controversy, is neutrinoless double beta decay. From the observation of the rate of double beta decay in ^{128}Te and ^{130}Te by Henneke *et al.*[7] it follows that the lepton number violating neutrinoless process must be present. A Majorana neutrino mass of about 10 eV would be needed to explain their findings. More recently, a remeasurement of these decay rates by Kirsten *et al.*[8] gives a result in good agreement with the expectation for a lepton number conserving two-neutrino process, the upper limit for neutrino mass being about 3 eV.

Finally, the result of the ITEP group[9] provides evidence for a finite neutrino rest mass, as derived from the upper energy of the ^3H beta decay, bracketed between $14 < m_\nu < 46$ eV. A serious problem, recently pointed out by Simpson[10], may cloud this experimental result. It has to do with the spectrometer resolution function derived from a calibration line (the M-shell internal conversion lines of the 20 keV gamma transition in ^{169}Tm). Owing to the finite width of the electronic M levels of about 15 eV[11], it appears that the ^3H spectrum was deconvoluted with a resolution function that is wider than appropriate. The effect on the spectral shape and endpoint would be such that the evidence for a finite neutrino mass would be weaker or even disappear. A final verdict has to await a careful reevaluation of the width of the calibration line.

As a conclusion, there does not exist at present confirmed and convincing experimental evidence for neutrino mixing and finite neutrino mass.

Let me then review the subject of neutrino oscillations. Following general practice[1] we write the weak-interaction neutrino state $\nu_\ell (\ell = e,\mu,\tau)$ as a superposition of mass-eigenstate neutrinos $\nu_i (i = 1,2,3)$,

$$\nu_\ell = \sum_i U_{\ell i} \nu_i .$$

The expansion coefficients $U_{\ell i}$ which are the elements of a unitary matrix represent the mixing amplitudes.

The dynamics of a mixed neutrino state follows by considering the evolution in time of a mass eigenstate $\nu_i(t)$,

$$\nu_i(i) \sim e^{-iE_i t}; \quad E_i = \sqrt{p^2 + m_i^2} .$$

Assuming that at least one of the masses m_i of the pure states ν_i is non-zero, neutrino oscillations will occur.

For the simplified case of 2 neutrinos, ν_{ℓ_1} and ν_{ℓ_2}, the matrix $U_{\ell i}$ contains only one independent parameter, the mixing angle θ. The probability for appearance of a new state ℓ_2 in a detector a distance L (m) away from the source emitting neutrinos in the state ℓ_1 is given by

$$P(\nu_{\ell_1} \to \nu_{\ell_2}) \sim \frac{\sin^2 2\theta}{2} \left[1 - \cos(2.53\ \Delta m^2 (L/E_\nu)) \right] .$$

The quantities $\sin^2 2\theta$ and $\Delta m^2 \equiv |m_1^2 - m_2^2|$ (eV^2) are the two parameters characterizing neutrino oscillation in this 2 neutrino model; E_ν (MeV) is the neutrino energy. Similarly, the probability of disappearance of a state ℓ_1 is given by $P(\nu_{\ell_1} \to X) \sim 1 - P(\nu_{\ell_1} \to \nu_{\ell_2})$. The oscillations are characterized by an oscillation length $\Lambda(m) = 2.48 E_\nu (MeV)/\Delta m^2 (eV^2)$.

We now review the experimental results on oscillations in the appearance channels $\nu_\mu \to \nu_e$ and $\nu_\mu \to \nu_\tau$. No evidence for oscillations has been reported and the current best upper limits of the parameters Δm^2 and $\sin^2 2\theta$ are shown by the solid lines in Fig. 1. The curves which are 90% confidence limit contour lines have been composed from the data referred to by the label. The regions to the right of the curves are excluded. A detailed list of references is contained in the recent review by Baltay[12].

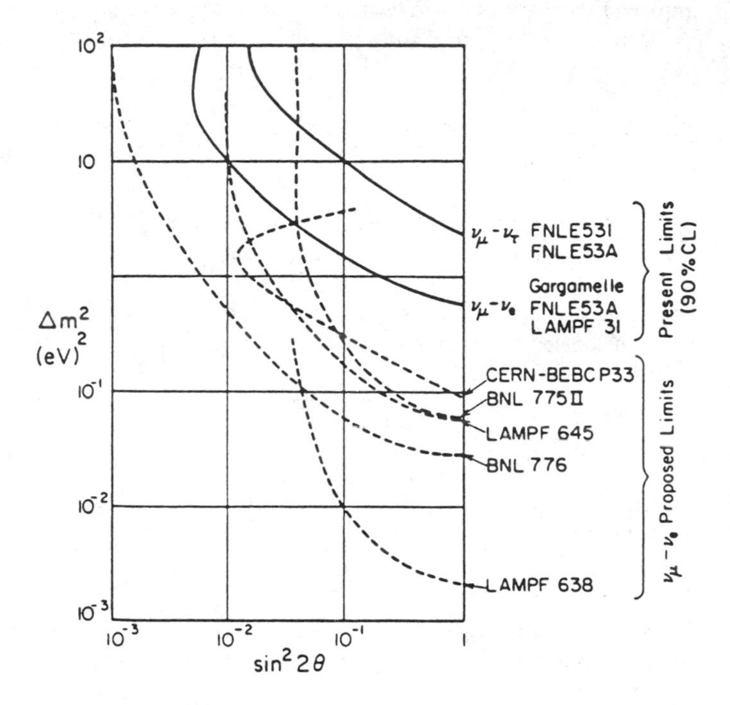

Fig. 1. Limits for neutrino oscillations $\nu_\mu \rightarrow \nu_e$ and $\nu_\mu \rightarrow \nu_\tau$. The solid curves represent current experimental limits at 90% c.ℓ. The dashed curves illustrate forthcoming experimental limits.

As to forthcoming results, several experiments have been proposed with the aim of exploring the region of smaller Δm^2 and $\sin^2 2\theta$. The sensitivities expected to be reached in these experiments are shown in Fig. 1 by dashed lines (the experiments are identified by Laboratory and proposal number[13]). It appears that an improvement in sensitivity of more than an order of magnitude should be attainable in the next 2-3 years.

Disappearance experiments have been conducted at high energy accelerators as well as with the help of fission reactors. Their results, in terms of the two parameters Δm^2 and $\sin^2 2\theta$ are summarized in Fig. 2. By far the most stringent limits come from a recent $\bar{\nu}_e \rightarrow X$ experiment[3] at the Gösgen power reactor. I shall briefly describe this work here.

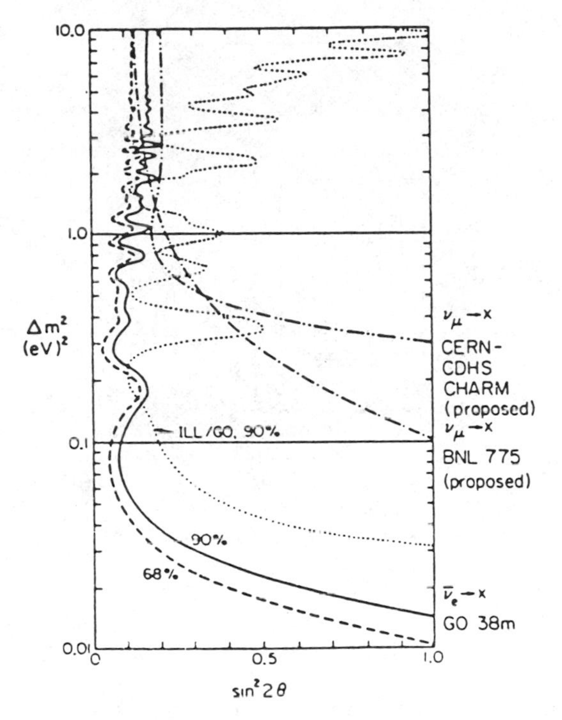

Fig. 2. Limits for neutrino oscillations $\nu_\mu \to X$ and $\nu_e \to X$. The solid curve represents current experimental limits at 90% c.ℓ. from the Gösgen reactor experiment. Limits obtained from the ratio of the ILL (8.7 m) and Gösgen (L = 38 m) data are shown by the dotted curves labelled ILL/GO. Forthcoming experimental limits are shown by broken lines.

The Caltech-Munich-SIN group[3] has recently completed a measurement of the neutrino spectrum at a distance of 38 m from the core of the 2800 MW reactor at Gösgen, Switzerland. The set up of the experiment is sketched in Fig. 3. The neutrinos were detected by the reaction $\bar{\nu}_e p - e^+ n$ using a composite liquid scintillation detector and ^3He multiwire proportional chambers. A time correlated e^+, n event constituted a valid signature.

Pulse shape discrimination in the scintillation counter has proofed to be a powerful technique to eliminate correlated neutron background events. Cosmic ray induced fast neutrons recoiling on protons in the liquid scintillation counter can give rise to scintillation counter trigger, followed, after a thermalization period,

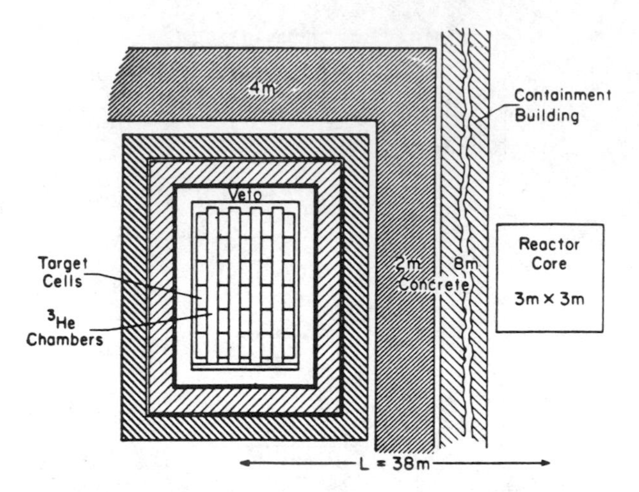

Fig. 3. Experimental set up of the Neutrino Detector at the Gösgen Reactor. (The drawing is not to scale).

by a neutron capture signal in the ^{3}He counter. The effect of the pulse shape discrimination is shown in Fig. 4. The positron peak corresponding to a short pulse-decay time is clearly visible.

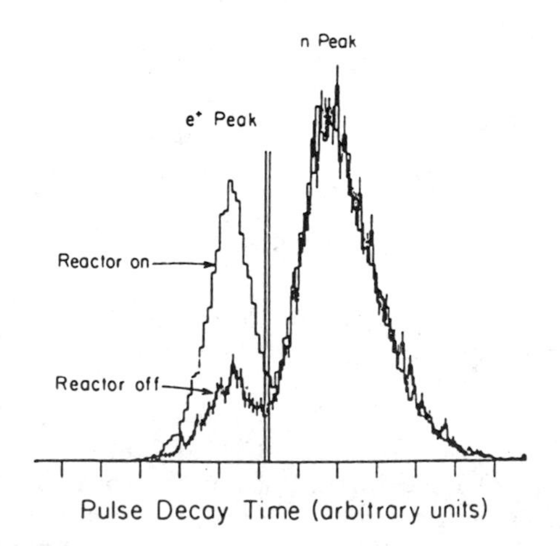

Fig. 4. Pulse shape discriminator spectrum in the liquid scintillation counter. the e^+ peak is the neutrino signal.

The neutron peaks for reactor on and reactor off have the same height, indicating that there are no reactor associated neutrons.

The signal-to-noise ratio in this experiment was considerably better than in the earlier Grenoble experiment[14]. This was achieved by introducing position correlation cuts for the positron and neutron events.

About 11,000 neutrino induced events were recorded in a six-months reactor-on period. Background was recorded during a one-month reactor-off period. The measured positron spectrum was compared with that expected for no oscillations. The latter was obtained from the on-line beta spectroscopic measurements at the Laue-Langevin reactor by Schreckenback *et al.*[15] studying ^{235}U and ^{239}Pu fission targets. These two isotopes account for about 89% of the total fission energy at the Gösgen reactor. The remaining 11% are due to fission of ^{238}U and ^{241}Pu. The calculations of Ref. 16 were used to evaluate the contribution to the beta spectra from ^{238}U and ^{241}Pu. The variation in time of the contributions of each fissioning isotope is well known and was taken into account. The calculated spectra to which the experimental spectrum was compared were obtained by multiplying the no-oscillation spectrum with the oscillation functions $P(E_\nu, L, \Delta m^2, \theta)$ and integrating over detector and core dimensions.

Figure 5 shows the ratio of the observed yield to the yield expected for the no-oscillation case, as a function of $L/E_{\bar{\nu}}$, together with the best fit for $\Delta m^2 = 0$ and for two possible $\Delta m^2 \neq 0$ solutions. A systematic χ^2 test to all possible values of Δm^2 and $\sin^2 2\theta$, including those shown in Fig. 5, followed by a maximum likelihood ratio test resulted in the 68% and 90% confidence limit contour lines displayed in Fig. 2. The same procedure was also applied to the ratio of the data of Ref. 14 taken at L = 8.7 m to the present data (L = 38 m) and the resulting contour lines are also shown in Fig. 2. This ratio test is somewhat less restrictive because of the moderate statistical accuracy of the data of Ref. 14.

It must be concluded from this experiment that there are no neutrino oscillations with parameters larger than those given by the curves GO in Fig. 2. For large mass parameters it is found that $\sin^2 2\theta \leq 0.17$ (90% c.ℓ.), and for maximum mixing the limit is $\Delta m^2 < 0.016\ eV^2$ (90% c.ℓ.). The results are in clear disagreement with those by Reines *et al.*[2] For the integral yield it is found that the experimental value divided by the predicted no-oscillation value is 1.05 ± 0.05 (68% c.ℓ.).

We note that similar experiments are in progress elsewhere[17,18].

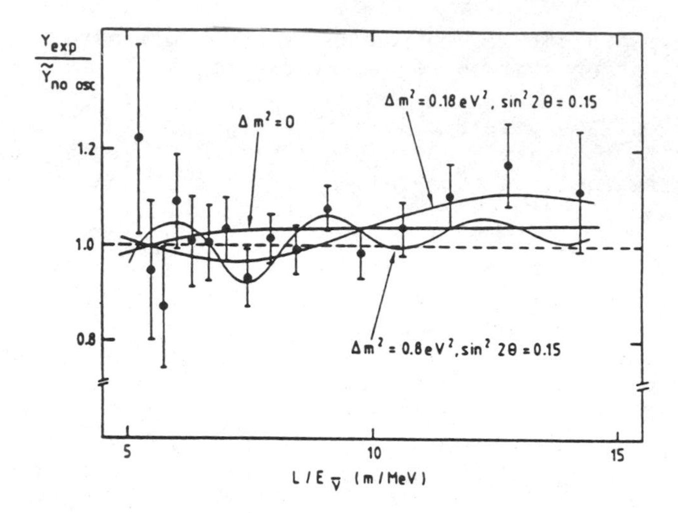

Fig. 5. Ratio of experimental to predicted (for no oscillations) positron spectra. The errors of the data points shown are statistical. Oscillation functions consistent with the data are shown for various sets of oscillation parameters.

At present the Caltech-Munich-SIN group is collecting data at a position of L = 46 m. It will then be possible to compare the neutrino spectra at 38 m and 46 m, allowing a ratio test independent of assumptions for the spectral shape.

Recent searches[19] for particle-antiparticle transitions $\nu_\mu \rightarrow \bar{\nu}_e$ and $\nu_e \rightarrow \bar{\nu}_e$ revealed no evidence having oscillation parameters Δm^2 (full mixing) larger than 0.7 eV^2 and 7 eV^2, respectively.

Finally, the important question should be addressed where, in the Δm^2 vs. $\sin^2 2\theta$ plane, should we continue to search for oscillations. Unfortunately there is no guidance from theory whatsoever. As to the mixing angle, we can state that present limits are smaller than the Cabibbo angle. Figure 6 shows these limits, together with other possible dimensional guesses (lepton mass ratios). If the solar experiments are indeed telling us that neutrinos oscillate with large mixing angles, the Δm^2 values must lie between 10^{-2} and 10^{-10} eV^2, a region increasingly more difficult to explore.

$n\bar{n}$ OSCILLATIONS

As in the case of neutrinos, the neutron-antineutron system can exhibit oscillations if the neutrons and antineutrons (n and $\bar{n}$)

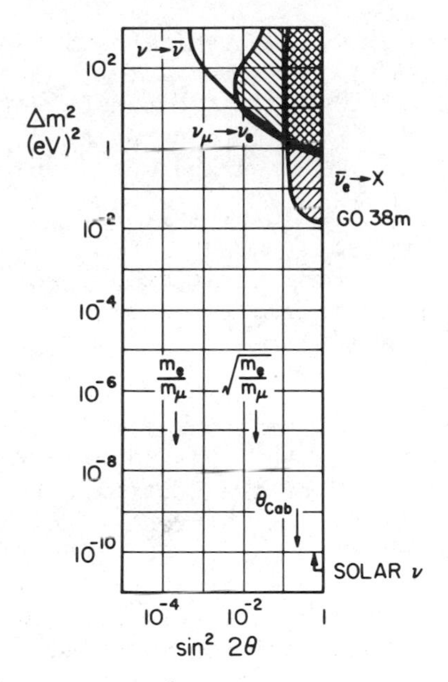

Fig. 6. Expanded Δm^2 vs $\sin^2 2\Theta$ plane showing current experimental limits and somedimensional guesses.

are not pure states but, instead, superpositions of mass eigenstates (n_1 and n_2). The mass matrix of the $n\bar{n}$ system is

$$\begin{pmatrix} E_o & \delta m \\ \delta m & E_o \end{pmatrix},$$

where E_o is the neutron (or antineutron) mass and δm is the transition mass which is related to the free neutron mixing time τ by $\tau = 1/\delta m$ ($\hbar = 1$). The mixing amplitude is unity, and the eigenvalues of the mass matrix are $E_{1,2} = E_o \pm \delta m$.

The time evolution of the mass eigenstates gives rise to transitions from n to $\bar{n}$, with a probability given by

$$P_{\bar{n}}(t) = \tfrac{1}{2}(1 - \cos 2\,\delta m\, t) \approx (\delta m\, t)^2 ,$$

for times short compared to the free neutron lifetime. The mixing time τ can be obtained by measuring the number, $N_{\bar{n}}$, of $\bar{n}$ states formed after a time t, $N_{\bar{n}} \approx N_n\,(t/\tau)^2$, always assuming that the neutrons are free particles.

An estimate of a lower limit of the mixing time τ has been obtained by considering matter stability. An experimental limit for $n\bar{n}$ decay of about $\tau_{\Delta B=2} > 10^{30}$ y can be set from the absence of events with $E \sim 2$ GeV in proton decay experiments. Relations between $\tau_{\Delta B=2}$ and τ have been worked out[20] by Mohapatra and Marshak, Kuz'min *et al.*, and others leading to lower limits for τ between $(0.3 - 2) \times 10^7$ sec. With $\tau > 2 \times 10^7$ sec it follows that $\delta m < 3 \times 10^{-23}$ eV. As proton decay experiments provide better limits on $\tau_{\Delta B=2}$, the limit for τ should be raised by a factor of $\sqrt{\tau_{\Delta B=2}(y)/10^{30} y}$.

The previously made assumption of a free neutron is, of course, not justified. The problem thus has to be reformulated taking into account the interaction ΔE of the neutron with external fields. The neutron magnetic dipole moment interacting with the earth magnetic field yields an interaction energy $\Delta E = \mu B \sim 10^{-11}$ eV. Although this interaction can be reduced by shielding of the magnetic field to, say, 10^{-14} eV, it remains large compared to the mentioned estimated limit for the free transition mass δm. As a consequence, the admixture between n and $\bar{n}$ states is reduced from unity to the value $\delta m/\Delta E \gtrsim 3 \times 10^{-9}$. The probability for a transition $n \to \bar{n}$ becomes

$$P_{\bar{n}}(t) = \left(\frac{\delta m}{\Delta E}\right)^2 \frac{1}{2} (1 - \cos\Delta E t).$$

The amplitude of $n\bar{n}$ oscillations thus is less than 10^{-17}.

For short observation times, $t < 10^{-2}$ sec, we have $\Delta E t \ll 1$, and the "quasi free" transition probability is

$$P_{\bar{n}}(t) \simeq (\delta m t)^2 = (t/\tau)^2 \gtrsim 10^{-18} .$$

We see from this rough estimate that a meaningful experiment with a neutron beam, and an interaction time of 10^{-2} sec, requires that more than 10^{18} neutrons enter the decay section of the apparatus.

Searches for $n\bar{n}$ oscillations using reactor neutrons have been carried out at the reactor of the ILL in Grenoble [21]. A recent experiment by a CERN - ILL - Padova - REHL - Sussex collaboration[21] used a cold (25°K) neutron beam with an intensity of 1.5×10^9 neutrons/second. The flight time in the 4.5 m drift section was 0.03 sec. A 60 day run with reactor on and reactor off periods produced no $n\bar{n}$ annihilation events, giving a limit of $\tau > 10^6$ sec. The magnetic field was shielded to better than 10^{-3} Gauss. The composite detector adjacent to the ^{6}LiF target consisted of streamer and scintillation counters, allowing a reconstruction of the $n\bar{n}$ vertex.

The present status of the neutron oscillation efforts is summarized in Fig. 7. Most of the predictions[20] are based on the lifetime for matter stability, $\tau_{\Delta B=2} = 10^{30}$y. The experimental limits of the Grenoble experiments[21] are shown, as well as reported sensitivities of several proposed experiments[22]. At present, only the ILL III and the Pavia experiment are funded. It appears that a limit for τ of about 1-2 x 10^8 sec is achievable in the near future.

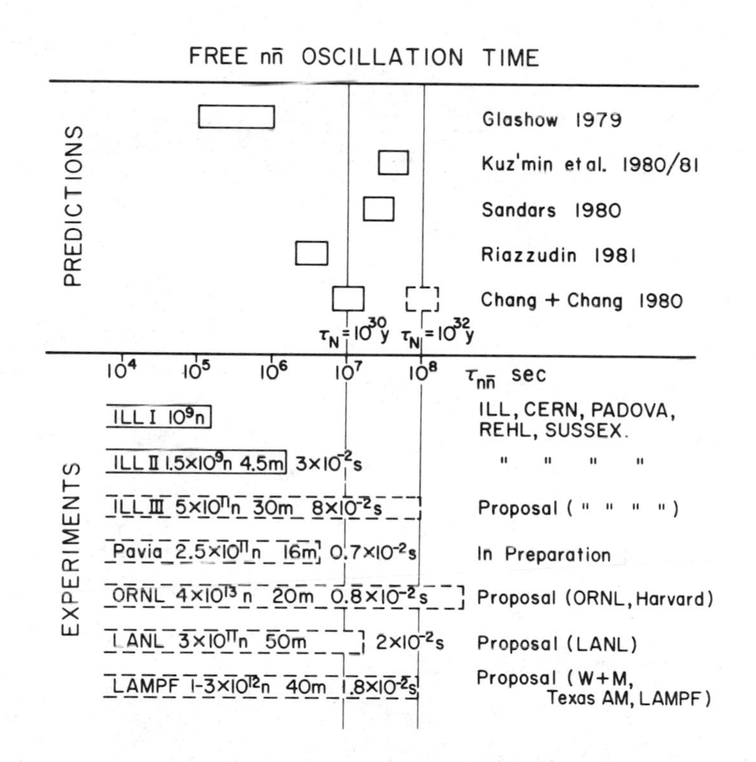

Fig. 7. Overview of some predicted limits as well as experimentally established and proposed limits for the free neutron mixing time τ. A complete list of calculated values can be found in G. Fidecaro[21], with references given there, as well as under Ref. 20. The ILL results are described in Ref. 21 and the proposals are listed under Ref. 22.

In comparison, matter stability experiments are now approaching comparable sensitivities, inasmuch as a lifetime limit agains $n\bar{n}$ annihilation in matter of 10^{32} is equivalent to a free neutron mixing time of $\tau \geq (0.3 - 2) \times 10^8$ sec.

REFERENCES

1. See the review by P.H. Frampton and P. Vogel, Phys. Reports 82, 339 (1982).
2. F. Reines *et al.*, Phys. Rev. Lett. 45, 1307 (1980).
3. J.-L. Vuilleumier *et al.*, Phys. Lett. 114B, 298 (1982).
4. See Ref. 1, p. 372.
5. R.J. Loveless, Neutrino 82, p. 89, A. Frenkel, editor, Budapest 1982.
6. See review by J. Bahcall *et al.*, Revs. Mod. Phys. 54, 767 (1982), and recent results on ^{7}Be(p,γ) by B.W. Fillippone, Phys. Rev. Lett. 50, 412 (1983).
7. E. Hennecke *et al.*, Phys. Rev. C 11, 1378 (1975).
8. T. Kirsten *et al.*, Phys. Rev. Lett. 50, 474 (1983).
9. V.A. Lubimov *et al.*, Phys. Lett. 94B, 266 (1980).
10. J.J. Simpson, paper presented at ICOMAN 83, Frascati, Italy, January 1983.
11. O. Keski - Rahkonen and M.O. Krause, At. Nucl. Data Tables 14, 139 (1974).
12. C. Baltay, Neutrino 81, Vol. II, p. 295, Univ. Hawaii, 1982.
13. For CERN Proposals, see A.L. Grant, Neutrino 81, Vol. II, p. 214; M. Murtagh *et al.*, Brookhaven Proposal E775; A. Pevsner *et al.*, Brookhaven Proposal E776; T. Romanowsky *et al.*, LAMPF Proposal 645; T. Dombeck *et al.*, Los Alamos Proposal 638.
14. H. Kwon *et al.*, Phys. Rev. D 24, 1097 (1981).
15. K. Schreckenbach *et al.*, Phys. Lett. 99B, 251 (1981).
16. P. Vogel *et al.*, Phys. Rev. C 24, 1543 (1981).
17. M. Mandelkern, Neutrino 81, Vol. II. p. 203 (1981).
18. J.F. Cavaignac *et al.*, Raport Annuel 1980, ISN 81.01, p. 71.
19. A.M. Cooper *et al.*, Phys. Lett. B 112, 97 (1982).
20. R.N. Mohapatra and R.E. Marshak, Phys. Rev. Lett. 44, 1316 (1980), 44 1644(E) (1980). V.A. Kuz'min and M.E. Shapashnikon, JETP Lett. 32, 82 (1980). D.G. Chetyrkin, M.V. Kazarnovsky, and V.A. Kuz'min, Phys. Lett. 99B, 358 (1981). L.N. Chang and N.P. Chang, Phys. Rev. Lett 45, 1540 (1980), Phys. Lett. 92B,103(1980). T.K. Ku and S.T. Low, Phys. Rev. Lett. 45, 93 (1980). Riazzudin, Phys. Rev. D 25, 885 (1982). P.G.H. Sandars, J. Phys. G 6 L 616 (1980).
21. G. Fidecaro, Neutrino 81, Vol. I, p. 264, Univ. Hawaii, 1982. M. Baldo-Ceolin, Neutrino 82, Suppl. p. 39. A. Frenkel and L. Jenik, ed, Budapest 1982.
22. NADIR, Pavia Reactor Proposal, 1981. H.R. Anderson, Neutrino 81, LANL Proposal, 1981. R. Ellis *et al.*, LAMPF Proposal 647 (1981). G.R. Young *et al.*, Oak Ridge Proposal, 1981.

NEUTRINO MASS EXPERIMENTS

B. Jonson[1], J.U. Andersen[2], G.J. Beyer[1*], G. Charpak[1], A. De Rujula[1], B. Elbek[3], H.A. Gustafsson[4], P.G. Hansen[2], P. Knudsen[3], E. Laegsgaard[2], J. Pedersen[3] and H.L. Ravn[1]
and
The ISOLDE Collaboration

1. INTRODUCTION

The current status of experiments to determine the mass of the different kinds of neutrinos is given in Table 1. The most stringent results are available for the electron antineutrino mass, results that are coming from a series of successive refinements of the experimental techniques to measure the shape near the upper end-point of the β^- spectrum from the tritium decay. Further improvements in this technique are to be expected; the striking conclusion reached by Lyubimov et al. $m_{\bar{\nu}_e} \cong 35$ eV has redirected much effort into ^3H experiments.

Little attention has been paid to the mass of the electron neutrino, and for good reasons. The combination of the competition

1) CERN, Geneva, Switzerland.
2) Institute of Physics, University of Aarhus, Aarhus, Denmark.
3) Tandem Accelerator, Niels Bohr Institute, Risø Denmark.
4) Institute of Physics, Technical University of Lund, Lund, Sweden.
*) Visitor from Zentralinstitut für Kernforschung, Rossendord bei Dresden, German Democratic Republic.

Table 1 : Experimental Neutrino Mass Limits

	Mass	Conf. level (%)	Technique	Ref.
$m_{\bar{\nu}_e}$	< 55 eV	90	^3H, mag. spec.	1
	$14<m_\nu<46$ eV	99	^3H, mag. spec.	2
	< 65 eV	95	^3H, sol.st.det.	3
m_{ν_e}	< 4100 eV	67	^{22}Na, β^+	4
	< 500 eV	90	^{193}Pt, IBEC	5,6
m_{ν_μ}	< 570 keV	90	$\pi^+ \to \mu^+ \nu_\mu$	7
	< 500 keV	90		8
m_{ν_τ}	< 250 MeV	95	$\tau \to e\, \bar{\nu}_e \nu_\tau$	9

between electron-capture (EC) and β^+ decay and the strong Z dependence favouring the EC process for heavy atoms and at low energies exclude that a β^+ case anywhere near comparable to the tritium case can exist.

The present paper is concerned with the development [10] of a new technique based on the observation of internal bremsstrahlung from EC processes (IBEC). Experiments of this kind are capable of providing greatly improved limits on the electron-neutrino mass and, more important, it is probable but not certain that they could be developed to become superior to the tritium experiments.

The method to determine the electron neutrino mass from the IBEC process was suggested by De Rújula [11] about two years ago. The idea is to study the shape near the end-point of the continuous internal bremsstrahlung spectrum. The spectrum is modified by a finite neutrino mass in a way that is completely analogous to that for the tritium electron spectrum : it has a horizontal tangent if the mass of the neutrino is zero and would be modified to have a vertical tangent if $m_\nu \neq 0$.

2. THE THEORY OF IBEC

The IBEC radiation is emitted in beta decay by the capture of bound atomic electrons according to the scheme :

$$A_Z + e^-_{n,\ell,j} \rightarrow A_{Z-1} + \nu_e + \gamma_{IB}, \tag{1}$$

where n,ℓ,j are the quantum numbers of the bound electrons. This process has a three-body final state, and the spectrum of emitted photons is continuous with a maximum energy

$$k^0_{max} = Q_{EC} - B[n,\ell,j]; \quad m_\nu = 0 \quad , \tag{2}$$

where Q_{EC} is the energy available for EC decay and $B[n,\ell,j]$ is the binding energy of the captured electron.

For Q_{EC} large, compared with atomic binding energies, the dominant contribution to the IBEC spectrum arises from capture from the S-states. The shape of the S-wave IBEC spectrum was first derived by Møller [12] and by Morrison and Schiff [13]. The shape is very simple and is found to be proportional to a universal three-body phase-space factor (daughter atom and two light particles) divided by k :

$$\frac{d\omega}{dk} \sim \phi(k) = k(k^0_{max}-k)\ [(k^0_{max}-k)^2 - m^2_\nu]^{1/2}, \tag{3}$$

where k is the photon energy and k^0_{max} its maximum allowed value for vanishing neutrino mass. The ratio of the radiative-to-non-radiative S capture is of the order of 10^{-4} and is proportional to $k^{0\,2}_{max}$. At low value for k^0_{max}, which is clearly most favourable for a neutrino-mass hunt, the S-wave IBEC intensities therefore become extremely low. The reason why an IBEC experiment may still compete successfully with the favourable tritium decay is due to the appearance of another mechanism : at low energies the IBEC spectrum arises mainly from P capture. This was first understood by Glauber and Martin [14]. In the P-wave IBEC the capture takes place from a virtual S-state associated with a radiative $P \rightarrow S$ electromagnetic transition. The resonant nature of the process leads to important enhancements of the photon intensity at low energies and, in particular, close to the resonance energies corresponding to $P \rightarrow S$ transitions, i.e. the normal atomic x-rays. The continuous nP IBEC spectrum differs, in fact, from an ordinary nS non-radiative capture followed by the radiation of a characteristic x-ray only in a relaxation of the requirement of energy conservation in the intermediate state. In the immediate neighbourhood of the line, the process is identical to the real, resonant emission of x-rays. On the other hand, the P-state intensity remains high also for energies that are considerably different from the resonance energies. The intensity observed at the off-resonance energies arises in the nP IBEC process, and there several virtual n'S-states contribute significantly to the photon spectrum.

The shape of the nP IBEC spectrum is, as demonstrated by Glauber and Martin [14], proportional to the product of $\phi(k)$ [eq.3] and a matrix element $Q^2_{nP}(k)$:

$$\frac{d\omega}{dk} \sim \phi(k) \cdot Q^2_{nP}(k) \; . \tag{4}$$

The Q_{nP} function was discussed in detail in ref. 14 for energies above the 1S x-ray pole, $k > B_{1S}-B_{2P}$, and later extended by De Rùjula [11] to the region below the pole. We show as example $Q^2_{2P_{3/2}}$ and $Q^2_{3P_{3/2}}$ functions for the case ^{193}Pt in Fig. 1.

The EC Q-value has nuclear origin and varies for different isotopes of the same element, whilst the atomic binding energies are essentially constant.[The challenge is thus to find nuclei with Q-values giving the maximum profit for the resonance enhancement.

The ideal case would be one where Q has a value just below the S binding energy of one of the shells.]

The two cases that best meet this requirement are the isotopes ^{193}Pt and ^{163}Ho. The present results for them are summarized below.

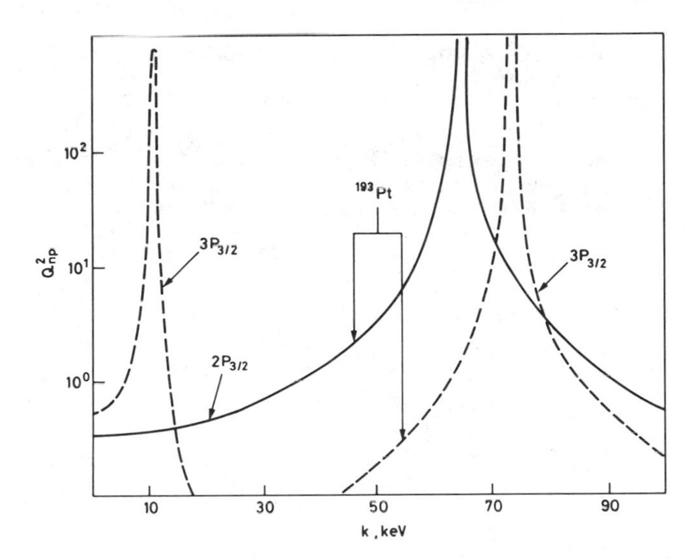

Fig. 1 : The $Q_2^2 P_{3/2}$ and $Q_2^2 P_{3/2}$ functions for the element Pt calculated in the pole approximation [eqs. (10.5b) and 10.6b) in ref.11]. The end-points of the respective IBEC spectra are indicated.

3. STUDIES OF THE EC DECAY OF ^{193}Pt AND ^{163}Ho

Wu and collaborators [15] have used the isotope ^{131}Cs to show that the internal bremsstrahlung spectrum (Q_{EC} = 355 keV) has a small excess of intensity at low energy, and that this part of the spectrum is not coincident with K x-rays. These observations are consistent with the interpretation of the low-energy part as arising from 2p, 3p radiative capture, as demanded by the Glauber-Martin theory. For a more exacting check it is necessary to go to a case where the Q-value is much closer to the resonance points.

3.1 The Case of ^{193}Pt

The isotope ^{193}Pt discovered by Hopke and Naumann [16], is ideally suited for a test of the IBEC theory.It decays by a ground-state transition with a Q-value of 56.3 ± 0.3 keV [5,6]. The K binding energy in Ir is 76.1 keV which means that anly capture from the L and higher shells is possible. The IBEC spectrum will be dominated by the K pole as illustrated in Fig. 1.

One of the main experimental problems in the ^{193}Pt experiment is the preparation of a radio-chemically clean source. A complete outline of the source preparation is described in ref. 17.

The IBEC experiment was performed with two x-ray detectors, one 2 cm^2 intrinsic Ge detector for the bremsstrahlung and a high-resolution Si(Li) detector for the x-rays. The source was mounted in a small hole at the centre of a 1 mm thick Cd plate and placed between the two detectors. The Cd plate served as a barrier to cross talk between the two detectors.

The detailed test of the IBEC theory made in the ^{193}Pt experiment is described in ref. 18. Here we shall confine our discussion to the 2p component which was obtained from the data by demanding a coincidence condition with L X-rays from Ir. The observed spectrum is shown in Fig. 2 and is put on an absolute scale on the basis of the number of L X-ray gates. The curve in the figure is the calculated 2p spectrum and we find a quantitative agreement.

Finally we examine the shape in more detail by dividing out the phase-space factor. The result (Fig. 3) coincides exactly with the calculation for 2p capture; the 2s capture would correspond to a constant shape-factor plot.

In short then, the experiments are in excellent agreement with what is basically a non-relativistic hydrogenic-atom theory; improvements in theory or experiment can only make the agreement

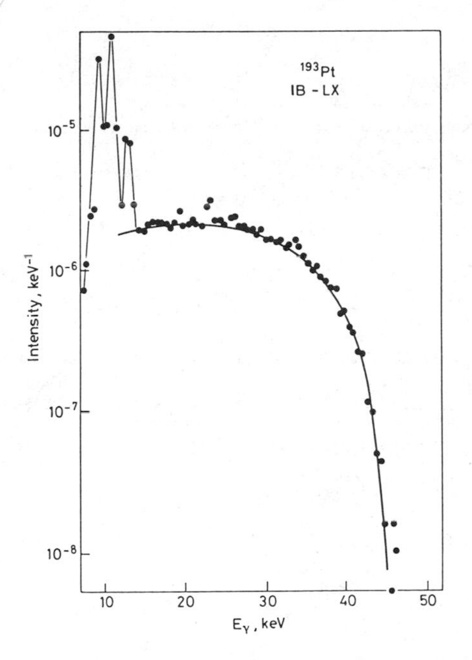

Fig. 2 : Internal bremsstrahlung from ^{193}Pt observed in coincidence with (all) L x-rays. The spectrum has been corrected for randoms but, owing to errors in the subtraction, weak L x-rays remain. The lines at 23 and 26 keV arise from fluorescent excitation of the Cd source holder. The spectrum has been brought to an absolute scale on the basis of the number of L x-rays gates and it has been corrected for the tail of the line but not for finite resolution. The theoretical curve is the sum of the $2p_{1/2}$ and $2p_{3/2}$ components, also given on an absolute scale.

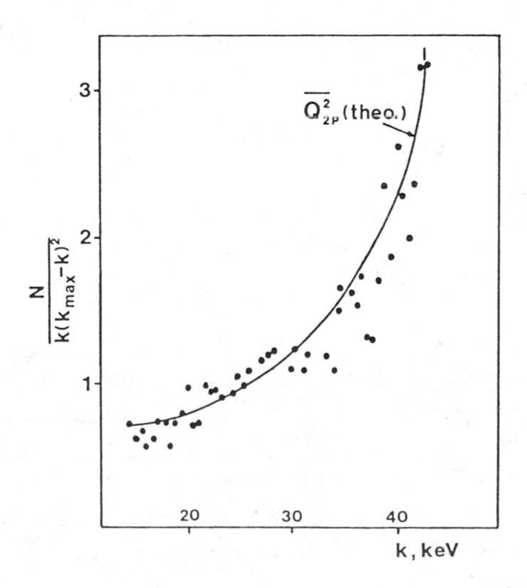

Fig.3 : Shape-factor plot of the coincident 2p inner-bremsstrahlung (IB) spectrum from Fig. 2 and the corresponding theoretical curve. (The 2s bremsstrahlung would in this representation be constant - 0.01.)

worse. The main reason for our being willing to forgo more detailed atomic calculations, at least at the present stage is, however, that our intended application involves a kinematically determined and very local effect, as can be seen from an analysis of the influence of the neutrino mass.

3.2 A Limit on the Neutrino Mass

The spectrum in coincidence with the 9.14 keV LX-rays in Fr which is pure L_{III} $M_{IV,V}$ (leaving aside a small correction for Coster-Kronig transitions) represents pure $2P_{3/2}$ radiative capture. If these data are transformed in a manner similar to the usual Kurie plot, we obtain the result shown in Fig. 4.

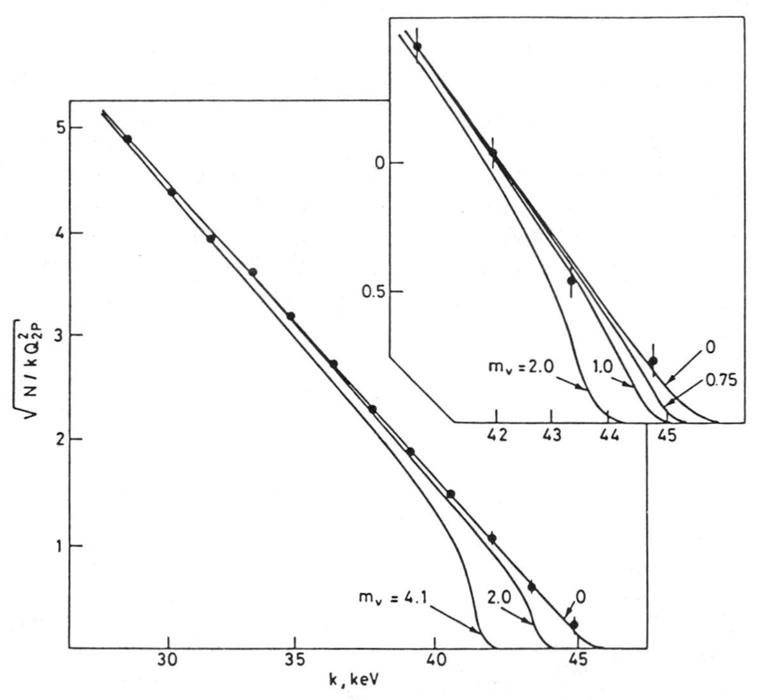

Fig. 4 : The high-energy part of the IB spectrum in coincidence with the 9.1 keV L x-ray. The spectrum, which thus represents essentially the $2p_{3/2}$ component. has been transformed similarly to a "Kurie plot". Theoretical curves are shown for different assumed values of the neutrino mass; the curves have been corrected for the detector resolution of 0.50 keV.

As this experiment is essentially statistics limited one finds from a simple analysis that it sets an upper limit of 500 eV for a 90% confidence level. This limit replaces the previous one of 4100 eV (67% confidence level) set by Beck and Daniel [4].

It should be relatively easy to improve this experiment further, but it would hardly be worth the effort as another radioactive isotope, ^{163}Ho, offers much more promise for a precise determination of the electron-neutrino mass.

3.3 The Case of ^{163}Ho

The most promising case for an IBEC experiment, capable of competing in precision with the ^{3}H experiments (Table 1), is ^{163}Ho. This isotope was discovered by Naumann et al. [18,19] and recently its M capture half-life has been measured to be 40,000 ± 12,000 years. A detailed account for the half-life determination is given in ref. 10 and we give here only a summary of the main experimental steps.

The crucial problem in a ^{163}Ho experiment is to prepare a suitable source. The bremsstrahlung is weak and with the very long half-life for ^{163}Ho it is necessary to make a decontamination from other elements and isotopes by a factor 10^{12}-10^{14}. For the M half-life measurement it has to contain a known number of atoms, preferably in the form of a mono-atomic layer so that one can count the M shell radiations with energies just above 1 keV. This objective was reached by a combination of mass separation, radiochemistry and highly sensitive ion-beam analysis.

The radioactivity was produced in the ISOLDE Facility connected to the CERN Synchro-Cyclotron. A hot tantalum target of 122 g/cm^2 was bombarded with a 2.4 μA beam of 600 MeV protons; rare-earth elements evaporating from the target were ionized by surface ionization and mass separated. The sample collected at mass 163 consisted predominantly of the elements holmium to lutetium (the yield is increased by collecting also the isobars, which decay to ^{163}Ho). The collection took place in a Faraday cup; from the integrated current the number of atoms on the collector foil was calculated to be 1.36×10^{12}. After some weeks of cooling, the main contaminants in the sample (determined by γ spectroscopy) were ^{147}Gd (1.4×10^{12} atoms collected as an oxide side band), and ^{167}Tm and ^{169}Yb (both about 4×10^{9} atoms from "tails" in the separator). A further purification was obtained by a complex procedure based on wet chemistry. The collector foil was dissolved in the presence of 3 mg of lanthanum carrier and microgram amounts of other rare earths, except holmium. A small amount of short-lived ^{160}Ho was added to permit a monitoring of the chemical yields. The separation was carried out in standard precipitation steps followed by ion-exchange chromatography on an Aminex A5 column with 0.11 M α-hydroxyisobutylic acid as the eluant. The holmium fraction was cleaned twice, again by the same procedure. From the shape of the elution curves (see Fig. 5) it was estimated that each step gave a decontamination of a factor of 10^{3} from the heavier rare earths (such as Tm, Yb) and considerably more from the lighter ones. Thus the total decontamination offered by the combination of mass separation and chemistry was of the order of 10^{14}-10^{15}. The yield in the chemical procedure was 44%.

Two thin sources were prepared from this stock of radioactivity by vacuum evaporation from a rhenium ribbon at 1800° C. The first (source A) was made before the ^{160}Ho tracer had decayed, so that the yield could be determined : the result of 49% corresponds to 1.9×10^{13} atoms in the central part of a 64 cm^{2} tantalum backing. A second thin source (source B) was prepared on a 0.1 mm beryllium foil.

In order to obtain an independent measurement of the number of Ho atoms, the contents of source B were obtained by means of a

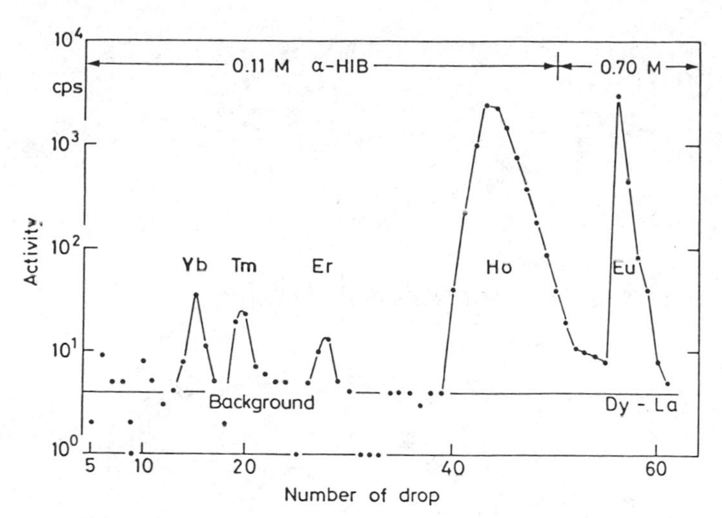

Fig. 5 : Ion-exchange chromatogram for the first holmium separation. Note that the tailings, of which only europium from a $^{147}Eu^{16}O$ side band is seen, have been eluted with a more concentrated solution of α-hydrosyisobutylic acid. Only the three drops at the peak of the holmium curve were used in subsequent separation.

highly sensitive surface-analysis technique. The elastic Coulomb scattering of α particles in back angles offers a probe for surface layers and a good mass-resolving power arising from the kinematic energy loss in the collision. The experiment carried out with a beam of 3.5 MeV α particles from the Aarhus single-stage Van De Graaff accelrator, shows (Fig. 6) a well-resolved peak at the holmium mass position with an intensity (integrated over the source) of 5.0×10^{13} atoms. This agrees well with the value 4.5×10^{14} obtained from the Faraday-cup and chemical-yield data. Thus a consistent scale for the number of ^{163}Ho atoms has been obtained by two entirely independent techniques.

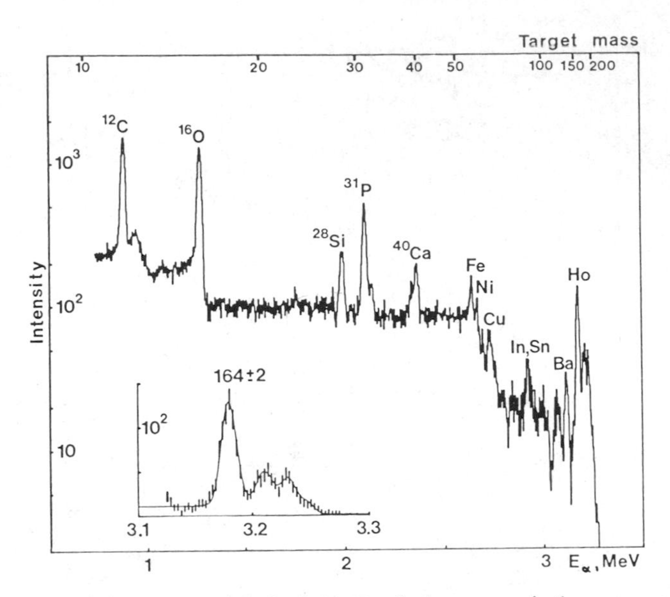

Fig. 6 : Energy spectrum of 3.5 MeV alpha particles scattered from the centre of foil B through 170°. Surface impurities such as holmium appear as clearly resolved peaks, whilst bulk impurities give rise to a step frunction which reflects the energy loss of the alpha particle inside the target. The energy scale was calibrated by scattering the beam from a thin target of gold. The inset shows that the high-energy region is dominated by a peak at mass 164 ± 2 with weaker peaks around 183 and 203, presumably Ho, Ta/Re and Au/Pb, respectively. The intensity of the holmium peak, integrated over the area of the foil, gives an independent check on the number of atoms in the source.

The sources were counted in a 10 × 10 cm^2 multi-wire proportional chamber filled with an argon-methane mixture. Source B was also counted with a single-wire proportional counter filled with helium and methane (10%) for the measurement of electrons, and filled with argon and methane for the measurement of M x-rays. For the second measurement the Auger electrons were absorbed in a thin mylar foil. The results of these measurements were consistent and resulted in a value $t^M_{1/2} = (4.0 \pm 1.2) \times 10^4$ years. This value has recently been confirmed in an independent experiment performed by Yasumi et al. [19]. The calculated total half-life for ^{163}Ho is $(7 \pm 2) \times 10^3$ years.

It is possible to convert the measured M-capture transition probability into an estimate of the phase-space factor for the ^{163}Ho decay by capture for the M_I shell

$$\phi(M_I) = [Q-E(M_I)]\left\{[Q-E(M_I)]^2 - m_\nu^2c^4\right\}^{1/2} = (0.53 \pm 0.10\ \text{keV})^2 \quad ,$$

where $E(M_I)$ is the M_I binding energy and m_ν is the electron-neutrino rest mass. The details of this calculation have been given elsewhere [5]. Essentially, it relies on the fact that the only other unknown quantity is the nuclear matrix element, which owing to a fortunate coincidence can be fixed precisely by reference*) to the corresponding transition in the nucleus ^{163}Ho. Assuming $m_\nu = 0$ and taking the known value $E(M_I) = 2.05$ keV, we thus find $Q = 2.58 \pm 0.10$ keV.

A measurement of the ^{163}Ho-^{163}Dy mass difference (Q in eq. 2) can also be made in a nuclear reaction experiment. The reaction chain

$$^{163}\text{Dy}(d,t)^{162}\text{Dy}(^3\text{He},d)^{163}\text{Ho}$$

was used in ref. 10 and a value of $Q = 2.3 \pm 1.0$ keV has been obtained. Further improvements in the precision are possible. A combination of this result with the value of $\phi(M_I)$ given above results in a limit [10] of 1.3 keV in m_{ν_e}.

4. CONCLUDING REMARKS

The theory of resonance enhanced IBEC has been confirmed in the ^{193}Pt experiment. The ^{163}Ho experiment gives a Q-value only 530 eV away from the M_I resonance point which makes ^{163}Ho the present best candidate for an electron-neutrino mass experiment. The

---

*) The essential difference originates in the so-called pairing correction factor $\mu_p^2\mu_n^2$ which was calculated theoretically for the two cases.

next phase of these experiments will follow two main lines : (i) to measure the IB in coincidence with Auger electrons, (ii) to make total absorption spectrometry based on solid-state detectors. The second, calorimetric technique would be especially interesting since it would exploit the large enhancement of the electron-ejection process [20].

For the anti-neutrino experiments many groups are planning refined studies of the ^3H decay. One may expect to have an independent check of the ITEP result [2] within the next year.

REFERENCES

[1] K.E. Bergkvist, Nucl. Phys. B39 (1972 317; Phys. Scripta 4 (1971) 23.

[2] V.A. Lyubimov, E.G. Novikov, V.Z. Nozik, E.F. Tretyakov and V.S. Kosik, Phys. Lett. 94B (1980) 266.

[3] J.J. Simpson, Phys. Rev. D23 (1981) 649.

[4] E. Beck and Daniel. Z. Phys. 216 (1968) 229.

[5] P.G. Hansen, J.U. Andersen, D.F. Andersen, G.I. Beyer, G. Charpak, A. De Rújula, B. Elbeck, H.Å. Gustafsson, B. Jonson, P. Knudsen, E. Laegesgaard, J. Pedersen and H.L. Ravn, Eighth Int. Workshop on Weak Interactions and Neutrinos, Javea, Spain 1982.

[6] B. Jonson, J.U. Andersen, G.J. Beyer, G. Charpak, A. De Rújula, B. Elbeck, H.Å. Gustafsson, P.G. Hansen, P. Knudsen, E. Laegesgaard, J. Pedersen and H.L. Ravn, Nucl.Phys. A396 (1983) 479c.

[7] M. Daum, G.H. Eaton, R. Frosch, H. Hirschmann, J. McCulloch, R.C. Minehart and E. Steiner, Phys. Lett. 74B (1978) 126.

[8] H.B. Anderhub, J. Boecklin, H. Hofer, F. Kothermann, P. Le Coultre, D. Makoviecki, H.W. Reist, B. Sapp and P.G. Seiler, Phys. Lett. 114B (1982) 76.

[9] W. Bacino, T. Ferguson, L. Nodulman, W.E. Slater, H.K. Ticho, A. Diamant-Berger, G. Donaldson, M. Duro. A. Hall, G. Irwin, J. Kirkby, F. Merritt, S. Wojciki, R. Burns, P. Condon and P. Cowell, Phys. Rev. Lett. 42 (1979) 749.

[10] J.U. Andersen, G.J. Beyer, G. Charpak, A. De Rújula, B. Elbek, H.Å. Gustafsson, P.G. Hansen, B. Jonson, P. Knudsen, E. Laegsgaard, J. Pedersen and H.L. Ravn, Phys. Lett. 113B (1982) 72.

[11] A. De Rújula, Nucl. Phys. B188 (1981) 414.

[12] C. Møller, Phys. Z. Sowjetunion 11 (1937) 9; Phys. Rev. 51 (1937) 84.

[13] P. Morrison and L.I. Schiff, Phys. Rev. 58 (1940) 24.

[14] R.J. Glauber and P.C. Martin, Phys. Rev. 104 (1956) 158; P.C. Martin and R.J. Glauber, Phys. Rev. 109 (1958) 1307.

[15] M.H. Biavati, S.J. Nassif and C.S. Wu, Phys. Rev. 125 (1962) 1364.

[16] P.K. Hopke and R.A. Naumann, Phys. Rev. 185 (1969) 1565.

[17] H.L. Ravn, J.U. Andersen, G.J. Beyer, G. Charpak, A. De Rújula, B. Elbeck, H.Å. Gustafsson, P.G. Hansen, B. Jonson, P. Knudsen, E. Laegsgaard and J. Pedersen, in proceedings of Neutrino Mass Workshop, Telemark Lodge, 1982.

[18] A. De Rújula, H.Å. Gustafsson, P.C. Hansen, B. Jonson, H.L. Ravn, to be published.

[19] S. Yasumi, G. Rajasekaran, M. Ando, F. Ochiai, H. Ikeda, T. Ohta, P.M. Stefan, M. Maruyama, N. Hashimoto, M. Fusioka, K. Ishii, T. Shinozuka, K. Sera, T. Omori, G. Izawa, M. Yagi, K. Masumoto and K. Shima, Phys. Lett. in press.

[20] A. De Rújula, M. Lusignoli, Phys. Lett. B118 (1982) 429.

A DECADE OF TESTS OF GRAND UNIFIED THEORIES*

Alfred K. Mann

Department of Physics
University of Pennsylvania
Philadelphia, PA 19104

ABSTRACT

This paper considers from an empirical point of view: (i) the agreement between experiment and the grand unified theory (SU(5)) prediction of $\sin^2\theta_W$; (ii) the results of the most recent searches for baryon number and lepton number non-conservation; (iii) possible lepton structure; and (iv) possible electric charge non-conserving processes.

INTRODUCTION

One of the exhilarating trends in today's physics is the apparent formation of a symbiotic relationship between the cosmology of the expanding universe and the attempts to develop an unified theory of the strong, weak and electromagnetic interactions. The "big bang" cosmology is essentially founded on the synthesis of a few fundamental data: these are principally the expansion (Hubble) parameter; the 2.7 °K black body radiation; the hydrogen and helium abundances in the universe; the baryon asymmetry in the universe, $(n_B - n_{\bar{B}})/n_\gamma \neq 0$; and, perhaps, the mechanism of nucleosynthesis. Attempts at unified theories build upon this synthesis and upon an additional small number of data. Specifically, at present, the accomplishments of such attempts are: a rationalization of the equality of electric charges of the electron and proton; and, in the minimal SU(5) model, a precise calculation of the value of the basic constant of the electroweak theory, $\sin^2\theta_W$.

There are also, on a somewhat less substantial level, predictions of grand unified theories concerning small violations of baryon number and lepton number, and the possible existence of massive magnetic monopoles. Furthermore, such theories seem able to accommodate more or less naturally the baryon asymmetry in the universe and a concomitant CP-violation, as well as any reasonable number of flavor families.

One reaches, however, areas of still greater weakness in these theories when certain other basic questions are considered. As examples, the questions of neutrino mass, of possible quark or lepton structure, and of the possible nonconservation of electric charge, among others, are not directly addressed.

It is the purpose of this talk to consider from an empirical point of view a few of the accomplishments, predictions and possible inadequacies of grand unified theories. In particular, it is worthwhile to (i) reconsider the agreement between theory and experiment on $\sin^2\theta_W$, (ii) look at the results of the most recent searches for baryon number and lepton number nonconservation, (iii) discuss possible lepton structure, and (iv) speculate on possible electric charge nonconserving processes.

(i) Theory and Experiment: $\sin^2\theta_W$

It is well known[1] that minimal SU(5) yields

$$(\sin^2\theta_W)_{SU(5)} = 0.214 + 0.006 \ln (0.16 \text{ GeV}/\Lambda_{\overline{MS}}) = 0.214 {}^{+0.004}_{-0.003} \text{ for } \Lambda_{\overline{MS}} = 0.16 {}^{+0.10}_{-0.08} \tag{1}$$

and that numerically the fundamental quantities $\Lambda_{\overline{MS}}$, $\sin^2\theta_W$, M_W, M_Z, M_X and τ_N are related as in Table I, assuming only one Higgs multiplet, N_H.

For $N_H \to N_H + 1$:

$$\sin^2\theta_W \to \sin^2\theta_W + 0.004$$
$$M_W \to M_W - 0.80 \text{ GeV}$$
$$\tau_N \to \tau_N (0.17)$$

Experimentally, the precise prediction of $\sin^2\theta_W$ may be headed for trouble. The recently determined lower limit of $\sim 10^{31}$ yr

Table I. Relationships between certain fundamental quantities in the minimal SU(5) theory. $\Lambda_{\overline{MS}}$ is the QCD mass scale, M_X is the grand unified theory mass scale, and τ_N is the nucleon lifetime. From Marciano and Sirlin, reference 1.

$\Lambda_{\overline{MS}}$ (GeV)	$\sin^2\theta_W$	M_W (GeV)	M_Z (GeV)	M_X (10^{14} GeV)	τ_N (yr)
0.1	0.216	82.8	93.6	1.4	$4 \times 10^{(28\pm1)}$
0.2	0.212	83.6	94.2	2.7	$5 \times 10^{(29\pm1)}$
0.3	0.210	84.0	94.6	4.1	$3 \times 10^{(30\pm1)}$
0.4	0.208	84.4	94.9	5.5	$9 \times 10^{(30\pm1)}$

on τ_N from the IMB experiment[2] corresponds to a value of $\Lambda_{\overline{MS}}$ that greatly exceeds the value obtained from present experiments[3], and is inconsistent as well with the present experimental value of $\sin^2\theta_W$ [4].

It should also be noted that the world average of $\sin^2\theta_W$ is dominated largely by two experiments that exhibit the smallest errors: polarized electron-deuteron inelastic scattering[5], and inelastic neutrino scattering[6]. There is, however, evidence accumulating which suggests that the value of $\sin^2\theta_W$ may be significantly larger[7] than the value obtained from those two experiments. That evidence is still insufficient to lead to a firm conclusion but it does indicate the need for caution in taking for granted the apparent agreement of $(\sin^2\theta_W)_{SU(5)}$ and $(\sin^2\theta_W)_{Expt}$. From Table I it is clear that an experimental value of $\sin^2\theta_W$ greater than, say, 0.22 would cast serious doubt on the validity of minimal SU(5) when taken in conjunction with the present experimental value of $\Lambda_{\overline{MS}}$ and the limit on τ_N.

We may expect this situation to be clarified in the next few years by new data on neutrino scattering, both inelastic[8] and elastic $[\nu_\mu(\bar{\nu}_\mu) + e^- \rightarrow \nu_\mu(\bar{\nu}_\mu) + e^-$ and $\nu_\mu(\bar{\nu}_\mu) + p \rightarrow \nu_\mu(\bar{\nu}_\mu) + p]$[9], and, of course, by precise measurements of the intermediate vector boson masses[10].

(iia) Baryon Instability

We have noted the recent lower limit on τ_N from the IMB experiment. It is useful to consider that experiment in somewhat greater detail. In 80 days of data-taking time, the IMB experiment with a fiducial mass of about 3.5 metric kilotons found about 60 cosmic ray neutrino-induced interactions in the detector, and set a lower limit on the proton lifetime (based on the decay mode $p \rightarrow e^+ + \pi^\circ$) of roughly 10^{31} yr. These data suggest that (a) an improvement in that limit by another order of magnitude will require several ($\approx$3) more years of data-taking, and (b) the approximately 600 neutrino interactions observed in that period will require a rejection factor of about 10^3 to achieve that lower limit. Of course, if protons do in fact decay with a lifetime less than 10^{32} yr, and if the decay mode $p \rightarrow e^+ + \pi^\circ$ is not strongly suppressed, the IMB experiment is very likely to observe that decay.

On the horizon are also several more proton decay experiments[11] some of which use water Cherenkov detectors, while others are fine-grained modular electronic detectors. All are proposed to have a mass of one metric kiloton or less. In view of the IMB results and prospects, and of the summary of possibilities in Fig. 1, it would appear that at least one detector of fiducial mass 10 metric kilotons or greater needs to be considered seriously, if we wish to reach the lifetime limit of 10^{33} yr or wish to study proton decay in detail once it is found at a shorter lifetime.

The attractiveness of grand unified theories and the difficulty of testing them require, in my opinion, that we carry out the search for nucleon instability to the limit imposed by technical feasibility. This appears to be of the order of a few times 10^{33} yr, which value if reached without a positive result would necessarily lead to a critical rethinking of the foundations of grand unified theories and perhaps of the cosmology associated with their foundations. Clearly, this will need at least a decade of further experimentation.

(iib) Neutrino Oscillations[12]

There is another area in which we are faced with the need to

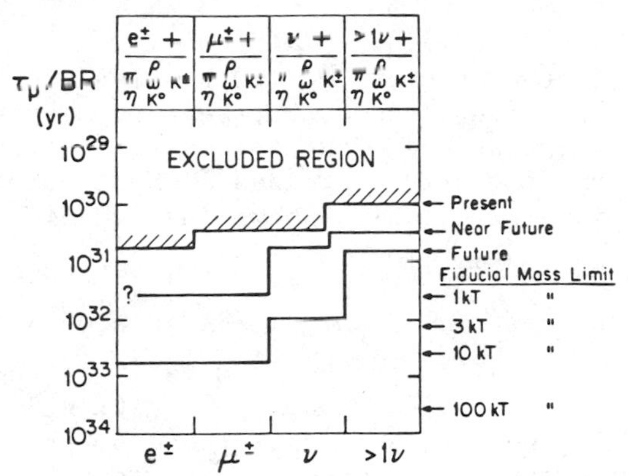

Fig. 1. Approximate expectations for present and future experimental limits on the nucleon lifetime for various decay modes. The limits shown would result from either (1) fewer than five decay events per year or (2) fewer decay events, than neutrino background events, assuming 100% of the decays go into the given channel. Detectors are assumed to have a 33% detection efficiency for decay events, after cuts to remove the neutrino background events. The curve labelled "Present" is for experiments which have been in operation for some time, and are characterized by minimal rejection of the neutrino-induced background (assumed to consist of $\nu_\mu : \nu_e = 2:1$). The curve labelled "Near Future" refers to the expected results from the water Cherenkov experiments and the Frejus tunnel calorimeter. The "Future" curve refers to expectations from fine-grained detectors with $<$ 10 kton fiducial mass and 100 times better background rejection for the electron and muon modes than present experiments. The fiducial-mass limits indicated show what could be achieved with a one-year exposure on the basis of the nucleon content of a detector alone. Taken from D. S. Ayres et al., Proceedings of the DPF Summer Study on Particle Physics and Future Facilities, Snowmass, Colorado, June, 1982, p. 581.

search for a very rare process that is outside the present standard electroweak theory. Here the phenomena--involving non-zero neutrino mass and violation of lepton number conservation--are qualitatively well-defined as neutrino oscillations and double beta-decay, but highly uncertain quantitatively, and, like nucleon decay, may not exist at all.

The present status of neutrino oscillation experiments of the "disappearance" type, i.e., those that seek to measure as a function of distance the disappearance of neutrinos of a given flavor from a neutrino beam that originally contained a known quantity of that flavor, is given in Fig. 2, which shows the two-dimensional space of $\Delta m^2 \equiv |m_i^2 - m_j^2|$ and $\sin^2 2\alpha$, where m_i and m_j represent the masses of possible eigenstate neutrinos, and $\sin^2 2\alpha$ is a measure of the neutrino mixing. In Fig. 2 one sees how large is the $\Delta m^2 - \sin^2 2\alpha$ space that we are more or less capable of testing, and also how relatively small is the area so far excluded by experiment. There is virtually no theoretical guidance as to where to look.

Improvements in the neutrino oscillation data from accelerators are likely to come in the next few years as indicated in Fig. 2, and probably also in the data from reactors. But the most significant expansion of the excluded area in Fig. 2 is likely to follow from experiments using cosmic ray neutrinos as the source and the diameter of the earth as the distance traversed[13], and also from experiments using solar neutrinos as the source and the sun-earth separation as the distance traversed[14]. The latter experiments are capable of searching in the regions suggested in Fig. 2.

To save time, I will forego discussion of neutrino oscillation experiments of the "appearance" type, and of double beta-decay experiments. New double beta-decay experiments, which are treated at length elsewhere[15], will place new strains on experimental techniques but should improve significantly (orders of magnitude) the present limits on neutrinoless double beta-decay and therefore on Majorana neutrino mass and lepton non-conservation contributions to that process.

In short, searches for lepton non-conservation and efforts to measure neutral lepton mass are of comparable interest to but no more tractable than the search for baryon non-conservation. They will therefore also persist for much of the coming decade.

(iii) Structure of Leptons

Still another basic question concerning leptons which is beyond

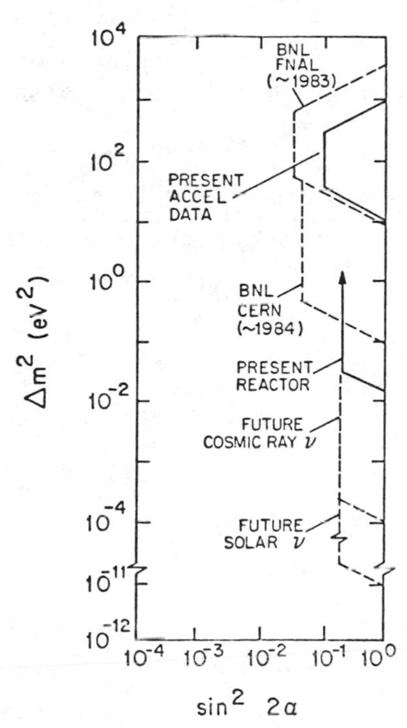

Fig. 2. Plot of the generalized mass parameter $\Delta m^2(eV^2)$ vs the generalized mixing strength parameter $\sin^2 2\alpha$ showing approximate limits of present and future neutrino oscillation data from experiments that measure the disappearance of neutrinos of a given flavor from an incident beam. Note abrupt change in scale of the ordinate.

present theory is that of their possible internal structure. At present, when quarks and leptons are joined in flavor families, it is especially interesting to search for possible structure in the more accessible leptons.

For the charged leptons there are excellent upper limits on the spatial extent of possible structure, i.e., on the magnitude of the smallest distance at which quantum electrodynamics of point charged particles is valid. These are mostly due to precise experiments done at Petra[16], which yield a limiting distance of about 10^{-16} cm for the electron, and slightly larger limits for the muon and tau.

For neutrinos, the limit is less satisfactory. The magnetic moment of $\nu_e(\nu_\mu)$ is known to be less than about 10^{-9} (10^{-8}) Bohr magneton from ν_e + e scattering (pion production by ν_μ). The charge radius of $\nu_e(\nu_\mu)$ is $<4 \times 10^{-15}$ $(<10^{-15})$ cm from the same sources. Observe, however, that non-zero neutrino mass or electromagnetic moments do not imply internal structure of the neutrino any more than they do for charged leptons. Nevertheless, it is of interest to continue to search for a non-zero neutrino magnetic moment by comparing data from high precision neutrino scattering experiments with the electroweak theory, e.g., the y $(\equiv E_e/E_\nu)$-distributions from $\nu_\mu(\bar{\nu}_\mu) + e^- \rightarrow \nu_\mu(\bar{\nu}_\mu) + e^-$ and the q^2-distributions from $\nu_\mu(\bar{\nu}_\mu) + p \rightarrow \nu_\mu(\bar{\nu}_\mu) + p$.

A somewhat different limit on neutrino structure is provided by a comparison of the q^2-dependence of the nucleon structure function $F_2(x,q^2)$ obtained from charged current neutrino-nucleon and muon-nucleon deep inelastic scattering experiments[17]. This procedure essentially searches for structure at the ν_μ-μ-W vertex in the former process relative to the (known absence of) structure at the μ-μ-γ(Z) vertex of the latter process. The upper limit on internal structure of ν_μ that is found from data with q^2-values up to about 200 GeV^2 is of the order of 10^{-15} cm. We may hope to improve that limit substantially when it becomes possible to compare $e^- + p \rightarrow e^- + X$ with $e^- + p \rightarrow \nu_e + X$ at q^2-values in the vicinity of 10^4 GeV^2.

These limits on lepton structure are still one or two orders of magnitude larger than the effective structure that arises from radiative corrections due to the exchange of photons and neutral intermediate vector bosons. This effective structure is of order $M_Z/\sqrt{\alpha} \approx 10^{-17}$ cm, below which it will be very difficult, if not impossible, to search for intrinsic structure of either charged or neutral leptons. Accordingly, with this goal in sight, experiments to search for lepton structure will undoubtedly be pursued in the coming decade.

(iv) Charge Conservation and Photon Mass (with H. Primakoff)

Once we call into question the conservation of baryons and leptons we are led to wonder still another time about the sacrosanct conservation of electric charge. In the standard model the symmetry of $SU(3)_C$ is unbroken, but the symmetry of $SU(2)_L \times U(1)$ is thought to be spontaneously broken, albeit by some mechanism still undemonstrated. U(1) itself is gauge invariant which, as is well-known, implies the conservation of electric charge and also zero mass of the photon. It is of interest to see how well these conditions are satisfied empirically.

In Table II are shown the experimental limits on the lifetime of the nucleon in baryon number violating but electric charge conserving processes, e.g., $p \to e^+ + \pi^\circ$, and in baryon number conserving but electric charge nonconserving processes, e.g., $^{87}Rb \to {}^{87m}Sr$ + neutrals. Also in Table II is a limit on transitions involving lepton number conservation but charge nonconservation, e.g., $e \to \nu_e + \gamma$, which is quantitatively similar to the corresponding limit obtained with baryons. One sees from Table II that the limit on charge non-conservation is eight or nine orders of magnitude worse than the limit on baryon non-conservation.

Table II. Comparison of limits on baryon non-conservation and electric charge non-conservation.

Transition Type	Example	Limit (yr)	Reference
$\Delta B \neq 0$; $\Delta Q = 0$	$p \to e^+ + \pi^\circ$	$>10^{31}$	2
$\Delta B = 0$; $\Delta Q \neq 0$	$^{87}Rb \to {}^{87m}Sr$ + neutrals	$>10^{23}$	18
$\Delta L = 0$; $\Delta Q \neq 0$	$e \to \nu_e + \gamma$	$>10^{22}$	19

It is difficult to see how to improve the experimental limit on charge non-conservation by very much. Study of elementary particle or nuclear processes is unlikely to yield more than about one or two orders of magnitude improvement. The limit obtained from astrophysical data[20] is extracted by considering the net flow of electric charge to the negatively charged earth from all sources, including such charge non-conserving processes as those in Table II. (Note that those are $\Delta Q = +1$ transitions.) The current carried by atmospheric electricity dominates all other currents to and from the earth. That current is less than or about 10^3 amp or 3×10^{10} coul/yr. If, for example, electrons in the earth were decaying by the processes $e \to \nu_e + \gamma$ or $e \to \nu_x + \bar{\nu}_x + \nu_e$ at a rate τ^{-1}, then the resultant rate of change of charge would be $N_e e/\tau \approx 10^{32}$ coul/τ(yr), where N_e is the number of electrons in the earth. Setting the two currents equal yields $\tau \gtrsim 10^{22}$ yr. Thus the astrophysical limit is essentially constrained by the magnitude of the charge carried by atmospheric electricity to the value in Table II.

In an attempt to improve the astrophysical limit one might treat the earth and its atmosphere as a single neutral electrical entity and consider the force of attraction between the earth and its moon. Assuming that $\Delta Q = +1$ processes have been occurring in both bodies since the formation of the solar system, the ratio of the electrical to the gravitational force between them would be simply

$$\frac{F_e}{F_g} = \frac{e^2}{Gm_N^2}\left(\frac{T_s}{\tau}\right)^2 \tag{2}$$

where G is the gravitational constant, m_N is the mass of the nucleon, τ is the reciprocal of the rate of the $\Delta Q \neq 0$ process and T_s is the age of the solar system. If the uncertainty in the calculation of F_g were taken as the limit on F_e, then even if the left-hand side of eq (2) were unity, the limit on τ would be of order 10^{27} yr.

Alas, the elegance of eq (2) is probably rendered useless by the individual currents in the solar wind which are each of order 10^{10} amp. If roughly 10^{-7} of this solar wind current were to penetrate the magnetopause, it would neutralize any charge build-up on the earth plus atmosphere which is less than that corresponding to a decay rate of $1/10^{22}$ yr.

It is interesting to speculate about the process $e \to \nu_e + \gamma$ in a little more detail. The analog of the electromagnetic interaction, $\sqrt{\alpha_{ee}}\, A_\mu\, \bar{\Psi}_e\, \gamma_\mu\, \Psi_e$, might be an interaction of the form $\sqrt{\alpha_{e\mu}}\, A_\mu\, \bar{\Psi}_e\, \gamma_\mu\, \Psi_{\nu_e}$ which would yield in lowest order $\Gamma(e \to \nu_e + \gamma)$, and in next order a nonzero value of m_γ/m_e. One finds in a crude, arbitrarily cut off calculation

$$\frac{m_\gamma}{m_e} \approx \sqrt{\alpha_{e\nu}} \approx \sqrt{\frac{\hbar\Gamma(e \to \nu_e + \gamma)}{m_e c^2}} \tag{3}$$

which gives for $\Gamma(e \to \nu_e + \gamma) = 1/10^{22}$ yr

$$\frac{m_\gamma}{m_e} \leq 10^{-25} \tag{4}$$

This value, when compared with the value $m_\gamma/m_e \leq 10^{-21}$ that is obtained from direct measurements[21], suggests that charge non-conserving processes, e.g., $e \to \nu_e + \gamma$, with rate $\lesssim 10^{-22}$ yr^{-1} (see Table II) are not ruled out by the experimental limit on m_γ/m_e.

A final speculation in this connection follows from the possibility that gauge invariance is broken in some unknown way by the weakest of all known forces, gravity. This gives rise to the amusing numerical coincidence that

$$\alpha_{e\nu} \approx G\, m_e\, m_\nu/\hbar c \tag{5}$$

$$\approx 10^{-50} \text{ , if one takes } m_\nu = 1 \text{ eV.}$$

SUMMARY

It seems clear that the preoccupation of much of the 1980's will be the consolidation of the important experimental and theoretical accomplishments of the previous decade. There is and will continue to be increasing effort, however, to search beyond the present theory in domains of particle physics and astrophysics that present a different challenge. This renewed effort has been stimulated by the possibility of grand unification and the progress in grand unified theories. To the extent that we are able to do so, it is necessary to find either a firm experimental footing for these theories or to repudiate them experimentally. This will be another important stream of research in the 1980's.

REFERENCES

* Work supported in part by U. S. Department of Energy.

1. H. Georgi and S. Glashow, Phys. Rev. Lett. 32, 438 (1974); H. Georgi, H. Quinn and S. Weinberg, Phys. Rev. Lett. 33, 457 (1974); W. Marciano and A. Sirlin, Phys. Rev. Lett. 46, 163 (1981); P. Langacker, Phys. Repts. 72, 185 (1981).

2. R. M. Bionta et al., Phys. Rev. Lett. 51, 27 (1983).

3. See, for example, A. Ali and F. Barreiro, Phys. Lett. 118B, 155 (1982).

4. J. E. Kim, P. Langacker, M. Levine, and H. H. Williams, Rev. Mod. Phys. 53, 211 (1981).

5. C. Y. Prescott et al., Phys. Lett. B84, 524 (1979).

6. See reference 4 for the relevant discussion and references.

7. See, for example, talk given by P. Rapidis in the Proceedings of the Europhysics Study Conference on Electroweak Effects at High Energies, Erice, Sicily, February, 1983 (ed. by H. Newman).

8. H. E. Fisk, Bulletin of the APS 28, 643 (1983).

9. S. M. Heagy, Bulletin of the APS 28, 643 (1983).

10. G. Arnison et al., CERN-EP/83-73, June, 1983, submitted to Phys. Lett. B.

11. See, for example, the summary in A. K. Mann, in Science Underground, Los Alamos, 1982, AIP Conference Proceedings No. 96, (ed. by M. M. Nieto, W. C. Haxton, C. M. Hoffman, E. W. Kolb, V. D. Sandberg and J. W. Toevs), p. 16.

12. A. K. Mann and H. Primakoff, Phys. Rev. D 15, 655 (1977); B. Pontecorvo and S. M. Bilenky, Phys. Repts. 41C, 225 (1978).

13. D. S. Ayres, Jr., et al., Proceedings of the 1982 DPF Summer Study on Elementary Particle Physics and Future Facilities,

Snowmass, Colorado, 1982, (ed. by R. Donaldson, R. Gustafson, and F. Paife), p. 590.

14. R. Davis, B. T. Cleveland and J. K Rowley, in Science Underground, Los Alamos, 1982, AIP Conference Proceedings No. 96, (ed. by M. M. Nieto, W. C. Haxton, C. M. Hoffman, E. W. Kolb, V. D. Sandberg and J. W. Toevs), p. 2.

15. D. Caldwell, Proceedings of the 1982 DPF Summer Study on Elementary Particle Physics and Future Facilities, Snowmass, Colorado, 1982, (ed. by R. Donaldson, R. Gustafson, and F. Paife), p. 600.

16. See, for example, A. Böhm, in Proc. XX International Conference on High Energy Physics, Madison, Wisconsin, 1980 AIP Conference Proceedings No. 68, (ed. by L. Durand and L. G. Pondrom).

17. A. K. Mann, Phys. Rev. D 23, 1609 (1981).

18. A. W. Sunyar and M. Goldhaber, Phys. Rev. 120, 871 (1960); E. B. Norman and A. G. Seamster, Phys. Rev. Lett. 43, 1226 (1979); S. C. Vaidya et al., Phys. Rev. D 27, 486 (1983).

19. M. K. Moe and F. Reines, Phys. Rev. 104B, 992 (1965); R. I. Steinberg et al., Phys. Rev. D 12, 2582 (1975); E. L. Kovalchuk, A. A. Pomansky and A. A. Smolnikov, in Neutrino '79, Bergen, Norway, 1979, (ed. by A. Haatuft and C. Jarlskog), Univ. of Bergen, 1980, Vol. 2, p. 650.

20. A. A. Pomansky, in Neutrino '76, Aachen, W. Germany, 1976, (ed. by H. Faissner, H. Reithler, P. Zerwas), Vieweg, Branschweig, W. Germany, 1977, p. 671; M. Visser, Phys. Rev. D 24, 2542 (1981).

21. A. S. Goldhaber and M. M. Nieto, Rev. Mod. Phys. 43, 277 (1971).

PHENOMENOLOGICAL SUPERGRAVITY

D.V. Nanopoulos

CERN

Geneva

1. WHY SUPERGRAVITY?

Grand unified theories (GUTs), despite all their successes and aesthetic appeal, seem to be problematic[1]. Their main defect is the gauge hierarchy problem. It seems very difficult to create, and then keep intact to all orders in perturbation theory, the energy gap of thirteen decades that exists between the electroweak scale $M_W(\sim 100$ GeV) and the GUT scale $M_X(\gtrsim O(10^{15}$ GeV)):

$$\frac{M_W}{M_X} \leqq O(10^{-13}) \tag{1}$$

Supersymmetry (SUSY), or Fermion-Boson symmetry[2], solves at least the technical aspect of this problem automatically[3]. Given (1) at the tree level, SUSY theories are able to keep it intact to all orders in perturbation theory.

In supersymmetric theories, fermions and bosons are sharing the same multiplet (supermultiplet), and thus new relations between fermion and boson masses, as well as between gauge and scalar self-couplings, are enforced. This is analogous to GUTs where, by putting together quarks and leptons, we get relations between their masses, as well as relations between different gauge coupling constants[1]. In the case of SUSY gauge theories, because of their high symmetry, a lot of "miraculous" cancellations take place and render the theory much more ultra-violet convergent than usual gauge theories. A quantitative expression of this behaviour is shown by the so-called non-renormalization theorems[4]. There is wave function renormalization for each chiral and gauge supermultiplet and gauge coupling renormalization

BUT no separate renormalization of masses, scalar and Yukawa couplings is needed. The non-renormalization theorems[4] go beyond the fact that parameters related by SUSY are expected to be renormalized by the same constant. They say that some parameters are not renormalized at all: THE SET IT AND FORGET IT principle. It is exactly this property of SUSY theories that solves the technical aspect of the gauge hierarchy problem, as mentioned before[3]. Also in SUSY theories, the technical aspect of the strong CP problem is solved automatically[5]. The parameter θ_{QCD} which characterizes the magnitude of the effect ($\theta_{QCD} < 10^{-9}$ experimentally), if set equal to zero, will stay zero, if SUSY is good, because the fermion mass matrix is not renormalized[5]. When SUSY is spontaneously broken, θ_{QCD} acquires a small, finite, acceptable value[5]. Furthermore, if SUSY is exact, no vacuum renormalization is needed[6], a property that eventually may shed light on the absence of the cosmological constant.

Clearly, SUSY cannot be an exact symmetry in Nature, since then all members of a supermultiplet will have the same mass, which is experimentally excluded. For example, there is no scalar electron (selectron) with mass of 0.5 MeV! Actually, present experimental limits[7] put the masses of any charged SUSY particle above O(20 GeV). SUSY needs to be broken. We may consider spontaneous, or explicit but soft breaking, SB or ESB respectively. Arbitrary ESB is ad hoc and problematic (see below). SB of global SUSY is associated with the existence of a Goldstone fermion, the goldstino, ψ_g. If g_{bf} denotes the coupling of the goldstino ψ_g to a supermultiplet that contains a boson b and a fermion f, then the boson-fermion mass splitting is given by a Goldberger-Treiman type relation:

$$m_b^2 - m_f^2 = g_{bf} M_S^2 \tag{2}$$

where M_S denotes the SUSY breaking scale. A priori, g_{bf} and M_S are arbitrary parameters, but they have to satisfy certain physical constraints: (i) absence of anything "new" below 20 GeV[7] imposes a lower bound, at least for the charged SUSY partners of the "observed" particles:

$$g_{bf} M_S^2 \geq O\left((20 \text{ GeV})^2\right) \tag{3}$$

(ii) absence of radiative corrections to the mass of the electroweak Higgs (E-Higgs) larger than $O(M_W)$, i.e., solution of the gauge hierarchy problem (technical aspect), imposes an upper bound:

$$g_{bf} M_S^2 \leq O\left((M_W)^2\right) \tag{4}$$

Clearly, the most naive way to satisfy (3) and (4) is to identify g_{bf} with some gauge coupling and M_S with M_W:

$$\begin{aligned} g_{bf} &\sim g \\ M_S &\sim O(M_W) \end{aligned} \tag{5}$$

This approach has been tried for many years now by Fayet, with rather limited success[8]. The Fayet-type models[8] suffer from incurable diseases, such as broken colour and/or electromagnetism, and/or Adler-Bell-Jackiw anomalies, and/or dangerous flavour-changing neutral currents, etc.[9]. We need something better. The other way is to diminish g_{bf} with an approximate increase of M_S, such that (3) and (4) are still satisfied. This approach, instigated by the work of Barbieri, Ferrara and myself[10], and followed by several other groups[11], seems to be more successful. The idea is very simple[10]. One has to find a natural way, so that the effective goldstino coupling to "ordinary" matter (quarks, leptons, E-Higgs, $W^{\pm}$, Z, γ, g) and their SUSY partners is extremely weak. This tiny effective coupling can be arranged at the tree level[10], or if absent at the tree level, may be induced, after some gymnastics, at some number of loops by radiative corrections[11]. In any case, in a large class of this type of models, one gets the following relations, as discussed in Ref. 10):

$$\begin{aligned} g_{bf} &\sim \frac{M_W}{M_{Pl}} \\ M_S^2 &\sim M_W M_{Pl} \sim (10^{10}\ \text{GeV})^2 \end{aligned} \tag{6}$$

which automatically satisfy the constraints (3) and (4). In other words, despite the fact that the whole theory is characterized by a SUSY breaking scale $M_S \sim 10^{10}$ GeV, the "low energy" world ($\equiv$ "ordinary" matter plus SUSY partners) is feeling an effective SUSY breaking

$$(M_S)_{eff} \sim O(M_W) << M_S \tag{7}$$

Thanks to the "magic" properties[4] of SUSY theories, this decoupling has been shown to persist to all orders in perturbation theory[10],[12].

Up to now, we have considered global SUSY gauge theories and have shown that the construction of "realistic"(?), phenomenologically accepted models[10],[11], has enforced upon us Eq. (6). This fact is extremely important because the supergravitational (SUGAR)[13] coupling between the goldstino and "ordinary" matter ($m \sim M_W$)

$$(g_{bf})_{SUGAR} \sim \sqrt{G_N}\, m \simeq \frac{M_W}{M_{P\ell}} \tag{8}$$

is exactly of the same order of magnitude as the corresponding non-gravitational coupling, given by Eq. (6). In other words, (super) gravitational effects cannot be neglected anymore[14),15),16)].

We may go even further and assume that non-gravitational g_{bf}'s are altogether absent and only those induced by supergravity remain. This is a spectacular idea, because it is not only very exciting that gravity, for the first time, plays a major role in low energy physics, but also because the resulting models are the simplest of all the realistic models of broken SUSY.

If we follow this line of thought, then (8) and (4) set an upper limit[15)] on M_S:

$$M_S^2 \leq O(M_W M_{P\ell}) \tag{9}$$

while (8) and (3) set a lower limit on M_S:

$$M_S^2 \gtrsim O\left(\left(\frac{20\,GeV}{M_W}\right)^2 M_W M_{P\ell}\right) \tag{10}$$

More or less in SUGAR-type models, M_S is determined uniquely from physics constraints to be

$$M_S^2 \sim O(M_W M_{P\ell}) \tag{11}$$

The coincidence between (6) and (8), (11) is rather amazing. Intuitively, one can easily understand the need to move from global to local SUSY or supergravity[13)]. Consider the one-loop correction to the E-Higgs mass due to the graviton

$$\delta m_H^2 \sim \frac{\Lambda^4}{M_{P\ell}^2} \tag{12}$$

This is catastrophic because the natural scale for the cut-off Λ is $M_{P\ell}$, which, through (12), implies electroweak breaking at the Planck scale! We need either extremely accurate fine tuning or unavoidable cancellation of the graviton (boson) loop by a fermion loop, i.e., supersymmetric gravity, thus SUPERGRAVITY[13)]. The needed spin 3/2 fermion sits on the same supermultiplet with the graviton (spin 2) and it is called gravitino. As the graviton may be considered the "gauge boson" of the Poincaré algebra, the gravitino is the "gauge fermion" of local SUSY. After SB of local SUSY, the gravitino "eats" the goldstino, thus becoming massive[17]:

$$m_{3/2} \sim \sqrt{G_N}\, M_S^2 \sim \frac{M_S^2}{M_{P\ell}} \tag{13}$$

which implies, by using (11) [10]:

$$m_{3/2} \sim O(M_W) \tag{14}$$

One may argue that, because of the anticommutation relation, which in standard notation[2] reads

$$\{ Q_\alpha, \bar{Q}_\beta \} = -2(\gamma_\mu)_{\alpha\beta} P_\mu$$

global SUSY plus gravity implies automatically local SUSY or supergravity, so what's new? Well, that is correct, but it was thought for a long time that, as in the case of ordinary gravity, the effects of supergravity, at least for low energy physics, would be negligible. We have proved here that this is not necessarily the case. Only if the Fayet-type models[8] were correct, would (8) indeed be much smaller than (5) and SUGAR effects would be negligible. Nevertheless, Fayet-type models[8] do not work[9], thus supergravity enters the "low energy" physics world in full strength and glory.

2. PHYSICAL STRUCTURE OF SIMPLE (N = 1) SUPERGRAVITY

We are then led to consider local SUSY gauge theories[18]. The effective theory below the Planck scale must be[19] N = 1 supergravity. The restriction to N = 1 follows from the apparent left-right asymmetry of the "known" gauge interactions. Since we are dealing with local SUSY, the breaking of SUSY must be spontaneous, not explicit, if Lorentz invariance or unitarity are not to be violated. It is remarkable that the effective theory below $M_{P\ell}$ has been uniquely determined[19] to be a spontaneously broken N = 1 local SUSY gauge theory[18].

We start with a reminder of the structure of N = 1 supergravity actions[18] containing gauge and matter fields (if not explicitly stated, we use natural units $k^2 \equiv 8\pi G_N = (8\pi/M_{P\ell}^2) \equiv 1/M^2 = 1$)

$$A = \int d^4x\, d^4\theta\, E \Big\{ \Phi(\varphi, \bar{\varphi} e^{2V}) + \mathrm{Re}[R^{-1} g(\varphi)] + \mathrm{Re}[R^{-1} f_{\alpha\beta}(\varphi)\, W_a^\alpha\, \varepsilon^{ab}\, W_b^\beta] \Big\} \tag{15}$$

where E is the superspace determinant, Φ is an arbitrary real function of the chiral superfields ϕ and their complex conjugates $\bar{\phi}$, V is the gauge vector supermultiplet, R is the chiral scalar curvature superfield, g is the chiral superpotential, $f_{\alpha\beta}$ is another chiral function of the chiral superfields ϕ, and W^a_α is a gauge-covariant chiral superfield containing the gauge field strength. In addition to all the obvious general co-ordinate transformations, local supersymmetry and gauge invariance, the action (15) is also invariant[18] under the transformations

$$\begin{aligned} J \equiv 3 \ln\left(-\tfrac{1}{3}\Phi\right) &\rightarrow J + K(\phi) + K^*(\bar{\phi}) \\ g(\phi) &\rightarrow e^{K(\phi)} g(\phi) \end{aligned} \tag{16}$$

These transformations can be related to a description of the chiral superfields ϕ as co-ordinates on a Kähler manifold with Kähler potential Φ, and the transformations (16) are known as Kähler gauge transformations[20]. One particular manifestation of this Kähler gauge symmetry is in the effective scalar potential

$$V = -\exp(-G)\left(3 + G'_i G''^{i}_{j}{}^{-1} G'^{j}\right) + (\text{gauge terms}) \tag{17}$$

where

$$G \equiv J - \ln\left(\tfrac{1}{4}|g|^2\right) \tag{18}$$

which is clearly invariant under the transformations (16). In general, the action depends on a real function

$$\hat{\Phi} \equiv \Phi / |g(\phi)|^{2/3} \tag{19}$$

and on the chiral function $f_{\alpha\beta}(\phi)$. The most familiar forms of these functions are $J = -\phi\bar{\phi}/2$, giving canonical kinetic energy terms for the chiral superfields, $g(\phi)$ a cubic polynomial giving renormalizable matter interactions of dimension ≤ 4, and $f_{\alpha\beta} = \delta_{\alpha\beta}$. We expect that more complicated functions will contain terms $O(\phi/M_{P\ell})^n$ relative to these canonical leading terms.

Ellis, Tamvakis and myself have suggested[19] interpreting Eq. (15) as an effective action suitable for describing particle interactions at energies $\ll M_{P\ell}$ just as chiral $SU(N) \times SU(N)$ Lagrangians were suitable for describing hadronic interactions at energies $\ll 1$ GeV. In much the same way as we know that physics gets complicated at $E = 1$ GeV, with many new hadronic degrees of freedom having masses of this order, we also expect many new "elementary particles" to exist with masses $O(M_{P\ell})$. It may well be that all the known light "elementary particles", as well as these heavy ones, are actually composite, and that at energies $\gg M_{P\ell}$ a simple preonic picture will emerge, analogously to the economical description of high-energy hadronic interactions in terms of quarks and gluons. It may even be that these preonic constituents are themselves ingredients in an extended supergravity theory[21]. But let us ignore these speculations for the moment and return to our pedestrian phenomenological interpretation of the action (15).

The well-known rules of phenomenological Lagrangians[22] are that one should write down all possible interactions consistent with the conjectured symmetries [e.g., chiral $SU(2) \times SU(2)$], and only place absolute belief in predictions which are independent of the general form of the Lagrangian (e.g., $\pi\pi$ scattering lengths). These are the reliable results which could also be obtained using current algebra arguments. It does not make sense to calculate strong interaction radiative corrections (read: supergravity loop corrections) to these unimpeachable predictions: these are ambiguous until we know what happens at the 1 GeV scale (read: $M_{P\ell}$), and our ignorance can be subsumed in the general form of the phenomenological Lagrangian, in which any and all possible terms are present a priori (read: non-trivial J, non-polynomial g and $f_{\alpha\beta}$). On the other hand, non-strong interaction radiative corrections can often be computed meaningfully (e.g., the $\pi^+-\pi^o$ mass difference, large numbers of pseudo-Goldstone boson masses in extended technicolour theories). Similarly, it makes sense to compute matter interaction (gauge, Yukawa, Higgs) corrections to the tree-level predictions of the effective action (15).

Since the supergravity action is non-renormalizable, and since both the Φ and $f_{\alpha\beta}$ terms in the action (15) have a $\int d^4\theta$ form, we expect general variants of them to be generated by loop corrections. Presumably, radiative corrections maintain the essential geometry of the Kähler manifold[20]. Therefore, we expect loop corrections to fall into the class of Kähler gauge transformations (16). The only analogous transformation allowed in a conventional renormalizable theory is K = constant, corresponding to a wave function renormalization. In our case, more general gauge functions $K(\phi)$ might appear.

In $N = 1$ SB local SUSY gauge models, called for abbreviation supergravity or SUGAR models, one usually distinguishes two sectors: (i) the "observable" sector containing quarks, leptons, Higgs and gauge bosons of electroweak and GUT types, as well as their SUSY partners; (ii) the "hidden" sector containing at least the goldstino and its SUSY associate. The "observable" and "hidden" sectors both couple to supergravity, but not directly to each other. In other words, ignoring supergravity, physics in the "observable" sector would be completely supersymmetric. A realization of this programme occurs, for example, by splitting the superpotential of the theory into the sum of two terms

$$g(\varphi_i) = h(Z_i) + f(\chi_i) \tag{20}$$

where χ_i are the "observable" fields and Z_i the "hidden" fields. The f - part of the superpotential has to do with the "observed" sector and contains most of the physics, while the h - part of the superpotential has to do with SB of local SUSY and the vanishing of the cosmological constant. The scalar fields of the hidden sector typically have vev's of $M_{P\ell}$ which cause SUSY to be SB at a scale M_S determined by the parameters in the h - part. We may choose these parameters to be such that (11), or equivalently (14), are satisfied, as well as cancelling the cosmological constant at the same time ("superHiggs effect")[18]. A celebrated example is the Polonyi potential[23]:

$$h(Z) = M_S^2 (Z + B) \tag{21}$$

which, in the absence of other fields, and for $B = (2-\sqrt{3})M$ and $\langle Z \rangle = (\sqrt{3}-1)M$, $[M \equiv (M_{P\ell}/\sqrt{8\pi}) \simeq 2.5 \cdot 10^{18}$ GeV being the appropriate supergravity scale (superPlanck mass)], implies SB of local SUSY at M_S and vanishing cosmological constant. The communication between the two sectors is mediated by the auxiliary fields of the SUGAR multiplet and takes place at tree-level in a model-independent way. Exchange of gravitons or gravitinos plays no role at this level. The elimination of these auxiliary fields produces non-renormalizable interactions between the two sectors. These non-renormalizable interactions include the ones between the hidden fields Z_i, usually taken to be neutral under all gauge symmetries, and gauge fields.

All these effects have been summed into an effective scalar potential[18] [(17), but with canonical kinetic energy terms for the chiral superfields].

$$V(\varphi) = \exp\left(\frac{1}{M^2}\sum_i \varphi_i \bar{\varphi}_i\right)\left\{\sum_i \left|\frac{\partial g(\varphi)}{\partial \varphi_i} + \frac{\bar{\varphi}_i\, g(\varphi)}{M^2}\right|^2 - \frac{3\,|g(\varphi)|^2}{M^2}\right\} + \frac{1}{2}\sum_a D_a^2(\varphi,\bar{\varphi}) \tag{22}$$

This mess involves the scalar fields ϕ_i, the superpotential g and the D-terms, which have their usual global SUSY form. Incidentally, by comparing (22) with the global SUSY potential

$$V(\varphi) = \sum_i \left| \frac{\partial g(\varphi)}{\partial \varphi_i} \right|^2 + \frac{1}{2} \sum_a D_a^2 (\varphi, \bar{\varphi})$$

we notice the richer structure of the SUGAR potential [(22)] and thus we anticipate that more physics will come out from it. When the scalar components of the hidden fields are replaced by their vev's, all superheavy fields are integrated out, and expanding in powers of $1/M$, the resulting effective theory, just below M_X, for the light observable fields, will contain both the usual SUSY terms and a soft SUSY breaking piece[15),24)-26)]:

$$\mathcal{L}_{SOFT} = \sum_i m_i^2 |\chi_i|^2 + m_{3/2} \sum_n (A_n f_n + h.c.) - \sum_{\alpha=1}^{3} \left(\frac{1}{2} M_\alpha \lambda_\alpha \lambda_\alpha + h.c. \right) \tag{23}$$

where λ_α is a gauge fermion [gluino (α=3), wino (α=2) or bino (α=1)], f_n is any term in f and χ_i is any scalar (Higgs, squark or slepton). The A_n's [25)] are expected to be of order one, while m_i and M_α should be in general of order $m_{3/2}$ [15),19),27)]. Actually, when corrections[26)] from integrating out superheavy fields are ignored, the following relations hold[15)]:

$$m_i^2 = m_{3/2}^2 \tag{24}$$

for every scalar field, and[24),25)]

$$A_n = A - 3 + d_n \tag{25}$$

where A is a universal number[25)] (assumed real) and d_n is the number of fields multiplied together in term n. If we imagine that the low energy theory is embedded in a GUT model at some GUT scale below the Planck mass, then all gaugino masses are equal at the GUT scale, so that only one single parameter M_o is needed [M_o _in principle_ may be of $O(m_{3/2})$]:

$$M_\alpha(M_X) = M_0 \quad , \quad \alpha = 1,2,3 \tag{26}$$

while at lower energies M_α evolves in a manner identical for the gauge couplings:

$$\frac{M_\alpha(\mu)}{M_0} = \frac{\tilde{\alpha}_\alpha(\mu)}{\alpha_G} \quad , \quad \alpha = 1,2,3 \tag{26'}$$

with $\tilde{\alpha}_\alpha$, α_G the usual SU(3), SU(2), U(1), GUT fine structure constants.

In the following, we shall assume that corrections[26)] from integrating out superheavy fields are not very important and proceed with the simplified soft SUSY breaking piece, just below M_X,

$$\begin{aligned} \mathcal{L}_{SOFT} = {} & m_{3/2}^2 \sum_i |x_i|^2 + m_{3/2} \sum_n (A-3+d_n) f_n + h.c.] \\ & - \frac{1}{2} M_0 \sum_\alpha \lambda_\alpha \lambda_\alpha + h.c. \end{aligned} \tag{27}$$

The soft operators in (23) or (27) are determined by the couplings of the hidden fields and by radiative corrections at the Planck scale, including those due to gravity. In spite of this, the very interesting thing is that sometimes it is possible to give the form of these operators without making detailed assumptions about either the hidden sector or the effects of gravitational radiative corrections. For example, in (27), the form of the hidden sector enters only through the three parameters $m_{3/2}$, A and M_0. The non-renormalizable interactions of the hidden fields with gauge fields, discussed before, will generally lead to soft Majorana masses, M_0, for the gauginos [$f_{\alpha\beta} \neq \delta_{\alpha\beta}$ in (15)], while the non-renormalizable interactions between the "observable" and "hidden" sectors, commented before, will produce the soft operators in the scalar potential [first two terms in (23) or (24)].

In a nutshell, the SB of SUSY in SUSY gauge theories coupled to N = 1 SUGAR leads to an effective theory below the Planck scale in which global SUSY is explicitly broken by a constrained set of soft operators, at an effective scale:

$$(M_S)_{eff} \simeq m_{3/2} \tag{28}$$

As discussed before, physics constraints [see (7)] impose the condition $m_{3/2} \sim O(M_W)$ [see (14)], which creates a new hierarchy problem. Since we are dealing with gravitational phenomena, naively, the natural mass scale for the gravitino is $M_{P\ell}$ and not M_W. I call this the supergravity hierarchy problem, or the SUGAR hierarchy problem. On the other hand, it should be emphasized that the automatic soft breaking of global SUSY that is provided in the SUGAR framework not only splits "low energy" supermultiplets in the right way [all scalars and gauginos getting masses $O(m_{3/2})$], BUT it has the correct form to pass unscathed through all the traps set out by low energy phenomenology.

In the physics applications which follow, we shall make extensive use of two main characteristics of the general framework discussed above. First, since we are dealing with an effective theory (the $N = 1$ SUGAR action is non-renormalizable), the superpotential g is not anymore necessarily constrained by renormalizability to be at most cubic, but it may contain any higher powers, suitably scaled, by inverse powers of $M_{P\ell}$, the natural cut-off of the theory[19]. Secondly, because of the non-renormalization theorems[4] of SUSY (SET IT AND FORGET IT principle), we may set, as we wish, certain parameters equal to zero, even if no symmetry implies that - a very different situation from ordinary gauge theories. Here, no apologies are needed. As explained in detail before, most of the physics is contained in the "observable" sector superpotential $f(\chi_i)$ [see (20)]. Here we shall assume that, in one way or another, the "hidden" sector has played its role, as discussed previously, and we shall concentrate on the form of $f(\chi_i)$. We follow the natural (cosmic) evolution of things starting at energies below $M_{P\ell}$ and "coming down" to M_W. So we distinguish physics around the GUT scale (M_X) and physics around the electroweak (E-W) scale (M_W).

All physics from $M_{P\ell}$ down to (and including) low energies should emerge from such a programme. We will show next that this is indeed possible.

3. PHYSICS WITH SIMPLE (N = 1) SUPERGRAVITY

A. Physics Around the GUT Scale (M_X)

The superhigh energy regime ($\sim 10^{16}$ GeV) is the theorists' paradise. There is a lot of freedom in building models, even though the constraints both from particle physics and cosmology become tighter and tighter. For definiteness, simplicity, and out of habit, we shall take as our prototype GUT an SU(5) type model[1]. All GUT physics information will be contained in f_{GUT}, the GUT part of the "observable sector" superpotential. There is no consensus about the definite form of this superpotential, but it should unavoidably contain a piece (f_I) that breaks SU(5) down to SU(3) × SU(2) × U(1) and if possible, a piece (f_{II}), providing some explanation about the tree-level gauge hierarchy problem, so we write:

$$f_{GUT} = f_I + f_{II} \tag{29}$$

For example, we may take[28)]

$$f_I = \frac{a_1}{M} X^4 + \frac{a_2}{M^2} X^2 \, Tr(\Sigma^3) \tag{30}$$

and[29)]

$$\begin{aligned} f_{II} = & \; \bar{\Theta} H \left(\lambda_1 \frac{\Sigma^2}{M} + \lambda_2 \frac{\Sigma^3}{M^2} + \cdots \right) \\ & + \bar{H} \Theta \left(\lambda_1' \frac{\Sigma^2}{M} + \lambda_2' \frac{\Sigma^3}{M^2} + \cdots \right) + M_\Theta \bar{\Theta} \Theta \end{aligned} \tag{31}$$

where $X = \underline{1}$, $\Sigma = \underline{24}$, $\overset{(-)}{\theta} = \overset{(-)}{\underline{50}}$, $\overset{(-)}{H} = \overset{(-)}{\underline{5}}$ are chiral superfields of SU(5). The Higgs fields H and $\bar{H}$ couple to quark and lepton fields in the usual way. All components of θ and $\bar{\theta}$ have a bare mass M_θ (which is taken to be of order M_X or larger), and so remain heavy after SU(5) breaks to SU(3) × SU(2) × U(1). After minimizing the potential, obtained by plunging into (22) the sum of f_I and f_{II} as given by (30) and (31), we get zero vev's for $\overset{(-)}{H}$ and $\overset{(-)}{\theta}$ but non-zero ones for

$$\begin{aligned} \langle X \rangle &= \left(\frac{m_{3/2}}{M} \right)^{3/8} M \\ \langle \Sigma \rangle &= \left(\frac{m_{3/2}}{M} \right)^{1/4} M \end{aligned} \tag{32}$$

Furthermore, we find[28)] that the SU(3) × SU(2) × U(1) symmetric minimum is the lowest one for all values of a_1 and a_2, with a value[28)]

$$V_{eff} \simeq - \left(\frac{m_{3/2}}{M} \right)^{5/2} M^4 \tag{33}$$

What do these results mean? First, since the vev of Σ sets the scale of SU(5) breaking, we find that the GUT scale M_X satisfies[28)]

$$M_X^4 \simeq O(m_{3/2} M^3) \tag{34}$$

which is a highly successful relation. Using as an input the <u>non-hierarchical</u> and easy to explain ratio $(M_X/M) \sim 10^{-2}$-10^{-4}, we obtain that $m_{3/2} \sim O(100\ \text{GeV})$! More generally, relations of the form $M_X^{2p-2} \simeq O(m_{3/2} M^{2p-3})$ with $p \geq 3$, are also possible[28)] by suitably modifying the exponents in (30). The supergravity hierarchy problem has been solved in a rather simple way.

Secondly, the $SU(3) \times SU(2) \times U(1)$ symmetric minimum is lower in energy density than the SU(5) symmetric minimum $X = \Sigma = 0$ by an amount $(m_{3/2} / M)^{5/2} M^4$. Thirdly, the barrier between these two minima is never larger than $(m_{3/2} / M)^{5/2} M^4$, the same as the splitting between the states. Why this is so can be seen by noting that if we replace X by its vev (32) in (30), the effective renormalizable self-coupling of Σ is 10^{-12} $\text{tr}(\Sigma^3)$. Thus we have generated[28)] a small renormalizable coupling for Σ from our starting point of only non-renormalizable interactions among X and Σ. This small coupling suppresses the barrier between the SU(5) and the $SU(3) \times$ $\times\ SU(2) \times U(1)$ phases. The consequences of this suppression for supercosmology[30)] are difficult to overestimate. Simply, it now makes possible the transition from the SU(5) to the $SU(3) \times SU(2) \times U(1)$ phase at temperatures $T \sim 10^{10}$ GeV, which was previously blocked, since the barrier between the two phases was of the order of $(M_X)^4$. Incidentally, in this picture, the number density of GUT monopoles is naturally suppressed[30)] below its present experimental upper bound.

It should be clear that the basic result - small renormalizable couplings arising from non-renormalizable ones suppressed only by inverse powers of M - is quite general and does not depend on the detailed form of the superpotential (f_I) [28)]. The main characteristics of these types of models[28),31)] are that they provide relations of the type (34); they make possible "delayed" SU(5) to $SU(3) \times$ $\times\ SU(2) \times U(1)$ phase transitions at $T \sim 10^{10}$ GeV, and they contain[28),31)] more "light" particles than the ones in the minimal SUSY $SU(3) \times SU(2) \times U(1)$ model. This last fact may sound dangerous when calculating M_X, $\sin^2\theta_{E-W}$ and m_b/m_τ, since in general an arbitrary increase of "light" stuff gives an out-of-hand increase[32),33)], and thus experimentally unacceptable values for, the above-mentioned quantities. A more careful analysis[34)] of these cosmological acceptable models (CAM)[34)] shows that they make predictions as successful (for $\sin^2\theta_{E-W}$, m_b/m_τ,...) as <u>at least</u> the ones[32),33)] of the phenomenologically acceptable minimal type models (MIMO). For a detailed, thorough phenomenological analysis of CAMs, see Ref. 34).

Next, we discuss[29)] physics related to f_{II} as given in (31). The vev of Σ does not only break SU(5) to SU(3) × SU(2) × U(1) but also provides a mass term which mixes the colour triplets in H and $\bar{H}$ with those in θ and $\bar{\theta}$. However, there is no weak doublet in the 50, and so the weak doublets in H and $\bar{H}$ remain massless. The colour triplets will have a mass matrix[29)]

$$\begin{pmatrix} 0 & \sim \frac{M_X^2}{M} \\ \sim \frac{M_X^2}{M} & M_\theta \end{pmatrix} \tag{35}$$

where M_θ should be of order M_X or larger ($\leq M$), to avoid having particles from θ and $\bar{\theta}$ influencing the renormalization group equations at scales below M_X (or even M). The eigenvalues of this mass matrix are $O(M_\theta)$ and $O(M_X^4/M_\theta M^2)$; this latter eigenvalue is about 10^{10} GeV for $M_X \sim O(10^{16}$ GeV) and $M_\theta \sim O(M)$. In this case, the Higgs colour triplet can be used to generate[28),30)] the baryon number of the Universe after the SU(5) to SU(3) × × SU(2) × U(1) transition which, as discussed earlier, occurs at temperatures $T \sim 10^{10}$ GeV in CAMs[34)]. It is remarkable that $O(10^{10}$ GeV) is the lower bound[35)] allowed for colour triplet Higgs masses from present limits on proton decay ($\tau_p > 10^{31}$ years). If indeed there are 10^{10} GeV Higgs triplets, then protons should decay predominantly[36)] to $\bar{\nu}_\mu K, \mu K$ with a lifetime $\sim O(10^{31}$ years).

The role of supergravity in this natural explanation[29)] of the Higgs triplet-doublet mass splitting ($\equiv$ tree-level gauge hierarchy problem) is fundamental, in several aspects. The same kind of explanation had been suggested before[37)] in the framework of renormalizable global SUSY GUTs, where Σ^2 in (31) was replaced by a 75 of SU(5) and higher than two powers of Σ were absent[37)]. Unfortunately, the use of 75 drastically conflicts with cosmological scenarios[28)-31)] based on SUSY GUTs. The barrier between the SU(5) and SU(3) × SU(2) × U(1) phases is impossible to overcome unless most of the 75 is very light ($\sim M_W$). But then, all hell breaks loose. A light 75 makes the gauge coupling in the SU(5) phase decrease at lower energies so there is no phase transition at all[28),30)]. Furthermore, the presence of these new light particles in the SU(3) × SU(2) × U(1) phase changes the renormalization group equations, and prevents perturbative unification. On the contrary, in SUGAR theories, since we may use non-renormalizable terms, we may replace the fundamental 75 by an "effective" 75 contained in Σ^2. Unlike a light 75, a light 24 neither makes the SU(5) gauge coupling decrease at energies below M_X, nor upsets perturbative unification[29)]. The previously mentioned cosmological scenarios[28)-31)] can proceed without modification. In addition, SUSY non-renormalization theorems[4)] ensure the stability of the triplet-doublet splitting to all orders in perturbation theory. Since the only modifications of the theory are at the GUT scale M_X, it seems that we have got[29)] a harmless and elegant solution of the tree-level, and for that matter, to all orders in perturbation theory, gauge hierarchy problem.

SUGAR models give good physics at the GUT scale - unique, cosmologically acceptable breaking of SU(5) to SU(3) × SU(2) × U(1), with an explanation of the smallness of the gravitino mass[28)] [(34)-like relations], and a natural explanation[29)] of the Higgs triplet-doublet splitting, cosmologically fitted and general enough. We believe that even if the very specific form of f_I in (30) may change, then f_{II} as given by (31) (or its obvious generalization to other GUT models) will be always a useful part of the f_{GUT}.

After finding plausible explanations for the SUGAR hierarchy problem [gravitino mass ∿ O(100 GeV)], the tree-level and higher orders gauge hierarchy problem (triplet-doublet Higgs splitting), it is time to explain the last gauge hierarchy problem (1), i.e., why does $M_W/M_X \leqq O(10^{-13})$? This problem brings us naturally to our next subject.

B. Physics Around the Electroweak (E-W) Scale (M_W)

Although there is no consensus on the best way to incorporate grand unification in SUGAR models, a unique minimal low energy model has recently emerged[38)-40)]. In this model, the physics of the TeV scale is described by an effective SU(3) × SU(2) × U(1) gauge theory, in which the breaking of weak interaction gauge symmetry is induced by renormalization group scaling of the Higgs $(\text{mass})^2$ operators[19)]. Much of the attractiveness of this model stems from the fact that no gauge symmetries or fields beyond those required in any low energy SUSY theory are included. Sometimes, it may happen, as is the case of Cosmological Acceptable Models[34)] (CAMs), that there are GUT relics which are light ($\sim M_W$), but they do not seem to play any fundamental role at low energies, so we may neglect them in our present discussion. Furthermore, adding random chiral superfields to the low energy theory may be problematic. For example, the presence of a gauge singlet superfield coupled to the Higgs doublets and added to trigger SU(2) × U(1) breaking[41),24),27)], usually (but not always[42)]) destroys[43)] any hope of understanding the gauge hierarchy problem; the reason being[43)] that in a GUT theory, the gauge singlet does not only couple to the Higgs doublets but also to their associate, superheavy colour triplets. Then we have to try hard[42)] to avoid 10^{10} GeV Higgs doublet masses, generated by[43)] one-loop effects involving colour triplets. Something smells fishy.

We focus then on the standard low energy SU(3) × SU(2) × U(1) gauge group, containing three generations of quarks and leptons, along with two Higgs doublets, as chiral superfields. The low energy effective superpotential (f_{LES}) of the model consists only of the usual Yukawa couplings of quark and lepton superfields to the Higgs superfields, along, in general, with a mass term coupling the two Higgs doublets, H_1 and H_2. Explicitly, in a standard notation:

$$f_{LES} = h_{ij} U_i^c Q_j H_2 + \tilde{h}_{ij} D_i^c Q_j H_1 + f_{ij} L_i E_j^c H_1 + m_4 H_1 H_2 \tag{36}$$

where a summation over generation indices (i,j) is understood and $Q(U^c)$ denote generically quark doublets (charge -2/3 antiquark singlet) superfields, while $L(E^c)$ refers to lepton doublets (charge -1 antilepton singlet) superfields. With the exceptions of the top quark Yukawa coupling and the mass parameter m_4, which in principle may be of order $O(m_{3/2})$, all other parameters appearing in (36) contribute to the masses of the observed quarks and leptons and are known to be small. Neglecting these small couplings, the effective Low Energy Potential (V_{LEP}) can be written as [see (22), (23) and (27)]:

$$\begin{aligned} V_{LEP} = {} & \sum_{i=1}^{3} \left\{ m_{L_i}^2 |L_i|^2 + m_{E_i}^2 |E_i^c|^2 + m_{Q_i}^2 |Q_i|^2 + m_{U_i}^2 |U_i^c|^2 + m_{D_i}^2 |D_i^c|^2 \right\} \\ & + m_1^2 |H_1|^2 + m_2^2 |H_2|^2 + A h_t m_{3/2} (U_3^c Q_3 H_2 + h.c.) \\ & + B m_{3/2} m_4 (H_1 H_2 + h.c.) + h_t m_4 (H_1^+ Q_3 U_3^c + h.c.) \\ & + h_t^2 \left(|Q_3|^2 |U_3^c|^2 + |Q_3|^2 |H_2|^2 + |U_3^c|^2 |H_2|^2 \right) \\ & + \text{"D-terms"} \end{aligned} \tag{37}$$

The effective parameters appearing in (37) take, at large scales ($\sim M_X$), the values

$$m_1^2(M_X) = m_2^2(M_X) = m_{3/2}^2 + m_4^2(M_X)$$

$$m_{Q_i}^2(M_X) = m_{U_i}^2(M_X) = m_{D_i}^2(M_X) = m_{L_i}^2(M_X) = m_{E_i}^2(M_X) = m_{3/2}^2 \quad (i = 1,2,3) \tag{38}$$

$$A(M_X) = A$$

$$B(M_X) = A-1$$

as dictated by (27). It should be stressed once more that the boundary conditions (38) are exact, if we only neglect corrections at the Planck scale, ignore the scaling of parameters from M to M_X, and pay no attention to corrections[26] at the GUT scale. All these effects are expected to be small and it is assumed that they do not seriously disturb (38) and the picture hereafter.

It is apparent from (37) that SUGAR models can easily succeed in giving weak interaction scale masses ($m_{3/2} \sim M_W$) to squarks, sleptons and gauginos [see (27)]. Alas, SUGAR models also give large positive (mass)^2 to the Higgs doublets, thus making the breaking of $SU(2) \times U(1)$ difficult. One way to overcome this difficulty is the introduction[41),24),27)] of a gauge singlet coupled to H_1 and H_2, but, as mentioned above, with disastrous effects[43)] for the gauge hierarchy. A particularly simple solution to the $SU(2) \times U(1)$ breaking relies upon the fact that the boundary conditions (38) need be satisfied only at M_X (or M), and that large renormalization group scaling effects can produce a negative value for m_H^2 at low energies[19)]. The full set of renormalization group equations for the parameters in V_{LEP} (37) has been written elsewhere[44)]. Here we concentrate on the most interesting equation, the one for the mass-squared of the Higgs ($m_2{}^2$), which gives mass to the top quark:

$$\mu \frac{\partial}{\partial\mu} \begin{bmatrix} m_2^2 \\ m_{U_3}^2 \\ m_{Q_3}^2 \end{bmatrix} = \frac{h_t^2}{8\pi^2} \begin{bmatrix} 3 & 3 & 3 \\ 2 & 2 & 2 \\ 1 & 1 & 1 \end{bmatrix} \begin{bmatrix} m_2^2 \\ m_{U_3}^2 \\ m_{Q_3}^2 \end{bmatrix} + \frac{|A|^2 h_t^2 m_{3/2}^2}{8\pi^2} \begin{bmatrix} 3 \\ 2 \\ 1 \end{bmatrix} - \frac{h_t^2 m_4^2}{8\pi^2} \begin{bmatrix} 0 \\ 2 \\ 1 \end{bmatrix} - \frac{8\alpha_3}{3\pi} M_3^2 \begin{bmatrix} 0 \\ 1 \\ 1 \end{bmatrix} \tag{39}$$

where we have neglected gauge couplings other than the "coloured" one, $\alpha_3(\equiv g_3^2/4\pi)$, M_3 is the gluino mass [see (26) and (27)], and Yukawa couplings other than h_t, for the top, have been dropped. The physics content of (39) is apparent. Since μ is decreasing (we come from high energies down to low energies), the sign of the first two terms in (39) is such as to make all m^2_{2,U_3,Q_3} smaller at low energy, with the decrease of $m_2{}^2$ becoming more pronounced because of the 3:2:1 weighting. On the other hand, the sign of the last two terms in (39) is such as to make $m^2_{U_3}$ and $m^2_{Q_3}$ (the squark masses) larger at low energy, but have no direct effect on $m_2{}^2$ [notice the "zeros" in the corresponding matrices in (39)]. Indirectly though, the net effect on $m_2{}^2$ of the last two terms in (39) is to enhance further its decrease at low energies, by increasing $m^2_{U_3}$ and $m^2_{Q_3}$, which then drive down $m_2{}^2$ via the first two terms of (39). This is exactly what we are after! We want large ($\sim M_W^2$) and positive squarks and slepton $(\text{masses})^2$, BUT negative Higgs $(\text{mass})^2$ to trigger $SU(2) \times U(1)$ breaking. The ways of obtaining negative Higgs $(\text{mass})^2$ now become clear [see (39)]. We have to use either a large top Yukawa coupling (h_t), or large A, or large m_4, or a fourth generation to provide large Yukawa couplings, or some suitable, physically plausible combination of the above possibilities. There are pros and cons for every one of the above situations. In the case of large h_t, a lower bound on the mass of the top quark is set[19),38)-40)]:

$$m_t > O(60 \text{ GeV}) \tag{40}$$

which some people may find uncomfortable. We may avoid a large h_t by moving it into the large $A(>3)$ regime[38),45)]. The price, though, is high[38),45)]. The phenomenologically acceptable vacuum becomes unstable against tunnelling into a vacuum in which all gauge symmetries, including colour and electromagnetism, are broken. We must[38),45)] then arrange things in such a way that the lifetime for this vacuum decay process is greater than the age of the Universe. Some people, not without reason, may find this possibility dreadful. We may avoid large h_t and/or large A by using[34)] non-vanishing $m_4(\sim m_{3/2})$ where a rather satisfactory picture then emerges[34)]. Some people may object here to the basic assumption of large $m_4(\sim m_{3/2})$, since in the case of natural triplet-doublet Higgs splitting-type models[29)] [see (31)], m_4 has a tendency to be small, if not zero, even though other sources of m_4 may be available.

Finally, we come to the possibility of a fourth generation which, suitably weighed, may help us to avoid large h_t, A, or m_4. The problem here is that low energy phenomenology (evolution of coupling constants, $m_b/m_\tau, \ldots$) [32)-34)] as well as firm cosmological results like nucleosynthesis (especially ^{4}He abundance)[46)], may suffer almost unacceptable modifications. Furthermore, one has to watch out for the mass of the fourth generation charged lepton, since it is going to behave like m_2^2 in (39), and thus $m^2_{L_4,E_4}$ may easily go negative, breaking electromagnetic gauge invariance.

Whatever mechanism (if any) turns out to be correct, it is rather remarkable that in SUGAR-type models, there is a simple explanation of the breaking of $SU(2) \times U(1)$ and of the non-breaking of $SU(3) \times U(1)_{E-M}$. Furthermore, for the first time, we have a simple explanation of why $M_W \lll M_X$ (or M), i.e., a simple solution of the cumbersome gauge hierarchy problem. Starting with a positive Higgs $(\text{mass})^2$, of order $m^2_{3/2}$ at M_X, and noticing that [see (39)] the evolution with μ^2 of the Higgs $(\text{mass})^2$ is very slow (logarithmic), it is not surprising that we have to come down a long way in the energy scale, before the Higgs $(\text{mass})^2$ turns negative and is thus able to trigger $SU(2) \times U(1)$ breaking. For example, in a class of models[38)] characterized by "small" gravitino masses ($\ll O(M_W)$) and by a Coleman-Weinberg-type[47)] radiative $SU(2) \times U(1)$ breaking, occurring naturally, we get[48)], by dimensional transmutation

$$M_W \simeq \Lambda_{QCD} \cdot \frac{g_2(M_W)}{2\sqrt{}} \exp\left(\frac{2\pi}{\alpha_G} F\left(\frac{h_t^2(M_X)}{\alpha_G}; A\right)\right) \tag{41}$$

where g_2 denotes the $SU(2)$ gauge coupling constant, α_G is the GUT fine structure constant and F is a rather involved function of its indicated variables. Using standard values for $g_2(M_W)$ (~ 0.67), α_G($\sim 1/25$) and reasonable values for $h_t(M_X)$ (0.2-0.3), $A(M_X)$ (2-3), (41) gives[48)]

$$M_W \simeq (300\text{-}600)\,\Lambda_{QCD} \tag{42}$$

a rather remarkable equation from many points of view. It does not only give $M_W \lll M_X$ as was required by the gauge hierarchy, but it also provides a new and successful relation between the scale of electroweak unification M_W($\sim$ 80 GeV) and the fundamental scale of strong interactions, Λ_{QCD} ($\sim$ 0.15-0.3 GeV) in terms of dimensionless parameters. Clearly, this occurs because the rapid final stages of evolution of m_2^2 are driven by the increases in t-quark Yukawa coupling, and more importantly, in the squark masses which occur when $g_3^2/4\pi$ becomes large. It is only in the SUSY Coleman-Weinberg scenario[38),34)] that the weak interaction scale is related to that of strong interactions. This contrasts with what usually happens in weak gauge symmetry breaking in SUGAR models[39)-40)] where M_W

is connected to $m_{3/2}$, but it is not directly related to the strong interaction scale.

Another very amazing fact is that the values of the parameters of the low energy world seem to co-operate with us. Since quarks are feeling strong interactions, (39) tells us that quarks may enjoy large masses (Yukawa couplings) without making squark $(\text{masses})^2$ negative, because of the last term $(\sim \alpha_3)$, which easily balances off large Yukawa couplings, without any sweat. On the other hand, since leptons are not feeling strong interactions, the balance-off between the weak gauge couplings and large Yukawa coupling becomes extremely delicate and could be problematic. How nice that for all three generations, leptons and down quarks weigh less than 5 GeV and especially for the third generation that the top quark (t) is heavier than the bottom quark (b). An inverse situation would be disastrous, because in any reasonable GUT, a very heavy b quark would mean a very heavy τ lepton, thus making electromagnetic gauge invariance tremble in such SUGAR-type schemes. I will not go any further into the esoterics of this type of $SU(2) \times U(1)$ breaking models, since a rather thorough and detailed exposé of these types of theories and of their phenomenological consequences is now available[34]. It should be stressed that things are now very constrained, as we see from the Table, taken from Ref. 34), where the whole low energy spectrum is worked out, in terms of very few parameters, $m_{3/2}$, $A(M_X)$, $\xi \equiv M_0/m_{3/2}$ [see (26)] and $m_4(M_X)$. Eventually, with more theoretical insight, we hope to determine even these very few parameters, thus predicting uniquely the low energy spectrum. For example, we have already discussed ways of determining $m_{3/2}$ [see (34)], while some people may favour $A(M_X) = 3$ as a natural solution[49] to the absence of the cosmological constant problem, etc. Among other interesting things contained in the Table, the existence of a very light ($\sim$ [3-6) GeV]) neutral Higgs, with the usual Yukawa couplings to matter, should not escape our attention. Since such a particle is a common feature of a large class of models[38],[34], a search in the $\Upsilon \to H^0 + \gamma$ channel, which is expected to be a few per cent of the $\Upsilon \to \mu^+\mu^-$ decay, may turn out to be very fruitful.

There are other, very interesting features concerning low energy phenomenology stemming out from the general form of V_{LEP} [(37) and (38)] in SUGAR models. Very tight constraints coming from natural suppression of flavour changing neutral currents[50] (FCNC), absence of large corrections to $(g-2)$[51] and ρ [39],[52] $(\equiv (M_W/M_Z\cos\theta)^2)$ as well as to θ_{QCD}[5], which have been the nemesis[50)-52),5] of SUSY models with arbitrary and explicit soft SUSY breaking, are satisfied in SUGAR models. The highly-constrained set of soft SUSY operators (27) in SUGAR models fits the bill[19],[39]. Concerning FCNC, (38) guarantees the super-GIM mechanism, since the mass matrices for the quarks and leptons are diagonalized by the same transformation that renders the mass matrices for their scalar partners and gluino couplings generation diagonal. Despite the fact that this property does

TABLE

PARTICLE SPECTRUM [34)]

		CAM	MIM	CAM	MIM	CAM	MIM
	$A(M_X)$	3	3	2.8	2.8	2.0	1.6
	$m_{\frac{3}{2}}$	15	15	15	15	15	15
	ξ	2.8	2.2	3.2	3.1	3.5	1.9
	$m_4(M_X)$	15	17	16	18	11	7
	top	25	25	35	35	50	50
All families	$(\text{sleptons})_L$	29	27	32	36	35	25
	$(\text{sleptons})_R$	21	20	22	23	23	19
1st and 2nd families	$(\text{squark})_L$	58	77	66	108	72	67
	$(\text{squark})_R$	54	74	61	104	67	65
		54	74	60	103	66	65
3rd family	$(\text{sbottom})_L$	58	76	64	106	68	66
	$(\text{sbottom})_R$	54	74	60	104	66	65
	$(\text{stop})_L$	81	96	95	132	112	106
	$(\text{stop})_R$	26	54	23	78	21	37
	Charged Higgses	96	93	95	94	88	83
	Neutral Higgses	106	104	105	105	100	95
		3	3	4.4	5.3	6	5
	"Axion"	51	46	49	48	35	19
	Gluinos	42	84	47	118	52	72
	Photino	11	4.6	9.4	7.3	4	3
	HW, WH-inos	89	87	87	94	84	90
		78	82	79	79	82	75
	HZ, HZ-inos	99	108	99	116	101	106
		92	85	91	80	88	83
	Axino	26	23	24	24	17	9

Physical mass spectrum of the cosmologically acceptable model (CAM) and the minimal model (MIM) corresponding to the same gravitino mass $m_{3/2} \simeq 15$ GeV for top quark masses equal to 25 GeV, 35 GeV and 50 GeV respectively. ξ denotes the ratio of the gaugino to the gravitino mass at M_X. All masses are in GeV units. The light neutral Higgs gets its mass via radiative corrections.

not survive, in general, after renormalization, it has been shown[53),54)] that these effects are controllable. Furthermore, the Buras stringent upper bound[55)] on the top quark mass [< O(40 GeV)], coming from kaon phenomenology (K_L-K_S and $K_L \to \mu^+\mu^-$ systems), is avoided[53)] in SUGAR models. There are a lot of cancellations between ordinary and SUSY contributions in K processes[56)], such that the top quark mass may be stretched up to 100 GeV without problem[53)]. That sounds very satisfactory, especially for SUGAR models[38)-40)] that do need a large top quark mass for $SU(2) \times U(1)$ breaking. It looks like a self-service situation. Similar comments apply in the case of $(g-2)_\mu$ or ρ, where it has been shown that SUGAR model contributions are acceptable[34),39),52)]. Typical values for SUGAR contributions are[34)] $|\Delta(g-2)_\mu| \lesssim (3 \cdot 10^{-9})$ and[39),52)] $\Delta\rho \leqq 0.01$, which compare favourably with the present experimental upper bounds of $(4 \cdot 10^{-8})$ and (0.03) respectively, but are large enough to be interesting. Better experimental bounds, especially on $\Delta\rho$, could be revealing.

Concerning θ_{QCD}, it has been shown[19),57)] that in SUGAR-type models, we not only understand the non-renormalization[5)] of θ, but we also understand[19),57)] why θ is zero or small ($< 10^{-9}$) to start with. This fact is related with our freedom, discussed before, to use non-minimal ($\neq \delta_{\alpha\beta}$) $f_{\alpha\beta}$ in (15). It has already been observed[18)] that the gauge kinetic term in (15), as well as giving rise to the canonical

$$-\frac{1}{4} F^\alpha_{\mu\nu} F^\beta_{\mu\nu} (\mathrm{Re}\, f_{\alpha\beta}) \tag{43}$$

could also yield the CP-violating θ vacuum term

$$\varepsilon^{\mu\nu\rho\sigma} F^\alpha_{\mu\nu} F^\beta_{\rho\sigma} (\mathrm{Im}\, f_{\alpha\beta}) \tag{44}$$

We know that θ_{QCD} is $< O(10^{-9})$ experimentally, and that θ is not renormalized in a supersymmetric theory[5)]. It is finitely renormalized when supersymmetry is broken, but[5)] this is plausibly only by an amount $\delta\theta = O(10^{-16})$ in the popular Kobayashi-Maskawa model[58)]. Thus we see that θ should be less than $O(10^{-9})$ in a supersymmetric GUT and may be very small. An attractive hypothesis is that $f_{\alpha\beta}$ is a function with only real coefficients as found in extended supergravities[59)]. In this case, $\mathrm{Im} f_{\alpha\beta} = 0$ when $\langle 0|\phi|0\rangle = 0$, and the theory is CP-invariant in the gauge sector. If some of the ϕ then acquire complex vacuum expectation values, they will induce a non-zero value of $\mathrm{Im} f_{\alpha\beta}$ and hence violate CP spontaneously in the gauge sector, which is a new twist on an old proposal[60)]. If the moduli of some of these complex $\langle 0|\phi|0\rangle$ were $O(M_{P\ell})$, then the effective θ parameter would be $O(1)$ which is phenomenologically unacceptable. However, it is easy to imagine scenarios where θ is much smaller. For example, if $f_{\alpha\beta} = \delta_{\alpha\beta} + O(\phi^2/m_{P\ell}^2)$ and the culprit $\langle 0|\phi|0\rangle =$ $= O(m_X)$, then

$$\theta = O\left(\left(\frac{M_X}{M_{P\ell}}\right)^2 \times \text{ some small (?) angle}\right) \lesssim O(10^{-7}) \tag{45}$$

and the phenomenological constraint on θ_{QCD} could easily be respected. If the only complex $\langle 0|\phi|0\rangle$ were $O(m_W)$, or if all the $\langle 0|\phi|0\rangle$ were real as in all supersymmetric GUTs proposed to date, then the bare $\theta = 0$. Hence supergravity offers the other half of an answer to the θ vacuum problem. It should be stressed that low energy supergravity models have new sources[39),57),61),62)] of CP violation beyond the standard model. Unfortunately, they do not shed more light on the smallness of the observed CP violation in the K system[61)]. On the other hand, potential large contributions[61),57)] to ε', θ_{QCD} and to the Dipole Electric Moment Of the Neutron (DEMON) put, in general, severe constraints[57),61)] on these new possible CP-violating phases. Actually, it seems almost unavoidable[5),39),57),62)] that in SUGAR models the DEMON should be near, but not above, the present experimental upper bound[63)] of $6 \cdot 10^{-25}$ e.cm, a rather drastic and experimentally testable prediction, in sharp contrast with the standard model prediction $O(10^{-30}$ e.cm). We may know soon.

Turning now to baryon decay, an interaction of the form[19)]

$$f \ni \frac{\lambda}{M} \bar{F}TTT \tag{46}$$

where $\bar{F}$ is a $\underline{\bar{5}}$ of matter (quark + lepton) chiral superfields in SU(5), T is a $\underline{10}$ of matter superfields and λ is some generic Yukawa coupling, could replace the Higgs exchange in the Weinberg-Sakai-Yanagida[64)] loop diagram for baryon decay. The magnitude of the diagram with (46) relative to the conventional Higgs diagram is:

$$\left(\lambda/M\right) \Big/ \left(\frac{\lambda^2}{M_{H_3}}\right) \simeq \left(\frac{M_{H_3}}{\lambda M}\right) \tag{47}$$

The ratio (47) could easily be >1, making a non-renormalizable superpotential interaction the dominant contribution to proton decay. A careful analysis of SUGAR-induced baryon decay shows[65)], surprisingly enough, that the expected hierarchy of decay mode is similar to that[32] coming from conventional minimal SUSY GUTs. One might have wrongly expected that no hard and fast predictions could be made about gravitationally-induced baryon decay modes. Anyway, this mechanism could give observable baryon decay even if the GUT mass $M_X \simeq M$.

Incidentally, similar terms like (46) have been considered[66] in efforts to explain the "lightness" of the first two generations of quarks and leptons. One replaces[66] direct Yukawa couplings for the first two generations with (very schematically):

$$f \ni \frac{\tilde{\lambda}_2}{M} \bar{H} \Sigma T_2 \bar{F}_2 + \frac{\tilde{\lambda}_2'}{M} H T_2 T_2 \Sigma + \frac{\tilde{\lambda}_1}{M^2} \bar{H} \Sigma^2 T_1 \bar{F}_1 + \frac{\tilde{\lambda}_1'}{M^2} H T_1 T_1 \Sigma^2 + \ldots \tag{48}$$

which not only repairs[19),66] wrong relations like $m_d(M_X) \simeq m_e(M_X)$, very difficult to correct[67] in conventional SUSY GUTs, <u>but</u> also provides reasonalbe masses for the first two generations. Indeed, it follows from (48) that the second generation is getting masses $(M_X/M)M_W \sim (0.1\text{-}1\ \mathrm{GeV})$, while the first generation masses are $(M_X/M)^2 M_W \sim (1\text{-}10\ \mathrm{MeV})$, exactly what was ordered[66]. It is amazing that in SUGAR models, by increasing $M_X (\sim 10^{16}\ \mathrm{GeV})$, relative to its ordinary GUT value $(10^{14}\ \mathrm{GeV})$, and by decreasing $M_{P\ell}$, what is relevant is the superPlanck scale $M(\sim 10^{18}\ \mathrm{GeV})$, the highly-desired ratio $(M_X/M) \sim 10^{-2}$, appears naturally. It seems now, for the first time, that gravitational interactions may be responsible for the masses of at least the first two generations. Once more, non-renormalizable interactions contained in SUGAR models provide a simple solution[66] to another hierarchy problem, the <u>fermion mass hierarchy problem</u>.

4. FINAL REMARKS

We have shown that gravitational effects, as contained in SUGAR theories, cannot be neglected anymore in the regime of particle physics. On the contrary, it may be that supergravitational effects are really responsible: <u>for</u> the SU(5) breaking at M_X with an automatic triplet-doublet Higgs splitting, <u>for</u> the $SU(2) \times U(1)$ breaking [and $SU(3) \times U(1)_{E-M}$ <u>non-breaking</u>] at M_W, naturally exquisitely smaller than M_X, <u>for</u> the "constrained" soft SUSY breaking at $m_{3/2}$, hierarchically smaller than M in a natural way, <u>for</u> definite, at present experimentally acceptable departures from the "standard" low energy phenomenology [like the DEMON, or $\Delta\rho$, with values below, but not far from, their present experimental upper bounds, or the existence of very light ($< O(10\ GeV)$) neutral Higgs bosons], as well as a rather well-defined low-energy SUSY spectrum, for observable baryon decay even if $M_X \simeq M$, and for the light fermion masses of the first two generations. Furthermore, supergravity theories may provide, for the first time, a problem-free cosmological scenario, from primordial inflation[68] through GUT phase transitions[28)-31] to baryon and nucleosynthesis, ostracizing troublesome particles such as GUT monopoles[28,30], gravitinos[69], Polonyi fields[70] or other SUSY relics[71].

Putting the whole thing together, it becomes apparent that spontaneously broken $N = 1$ local SUSY gauge theories, with their prosperous and appropriate structure, may well serve as an effective theory describing all physics from $M_{P\ell}$ down to (and including) low energies, with well-defined and rich experimental consequences. What's next, then? Well, we really have to understand where this highly successful theory comes from. There are reasons to believe[21] that $N = 8$ extended SUPERGRAVITY[13] may provide the fundamental theory. But this next move asks for a deep understanding of physics at Planck energies, which is as exciting as it is difficult, taking into account that even QUANTUM MECHANICS may need modification[72], if quantum gravitational effects have to be considered seriously.

ACKNOWLEDGEMENTS

Many thanks to Harvey Newman for organizing a very stimulating meeting. I should like to thank J. Prentki, A. Lahanas and K. Tamvakis for reading the manuscript and for their valuable comments. I should also like to express my gratitude to my wife Myrto for her patience and tolerance while this review was being written during vacations.

REFERENCES

1. For reviews, see:
D.V. Nanopoulos, Ecole d'Eté de Physique des Particules, Gif-sur-Yvette, 1980 (IN2P3, Paris, 1980), p. 1;
J. Ellis, in "Gauge Theories and Experiments at High Energies", K.C. Bowler and D.G. Sutherland, eds., (Scottish Universities Summer School in Physics, Edinburgh, 1981), p. 201; and CERN preprint TH.3174 (1981), published in the Proceedings of the 1981 Les Houches Summer School;
P. Langacker, Phys. Rep. 72C:185 (1981), and Proceedings of the 1981 International Symposium on Lepton and Photon Interactions at High Energies, W. Pfeil, ed., Bonn University (1981), p. 823.

2. For reviews, see:
P. Fayet and S. Ferrara, Phys. Rep. 32:251 (1977);
B. Zumino, University of California Berkeley preprint UCB/PTH 83-2 (1983);
D.V. Nanopoulos, in "Grand Unification and Physical Supersymmetry", lectures given at the XIIIth GIFT International Seminar on Theoretical Physics, Girona, Spain, 1982, J.A. Grifols, A. Mendez and A. Ferrando, eds., World Scientific Pub. Co., Singapore (1983).
3. L. Maiani, Proceedings of the Summer School of Gif-sur-Yvette, (IN2P3, Paris, 1980), p. 3;
E. Witten, Nucl. Phys. B188:513 (1981).
4. J. Wess and B. Zumino, Phys. Lett. 49B:52 (1974);
J. Iliopoulos and B. Zumino, Nucl. Phys. B76:310 (1974);
S. Ferrara, J. Iliopoulos and B. Zumino, Nucl. Phys. B77:413 (1974);
M.T. Grisaru, W. Siegel and M. Roček, Nucl. Phys. B159:420 (1979).
5. J. Ellis, S. Ferrara and D.V. Nanopoulos, Phys. Lett. 114B:231 (1982).
6. B. Zumino, Nucl. Phys. B89:535 (1975).
7. For reviews, see:
A. Savoy-Navarro, in Proceedings of the "Supersymmetry versus Experiment" Workshop, D.V. Nanopoulos and A. Savoy-Navarro. eds., appearing soon as a Physics Report;
K.H. Lau, SLAC-PUB-3001 (1982);
J.G. Branson, MIT Technical Report #133 (1983).
8. P. Fayet, Phys. Lett. 69B:489 (1977); Phys. Lett. 84B:416 (1979), and innumerable talks.
9. See, for example:
G.R. Farrar and S. Weinberg, Phys. Rev. D27:2732 (1983).
10. R. Barbieri, S. Ferrara and D.V. Nanopoulos, Zeit. f. Phys. C13:267 (1982); Phys. Lett. 116B:16 (1982).
11. J. Ellis, L. Ibañez and G.G. Ross, Phys. Lett. 113B:983 (1982); Nucl. Phys. B221:29 (1983);
L. Alvarez-Gaumé, M. Claudson and M. Wise, Nucl. Phys. B207:16 (1982);
M. Dine and W. Fischler, Nucl. Phys. B204:346 (1982);
T. Banks and V. Kaplunovsky, Nucl. Phys. B206:45 (1982); Nucl. Phys. B211:529 (1983);
S. Dimopoulos and S. Raby, Nucl. Phys. B219:479 (1983).
12. J. Polchinski and L. Susskind, Phys. Rev. D26:3661 (1982).
13. For a review, see:
P. van Nieuwenhuizen, Phys. Rep. 68C:189 (1981).
14. R. Barbieri, S. Ferrara, D.V. Nanopoulos and K. Stelle, Phys. Lett. 113B:219 (1982).
15. J. Ellis and D.V. Nanopoulos, Phys. Lett. 116B:133 (1982).
16. S. Weinberg, Phys. Rev. Lett. 48:1776 (1982);
A.H. Chamseddine, R. Arnowitt and P. Nath, Phys. Rev. Lett. 49:970 (1982).
17. S. Deser and B. Zumino, Phys. Rev. Lett. 38:1433 (1977).

18. E. Cremmer, B. Julia, J. Scherk, S. Ferrara, L. Girardello and P. van Nieuwenhuizen, Phys. Lett. 79B:231 (1978); Nucl. Phys. B147:105 (1979);
E. Cremmer, S. Ferrara, L. Girardello and A. Van Proeyen, Phys. Lett. 116B:231 (1982); Nucl. Phys. B212:413 (1983).
19. J. Ellis, D.V. Nanopoulos and K. Tamvakis, Phys. Lett. 121B:123 (1983).
20. B. Zumino, Phys. Lett. 87B:203 (1979).
21. E. Cremmer and B. Julia, Nucl. Phys. B159:141 (1979);
J. Ellis, M.K. Gaillard and B. Zumino, Phys. Lett. 94B:343 (1980).
22. S. Weinberg, Phys. Rev. 166:1568 (1968);
S. Coleman, J. Wess and B. Zumino, Phys. Rev. 177:2239 (1968);
C. Callan, S. Coleman, J. Wess and B. Zumino, Phys. Rev. 177:2247 (1968).
23. J. Polonyi, Budapest preprint KFKI-1977-93 (1977).
24. R. Barbieri, S. Ferrara and C.A. Savoy, Phys. Lett. 119B:343 (1982).
25. H.P. Nilles, M. Srednicki and D. Wyler, Phys. Lett. 120B:346 (1982).
26. L. Hall, J. Lykken and S. Weinberg, Phys. Rev. D27:2359 (1983).
27. E. Cremmer, P. Fayet and L. Girardello, Phys. Lett. 122B:41 (1983).
28. D.V. Nanopoulos, K.A. Olive, M. Srednicki and K. Tamvakis, Phys. Lett. 124B:171 (1983).
29. C. Kounnas, D.V. Nanopoulos, M. Srednicki and M. Quiros, Phys. Lett. 127B:82 (1983).
30. D.V. Nanopoulos and K. Tamvakis, Phys. Lett. 110B:449 (1982);
M. Srednicki, Nucl. Phys. B202:327 (1982); Nucl. Phys. B206:139 (1982);
D.V. Nanopoulos, K.A. Olive and K. Tamvakis, Phys. Lett. 115B:15 (1982).
31. C. Kounnas, J. Leon and M. Quiros, Phys. Lett. 129B:67 (1983);
C. Kounnas, D.V. Nanopoulos and M. Quiros, CERN preprint TH.3573 (1983).
32. J. Ellis, D.V. Nanopoulos and S. Rudaz, Nucl. Phys. B202:43 (1982).
33. D.V. Nanopoulos and D.A. Ross, Phys. Lett. 118B:99 (1982).
34. C. Kounnas, A.B. Lahanas, D.V. Nanopoulos and M. Quiros, CERN preprints TH.3651 and TH.3657 (1983).
35. J. Ellis, M.K. Gaillard and D.V. Nanopoulos, Phys. Lett. 80B:360 (1979).
36. D.V. Nanopoulos and K. Tamvakis, in Ref. 30; Phys. Lett. 113B:151 (1982); Phys. Lett. 114B:235 (1982).
37. A. Masiero, D.V. Nanopoulos, K. Tamvakis and T. Yanagida, Phys. Lett. 115B:380 (1982);
B. Grinstein, Nucl. Phys. B206:387 (1982).
38. J. Ellis, J. Hagelin, D.V. Nanopoulos and K. Tamvakis, Phys. Lett. 125B:275 (1983).
39. L. Alvarez-Gaumé, J. Polchinski and M. Wise, Nucl. Phys. B221:495 (1983).

40. L. Ibañez and C. Lopez, Phys. Lett. 126B:54 (1983).
41. A.H. Chamseddine, R. Arnowitt and P. Nath, in Ref. 16).
42. S. Ferrara, D.V. Nanopoulos and C.A. Savoy, Phys. Lett. 123B:214 (1983).
43. H.P. Nilles, M. Srednicki and D. Wyler, Phys. Lett. 124B:337 (1983);
A.B. Lahanas, Phys. Lett. 124B:341 (1983).
44. K. Inoue, A. Kakuto, H. Komatsu and S. Takeshita, Prog. Th. Phys. 68:927 (1982).
45. M. Claudson, L.J. Hall and I. Hinchliffe, LBL preprint 15948 (1983).
46. K.A. Olive, D.N. Schramm, G. Steigman and J. Yang, Ap. J. 246:557 (1981).
47. S. Coleman and E. Weinberg, Phys. Rev. D7:1888 (1973).
48. J. Ellis, J. Hagelin, D.V. Nanopoulos and K. Tamvakis, to be published.
49. E. Cremmer, S. Ferrara, C. Kounnas and D.V. Nanopoulos, CERN preprint TH.3667 (1983).
50. J. Ellis and D.V. Nanopoulos, Phys. Lett. 110B:44 (1982);
R. Barbieri and R. Gatto, Phys. Lett. 110B:211 (1982).
51. J. Ellis, J. Hagelin and D.V. Nanopoulos, Phys. Lett. 116B:283 (1982);
R. Barbieri and L. Maiani, Phys. Lett. 117B:203 (1982);
J.A. Grifols and A. Mendez, Phys. Rev. D26:1809 (1982).
52. R. Barbieri and L. Maiani, Rome preprint (1983).
53. A.B. Lahanas and D.V. Nanopoulos, CERN preprint TH.3588 (1983).
54. J.F. Donoghue, H.P. Nilles and D. Wyler, CERN preprint TH.3583 (1983).
55. A.J. Buras, Phys. Rev. Lett. 46:1354 (1981).
56. T. Inami and C.S. Lim, Nucl. Phys. B207:533 (1982).
57. D.V. Nanopoulos and M. Srednicki, Phys. Lett. 128B:61 (1983).
58. J. Ellis and M.K. Gaillard, Nucl. Phys. B150:141 (1979).
59. M.T. Grisaru, M. Roček and A. Karlhede, Phys. Lett. 120B:110 (1983).
60. T.D. Lee, Phys. Rep. 9:143 (1973).
61. F. del Aguila, J.A. Grifols, A. Mendez, D.V. Nanopoulos and M. Srednicki, Phys. Lett. 129B:77 (1983).
62. W. Büchmuller and D. Wyler, Phys. Lett. 121B:321 (1983);
F. del Aguila, M.B. Gavela, J.A. Grifols and A. Mendez, Phys. Lett. 126B:71 (1983);
J. Polchinski and M. Wise, Phys. Lett. 125B:393 (1983).
63. I.S. Altarev et al., Phys. Lett. 102B:13 (1981);
W.B. Dress et al., Phys. Rev. D15:9 (1977);
N.F. Ramsey, Phys. Rep. 43:409 (1978).
64. S. Weinberg, Phys. Rev. D26:187 (1982);
N. Sakai and T. Yanagida, Nucl. Phys. B197:533 (1982).
65. J. Ellis, J. Hagelin, D.V. Nanopoulos and K. Tamvakis, Phys. Lett. 124B:484 (1983).

66. D.V. Nanopoulos and M. Srednicki, Phys. Lett. 124B:37 (1983).
67. L. Ibañez, Phys. Lett. 117B:403 (1982);
A. Masiero, D.V. Nanopoulos and K. Tamvakis, Phys. Lett. 126B:337 (1983).
68. J. Ellis, D.V. Nanopoulos, K.A. Olive and K. Tamvakis, Nucl. Phys. B221:524 (1983);
D.V. Nanopoulos, K.A. Olive, M. Srednicki and K. Tamvakis, Phys. Lett. 123B:41 (1983);
D.V. Nanopoulos, K.A. Olive and M. Srednicki, Phys. Lett. 127B:30 (1983);
G. Gelmini, D.V. Nanopoulos and K.A. Olive, CERN preprint TH.3629 (1983).
69. J. Ellis, A.D. Linde and D.V. Nanopoulos, Phys. Lett. 118B:59 (1982).
70. D.V. Nanopoulos and M. Srednicki, CERN preprint TH.3673 (1983).
71. J. Ellis, J. Hagelin, D.V. Nanopoulos, K.A. Olive and M. Srednicki, SLAC-PUB-3171 (1983).
72. J. Ellis, J. Hagelin, D.V. Nanopoulos and M. Srednicki, CERN preprint TH.3619/SLAC-PUB-3134 (1983).

RECENT DEVELOPMENTS IN N = 8 SUPERGRAVITY

H. Nicolai

CERN - Geneva, Switzerland

ABSTRACT

Some recent developments in (gauged) N = 8 supergravity are reviewed. Particular emphasis is placed on the connection between this theory and N = 1 supergravity in eleven dimensions and on the mechanism of spontaneous compactification by which spontaneously broken versions of supergravity are generated.

1. INTRODUCTION

During the past two years, we have witnessed a dramatic increase in "public interest" in supersymmetry and supergravity[1]. This interest cannot as yet be based on any solid experimental fact; there is no evidence that would force us to abandon the presently very successful standard model of strong and electro-weak interactions in favour of some supersymmetric theory. Nevertheless, it has become clear that supersymmetry has many attractive features to offer and that it will perhaps play a crucial role in the ultimate unification of fundamental particle interactions. This unification constitutes one of the most ambitious endeavors in theoretical physics because it necessarily involves the extrapolation in energy over many orders of magnitude where no experimental information is expected to become available. The main reason why supersymmetry and supergravity are so attractive is that, at the present time, no other candidate theories exist which may enable us to simultaneously solve the three outstanding problems of modern elementary particle physics. These are:

i) the unification of gravity with the other fundamental interactions and the construction of a consistent, i.e., finite or renormalizable, theory of quantum gravity;

ii) the explanation of the current proliferation of fundamental fields and coupling constants ("who ordered the muon?"); and

iii) the hierarchy problem in its most fundamental form: why are the relevant mass scales of the standard model so tiny in comparison with the Planck mass of 10^{19} GeV, or why is the gravitational force so much weaker than the other fundamental forces?

Supersymmetry describes bosons and fermions in a unified way as partners in a supermultiplet which thus contains particles of different spin. These supermultiplets form irreducible representations of the basic superalgebra

$$\{Q^i_\alpha, \overline{Q}_{\dot{\beta}j}\} = 2\,\delta^i{}_j\, \sigma^\mu{}_{\alpha\dot{\beta}}\, P_\mu \tag{1.1}$$

If there are N independent supersymmetries, the indices i, j, ... assume the values 1, ..., N and the irreducible multiplets cover a range of spins of at least $\hbar$ N/4. For N > 4, it turns out that there are no "matter supermultiplets" any more and that all relevant multiplets contain spin-2 particles. Hence a unification with gravity is forced upon us. From the algebra (1.1), it is also easy to see that local supersymmetry implies gravity: the commutator of two space-time dependent supersymmetry transformations yields a space-time dependent translation which is nothing but a general coordinate transformation.

The maximally extended theory which does not contain spins higher than s = 2 is N = 8 supergravity[2)-4)]. This theory, which is invariant under N = 8 independent supersymmetries, describes the interactions of one graviton, eight gravitinos, 28 spin-1, 56 spin-1/2 and 70 spin-0 fields. There is an alternative way to characterize this theory. Namely, the number of supersymmetry generators also increases as one goes to higher space-time dimensions because spinors in higher dimensions have more components. One can show that supergravities can be constructed in only up to eleven dimensions[5)] and that N = 1 supergravity in eleven space-time dimensions[6)] corresponds to N = 8 supergravity in four dimensions. This can be easily understood because the 32 components of a spinor in eleven dimensions decompose into eight four-components spinors in four dimensions. One may regard the eleven-dimensional theory as more fundamental since it gives rise not only to N = 8 supergravity in four dimensions but also to various N ≤ 8 supergravity theories with and without spontaneous breaking of supersymmetry. In the general framework of Kaluza-Klein theories[7], the N = 8 theory is presumably obtained by "spontaneously compactifying"[8],[9] the eleven dimensional theory on the seven-sphere S^7 [10][11]. Only the low-energy modes are retained in this scheme, but one should realize that the full theory also contains an infinite tower of higher excited modes with masses of the order of the Planck mass and multiples thereof.

In this contribution, we will review some of the progress that has been recently made in understanding the basic mechanism of spontaneous compactification and its implications for supergravity. Although various aspects are not yet fully understood, it is now clear that the rich structure of supergravity theories may also lead to a better understanding of spontaneous compactification.

2. SPONTANEOUS COMPACTIFICATION OF ELEVEN-DIMENSIONAL SUPERGRAVITY

Eleven-dimensional supergravity was constructed in Ref. 6). It is based on the following multiplet of fields: an elfbein $E_M{}^A$ which describes ordinary gravity in eleven dimensions, a 32-component Majorana spinor Ψ_M which is the analog of the gravitino in four dimensions and an antisymmetric three-index tensor A_{MNP}. The latter is subject to the Maxwell-like Abelian gauge transformations

$$\delta A_{MNP} = \partial_{[M}\Lambda_{NP]} \tag{2.1}$$

which are necessary to balance the bosonic and fermionic physical degrees of freedom. Under local supersymmetry transformations of parameter ε, these fields tranform into each other according to[6]

$$\delta E_M{}^A = -\frac{i}{2}\bar{\varepsilon}\,\tilde{\Gamma}^A\,\Psi_M$$

$$\delta A_{MNP} = \frac{\sqrt{2}}{8}\bar{\varepsilon}\,\tilde{\Gamma}_{[MN}\Psi_{P]}$$

$$\delta\Psi_M = D_M\varepsilon + \frac{\sqrt{2}}{288}\,i\,(\tilde{\Gamma}_M{}^{NPQR} - 8\,\delta_M^N\,\tilde{\Gamma}^{PQR})\,\varepsilon\,F_{NPQR}{}^* \tag{2.2}$$

The eleven-dimensional $\tilde{\Gamma}$ matrices which appear in (2.2) are 32 × 32 matrices and may be represented as direct products of ordinary 4 × 4 γ matrices and 8 × 8 matrices Γ_m which generate the Clifford algebra in seven dimensions

*We adhere to the notation and conventions of Ref. 11).

$$\Gamma_\mu = \gamma_\mu \otimes 1 \qquad\qquad \tilde{\Gamma}_m = \gamma^5 \otimes \Gamma_m \tag{2.3}$$

Here and in what follows, we will always split eleven-dimensional indices $M = 1, \ldots, 11$ into four-dimensional ones $\mu, \ldots = 1, \ldots, 4$ and seven-dimensional ones $m, \ldots = 5, \ldots, 11$.

There is a unique Lagrangian which is left invariant under (2.2). However, we will need here only the bosonic field equations which follow from this Lagrangian. They are

$$R_{MN} - \frac{1}{2} g_{MN} = -\frac{1}{48} \{8 F_{MPQR} F_N{}^{PQR} - g_{MN} F^2\}$$

$$F^{MNPQ}{}_{;M} = -\frac{\sqrt{2}}{1152} \varepsilon^{M_1 \ldots M_8 NPQR} F_{M_1 \ldots M_4} F_{M_5 \ldots M_8} \tag{2.4}$$

Spontaneous compactification means that these equations admit solutions which "spontaneously" split eleven-dimensional space-time into a produce of four-dimensional space-time and some seven-dimensional "internal" manifold and thereby "explain" the four-dimensionality of space-time. That this is indeed the case, was first noted in Ref. 9). The second of Eqs. (2.4) is solved by

$$F^{\mu\nu\rho\sigma} = 3\sqrt{2}\, m\, e^{-1} \varepsilon^{\mu\nu\rho\sigma} \tag{2.5}$$

all other F's = 0

Inserting (2.5) into the right-hand side of the first of Eqs. (2.4), one easily sees that these become two independent equations, one for ordinary space-time

$$R_{\mu\nu} = 12\, m^2 g_{\mu\nu} \tag{2.6}$$

and another for the "internal" seven-dimensional manifold

$$R_{mn} = -6\, m^2 g_{mn} \tag{2.7}$$

Thus, the mass parameter m in (2.5) serves as an order parameter for spontaneous compactification, and the four-dimensionality of space-time is a consequence of the fact that the field strength F_{MNPQ} has four indices[9]!

At this point, the choice of four- and seven-dimensional manifolds is still quite arbitrary. Equations (2.6) and (2.7) only require them to be Einstein spaces.: a four-dimensional one with negative cosmological constant and a seven-dimensional one with positive cosmological constant. One can now further restrict the set of solutions by demanding that the resulting theory in four dimensions have eight supersymmetries[10]. To do so, one must analyze the transformation law of Ψ_m [see Eq. (2.2)] in the background (2.5), (2.6) and (2.7). Inserting (2.5) into $\delta\Psi_m$, one finds, using the representation (2.3),

$$\delta\Psi_m = (D_m - \frac{1}{2} m \Gamma_m) \varepsilon \tag{2.8}$$

where Γ_m acts trivially on four-dimensional Dirac indices. For supersymmetry transformation parameters ε which correspond to unbroken supersymmetries in four dimensions, the right-hand side of (2.8) must vanish. A necessary condition for this to happen is the integrability constraint

$$[D_m - \frac{1}{2} m \Gamma_m, D_n - \frac{1}{2} m \Gamma_n] \varepsilon =$$

$$= (+ \frac{1}{4} R_{mn}{}^{pq} (E) \Gamma_{pq} + \frac{m^2}{2} \Gamma_{mn}) \stackrel{!}{=}$$

$$\stackrel{!}{=} 0 \tag{2.9}$$

But (2.9) immediately implies

$$R_{mnpq} (E) = -m^2(g_{mp} g_{nq} - g_{mq} g_{np}) \tag{2.10}$$

and therefore the "internal" manifold is the maximally symmetric space in seven dimensions: the seven-sphere S^7. In a similar way, one can show that the evaluation of the integrability constraint on the four-dimensional manifold leads to

$$R_{\mu\nu\rho\sigma} (E) = 4m^2(g_{\mu\rho} g_{\nu\sigma} - g_{\mu\sigma} g_{\nu\rho}) \tag{2.11}$$

The curvature tensor (2.11) describes a four-dimensional anti de Sitter space. The integrability constraint thus singles out the unique choice of AdS $\times$ S^7 as the compactification that preserves eight supersymmetries. Conversely, one can also prove that this is indeed the only such compactification[11]. In order to exhibit the eight remaining supersymmetries explicitly, it is necessary to distinguish between four-dimensional co-ordinates x^μ. and seven-dimensional co-ordinates y^m. The supersymmetry transformation parameter ε which is a function of both x^μ and y^m may then be expanded according to

$$\varepsilon(x,y) = \varepsilon^I(x) \otimes \eta^I(y) + \ldots \tag{2.12}$$

where ε^I are four-dimensional and η^I seven-dimensional spinors, and the index I labels the independent supersymmetries. Inserting (2.12) into (2.8) and observing that the differential operator in (2.8) only acts on seven-dimensional spinor indices, one obtains

$$(D_m - \frac{1}{2} m \Gamma_m)\eta^I(y) = 0 \tag{2.13}$$

For S^7, there are precisely eight such covariantly constant spinors[10]. Furthermore, the natural metric on S^7 is SO(8) invariant, and therefore the label I = 1, ..., 8 is an SO(8) index. The four-dimensional parameters $\varepsilon^I(x)$ in (2.12) must therefore be identified with the eight local supersymmetry transformation parameters of N = 8 supergravity in four dimensions. The eight gravitino fields of the four-dimensional theory are obtained in an analogous manner by expanding the gravitino field Ψ_μ of the eleven-dimensional theory

$$\Psi_\mu(x,y) = \Psi_\mu{}^I(x) \otimes \eta^I(y) + \ldots \tag{2.14}$$

Equations (2.12) and (2.14) in conjunction with (2.2) guarantee that

$$\delta\Psi_\mu{}^I(x) = D_\mu\, \varepsilon^I(x) + \ldots \tag{2.15}$$

which is the expected transformation law[2)-4)].

It requires some more work to prove that the resulting four-dimensional theory is gauged N = 8 supergravity[4)]; this has been demonstrated in Refs. 10) and 11)*. For the proof, one must construct the analogous ansätze to (2.14) for the other fields as well and show that the spectrum of N = 8 supergravity emerges. This involves the identification of zero eigenmodes of certain differential operators on S^7 which is a non-trivial task. That the construction is actually possible may be traced to the unique mathematical properties of S^7.

3. SPONTANEOUS SYMMETRY BREAKING

One of the advantages of spontaneous compactification is that it allows us not only to derive gauged N = 8 supergravity from eleven-dimensional supergravity but also to construct spontaneously broken versions of supergravity. While (2.5), (2.10) and (2.11) characterize the only solution of the field equations (2.4) with eight local supersymmetries, there are other solutions of (2.4) which exhibit spontaneous breaking of supersymmetry. However, most of these spontaneously broken solutions cannot be related to gauged N = 8 supergravity will be briefly discussed.

The first solution of this type was exhibited in Ref. 12). It is characterized by the relations [cf. Eqs. (2.5), (2.10) and (2.11)]

*To be precise, the correspondence has only been established for the linearized theory so far but is expected to extend to all orders.

$$F^{\mu\nu\rho\sigma} = 2\sqrt{2}\, m\, e^{-1}\, \varepsilon^{\mu\nu\rho\sigma}$$

$$R_{\mu\nu\rho\sigma}(E) = \frac{10}{3} m^2 (g_{\mu\rho}\, g_{\nu\sigma} - g_{\mu\sigma}\, g_{\nu\rho})$$

$$R_{mnpq}(E) - m^2 (g_{mp}\, g_{nq} - g_{mq}\, g_{np}) \tag{3.1}$$

In addition, the field strength F_{mnpq} now has a non-vanishing expectation value given by

$$F_{mnpq} = \frac{\sqrt{2}}{6} m\, \varepsilon_{mnpqrst}\, \bar{\psi}\, \Gamma^{rst}\, \psi \tag{3.2}$$

where the covariantly constant spinor ψ satisfies the equation

$$(D_m + \frac{1}{2} m\, \Gamma_m)\, \psi(y) = 0 \tag{3.3}$$

Note that the sign in (3.3) is opposite to the one occurring in (2.13); therefore the parity of ψ is opposite to that of $\eta^I(y)$. One can now verify that (3.1) and (3.2) also solve the field equations (2.5)[12]. Since the quantity $\bar{\psi}\Gamma_{mnp}\psi$ is the torsion which "parallelizes" the seven-sphere, this solution is sometimes called the "parallelized solution."

An especially nice feature of this solution is that it may be interpreted as a spontaneously broken version of N = 8 supergravity in four dimensions [such an interpretation was first suggested in Ref. 13)]. One can show that it corresponds to a solution of N = 8 supergravity with a vacuum expectation value of the pseudoscalar fields B^{IJKL}. For suitably chosen ψ, this vacuum expectation value is explicitly given by[11]

$$\langle B^{IJKL} \rangle \sim a_{mnp}\, \Gamma^{mn}_{[IJ}\, \Gamma^{p}_{KL]} \tag{3.4}$$

where a_{mnp} are the octonionic structure constants. All supersymmetries are broken by this solution and the surviving gauge symmetry is the exceptional group G_2 [11),14)]. One can furthermore explicitly verify that the condition for a stationary point of the four-dimensional scalar field potential[4] is satisfied by the solution (3.4)[15]. Thus, the connection between the "parallelized solution" and the symmetry breaking of gauged N = 8 supergravity is rather well understood.

A second solution is obtained by "squashing" the seven-sphere[16]. There arw two Einstein metrics on S^7, one corresponding to the "round" S^7 and another one corresponding to a distorted S^7 [17]. The curvature tensor of the "squashed solution" is no longer given by (2.10), but since the equations of motion (2.2) only contain the Ricci-tensor and not the full curvature tensor, it is easy to see that the squashed S^7 still solves the equations of motion. The most interesting feature of this solution is that precisely one supersymmetry survives the squashing[16]. Accordingly, there is one covariantly constant spinor η with

$$(D_m - \frac{1}{2} m \Gamma_m) \eta(y) = 0 \qquad (3.5)$$

where D_m is now constructed from the squashed metric on S^7. It turns out that this spinor is not only covariantly constant, but actually constant[16),18)], i.e.,

$$\eta(y) = \text{constant} \qquad (3.6)$$

In contrast to the "parallelized" solution, it is not clear whether the squashed solution is interpretable as a spontaneously broken version of N = 8 supergravity. Whereas it has been demonstrated that the former gives vacuum expectation values to the 35 zero-mass pseudoscalars only, the latter will give vacuum expectation values not only to the 35 zero-mass scalars but also the higher excited modes[19]. One reason for this difference is that both unbroken and the parallelized solution live on the "round" S^7 which may be represented as the coset-space SO(8)/SO(7) whereas the squashed S^7 is the distance sphere in the eight-dimensional quarternionic projective space[17] which is metrically inequivalent to SO(8)/SO(7).

Finally, there is a third solution which combines squashing and parallelization[18),19)]. It is straightforwardly obtained by noting once more that the equations of motion only contain the Ricci tensor. All supersymmetries are broken for this solution [in fact, whenever $F_{mnpq} \neq 0$, all supersymmetries are broken[18]]. The most curious feature of this solution is that, owing to the constancy of η in (3.6), the parallelizing torsion is also constant[18]. This is vaguely reminiscent of group manifolds although there is no seven-dimensional semi-simple compact Lie group.

The constancy of the parallelizing torsion implies that the pseudoscalars of gauged N = 8 supergravity do <u>not</u> get vacuum expectation values in striking contrast to the parallelized solution without squashing. This is further evidence that squashed solutions should be viewed as spontaneously broken versions of the eleven-dimensional theory itself rather than of N = 8 supergravity. This conclusion is reinforced by the observation that the constant spinor (3.6) cannot be represented as a linear combination of the spinors $\eta^I(y)$ of Eq. (3.12) with constant coefficients: the unbroken supersymmetry associated with η is a linear combination of supersymmetries originating from higher modes in the expansion (2.12).

4. PHYSICS ?

The outstanding problem in the context of N = 8 supergravity is, of course, the question of whether this theory has anything to do with physics. On one hand, the theory should predict sufficiently many (almost) massless fermions, on the other hand the supersymmetries should be sufficiently broken in order not to be in blatant conflict with present day phenomenology. We have seen that the theory does admit spontaneous symmetry breaking, and that the different patterns of symmetry breaking are beautifully related to the unique mathematical properties of S^7. However, we do not know yet how many massless fermions the various broken versions predict; in principle, it could be that none are left. In this respect, only the squashed solution is in comparatively good shape. Since there is one supersymmetry left, the associated gravitino will pair up with the graviton to form a N = 1 supermultiplet. If the residual gauge symmetry coincides with the isometry group SO(5) × SO(3) of the squashed S^7 as was asserted in Ref. 16), there will be 13 = 10 + 3 massless gauge bosons, and N = 1 supersymmetry then predicts 13 massless spin-1/2 partners. One can furthermore show that these are the only massless states of this solution[19]: there are no massless pseudoscalars (essentially because the covariantly constant spinor of opposite parity is not available) and since N = 1 supersymmetry would require such pseudoscalars as partners of scalars and matter spin-1/2 fields the latter must also be absent.

Even if there are no massless spin-1/2 fermions left, this is not necessarily disastrous. It has been pointed out in Ref. 11) that, in the spontaneously broken versions, one cannot a priori rule out the possibility that some of the previously massive states become (almost) massless. If this were true, higher modes would not only affect the ultra-violet behavior[20] but also have important implications for low energy physics.

Another possibility is that the physical particles which we observe are bound states[21]. This possibility also suggests itself in view of the fact that conventional Kaluza-Klein theories only predict vector-like fermions. In Ref. 3), it was shown that unbroken N = 8 supergravity possesses a "fake" local chiral SU(8) invariance which could become dynamical at the quantum level. Since gauged N = 8 supergravity also has this local SU(8) invariance[4], one may conjecture that some chiral subgroup thereof survives the symmetry breaking and that this residual group should be identified with one of the usual GUT groups.

It remains to be seen which of these contending scenarios eventually emerges as the correct one. Since supergravity confronts us with unprecedented complexities, we are barely beginning to understand its implications, and much work remains to be done.

REFERENCES

1) See, e.g.,
P. van Nieuwenhuizen, Phys. Rep. 68 (1981) 189.
2) B. de Wit and D.Z. Freedman, Nucl. Phys. B130 (1977) 105;
B. de Wit, Nucl. Phys. B158 (1979) 189.
3) E. Cremmer and B. Julia, Phys. Lett. 80B (1978) 48; Nucl. Phys. B159 (1979) 141.
4) B. de Wit and H. Nicolai, Phys. Lett. 108B (1981) 285; Nucl. Phys. B208 (1982) 323.
5) W. Nahm, Nucl. Phys. B135 (1978) 149.
6) E. Cremmer, B. Julia and J. Scherk, Phys. Lett. 76B (1978) 409.
7) T. Kaluza, Sigzungsber. Preus. Akad. Wiss. Berlin KI (1921) 966;
O. Klein, Z. Phys. 37 (1926) 895.
8) E. Cremmer and J. Scherk, Nucl. Phys. B103 (1976) 399.
9) P. Freund and M. Rubin, Phys. Lett. 97B (1980) 233.
10) M.J. Duff and C. Pope, preprint ICTP/82/83-7, to appear in "Supergravity'82," edited by S. Ferrara, J. Taylor and P. van Nieuwenhuizen.
11) B. Biran, B. de Wit, F. Englert and H. Nicolai, CERN preprint TH.3489 (1982).
12) F. Englert, Phys. Lett. 119B (1982) 339.
13) M.J. Duff, CERN preprint Th.3451 (1982).
14) R. D'Auria, P. Fré and P. van Nieuwenhuizen, CERN preprint TH.3453 (1982).
15) B. Biran, B. de Wit, F. Englert and H. Nicolai, in preparation.
16) M. Awada, M.J. Duff and C. Pope, preprint ICTP/82-82/4 (1982).
17) J. Bourguignon and A. Karcher, Ann. Sci. d'Ecole Normale Sup. 11 (1978) 71.
18) F. Englert, M. Rooman and P. Spindel, preprint Brussels university (1983).
19) H. Nicolai and P. van Nieuwenhuizen, to be published.
20) M.J. Duff, in "Supergravity '81," edited by S. Ferrara and J. Taylor, Cambridge Univ. Press.
21) J. Ellis, M.K. Gaillard and B. Zumino, Phys. Lett. 94B (1980) 343.

THE NEED FOR AN INTERMEDIATE MASS SCALE IN GUTS

Qaisar Shafi

NASA/Goddard Space Flight Center
Greenbelt, MD 20771 and International
Center for Theoretical Physics
Trieste, Italy

Abstract

The minimal SU(5) GUT model fails to resolve the strong CP problem, suffers from the cosmological monopole problem, sheds no light on the nature of the "dark" mass in the universe, and predicts an unacceptably low value for the baryon asymmetry. All these problems can be overcome in one fell swoop in suitable grand unified axion models with an intermediate mass scale of about 10^{11}-10^{12} GeV. An example based on the gauge group SO(10) is presented. Among other things, it predicts that the axions comprise the "dark" mass in the universe, and that there exists a galactic monopole flux of 10^{-8} - 10^{-7} cm^{-2} yr^{-1}. Other topics that are briefly discussed include proton decay, family symmetry, neutrino masses and the gauge hierarchy problem.

Despite the remarkable successes enjoyed by the standard SU(3)xSU(2)xU(1) gauge model in describing electroweak and strong interaction phenomenon at present energies, there are good reasons to suspect that the model is only a part of a more complete theory. Let me point out some of them:

i) The model involves three independent gauge couplings g_3, g_2 and g_1, associated with SU(3), SU(2) and U(1) respectively. If the couplings could be related, $\sin^2\theta_W$ can be predicted.

ii) Charge quantization (in units of e/3, where e is the electron charge) is put in by hand.

iii) Model allows, in principle, leptons with fractional charges or color triplet fermions with integer charges. Such particles are not found.

iv) The model does not explain why $\bar{\theta}_{QCD} < 10^{-9}$.

v) The origin of fermion masses, mixing angles etc. is left unexplained.

vi) The standard model fails to shed light on several important problems in cosmology, such as the origin of baryon asymmetry in the universe, the nature of the dark mass in the universe, etc.

A promising approach for resolving at least some of these questions is offered by grand unified theories (or GUTS, for short). The basic idea is to embed the standard model in a larger gauge group.[1] The simplest GUT model is based on SU(5) which is a rank four group.[2] This model nicely takes care of points (i), (ii) and (iii) listed above. It makes some other interesting predictions such as the occurence of baryon number violating processes and superheavy ($\sim 10^{16}$ GeV) magnetic monopoles. But there are problems with the minimal model. Below I list a few of them, some taken from particle physics and some others from cosmology.

Particle Physics	SU(5)	Suggested Cures Include
Strong CP problem ($\bar{\Theta}_{QCD} < 10^{-9}$)	U(1) global chiral symmetry; but does not work(see later)	GUTS with an intermediate mass scale; Supersymmetry
Gauge Hierarchy problem ($m_W/m_X \sim 10^{-13}$)	Fine tune to each order in the perturbation expansion	Supersymmetry
Fermion masses; Mixing Angles	$m_b \simeq 3m_\tau$; Some bad relations or no predictions	Family Symmetry; Kaluza-Klein approach (e.g. N=8 Supergravity)
Higgs sector	Largely arbitrary	Dynamical symmetry breaking; Kaluza-Klein

Cosmology	SU(5)	Suggested cures include
n_b/n_γ (baryon asymmetry)	Orders of magnitude too low	Extended higgs sector; Larger GUTS
Dark mass (Missing mass problem)	No non-baryonic candidate	Larger GUTS with massive neutrinos; axions, photino, gravitino, higgsino.
Primordial magnetic monopoles	Too many	Inflation (but cannot be implemented in SU(5)); Larger GUTS with an intermediate mass scale;
$\delta\rho/\rho$ (density perturbations in the very early universe)	Assuming inflation get $\delta\rho/\rho >> 10^{-4}$ which is unacceptable	Extended structures (e.g. strings) from larger GUTS; Gravity
Cosmological constant $\Lambda \simeq 0$.	Fine tune	Kaluza-Klein

My intention here is to argue that at least some of these problems (in particular the strong CP and the cosmological monopole problems) can be nicely resolved by introducing an intermediate scale of about 10^{11}-10^{12} GeV in grand unification theories. Clearly, this entails going beyond SU(5).

Strong CP problem and the Peccei-Quinn Mechanism

The strong CP problem arises because non-perturbative QCD effects force one to add to the standard SU(3)xSU(2)xU(1) Lagrangian an extra term ΔL given by[3]

$$\Delta L = \bar{\theta} \frac{g^2}{32\pi^2} F^a_{\mu\nu} \tilde{F}^a_{\mu\nu}$$

where $\bar{\theta} = \theta + \arg \det M$. Here θ is the angle that characterizes the QCD vacuum and M is the quark mass matrix in the weak eigenstate basis. Clearly, the extra term ΔL violates CP and its presence induces an electric dipole moment of the neutron, current experimental upper limits on which require that $\bar{\theta} < 10^{-9}$. How $\bar{\theta}$ happens to have such a small value is not explained by the stardard SU(3)xSU(2)xU(1) gauge model and is referred to as the strong CP problem.

The small value of $\bar{\theta}$ is most naturally understood, as we shall see, in those models that possess a spontaneously broken global chiral U(1) (Peccei-Quinn) symmetry.[4] Briefly, the Peccei-Quinn mechanism works as follows. Under U(1) transformations not only the fermion fields but also the scalar fields transform suitably, such that the classical Lagrangian is U(1) invariant. The U(1) symmetry, however, is explicitly broken by QCD instanton effects, with the consequence that $\bar{\theta}$ also transforms under U(1) rotations, i.e., $\bar{\theta}$ becomes one of the dynamical variables. The potential energy contains a term proportional to

$$\Lambda^4 (1-\cos\bar{\theta}), \quad \Lambda \simeq 100\text{MeV is the QCD scale,}$$

which is minimized for $\langle\bar{\theta}\rangle = o$. The strong CP problem is no more!

The spontaneous breaking of the U(1) symmetry (broken only by QCD instanton effects) leads to the appearance of a pseudo-Goldstone boson known as the axion.[5] The axion has a mass $m_a \sim f_\pi m_\pi / f_A$, where f_A is the dominant U(1) breaking scale. The important question now is: What is f_A?

The axion is consistent with all known laboratory constraints for $f_A \gtrsim 1$ TeV. A more stringent constraint on f_A comes from astrophysical considerations. In order that the power radiated in axions by the helium core of a red giant be not too excessive, one requires that $f_A \gtrsim 10^8$ GeV.[6] So we conclude that the axion must be light ($\lesssim 10^{-1}$ eV), and also weakly coupled (its couplings to ordinary matter should be suppressed by inverse powers of f_A). Remarkably enough, there is even an upper bound on f_A which comes from cosmology.[7] In order to derive it, we must briefly consider axion production and their subsequent evolution in the early universe.

Primordial Axions

At $T \sim f_A$, the U(1) symmetry is spontaneously broken by $\langle\phi\rangle \sim f_A e^{i\theta}$, with $\bar{\theta}$ taking some value between 0 and 2π. The axion field is defined to be $\phi_A \equiv f_A \bar{\theta}$. For $\Lambda < T < f_A$, $L(\phi_A) \sim (\partial_\mu \phi_A)^2$ so that the axion field essentially behaves as a free massless scalar field. For $T \lesssim \Lambda$, the QCD instanton effects introduce the term $\Lambda^4(1-\cos\bar{\theta})$ and the field ϕ_A starts to perform damped oscillations about $\langle\bar{\theta}\rangle=0$. These oscillations produce a coherent state of axions at rest. They turn out to be non-relativistic even though they are produced at $T \sim \Lambda \gg m_a$! The energy density in the axion gas decreases as R^{-3}, whereas the radiation energy density falls off as R^{-4}. Imposing the requirement

$$\rho_a \equiv \rho_{axions} \lesssim (1\text{-}10)\,\rho_c \quad (\rho_c \sim 2 \times 10^{-29}\ \text{g cm}^{-3}$$

is the critical energy density)

one obtains the promised upper bound on f_A,[7]

$$f_A \lesssim 10^{11}\text{-}10^{12}\ \text{GeV}$$

Thus, the Peccei-Quinn mechanism can be satisfactorily implemented only if we are prepared to introduce an intermediate mass scale in GUTS. We must go beyond SU(5).

Before discussing other problems from our list, we must consider another constraint on axion models that arises from cosmological considerations. In the effective field theory describing physics at ordinary energies, the Peccei-Quinn U(1) symmetry is realized non-linearly. Under a U(1) transformation $e^{i\psi Q}$,

$$\bar{\theta} \rightarrow \bar{\theta} - 12\psi$$

provided there are three fermion familes. For $\psi = n\pi/6$, $\bar{\theta} \rightarrow \bar{\theta} + 2\pi n$ which is an identity transformation. It follows that there is a discrete subgroup of U(1), consisting actually of six distinct elements, which is not broken by QCD effects, but which is spontaneously broken by higgs vacuum expectation values. The spontaneous breaking of the discrete symmetry implies the existence of topologically stable domain walls which are cosmologically unacceptable[8]. Thus, we are confronted with a domain wall problem.

Let us summarize what we have learnt so far. In order to satisfactorily implement the Peccei-Quinn mechanism, we must ensure that

1) the spontaneous breaking scale f_A of U(1) satisfies

$$10^8 \text{ GeV} \lesssim f_A \lesssim 10^{12} \text{ GeV}$$

2) there are no topologically stable domain walls.

Several remarks are now in order:

A) It has already been mentioned that (1) can be taken care of by going to GUTS larger than SU(5), e.g., SO(10).

B) The resolution of (2) necessarily involves the introduction of new fermions that transform under real representations of the gauge group.[9,10] The vacuum structure of the theory can then be made topologically trivial, and the domain wall problem is avoided. In some models the U(1) symmetry prevents the additional fermions from acquiring huge masses through direct coupling to higgs that acquire large vacuum expectation values. Radiatively acquired masses, in two or three loops, then make these fermions relatively light, of order 10^2-10^3 GeV. It is important to look for such fermions in the next generation of high energy machines. A characteristic signature would be their V+A couplings to the W bosons.

C) An elegant resolution of (2) involves embedding the unbroken discrete elements of U(1) in a continuous symmetry which is most naturally identified with a family symmetry[9]. The family symmetry could either be global or local.

D) Suppose it is global and also spontaneously broken.[11] This then implies the existence of goldstone bosons called familons. These objects can be looked for in rare decays such as

$$\mu^- \to e^- + f \text{ (familon)}$$

$$K^+ \to \pi^+ + f$$

One expects

$$\frac{\Gamma(\bar{\mu} \to \bar{e} + f)}{\Gamma(\mu^- \to e^- + \nu_\mu + \bar{\nu}_e)} \sim 2.5 \times 10^{14} \left(\frac{\text{GeV}}{F}\right)^2 \text{ etc.,}$$

where F is the relevant symmetry breaking scale. Typically, $F \sim 10^{11}$-10^{12} GeV.

E) For $f_A \sim 10^{11} - 10^{12}$ GeV, $\rho_a \sim (1\text{-}10)\ \rho_c$. A new cosmological scenario where axions provide the dark matter in the universe has recently been constructed. Axion models predict the existence of topologically unstable extended structures called "walls bounded by strings". Fluctuations ($\delta\rho/\rho$) in the axion field energy density produced by these structures may cause the appearance of "axion clumps" with masses $\sim 10^6 M_\Theta$.[12] These objects would then form the "building blocks" for a clustering hierarchy theory of galaxy and supercluster formation on length scales up to 10 Mpc and mass $\sim 10^{15} M_\Theta$.[12,13] They also provide the seed potential wells needed for galaxy formation.

Thus, in axion models we may not have to postulate an arbitrary spectrum of initial density perburbations $\delta\rho/\rho$. The latter may come naturally and causally from the physics of the U(1) symmetry breaking which produces the axions to begin with. Another problem on our list can therefore been taken care of!

Next let me discuss a specific SO(10) model which satisfactorily implements the Peccei-Quinn mechanism and also possesses other interesting features.

An SO(10) Model

Consider the following breaking of SO(10)[14]

$$SO(10) \xrightarrow[M_X]{} SU(4)_c \times SU(2) \times U(1) \xrightarrow[M_C]{} SU(3) \times SU(2) \times U(1)_y \xrightarrow[M_W]{} SU(3) \times U(1)_{em}$$

The first step in the symmetry breaking can be achieved by a combination of a real $\underline{45}'$ and a real $\underline{54}$, both with PQ charge zero. The second breaking requires $\underline{126}$, $\underline{45}$ and $\underline{16}$ of Higgs fields, with PQ charges 2, 4, and zero respectively. The U(1) symmetry is also broken at this stage. Finally, a $\underline{10}$ plet of Higgs field with PQ charge -2 can achieve the last step in the symmetry breaking. The fermion content of the model is given by

$$\psi_{16}^{(i)}\ (i=1,2,3),\ \psi_{10}^{(\alpha)}\ (\alpha=1,2)$$

where the subscripts denote the dimension of the SO(10) representation to which the various fields belong. The U(1) transformation properties of the fermion fields are:

$$\psi_{16}^{(i)} \to e^{i\theta}\, \psi_{16}^{(i)}$$

$$\psi_{10}^{(\alpha)} \to e^{-2i\theta}\, \psi_{10}^{(\alpha)}$$

and are chosen so that the residual, discrete PQ symmetry coincides with the center Z_4 of SO(10). Had we not included $\psi_{10}^{(\alpha)}$ (α=1,2), the residual PQ symmetry would be Z_{12} which, of course, is too large to be embedded in Z_4.

We next use the one loop renormalization group equations for the various couplings constants to calculate M_x and M_c in terms of $\sin^2\theta_w(M_w)$ and $\alpha_s(M_w)$. We have included the following Higgs contributions. Between M_c and M_x we include the $(\overline{10},1,-1)$ component of $\underline{126}$, the (15,1,0) component of $\underline{45}$ and the (1,2,½) component of $\underline{10}$. Between M_w and M_c we include only the Weinberg-Salam doublet. The fermions in the $\underline{10}$ contribute to the renormalization group equations between M_c and M_x. The results for M_x and M_c are shown in Table I.

Table I

$\sin^2\theta_w(M_w)$	$\alpha_s^{-1}(M_w)$	M_c(GeV)	M_x(GeV)
0.22	7.5	3.5×10^{11}	2×10^{15}
0.22	8.0	6.2×10^{11}	1.3×10^{15}
0.22	9.0	1.9×10^{12}	5.7×10^{14}
0.23	7.5	3.7×10^{9}	2.6×10^{15}
0.23	8.0	6.6×10^{9}	1.7×10^{15}
0.23	9.0	2.0×10^{10}	7.2×10^{14}

Table I: M_c and M_x as functions of $\sin^2\theta_w(M_w)$ and $\alpha_s(M_w)$ with $SU(4)_c \times SU(2) \times U(1)$ as the intermediate symmetry group. The U(1) Peccei-Quinn symmetry is broken at scale M_c.

For the sake of definiteness, from now on we concentrate on the first possibility in Table I i.e. $M_c \sim 3.5\times10^{11}$ GeV and $M_c \sim 2\times10^{15}$ GeV, which corresponds to $\sin^2\theta_w \simeq 0.22$ and $\alpha_s(M_w) \simeq 0.13$. It follows from our previous considerations that $\rho_{axions} \sim \rho_c$.

Let us next see how the SO(10) axion model takes care of some of the other problems on our list. Consider first the cosmological monopole problem in the context of SU(5).

For $T > 10^{15}$ GeV, the expectation value of the Higgs 24-plet ϕ is zero so that the SU(5) gauge symmetry is restored. As the universe expands and cools below $T \sim 10^{15}$ GeV, $\langle\phi\rangle$ starts to acquire a non zero vacuum expectation value. Assuming a weakly first order phase transition, the Higgs field is rapidly quenched. Monopoles are produced during this phase transition as topological knots (Kibble mechanism). One estimates the initial relative monopole number density to be

$$r_{in} = (n/T^3)_{in} \gtrsim 10^{-10}$$

where n denotes the number density of the monopoles (and antimonopoles). Subsequent monopole - antimonopole annihilation cannot reduce this much below[15]

$$r_f \sim 10^{-10}$$

Thus, one expects to find roughly one monopole per baryon. Needless to say, this is cosmologically unacceptable and is referred to as the cosmological monopole problem.

Many attempts have been made to overcome the problem in the context of SU(5). Let me list a few of them:

1) Assume that the phase transition from SU(5) to SU(3)xSU(2)xU(1) is strongly first order, i.e., it proceeds only after a certain amount of supercooling. The parameters of the SU(5) higgs potential can be adjusted in order to achieve this. Unfortunately, this still probably leads to the production of an unacceptably large number density of monopoles through the Kibble mechanism.

2) Inflationary Scenario: Inflation presumably can overcome the monopole problem. However, as I mentioned earlier on, the scenario cannot be implemented in SU(5)[16] (nor for that matter in any of the known GUTS).

3) One possible way of overcoming the monopole problem in SU(5) is to add extra higgs field to the system and arrange parameters in such a way that the symmetry breaking pattern of SU(5) in the very early universe is very different from what is suggested by the simplest version[17]. For instance, one could envisage the following scenario

$$SU(5) \to SU(3) \to SU(3) \times SU(2) \times U(1) \to SU(3) \times U(1)_{em}$$

This model predicts essentially zero monopoles in the present universe. Although perfectly logical, I do not regard this resolution of the monopole problem as being particularly attractive. However, it can only be excluded by looking for GUT monopoles and finding one!

I shall now argue that the inevitable existence of an intermediate mass scale in axion models can be exploited to resolve the cosmological monopole problem[18]. To be specific consider the SO(10) axion model:

$$SO(10) \xrightarrow[M_x \sim 10^{15}\,GeV]{} SU(4)_c \times SU(2) \times U(1) \xrightarrow[M_c \sim 3.5\times 10^{11}\,GeV]{} SU(3) \times SU(2) \times U(1)_y$$

Monopoles are produced at the phase transition where the SO(10) symmetry breaks down to $SU(4)_c \times SU(2) \times U(1)$. This transition takes place at a critical temperature $T_c \sim 10^{15}$ GeV. We will assume that the initial relative monopole density

$$r_{in} \gtrsim 10^{-10}$$

Subsequent monopole - antimonopole annihilation reduces the relative density, as previously discussed, to

$$r_f \sim 10^{-10}$$

at temperatures of order 10^{12} GeV.

The parameters of the theory can be chosen so that the zero temperature effective potential for the breaking of $SU(4)_c \times SU(2) \times U(1)$ down to $SU(3) \times SU(2) \times U(1)_y$ is of the Coleman-Weinberg type[19]. In this case, as the universe cools below a critical temperature $T_{c1} \sim M_c$, the $SU(4)_c \times SU(2) \times U(1)$ phase becomes metastable. The vacuum energy density of this phase soon dominates over the radiation energy density, and the universe enters an exponentially expanding de Sitter state. Gravitational and thermal effects destabilize the $SU(4)_c \times SU(2) \times U(1)$ phase at a temperature T_{c2} of order the Hawking temperature T_H of this phase

$$T_{c2} \sim T_H = \frac{H}{2\pi} \sim \frac{M_c^2}{M_{PL}} \sim 10^4 \text{ GeV}$$

Here H is the Hubble constant of the de Sitter state and $M_{PL} \simeq 1.2 \times 10^{19}$ GeV is the Planck mass. The transition to the $SU(3) \times SU(2) \times U(1)_y$ phase is rapidly completed at T_{c2} and the latent heat is released. The Universe reheats to a temperature T_R of order $(1-3) \times 10^{11}$ GeV and the relative monopole density is diluted to[18]

$$r(T_R) \sim 10^{-10} \left(\frac{T_{c_2}}{T_R}\right)^3 \sim 10^{-32} - 10^{-31}$$

This value for r is consistent with the cosmological bounds from nucleosynthesis and from the observed values of the Hubble constant and the deceleration paramter. The cosmological monopole problem is therefore resolved. Moreover, the predicted monopole density is just about enough to sustain an anisotropic galactic monopole flux at the level of the Parker bound for 10^{10} years. Thus, a monopole flux of[18]

$$F \sim (10^{-8} - 10^{-7})\ \mathrm{cm}^{-2}\ \mathrm{yr}^{-1}$$

may still exist in our galaxy.

Note that the predicted monopole flux is compatible with a recent upper bound on it derived from considerations of observational limits on the diffuse ultraviolet and X-ray background[20]. This latter bound happens to coincide with well known bounds on 10^{16} GeV mass monopoles obtained in ref (21).

The resolution of the cosmological monopole problem in the manner described above implies that the observed baryon asymmetry in the universe gets created after completion of the intermediate phase transition. One must therefore require that there exist Higgs bosons with masses of order 10^{11} GeV which have $\Delta B \neq 0$ decay modes. The out of equilibrium decay of these bosons can create the observed baryon asymmetry.

Thus, the SO(10) axion model with an intermediate mass scale of about 10^{11} - 10^{12} GeV is able to overcome the problems explicitly listed in the abstract of the talk. Let me briefly discuss two other topics in the context of this model. First consider proton decay. The requirement that there exist higgs bosons with masses of order 10^{11} GeV and with $\Delta B \neq 0$ couplings to ordinary fermions suggests that nucleon decay mediated by these bosons competes with, and perhaps even dominates, the usual gauge boson mediated decays. Thus, processes like $p \rightarrow \mu^+ K^0$ may dominate over the usual decay mode of the proton such as $p \rightarrow e^+ \pi^0$.

Finally, note that B-L is spontaneously broken in this SO(10) model at the intermediate scale of about 10^{11} GeV. Following ref (22), this implies that the tau neutrino has a mass

$$m_{\nu_\tau} \sim \frac{m_t^2}{M_{B-L}} \gtrsim \text{several (eV), for } m_t \gtrsim 20 \text{ GeV}.$$

We thus have the intriguing possibility that both axions and neutrinos contribute significantly to the energy density of the universe.

Gauge Hierarchy Problem and Supersymmetry

The inability of the standard GUTS to explain small mass

ratios such as $m_W/m_X \sim 10^{-13}$ etc. in a satisfactory manner is referred to as the gauge hierarchy problem. For instance, in SU(5) $m_W/m_X \sim 10^{-13}$ is obtained only if the parameters of the theory are fine tuned to one part in 10^{26}. Moreover, ordinary perturbation theory does not respect fine tuning, so that adjustments to the parameters must be made in each order of the expansion. The last problem perhaps can be overcome if m_W/m_X happens to lie near a fixed point[23].

It was hoped that supersymmetry (SUSY) may overcome the gauge hierarchy problem. However, this is not borne out by recent calculations. Fine tuning remains part and parcel also of SUSY GUTS. This seems to hold both for global and N=1 local SUSY GUTS. Moreover, more often than not, the merger of cosmology and SUSY leads to unacceptable consequences.[24] Although one would like to think that an attractive idea like supersymmetry should play a role in particle physics, this has not yet been satisfactorily realized.

Concluding Remarks

The presence in GUTS of an intermediate mass scale of about 10^{11} - 10^{12} GeV can help resolve a number of apparently unrelated problems. We must be prepared to go beyond (SU(5). One can think of at least two ways of achieving this. Either by going to "standard" larger GUTS such as SO(10), or by attempting to combine SU(5) with a family symmetry. An example incorporating the first possibility is readily constructed and is described in the text. An elegant example utilizing the second possibility remains to be found. Finally, local supersymmetric GUTS also require an intermediate scale of about 10^{11} GeV[25]. Is there any connection between these two intermediate scales?

Acknowledgement

I would like to thank Abdus Salam for arousing my interest in the subject and Harvey Newman for arranging an extremely nice conference in Erice.

References

1. J. C. Pati, Abdus Salam, Phys. Rev. D10 (1974) 274.

2. H. Georgi, S. L. Glashow, Phys. Rev. Lett. 32 (1974) 438.

3. G. 't. Hooft, Phys. Rev. Lett. 37 (1976) 8;
   C. Callan, R. Dashen, D. Gross, Phys. Lett. 63B (1976) 334;
   R. Jackiw, C. Rebbi, Phys. Rev. Lett. 37 (1976) 1977

4. R. Peccei, H. Quinn, Phys. Rev. Lett. 38 (1977) 1440.

5. S. Weinberg, Phys. Rev. Lett. 40 (1978) 223;
F. Wilczek, Phys. Rev. Lett. 46 (1978) 279.

6. D. Dicus et al, Phys. Rev. D18 (1978) 1829.

7. J. Preskill, M. Wise, F. Wilczek, Phys. Lett. 120B (1983) 127;
F. Abbott, P. Sikivic, Phys. Lett. 120B (1983) 133;
M. Dine, W. Fischler, Phys. Lett. 120B (1983) 137.

8. P. Sikivie, Phys. Rev. Lett. 48 (1982) 1156.

9. G. Lazarides, Q. Shafi, Phys. Lett. 115B (1982) 21.

10. H. Georgi, M. Wise, Phys. Lett. 116B (1982) 123;
S. M. Barr, D. Reiss, A. Zee, Phys. Lett. 116B (1982) 227;
B. Grossman, Rockefeller, preprint, 1982.

11. F. Wilczek, Phys. Rev. Lett. 49 (1982) 1549.
Also see D. Reiss, Phys. Lett. 109B (1982) 365.

12. F. Stecker, Q. Shafi, Phys. Rev. Lett. 50 (1983) 928.

13. A clustering scenario for galaxy formation with axions has also been discussed by M. Turner, F. Wilczek, T. Zee, Univ. of Washington preprint, Feb. 1983.

14. R. Holman, G. Lazarides, Q. Shafi, Phys. Rev. D27 (1983) 995.

15. See, for instance, J. Preskill, Proceedings of the Nuffield Workshop on the Very Early Universe (June 1982), and references therein.

16. See proceedings of the Nuffield Workshop, Cambridge, June 1982.

17. P. Langacker, S. P. Pi, Phys. Rev. Lett. 45 (1980) 1.

18. G. Lazarides, Q. Shafi, Phy. Lett. B, April 21, 1983.

19. S. Coleman, E. Weinberg, Phys. Rev. D 7 (1983) 1888.

20. V. Rubakov, V. Kuzmin, Trieste preprint, Feb. 1983.

21. G. Lazarides, Q. Shafi, T. Walsh, Phys. Lett. 100B (1981) 21.
Also see F. Bogodan, G. Parker, M. Turner, Phys. Rev. D26 (1982) 1296.

22. M. Gell-Mann, P. Ramond, R. Slansky, Rev. Mod. Phys. 1979
C. Wetterich, Nucl. Phys. B. 187 (1981) 343.

23. C. Wetterich, CERN preprint (1983).

24. See, for instance, G. Lazarides, Q. Shafi, Phys. Rev. D., May 15, 1983.

25. R. Barbieri, S. Ferrara, C. Savoy, Phys. Lett. 119B (1982) 343;
R. Arnowitt, A. Chemseddine, P. Nath, Phys. Rev. Lett. 49 (1982) 970;
J. Ellis, D. V. Nanopoulos, K. Tamvakis, CERN preprint TH3418 (1982);
L. Hall, J. Lykken, S. Weinberg, Univ. of Texas preprint 1983.

CAN SUPERSYMMETRY BE CONNECTED TO OBSERVATION?

Murray Gell-Mann

California Institute of Technology

452-48, Pasadena, California 91125

The subject of this article is supersymmetry[1] and supergravity[2]. I shall comment on what I believe are the main approaches to the problem of relating supergravity to observation. I shall also discuss briefly an idea of my own, published here for the first time; although it does not seem to work, I think it is amusing and maybe someone else will be able to do something with it.

Supersymmetry is a very clever way of associating fundamental fields of different spins in the same supermultiplet. In this way the different fundamental fields, or haplons (from the Greek word for simple or single) can be thought of as components of a single field.

There are two other proposals for associating fields of different spin. One of them[3], which I shall not pursue here, is discussed in some very interesting, highly speculative work of Kevin Cahill and of John Ward, who have arrived independently at somewhat similar conclusions. They relate different integral spins to one another and different half-integral spins to one another, while supersymmetry directly connects integral and half-integral spins.

Another way to connect integral spins together is the method of Kaluza and Klein[4]. It involves associating other spatial dimensions with the four dimensions of our space time and using these supplementary dimensions, assuming they are wrapped up in a little ball, to obtain other fields out of the metric $g_{\mu\nu}$. If i and j refer to the supplementary dimensions and μ and ν refer to the four dimensions with which we are familiar, then the $g_{\mu i}$ and g_{ij} metric tensor components represent vectors and scalars respectively, as far as four-dimensional space time is concerned. The Kaluza-Klein idea, which goes back some sixty years, is now popular again, after a long period of neglect. It can be and is being used in connection with supersymmetry[5]. The methods of Cahill and Ward might possibly be compatible with supersymmetry as well.

Well, what is supersymmetry? Let me review it. Supersymmetry comes in eight different kinds, and the simplest one is the so-called $N=1$ supersymmetry. One has a spinor-raising operator and a spinor-lowering operator that change the z-component of angular momentum of states by half a unit. The representations of this supersymmetry for massless states involve two adjacent spins. One can have for example $J_z=+1$ associated with $J_z=+½$. This is a representation of ordinary supersymmetry, and by CPT one also has to add the states $J_z=-½$ and $J_z=-1$. The result is a massless spin 1 vector and a massless Majorana spinor. There can be a set of these, belonging to the adjoint representation of a gauge group:

		# states	
$J_z =$	+1	1	
	+½	1	
	0		×(adjoint of a gauge group)
	−½	1	
	-1	1	

This means that the spin 1 particles can be Yang-Mills particles (for example, for SU(3) one would have eight vectors and eight Majorana spinors). One can also have other representations, for example $J_z=+½$ and 0 accompanied by $J_z=-½$ and 0. Here one has a scalar, a pseudoscalar and a Majorana spinor belonging to any representation of the symmetry group:

		#states	
$J_z =$	½	1	
	0	1+1	×(representation of the gauge group)
	−½	1	

If supersymmetry is exact, it must be spontaneously violated, as discussed further on. The violation produces a "supergap" in mass between each particle and its "superpartner," which has spin differing by one-half unit. That gap could be enormous, say of order 10^{19} GeV, or it could be relatively small, say less than 10^3 GeV, or it could be somewhere in between. If it is small, then superpartners could play a very important role in the results of accelerator experiments and our standard theory at moderate energies would be replaced by a "supertheory."

Several attempts have been made over the last few years to use the above supermultiplets for the weak and electromagnetic interactions, for the strong color interaction, and for unified Yang-Mills theories. At the very beginning of this period, in the early work of Fayet[6], it was hoped that the spin 1 particles would be accompanied by spin ½ particles that one could use, and that the spin ½ particles would be accompanied by spin zero particles that one could use, or vice versa. This turned out not to be true, and it is necessary to add new particles with different spins to accompany everything that is known. With the O'Raifertaigh[7] mechanism, as opposed to the old Fayet[8] mechanism for spontaneous violation of supersymmetry, it is now necessary to add on, in addition, a large number of spin ½ and spin 0 particles that are superpartners of each other, which nobody wanted at all. With this long list, though, one can elevate the usual theory of the weak, electromagnetic and strong interactions (or a typical unified Yang-Mills theory) into a supertheory. In such a supertheory, there is a large number of unwanted particles of spin ½ and spin 0, which one must push up to higher energy where they could not have been seen yet. That can be arranged, with whatever supergap is needed.

The supertheory has some advantages, as one gains control of radiative corrections to certain quantities that are known to be small. When quantities are small, one would like not to have them subject to arbitrary corrections from renormalization. One would like to be able to calculate radiative corrections to them in principle, and show they are small too. That can be done in supertheories[9]. For example, the ratio of the mass scale for the violation of SU(2)×U(1) to that for the violation of the unified Yang-Mills symmetry, which is around 10^2 GeV/10^{15} GeV = 10^{-13}, does not get arbitrary corrections. Neither does the CP violating angle in QCD. The smallness of the breaking of weak interaction symmetry compared to the largeness of the breaking of the unified Yang-Mills symmetry comes about because the Higgs bosons for the weak interactions have nearly zero mass on the scale of the unified Yang-Mills theory. Why is the mass of the Higgs boson small? What may happen is the following: In a supertheory the Higgs bosons are superpartners of spin ½ particles, which have their masses tied down to relatively small values by approximate chiral invariance. The Higgs bosons and their superpartners have their masses connected by supersymmetry, and in this way one can keep Higgs boson masses near zero if they are originally set to a small value.

In unified Yang-Mills theories there are several very small numbers. These are the ratio between the mass of the intermediate bosons and the unification mass:

$$\frac{M_{IB}}{M_{UNIF}} \approx 10^{-13} \ ,$$

the ratio between the renormalization group invariant mass in QCD, of about 200-250 MeV, and the unification mass:

$$\frac{M_{QCD}}{M_{UNIF}} \approx 10^{-16} \ ,$$

and the ratios between the fermion masses and the unification mass:

$$\frac{M_{FERM}}{M_{UNIF}} \approx 10^{-18} \ to \ 10^{-13} \ ,$$

10^{-18} for the electron to 10^{-13}, perhaps, for the t-quark, if it exists. The middle number, M_{QCD}/M_{UNIF}, is usually thought of as being exp(-40) or so, where the 40 is some group theory constant of the order of 1, divided by something like $\frac{1}{40}$, which is the dimensionless coupling constant of the unified Yang-Mills theory. Why is the coupling constant small, say $\frac{1}{40}$? Nobody has the vaguest idea. (In fact, this is the modern version of the old mystery of $\frac{1}{137}$.)

Anyway, M_{QCD}/M_{UNIF} and the other numbers are dialed to small values. What is accomplished in the supertheories is that the dialed numbers remain small. It is unexplained why they are dialed to small values, and in most cases it is also unexplained why they are all similarly small, which in my view is the most mysterious thing of all.

The gains are modest, and whether it is worth introducing these supertheories with a relatively small supergap only time will tell. It is an experimental matter because experimentalists will soon be looking for the superpartners of the known particles. The experimentalists are, of course, delighted with supertheories, because they have made the "desert" bloom; in fact they have turned it into a jungle!

Supersymmetry is also applied, of course, to Einstein's theory of gravitation. In that theory we have a massless graviton with two degrees of freedom $J_z = \pm 2$ obeying an infinitely nonlinear equation. In Feynman diagram notation:

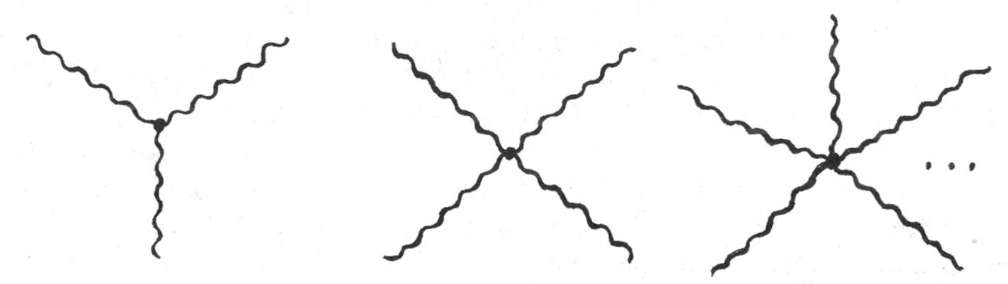

One can have any number of graviton lines meeting at a point. Gravity has Newton's constant as a dimensional parameter. The square root of Newton's constant, κ, is proportional to the inverse of a mass, the Planck mass M_p, such that we get a dimensionless constant of order unity:

$$\frac{\kappa^2 M_p^2}{4\pi\hbar c} \approx 1, \quad M_p \approx 2 \times 10^{19}\, GeV \sim 4 \times 10^{-5} gm.$$

The corresponding Compton wave length is around $10^{-33} cm$, and is called the Planck length.

The quantum theory of gravity is actually not very singular. It is finite in one loop[10]. In two loops, there is a candidate for a divergence [11], but no theorist has yet had the strength to calculate the coefficient and see if it is non-zero. So the theory may not be singular at all, or if it is singular it is very slightly so, with mild divergences in two loops and so on. When you couple gravity to matter, however, whether it is scalar matter, spinor matter, or vector matter, then you get enormous singularities[12], tremendous divergences in the perturbation expansion, and we no longer know how to handle the theory. That is what is meant by the statement that quantum gravity gives difficulties.

Let us now look at supergravity[2]. Here the graviton belongs to a supermultiplet, and one has spin 2 together with spin $\frac{3}{2}$, both massless. The spin $\frac{3}{2}$ particle is called a gravitino. In supergravity one has a generalization of Einstein's equation in which the graviton and the gravitino obey coupled nonlinear equations. If we want supersymmetry to be an exact symmetry of nature, it must be spontaneously violated, since no degeneracy between fermions and bosons is observed in nature. When one has spontaneous violation of an ordinary continuous symmetry, one always has massless spin-zero particles. Since supersymmetry is a spinor symmetry, the corresponding zero mass particle is a spinor particle, a Goldstone fermion, commonly referred to as a goldstino. This spin ½ Goldstone particle can be found among the haplons of the theory, can be composite, or can be introduced separately, as long as it is there. The gravitino acquires mass by eating the goldstino. (A spin $\frac{3}{2}$ particle, in order to get mass, needs 4 degrees of freedom corresponding to $J_z = \frac{3}{2}, \frac{1}{2}, -\frac{1}{2}, -\frac{3}{2}$, whereas a massless spin $\frac{3}{2}$ particle has only $J_z = \frac{3}{2}, -\frac{3}{2}$.)

What about supergravity and its divergences? Pure supergravity is free of divergences both in one-loop and two-loops[13,14], although there is a candidate for an infinity in three loops, so again it is not a very singular theory and it might even be finite for all we know. If supergravity is coupled to external supermatter, however, one gets terrible divergences in perturbation theory[15], and nobody knows whether the theory makes sense or how to calculate with it.

Many people are now playing the game of ignoring this difficulty with infinities, introducing an effective cutoff at the Planck mass and then considering the corrections to supertheories of weak and electromagnetic, strong, and especially unified Yang-Mills interactions, as a result of coupling to $N=1$ supergravity. They look at all the terms that could possibly come in, classify them according to powers of the cutoff and powers of the Planck mass, and calculate rough orders of magnitude for the coefficients. There may be large numerical coefficients that are gotten wrong, or some might be zero for reasons that nobody knows today, and therefore these estimates are not very reliable. Nevertheless it is an amusing thing to do and it is done, for example, in a recent paper by Ellis, Nanopoulos and Tamvakis[16], among other papers by other authors on the same subject. They find significant corrections to all sorts of important phenomena. In fact, if their work is to be taken seriously with the cutoff, it means that all the standard predictions at low energies that come from very high energies are wrong. The great boom in experimental testing of unified Yang-Mills theories was due to the possibility of doing experiments at moderately low energies, on proton decay, on neutrino masses, on monopoles, even though they involved the extraordinarily high energy phenomena of unified Yang-Mills theories at, say 10^{15} or 10^{16} GeV. What these people have found by turning the unified theory into a supertheory, coupling it to $N=1$ supergravity with a cutoff, and estimating the corrections coming from virtual supergravity is that all the low energy predictions are altered. For example, proton decay has a different rate, which one cannot calculate very well, and the decay into K-particles should be much more important than the decay into pions. These superpeople, therefore, welcome results on low rates or low upper limits for pion decay modes of the proton, since they want K-decay instead. Likewise the charged lepton and quark masses can be altered, so that calculations of these masses from Higgs breaking need corrections. The neutrino masses may be altered as well. In general these attempts at making the unified Yang-Mills theory into a unified Yang-Mills supertheory and coupling it to $N=1$ supergravity introduce a wild card into the whole subject. I think it is very important to remember this.

We have said that when one couples $N=1$ supergravity to $N=1$ supermatter one finds huge infinities in perturbation theory. This does not necessarily mean that there is no non-perturbative way of calculating that would not have infinities, but, after thirty years of trying, theorists have never found one. Therefore, it is unlikely that it could be done here, although the possibility is not eliminated logically.

There is a way to couple matter to gravity in a very nonsingular manner, although it may not be completely free of infinities, and that is to go to higher supergravity theories using "extended" supersymmetry with $N>1$. One gets a supermultiplet of the following kind:

		#states
$J_z =$	$+2$	1
	$+\frac{3}{2}$	N
	$+1$	$\frac{N(N-1)}{2}$
	$+\frac{1}{2}$	$\frac{N(N-1)(N-2)}{6}$
	.	.
	.	.
	.	.

The index N denotes the number of supersymmetries and runs from 1 to 8. We have one graviton, N gravitinos that eat the N goldstinos, $\frac{N(N-1)}{2}$ spin one bosons that can gauge SO(N), and so on. It appears desirable, although perhaps not absolutely necessary, to make the list terminate with $J_z \geqslant 2$ and thus to confine N between 1 and 8. Nobody knows how to write interacting theories for spins higher than 2, and when we include the CPT conjugate representation we would have more than one graviton, more than N gravitinos, and so forth, if we took $N>8$.

In the N-extended supergravities one gets for the haplon fields nonlinear equations, which are generalizations of Einstein's equation. If one gauges the SO(N), one gets a theory with two coupling constants, the gravitational coupling κ and the dimensionless self-coupling e for the supermultiplet. The largest supergravity theory we know, $N=8$ supergravity, contains 1 graviton, 8 gravitinos, 28 potential gauge bosons, 56 Majorana spinors, 35 scalars, and 35 pseudoscalars.

In $N=8$ supersymmetry there is no problem with external supermatter, since there is no room for it. The supermultiplet goes from $J_z=+2$ to $J_z=-2$ and is not doubled for CPT conjugation since it is already self-conjugate. Everything in the world has been put into this one supermultiplet; there are only these haplons in the world, and they are all superpartners of the graviton. The theory is highly non-singular, although there may be infinities. The various theorists who work on these matters dispute among themselves whether the first threatened infinity comes at three loops or at seven; and of course the theory could be finite.

If we believe in an effective $N=1$ super-Yang-Mills theory with a relatively small supergap coupled to $N=1$ supergravity with an effective cutoff, we may suppose that seven of the eight supersymmetries are broken up near the Planck mass, and one of them is left to come down and be broken at a much lower energy.

What are the symmetries of $N=8$ supergravity? It depends on whether we have a dimensionless self-coupling or not. If the dimensionless self-coupling is omitted (e=0)[17], there is a global SU(8) symmetry. This SU(8) is chiral, that is, particles spinning one way and particles spinning the other way transform under different representations of the SU(8):

		SU(8) reps.
$J_z =$	$+2$	$\underset{\sim}{1}$
	$+\frac{3}{2}$	$\underset{\sim}{8}$
	$+1$	$\underset{\sim}{28}$
	$+\frac{1}{2}$	$\underset{\sim}{56}$
	0	$\underset{\sim}{70}$
	$-\frac{1}{2}$	$\underset{\sim}{\overline{56}}$
	-1	$\underset{\sim}{\overline{28}}$
	$-\frac{3}{2}$	$\underset{\sim}{\overline{8}}$
	-2	$\underset{\sim}{1}$

If we introduce a non-zero value of the dimensionless gauge-coupling constant[18], the symmetry is reduced to a local SO(8):

		SO(8) reps.
$J_z =$	$+2$	$\underset{\sim}{1}$
	$+\frac{3}{2}$	$\underset{\sim}{8}$
	$+1$	$\underset{\sim}{28}$
	$+\frac{1}{2}$	$\underset{\sim}{56}$
	0	$\underset{\sim}{35} + \underset{\sim}{35}'$
	$-\frac{1}{2}$	$\underset{\sim}{56}$
	-1	$\underset{\sim}{28}$
	$-\frac{3}{2}$	$\underset{\sim}{8}$
	-2	$\underset{\sim}{1}$

We have lost the chiral part of the symmetry.

Can the spin-0 fields in the gauged $N \geqslant 4$ supergravities be utilized to break symmetries by having an expected value in the vacuum? Well, the scalar potentials in these theories have been constructed, and they are always unbounded from below, that is to say, they go on down toward a negatively infinite energy density. It therefore seemed that a stable vacuum would not exist. Breitenlohner and Freedman[19], however, have shown that this is not the case if one exploits kinetic energy effects in the theory. The scalar fields can acquire stable vacuum expectation values, and can break both the SO(N) symmetry and supersymmetry.

How about the identification with the real world? I worked on that problem before supergravity, even $N=1$ supergravity, was written down. Back in 1974 Yuval Ne'eman and I[20] looked at the SO(8) group, assuming it was gauged, and at the list of haplons given above. We asked whether we could compare those in any way with experiment. The group SO(8) is too small to include SU(3)×SU(2)×U(1), so we gave up right away on explaining the intermediate vector bosons as fundamental particles. However, SO(8) does include SU(3)×U(1)×U(1), so there is enough room for gluons and photons among the spin one particles. The spin ½ fermions are present in the right numbers, but the quantum numbers are not the correct ones. If one identifies the SU(3) with color SU(3), the fundamental $\underset{\sim}{8}$ of spin $\frac{3}{2}$ is broken down to:

$$\underset{\sim}{8} \rightarrow \underset{\sim}{3} + \underset{\sim}{\overline{3}} + \underset{\sim}{1} + \underset{\sim}{1}$$

under SU(3). One can assign two charges, a charge for the $\underset{\sim}{3}$ and a charge for the $\underset{\sim}{1}$, say Q and Q'; then the $\underset{\sim}{\overline{3}}$ and the other $\underset{\sim}{1}$ have charges -Q and $-Q'$:

$$\underset{\underset{\sim}{3}}{Q\,Q\,Q} \qquad \underset{\underset{\sim}{1}}{Q'} \qquad \underset{\underset{\sim}{\overline{3}}}{-Q\;-Q\;-Q} \qquad \underset{\underset{\sim}{1}}{-Q'}$$

The most convenient values for comparing the spin ½ particles with observation are obtained if one takes $Q = -\frac{1}{3}$ and $Q' = 0$. In this case one obtains a very interesting list of particles of spin ½, including some lepton-like particles with charges -1 and 0 and quark-like particles with charges $-\frac{1}{3}$ and $\frac{2}{3}$. One gets, however, only two families of quarks and one family of leptons, and many particles are wasted on $\underset{\sim}{6}$'s and $\underset{\sim}{8}$'s of color, which have not been seen so far. Some fermions come out right, but one would have to interpret, say, the electron and neutrino as fundamental but the muon and tau as composites, the u, d, c, and s quarks are fundamental but the t and b as composites; and that is so ugly that nobody has ever adopted such a scheme.

Recently, in 1980, I found another way to look at it, which also doesn't work, but is quite an amusing curiosity. After a few years of playing with this and not knowing what to do with it, I have decided to put it forward. What I noticed is the following.

Suppose we take the 56 spin ½ particles and we use up 8 of them as goldstinos, which are eaten by the gravitinos. This then leaves 48 particles. Effectively (a group theorist can see this immediately!) one is reducing the symmetry from SO(8) to SO(7), and correspondingly the $\underset{\sim}{56}$ breaks up into $\underset{\sim}{8} + \underset{\sim}{48}$ of SO(7), which contains $SU(3)\times U(1)$. Now, 48 particles are just as many as the left-handed quarks and leptons that we see [three families of 15 particles each, usually called a $\overline{10} + 5$ of SU(5)] plus three left-handed anti-neutrinos, which may perfectly well be there. (They would not have weak interactions and we would not see them.) Thus the haplons with spin ½ agree in number the quarks and leptons that we would like to have there, provided we include the left-handed anti-neutrinos. Now what is the discrepancy in quantum number between what we have and what we want? Instead of the u, c, and t quarks with charge $\frac{2}{3}$, we get quarks with a U(1) quantum number of ½, and instead of having three triplets of SU(3), we get an $\underset{\sim}{8}$ and a $\underset{\sim}{1}$:

	U(1)	SU(3)
$\{u,c,t\}$	½	$\underset{\sim}{8} + \underset{\sim}{1}$

Instead of the d, s, and b quarks with charge $-\frac{1}{3}$, one gets particles with a U(1) quantum number of $-\frac{1}{6}$, and instead of three triplets one gets a $\underset{\sim}{6}$ and a $\underset{\sim}{\bar{3}}$ of SU(3):

	U(1)	SU(3)
$\{d,s,b\}$	$-\frac{1}{6}$	$\underset{\sim}{\bar{3}} + \underset{\sim}{6}$

For the charged leptons, the e, μ, τ of charge -1, one gets a triplet of SU(3) with a U(1) charge of $-\frac{5}{6}$.

	U(1)	SU(3)
$\{e,\mu,\tau\}$	$-\frac{5}{6}$	$\underset{\sim}{3}$,

finally for the neutrinos one gets a U(1) charge of $-\frac{1}{6}$ and a $\underset{\sim}{\bar{3}}$ of SU(3):

	U(1)	SU(3)
$\{\nu_e, \nu_\mu, \nu_\tau\}$	$-\frac{1}{6}$	$\underset{\sim}{\bar{3}}$

This discrepancy between what one finds and what one would like to find can be formulated in a very elegant way, which does not, however, get rid of the discrepancy. Suppose we introduce an $SU^f(3)$ of family, and suppose we assign the u, c, and t quarks to an antitriplet $\underset{\sim}{\bar{3}}$ of family, the d, s, and b quarks to a triplet $\underset{\sim}{3}$, the leptons e, μ, and τ to a triplet $\underset{\sim}{3}$ and the neutrinos ν_e, ν_μ, and ν_τ to an antitriplet $\underset{\sim}{\bar{3}}$. Then the assignments would be:

	$SU^c(3)$, $SU^f(3)$
$\{u,c,t\}$	$(\underset{\sim}{3}, \underset{\sim}{\bar{3}})$
$\{d,s,b\}$	$(\underset{\sim}{3}, \underset{\sim}{3})$
$\{e,\mu,\tau\}$	$(\underset{\sim}{1}, \underset{\sim}{3})$
$\{\nu_e,\nu_\mu,\nu_\tau\}$	$(1, \underset{\sim}{\bar{3}})$

Now we look at the list of haplons in the theory, and we see what is going on. The SU(3) generators in the theory must be the sum of the generators of the two SU(3)'s. Using convenient jargon, we say that the SU(3) of the theory is the diagonal SU(3) of $SU^c(3) \times SU^f(3)$. In that case the $\underset{\sim}{3}$ and $\underset{\sim}{\bar{3}}$ of the $\{u,c,t\}$ would make an $\underset{\sim}{8} + \underset{\sim}{1}$ of the composite SU(3), and that is what we have. In addition, the U(1) charge in the theory is the electric charge plus or minus $\frac{1}{6}$. We use minus $\frac{1}{6}$ when there is a $\underset{\sim}{\bar{3}}^f$ and plus $\frac{1}{6}$ when there is a $\underset{\sim}{3}^f$. We can describe these assignments by means of what Gregor Wentzel called a "spurion." This spurion carries $(\underset{\sim}{\bar{3}}^c, \underset{\sim}{3}^f)$ with charge $Q=\frac{1}{6}$ and $(\underset{\sim}{3}^c, \underset{\sim}{\bar{3}}^f)$ with $Q=-\frac{1}{6}$, and is a singlet under the diagonal SU(3). If we could somehow attach this spurion to the theory, we would get what we observe.

In this scheme, by the way, the weak interactions would violate $SU^f(3)$, but if family SU(3) breaks down to an SO(3) of family, the known intermediate bosons would be singlets under that SO(3). Whether all this is of any use I have absolutely no idea, but I thought I would mention it.

A more popular idea for how to compare N=8 supergravity with the observed particles is a different one. We abandon the identification of the haplons with the observed particles, except for the graviton. We say that the particles that we observe today are composites of haplons and that the haplons would be observed only at enormously high energies. All our renormalizable field theories with vector, spinor and scalar particles are simply phenomenological models, and the renormalizability is simply the rough independence of the cutoff, provided we lump the cutoff-dependent quantities into effective coupling constants and effective masses. This is the oldest view of renormalization, the view that prevailed thirty-five years ago, and many people are returning to it now. It would explain very nicely the enormous complexity our theories exhibit, specially in the Higgs sector, with large numbers of spinless bosons belonging to peculiar representations, with arbitrary self-couplings and arbitrary Yukawa couplings, following no obvious rules. This would all be explained because one would be dealing with lumped parameters depending on the real theory, say of supergravity, at higher energy. Of course we would not know how to calculate any of those numbers for a while, but at least it would explain why they exist. Even a unified Yang-Mills theory would be a phenomenological theory, depending on a real cutoff, but with most of the juice of the theory independent of the cutoff.

For example, Ellis, Gaillard and Zumino[21] suggested three years ago a particular picture, which contains, I think, some good ideas, although the proposal as a whole may not be right. They suggested dropping the dimensionless self-coupling and conjectured, following Cremmer and Julia[17], that there is a dynamical gauging of the global SU(8) invariance of the theory with composite gauge bosons doing this trick[22]. They also suggested that electric charge and color have pieces inside the SO(8) subalgebra of SU(8) and pieces outside SO(8). This is something we never discussed before; we always thought that charge and color would be found inside SO(8) and that therefore we must have symmetry between positive and negative charge and between color and anticolor. Of course, these symmetries apply when one changes the helicity from positive to negative, but for a given helicity one does not have these symmetries if one takes charge and color partially outside the SO(8). They proposed the following unsymmetrical assignment of the fundamental $\underset{\sim}{8}$: electric charges of $\{-\frac{1}{3}, -\frac{1}{3}, -\frac{1}{3}\}$, forming a triplet of color, $\{1\}$ forming a singlet of color, and $\{0,0,0,0\}$ all being singlets of color. In this way they were able to identify the first five, the triplet and two singlets, with the fundamental $\underset{\sim}{5}$ of SU(5), and the other three with a basic triplet of family SU(3), although they did not use that name.

$$\left.\begin{array}{llll} \text{charge} & -\frac{1}{3},-\frac{1}{3},-\frac{1}{3}, & 1, 0, & 0, 0, 0 \\ \text{color} & \underbrace{\qquad\underset{\sim}{3}\qquad\qquad\qquad \underset{\sim}{1},\underset{\sim}{1},}_{\underset{\sim}{5}\text{ of SU(5)}} & & \underbrace{\underset{\sim}{1},\underset{\sim}{1},\underset{\sim}{1}}_{\text{SU}^f(3)} \end{array}\right\} \text{for } J_z = \frac{3}{2}$$

The opposite helicity of course reverses all these things:

$$\left.\begin{array}{llll} \text{charge} & \underbrace{\frac{1}{3},\frac{1}{3},\frac{1}{3},} & -1, 0, & 0, 0, 0 \\ \text{color} & \underset{\sim}{\bar{3}}, & \underset{\sim}{1},\underset{\sim}{1}, & \underset{\sim}{1},\underset{\sim}{1},\underset{\sim}{1} \end{array}\right\} \text{for } J_z = -\frac{3}{2}$$

In this way overall CPT is correct for the theory, but within a given multiplet we have a lopsided situation. The rest of what they did involved some ad hoc assumptions, which finally led them to SU(5) with three families of $\underset{\sim}{5} + \underset{\sim}{\overline{10}}$ each. I am not sure they believe in those particular arguments any more.

The general idea might still be right, though, and it is a very exciting one. Electric charge and color are partly inside and partly outside the SO(8) subalgebra of SU(8), and the photon and gluons are supposed to be composites and dynamically gauged. Is this possible? There are theorems that say that exact gauge bosons cannot be composites[23], but Gaillard and Zumino have checked carefully, and they have found that the assumptions of those theorems are violated in this particular case, and therefore nobody can prove today that the suggestion is impossible. Of course, it may still be impossible, anyway. Maybe if the photon and the gluon are composite effective gauge bosons, one pays some small price: a little mass for the photon for instance, or a tiny mass for the gluon; a slight violation of electric charge perhaps, or a slight violation of color, or something. All this I believe should be carefully investigated.

Another thing that one might look at is a still larger group, a non-compact symmetry group. In the work of Cremmer and Julia, if one puts e=0 one has not only an SU(8) global symmetry, but also a non-linearly realized non-compact symmetry $E_{7(7)}$. Slansky and I are looking at this symmetry, and so are Gaillard and Zumino and no doubt others, to see whether what happens is that one gets unitary infinite-dimensional representations of $E_{7(7)}$. I have no reason to believe that this work will lead to anything useful, but I'm just mentioning it as something that theoreticians do these days to keep off the streets.

At this point I should issue a warning. If e=0, we have the non-compact non-linearly realized global symmetry $E_{7(7)}$, which contains the global SU(8) that might be spontaneously gauged. When $e \neq 0$ we have the SO(8) subgroup gauged. But in the formalism of Cremmer and Julia there is actually a bigger symmetry, the product group $E_{7(7)} \times SU(8)$, when $e=0$. However this product group appears to be a gauge artifact unsuitable for phenomenology. The same applies to the smaller product group $SU(8) \times SO(8)$ that appears formally when $e \neq 0$.

Now let me mention strings and extra dimensions. $N=8$ supergravity is not a unique theory, but falls into a class of theories that are more general. It is very important to remember this because $N=8$ supergravity itself might not work but one of these generalizations of it might. One way to get these theories is to look at superstrings. Superstrings are actually very old, much older than supergravity, and Neveu, Schwarz, and Ramond[24] worked on them around 1970. Neveu and Schwarz were looking at string theories for hadrons, and found a string theory of fermions and bosons with interactions. It was later proved in 1976 that restricting the spectrum of the Neveu-Schwarz model led to a string theory [25] without negative probabilities and without negative masses squared. It is a theory in 10 dimensions, however. Although this 10-dimensional Neveu-Schwarz string theory was originally conceived as a theory of hadrons, we have believed since 1972 or so that QCD with confined quarks and gluons is the theory of hadrons and that string theory is only a crude approximation to QCD. Scherk and Schwarz[26], in the middle seventies, revived the Neveu-Schwarz theory with a different interpretation, namely, supergravity and super-Yang-Mills supermatter plus higher mass particles. Quantized strings have the characteristic that there are particle states of increasing angular momentum along Regge trajectories, and Schwarz and Scherk altered the inverse slope of that Regge trajectory from around $(1 GeV)^2$ to around $(10^{19} GeV)^2$, the square of the Planck mass. In order to obtain a theory in four dimensions, one assumes that six of the ten dimensions are somehow spontaneously wrapped up into a little ball of some kind.

The topology of that ball is very important, but I will not discuss it here. The main point is the small size, which means that one cannot see those 6 extra dimensions in most experiments. They are supposed to be real, however, and it may be that at very high energies and in the very early moments of the universe these extra dimensions play a real physical role. Under ordinary conditions, they can still supply some observable symmetry of the elementary particles.

The zero mass sector of the theory turns out to be N=4 supergravity plus N=4 super-Yang-Mills supermatter. The excited states lie on Regge trajectories with masses squared in multiples of the Planck mass squared. In 1980, Green and Schwarz[26] discovered that back in the 1970's two other theories in 10 dimensions with bosons and fermions had been overlooked. Now that we have abandoned the connection with hadrons, we can tolerate theories with closed strings only. Remember that in this interpretation it is the closed strings that give the supergravity and the open strings that give the supermatter. In a 10-dimensional superstring theory with closed strings only you would have only supergravity and no external supermatter, and indeed, that is what was found. They found two versions of a new theory, and they believe they have now listed all possible theories. These two new versions have no open strings and therefore no external supermatter, and they both give N=8 supergravity as the zero mass sector in four dimensions, with excited states at the Planck mass or higher. One of these theories is actually more interesting than the other, and Schwarz and Green are working on it very hard. There may really be additional dimensions that play a physical role at very high energies and in the early moments of the universe. If not, if we want to roll up the extra dimensions into a little ball, shrink the ball to zero, and throw it away, we can do that too. In that case, we get modified 4-dimensional theories, because the extra dimensions can introduce symmetry violations, and we can end up with broken N=8 supergravity in 4 dimensions. That is why I said that even in 4 dimensions, N=8 supergravity is actually one of a class of theories, including various kinds of broken N=8 supergravity, which can be derived from theories with higher dimensions by throwing the higher dimensions away.

There is also a non-string way of using higher dimensions, less appealing because such theories are possibly very singular in perturbation theory, whereas the string theories are probably finite to all orders. One can take N=1 supergravity in 11 dimensions[27], or N=2 supergravity in 10- dimensions[28], and these also reduce in 4 dimensions to N=8 supergravity plus extra states coming from the rolled-up dimensions. This is another direction that theorists are exploring. It is related to the Kaluza-Klein idea of having higher dimensions with extra components of the metric giving vector and scalar fields. Some theorists are working on this approach and the plain Kaluza-Klein approach simultaneously, not knowing for sure whether they want supersymmetry or not. By the way, the two ten-dimensional string theories with only closed strings reduce, on restriction to zero mass in ten dimensions, to the two N=2 supergravity theories in 10 dimensions. Also, the N=1 supergravity theory in 11 dimensions reduces to one of the two N=2 supergravity theories in ten dimensions when the eleven dimensions are restricted to ten.

The last point I want to make is about the cosmological constant, which gives rise to what is perhaps the most important question in physics today. In Einstein's equation for gravity, on the right-hand side, one has the stress-energy-momentum tensor as the source of gravity. Einstein later considered the possibility of adding in another term, proportional to the metric tensor. This would be a kind of constant pressure and energy density throughout spacetime; call it λ:

$$R_{\mu\nu} - \tfrac{1}{2}R\, g_{\mu\nu} = 8\pi\kappa\Theta_{\mu\nu} + \lambda\, g_{\mu\nu}$$

Is there such a constant in Nature? If one looks at the large-scale behavior of matter in the universe, one sees that in the macroscopic sense this constant is very very small, perhaps exactly zero. In any case the astronomical value for λ comes out smaller than the natural dimensional value by a factor of 10^{-118} or so. This is one of the largest discrepancies in physics; even for astrophysics this is considered to be a large error! So one looks for some explanation of why, in natural units, λ is 0 or at least $\lesssim 10^{-118}$.

One question one must ask is whether the fundamental constant in the microscopic Einstein equation is really the same as the quantity that is measured macroscopically in astronomy, and it turns out that this is not at all clear.

Another point is that a λ problem appears in particle physics since, when gravity is around, the trick that we all learned in kindergarten of subtracting the vacuum energy in field theory doesn't work anymore. It is a swindle that one can get away with as long as gravity is not watching, but if gravity is watching the vacuum energy enters Einstein's equation as a $\lambda\, g_{\mu\nu}$ term and cannot be thrown away. One of the beautiful things about supersymmetry is that unbroken supersymmetry, together with a symmetry called R-invariance, makes this vacuum energy zero to all orders[29]. The boson zero-point energy cancels against the fermion zero-point energy, and all the radiative corrections cancel too. However, when one breaks supersymmetry spontaneously, there is once again a vacuum energy. The argument is that supersymmetry acting on the vacuum is no longer zero if supersymmetry is spontaneously broken: $S|0\rangle \neq 0$. Therefore, the expected value in the vacuum of $S^{+}S$ is greater than zero: $\langle 0|S^{+}S|0\rangle > 0$. But $S^{+}S$ is the energy in supersymmetry theory, and so the expected value of the energy is greater than zero. In supergravity, if one introduces a dimensionless self-coupling e, then one gets a cosmological constant of magnitude e-squared times the Planck mass squared, and of the opposite sign from that given by the spontaneous breaking. One might conceive of finding some principle that would allow these to cancel against each other; nobody has ever done that in a fully convincing way, but people have found moderately elegant ways of doing so[30]. However, there is still another effect suggested by Hawking and his collaborators in Cambridge[31].

The particle physics community has paid insufficient attention to what Hawking and his friends are doing, and Hawking and his friends have not fully succeeded in explaining their point of view to most particle theorists. As a result there has been insufficient comparison of the two approaches with each other. What Hawking points out is that there is a tendency in quantum gravity for spacetime to curl up into tiny bubbles the size of the Planck length. If that happens, what we observe macroscopically is an average over all that foam, or whatever one wants to call it; the astronomical value of λ is not the same as the value of λ in the fundamental particle physics equations; and we cannot discuss the relation between them without doing a lot of work.

On a more elementary level, we may remark that if we take the trace of Einstein's equation, we find that the curvature scalar R has a λ term and also a term in Θ^{μ}_{μ}, the trace of the stress-energy-momentum-tensor. This is formally zero in our equations for massless fields, but there can be anomalies that are not zero. These can be gravitational anomalies and also anomalies from strong, weak, and other interactions. Expected values of these anomalies can disturb the relation between λ and the macroscopic R. It is very important to go over this whole field, including the observations of the Cambridge people, and really straighten out the relation between what is observed astronomically and the parameter λ that occurs in the equation.

There is also the extraordinarily interesting idea formulated originally by Guth and Tye[32], and then corrected by Albrecht and Steinhardt[33] and, in the Soviet Union, by Linde. This is the idea of a universe exploding during its early moments. This theory makes use of a change in the vacuum state in particle physics. During the exploding phase, and in the later phase of reheating and ordinary expansion, there are different vacua, with different values of the particle physics energy. The symmetry may change, for example, from SU(4) x U(1) to SU(3) x SU(2) x U(1). This can give a change in the expected value of the energy and therefore a change in the cosmological constant, so that in the early "inflationary period," or exploding period, there is one kind of cosmological constant corresponding to explosion, and afterwards during the period of reheating and ordinary expansion, in which we still live, there would be another value of the cosmological constant, mysteriously very close to zero. This theory of the "inflationary" or exploding universe is very beautiful because it dilutes out monopoles; it dilutes out asymptotic curvature of the universe, and thus explains why the universe is asymptotically nearly flat, in other words, why the density of matter in the universe is so close to the critical density; it dilutes out inhomogeneities and anisotropies in the universe; and it explains the so-called horizon paradox, how the different parts of the universe know that they are to behave similarly when it looks, without the explosion or inflation, as if in the early moments they would have been outside one another's light cone and not known about one another. All of these puzzles are explained by this very beautiful idea, but let's remember that while the change of vacuum can perhaps be explained by particle physics, the fact that in the second phase the observed cosmological constant is near zero is not explained. That puzzle remains and must be resolved.

References

1. Yu. A. Gol'fand and E. P. Lickhtman, JETP Letters 13; 323 (1971).
   J. Wess and B. Zumino, Nucl. Phys. B70; 39 (1974).
2. S. Ferrara, D.Z. Freedman and P. van Nieuwenhuizen, Phys. Rev. D13; 3214 (1976). 3214;
   S. Deser and B. Zumino, Phys. Lett. 62B; 335 (1976).
3. K. Cahill, Phys. Rev. D26; 1916 (1982).
   J. C. Ward, Proc. Nat. Acad. USA 75; 2568 (1978).
4. The Kaluza, Sitzungsber. Preuss. Acad. Wiss. Berlin Math. Phys. K1; 966 (1921).
   O. Klein, Z. Phys. 37; 895 (1936).
5. For a review, see A. Salam and J. Strathdee, Ann. Phys. 141; 316 (1982).
6. P. Fayet, Nucl. Phys. B90; 104 (1975).
7. L. O'Raifertaigh, Nucl. Phys. B96; 331 (1975).
8. P. Fayet and J. Iliopoulos, Phys. Lett. 51B; 461 (1974).
9. M. T. Grisaru, M. Rocek and W. Siegel, Nucl. Phys. B159; 429 (1979).
10. G. 't Hooft and M. Veltman, Ann. Inst. H. Poincaré 20; 69 (1974).
11. P. van Nieuwenhuizen and C. C. Wu, J. Math. Phys. 18; 182 (1977).
12. G. 't Hooft and M. Veltman, Ref. [10];
    S. Deser, H. S. Tsao and P. van Nieuwenhuizen, Phys. Rev. D10; 3337 (1974), and references therein.
13. M. T. Grisaru, P. van Nieuwenhuizen and J.A.M. Vermaseren, Phys. Rev. Lett. 37; 1662 (1976).
14. M. T. Grisaru, Phys. Lett. B66; 75 (1977);
    E. T. Tomboulis, Phys. Lett. 67B; 417 (1977).
15. P. van Nieuwenhuizen and J. A. M. Vermaseren, Phys. Lett. 65B; 263 (1976).
16. J. Ellis, D. V. Nanopoulos and K. Tamvakis, Phys. Lett. 121B; 123 (1983).
17. E. Cremmer and B. Julia, Nucl. Phys. B159; 141 (1979).
18. B. de Wit and H. Nicolai, Nucl. Phys. B208; 323 (1982).
19. P. Breitenlohner and D. Z. Freedman, Phys. Lett. 115B; 197 (1982).
20. M. Gell-Mann, Bull. Am Phys. Soc. 22; 541 (1977).
21. J. Ellis, M. K. Gaillard and B. Zumino, Phys. Lett. 94B; 343 (1980).
22. A. D'Adda, P. Di Vecchia and M. Luscher, Nucl. Phys. B146; 63 (1978).
23. S. Weinberg and E. Witten, Phys. Lett. 96B; 59 (1980).

24. A. Neveu and J. H. Schwarz, Nucl. Phys. B31; 86 (1971);
P. Ramond, Phys. Rev. D3; 2415 (1971).

25. F. Gliozzi, J. Scherk and D. Olive, Nucl. Phys. B122; 253 (1977).

26. M. Green and J. H. Schwarz, Phys. Lett. 109B; 444 (1982).

27. E. Cremmer, B. Julia and J. Scherk, Phys. Lett. 76B; 409 (1978).d

28. J. H. Schwarz, Caltech Preprint CALT-68-1016, to be published in Nucl. Phys. B;
P. S. Howe and P. C. West, King's College Preprint, 1983.

29. B. Zumino, Nucl. Phys. B89; 535 (1975).

30. S. Deser and B. Zumino, Phys. Rev. Lett. 38; 1433 (1977).

31. For a review see S. W. Hawking in "General Relativity. An Einstein Centennial Survey," eds. S. W. Hawking and W. Israel (Cambridge University Press, 1979).

32. A. Guth and S. H. H. Tye, Phys. Rev. Lett. 44; 31 (1980); erratum ibid. 44; 963 (1980);
A. Guth, Phys. Rev. D23; 347 (1981).

33. A. Albrecht and P. Steinhardt, Phys. Rev. Lett. 48; 1220 (1982);
A. D. Linde, Phys. Lett. 108B; 309 (1982).

COMPOSITENESS, γ-W MIXING, AND ELECTROWEAK UNIFICATION

Pisin Chen

Department of Physics
University of California
Los Angeles, CA 90024

MEMORIUM

Today I am here with mixed feelings. It is a great honor for me to receive the J. J. Sakurai Memorial Fellowship and to speak to you. Yet the sad fact is that Sakurai is no longer with us. As the last student of Professor Sakurai before his untimely death, I would like to report to you his approach to the electroweak interactions, and in particular to present our recent study on the electroweak unification in the context of composite models.

The main ideas in this talk were raised by Sakurai last September during the time he went back to California. After he passed away, the work was continued by me in collaboration with F. M. Renard.[1] Therefore I would like to dedicate this talk to Sakurai with the dearest memory of him.

If one traces Sakurai's approaches toward strong, weak and EM interactions, one easily sees parallel developments between strong and EM, and weak and EM interactions.[2] The γ-W mixing model of electroweak interactions of the seventies is a metaphor of the vector dominance model (VDM) of EM and strong interactions of the sixties. While the "Q^2-Duality" of the early seventies[3] in QCD has been extended to QHD (quantum haplodynamics) in the early eighties.[4]

It is ironic that Sakurai was a pioneer in constructing a model of hadronic physics based on Yang-Mills theory,[5] a brilliant accomplishment despite the fact that it was later realized to be incorrect because of an identification of the global flavor symmetry with the local gauge symmetry (as a result the $\rho^{0,\pm}$ and ω mesons were treated as gauge-bosons). Again, an (anti-) parallel development in his approach to the weak interactions can be traced. When the majority in our community adopted the standard model, he[2] (and Bjorken[6]) argued that the $SU(2)_L$ symmetry in weak interactions could be equally well treated as a global rather than a local symmetry. At the time (1978) when this alternative was raised, it was a bit wild. Why do we need to depart from the gauge theory which not only offers aesthetic beauty theoretically but also agrees with all the facts experimentally? It was not until the composite models of leptons and quarks[7] aroused lots of attention that this alternative was taken seriously.[8] But will the history repeat itself?

It is much too early to answer this question. The fact that even QCD is not entirely confirmed to be "the" theory of strong interaction tells us that the QHD, if any, still has a long way to go. But even without detailed knowledge of the underlying dynamics, one still can shed some light on this question by appealing to general principles. This is what I am going to do later.

In closing this memorium let me say that the legacy that Sakurai left for us seems to be the following: In searching for the ultimate theory of elementary particle interactions, we need to be careful to distinguish what has been firmly established based on experimental facts from what is purely speculation. Since nature is often subtler than we expect, such an effort helps to release us from theoretical prejudices.

INTRODUCTION

In the γ-W mixing approach[2] to the electroweak interaction there is an extra degree of freedom associated with the mixing strength λ of the γ-W junction. As a result the "unification" between EM and weak interactions becomes less restrictive. Namely, in the standard model we have

$$\frac{e^2}{g^2} = \sin^2\theta_W , \tag{1}$$

whereas the Hung-Sakurai model is replaced by

$$\lambda \frac{e}{g} = \sin^2\theta_W . \tag{2}$$

There is no a priori reason to insist on relation (1), since with Eq. (2) the Hung-Sakurai model reproduces all the presently observable low-energy phenomenology. However, if one insists on Eq. (1), then Hung and Sakurai were able to fix this extra degree of freedom by imposing the condition of "asymptotic SU(2) × U(1) symmetry" on the neutral current Hamiltonian. What they got was a relation called the "unification condition:"

$$\lambda = \frac{e}{g} \; . \tag{3}$$

With this condition the relation between $W^{\pm}$ and Z masses becomes the same as in the standard model. This is not very surprising since the standard model also has the property of asymptotic SU(2) × U(1) gauge symmetry. What is surprising is that when one asks for good high energy behavior of $f\bar{f} \to W^+W^-$, one arrives at Eq. (3) again. Is it a coincidence that two independent requirements give us the same result? Why do we always fall back to the standard model?

In the composite models with composite $W^{\pm}$ and Z, γ-W mixing scheme is commonly imposed[8] to ensure the electroweak phenomenology. In this context, is the "unification condition" (Eq. (3)) still OK? Potential complexities may arise when $s \gg M_W^2$. We expect to see, in composite models, exotic states like e*, e** ... etc. when $s \gtrsim \Lambda^2$, where Λ is the mass scale of the underlying composite dynamics. As a result the asymptotic picture may not be as clear anymore.

But if $\Lambda \gg M_W$, and if we limit ourselves to the energy range,

$$\Lambda^2 \gtrsim s > 4M_W^2 \; , \tag{4}$$

then the study of "asymptotic SU(2) × U(1) symmetry" and "good high energy behavior of $f\bar{f} \to W^+W^-$" can be carried out without unwanted complexities.

The outcome is a set of constraints on the W form factors (which we shall call "evolution conditions") required of any composite (W) model. The evolution conditions imply:

1. EM and weak interactions can be "unified" in non-standard ways even though all low-energy observed quantities agree with the standard model.
2. At low-energy, electroweak theory can be non-standard. When the energy increases, the effective theory should evolve in such a way that it looks more and more like a gauge theory.

Before going into details, we briefly review the idea of γ-W mixing and discuss the composite mass scale Λ.

THE γ-W MIXING FORMALISM AND COMPOSITE MASS SCALE

The γ-W Mixing

The basic assumptions of the Hung-Sakurai model are the following:

1. There exists a global $SU(2)_L$ symmetry, and $\vec{W}$ is a triplet under it. The free Lagrangian is

$$L_o = -\frac{1}{4}\vec{W}_{\mu\nu}\cdot\vec{W}^{\mu\nu} + \frac{1}{2}M_W^{\ 2}\,\vec{W}_\mu\cdot\vec{W}^\mu \,, \qquad (5)$$

where $\vec{W}_{\mu\nu} \equiv \partial_\mu\vec{W}_\nu - \partial_\nu\vec{W}_\mu$.

2. There exists a mixing between the photon and the W^3. The Lagrangians for the EM field and the mixing are

$$L_{em} + L_{\gamma W} = -\frac{1}{4}F_{\mu\nu}F^{\mu\nu} - \frac{1}{4}\lambda\,(F^{\mu\nu}W_{\mu\nu}^{\ 3} + W_{\mu\nu}^{\ 3}F^{\mu\nu}) \,. \qquad (6)$$

Using the "Propagator Formalism" we can diagonalize the mass matrix and identify the poles as the physical photon and $W^3(Z)$ masses. The results are

$$m_\gamma = 0 \quad ; \quad J_\mu^{\ em} \text{ unchanged} \,,$$

$$M_Z^{\ 2} = \frac{M_W^{\ 2}}{1-\lambda^2} \quad ; \quad J_\mu^{\ Z} = \frac{1}{\sqrt{1-\lambda^2}}\,[J_\mu^{\ 3} - \lambda J_\mu^{\ em}] \,. \qquad (7)$$

The Hamiltonians for charged and neutral currents are

$$H^{CC} = \frac{1}{2}\,\frac{g^2}{M_W^{\ 2}-s}\,J_\mu^{\ (+)}\cdot J^{\mu(-)} \qquad (8)$$

and

$$H^{NC} = \frac{1}{2}\left[-\frac{J_\mu^{\ em}J^{\mu em}}{s} + \frac{J_\mu^{\ Z}J^{\mu Z}}{M_W^{\ 2}-s}\right] \,. \qquad (9)$$

Define $J_\mu^{\ 3} \equiv gI_\mu^{\ 3}$, then $J_\mu^{\ em}$ and $J_\mu^{\ Z}$ are related by

$$J_\mu^{em} = e[I_\mu^3 + Y_\mu] \equiv eQ_\mu ,$$

where Y_μ : U(1) hypercharge current and

$$J_\mu^Z = \frac{g}{\sqrt{1-\lambda^2}} \left[I_\mu^3 - \frac{e\lambda}{g} (I_\mu^3 + Y_\mu) \right] = \frac{g}{\sqrt{1-\lambda^2}} \left[\left(1 - \frac{e\lambda}{g} \right) I_\mu^3 - \frac{e\lambda}{g} Y_\mu \right] . \tag{10}$$

When related to experimental measurable quantities:

$$\frac{8G_F}{\sqrt{2}} = \frac{g^2}{M_W^2} , \quad \sin^2\theta_W = \frac{e\lambda}{g} , \quad \rho = 1 \tag{11}$$

and defining our free parameter as

$$F \equiv \frac{g\lambda}{e} , \tag{12}$$

we get

$$M_Z^2 = \frac{M_W^2}{1 - F \sin^2\theta_W} . \tag{13}$$

The "unification condition" corresponds to $F = 1$, or $\lambda = e/g$. To achieve this, as we have stated, we impose

1. "Asymptotic SU(2) × U(1) symmetry." This means that there should be no $I_\mu^3 Y^\mu$ term in $H^{NC}(s \to \infty)$, or
2. "Good high energy behavior of $\nu\bar{\nu} \to W^+W^-$ and $e^+e^- \to W^+W^-$." This is actually part of a general theorem by Llewellyn-Smith and Cornwall et al. which says that good high energy behavior of the fW, $f\bar{f}$, $\phi\phi$ and ϕW processes necessarily ensures the Yang-Mills structure of the theory.

The Composite Mass Scale Λ

It is argued from various constraints[10] that the mass scale Λ in composite models should be very large ($\Lambda \gg M_W$). In the context of γ-W mixing, how does the largeness of the mass scale come about? Recall that in hadron physics the γ-ρ mixing strength is (see Fig. 1a)

$$\lambda^2_{\gamma\rho} \simeq \frac{3\alpha}{2\pi} \sim \frac{1}{300} , \tag{14}$$

whereas in composite models (Fig. 1b)

$$\lambda^2_{\gamma W} = \sin^2\theta_W \simeq \frac{1}{4} . \tag{15}$$

How can we explain the difference? The Q^2-duality sum rule[4] says

$$\lambda^2 = \frac{\alpha}{3\pi} N\tilde{e}^2 \left(\frac{\Delta m^2}{m^2} \right) \tag{16}$$

where N stands for the number of "colors" in the underlying theory, and $\tilde{e}$ is the charge of the constituent fermions (quarks in QCD or haplons in QHD). m^2 and Δm^2 are the vector bound state mass-squared and the mass-squared difference between the ground and excited states of the vector particle, respectively. For the ρ meson in QCD: $N_c = 3$, $\tilde{e}_c^{\,2} = 1/2$, $\Delta m^2 = m^2_{\rho'} - m_\rho^{\,2} \simeq 3m_\rho^{\,2}$, therefore $\lambda^2_{\gamma\rho} \simeq 1/300$. On the other hand, for the W in QHD, e.g. in the Fritzsch-Mandelbaum model,[8] if $N_H = 3$ with $\tilde{e}_H^{\,2} = 1/2$, then $\lambda^2_{\gamma W} \simeq 1/4$ if $M_{W'} \simeq 10\, M_W$. Thus we find it necessary to have the mass scale

$$\Lambda \sim \Delta m \simeq 1 \text{ TeV} \tag{17}$$

in order that γ and W^3 be strongly mixed.

THE EVOLUTION CONDITION

In composite models the electroweak interaction is an effective one; thus the couplings are quite unrestricted. The most general CPT invariant forms of the γW^+W^- vertex[11] are:

$$I_1^{\,\mu} = \varepsilon^+\cdot\varepsilon^-(k^--k^+)^\mu - \varepsilon^+\cdot k^-\varepsilon^{-\mu} + \varepsilon^-\cdot k^+\varepsilon^{+\mu} ,$$

$$I_2^{\,\mu} = \varepsilon^-\cdot k^+\varepsilon^{+\mu} - \varepsilon^+\cdot k^-\varepsilon^{-\mu} , \tag{18}$$

$$I_3^{\,\mu} = \varepsilon^-\cdot k^+\varepsilon^+\cdot k^-(k^--k^+)^\mu ,$$

where $k^\pm$ are the four-momentum, and $\varepsilon^\pm$ are the polarizations of $W^\pm$, respectively. Notice that $I_3^{\,\mu}$ is associated with the $W^\pm$ quadrupole moment and is known[9,12] to give an incurably bad high energy behavior unless it is attached to a vanishing form factor. Thus we ignore

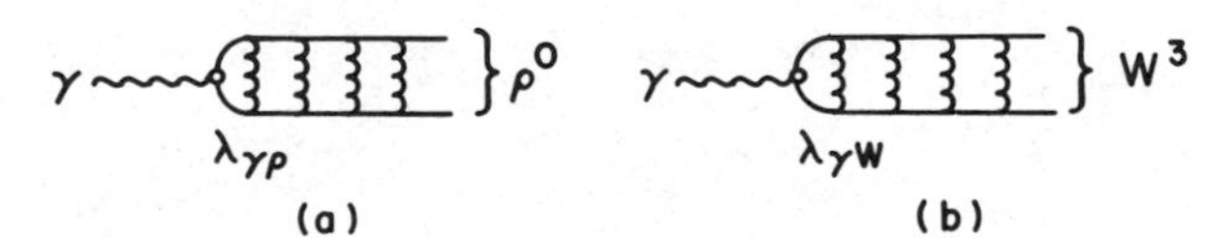

Fig. 1

this term in the following discussion. There arises four independent form factors when we express J_μ^{em} and J_μ^Z in terms of $I_{1\mu}$ and $I_{2\mu}$:

$$\begin{aligned} J_\mu^{em} &= e[I_{1\mu}F_1^{em}(s) + I_{2\mu}F_2^{em}(s)] \quad , \\ J_\mu^Z &= g[I_{1\mu}F_1^3(s) + I_{2\mu}F_2^3(s)] \quad . \end{aligned} \tag{19}$$

The normalization condition (at s=0) can be fixed by the "minimal substitution scheme,"[2] i.e. with

$$\begin{aligned} \partial_\mu &\rightarrow \partial_\mu - ieQA_\mu \qquad \text{for EM couplings} \quad , \\ \partial_\mu &\rightarrow \partial_\mu - ig\vec{T}\cdot\vec{W}_\mu \qquad \text{for weak couplings} \quad . \end{aligned} \tag{20}$$

Then from L_o (Eq. (5)),

$$\begin{cases} F_1^{em}(0) = Q_W = 1 \\ F_2^{em}(0) \quad \text{no restriction} \end{cases} \tag{21}$$

and from $L_{em} + L_{\gamma W}$ (Eq. (6)) we get

$$\begin{cases} F_2^{em}(0) = \dfrac{\lambda g}{e} \equiv F \\ F_1^3(0) = F_2^3(0) = 1 \end{cases} \tag{22}$$

After mixing

$$J_\mu^Z = \frac{1}{\sqrt{1-\lambda^2}}\,[J_\mu^3 - \lambda J_\mu^{em}]$$

$$= \frac{g}{\sqrt{1-\lambda^2}} \left\{ I_{1\mu}\left[F_1{}^3(s) - \frac{e\lambda}{g} F_1{}^{em}(s) \right] \right.$$

$$\left. + I_{2\mu}\left[F_2{}^3(s) - \frac{e\lambda}{g} F_2{}^{em}(s) \right]\right\} . \tag{23}$$

If all F(s) are unity this expression becomes again the one of Hung-Sakurai. Now we look for $f\bar{f} \rightarrow W^+W^-$ scattering amplitudes ($m_e \simeq 0$). For $\nu\bar{\nu} \rightarrow W^+W^-$ (Fig. 2)

$$\begin{aligned} R_{fi} &= - \frac{g^2[F_e(t)]^2}{4t} \bar{v}(p') \not{\epsilon}^-[\not{k}^- - \not{p}'] \not{\epsilon}^+(1-\gamma_5) u(p) \\ &+ \frac{g^2 F_\nu{}^3(s)}{4(1-\lambda^2)(s-M_Z{}^2)} \bar{v}(p')\left\{\left[F_1{}^3(s) - \frac{e\lambda}{g} F_1{}^{em}(s) \right] \not{I}_1 \right. \\ &\left. + \left[F_2{}^3(s) - \frac{e\lambda}{g} F_2{}^{3m}(s) \right] \not{I}_2 \right\} (1-\gamma_5) u(p) . \end{aligned} \tag{24}$$

For $e^+e^- \rightarrow W^+W^-$ (Fig. 3)

$$\begin{aligned} R_{fi} &= - \frac{g^2[F_\nu(t)]^2}{4t} \bar{v}(p') \not{\epsilon}^+(\not{k}^+ - \not{p}^-) \not{\epsilon}^-(1-\gamma_5) u(p) \\ &+ \frac{e^2 F_e{}^{em}(s)}{s} \bar{v}(p')\left\{ F_1{}^{em}(s) \not{I}_1 + F_2{}^{em}(s) \not{I}_2 \right\} u(p) \end{aligned}$$

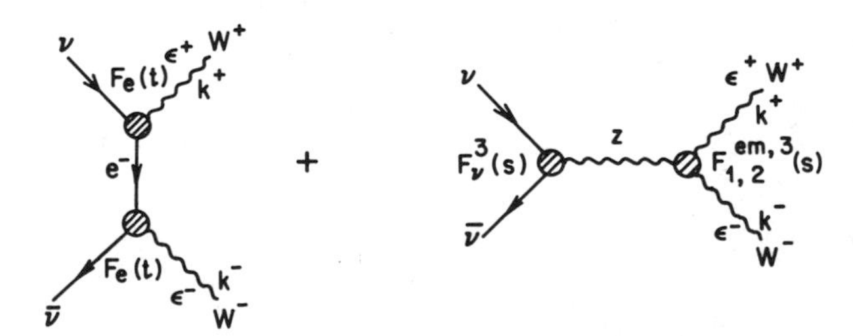

Fig. 2

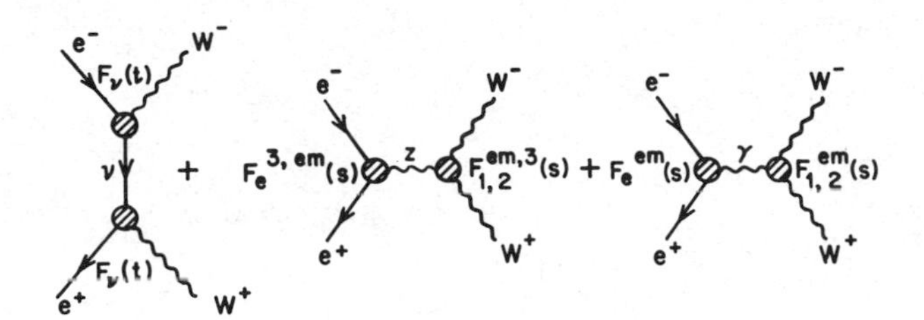

Fig. 3

$$+ \frac{g^2}{(1-\lambda^2)(s-M_Z^2)} \bar{v}(p')\left\{ -\frac{1}{4}(1+\gamma_5) F_e^{\ 3}(s) - \frac{e\lambda}{g} F_e^{\ em}(s) \right\}$$

$$\times \left\{ [F_1^{\ 3}(s) - \frac{e\lambda}{g} F_1^{\ em}(s)] \not{I}_1 + [F_2^{\ 3}(s) - \frac{e\lambda}{g} F_2^{\ em}(s)] \not{I}_2 \right\}$$

$$\times u(p) \tag{25}$$

with

$$\frac{1}{2} gF_f^{\ 3}(s)\ \bar{f}\ \gamma^\mu(1-\gamma_5)\ W^3 f \quad \text{for} \quad W^3 f\bar{f} \text{ couplings}$$

and

$$eF_f^{\ em}(s)\ \bar{f}\ \gamma^\mu A_\mu f \quad \text{for} \quad \gamma f\bar{f} \text{ couplings .}$$

These amplitudes presumably will have bad high energy behaviors when one (to order $\sqrt{s}$) or both (to order s) polarizations $\varepsilon^\pm$ are longitudinal. At high energy ($s >> M_W^2$), the amplitudes behave as follows:

(i) For the $1-\gamma_5$ part
to order s:

$$[F_e(t)]^2 - \frac{F_\nu^{\ 3}(s)}{1-\lambda^2} \left\{ F_2^{\ 3}(s) - \frac{e\lambda}{g} F_2^{\ em}(s) \right\} ,$$

$$(\text{from } \nu\bar{\nu} \to W^+W^-) \tag{26}$$

$$[F_\nu(t)]^2 - \frac{F_e^{\,3}(s)}{1-\lambda^2}\left\{ F_2^{\,3}(s) - \frac{e\lambda}{g} F_2^{\,em}(s) \right\} ,$$

$$\text{(from } e^+e^- \to W^+W^-) \qquad (27)$$

to order $\sqrt{s}$:

$$[F_e(t)]^2 - \frac{F_\nu^{\,2}(s)}{2(1-\lambda^2)}\left\{ F_1^{\,3}(s) + F_2^{\,3}(s) - \frac{e\lambda}{g}[F_1^{\,em}(s) + F_2^{\,em}(s)] \right\} \quad \text{(from } \nu\bar{\nu} \to W^+W^-) \qquad (28)$$

$$[F_\nu(t)]^2 - \frac{F_e^{\,2}(s)}{2(1-\lambda^2)}\left\{ F_1^{\,3}(s) + F_2^{\,3}(s) - \frac{e\lambda}{g}[F_1^{\,em}(s) + F_2^{\,3m}(s)] \right\} \quad \text{(from } e^+e^- \to W^+W^-) \ . \qquad (29)$$

(ii) For the vector part (from $e^+e^- \to W^+W^-$ only) to order s:

$$\frac{eF_e^{\,em}(s)}{1-\lambda^2}\{eF_2^{\,em}(s) - \lambda g F_2^{\,3}(s)\} , \qquad (30)$$

to order $\sqrt{s}$:

$$\frac{eF_e^{\,em}(s)}{1-\lambda^2}\{e[F_1^{\,em}(s) + F_2^{\,em}(s)] - \lambda g[F_1^{\,3}(s) + F_2^{\,3}(s)]\} \ . \qquad (31)$$

One can easily check that with all form factors equal to one, all the above coefficients will vanish if $e\lambda/g = 1$. This is the Hung-Sakurai result.

The existence of t-dependent form factors deserves a special discussion. Note that by subtracting Eq. (26) from Eq. (28), and Eq. (27) from Eq. (29), we get t-independent constraints:

$$F_1^{\,3}(s) - F_2^{\,3}(s) - \frac{e\lambda}{g}[F_1^{\,em}(s) - F_2^{\,em}(s)] = 0$$

$$\text{(from } \nu\bar{\nu} \to W^+W^-) \qquad (32)$$

and

$$eF_2^{em}(s) - \lambda g F_2^{3}(s) = eF_1^{em}(s) - \lambda g F_1^{3}(s) = 0$$
$$(\text{from } e^+e^- \to W^+W^-) \ . \qquad (33)$$

If we take a non-trivial condition where $F_e^{em}(s)$ and $F_\nu^{3}(s)$ are not necessarily vanishing, then one has the solutions

$$F_1^{em}(s) = F_2^{em}(s) \ , \quad F_1^{3}(s) = F_2^{3}(s) \ ,$$

and

$$\frac{\lambda g}{e} = \frac{F_1^{em}(s)}{F_1^{3}(s)} = \frac{F_2^{em}(s)}{F_2^{3}(s)} \equiv F \ . \qquad (34)$$

Now we turn back to the coefficients of the $1-\gamma_5$ parts. Notice that these t-dependent constraints cannot be satisfied without additional contributions if $F(t) \neq 1$. So either we restrict our discussion to the "t = 0 window," or we add effective "contact terms" to assure the good high energy behavior.

In the first case, at $t = 0$ we let $F_e(t=0) = F_\nu(t=0) = 1$, then good high energy behavior requires that

$$F_1^{3}(s) \cdot F_\nu^{3}(s) = F_1^{3}(s) \cdot F_e^{3}(s) = 1 \ . \qquad (35)$$

We see that even if $F_\nu(s) \simeq F_e^{e}(s) \simeq F_1^{3}(s) \simeq 1$ we are still left with an undetermined $F_1^{em}(s) = F_2^{em}(s)$ asymptotic value. Thus our free parameter $\lambda g/e = F$ is still undetermined.

This "t=0 window" can be compared to the "low $p_\perp$ window" in hadronic processes where the substructure effects do not directly show up. This is in contrast to the "large $p_\perp$" processes where hard scatterings of the subconstituents do occur (e.g. the jets).

In the second case, we can add contact terms (Fig. 4) like

$$H(s,t) \ \bar{v}(p')[\not{l}_1 + \not{l}_2](1-\gamma_5) \ u(p) \qquad (36)$$

such that the t-dependent constraints are automatically satisfied. Notice that the t-independent constraints are not modified by this additional term.

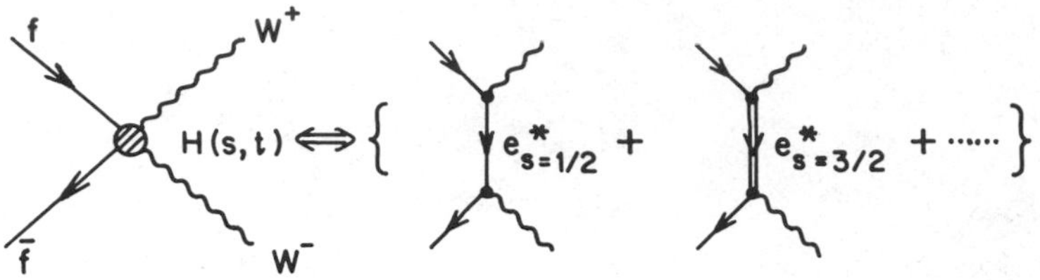

Fig. 4

It seems likely that a general solution with arbitrary s and t can be achieved by simply adding the contact terms to the scattering amplitude. However, the contact terms may actually simulate the complexities due to excited states like e*, e**, ... and spins 1/2, 3/2, ... etc., which we have assumed to neglect. Actually, since the t-independent constraints together with the t = 0 constraint allow another self-consistent solution, which is not modified by t-dependent considerations, it is reasonable to believe that they form a closed set of properties which is independent of the remaining unknowns. This set of constraints are called the "evolution condition" (i.e. Eq. (34) and Eq. (35)). Given initial (i.e. the normalization) conditions (Eq. (21) and Eq. (22)) at s = 0 in a particular model, the W form factors should evolve in such a way that at high energy ($4M_W^2 \ll s \lesssim \Lambda^2$) the evolution condition is satisfied.

In more realistic considerations, there should be additional bound states other than the $\vec{W}$'s formed by the same subconstituents, for example, the isosinglet weak-boson W^o. It is shown[1] that with some generalizations, the evolution condition still holds.

CONSEQUENCES FROM THE CONSTRAINTS

Now we discuss some of the consequences of the evolution condition.

(i) Low Energy Phenomenology

As mentioned in Eq. (11), the following quantities

$$G_F = \frac{g^2}{4\sqrt{2}\, M_W^2}\ , \quad \sin^2\theta_W = \frac{e\lambda}{g} = \frac{\lambda^2}{F}\ , \quad \rho = 1 \tag{37}$$

are satisfied for any value of F. However

$$M_W^2 = F\left(\frac{e^2}{4\sqrt{2}\, G_F \sin^2\theta_W}\right), \quad M_Z^2 = \frac{M_W^2}{1 - F \sin^2\theta_W}, \tag{38}$$

may differ from that of the standard model if $F \neq 1$.

(ii) $W^\pm$ Magnetic Moment

At low energy ($s \simeq 4M_W^2 \ll \Lambda^2$), from the equation of motion of $W^\pm$ we can get

$$\mu_W = \frac{e}{2M_W}\,[1 + F_2^{em}(0)] = \frac{e}{2M_W}\left[1 + \frac{\lambda g}{e}\right]. \tag{39}$$

Recall that in the standard model

$$\mu_W = \frac{e}{M_W}.$$

Again, if $F = \frac{\lambda g}{e} \neq 1$, the magnetic moment will be affected.

(iii) The F Parameter

Let us observe that the F parameter is actually

$$F = \frac{\gamma \text{ form factor}}{W^3 \text{ form factor}} \text{ at large value of } s.$$

It is possible that this value is independent of s. In any case, when the energy range of our interest is $4M_W^2 \lesssim s \lesssim \Lambda^2$, for definiteness we can take

$$F \equiv \frac{F_1^{em}(\Lambda^2)}{F_1^3(\Lambda^2)}. \tag{40}$$

Moreover, if there is a hard core inside the composite W (also e and ν) with size $1/\Lambda$, then it is perfectly possible that the form factors approach non-zero values asymptotically. Then F = constant.

(iv) The "Evolution" of Currents

At high energy,

$$J_\mu^{em} \to eF_1^{em}(s)[I_{1\mu} + I_{2\mu}] ,$$

$$J_\mu^{3} \to \frac{g}{F} F_1^{em}(s)[I_{1\mu} + I_{2\mu}] , \qquad (41)$$

$$J_\mu^{Z} \to \frac{e}{\sqrt{F}\, \sin\theta_W} \sqrt{1 - F \sin^2\theta_W} \cdot F_1^{em}(s)[I_{1\mu} + I_{2\mu}] .$$

So the effective theory of electroweak interaction is similar to the standard model at $s >> 4M_W^2$. Actually, a close look reveals that it is not precisely the standard model. Rather, it is a "rotated" standard model with $\sin\theta_W$ replaced by $\sqrt{F}\, \sin\theta_W$.

CONCLUDING REMARKS

To conclude, I shall emphasize that once $\vec{W}$'s are composite there is no "unification" between EM and weak interactions in the Weinberg-Salam sense. Even if the "unification condition" $F = \lambda g/e = 1$ is satisfied due to dynamical reasons via γ-W mixing strength λ, not all features become the same as in the standard model.

$$M_Z^2 = \frac{M_W^2}{1 - \sin^2\theta_W} ,$$

but form factor effects of W can be present.

Finally, from $\mu_W = \frac{e}{2M_W}[1+F]$ it is reasonable to say that

$$(F-1) \sim \frac{M_W}{\Lambda} \lesssim 10\% . \qquad (42)$$

So instead of radiative corrections, we argue that F-1 gives rise to a similar shift to the W mass, i.e.

$$M_W^{QHD} = FM_W^{standard} \simeq 77+8 \text{ GeV} . \qquad (43)$$

REFERENCES

1. P. Chen and F. M. Renard, SLAC-PUB-3049 (January 1983).
2. P. Q. Hung and J. J. Sakurai, Nucl. Phys. B143, 81 (1978), 538 (1979).
3. A. Bramón, E. Etim and M. Greco, Phys. Lett. 41B, 609 (1972); J. J. Sakurai, Phys. Lett. 46B, 209 (1973).
4. P. Chen and J. J. Sakurai, Phys. Lett. 110B, 481 (1982).
5. J. J. Sakurai, Ann. Phys. 11, 1 (1960).
6. J. D. Bjorken, Phys. Rev. D19, 335 (1979).
7. For a review see: M. E. Peskin, Proc. 1981 International Symposium on Lepton and Photon Interactions, Bonn, ed. W. Pfeil; and H. Harari, Lectures given at SLAC Summer Institute, August 1982, to be published.
8. L. Abbott and E. Farhi, Phys. Lett. 101B, 69 (1981); H. Fritzsch and G. Mendelbaum, Phys. Lett. 102B, 319 (1981).
9. C. H. Llewellyn-Smith, Phys. Lett. 46B, 233 (1973); J. M. Cornwall, D. Levin and G. Tiktopoulos, Phys. Rev. D10, 1145 (1974).
10. See, for example, L. Lyons, Oxford University Report 52/82 (1982).
11. F. M. Renard, "Basis of e^+e^- Collisions," ed. Frontiéres (1981).
12. K. J. F. Gaemers and G. J. Gounaris, Z. Phys. C1, 259 (1979).

THE STRUCTURE OF FUNDAMENTAL INTERACTIONS

Roberto D. Peccei

Max-Planck-Institut für Physik und Astrophysik
- Werner Heisenberg Institut für Physik -
Munich, Fed. Rep. Germany

ABSTRACT

I describe both the achievements as well as the unsolved puzzles of the modern gauge theories of weak, strong and electromagnetic interactions. This is followed by a discussion of the various speculative attempts being pursued to remedy some of the physical limitations of the standard model. In particular, some of the successes and shortcomings of grand unified theories, without or with supersymmetry, and of composite models, with quarks and leptons as quasi Goldstone fermions, are exhibited.

In the academic year 1981/82 we were very fortunate, at the Max-Planck-Institut in Munich, to have had J.J. Sakurai visit us, as a Senior Alexander von Humboldt fellow. This opportunity allowed me, to get to know him better as a physicist and as a man. It is, therefore, with great sadness, and with trepidation for my own inadequacy, that I have undertaken the task to prepare a lecture in honor of his memory. My only consolation is that I have at least tried - albeit, I am sure, not totally successfully - to overview high energy physics guided by his critical and discerning point of view. Indeed, his legacy for me will remain his frankness and open attitude towards physics and life.

I begin by looking at what J.J. Sakurai would have called orthodoxy:

WHAT WE THINK WE KNOW

QCD as the theory of the strong interactions

In the early and mid 1960's, three observations were made which turned out to be of crucial importance for the establishment of QCD as the theory of the strong interactions:

(1) Quarks, as subcomponents of the hadrons, were suggested by Gell Mann and Zweig[1]. Even though no clear understanding existed for the quark binding dynamics, different flavor varieties of quarks could account for the charge and strangeness properties of the then observed hadrons.

(2) Besides flavor, it was found that quarks needed to possess a further attribute - color[2] - so that properly antisymmetric s-wave nonrelativistic wavefunctions could be written down for baryons. Again, even though the nonrelativistic reasoning that led to color was suspect, the inference of the existence of color proved to be correct. This inference received its first real test with the predictions concerning R and $\pi^\circ \to 2\gamma$[3].

(3) The Hamiltonian of the strong interactions was discovered to exhibit an approximate chiral SU(2) x SU(2) symmetry, with chirality being spontaneously broken[4]. This observation allowed for the calculation of threshold properties of hadronic processes involving the (approximately) Goldstone pions, either through current algebra techniques[5] or via effective chiral Lagrangians[6]. The approximate chiral symmetry of H_{strong} implied that the forces among quarks must be vectorial.

QCD, being a nonabelian gauge theory based on the group $SU(3)_c$, naturally uses the above points. The intercharge of color vector gluons among the quarks, presumably, provides the binding force which holds quarks together to make hadrons. I should remark here that Sakurai[7], in his famous Annals of Physics paper, was the first person to suggest that the strong interactions may be based on a non-Abelian gauge theory. He thought, however, that the role of the gluons was played by the vector mesons.

QCD, of course, does more than encapsulate in a neat form the knowledge acquired in the past. It is a predictive theory, which at short distances exhibits asymptotic freedom[8] and which is presumed to confine at large distances. The asymptotic freedom of QCD allows one to use perturbation theory in limited domains. In practice, however, theorists tend to forget these natural boundaries, so that these days all curves drawn through experimental data are always labelled QCD predictions ! One should be aware, however, that these predictions range in reliability from those based on impeccable theory to those based on lore, legend and magic. On this scale, at the upper end one would find predictions for R and for the q^2-dependence of the moments of the deep inelastic structure functions, while at the lower end would appear specific z distributions for quark fragmentation functions.

To my mind QCD still has to pass certain crucial tests, which on the whole are not experimental, before it can be truly thought off as "the theory" of the strong interactions. Although, it is far from clear what would replace QCD, if it failed ! To wit, one must:

(1) Calculate the hadronic spectrum

(2) Calculate, or at least be able to get the general features right of soft hadronic processes (e.g. πN scattering at 14.7 GeV).

(3) Derive the nuclear force, and hence understand nuclear physics, from first principles.

Some remarkable early results on point (1) above have been achieved recently, using lattice Monte Carlo techniques[9]. However, one is still not totally sure whether these very nice results are totally free of possible lattice artifacts[10].

There are a few structural remarks that I want to make on QCD. If one neglects the masses of the light quarks u, d and stays only in this sector of the theory, it is clear that the QCD Lagrangian, at the classical level, is that of a scale invariant theory. Nevertheless, through quantum effects, QCD acquires a scale Λ_{QCD}, which is a purely dynamical scale. As a result, all u, d hadron composites have their mass proportional to Λ_{QCD}. One may define Λ_{QCD} as the scale in momentum space where the running coupling constant $\alpha_s(q^2) = \frac{g_3^2(q^2)}{4\pi}$ goes through unity. This scale can be obtained from deep inelastic experiments by tracing back, with the aid of the renormalization group, the observed logarithmic deviations from the parton model. I shall return later on again to this idea of theories with purely dynamically induced scales.

The approximate chiral SU(2) x SU(2) property of the strong Hamiltonian is seen in QCD because the mechanical masses of the u and d quarks are much less than Λ_{QCD}. Best estimates[11] put the sum of the u and d quark masses to about 10 MeV, while Λ_{QCD} is of the order of hundreds of MeV*. The formation of chiral breaking

*This statement is sloppy, since Λ_{QCD} depends on the renormalization scheme one has adopted. Nevertheless, for all schemes one has always $\Lambda_{QCD} \gg m_u, m_d$.

condensates of the type $\langle\bar{u}u\rangle \sim \Lambda^3_{QCD}$ cause the spontaneous breakdown of the SU(2) x SU(2) symmetry and the appearance of a nearly massless triplet of pions. This behaviour has been checked on the lattice by direct computation of a nonvanishing value for the pion decay constant, $F_\pi \sim \Lambda_{QCD}$ [12]. It provides for a rational expalnation of why the pion mass is so much lighter than that of any other hadron through the relation[13]

$$m_\pi^2 F_\pi^2 = \bar{m} \langle \bar{u}u \rangle \tag{1}$$

with

$$\bar{m} = ½(m_u + m_d) \tag{2}$$

I should remark that $\langle\bar{u}u\rangle$ is not the only condensate formed in QCD. One knows from the QCD sum rules of Shifman, Vainshtein and Zakharov[14] that a non-zero gluon condensate $\langle G^2\rangle = \langle G_a^{\mu\nu} G^a_{\mu\nu}\rangle$ exists. Its numerical value, extracted from the sum rules, has been also checked by some lattice work[15], although the actual calculations are quite tricky because one must know how to correctly subtract a perturbative background. The existence of a non-zero $\langle G^2\rangle$ is not associated with the breakdown of any symmetry* and hence it does not give rise to Goldstone boson. Rather, it is throught to be a signal of finement.

With regards to condensate formation in QCD, there is an amusing dynamical question that one can pose. QCD has baryon number as a global symmetry. If a condensate like $\langle u\ d\ d\ u\ d\ d\rangle$ formed in QCD then one would have spontaneous breakdown of baryon number and concomitant $n - \bar{n}$ oscillations. We know, experimentally, that this does not happen because the Goldstone boson associated with the $U(1)_B$ breakdown, χ, would allow the deuteron to disintegrate very rapidly into χ + pions. Nevertheless, it is theoretically interesting to show that a condensate like $\langle u\ d\ d\ u\ d\ d\rangle$ never forms in QCD. My guess is that this is probably because the six quark state u d d u d d finds it energetically favorable to regroup into two color singlet neutrons.

It appears to me that the extant problems in QCD are "only" calculational. If one assumes that confinement really works, we have no hint from either theory or experiment that QCD cannot be "the theory" of the strong interactions. The only real problem I know in QCD is that of strong CP[16]. However, this is more a problem of how quark masses are generated, than a problem of QCD itself. The strong

---

*Dilatation invariance is not a good symmetry, since it has anomalies.

CP problem arises because one can always add to the Lagrangian of the theory a term which violates both P and CP

$$L_{\text{Total}} = L_{\text{QCD}} + \theta G_a^{\mu\nu} \tilde{G}_{\mu\nu}^a \tag{3}$$

Present day bounds on the neutron dipole moment[17] imply that $\theta \lesssim 10^{-8} - 10^{-9}$[18]. Such a small number is difficult to understand and the suggestion was made[19] that perhaps θ was effectively zero, because of the presence of an extra chiral symmetry in the problem. Although the initial suggestion[19] for curing the strong CP problem probably does not work, since the standard axion it predicts[20] does not seem to exist[21], I feel that when we finally will understand the origin of quark masses, the problem of strong CP will also go quietly away.

Electroweak Interactions

As was the case for QCD also the modern gauge theory of electroweak interactions - the, so-called, standard model[22] - has its roots on observations made in the past. As a result of more than 30 years of experimentation and theoretical analysis one knew by the middle 1960's that low energy charged current weak interactions could be summarized by the current-current Fermi Lagrangian*

$$L_{\text{eff}} = \frac{G}{\sqrt{2}} J_+^\mu J_{-\mu} \tag{4}$$

where

$$J_+^\mu = \bar{\nu}_e \gamma^\mu (1-\gamma_5) e + \bar{\nu}_\mu \gamma^\mu (1-\gamma_5)\mu + \cos\theta_c \bar{u} \gamma^\mu (1-\gamma_5) d$$

$$+ \sin\theta_c \bar{u} \gamma^\mu (1-\gamma_5) s \tag{5a}$$

$$J_-^\mu = (J_+^\mu)^\dagger \tag{5b}$$

The above Lagrangian contains already a remarkable amount of information. To wit, one sees that the hadronic vector current is just the isospin current (CVC), that the interactions are maximal parity violating (V-A) and that there is universality of the strength of the charged current interactions, through the appearance of the Cabibbo angle.

---

*That this form applied also for nonleptonic interactions was a pure inference, given the difficulty of calculating these processes.

Today, roughly ten years after the discovery of the weak neutral currents[23], the Lagrangian (4) continues to be valid for the charged current interactions, except that these currents have further pieces containing heavy quarks. There is, however, an additional piece to L^{eff} which describes low energy neutral current weak interactions, and the total Lagrangian reads:

$$L_{eff} = \frac{G}{\sqrt{2}}(J^{\mu}_{+}J_{-\mu} + \rho J^{\mu}_{NC}J_{NC\mu}) \tag{6}$$

The neutral current is given by

$$J^{\mu}_{NC} = 2(J^{\mu}_{3} - \sin^2\theta_W J^{\mu}_{em}) \tag{7}$$

where J^{μ}_{3} is the third component of the weak $SU(2)_L$ current of the standard model[22]. Ignoring quark mixing, the charged currents are also given in terms of the $SU(2)_L$ currents by

$$J^{\mu}_{\pm} = 2(J^{\mu}_{1} \mp iJ^{\mu}_{2}) \tag{8}$$

When mixing is taken into account, these currents will contain a generalized unitary matrix of mixing angles (and phases), which, for three generations of quarks and leptons, was first written down by Kobayashi and Maskawa[24]. Experimentally one knows that the free parameters ρ and $\sin^2\theta_W$ take the values*

$$\rho = 0.992 \pm 0.017\,; \quad \sin^2\theta_W = 0.224 \pm 0.015 \tag{9}$$

The value $\rho = 1$ corresponds, in the standard model[22], to breaking the $SU(2)_L$ symmetry in the simplest possible way - via doublets of Higgs bosons.

Of course, the great excitement of 1983 has been the announced discovery of the W[26,27], which is supposed to mediate the charged current weak interactions. One finds[26], essentially without any model assumptions,

$$M_W > 73\,\text{GeV} \tag{10}$$

and, in a more model dependent way,

$$M_W = 81 \pm 5\,\text{GeV} \tag{11}$$

This latter value is in remarkable agreement with the prediction of the standard model, including radiative corrections[28]

$$M_W = 83 \pm 2.4\,\text{GeV} \tag{12}$$

---

*This are the results of the best fit to high statistics experiments, from Ref. 25, and do not include radiative corrections.

Certain remarks are in order:

1. The effective Lagrangian (6) is a prediction of the standard model[22], for $q^2 \ll M_W^2$. However, just because experiment agrees with the predictions of (6), it does not necessarily prove that the standard model is correct. Indeed as Bjorken and Hung and Sakurai[29] pointed out, one obtains (6) also by the much weaker assumption that the weak interactions are globally $SU(2)_L$ invariant and that there is γ -Z° mixing. Generally speaking, in models of this kind[31] one expects that there should be not only one set of vector bosons, $W^\pm$, Z, but a series of them. In that case one can show that in addition to (6) one expects a term

$$L_{add} = \frac{G}{\sqrt{2}} c J^\mu_{em} J_{em\mu} \tag{13}$$

Petra experiments give a fairly strong bound on c:[31]

$$c \lesssim 0.01 \tag{14}$$

but it is clearly of interest to continue probing in this direction.

2. The most crucial tests of the standard model[22], to my mind, concern the values predicted, after radiative corrections for the W and Z masses. As Sakurai himself used to point out[32], alternative models of the weak interactions have no particular reason for yielding exactly the values for M_W and M_Z predicted by the standard model. General vector dominance arguments[33] will favor values near those predicted by the standard model for W and Z, but one does not necessarily obtain the same values. I have given in Eq. (12) the radiatively corrected value expected for M_W. Similarly one has[28], for the neutral intermediate vector boson:

$$M_{z^0} = 93.8 \pm 2\text{GeV} \tag{15}$$

The errors in the predicted values are reflections of experimental errors in the determination of $\sin^2\theta_W$ and certain theoretical uncertainties - like the influence of yet unseen heavy quarks - in the radiative corrections. For instance, from the value[34] $\sin^2\theta_W^{exp} = 0.229 \pm 0.009$, which is the best fit result of high statistic eD and deep inelastic experiments, one obtains, after radiative corrections a value[35] for $\sin^2\theta_W\big|_R^\mu = 0.202 \pm 0.014$. Using the, self-consistent, definition

$$\sin^2\theta_W\big|_R^\mu = \frac{\pi\alpha}{\sqrt{2}G_\mu M_W^2} \tag{16}$$

with G_μ being the Fermi constant extracted from μ-decay, after having included the appropriate radiative corrections, and the above value for $\sin^2\theta_W\big|_R^\mu$, yields Eq. (12). It may perhaps be worthwhile to emphasize that the radiative corrections should be applied to each individual experiment, since the final result can depend on the cuts made. Thus, it is possible that, with further reanalysis, one may still be able to reduce somewhat the errors expected on the W and Z masses.

3. Although L_{eff} of Eq. (6) describes accurately all known leptonic and semileptonic processes, and provides the basis for nonleptonic calculations, there are still pieces in it totally untested, like $\nu - \nu$ scattering or $\nu_e e$ scattering. The above processes are unlikely to ever be tested directly. However, there is an aspect of L_{eff} which could be tested shortly and which is of unusual interest. This concerns whether or not there is a CP violating phase in the $\Delta S = 1$ part of L_{eff}. One knows, from the work of Kobayashi and Maskawa[24], that for three generations of quarks and leptons the mixing matrix in the charged current sector can contain a phase δ. If δ were zero, one would expect to find the CP violation parameter ϵ' in the kaon system to vanish. The observed CP violation in the kaon system would then be due to some new $\Delta S = 2$ interactions, yielding a nonzero mixing parameter ϵ. If CP violation, on the other hand, is purely due to the Kobayashi-Maskawa phase δ - the simplest possibility - then it turns out that both ϵ' and ϵ are proportional to $\sin\delta$ and that their ratio is calculable. One finds[36,37]

$$\frac{1}{50} \leqslant \left|\frac{\varepsilon'}{\varepsilon}\right| \lesssim \frac{1}{500} \tag{17}$$

where the large uncertainty in the result is due to difficulties in estimating a particular hadronic matrix element. Experiments now in progress at Brookhaven and Fermilab and proposed at CERN have the ability to measure $|\epsilon'/\epsilon|$ in the range of Eq. (17). These important experiments will be able to tell us if one needs to add or not additional pieces to L_{eff} of Eq. (6), to describe CP-violation phenomena.

I have said earlier that, apart from the still unsettled question of confinement, QCD appears to have no problems of principle. In contrast the standard model[22] has within it at least 6 mysteries. Mysteries are not necessarily theoretical problems. Rather they are things, which the theory itself cannot explain. The mystery list includes:

1. Why is parity violated ? [This is put in, in the standard model, by using asymmetric $SU(2)_L$ representations for the right and left handed fermions of the theory].

2. Why do quarks have 1/3 integer charges? [This is again fed in, in the standard model, by picking the values of the quark hypercharges appropriately].

3. Why are there distinct coupling constants for strong and electroweak effects ? [If the standard theory is $SU(3)_c \times SU(2)_L \times U(1)$, clearly the coupling constants g_3, g_2 and g_1 can be taken to be distinct].

4. Why is not the Higgs mass of $O(M_{Planck})$? [If radiative corrections were to be cut off by gravity one would expect this result. Formally, in the standard model, renormalizability saves the day].

5. Why are the masses of the quarks and leptons so strange? [Because these masses are proportional to arbitrary Yukawa coupling constants, they are uncalculable in the standard model].

6. Why do lepton and quarks come in replicas ? [Any number of families of quarks and leptons are allowed to exist in the electroweak theory[22]. Only requiring that $SU(3)_c$ be asymptotically free puts the mild restriction $N_f \lesssim 8$].

Attempts to answer these mysteries have lead theorists to speculate on possible physics beyond that of the standard picture. Most of these speculations start by assuming that QCD and the standard electroweak gauge theory are true and then proceed to extend these concepts. But perhaps one should remember, as Sakurai used to remind us[32], that only L_{eff} of Eq. (6) - and now also the existence of the W - has been checked experimentally. There is really no evidence at all for any of the theoretical speculations I shall discuss below. However, to distinguish them from theology we will need some experimental evidence in their favor, reasonably soon ($t \lesssim 10$ years ?). One should always keep in mind that, after all, physics is an experimental science.

SPECULATIONS

Essentially each of the mysteries I discussed in Sec. I has engendered its own set of theoretical speculations. I have tried to capture the flavor of the various speculative ideas in Table I, where mysteries are roughly paired up with the relevant speculations on how to resolve them. Several possible interconnections are also shown, with supersymmetry and beyond it supergravity overshadowing the lot.

I should point out that these speculative ideas do not only try to answer why certain of the mysteries of the standard model occur, but they also give rise to an impressive list of new phenomena and particles. I collected in Table II some of these predictions, indicating in italics the inevitably rich variety of new particles expected from supersymmetry alone.

Table I. Mysteries and Speculations

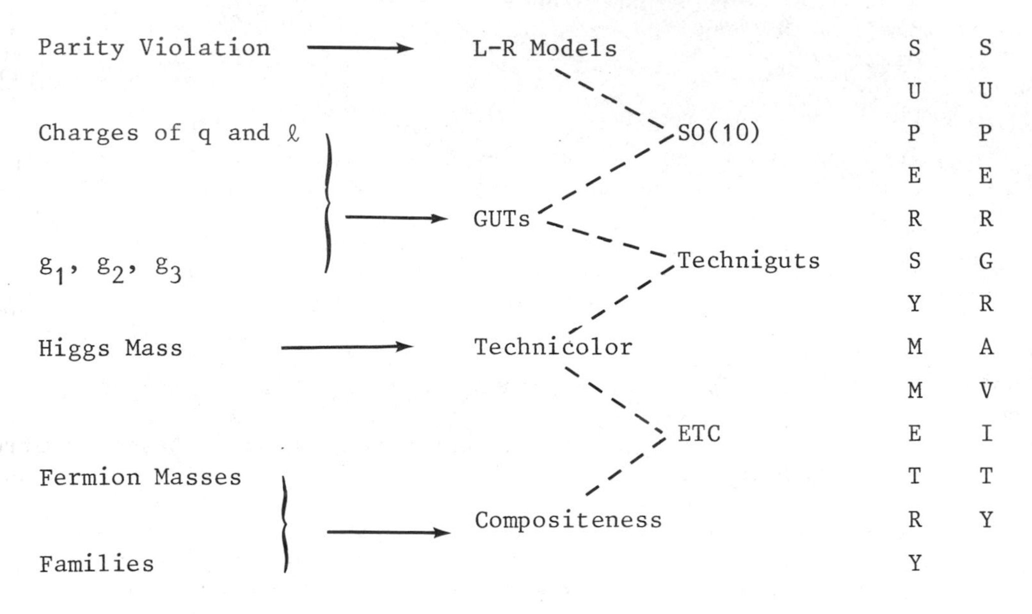

Table II. Some of the Physical Predictions of Speculations Beyond the Standard Model

Neutrino Masses		Neutrino Oscillations
Shiggs	Neutrinoless 2ß-decay	*Sleptons*
n-$\bar{n}$ oscillations	*Squarks*	Proton decay
Gluinos	Monopoles	*Photino*
Technipion	*Gravitino*	Technihadrons
Wino	Lepton and Quark Structure	*Zino*

There are two bothersome features about the speculations, apart from the fact that there is as yet no evidence for any of them:

1. The suggestions for solving the mysteries of the standard model, although interrelated in some ways, are in other ways almost orthogonal. Thus it may be that more than one of the above speculations need be true. This is bad since probabilities, unfortunately, do not add but multiply !

2. The physical predictions of Table II, although rich and varied, have very limited experimental windows where they can be tested. For instance, electron neutrino masses can be found if they are not much below 1 ev; the achievable limit for ν_u and ν_τ masses is probably around 200 kev and 80 Mev, respectively, with the latter limit coming from the process D. Similarly, it is probably hopeless to search for neutrinoless 2ß decay below $\tau_{2\beta} \simeq 10^{24}$ years, n-$\bar{n}$ oscillations with an oscillation time $t \gtrsim 10^8$ sec, and proton decay beyond $\tau_p \simeq 10^{33}$ years. If the monopole flux on earth is below the Parker limit[38], $\Phi_P = 10^{-15}$ cm^{-2} sec^{-1} st^{-1}, it will be probably be impossible to detect (This assumes that the Cabrera event[39] is a fluke). Earth neutrino oscillation experiments are hard put to go beyond $\Delta m^2 \simeq .01$ ev^2 and nothing will be ever seen if the mixing angles are very small. Finally, accelerator physics itself is reaching pretty natural barriers, with p$\bar{p}$ interactions at LEP probably yielding the highest energy ever likely to be achieved on earth.

This gloominess about realities should not deter one, however, from pursuing these speculations. After all, the mysteries are there and a complete theory must somehow be able to solve them ! I will, in the remainder of this section, go through some of the characteristics and difficulties of each of the principal lines of speculations, thereby providing a tour of what may be (or may not be !) the physics of the future.

a) L-R Model

Left-right symmetric models (L-R models[40]) are the simplest extension of the standard electroweak theory[22]. The gauge group instead of $SU(2)_L \times U(1)$ is taken to be $SU(2)_L \times SU(2)_R \times U(1)_{B-L}$. This has two advantages: 1) Fermions are treated in a manifestly L-R symmetry way. For instance $(\nu_e, e)_L$ is a doublet of $SU(2)_L$ and singlet of $SU(2)_R$ and the reversed assignment holds for $(\nu_e, e)_R$. Thus, under parity, the Lagrangian of the theory is invariant. Nevertheless, the parity asymmetry of the weak interactions arises because the vacuum does not respect the L-R symmetry, so that the vector bosons corresponding to $SU(2)_R$, W_R, are much heavier than those correspondingly to $SU(2)_L$, W_L.

2) Identifying the U(1) group with (B-L) transformations gives a much more physical interpretation of the weak hypercharge:

$$Y = T_{3R} + \tfrac{1}{2}(B-L) \tag{18}$$

In L-R models neutrino masses cannot be avoided. However, one can turn a possible source of embarrassment into a positive prediction if one assumes[41] that the right-handed neutrino has a large Majorana mass (since this Majorana mass breaks B-L, while Y is still supposed to be good, it must, according to (18) also break $SU(2)_R$. Whence $M_{\nu_R} \sim M_{W_R}$). Because of this large Majorana mass, the "see-saw" mechanism applies[42], and a heavy neutrino, of mass M_{W_R}, as well as a light neutrino, of mass $m_\nu \sim m_\ell^2/M_{W_R}$ appear. This is very nice, since the light neutrino, which has all the properties of the normally expected left-handed neutrino, is forced by the see-saw mechanism to have a naturally small mass, with respect to the charged lepton mass m_ℓ. I should note that the presence of Majorana neutrinos allows for neutrinoless double ß-decay, with a rate depending on the assumed large Majorana mass. Further, there is a connection between the amount of parity violation, proportional to M_{W_L}/M_{W_R}, and the mass of the neutrino[41]. As $M_{W_R} \to \infty$ one recovers the standard model with massless neutrinos.

L-R models also suggest, without having to go all the way to grand unification, that B-violating processes may exist. In particular $\Delta B = 2$ transitions[43], giving rise to n-$\bar{n}$ oscillations are possible in these models. Of course, the tell tale sign of L-R models is the fact that they contain 2 Z°'s and 2 $W^\pm$'s. If M_{W_R} is sufficiently light*, these models may give observable shifts from the precise predictions of the standard model, relating to the W and Z° masses, and experiments at the $p\bar{p}$ collider, but more likely at LEP or the SLC will see these. HERA, will of course, also provide an excellent setting for testing L-R models with a relative light M_{W_R}. If $M_{W_R} \gg M_{W_L}$, on the other hand, it will be difficult to see any difference between the predictions of the standard model and those of the L-R model. Further, one must then explain theoretically how the large hierarchy between M_{W_R} and M_{W_L} arises.

My own personal conclusion is that L-R models could well be true and, if W_R is sufficiently low, may even cause great excitement. However, they do not provide, per se, any deep elucidation of the structural problems of electroweak theories.

---

* Conservative bounds on M_{W_R}/M_{W_L} put this value around 3. More stringent, but more speculative bounds exist. For a discussion, see for example, Ref. 44.

b) GUTs

Grand Unified Theories involve fearsome (and fearless !) extrapolations, which, from a purely phenomenological point of view appear, on the surface, to be justified. The prescription is easy, and it was spelled out long ago by Georgi, Quinn and Weinberg[45]. If one takes the experimentally observed values for the coupling constant of QCD, g_3, and of those of the electroweak group, g_2 and g_1,*, which are related to e and $\sin^2\theta_W$, and lets these coupling constants evolve according to the renormalization group, one finds - rather amazingly - that they tend to come together around a scale $M_x \simeq 10^{14} - 10^{15}$ Gev. One can take two attitudes about this result; ignore it or try to explain it. The only sensible explanation for this triple coincidence is that beyond M_x, the gauge interactions of $SU(3)_c \times SU(2)_L \times U(1)$ unify into those of a larger group G, and that M_x itself is the scale at which this group suffers a spontaneous breakdown to the standard model group. This explanation, which was arrived at originally in a different way by Georgi and Glashow[46], solves at once two mysteries. Clearly it interelates naturally g_1, g_2 and g_3. Further if G is a simple group, since the charge must be one of its generators, it follows that the charges of fermions in the theory lie in particular ratios with respect to one another.

GUTs make some "minor" predictions, which depend more on the details of the particular grand unified theory. For instance, there is the well known prediction[47] for m_τ/m_b, which follows if one assumes a minimal Higgs structure in SU(5). Of a similar nature are predictions of radiatively induced neutrino masses in SO(10)[48]. The principal predictions of GUTs, however, are related to matter instability and with the precise value for $\sin^2\theta_W$. Let me discuss first the second point. If one assumes that, indeed, at some point M_x $SU(3)_c \times SU(2)_L \times U(1)$ unify into a larger group, one can determine M_x from the condition that, say g_3 and g_2 unify. Then, for g_1 to also unify at M_x requires a precise initial value for $\sin^2\theta_W$. Carrying this out for the simplest grand unified group, SU(5)[46], by doing a two loop renormalization group calculation, yields an impressive result. One predicts[49] from SU(5)

$$\sin^2\theta_W(M_W)|_{\overline{MS}} = 0.214 \tag{19}$$

This value must be compared to the experimental value extracted in the same $\overline{\text{MS}}$ renormalization scheme:

$$\sin^2\theta_W(M_W)\Big|_{\overline{MS}}^{\text{exp}} = 0.215 \pm 0.014 \tag{20}$$

*One must be careful to normalize g_1 appropriately.

and, clearly, the agreement is unbelievably good. However, as I will comment below, this agreement may have to be take with a grain of salt.

That GUTs lead to matter instability is easy to understand. In GUTs quarks and leptons sit in the same representations of the group G, and hence there is not an a priori reason why quark-lepton transitions involving $\Delta B \neq 0$ should not be allowed. The simplest $\Delta B \neq 0$ effective interactions induced by GUTs will contain a product of 4 fermion fields and hence must be scaled by $1/M_x^2$. Whence, it follows that the lifetime for proton, or bound neutron, decay scales with M_x^4. Protons are so long lived, only because the unification scale is so high ! Weinberg and Wilczek and Zee[51] showed, more precisely, that if one assumes that $SU(3)_c \times SU(2)_L \times U(1)$ is a good symmetry, then the Lorentz structure of the simplest $\Delta B \neq 0$ effective interaction (of dim 6) forces it to conserve B-L. Whence, one would expect that if the proton decays it will decay in modes like $p \to \pi^\circ e^+$ rather than $p \to e^-\pi^+\pi^+$.

The simple SU(5) model, which had such a nice prediction for $\sin^2\theta_W$, predicts for the unification scale[49]

$$M_x = \left(2.1 \,{}^{+1.7}_{-1.2}\right) \times 10^{14} \mathrm{GeV} \tag{21}$$

In turn, this leads to an estimate for the proton lifetime[28] of

$$\tau_p = 2 \times 10^{29\pm2} \mathrm{years} \tag{22}$$

where the large error is caused by uncertainties in M_x itself and in evaluating the relevant decay matrix elements. After some encouraging events reported by the Kolar Goldfield experiment[51] and the Mt. Blanc experiment[52], the prediction (22) has been put in jeopardy by very recent results of the IMB experiment[53]. They searched extensively for the decay mode $p \to e^+ \pi^\circ$, which should account for roughly 20% of the decays, and found no such events. This allows them to put a strong limit on τ_p for this mode of greater than 6.5×10^{31} years.

A plausible explanation for the nonobservation yet of proton decay (even though it is supposed to be near an available experimental window) is that perhaps the grand unified group is larger than SU(5), with a more complicated spontaneous breakdown. For instance, in SO(10)[54] the chain

$$SO(10) \underset{M_x}{\to} SU(4) \otimes SU(2)_L \otimes SU(2)_R \underset{M_c}{\to} SU(3)_c \otimes SU(2)_L \otimes SU(2)_R \otimes U(1)_{B-L}$$

$$\underset{M_{W_R}}{\to} SU(3)_c \otimes SU(2)_L \otimes U(1) \underset{M_{W_L}}{\to} SU(3)_c \otimes U(1)_{em} \tag{23}$$

has M_x larger than 10^{16} Gev. The practical question remains, however, that if proton decay is not observed there remain truly few tests of GUTs. GUTs, as 't Hooft[55] and Polyakov[56] showed, do predict the existence of heavy ($M > M_x$) magnetic monopoles, but except for Cabrera's event[39], there is no positive evidence on them yet. GUTs also allow for the establishment of a nonzero asymmetry between matter and antimatter in the universe[57], and calculations[58] give for it roughly the order of magnitude observed. Although I consider this point one of the nicest arguments in favor of GUTs, I must confess that, per se, it cannot really "prove" the existence of nontrivial physics at the GUT scale M_x.

Although GUTs solve two of the mysteries they, on the whole, have other theoretical problems of their own. GUTs are, in general, a one family solution ($\bar{5}$ + 10 in SU(5); 16 in SO(10)) and generations must be put in by hand. Fermion masses, except for a few relations in special cases, are still essentially arbitrary. The Higgs problem is exacerbated. There is now a plethora of Higgs and there is a hierarchy problem, namely the ratio M_x/M_W, even without gravity. In fact, it requires very fine tuning indeed to guarantee that, in the Higgs potential, parameters are such to allow for $M_x^2/M_W^2 = 10^{26}$! Finally, GUTs in their simplest form suffer from what may just be a philosophical problem - which is that of the great desert between $M_W \simeq 10^2$ Gev to $M_x \simeq 10^{15}$ Gev. Can it really be true that there is no physics going on for all these many orders of magnitude ?*

My conclusion is that GUTs are in very bad need of some positive experimental impetus. GUTs may indeed be true, but they are certainly not the answer to all problems. In this respect, it is a hopeful sign that they are able to coexist with other ideas - like compositeness or supersymmetry. In fact, in the past couple of years there has been spirited activity going on towards trying to build supersymmetric grand unified theories (SUSY-GUTs) and I want to comment briefly on this point here. The principal advantage of SUSY-GUTs[60] is that the hierarchy problem is considerably ameliorated. The point is that, in supersymmetric theories, once one fixes the tree parameters to give a hierarchy of breakings, the hierarchy will require no further fine tuning. Apart from this point, however, SUSY-GUTs tend to be less predictive than GUTs. Typically[61], the predicted $\sin^2\theta_W$ tebds to be bigger than that in GUTs and $M_x > 10^{16}$ Gev. However, the proton lifetime is predicted to be shorter (longer) than 10^{30} years depending on the presence (absence) of certain dimension 5 operators in the theory[62]. The principal prediction, of course, of SUSY-GUTs is that there should be

*In fact, as one can readily demonstrate[59], it is possible to populate the desert and still get unification, if one is careful about what representations and interactions one allows.

supersymmetric partners to all the low energy states we know. Where these Spartners (Squarks, Sleptons, etc.) are depends, unfortunately, crucially on how the supersymmetry is broken in the theory. In my opinion, SUSY-GUTS are not a reasonable trade off to solve the hierarchy problem. If, however, one could really show that they arise as some N = 1 approximation to an N = 8 supergravity theory[63], with a reasonable spectrum of states, then they would be extremely interesting. I hope we shall be able to know the answer to this question in the next 5 years or so.

c) Technicolor

Sometimes people think that the hierarchy problem of GUTs is due to the presence of unnatural large members. This is not necessarily so. For instance $M_x/M_W \simeq 10^{13}$ has no natural explanation in GUTs, but the larger number $M_x/\Lambda_{QCD} \simeq 10^{16}$ is perfectly natural. This is so, since the latter number can be immediately understood, given the initial condition $\alpha_3(M_x) = \alpha_{GUT}(M_x)$, from renormalization group arguments. M_W being the scale where $SU(2)_L \times U(1)$ breaks down, unfortunately appears to have no such renormalization group interpretation. This is not quite strictly true, because one knows[64] that in pure QCD the $<\bar{u}u>$ chiral breaking condensates that form - whose scale is set by Λ_{QCD} - also break $SU(2)_L \times U(1)$, giving a contribution to M_W of $0(\Lambda_{QCD})$. Hence, if M_W were of $0(\Lambda_{QCD})$ and not around 10^2 GeV, there would be no hierarchy problem.

Technicolor[64] provides a solution to the question of the naturalness of the scale M_W along these lines, by supposing that there exist another QCD-like theory - technicolor - with an intrinsic dynamical scale $\Lambda_{Tc} >> \Lambda_{QCD}$. The presence of condensates of the fermions of this theory - techniquarks - with nonzero $SU(2)_L \times U(1)$ quantum numbers, causes the spontaneous breakdown of the electroweak theory, without having to invoke explicit Higgs fields. Typically, one assumes there exist both T_R and T_L techniquarks, which are respectively singlets and doublets under $SU(2) \times U(1)$. Then the existence of a $<\bar{T}_L T_R>$ condensate ($<\bar{T}_L T_R> \sim \Lambda^3_{Tc}$) with Λ_{Tc} of 0 (1 Tev) breaks $SU(2)_L \times U(1)$ dynamically, yielding a W mass, $M_W \simeq e \Lambda_{Tc}$, of the right order of magnitude. Effectively, the $<\bar{T}_L T_R>$ condensates have replaced the Higgs field vacuum expectation value, $<\phi>$, of the standard model.

Technicolor is a very nice idea for generating the gauge boson masses. However, it cannot alone generate any fermion masses. This is its principal drawback, because one must then suppose that there exist further interactions, extended technicolor (or ETC)[65] which provide connections between the quarks and techniquarks. With ETC it is then possible to get fermion masses which scale like $m_f \sim \Lambda^3_{Tc}/M^2_{ETC}$, with M^2_{ETC} being the scale of the ETC gauge bosons. However, one is then faced with a list of problems. To mention a few:

1. There is no theoretical understanding of the origin of the mass scale M_{ETC}, although the tumbling scenario[66] is appealing.

2. To get reasonable fermion masses, M_{ETC} must be near $_{Tc}$ ($M_{ETC}/\Lambda_{Tc} \simeq 10$). This requirement is in conflict with bounds from the absence of flavor changing neutral currents[67].

3. In the theory there exist some low mass spin zero excitations, the, so-called, technipions, which should perhaps already have been observed experimentally[68].

My conclusion, especially in view of the very byzantine structures emerging from "semirealistic" ETC theories, is that, while technicolor may be an appealing way to solve the Higgs hierarchy problem (the origin of the M_W scale), it is unlikely that the origin of fermion masses and families will emerge out of an ETC theory.

d) Compositeness

L-R models, GUTs with or without supersymmetry, Technicolor and ETC never address squarely the question of families. To my mind this is the most deeply clothed mystery of particle physics - one to which we only have, at best, vague hints on how to resolve it. Parallel to the question of families is the whole issue of fermion masses. Although I cannot prove it as a theorem, since ETC provides a counter example, I believe that the mass spectrum of the quarks and leptons will only be able to be computed if the quarks and leptons are themselves composites, of yet more fundamental objects (preons). I also believe, but have no proof, that in compositeness will lie the key for the understanding of families. This said, I must add that of all the speculations I have discussed up to now, the one of compositeness of some (or all ?) the fundamental excitations of the standard model, is perhaps the one which is the most daring. Indeed, not only does one not have any evidence for compositeness, but there is very good evidence for the elementarity of, say, electrons and muons !

People who work on composite models, a fraternity which I have joined in the last couple of years, spend - with good reason - most of their time worrying about the following points:

1. From many different sources (g-2, absence of flavor changing neutral currents etc.) one knows[69] that leptons and quarks are rather elementary. If Λ_{comp} is a typical scale associated

with compositeness, then one finds that it must be at least as large as 1 Tev. What dynamics then allows for the formation of bound states - the quarks and leptons - whose mass is so much less than $\Lambda_{comp.}$?

$$\Lambda_{comp.} >> m_f \tag{24}$$

2. The dynamics of composite models must involve, at some stage, a strong coupling theory. Yet, because we are interested in essentially massless bound states, it is unlikely that the kind of non-relativistic argumentation that helped with QCD - another strong coupling theory - will be of any help here. The lesson of ground state baryons and color is unlikely to be repeated.

3. At some stage in model building one must decide what should be composite. A list, in an ascending order of radicalness, would include: Higgs bosons; quarks and leptons; W and Z bosons; gluons, photons and gravitons. Most "conservative" practitioners stop at the second level above.

4. A reason for families must be found, along with some dynamics that impedes their too rapid communication with one another. It almost appears as if one needs some "orthogonal axis" to the dynamics to allow for replications. Because the dynamics must satisfy Eq. (24), i.e. the quarks and leptons are effectively almost massless with respect to $\Lambda_{comp.}$, the Case-Gasiorowicz-Weinberg-Witten theorem[70] provides, at least, a reason why light higher spin excitations of quarks and leptons should not be formed. Although, no really appealing theoretical idea on families exists yet, our experimental colleagues may prove of great help here. The measurement of the total width of the Z° can shed some light in the number of generations, since each extra neutrino species adds approximately 180 Mev to the width*.

In the study of composite models, Eq. (24) and its concomitant dynamics have received the most attention. I should point out, parenthetically, that if weak bosons themselves are composite, as has been suggested by a growing list of authors[71], including Sakurai[32], then $\Lambda_{comp.}$ cannot be much greater than around a Tev, since there exists no dynamics known to make light composite vector bosons. Fermions, on the other hand, can be made light by making use of protective symmetries. Indeed, the situation is perhaps best described in a converse way. One

*This assumes, of course, that nearby compositeness has not distorted too much the standard picture.

can argue, on quite general grounds, that without such protective symmetries no light composite quarks and leptons could ever emerge ! The logic of this last remark is as follows. Composite confined quarks must be made of preon constituents which are themselves also confined. Thus it is very likely that the preon theory is also a non-Abelian gauge theory*. Such theories, assuming massless preons, have only one dynamical scale, $\Lambda_{comp.}$, where the appropriate running coupling constant, $\alpha_{comp.}$, goes through unity. Whence, as in QCD, all bound states are forced to have masses of $O(\Lambda_{comp.})$. The only exception can be states which, for symmetry reasons, are forced to have zero mass. If the symmetry is only approximately true, then the relevant states may pick up a small mass, with respect to $\Lambda_{comp.}$.

't Hooft[72] suggested a very simple way to obtain massless composite quarks and leptons. Namely, they could arise from an underlying preon theory which possessed an unbroken chiral symmetry. Furthermore, he discussed specific consistency conditions required to be satisfied, by the underlying theory and the spectrum of massless composite states, to guarantee that in the binding no chiral breakdown occured, to spoil the picture. Specifically, one has to check that all triangle anomalies[73] arising from the assumed global chiral symmetry matched, when computed either at the underlying or composite level. A variety of models have been found where this anomaly matching occurs. However, the roster of available models is drastically reduced if one adds as a further, rather physical, constraint, that the anomaly matching should also hold for subsets of the original theory. The specific requirement is that the model should still give no chirality breaking, if one removes from the theory a given preon, assumed to get a large mass, and all the bound states containing this massive preon. The most general analysis of models, where this "persistence mass" condition applies, is due to Bars[74]. Unfortunately, although some of the models found by him have very interesting characteristics, they do not appear to be as close to reality as one may wish. Perhaps the most interesting feature of this line of investigation is that, in trying to match anomalies, one, at times, is forced to have a repetition of certain bound states. This suggests a raison d'être for families, alas perhaps a little bit too mechanical.

I would like to end this lecture by discussing a new idea, developed at the Max-Planck-Institute in collaboration with Wilfried Buchmüller and Tsutomo Yanagida, for obtaining (approximately) massless composite quarks and leptons. I find it particularly appropriate to discuss our suggestion here since, as it will emerge, it can lead to models where the weak interactions are effective and one has an,

---

*After all, these are the only realistic theories one believes confine.

approximate, global $SU(2)_L$ symmetry, precisely as it was originally envisaged by Sakurai. Our idea can be very simply explained, and the analogy with pre-QCD strong interactions will certainly not escape the reader. Imagine having a confining preon theory which has the following properties:

(1) It is supersymmetric

(2) It has some (approximate) global symmetry G

(3) Condensates form in the theory, which spontaneously break G to some subgroup H.

Then, irrespective of the detailed preon dynamics, there will appear in the theory n (approximately) massless Goldstone bosons, with n being the dimension of the coset space G/H. Furthermore, because of the supersymmetry, these n Goldstone bosons will be joined by fermionic, and possibly other bosonic, partners to form an (approximately) massless supermultiplet. Our suggestion* is that the quarks and leptons which we see are precisely the quasi Goldstone fermions of some specific preon theory.

A technical remark is in order[77]. Although in the breakdown $G \rightarrow H$ there need appear always n Goldstone bosons, the number of quasi Goldstone fermions may range from $\frac{n}{2}$ $(\frac{n+1}{2})$, for n even (odd), to n. To restore the supersymmetry the theory then has also an appropriate number of quasi Goldstone bosons. Here I shall consider only the total doubling case, with n quasi Goldstone fermions and n quasi Goldstone bosons in addition to the expected n Goldstone bosons. In this total doubling case, these quasi states sit in "real" representations with respect to H and since quarks and leptons are complex with respect to $SU(2)_L \times U(1)$, we know that H cannot contain $SU(2)_L \times U(1)$ as a product group. (Although it may contain it as a subgroup).

Quarks and leptons are, however, "real" with respect to $SU(3)_c \times U(1)_{em}$. This suggested to us[76] that H could perhaps be this group. Since, furthermore, 9, which is the dimension of H, plus 15 which is the number of quarks and leptons of one family, adds to 24, it was easy to guess that G = SU(5). Let me restate the result. If one had a supersymmetric preon theory with a global SU(5) invariance which, because of condensate formation, spontaneously broke down to $SU(3)_c \times U(1)_{em}$ then necessarily there would exist in the spectrum

---

*The general idea of quasi Goldstone fermions was developed in work done in collaboration with S. Love, see Ref. 75. The identification of quarks and leptons with quasi Goldstone fermions is made in Ref. 76. A comprehensive discussion of the many ramifications of the idea is contained in Ref. 77.

of the theory 15 massless fermions with precisely the quantum number of the quarks and leptons of one family. This result is nontrivial since there is a good dynamical reason for having the m = 0 states independent of the scale of $\Lambda_{comp.}$, and, furthermore, these states turn out to have precisely the right set of quantum numbers. This double coincidence is quite rare in composite models, where one is always either trying to get rid of superfluous states and/or is trying to fix the dynamics so that it may be sensible.

The above example is, however, far from being a theory yet. First of all, as it is discussed in much more detail in ref. 77, there are many other group theoretical possibilities besides $SU(5) \rightarrow SU(3)_c \times U(1)_{em}$, which also lead to realistic models. Secondly, a barrage of difficult questions must be answered before any real progress can be made. I think it is useful here to at least list some of these crucial questions, even though our answers to them are still in a rather primitive stage. The last of these questions will lead us back to the work of Sakurai.

Among the questions that need to be answered are:

1. How does one give the bosonic partners of the quarks and leptons sufficient mass so that one does not enter into contradictions with experiment ? This clearly depends on how one does the supersymmetry breaking. However, it is a tricky question since the Goldstone bosons were the dynamical reason why the quasi Goldstone fermions were rendered massless originally ! It may be necessary that the fermions have some additional dynamical protection - of which $SU(2)_L \times U(1)$ could be an example[77] - for remaining light.

2. How do families originate ? Although there exist models where families naturally arise, like $E_7 \rightarrow SU(6) \times SU(3)$ or $E_7 \rightarrow SU(5)$ (with not total doubling), a suitable "orthogonal axis" may be provided by the index of the supersymmetry. Unfortunately, we do not even have toy versions for these latter models.

3. How do quarks and leptons get their small masses ? This is an extraordinary difficult question, probably not uncoupled to (2). Furthermore, in contrast to what happens in QCD, here it is necessary to break explicitly the supersymmetry, as well as the global symmetry G before getting masses. Only then the intuitively nice idea, that charged leptons get their mass because of electromagnetism, quarks get their mass because of color and electromagnetism and neutrinos remain massless, may be realized.

Finally, one should ask,

4. Where are the weak interactions ? It is possible that they may be embedded in the preon theory, much as we now think that pro-

tons have weak interactions because quarks do. Models of this kind are discussed in Ref. 77. These models, however as a rule, tend to contain more massless states than just the ordinary quarks and leptons. On the other hand, it could be that the weak interactions could arise as effective residual interactions among the quarks and leptons, due to the binding forces among their constituents. This last option is close in spirit to Sakurai's idea and I would like to close my discussion by describing some of our results along these lines in a bit of detail.

Because of the assumed quasi Goldstone nature of the quarks and leptons, it turns out that it is possible to write down an explicit expression for the interactions among these fields. This is analogous to what one did for pions in pre-QCD days. The fact that pions were the Goldstone bosons of the SU(2) x SU(2) breakdown allowed one to calculate, by effective Lagrangian techniques[6], their threshold scattering properties. Here too, it is possible to write down an effective Lagrangian for the interaction among the Goldstone and quasi Goldstone particles[76]. The only difference between this case and the pion case is that the threshold interactions among the pions are totally fixed, apart from the scale F_π, by group theory, while here the interactions involving the quasi Goldstone particles have considerable more freedom. This freedom arises because of the not strictly Goldstone nature of these states, and it reflects the underlying theory[77].

If one restricts oneself to just the residual interactions among the quasi Goldstone fermions, one finds[76,77] that the first nontrivial interaction is of the form of a current-current interaction:

$$L_{\text{eff}} = c_{ijkl} J^{\mu}_{ij} J_{kl\mu} + \text{h.c.} \tag{25}$$

where

$$J^{\mu}_{ij} = \overline{\Psi}_{iL} \gamma^{\mu} \Psi_{jL} \tag{26}$$

and the ψ_{iL} are left-handed quarks and lepton fields. Of course, for a given breakdown $G \to H$, L_{eff} will be H invariant. However, it may be also approximately invariant under some larger symmetry. That this may be the case follows since, as discussed above, in the construction of the effective Lagrangian for the quasi Goldstone fields one has more freedom than in the case of a Lagrangian for purely Goldstone fields.

We have checked that for the case of the breakdown $SU(5) \to SU(3)_c \times U(1)$ indeed one may choose parameters in the theory such that L_{eff} of Eq. (25) is approximately $SU(2)_L$ invariant[77]. Whence it follows that for such a theory, after γ-Z° mixing[29], one could recover precisely the standard Lagrangian of Eq. (6). This provides an explicit realization of weak interactions as residual physics, along Sakurai's lines. In fact, there is a much more natural model for achieving this than $SU(5) \rightarrow SU(3)_c \times U(1)_{em}$. If one considers the breakdown[77] $SU(6) \rightarrow SU(4) \times SU(2) \times U(1)$ then quark-lepton universality is immediate, and the approximate $SU(2)_L$ nature of the residual interactions is more easily obtained. Of course, in both these examples, the crucial question to determine is what underlying preon dynamics forces these approximate low energy symmetries on the theory. Note, in this respect that just because L_{eff} is of a current-current form that does not mean there is no W. L_{eff} is valid only for q^2 very small compared to the binding scale - which in this case must be related to M_W. At any rate much more work needs to be done before the viability of these kind of models can really be established.

CONCLUSIONS

I have gone from describing things we think we know to describing things which we really do not know. The boundary line between these regions can only be moved by experiment. Sakurai's legacy, perhaps, is that he taught us to be ready to question even our more cherished preconceptions. I would like to end this lecture in his memory by paraphrasing him:

"The only orthodoxy is facts"

REFERENCES

1. M. Gell Mann, Phys. Lett. 8 (1964) 214; G. Zweig, CERN preprints TH 401 and TH 412 (1964) (unpublished).
2. O.W. Greenberg, Phys. Rev. Lett. 13 (1964) 598; M. Han and Y. Nambu, Phys. Rev. 139 (1965) B1006.
3. W.A. Bardeen, H. Fritzsch and M. Gell Mann, in Scale and Conformal Symmetry in Hadron Physics, edited by R. Gatto (Wiley, New York 1973).
4. M. Gell Mann, Phys. Rev. 125 (1962) 1067; Physics 1 (1964) 63.
5. See for example the monograph by S.L. Adler and R.F. Dashen, Current Algebras (W.A. Benjamin, New York 1968).
6. S. Weinberg, Phys. Rev. 166 (1968) 1568.
7. J.J. Sakurai, Ann. Phys. (N.Y.) 11 (1960) 1.
8. D.J. Gross and F. Wilczek, Phys. Rev. Lett. 30 (1973) 1343; Phys. Rev. D8 (1973) 3633; H.D. Politzer, Phys. Rev. Lett. 30 (1973) 1346; G. 't Hooft (unpublished).
9. For a review, see for example, C. Rebbi, Proceedings of the XXI International Conference on High Energy Physics, Paris 1982.

10. P. Hasenfratz and I. Montvay, Phys. Rev. Lett. 50 (1983) 309.
11. J. Gasser and H. Leutwyler, Phys. Rept. 87C (1982) 77.
12. E. Marinari, G. Parisi and C. Rebbi, Phys. Rev. Lett. 47 (1981) 1795.
13. M. Gell Mann, R. Oakes and B. Renner, Phys. Rev. 175 (1968) 2195.
14. M.A. Shifman, A.I. Vainshtein and V.I. Zakharov, Nucl. Phys. B147 (1979) 385; ibid B147 (1979) 448.
15. A. D. Giacomo and G. Paffuti, Phys. Lett. 108B (1982) 327; J. Kripfganz, Phys. Lett. 101B (1981) 169; T. Banks, R. Horsley, H. Rubinstein and U. Wolff, Nucl. Phys. B190 (1982) 129.
16. G. 't Hooft, Phys. Rev. Lett. 37 (1976) 8; Phys. Rev. D14 (1976) 3432.
17. W.B. Dress et al., Phys. Rev. D15 (1977) 9; L.S. Altarev et al., Nucl. Phys. A341 (1980) 269; Phys. Lett. 102B (1981) 13.
18. V. Baluni, Phys. Rev. D19 (1979) 2227; R. Crewther, P. Di Vecchia, G. Veneziano and E. Witten, Phys. Lett. 89B (1979) 123.
19. R.D. Peccei and H.R. Quinn, Phys. Rev. Lett. 38 (1977) 1440; Phys. Rev. D16 (1977) 1791.
20. S. Weinberg, Phys. Rev. Lett. 40 (1978) 223; F. Wilczek, Phys. Rev. Lett. 40 (1978) 279.
21. R.D. Peccei in Neutrino 81, edited by R. Cence, E. Ma and A. Roberts (Univ. of Hawaii, Honolulu 1981).
22. S.L. Glashow, Nucl. Phys. 22 (1961) 579; S. Weinberg, Phys. Rev. Lett. 19 (1967) 1264; A. Salam in Elementary Particle Theory edited by N. Svartholm (Almquist and Wiksells, Stovkholm 1969).
23. F. J. Hasert et al., Phys. Lett. B46 (1973) 121, 138: A. Benvenuti et al., Phys. Rev. Lett. 32 (1974) 800.
24. M. Kobayashi and T. Maskawa, Prog. Theor. Phys. 49 (1973) 652.
25. J.E. Kim, P. Langacker, M. Levine and H.H. Williams, Rev. Mod. Phys. 53 (1981) 211.
26. G. Arnison et al., Phys. Lett. 122B (1983) 103.
27. M. Banner et al., Phys. Lett. 122B (1983) 476.
28. W. Marciano, in Proceedings of the 19th Orbis Scientiae Meeting, Coral Gables, Florida 1982.
29. J.D. Bjorken, Phys. Rev. D19 (1979) 335; P.Q. Hung and J.J. Sakurai, Nucl. Phys. B143 (1978) 81.
30. J.D. Bjorken ref. 29; G.J. Gounaris and D. Schildknecht, Zeit. Phys. C12 (1982) 571.
31. P. Söding, Proceedings of Neutrino 82, Balatonfüred, Hungary 1982.
32. J.J. Sakurai, Proceedings of the XII Internationale Universitätswochen für Kernphysik, Schladming, Austria 1982.
33. H. Fritzsch, D. Schildknecht and R. Kögeler, Phys. Lett. 114B (1982) 157.
34. J.E. Kim et al. Ref. 25; I. Liede and M. Roos, Nucl. Phys. B167 (1980) 3971.
35. W. Marciano and A. Sirlin, Nucl. Phys. B189 (1981) 442; C. Llewellyn Smith and J. Wheater, Phys. Lett. 105B (1981) 486.
36. F. Gilman and M. Wise, Phys. Lett. 83B (1979) 83.
37. B. Guberina and R.D. Peccei, Nucl. Phys. B163 (1980) 289.

38. E.N. Parker, Ap. J. 160 (1970) 383; M.S. Turner, E.N. Parker and T.J. Bogdan, Phys. Rev. D26 (1982) 1296.
39. B. Cabrera, Phys. Rev. Lett. 48 (1982) 1378.
40. J.C. Pati and A. Salam, Phys. Rev. D10 (1974) 275; R.N. Mohapatra and J.C. Pati, Phys. Rev. D11 (1975) 566, 2558; G. Senjanović and R.N. Mohapatra, Phys. Rev. D12 (1975) 1502.
41. R.N. Mohapatra and G. Senjanović, Phys. Rev. Lett. 44 (1980) 912.
42. T. Yanagida, in Proceedings of the Workshop on the Unified Theory and the Baryon Number of the Universe, KEK, 1979; M. Gell Mann, P. Ramond and R. Slansky, in Supergravity, Proceedings of the Supergravity Workshop at Stony Brook (North Holland, Amsterdam, 1980).
43. R.N. Mohapatra and R. Marshak, Phys. Lett. 91B (1980) 222; Phys. Rev. Lett. 44 (1980) 1316.
44. M.A. Beg, R. Budny, R.N. Mohapatra and A. Sirlin, Phys. Rev. Lett. 38 (1977) 1252; J. Maalampi, K. Mursula and M. Roos, Helsinki preprint HU-TFT-82-4 (1982); for a discussion see also, R.D. Peccei in Proceedings of the 1982 CERN School of Physics, Cambridge, England.
45. H. Georgi, H.R. Quinn and S. Weinberg, Phys. Rev. Lett. 33 (1974) 438.
46. H. Georgi and S.L. Glashow, Phys. Rev. Lett. 32 (1979) 438.
47. M.S. Chanowitz, J. Ellis and M.K. Gaillard, Nucl. Phys. B128 (1977) 506.
48. E.Witten, Phys. Lett. 91B (1980) 81.
49. W. Marciano, Ref. 28; W. Marciano and A. Sirlin, Ref. 35.
50. S. Weinberg, Phys. Rev. Lett. 43 (1979) 1566; F. Wilczek and A. Zee, Phys. Rev. Lett. 43 (1979) 1571.
51. M.R. Krishnaswamy et al., Phys. Lett. 115B (1982) 349.
52. G. Battistoni et al., Phys. Lett. 118B (1982) 461.
53. R.M. Bionta et al., Phys. Rev. Lett. to be published.
54. H. Georgi in Particles and Fields 74, edited by C. Carlson (AIP, New York 1975); H. Fritzsch and P. Minkowski, Ann. of Phys. (N.Y.) 93 (1975) 193.
55. G. 't Hooft, Nucl. Phys. B79 (1974) 276.
56. A.M. Polyakov, JETP Letters 20 (1974) 194.
57. M. Yoshimura, Phys. Rev. Lett. 41 (1978) 381; E42 (1979) 746; Phys. Lett. 88B (1979) 294; A. Ignatiev, N. Krasnikov, V. Kuzmin and A. Tavkhelidze, Phys. Lett. 76B (1978) 436.
58. E.W. Kolb and S. Wolfram, Nucl. Phys. B172 (1980) 244; J.M. Fry, K.A. Olive and M.S. Turner, Phys. Rev. D22 (1980) 2953, 2977; J. Ellis, M.K. Gaillard and D. Nanopoulos, Phys. Lett. 80B (1979) 360; E82B (1979) 464.
59. R.D. Peccei and T. Yanagida, Nucl. Phys. B212 (1983) 189.
60. S. Dimopoulos and H. Georgi, Nucl. Phys. B193 (1981) 150; N. Sakai, Z. Phys. C11 (1982) 153.
61. For a review, see for example the Proceedings of the Supersymmetry versus Experiment Workshop, CERN 1982, ed. D. Nanopoulos, A. Savoy Navarro and Ch. Tao .

62. N. Sakai and T. Yanagida, Nucl. Phys. B197(1982) 533.
63. For a review of recent progress see H. Nicolai, these Proceedings.
64. L. Susskind, Phys. Rev. D20 (1979) 2619; S. Weinberg, Phys. Rev. D13 (1976) 974; D19 (1979) 1277.
65. S. Dimopoulos and L. Susskind, Nucl. Phys. B155 (1979) 237; E. Eichten and K. Lane, Phys. Lett. 90B (1980) 125.
66. S. Raby, S. Dimopoulos and L. Susskind, Nucl. Phys. B169 (1980) 373.
67. J. Ellis, D.V. Nanopoulos and P. Sikivie, Phys. Lett. 101B (1981) 387.
68. A. Ali and M.A. Beg, Phys. Lett. 103B (1981) 376; J. Ellis in Proceedings of the 1981 Les Houches Summer School.
69. For a discussion see, R.D. Peccei, Proceedings of the Arctic School of Physics, Akäslompolo, Finland 1982.
70. K.M. Case and S. Gasierowicz, Phys. Rev. 125 (1962) 1055; S. Weinberg and E. Witten, Phys. Lett. 96B (1980) 59.
71. An incomplete list includes, H. Terazawa, Y. Chikashige and K. Akama, Phys. Rev. D15 (1977) 480; H. Harari and N. Seiberg, Phys. Lett. 98B (1981) 269; O.W. Greenberg and J. Sucher, Phys. Lett. 99B (1981) 339; L. Abbott and E. Fahri, Phys. Lett. 101B (1981) 69; Nucl. Phys. B189 (1981) 547; H. Fritzsch and G. Mandelbaum, Phys. Lett. 102B (1981) 113.
72. G. 't Hooft in Recent Developments in Gauge Theories, Cargèse Lectures 1979 (Plenum Press, New York 1980).
73. S. Adler, Phys. Rev. 177 (1979) 2426; J.S. Bell and R. Jackiw, Nuovo Cimento 60A (1969) 47.
74. I. Bars, Nucl. Phys. B208 (1982) 77.
75. W. Buchmüller, S.T. Love, R.D. Peccei and T. Yanagida, Phys. Lett. 115B (1982) 233.
76. W. Buchmüller, R.D. Peccei and T. Yanagida, Phys. Lett. 124B (1983) 67.
77. W. Buchmüller, R.D. Peccei and T. Yanagida, Max-Planck-Institut preprint MPI-PAE/PTh 28/83.

PARTICIPANTS

G. ALTARELLI

Ist. di Fisica "G. Marconi"
and INFN-Sezione de Roma
Piazzale A. Moro 2
1-00185 Roma
Italy

D. BARBER

DESY
Notkestrasse 85
2000 Hamburg-52
West Germany

G. BELLA

Tel-Aviv University
Dept. of Physics and Astronomy
Ramat-Aviv
Tel-Aviv
Israel

F. BERENDS

Universitat Leiden,
Thc Institut-Lorenz
Nieuwsteeg 18
2311 Fr. Leiden
The Netherlands

A. BOEHM

DESY
Notkestrasse 85
2000 Hamburg-52
West Germany

F. BOEHM

California Institute of Technology
Physics Dept. 161-33
Pasadena, CA 91125
U.S.A.

B. BORGIA

Ist. di Fisica "G. Marconi"
and INFN-Sezione di Roma
Piazzale A. Moro 2
1-00185 Roma
Italy

W. BRAUNSCHWEIG

I. Physikal. Inst.
Rhein.-Westf. Tech. Hochschule
Physikzentrum
D-5100 Aachen
West Germany

R. BROWN

Case Western Reserve University
Physics Dept.
Cleveland, Ohio 44106
U.S.A.

J. BUON

Laboratoire de l'Accelerateur
Lineaire
Centre d'Orsay
Batiment 200
F-91405 Orsay
France

B. CABRERA

Stanford University
High Energy Physics Dept.
Stanford, CA 94305
U.S.A.

G. CHADWICK

Stanford Linear Accelerator Center
P.O. Box 4349, Bin 94
Stanford, CA 94305
U.S.A.

L. L. CHAU

Brookhaven National Lab
Physics Department
Upton, Long Island
New York 11973
U.S.A.

P. CHEN

University of California (UCLA)
Physics Dept.
Los Angeles, CA 90024
U.S.A.

H. FRITZSCH	Max-Planck Institut fur Physik und Astrophysik Fohringer Ring 6, D-8 Munchen-40 Germany
K. GAEMERS	University of Amsterdam Dept. of High Energy Physics Valckenierstraat 65 1018 XE Amsterdam The Netherlands
R. GASTMANS	Katholieke University Leuven Inst. Voor Theor. Fysica Celestijnenlaan 200D B-3030 Heverlee Belgium
M. GELL-MANN	California Institute of Technology Physics Dept. 452-48 Pasadena, CA 91125 U.S.A.
C. GEWENIGER	Institute of High Energy Physics University of Heidelberg Schroderstrasse 90 D-6900 Heidelberg West Germany
G. GOGGI	Ist. di Fisica Nucleare Via Bassi 1-27100 Pavia Italy
D. GROOM	University of Utah Physics Dept. Salt Lake City, Utah 84112 U.S.A.
P. GROSSE-WEISMANN	DESY Notkestrasse 85 2000 Hamburg-52 West Germany
B. JONSON	CERN EP Division Geneva 23 Switzerland

H. KAPITZA	DESY Notkestrasse 85 2000 Hamburg-52 West Germany
K. KLEINKNECHT	Universitat Dortmund Institut fur Physik Postfach 500500 4600 Dortmund 50 West Germany
D. KREINICK	Cornell University CLEO Group Newman Lab. of Nucl. Studies Ithaca, New York 14853 U.S.A.
R. LANOU	Brown University Physics Dept. Providence, Rhode Island 02912 U.S.A.
J. LAYTER	University of California, Riverside Physics Dept. Riverside, CA 92521
F. MANDL	Inst. f. Hochenergiephysik (HEPHY) Oesterreich. Akad. d. Wissensch. Nikolsdorfergasse 18 A-1050 Wien Austria
A. MANN	University of Pennsylvania Physics Department Philadelphia, PA 19104 U.S.A.
R. MARSHAK	Virginia Polytechnic Institute and State University Physics Dept. Blacksburg, VA 24061 U.S.A.
H. MAXEINER	DESY Notkestrasse 85 2000 Hamburg-52 West Germany

K. MESS	DESY Notkestrasse 85 2000 Hamburg-52 West Germany
G. MIKENBERG	Weizmann Institute of Science Physics Department Rehovot Israel
P. MINKOWSKI	University of Berne Sidlerstrasse 5 CH-3012 Berne Switzerland
B. MONTAGUE	CERN EP Division CH-1211 Geneva 23 Switzerland
J. MORFIN	Fermi Nat. Accelerator Lab.(FNAL) P.O. Box 500 Batavia, IL 60510 U.S.A.
D. NANOPOULOS	CERN EP Division CH-1211 Geneva 23 Switzerland
B. NAROSKA	DESY Notkestrasse 85 2000 Hamburg-52 West Germany
H. NEWMAN	California Institute of Technology Physics Dept. 256-48 Pasadena, CA 91125 U.S.A.
H. NICOLAI	CERN EP Division CH-1211 Geneva 23 Switzerland

H. OGREN

Indiana University
Physics Dept.
Swain Hall W 117
Bloomington, IN 47401
U.S.A.

R. PECCEI

Max-Planck Institut fur Physik
und Astrophysik
Fohringer Ring 6
D-8 Muchen-40
Germany

M. PERL

Stanford Linear Accelerator Center
Physics Department
P.O. Box 4349
Stanford, CA 94305
U.S.A.

C. PRESCOTT

Stanford Linear Accelerator Center
Physics Department
P.O. Box 4349
Stanford, CA 94305
U.S.A.

P. RAPIDIS

Fermi Nat. Accelerator Lab
P.O. Box 500
Batavia, IL 60510
U.S.A.

R. RAU

DESY
Notkestrasse 85
2000 Hamburg-52
West Germany

M. ROLLIER

Universita Degli Studi DiMilano
Ist. di Scienze Fisiche,
"Aldo Pontremoli"
Via Celoria, 16
1-20133 Milano
Italy

M. ROOS

Helsinki University
Dept. of High Energy Physics
Siltavuorenpenger 20C
SF-00170
Helsinki 17
Finland

J. SALICIO	Grupo de Altas Energias Div. de Fisica Junta de Energia Nuclear (JEN) Avenida Complutense 22 Ciudad Universitaria Madrid 3 Spain
G. SALVINI	Ist. di Fisica "G. Marconi" and INFN-Sezione di Roma Piazalle A. Moro 2 1-00185 Roma Italy
D. SCHILDKNECHT	Fak. fur Phys., Universitat Bielefeld Postfach 8640 D-48 Bielefeld 1 West Germany
L. SEHGAL	RWTH Physikzentrum, Sommerfeldstrasse D-5100 Aachen West Germany
Q. SHAFI	6478 Gainer Street Annandale, VA 22003 U.S.A.
G. SMADJA	CEN Saclay, B.P. No. 2 Dept. de Phys. des Particules Elementaires F-91190 Gif-sur Yvette France
P. SORBA	LAPP Lab d'Annecy-le-Vieux de Phys. de Particules Chemin de Bellevue-BP 909 F-74019 Annecy-le-Vieux France
M. STEUER	CERN EP Division CH-1211 Geneva 23 Switzerland

R. STROYNOWSKI — California Institute of Technology
Physics Dept. 256-48
Pasadena, CA 91125
U.S.A.

Y. SUZUKI — Brookhaven National Lab.
Upton, L.I.
New York 11973
U.S.A.

K. WINTER — CERN
EP Division
CH-1211
Geneva 23
Switzerland

F. YNDURAIN — Universidad Autonoma de Madrid
Department di Fisica Teorica, C-Xl
Canto Blanco
Madrid 34
Spain

INDEX